PERSONAL WIRELESS COMMUNICATIONS

IFIP – The International Federation for Information Processing

IFIP was founded in 1960 under the auspices of UNESCO, following the First World Computer Congress held in Paris the previous year. An umbrella organization for societies working in information processing, IFIP's aim is two-fold: to support information processing within its member countries and to encourage technology transfer to developing nations. As its mission statement clearly states,

> IFIP's mission is to be the leading, truly international, apolitical organization which encourages and assists in the development, exploitation and application of information technology for the benefit of all people.

IFIP is a non-profitmaking organization, run almost solely by 2500 volunteers. It operates through a number of technical committees, which organize events and publications. IFIP's events range from an international congress to local seminars, but the most important are:

- The IFIP World Computer Congress, held every second year;
- Open conferences;
- Working conferences.

The flagship event is the IFIP World Computer Congress, at which both invited and contributed papers are presented. Contributed papers are rigorously refereed and the rejection rate is high.

As with the Congress, participation in the open conferences is open to all and papers may be invited or submitted. Again, submitted papers are stringently refereed.

The working conferences are structured differently. They are usually run by a working group and attendance is small and by invitation only. Their purpose is to create an atmosphere conducive to innovation and development. Refereeing is less rigorous and papers are subjected to extensive group discussion.

Publications arising from IFIP events vary. The papers presented at the IFIP World Computer Congress and at open conferences are published as conference proceedings, while the results of the working conferences are often published as collections of selected and edited papers.

Any national society whose primary activity is in information may apply to become a full member of IFIP, although full membership is restricted to one society per country. Full members are entitled to vote at the annual General Assembly. National societies preferring a less committed involvement may apply for associate or corresponding membership. Associate members enjoy the same benefits as full members, but without voting rights. Corresponding members are not represented in IFIP bodies. Affiliated membership is open to non-national societies, and individual and honorary membership schemes are also offered.

PERSONAL WIRELESS COMMUNICATIONS

The 12th IFIP International Conference on Personal Wireless Communications (PWC 2007), Prague, Czech Republic, September 2007

Edited by

Robert Bestak
Czech Technical University, Czech Republic

Boris Simak
Czech Technical University, Czech Republic

Ewa Kozlowska
Czech Technical University, Czech Republic

 Springer

Library of Congress Control Number: 2007932608

Personal Wireless Communications

Edited by R. Bestak, B. Simak, and E. Kozlowska

p. cm. (IFIP International Federation for Information Processing, a Springer Series in Computer Science)

ISSN: 1571-5736 / 1861-2288 (Internet)
ISBN: 13: 978-0-387-74158- 1
eISBN: 13: 978-0-387-74159 -8

Printed on acid-free paper

Copyright © 2007 by International Federation for Information Processing.
All rights reserved. This work may not be translated or copied in whole or in part without the written permission of the publisher (Springer Science+Business Media, LLC, 233 Spring Street, New York, NY 10013, USA), except for brief excerpts in connection with reviews or scholarly analysis. Use in connection with any form of information storage and retrieval, electronic adaptation, computer software, or by similar or dissimilar methodology now known or hereafter developed is forbidden.
The use in this publication of trade names, trademarks, service marks and similar terms, even if they are not identified as such, is not to be taken as an expression of opinion as to whether or not they are subject to proprietary rights.

9 8 7 6 5 4 3 2 1
springer.com

Preface

The international conference Personal Wireless Communications (PWC 2007) was the twelfth conference of its series aimed at stimulating technical exchange between researchers, practitioners and students interested in mobile computing and wireless networks. On behalf of the International Advisory Committee, it is our great pleasure to welcome you to the proceedings of the 2007 event.

Wireless communication faces dramatic changes. The wireless networks are expanding rapidly in subscribers, capability, coverage, and applications, and costs continue to decrease. Mobile devices are becoming ubiquitous with greatly expanded computing power and memory, improved displays, and wireless local and personal area connectivity. The PWC 2007 program covered a variety of research topics that are of current interest, starting with Ad-Hoc Networks, WiMAX, Heterogeneous Networks, Wireless Networking, QoS and Security, Sensor Networks, Multicast and Signal processing. This year we enriched PWC with a poster session covering diversity topics related to wireless networks (e.g., filters, current conveyors, etc.).

We would like to thank the International Advisory Committee members and the referees. Without their support, the program organization of this conference would not have been possible. We are also indebted to many individuals and organizations that made this conference possible (Czech Technical University, IFIP, ESTEC). In particular, we thank the members of the Organizing Committee for their help in all aspects of the organization of this conference.

We hope that the attendees enjoyed the conference in Prague, and found it a useful forum for the exchange of ideas and results and recent findings. We also hope that the attendees found time to enjoy Prague and to visit its major cultural monuments.

September 2007

Robert Bestak
Borys Simak

Organization

PWC'07 is organized by the department of Electrical Engineering, Czech Technical Univeristy in Praghue and is sponsored by IFIP WG 6.8.

General Co-chairs

Boris Simak, Czech Technical University, Czech Republic
Robert Bestak, Czech Technical University, Czech Republic

Steering committee

Robert Bestak, Czech Technical University, Czech Republic
Augusto Casaca, INESC, Portugal
Pedro Cuenca, Universidad de Castilla La Mancha, Spain
Luis Orozco-Barbosa, Universidad de Castilla La Mancha, Spain
Ramn Puigjaner, Universidad de las Islas Baleares, Spain
Guy Pujolle, University of Paris 6, France
Boris Simak, Czech Technical University, Czech Republic
Jan Slavik, TESTCOM, Czech Republic
Ivan Stojmenovic, University of Ottawa, Canada

Technical Programm Committee

Khaldoun Al Agha, Universite Paris-Sud, France
Raffaele Bruno, IIT-CNR, Italy
Augusto Casaca, INESC, Portugal
Amitabha Das, Nanyang Technological University, Singapore
Maros Dora, Budapest Polytechnic, Hungary
Rajit Gadh, UCLA, USA
Javier Garcia, Universidad Complutense de Madrid, Spain
Jorge Garcia, Universidad Politcnica de Cataluna, Spain
Silvia Giordano, ICA-DSC-EPFL, Switzerland
Phippe Godlewski, ENST Paris, France
Takeshi Hattori, Sophia University, Japan
Sonia Heemstra de Groot, Ericsson EuroLab, The Netherlands
Villy Baek Iversen, Technical University of Denmark, Denmark
Ousmane Kone, University of Toulouse, France
Houda Labiod, ENST Paris, France
Hyong W. Lee, Korea University, Korea

Pascal Lorenz, University of Haute Alsace, France
Zoubir Mammeri, University of Toulouse, France
Vicenzo Mancuso, University of Palermo, Italy
Pietro Manzoni, Universidad Politcnica de Valencia, Spain
Philippe Martins, ENST Paris, France
Antonin Mazalek, University of Defence in Brno, Czech Republic
Ali Miri, University of Ottawa, Canada
Yan Moret, Institut EERIE, France
Manuel Perez Malumbres, Universidad Miguel Hernndez, Spain
Samuel Pierre, Ecole Polytechnique du Montreal, Canada
Ramn Puigjaner, Universidad de las Islas Baleares, Spain
Fernando Ramirez, ITAM, Mexico
Pedro Ruiz, Universidad de Murcia, Spain
Debashis Saha, Indian Institute of Management (IIM) Calcutta, India
Jun-Bae Seo, ETRI, Korea
Jan Slavik, TESTCOM, Czech Republic
Ottio Spaniol, University of Technology of Aachen, Germany
Dirk Staehle, University of Wurzburg, Germany
Ivan Stojmenovic, University of Ottawa, Canada
Luis Villasenor, CICESE, Mexico
Zuzana Vranova, University of Defence in Brno, Czech Republic
Jozef Wozniak, Technical University of Gdansk, Poland

Organizing committee

Ivo Bazant, Czech Technical University, Czech Republic
Robert Bestak, Czech Technical University, Czech Republic
Ewa Kozlowska, Czech Technical University, Czech Republic
Marek Nevosad, Czech Technical University, Czech Republic
Jan Rudinsky, Czech Technical University, Czech Republic
Boris Simak, Czech Technical University, Czech Republic
Jan Slavik, TESTCOM, Czech Republic

Table of Contents

Ad-Hoc Networks

Adaptive Admission Control for Mobile Ad Hoc Networks based on a Cross-layer Design 1
María Canales, José Ramón Gállego, Ángela Hernández-Solana, Antonio Valdovinos

Analysis of a TDMA MAC Protocol for Wireless Ad Hoc Networks under Multipath Fading Channels 13
José Ramón Gállego, María Canales, Ángela Hernández-Solana, Antonio Valdovinos

A QoS Architecture Integrating Ad-Hoc and Infrastructure in Next Generation Networks 25
Marek Natkaniec, Janusz Gozdecki, Susana Sargento

Federating Personal Networks over Heterogeneous Ad-hoc Scenarios 38
Luis Sánchez, Jorge Lanza, Luis Muños

Analysis of Algorithms for Radial Basis Function Neural Network 54
Jiri Stastny, Vladislav Skorpil

Application of Path Duration Study in MultiHop Ad Hoc Networks 63
Alicia Triviño-Cabrera, Jorge García-de-la-Nava, Eduardo Casilari, Francisco J. González-Cañete

WiMAX

WiMAX Throughput Evaluation of Conventional Relaying 75
Pavel Mach, Robert Bestak

Study of the IEEE 802.16 Contention-based Request Mechanism 87
Jesús Delicado, Francisco M. Delicado, Luis Orozco-Barbosa

Transmission Performance of Flexible Relay-based Networks on The Purpose of Extending Network Coverage 99
Ardian Ulvan, Robert Bestak

Impact of Relay Stations Implementation on the Handover in WiMAX... 107
Zdenek Becvar

Optimization of Handover Mechanism in 802.16e using Fuzzy Logic 115
Ewa Kozlowska

Heterogeneous Networks

Implementation of component based wireless communication protocol
for SDR AT 123
    Junsik Kim, Sangchul Oh, Hongsoog Kim, Namhoon Park, Nam Kim

A Hybrid WLAN-Bluetooth Access Network Solution for a More
Efficient VoIP-Data & Video Traffic Management 133
    David Pérez, José Luis Valenzuela, Ángela Hernández, Antonio Valdovinos

Interference Aware Bluetooth Scatternet (Re)configuration Algorithm
IBLUEREA 145
    Tomasz Klajbor, Jozef Wozniak

Performance Studies of MPLS Based Integrated Architecture for
3G-WLAN Scenarios with QoS Provisioning 157
    Iti Saha Misra, Chandi Pani

VoIP-PSTN Interoperability by Asterisk and SS7 Signalling 169
    Jan Rudinsky

Experimental NGN Lab Testbed for Education and Research in Next
Generation Network Technologies 174
    Eugen Mikoczy, Pavol Podhradsky, Ivan Kotuliak, Juraj Matejka

Wireless Networking I

Enhancing IEEE 802.11n WLANs using group-orthogonal code-division
multiplex 184
    Guillem Femenias, Felip Riera-Palou

Saturation Throughput Analysis of IEEE 802.11g (ERP-OFDM) Networks 196
    Krzysztof Szczypiorski, Jozef Lubacz

A Tiered Mobility Management Solution for Cellular/WLAN Integrated
Networks with Low Handoff Delay 206
    Iti Saha Misra, Sibaram Khara, Debashis Saha

The Positioning of Base Station in Wireless Communication with
Genetic Approach 217
    Yong Seouk Choi, Kyung Soo Kim, Nam Kim

A Door Access Control System with Mobile Phones 230
    Tomomi Yamasaki, Toru Nakamura, Kensuke Baba, Hiroto Yasuura

UWB Radar: Vision through a wall 241
    Ondrej Sisma, Alain Gaugue, Christophe Liebe, Jean-Marc Ogier

Wireless Networking II

SDU Discard Function of RLC Protocol in UMTS 252
   Vit Krichl

Scheduling algorithms for 4G wireless networks 264
   Jaume Ramis, Loren Carrasco, Guillem Femenias, Felip Riera-Palou

Precise Time Synchronization over GPRS and EDGE 277
   Petr Kovar, Karol Molnar

GSM Base Station Subsystem Management Application 285
   Vit Novotny, Pavel Svoboda

Transport resources reservation in IMS frameworks: Terminal vs. PDF driven 294
   Antonio Cuevas, Jose I. Moreno, Hans Einsiedler

Wireless Networking III

Enhancing access control for mobile devices with an agnostic trust negotiation decision engine 304
   Daniel Díaz-Sánchez, Andrés Marín, Florina Almenárez

Wireless Technology in Medicine Applications 316
   Otto Dostal, Karel Slavicek

Wireless Military Communications - NNEC Enabler 325
   Miroslav Hopjan, Zuzana Vranova

Wired Core Network for Local and Premises Wireless Networks 332
   Jiri Vodrazka, Tomas Hubeny

Classification of Digital Modulations Mainly Used in Mobile Radio Networks by means of Spectrogram Analysis 341
   Anna Shklyaeva, Petr Kovar, David Kubanek

QoS and Security

Security Issues of Roaming in Wireless Networks 349
   Jaroslav Kadlec, Radek Kuchta, Radimir Vrba

Secure Networking with NAT Traversal for Enhanced Mobility 355
   Lubomir Cvrk, Vit Vrba

Simulation model of a user-manageable quality of service control method . 367
   Karol Molnar

Voice Quality Planning for NGN including Mobile Networks 376
Ivan Pravda, Jiri Vodrazka

Sensor Networks

A Light-weight Security Protocol for RFID System 384
Jung-Hyun Oh, Hyun-Seok Kim, Jin-Young Choi

Analog Digitized Data Logger with Wireless and Wired Communication Interface and RFID Features 396
Radek Kuchta, Radimir Vrba

Zigbee-Based Wireless Distance Measuring Sensor System 403
Ondrej Sajdl, Jaromir Zak, Radimir Vrba

A proposal of a Wireless Sensor Network Routing Protocol 410
Cláudia Barenco Abbas, Ricardo González, Nelson Cardenas, L. J. García Villalba

On the Accuracy of Weighted Proximity Based Localization in Wireless Sensor Networks 423
Peter Brida, Jan Duha, Marek Krasnovsky

Multicast

Multicast in UMTS: Adopting TCP-Friendliness 433
Antonios Alexiou, Christos Bouras, Andreas Papazois

Simulation of Large-Scale IPTV Systems for Fixed and Mobile Networks . 445
Radim Burget, Dan Komosny, Milan Simek

Multicast Feedback Control Protocol for Hierarchical Aggregation in Fixed and Mobile Networks 456
Dan Komosny, Radim Burget

Experiences of Any Source and Source Specific Multicast Implementation in Experimental Network 468
Milan Simek, Radim Burget, Dan Komosny

Signal Processing

Image compression in digital video broadcasting 477
Kamil Bodecek, Vit Novotny, Milan Brezina

Fast lifting wavelet transform and its implementation in Java 488
Jan Maly, Pavel Rajmic

New watermarking scheme for colour image 497
Petr Cika

Data hiding error concealment for JPEG2000 images 505
Milan Brezina, Kamil Bodecek, Milan Brezina

Optimized discrete wavelet transform to real-time digital signal processing 514
Jan Vlach, Pavel Rajmic, Jiri Prinosil, Josef Vyoral, Ivan Mica

Enhanced estimation of power spectral density of noise using the
wavelet transform 521
Petr Sysel, Zdenek Smékal

Face detection in image with complex background 533
Jiri Prinosil, Jan Vlach

Posters

RF Pure Current-Mode Filters using Current Mirrors and Inverters 545
Jan Jerabek, Kamil Vrba

Multifunction RF Filters Using OTA 557
Norbert Herencsar, Kamil Vrba

New Multifunctional Frequency Filter Working in Current-mode 569
Jaroslav Koton, Kamil Vrba

Second-Order Multifunction Filters with Current Operational Amplifiers . 578
David Kubanek, Kamil Vrba

Continuous - Time Active Filter Design Using Signal - Flow Graphs 585
Martin Minarcik, Kamil Vrba

The Design of Optical Routes Applications 595
Martin Kyselak, Miloslav Filka

Measurement and therapeutical system based on Universal Serial Bus ... 602
Lukas Palko

Transmitting Conditions of Cable Tree 610
Vaclav Krepelka, Miloslav Filka

Non-linear circuits with CCII+/- current conveyors 616
Jiri Misurec

Field Programmable Mixed-Signal Arrays (FPMA) Using Versatile
Current/Voltage Conveyor Structures 628
Jiri Stehlik, Daniel Becvar

Modeling and Design of a Novel Integrated Band-Pass Sigma-Delta
Modulator 637
Lukas Fujcik, Jiri Haze, Radimir Vrba, Jiri Forejtek, Pavel Zavoral, Roman Prokop, Linus Michaeli

Composite Materials for Electromagnetic Interference Shielding 649
Pavel Steffan, Jiri Stehlik, Radimir Vrba

Single Chip Potentiostat Measurement System 653
Pavel Steffan, Radimir Vrba

Portable and Precise Measurements with Interdigital Electrodes at
Wide Range of Conductivity 660
Jaromir Hubalek, Radek Kuchta

New Architecture of Network Elements 669
Vladislav Skorpil, Martin Kral

Author Index 679

Adaptive Admission Control for Mobile Ad Hoc Networks based on a Cross-layer Design[*]

María Canales, José Ramón Gállego, Ángela Hernández-Solana, and Antonio Valdovinos

Institute of Engineering in Aragón, I3A, University of Zaragoza.
C\ María de Luna, 3, 50.018, Zaragoza (Spain).
mcanales@unizar.es,jrgalleg@unizar.es,anhersol@unizar.es,toni@unizar.es

Abstract. Radio resource management and QoS provision in Mobile Ad hoc NETworks (MANETs) require the cooperation among different nodes and the design of distributed control mechanisms, imposed by the self-configuring and dynamic nature of these networks. In this context, in order to solve the tradeoff between QoS provision and an efficient resource utilization, a distributed admission control is required. This paper presents an adaptive admission procedure based on a cross-layer QoS Routing supported by an efficient end-to-end available bandwidth estimation. The proposed scheme has been designed to perform a flexible parameters configuration that allows to adapt the system response to the observed grade of mobility in the environment. The performance evaluation has shown the capability of the proposal to guarantee a soft-QoS provision thanks to a flexible resource management adapted to different scenarios.

1 Introduction

Nowadays applications heavily demand the fulfillment of their Quality of Service (QoS) requirements. In Mobile Ad hoc NETworks (MANETs), characterized by a self-organizing capacity that leads to a distributed operation and a great dynamism, the complexity to solve the tradeoff between an efficient resource utilization and the required resource reservation according to the application demands points to an efficient design of distributed admission mechanisms.

The role of routing in a cooperative environment, where the establishment of multihop routes is required, is especially relevant in order to define the admission criteria [1]. A routing decision based on a metric that reflects the resources availability allows to identify the system ability to provide the demanded quality of service. However, this measurement highly depends on the resource management performed by the Medium Access Control (MAC) level, which suggests the collaboration between both layers (cross-layer [2], [3]) as a promising solution.

---

[*] This work was financed by the Spanish Government (Project TEC2004-04529/TCM from MEC and FEDER), Gobierno de Aragón for WALQA Technology Park and the European IST Project PULSERS Phase II (IST - 027142)

Please use the following format when citing this chapter:

Canales, M., Ramón Gállego, J., Hernández-Solana, Á., Valdovinos, A., 2007, in IFIP International Federation for Information Processing, Volume 245, Personal Wireless Communications, eds. Simak, B., Bestak, R., Kozowska, E., (Boston: Springer), pp. 1–12.

In addition, the inherent dynamism of mobile ad hoc networks involves new difficulties that require more flexible admission mechanisms in order to provide a resource management adapted to the topology-variant scenarios.

In this context, this paper describes an adaptive cross-layer proposal that allows to perform a distributed admission procedure with a flexible reallocation that responds to the different grades of mobility of the environment, triggering accordingly the appropriate configuration mode. Therefore, the proposed architecture guarantees an end-to-end quality of service provision with an appropriate adaptation to the variability of MANETs.

The remaining of the paper is organized as follows. Section 2 presents the cross-layer architecture, detailing the basis of the MAC protocol and the designed admission procedure based on a proposed QoS Routing. The mechanisms to adapt the operation to a mobility scenario are shown in Section 3. The proposal scheme has been evaluated through simulations and results are shown in Section 4. Finally, some conclusions are provided in Section 5.

2 The cross-layer proposal

2.1 Efficient MAC protocol

In this proposal, a MAC TDMA layer based on the ADHOC MAC protocol has been considered [4]. ADHOC MAC works on a slot synchronous physical layer and implements a distributed access technique capable of dynamically establishing a reliable single-hop Basic broadcast CHannel (BCH) for each active terminal. Each BCH carries signaling information, including priorities, which distributes layer-two connectivity information and resources occupation to all the terminals. In response to the QoS demanded by multimedia applications, the MAC level efficiently allocates resources by exploiting the in-band signaling provided by the protocol. The reservation strategy is based on the use of the BCH capabilities to signal the request before the access, in such a way that collisions can be theoretically avoided (Book In Advance Scheme - BIAS). According to the demanded resources, a user signals its attempts to reserve in addition to the occupied resources. Upon reception of different requests for the same resource, each node signals the decision about the possible reservation (even if it is not the destination) according to a common criterion. The soliciting node receives this information in the corresponding BCHs in the next frame. If there are not collisions, the resource is effectively reserved – Fig. 1. Preemption can be also carried out in order to allocate resources for high priority services despite the lower priority ones. The rules used to resolve the conflicts are explained in detail in [4]. In order to overcome the problems stemmed from the dynamic nature of a realistic environment, where the actual interference of all the active users is considered, some modifications to the basic operation of the protocol have been carried out [5]. As a result, the designed MAC protocol becomes a stable and reliable support for the cross-layer architecture that provides a suitable estimation of the available bandwidth and allows to efficiently reserve the required resources.

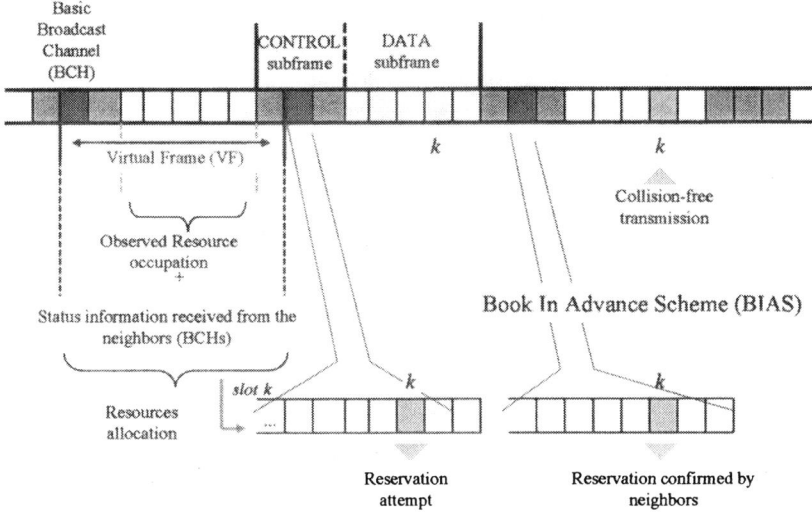

Fig. 1. Book in Advance Scheme (BIAS). Example of basic operation.

2.2 Admission control based on QoS Routing

A Distributed Admission Control (DAC) mechanism has been developed by defining a bandwidth metric required to establish a reliable path when trying to allocate a new application. According to this criterion, the routing protocol needs to find a connection path capable of satisfying the QoS requirements in terms of the demanded resources. The tradeoff between an efficient resource utilization and the required reservation that provides the desired QoS has a promising solution in the collaboration between the MAC level, responsible for the network resource management, and the routing protocol. The definition of a QoS metric that properly reflects the resource availability needs a correct estimation provided by the MAC level. In addition, in a TDMA structure, the end-to-end bandwidth cannot be estimated with the bottle-neck link (minimum local bandwidth), since links are not independent [6]. The hidden terminal problem imposes the disjoint property in three consecutive links requiring a global network perspective instead of a local vision [3], which can be achieved by including the end-to-end available bandwidth estimation in the routing process.

In this work, the Ad hoc On-demand Distance Vector (AODV) routing protocol [7] has been adapted to include BWC-FA (BandWidth Calculation - Forward Algorithm), a modified version of the proposal in [3], in order to measure the available bandwidth considering the whole path. The main features of this algorithm are described next. Additional information regarding slot availability is appended to the routing messages (request-RREQ, reply-RREP). During the discovery process, the proposed algorithm finds the available TDMA slots that can be used for transmitting in every link along the path so that these slots, if reserved, would be interference-free. The distributed bandwidth calculation is supported by the measurements in the MAC level reflected in the QoS routing metric, which defines the available end-to-end bandwidth in the partial path from the source to each intermediate node, that updates the metric and evaluates if the QoS constraints are met in order to forward the RREQ. The final value calculated in the destination node represents the maximum available bandwidth in the whole path. Repeated RREQs are not directly dropped in the destination in order to select the best path according to this final metric. The decision considers the path with the minimum number of hops that supports the highest bandwidth. During the reply phase, information regarding the specific available resources is appended in the RREP in order to appropriately select the necessary slots to cover the QoS demands. Upon selection, the BIAS mechanism of the ADHOC MAC protocol is performed to ensure their reliable reservation. A more detailed explanation of the algorithm can be found in [5].

The specific identification of the available resources in every link makes it possible an estimation of the end-to-end available bandwidth closer to the real resource occupation than the classical local link measurement of the bottle-neck. An admission decision according to this measurement and the reservation of the specified resources increase the probability of admitting connections with the guarantee of a fulfillment of the QoS requirements [5]. The improvement in the effective capacity is the result of the establishment of guaranteed virtual circuits.

3 Operation in a mobility scenario

3.1 Reallocation and readmission procedures

The DAC mechanism allows to better allocate resources for the demanding applications without congesting the network over its capacity. However, the variability in the network conditions leads to a different scenario from that when the active connections were admitted. The established paths in the new topology may be unviable to provide the expected QoS, degrading the corresponding admitted connections. This degradation can be due to two differentiated effects: the initially non-colliding applications can interfere each other leading to unexpected collisions reducing the experienced bandwidth, or the lost of connectivity between neighbors may affect active paths, which will suffer from broken links and the consequent packet dropping. In any case, it is required to initiate a readmission capable of ensuring the QoS provision demanded by the admitted applications. The reallocation criterion, however, should depend on the nature of the suffered degradation. Fig. 2 shows the proposed admission scheme. The implemented procedures are summarized in this paper, but they are fully described in [8]. In this figure, P_b represents the probability of blocking a new admission and P_d is the dropping probability of admitted connections. In addition, due to the distributed nature of the developed procedures, some degraded applications are not correctly identified and they are maintained in the system. This, in fact, is equivalent to dropping for the perspective of the end user. According to this assumption, the failure probability P_f also includes the completed connections that do not experience the required QoS.

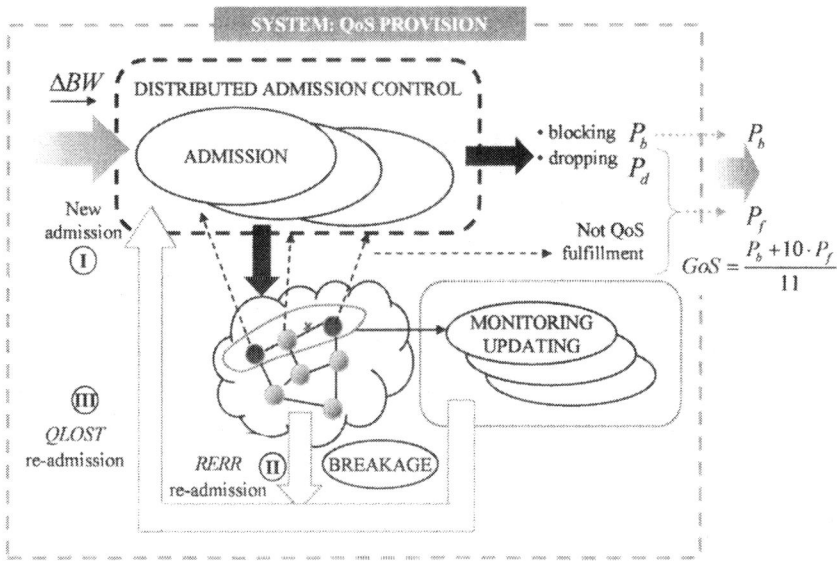

Fig. 2. Admission procedures. (I: new admissions, II-III: re-admission after breakage and QLOST)

As in the normal operation of the AODV, nodes react to broken links sending error messages that trigger a new discovery in the source of the connection. However, if the affected path requires certain QoS, a new attempt to allocate resources may fail due to the new topology which would lead to drop the connection increasing P_d. To reduce this probability, the QoS constraints are relaxed during the discovery including best-effort paths. If the connection is finally readmitted, a flexible resource reallocation or a new QoS readmission could finally provide the expected QoS, as it is explained next.

A proposed QoS monitoring process, based on the BWC-FA algorithm, is performed appending certain routing information piggy-backed in the DATA-ACK packets with a configurable updating period (t_{update}). The operation, similar to that in the RREQ-RREP exchange, allows to realize if the QoS constrains are not met anymore. If additional available bandwidth is found, new resources can be reserved without new discoveries, performing a flexible reallocation procedure. However, if the QoS cannot be maintained (after $n_{failupd}$ updates), a QLOST (QoS LOST) packet is sent to the source in order to trigger the discovery of a new path capable of satisfying the demanded bandwidth. This new discovery does not relax the routing constraints in order to find only a new QoS path. However, despite the degradation of the QoS, the former path is still viable to send traffic as best-effort, so it is maintained to avoid dropping packets in excess while discovering the new path. If it is found, the QoS is recovered. At the end, despite a temporary lost, certain soft-QoS is achieved allowing an acceptable performance. Nevertheless, if the changes in topology make unfeasible to reallocate the connection with the demanded QoS ($t_{qlost,max}$ seconds without the required bandwidth), it is dropped in order to free resources for other applications. The configuration of the specified parameters can modulate this operation in order to define the correct response of the system according to the scenario (Table 1).

Table 1. Admission – Readmission procedure

		QoS flows (QoS Routing - BWC-FA)			best-effort
		I	II	III	always
Admission criterion[a]		I	II	III	always
Discovery tries	$n_{disc,max}$	1		1	if $n_{pq} > 0$[b]
RREQs dropping		✓	-	✓	-
Waiting in dest.		t_{wait}	0	t_{wait}	0
Guard bandwidth		ΔBW	0	0	0
Available priority[c]		p_1	p_0	p_0	p_2
(preemptable resources[d])		(p_2)	(p_2, p_1)	(p_2, p_1)	-

[a] I: new connections, II: rediscovery after breakage y III: rediscovery after $QLOST$.
[b] Queued packets.
[c] Priority of the reserved resources if the connection is admitted. $p_2 < p_1 < p_0$
[d] Reserved resources that can be reallocated for the connection to allow its admission.

In order to ensure a satisfactory Grade of Service ($GoS = (P_b + 10 \cdot P_d)/11$ [9]), a new attempt to allocate resources for an admitted connection must be preferential. Therefore, a flexible priorization is performed in the MAC level in such a way that new applications cannot preempt resources reserved in virtual circuits, even during the best-effort operation (readmission after QLOST).

In addition to this basic readmission scheme, an additional parameter, ΔBW, can be configured to reduce the dropping probability. When a highly dynamic scenario leads to continuous degradations in the experienced QoS for the admitted connections, frequent new discoveries arise to readmit the disrupted ones. If the admission procedure allocates all the available resources, the variations in the admission conditions make it difficult to find sufficient alternative QoS paths to accommodate all these connections, which finally can increase the dropping probability. In order to provide extra free resources to flexibly reallocate the admitted connections when they are disrupted due to mobility, not only the demanded resources but also a guard bandwidth ΔBW are required during the initial admission. However, if admitted, only the necessary resources are effectively reserved. This ensures certain unused bandwidth available when required during a readmission.

3.2 Adaptive configuration

Despite the flexibility provided thanks to the reallocation and readmission procedures, the implemented mechanisms introduce control overhead that competes with the user data. In some cases, this occupation of priority resources can excessively reduce the bandwidth availability. In addition, when extra bandwidth is underutilized (function of ΔBW), this reduction can be even more representative. Despite the better grade of service, a reduction in the effective capacity does not represent a global improvement in the network performance. The benefit of a flexible resource management becomes significant when a high grade of mobility leads to continuous route breakages or bandwidth degradations that make the maintenance of the admitted QoS connections unfeasible. According to the analysis of several scenarios, two different operation modes have been selected:

- **Quasy-Static OPeration mode (QSOP):** In static or low mobility scenarios the control overhead degrades the effective capacity with a slight improvement in the GoS. Therefore, the QoS monitoring, the dropping capability and ΔBW are not configured.
- **Mobility OPeration mode (MOP):** A flexible resource management due to the whole readmission procedure (including ΔBW) allows to improve the grade of service without a significant reduction in the effective capacity. The appropriate configuration has been selected thanks to an exhaustive analysis of the application scenarios (section 4): $n_{failupd} = 3$, $t_{update} = 0.3$ sec. and $t_{qlost,max} = 3$ sec.

Therefore, an appropriate configuration is highly dependent on the grade of mobility, which makes it difficult to define the adequate readmission parameters without a previous knowledge of the application scenario. In order to carry out a correct operation in a dynamic scenario it could be more interesting to adaptively configure the response of the system. To this purpose, an adaptation mechanism based on the estimation of the grade of mobility in the environment is proposed.

Estimation of the grade of mobility

The selected MAC protocol bases its operation on a correct measurement of the stable neighbors. The maintenance of this connectivity information allows to realize the existence of link disconnections, which are more likely to happen with a higher grade of mobility. Thus, the variability of the number of neighbors that the MAC level observes can be a good approximation of the dynamism in the surrounding area. This local measurement can represent the mobility in the network in deployment scenarios with geographically uniform movement, such as those modeled through the Random WayPoint (RWP). A metric according to this idea is defined as next:

$$var_{NB,i}^{filt}(t) = \alpha \cdot var_{NB,i}^{filt}(t - \Delta t) + (1 - \alpha) \cdot var_{NB,i}(t) \quad (1)$$

$$var_{NB,i}(t) = \left| \frac{NB_i^{est}(t) - NB_i^{est}(t - \Delta t)}{\Delta t} \right| \quad (2)$$

where NB_i^{est} is the number of stable neighbors for node i, $var_{NB,i}(t)$ is the temporal variation of this measurement and the metric $var_{NB,i}^{filt}$ is the result of filtering this variation in order to estimate the mean value of $var_{NB,i}$, which is directly related to the grade of mobility. Δt has been selected to 1 sec. The parameter α should solve the tradeoff between determining an accurate estimation of the average value of $var_{NB,i}$ and a rapid response to its sudden variations. Due to the oscillations in the measurement, an stable configuration must be performed according to two defined thresholds (hysteresis):

- th_1: $var_{NB,i}^{filt}$ to trigger the MOP configuration.
- th_2: $var_{NB,i}^{filt}$ to change from the MOP to the QSOP configuration.

The value th_1 must be high enough to correctly identify the mobility environment avoiding the full readmission procedure in quasy-static scenarios, and it is related to the expected average $var_{NB,i}$ for the minimum mobility that requires a MOP configuration. The value th_2 tries to maintain an stable configuration of the readmission parameters in a mobility scenario without considering the metric variations, which must be absorbed by $(th_1 - th_2)$.

4 Performance evaluation

The proposed architecture (cross-layer QoS Routing - QoSR-CL) has been evaluated by means of simulation. With this purpose, we have built up an event-driven simulator in C++ which implements all the required functionalities. A set of 50 nodes are randomly positioned within a square area of 2 Km². Terminals follow a Modified Random WayPoint mobility model (MRWP), whose parameters are defined in Table 2. Speeds are measured in Km/h, time in sec. and distances in meters (topology grid: X_{max}, Y_{max}, equal to 1400 m, and $D_{max} = \sqrt{(X_{max}^2 + Y_{max}^2)}$). The additional parameters β_{max} and R_{max} define the maximum variation from the current direction and the maximum distance to the next destination for a new movement, which allows to follow a path closer to the natural movement. The simulation length has been selected in order to guarantee a stable topology distribution of the terminals according to the different mobility scenarios.

Table 2. Mobility scenarios. MRWP model.

Random Waypoint				MRWP	
V_{min}	V_{max}	T_{pausa}	V_m	β_{max}	R_{max}
1	3	2	~2	$\frac{\pi}{12}$	$0.1 \times D_{max}$
3	20	2	~9	$\frac{\pi}{12}$	$0.1 \times D_{max}$
10	40	2	~20	$\frac{\pi}{12}$	$0.1 \times D_{max}$

The connectivity among terminals is determined by the ability of decoding the BCH transmissions according to the received SIR, considering a transmitted power of 20 dBm, a Kammerman propagation model [5] and a minimum decoding threshold SIR_{th} of 5 dB. The sensing threshold CS_{th} is set to 1 dB.

An additional margin of 3.5 dB is selected to define the stable neighbors from the MAC level perspective [5]. Connections between different pairs of nodes are generated according to a Poisson process with rate [connections/sec.] ranging according to the simulated offered load and the mean connection duration (50 sec). QoS flows are generated as CBR sources demanding a bandwidth of 128 kbps (2 TDMA slots). Packets are early discarded to avoid an excessive delay. The tolerable packet loss P_{loss} is set to 3%, which defines the correctly dispatched traffic as the occupied bandwidth by the number of connections that ends with a throughput higher than 97%.

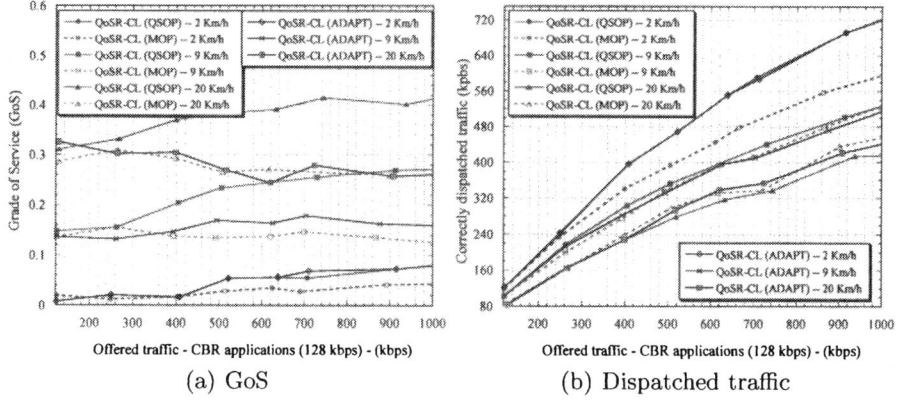

(a) GoS (b) Dispatched traffic

Fig. 3. Operation modes {QSOP, MOP and adaptive ($th_1 = 0.09$, $th_2 = 0.03$)} for different scenarios.

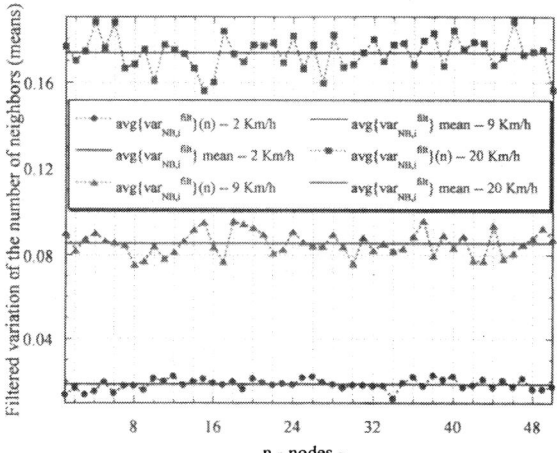

Fig. 4. Average filtered variation of the estimated number of neighbors. $\alpha = 0.99$.

The evaluation of the proposal in mobility conditions has shown the difficulties to guarantee the QoS in this dynamic scenario. As the average node speed increases, the number of correctly dispatched connections falls due to a growing packet dropping probability, since frequent route breakages or emerging interference disrupt the initially established virtual circuits. However, the correct configuration of the full readmission procedure (MOP operation) allows to improve the global performance in terms of grade of service (Fig. 3(a)) although only in scenarios with higher mobility (20 Km/h) the control overhead can be compensated with an increase in the effective capacity (Fig. 3(b)), thanks to a reduction of the failure probability (P_f). At lower speeds, P_f is not so significant and it cannot be extremely reduced, thus the increasing blocking probability due to both the overhead and the unused resources (ΔBW) has a negative effect over the effective capacity.

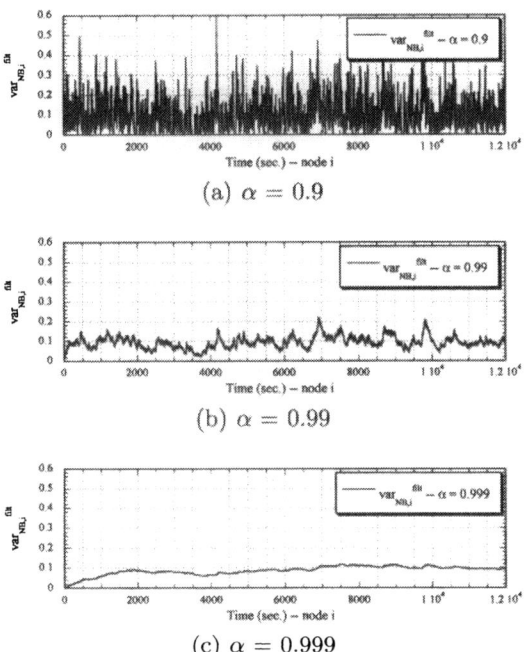

(a) $\alpha = 0.9$

(b) $\alpha = 0.99$

(c) $\alpha = 0.999$

Fig. 5. Evolution of $var_{NB,i}^{filt}$ with different values of parameter α.

In order to guarantee a correct operation in any scenario, adaptively configuring the operation mode (QSOP or MOP), the mobility metric $var_{NB,i}^{filt}$ is defined. As it is shown in Fig. 4, this estimation provides a clear differentiation among scenarios. The selection of $\alpha = 0.99$ in Fig. 4 comes from the observed results represented in Fig. 5. Lower values (Fig. 5(a)) generate a more fluctuating metric that can complicate the desired differentiation. On the contrary, an excessive memory (Fig. 5(c)), despite the better average estimation, implies a slower convergence inadequate to adapt the system response to potential changes in the mobility environment. The selected value provides a good approximation to the average $var_{NB,i}$ with a rapid adaptation, as it is observed in Fig. 5(b). Due to the metric oscillation, the adaptation is made according to a pair of thresholds (th_1 and th_2).

Results observed at 9 Km/h show this grade of mobility as an inflection point between both operation modes (QSOP, MOP). At lower speeds, the overhead of the MOP configuration cannot be assumed whereas, at a higher mobility, the global performance is improved thanks to the more flexible resource management. The mean and standard deviation of $var_{NB,i}^{filt}$ in the scenario of 9 Km/h (0.0865 and 0.035 respectively) have been considered as a reference. Finally, after the evaluation of different configurations, results have shown the convenience of a selection $\{th_1 = 0.09 - th_2 = 0.03\}$. The resulting ratio of time that corresponds to a MOP configuration is 0, 0.86 and 0.98 for 2, 9 and 20 Km/h respectively. As it is shown in Fig. 3, this provides an adaptation near the best performance in any mobility scenario.

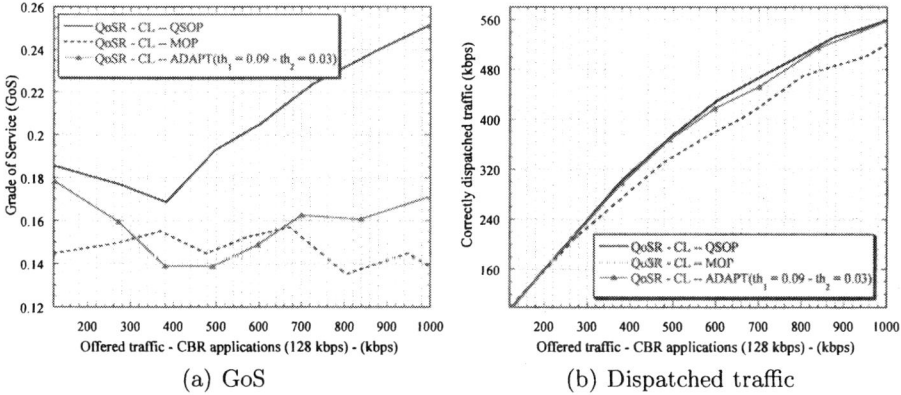

Fig. 6. Performance in a scenario with variable grade of mobility.

Obviously, a previous configuration adapted to the specific grade of mobility provides a more adjusted definition of the readmission parameters. However, an adaptive solution allows the system to auto-configure the procedure without a prior knowledge of the mobility scenario. In addition, variable mobility conditions can make it difficult to properly define the best operation mode. In order to evaluate the proposed solution in a mobility variant environment, a more dynamic scenario has been analyzed. The mobility pattern follows the MRWP model with three consecutive grades of mobility (2, 9 and 20 Km/h).

The proposed mobility pattern could suggest a MOP configuration as appropriate due to the highest percentage of high mobility (9 and 20 Km/h). However, despite the GoS improvement (Fig. 6(a)) the representative reduction on the effective capacity shows the impact of an incorrect configuration during the quasy-static mobility pattern (Fig. 6(b)). On the contrary, the adaptive readmission procedure allows to obtain a grade of service close to that provided by the MOP configuration thanks to the flexible reallocation (Fig. 6(a)) maintaining the effective capacity, evaluated as the correctly dispatched traffic, as it is shown in Fig. 6(b).

5 Conclusions

A distributed admission control designed for MANETs has been evaluated in different mobility scenarios. A cross-layer architecture based on a QoS Routing allows to allocate different applications according to their demanded end-to-end bandwidth establishing virtual circuits that guarantee the QoS provision. However, the dynamic nature of the environment leads to changes in the topology that modify the admission conditions, which has motivated the proposal of a flexible reallocation scheme. Results have revealed the difficulties to properly configure this mechanism in different mobility conditions. A proposed mobility estimation based on the measurement of the neighbors variability allows to implement an adaptive parametrization that performs a flexible management capable of guaranteeing a reasonable performance in different scenarios.

References

1. Chen, S., Nahrstedt, K.: An overview of quality-of-service routing for the next generation high-speed networks: Problems and solutions. IEEE Network Magazine (1998) 64–79
2. Zhou, B., Marshall, A., Wu, J., Lee, T.H., Liu, J.: A cross-layer route discovery framework for mobile ad hoc network. EURASIP Journal on Wireless Communications and Networking 5 (2005) 645–660
3. Zhu, C., Corson, M.: QoS routing for mobile ad hoc networks. In: Proc. IEEE INFOCOM'02. Volume 2., New York, USA (2002) 958–967
4. Gállego, J.R., Canales, M., Hernández-Solana, A., Campelli, L., Cesana, M., Valdovinos, A.: Performance evaluation of point-to-point scheduling strategies for the ADHOC MAC protocol. In: Proc. of WPMC'05, Aalborg, Denmark (2005) 1380–1384
5. Canales, M., Gállego, J.R., Hernández-Solana, A., Valdovinos, A.: Performance evaluation of cross-layer routing for QoS support in mobile ad hoc networks. Springer LNCS (IFIP PWC'06) **4217** (2006) 322–333
6. Chen, T.W., Tsai, J., Gerla, M.: QoS routing performance in multihop, multimedia, wireless networks. In: Proc. of IEEE ICUPC'97. Volume 2., San Diego, CA. USA (1997) 557–561
7. Perkins, C., Belding-Royer, E., Das, S.: Ad hoc on-demand distance vector (AODV) routing. Experimental RFC 3561, The IETF Network Working Group (2003) http://www.ietf.org/rfc/rfc3561.txt.
8. Canales, M., Gállego, J.R., Hernández-Solana, A., Valdovinos, A.: Performance analysis of cross-layer QoS routing for mobile ad hoc networks. In: Proc. of WPMC'06, San Diego, USA (2006) 946–950
9. Dimitriou, N., Tafazolli, R., Sfikas, G.: Quality of service for multimedia CDMA. IEEE Communications Magazine **38**(7) (2000) 88–94

Analysis of a TDMA MAC Protocol for Wireless Ad Hoc Networks under Multipath Fading Channels*

José Ramón Gállego, María Canales, Ángela Hernández-Solana, and Antonio Valdovinos

Institute of Engineering in Aragón, I3A, University of Zaragoza.
C\ María de Luna, 3, 50.018, Zaragoza (Spain).
jrgalleg@unizar.es,mcanales@unizar.es,anhersol@unizar.es,toni@unizar.es

Abstract. The behavior of MAC protocols for wireless ad hoc networks is conditioned by multipath fading, especially with regard to the links stability and the differences between channel errors and collisions due to interference. In this paper, we evaluate the impact of multipath fading on the performance of a TDMA proposal designed to provide QoS in wireless ad hoc networks. It lies in a frame subdivision that consists of a broadcast control phase and an information phase. We propose modifications to guarantee the reliability and efficiency of both broadcast control and data services.

1 Introduction

QoS provision in wireless ad hoc networks requires the support of efficient MAC protocols to solve the tradeoff between resource reservation and channel utilization. In this context, several synchronous approaches have been proposed in order to provide an efficient TDMA access capable of providing resource reservation [1-5]. The basic idea behind most of them lies in a reservation cycle with a given number of control mini-slots that allows a node to reserve a conflict-free information slot [1-4]. ADHOC MAC [5] is a proposal that does not require a reservation cycle based on mini-slots to provide access to a broadcast or point-to-point TDMA slot. Using slots of the same size in the whole frame partially alleviates the complexity of the synchronization. It implements a distributed access technique capable of establishing a reliable single-hop Basic broadcast Channel (BCH) for each active terminal. Each BCH carries signaling information that distributes layer-two connectivity information to all the terminals. Upon the basis of the ADHOC MAC protocol, we have proposed a MAC frame structure that consists of two subframes: A control subframe where terminals contend for a BCH in order to access the system and a data subframe where active terminals can allocate data communications in a contention-free situation [6].

* This work was financed by the Spanish Government (Project TEC2004-04529/TCM from MEC and FEDER), Gobierno de Aragón for WALQA Technology Park and the European IST Project PULSERS Phase II (IST - 027142)

Please use the following format when citing this chapter:

Ramón Gállego, J., Canales, M., Hernández-Solana, Á., Valdovinos, A., 2007, in IFIP International Federation for Information Processing, Volume 245, Personal Wireless Communications, eds. Simak, B., Bestak, R., Kozowska, E., (Boston: Springer), pp. 13-24.

In the evaluation of MAC protocols for wireless ad hoc networks it is often assumed that transmissions that are not sensed do not contribute to the total level of interference a terminal suffers (*Protocol Model* – [7]). Under these conditions, totally collision-free transmissions are possible with the appropriate MAC signaling. However, when the actual interference produced by all the transmitting terminals is taken into account (*Physical Model* – [7]), some assumptions carried out under the *Protocol Model* are no longer valid and collisions can still occur. Moreover, the impact of multipath fading hardly ever is taken into account [8], but it highly conditions the behavior of MAC protocols, especially with regard to the links stability and the differences between errors due to channel conditions and collisions due to interference.

In this paper, we analyze the impact of multipath fading on the performance of the proposed MAC structure. We propose solutions to guarantee the reliability of the protocol in this scenario, including mechanisms to provide link stability and power control procedures that highly increase the spatial reuse. The remaining of the paper is organized as follows. In Section 2 we provide an overview of the considered MAC structure. In Section 3 we show the effect of fading in the protocol operation and the required modifications. In Section 4, simulations results of the proposal are presented. Finally, in Section 5 some conclusions are provided.

2 MAC Protocol Structure and Operation

ADHOC MAC operates with a time slotted structure, where slots are grouped into virtual frames (VF) of length N, and no frame alignment is needed. For a correct operation, the ADHOC MAC needs that each active terminal has assigned a Basic CHannel (BCH), corresponding to a slot in the VF, which is a reliable single hop broadcast channel.

In the BCH, each terminal broadcasts information about the status of the channel it perceives. The BCH contains a control field, the Frame Information (FI). The FI is an N-elements vector specifying the status of the N slots preceding the transmission of the terminal itself. The slot status can be either BUSY if a packet has been correctly received or transmitted by the terminal or FREE otherwise. In the case of a BUSY slot, the FI also contains the identity of the transmitting terminal. Based on the received FIs, each terminal marks a slot, say slot k, either as RESERVED or AVAILABLE. It is RESERVED if slot k-N is coded as BUSY in at least one of the FIs received in the slots from k-N to k-1, and AVAILABLE, otherwise. If a slot is AVAILABLE, it can be used for new access attempts. Upon accessing an AVAILABLE slot, terminal j will recognize, after N slots (a frame), its transmission either *successful*, if the slot is coded as 'BUSY by terminal j' in all the received FIs, or *failed*, otherwise.

Once a terminal has acquired a BCH, it can establish additional data communications with all its neighbors by exploiting this distributed signaling provided by the FIs. In order to efficiently manage the available resources we consider a frame subdivision into N_{BCH} slots for new accesses and N_{Data} slots for data links among the active terminals. In [9] we propose an adaptive scheme for this subdivision, but in the remaining of the paper, we will consider a static subdivision. In order to establish point-to-point data communications in the data subframe avoiding the exposed terminal problem, each entry of the FI encloses a Point-To-Point (PTP) flag. A terminal sets the PTP flag of a given slot ON if the packet received in the slot is a broadcast one or the terminal itself is the destination. A point-to-point communication can use all the AVAILABLE slots. Further on, even some RESERVED slots can be used if it is satisfied that the PTP flag is signaled OFF in all the received FIs and the FI received from the destination terminal signals the slot as FREE. The whole set of slots that a terminal can select for PTP connections to a given destination will be referred as AVAILABLE PTP slots for that destination.

When dealing with multimedia applications, in response to the demanded QoS, the MAC level must efficiently allocate resources for several differentiated services. To this purpose, we have proposed a reservation scheme [6] to handle the access to the N_{Data} slots by exploiting the in-band signaling provided by the BCH that can provide user differentiation through the use of priorities. The basis of this strategy (Book In Advance Scheme - BIAS) relies on the use of the BCH capabilities to signal the request before the access, in such a way that collisions can be theoretically avoided. The details can be found in [6].

3 Operation under Fading Channels

The features of the described protocol guarantee under a simplified *Protocol Model* that collisions cannot occur once a slot is reserved. However, considering a *Physical Model*, where the interference produced by all the transmitting terminals is taken into account, a transmission is only considered successful if (1) is satisfied:

$$SIR_{rx,i,j}^k = \frac{P_{tx,i}^k \cdot h_{i,j}}{\sum_{\substack{n \in N_{tx}^k \\ n \neq i}} P_{tx,n}^k \cdot h_{n,j} + P_n} > SIR_{th} \qquad (1)$$

where $SIR_{rx,i,j}^k$ is the SIR (Signal to Interference Ratio) received by terminal j from terminal i in slot k, $P_{tx,i}^k$ is the power transmitted by user i in slot k, $h_{i,j}$ represents the propagation channel between users i and j, N_{tx}^k is the set of transmitting terminals in slot k, P_n is the thermal noise and SIR_{th} is the minimum required SIR to decode the information.

In addition to the path loss, dependent on the distance between transmitter and receiver ($L_{i,j}$), the propagation channel $h_{i,j}$ includes two random variations in a mobile environment:

- *Slow fading* or shadowing, due to the terrain variability and often modeled through a lognormal random variable added to the path loss.

- *Fast fading* due to multipath propagation, characterized by a Rayleigh distribution when there is not line of sight and with a spectral variation measured through the Doppler spectrum.

Thus, the collision-free property of the reserved slots, theoretically provided in an ideal scenario, cannot be ensured any more due to both channel variations in the received signal and the interference of distant nodes, that can make the received SIR go down the SIR_{th} in a reserved slot. In order to identify the slots where some power can be sensed, although the terminal is not capable of decoding any information, we include in the FI an additional slot, DIRTY. A terminal cannot transmit in a slot that the potential receiver signals as DIRTY, assuming that the reservation may fail due to interference.

3.1 Solutions for the broadcast control service

The maintenance of the BCH in the ADHOC MAC protocol requires the transmitting terminal i to receive the acknowledgment (slot 'BUSY by terminal i') in all the received FIs from its neighbors. According to this constraint, errors due to channel conditions can induce a high variability in the network activity as a consequence of nodes trying to reallocate a BCH. The control information carried in the BCH is the basis to perform an appropriate resource allocation and to maintain updated connectivity information, thus this variability makes it more difficult to perform the resource management.

In order to guarantee the correct operation of the broadcast service and to properly choose the nodes suitable to establish reliable point-to-point data links, a set of stable neighbors must be defined. With this purpose, two main problems must be solved. First of all, a criterion to consider a terminal as a stable neighbor is required. Then, when errors in the received packets occur, a mechanism to differentiate a collision due to an interference increase from adverse channel conditions must be provided.

Definition of stable neighbors

A stable neighbor should guarantee a packet error rate in the communications below certain threshold. In order to fulfill this requirement, the transmissions must be provided with a security margin to overcome the signal and interference variations. An additional threshold P_{rx-min} is used to verify the reliability of a link. This value is considered as the minimum received power to provide a margin of ΔSIR dB over SIR_{th} in the absence of any interference. Since this ΔSIR margin must absorb signal variations due to channel conditions with regard to $\overline{P_{rx}}$ (mean value of the received power), this value has to be known to determine the reliability of a neighbor. Thus, before considering a terminal as a stable neighbor, the received power in the BCH must be averaged for a given number of frames. With this purpose, we consider a first order IIR filter:

$$\overline{P_{rx}(n)} = \alpha \cdot \overline{P_{rx}(n-1)} + (1-\alpha) \cdot P_{rx}(n) \qquad (2)$$

where $\overline{P_{rx}(n)}$ is the mean signal power calculated in frame n, $P_{rx}(n)$ the power received in frame n and α the parameter that determines the tradeoff between filtering of fast fading and tracking of slow variations in the mean level of the received signal. The information about the useful received signal P_{rx} can be obtained from the estimated SIR [10] and the total received power.

Thus, we consider that a terminal is a stable neighbor when $\overline{P_{rx}(n)} > P_{rx,min}$ in the BCH. Since the estimation of the received power varies around the real mean level, those nodes with a mean power around $P_{rx,min}$ can enter in an oscillating dynamic 'stable non stable' that can complicate the establishment of links with them. In order to avoid this situation, an hysteresis mechanism is incorporated, in such a way that once $\overline{P_{rx}(n)} > P_{rx,min}$ and the terminal is considered as a stable neighbor, the estimation must be ΔP_{est} dB lower than $P_{rx,min}$ to stop being a stable neighbor $(\overline{P_{rx}(n)} < P_{rx,min} - \Delta P_{est})$.

Collision Detection

Fading effects can lead to receive a packet from an stable neighbor with $SIR < SIR_{th}$ even when the total level of interference is zero. According to the rules of the protocol, an error is always understood as a collision and a rescheduling of the BCH in a different slot is needed. However, when errors are caused by fading, the reallocation of the transmission does not solve the problem, since while the fading lasts, the quality of the received signal is low in any slot. In fact, continuous reschedulings due to channel errors lead to an important increase in the network instability. As a consequence, collision detection cannot be carried out through instantaneous values, but mean SIR values. We consider that a terminal suffers a collision if the average SIR goes down while the average received power keeps constant, i.e., if there is an increase in the mean level of interference.

The mean value of the received signal to interference ratio, $\overline{SIR}$, is obtained with the same filter as (2), applied over the estimated SIR values on a frame by frame basis. Then, when an error in the reception of the BCH from stable neighbor occurs (i.e., $\overline{P_{rx}} > P_{rx,min}$), it is checked whether $\overline{SIR} < SIR_{collision}$. Only if both conditions are fulfilled, the receiver will treat this error as a collision and will report it to the transmitter so that this one reallocates the transmission, as it is shown in figure 1. In order to differentiate a transmission error from the report of a collision (DIRTY status), an additional status in the FI is required in order to notify a negative ACK that does not require a slot reallocation. We name that status as ERROR, in such a way that terminal i will maintain its slot BCH as long as it observes the slot as 'BUSY or ERROR by terminal i' in all the received FIs from its stable neighbors.

The election of the specific value for $SIR_{collision}$ implies a tradeoff in the protocol design. High values, near the SIR that all the stable neighbors must guarantee when there is no interference ($SIR_{th} + \Delta SIR = P_{rx,min} - P_n$ in dB) lead to mistakes between channel errors and collisions. On the other hand, low values near SIR_{th} can disguise the existence of a collision, leading to keep on transmitting in a slot where the interference level has notably grown, with the subsequent increase in the packet error rate.

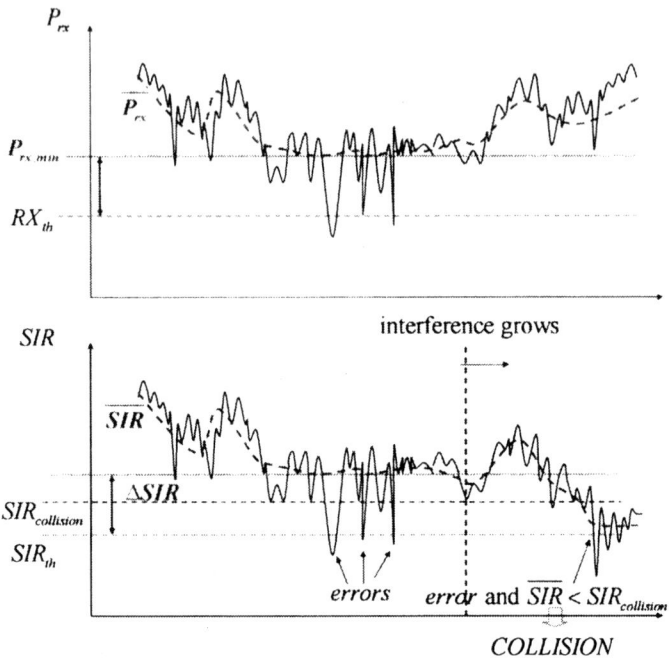

Fig. 1. Example of collision detection under fading.

3.2 Solutions for the point to point data service

We consider that point to point data communications can only be established between stable neighbors in order to guarantee the reliability of the dispatched connections. In these transmissions, differentiation between channel errors and collisions is also required. Therefore, in order to detect a collision in a data slot k, a terminal must calculate the average SIR in that slot, $\overline{SIR_{data,k}}$, in the same way as with the BCH. Thus, when there is an error in a data packet and $\overline{SIR_{data,k}} < SIR_{collision}$ a collision will be reported (DIRTY in the FI), while if $\overline{SIR_{data,k}} > SIR_{collision}$ only a negative acknowledgment will be sent (ERROR in the FI).

Power control has been shown to increase spatial reuse leading to an improvement in the network capacity [7]. The signaling information distributed by all the neighbors in their FI at a fixed transmission power, P_{tx-max}, supplies the basis to implement power control procedures in our proposed MAC structure [11]. This power control mechanism is based on estimating the propagation channel from the received power in the BCH of the destination terminal and, assuming symmetrical links, using this estimation to calculate the required transmission power. Since both terminals transmit in the same carrier frequency and with the same bandwidth, we can consider the bidirectionality to be valid. However, under time varying multipath channels, this scheme presents a main problem, consisting of the validity of the estimation carried out in a specific time to be applied in a later time. As terminal speeds grows, the previous estimation is less correlated with the actual channel, making unviable this mechanism. Therefore, the chosen alternative to solve these problems lies in carrying out the power control according to the mean value of the channel loss, in such a way that the transmission power must provide an additional margin to protect the signal against the fast fading. Moreover, this mean value can be directly obtained from the mean received power, $\overline{P_{rx}}$, already calculated for the management of the stable neighbors.

A slot can be used to establish a connection if the receiver signals the slot as FREE and there is not any receiving terminal in the surroundings of the transmitter. The adjustment of the transmission power allows to reduce the interference over distant connections, thus reducing the number of DIRTY slots, and consequently increasing the number of accessible FREE slots, which can lead to an improvement in the network capacity.

We assume that terminal i wants to establish a connection with terminal j in slot k, signaled as FREE by j. According to the estimated information about the channel loss, $\overline{h_{i,j}}$, terminal i can set its transmission power $P_{tx,i}^k$, on a frame by frame basis, to provide a target SIR (SIR_{tar}), which must guarantee a margin to overcome variations in the received signal and in the interference level:

$$P_{tx,i}^k = \begin{cases} \frac{SIR_{tar} \cdot P_n}{\overline{h_{i,j}}} & \text{if } P_{tx,i}^k \leq P_{tx-max}, \\ P_{tx-max} & \text{if } P_{tx,i}^k > P_{tx-max}. \end{cases} \quad (3)$$

where the mean channel loss from i to j, $\overline{h_{i,j}}$ is estimated by means of the mean received power from j to i, $\overline{P_{rx,j,i}}$: $\overline{h_{i,j}} = \overline{h_{j,i}} = \frac{\overline{P_{rx,j,i}}}{P_{tx-max}}$.

A more sophisticated procedure to increase the spatial reuse can be carried out by including in the FI additional information about the mean level of interference that a terminal estimates in each slot, $\hat{P}_{int}^k$, at the expense of increasing the control information sent in the BCH. In that case, the transmitter can access slot k regardless of being DIRTY or BUSY if:

$$P_{tx,i}^k = \frac{SIR_{tar} \cdot \left(P_n + \overline{\hat{P}_{int}^k}\right)}{\overline{h_{i,j}}} \leq P_{tx-max} \quad (4)$$

4 Performance Evaluation

The impact of multipath fading in the proposed MAC architecture has been evaluated by means of simulation. With this purpose, we have built up an event driven simulator in C++ which implements all the required functionalities. 400 terminals are randomly positioned within a square area of 1 Km^2. The considered mobility pattern is based on the well-known *Random Way-Point* (RWP) mobility model. The RWP is modified in such a way that each new destination of the model is not completely random, but it must not suppose a change in the current direction of the terminal higher that some specific angle ($\pi/12$ in the simulations). Each terminal generates PTP data communications according to a Poisson process with intensity X (PTP connections/s). The source of each point-to-point communication is randomly chosen among the users with an active BCH, while the destination is randomly chosen among the source's neighbors. The duration of each communication is exponentially distributed with mean D = 50 (frames). Likewise, each connection generates 1 packet per frame (CBR). X and D define the point-to-point offered traffic by each terminal. Each frame consists of 75 slots ($N_{BCH} = 50$, $N_{Data} = 25$), transmitted each 20 ms at 11 Mbps. SIR_{th} is set to 5 dB, P_n is −103 dBm and CS_{th} (Carrier Sense Threshold) is −102 dBm. Path loss is $L = -128.1 - 37.6 \cdot \log_{10}(d)$ dB (d in Km). The maximum transmission power, P_{tx-max}, is set to −7.5 dBm, which provides a maximum transmission range of 100 m. Log-normally distributed shadowing with standard deviation of 6 dB is also included. Two multi-path fading environments (pedestrian and vehicular - [12]) have been considered.

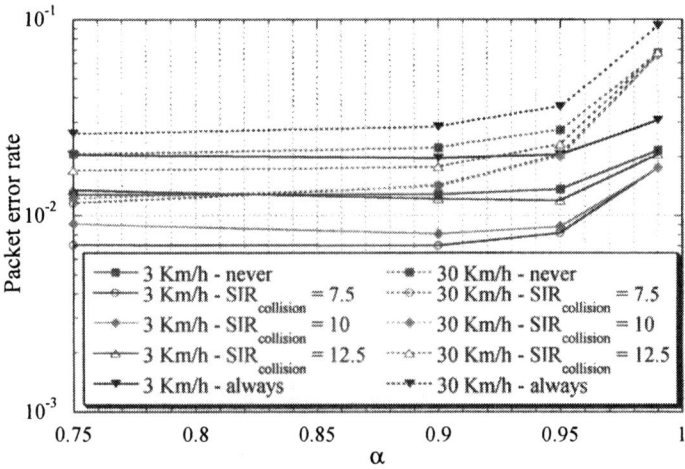

Fig. 2. Packet error rate for the stable neighbors in the broadcast service. $\Delta SIR = 10$ dB. Different values of $SIR_{collision}$.

Figure 2 shows the packet error rate for the broadcast service according to α (filtering of the received power) for different values of $SIR_{collision}$ with a margin against fading and interference of 10 dB (ΔSIR) and $\Delta P_{est} = 3$ dB. This error rate is defined as the ratio between wrong and total receptions in stable neighbors. Two extreme situations regarding collision detection are included: **never** detecting collision, regardless of the estimated SIR and **always** that an error is detected reporting a collision. High values for α (0.99), lead to an important increase in the error rate: This rate is related to the errors in stable neighbors, whose definition is given by the filtering of the received power. With high values for α, the estimated average power $\overline{P_{rx}}$, not only filters fast fading, but also the changes in channel loss due to terminals movement or shadowing, in such a way that a neighbor can be considered as stable despite not providing the required margin. With regard to $SIR_{collision}$, the extreme cases (always and never) provide the worst performance. If all errors are treated as collisions, the frequent reallocations of the BCH lead to an scenario of permanent contention for the resources, which increases access collisions and as a result, the error rate. When collisions are not detected, transmissions in slots suffering high levels of interference can persist, also increasing the error rate. In this scenario, a $SIR_{collision}$ of 7.5 or 10 dB provides a good tradeoff for both 3 and 30 km/h.

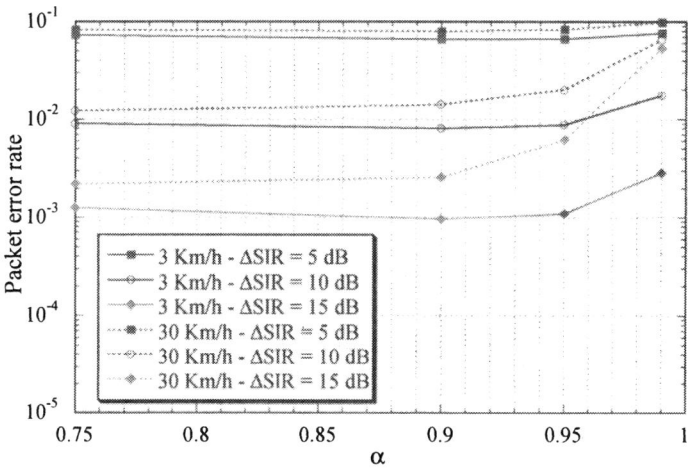

Fig. 3. Packet error rate for the stable neighbors in the broadcast service. Different values of ΔSIR.

Table 1. Mean duration of stable links (sec).

α	Speed	
	3 Km/h	30 Km/h
0.75	3.2	3.2
0.9	9.6	6.1
0.95	16.3	7.6
0.99	32.1	13.8

With an analogous analysis for ΔSIR of 5 and 15 dB, $SIR_{collision}$ values of 7.5 and 12.5 dB also provide a good tradeoff with regard to the collision detection. Figure 3 shows similar results, in this case for different values of ΔSIR (5, 10 and 15 dB), fixing the $SIR_{collision}$ that provides best performance in each case (7.5, 10 and 12.5 dB). Logically, the higher ΔSIR, the lower the error rate. However, an increase in ΔSIR implies a reduction in the *effective* coverage area, and consequently, in the number of terminals allowed to establish data communications. Thus, excessively high values for ΔSIR can lead to a higher probability of unconnected network. In the evaluation of point to point data services, a $\Delta SIR = 10$ dB has been chosen as a tradeoff between effective coverage area and reliability (error rate).

Regarding the election of α, in addition to the error rate, the capacity of reliably determining the set of stable neighbors is also important in order to establish efficient data communications. Table 1 shows mean duration of stable links for different values of α. The estimation of $\overline{P_{rx}}$ is more resistant against channel variations as α grows, and therefore, the duration of stable links is higher. However, as it is seen in figures 2 and 3, $\alpha = 0.99$ or even 0.95 for high speeds lead to wrong estimations in the stable links that make error rate grow. Therefore, we choose $\alpha = 0.9$ in order to evaluate data services.

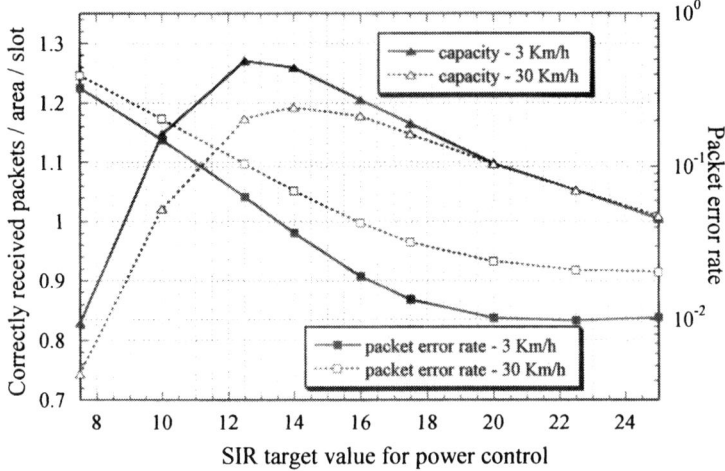

Fig. 4. Power control behavior for different SIR_{tar} values. Mean PTP offered traffic = 20 connections/terminal.

Figure 4 shows the capacity and the error rate for point to point data services obtained for different SIR_{tar} values when only FREE slots are used to access. The capacity is normalized to the number of data slots (N_{Data}) and the maximum coverage area (range: 100 m). Power control requires a tradeoff between reliability and capacity. The error rate notably decreases as the SIR_{tar} grows, specially up to 18 or 20 dB, since the available margin is higher. However, the maximum capacity is obtained around $SIR_{tar} = 12$ dB, and decreases for higher vales since the spatial reuse diminishes. We consider a tradeoff value of 17.5 dB.

Figure 5 compares the performance without power control with two different alternatives: basic mode with access only in the FREE slots (*Basic*) and access also in the BUSY or DIRTY slots according to the interference information distributed in the FIs (*Interference-Aware*), all of them with a mean terminal speed of 30 Km/h. Power control techniques involve a slight increase in the packet error rate, since the margin against fading and interference is reduced with regard to the situation where each terminal transmits at the maximum available power. However, in spite of this fact, the number of connections correctly dispatched is significantly higher thanks to the increase in the spatial reuse, specially when interference information ($\hat{P}_{int}^{k}$) is used.

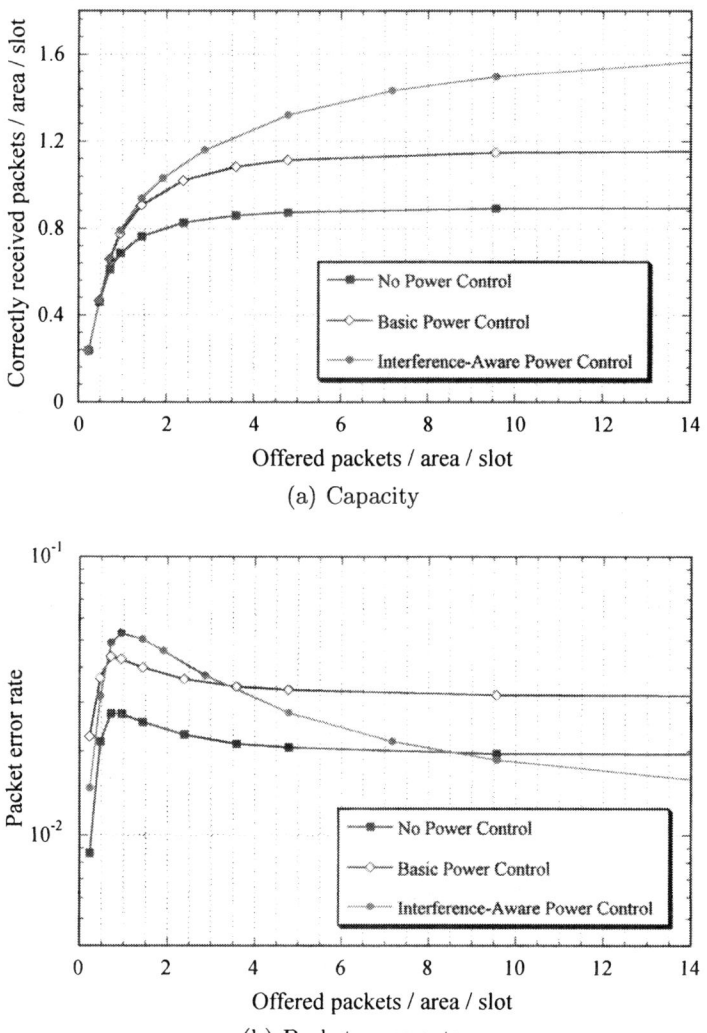

Fig. 5. Capacity and packet error rate for point-to-point data services with different strategies. $SIR_{tar} = 17.5$ dB for power control.

5 Conclusions

In this paper, we have analyzed the impact of multipath fading on the performance of a MAC proposal for QoS provisioning in wireless ad hoc networks based on a TDMA access. We have proposed solutions to guarantee the reliability and stability of both broadcast control and data services, thanks to the definition of a set of stable neighbors based on the average received power. Differentiation between errors due to channel conditions from collisions due to an interference increase has been provided. Finally, the validity of power control procedures to increase the network capacity has been shown.

References

1. Tang, Z., Garca-Luna-Aceves, J.J.: A protocol for topology-dependent transmission scheduling in wireless networks. In: Proc. of IEEE WCNC'99. Volume 3., New Orleans, USA (1999) 1333–1337
2. Zhu, C., Corson, M.S.: A Five-Phase Reservation Protocol (FPRP) for mobile ad hoc networks. Wireless Networks (WINET) **7**(4) (2001) 371–384
3. Fang, J.C., Kondylis, G.D.: A synchronous, reservation based medium access control protocol for multihop wireless networks. In: Proceedings of IEEE WCNC'03. Volume 2., New Orleans, USA (March 2003) 994–998
4. Ahn, C.W., Kang, C.G., Cho, Y.Z.: Soft reservation multiple access with priority assignment (SRMA/PA): A distributed MAC protocol for QoS-guaranteed integrated services in mobile ad-hoc networks. IEICE Transactions on Communications **E86-B**(1) (January 2003) 50–59
5. Borgonovo, F., Capone, A., Cesana, M., Fratta, L.: ADHOC MAC: a new MAC architecture for ad hoc networks providing efficient and reliable point-to-point and broadcast services. Wireless Networks (WINET) **10**(4) (2004) 359–366
6. Gállego, J.R., Canales, M., Hernández-Solana, A., Campelli, L., Cesana, M., Valdovinos, A.: Performance evaluation of point-to-point scheduling strategies for the ADHOC MAC protocol. In: Proc. of WPMC'05, Aalborg, Denmark (2005) 1380–1384
7. Gupta, P., Kumar, P.R.: The capacity of wireless networks. IEEE Transactions on Information Theory **46**(2) (2000) 388–404
8. Mullen, J., Huang, H.: Impact of multipath fading in wireless ad hoc networks. In: Proc. of ACM PE-WASUN'05, Montreal, Canada (2005) 181–188
9. Gállego, J.R., Campelli, L., Cesana, M., Capone, A., Borgonovo, F., Hernández-Solana, A., Canales, M., Valdovinos, A.: Efficient bandwidth allocation for basic broadcast and point-to-point services in the ADHOC MAC protocol. In: Proc. of IFIP PWC'05, Colmar, France (2005) 87–98
10. Lau, F., Tam, W.: Novel SIR-estimation-based power control in a CDMA mobile radio system under multipath environment. IEEE Transactions on Vehicular Technology **50**(1) (2001) 314–320
11. Gállego, J.R., Canales, M., Hernández-Solana, A., Valdovinos, A.: Performance analysis of an interference-aware MAC protocol with power control for wireless ad hoc networks. In: Proc. of IEEE PIMRC'06, Helsinki, Finland (2006)
12. 3GPP: Universal mobile telecommunications system (umts); selection procedures for the choice of radio transmission technologies of the umts (umts 30.03 version 3.2.0). Technical Report 101 112 V3.2.0, 3GPP (1998-04)

A QoS Architecture Integrating Ad-Hoc and Infrastructure in Next Generation Networks

Marek Natkaniec[1], Janusz Gozdecki[1], Susana Sargento[2]

[1] Department of Telecommunications AGH University of Science and Technology,
Al. Mickiewicza 30, Cracow, Poland
[2] Instituto de Telecomunicações, Universidade de Aveiro, Portugal
{natkaniec@kt.agh.edu.pl, gozdecki@kt.agh.edu.pl, ssargento@det.ua.pt}

Abstract. This paper proposes the complete QoS architecture for integration of ad-hoc with infrastructure networks. The technology, service differentiation mechanisms, and signaling protocols are discussed. The modules required in the network elements and its integration to provide end-to-end QoS in mobile ad-hoc networks are presented. The proposed solution is based on the SWAN model with some extensions to provide L2 differentiation for four traffic classes and supports the integration with infrastructure networks. The deployed hierarchical architecture guarantees scalability and make possible per-flow resource management in wireless access where scarce radio resources should be managed effectively, and per-aggregate traffic management using a DiffServ model in the core.

Keywords: Ad-hoc networks, Infrastructure networks, Quality of Service, Integration, Service Differentiation, Signaling protocols

1 Introduction

Wireless Local Area Networks (WLANs) are one of the fastest growing areas of modern telecommunications today. They can be installed in places that are very difficult to wire as, for example, trading floors, manufacturing facilities, warehouses or historical buildings. WLANs are being widely implemented in many venues from markets and airports to retail, manufacturing, hospitals and corporate environments; they are beginning to be available in public spaces such as schools, hotels, restaurants, malls and shops. This technology offers the highest level of performance and capability features among other local wireless solutions. WLANs play a very important role in the network architecture as a provider of easy and unconstrained access to the wired infrastructure. Currently ad-hoc networking is becoming a promising solution to increase the radio coverage of broadband wireless systems, extending coverage of hotspots. This business strategy is profitable for both the provider and the user. Since a radio range is strongly affected in closed spaces or in areas with dense radio interferences, the resilience provoked by the multi-hop characteristics of mobile ad-hoc networks makes these especially appropriate to provide increased radio coverage with low cost and easy deployment. Therefore, ad-hoc networks play an increasing role in network access.

Ad-hoc networks differ clearly from the traditional cable infrastructure. This kind of networks is characterized by very dynamic changes of topologies and hence their design requires special attention. To support the present users and service requirements, the ad-hoc network needs to support differentiated QoS, which is a major challenge. The protocols assuring QoS in ad-hoc networks need to operate in a distributed way along the ad-hoc nodes, with proper mechanisms for reacting in a responsive way to any changes (e.g., topology, new sessions, congestion). There are some solutions for QoS support in ad-hoc networks: SWAN (Stateless Wireless Ad-hoc Networks) [4], INSIGNIA [6], FQMM (Flexible Quality of service Model for Mobile ad-hoc networks) [5], and DS-SWAN (Differentiated Services-SWAN) [17]. DS-SWAN supports end-to-end QoS in ad-hoc networks connected to fixed DiffServ domains. The QoS proposal addressing the integration of ad-hoc networks and infrastructure networks was also deployed within the confines of Daidalos I project [1] and published in [7]. The work presented in this paper continues the concept of ad-hoc and infrastructure networks integration with complete end-to-end QoS support for four traffic classes. Following Daidalos I architecture, the QoS model in the wireless ad-hoc network is based on the extended SWAN approach. The QoS solution developed in Daidalos I was simplified by exploiting IEEE 802.11e technology, therefore introducing QoS differentiation at layer 2. The ad-hoc QoS mechanism is integrated with the NSIS signaling protocols suite [10] in the infrastructure network, which provides more flexibility in end-to-end signaling supporting different resource management models, and makes possible to set up bidirectional reservation. For out-of-path QoS signaling in infrastructure network a Diameter [11] protocol is used. Due to build-in security mechanisms it can be used for both inter and intra administrative domain signalling.

The paper is organized as follows. The proposed QoS architecture as well as the mobile node and gateway node schemes are presented in Chapter II. Chapter III contains the details about the SWAN protocol extensions for ad-hoc network integration including: signaling and dynamic regulation process, QoS differentiation and MAC layer measurements. The signaling protocols used in infrastructure are presented in Chapter IV. At the end of the paper, the conclusions are showed.

2 Network QoS Architecture

The QoS architecture presented in this paper is the continuation of work provided in IST Daidalos project [1] on QoS architecture in Next Generation Networks (NGN). The framework of NGN is recently under study by the most important standardization organization including International Telecommunication Union – Telecommunication Standardization Sector (ITU-T) Study Group 13 (SG13) and Focus Group NGN (FGNGN), and European Telecommunications Standard Institute (ETSI) TISPAN [8] initiative. Currently NGN standards groups become tightly coupled with mobile groups – SG13 and FGNGN with SG19, and TISPAN with 3GPP. The results of joint efforts can be found e.g. in [10], where integration of 3GPP QoS architecture in NGN is proposed.

In comparison to Daidalos I the main changes in the QoS architecture are:
1. supporting new signalling protocols, adding more flexibility in session setup,
2. support of Local Mobility Management,
3. tighter integration of broadcast and multicast,
4. integration of terminal multihoming,
5. new QoS solutions for the access network including ad-hoc.

In this paper the focus is put on QoS mechanisms in ad-hoc networks and integration of ad-hoc with infrastructure.

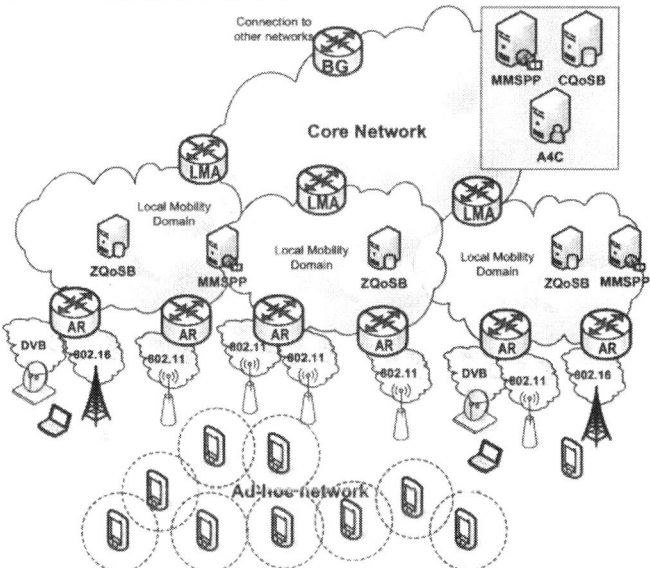

Fig. 1. QoS hierarchical architecture of a complete administrative domain.

In Fig. 1. the QoS architecture of a complete administrative domain is presented. The general QoS Daidalos II architecture follows the Daidalos I hierarchical network, where three main levels are distinguished. The top level makes up a core network, responsible for inter operator domains resource management and interconnecting second level Local Mobility Domains (LMD). The low level consists of access networks built in different technologies. The most important access network types from Daidalos point of view are IEEE 802.11 including ad-hoc and infrastructure mode, 802.16, DVB, and WCDMA.

In the Daidalos architecture, QoS resource management is divided into several areas, according to the structure of the network. The hierarchical architecture guarantees scalability and makes possible per-flow resource management in wireless access where scarce radio resources should be managed effectively. Whilst in the core network per aggregate traffic management using DiffServ [12] model of resource management is provided.

The mobility scheme deployed in the project is very important for QoS architecture due to its impact on QoS network structure. The mobility management is divided on

Global Mobility Management where MIPv6 is used and Local Mobility Management. The protocol supporting local mobility LMP (Local Mobility Protocol) is designed in Daidalos, and is based on concepts proposed by IETF NetLMM [13], a hierarchical mobile IP [14] and the IEEE 802.21 proposed standard [15]. The main mechanisms of LMP are implemented in the network, but some of L2 mechanisms, like IEEE 802.21 signalling, have to be deployed in mobile node (MN). To support QoS in LMD (Local Mobility Domain) the QoS solution must be tightly coupled with LMP.

Inside LMD a hand-over between access routers (ARs) does not require the change of CoA in MN – from network layer point of view the MN does not perform any L3 address reconfiguration of interface. The MN only re-associates to a new access point (AP) with the same IP address. The entity controlling this process in LMD is LMA (Local Mobility Anchor). LMD can integrate several heterogeneous access networks. Although from point of view of end-to-end signalling the session does not change, the QoS on links between LMA and ARs must be maintained; QoS in access network has also to be appropriately controlled. The main entity managing QoS in LMD is the ZQoSBr (Zone QoS Broker), which controls all routers in the domain. ZQoSBr is also a policy enforcement point of A4C (Authentication, Authorization, Accounting, Auditing and Charging) subsystem, controlling access to the network by QoS mechanisms deployed in ARs. In ARs there are interfaces responsible for QoS management in access networks (AN). To unify the access to different access technology mechanism the RAL (Radio Access Layer) was designed.

The entity responsible for QoS requirements signalling for application using SIP protocol is MMSPP (Multimedia Service Provisioning Proxy). MMSPP is responsible for extracting QoS requirements from QoS extended SIP protocol and signaling application requests to ZQoSBr.

To really make benefits of ad-hoc network, the integration with infrastructure and end-to-end QoS resource management is required. In Daidalos II architecture the interoperation between ad-hoc and infrastructure solution is provided in AR and MN. The physical interconnection between ad-hoc and infrastructure is performed in the AR. The interaction between ad-hoc and mechanisms independent of access network technology are performed in the QoS Client (QoSC) module in the MN. QoSC provides general QoS API for applications to reserve network QoS resources for application data flows, which is independent of access network technologies.

2.1 Ad-hoc QoS Architecture

The ad-hoc mobile node has a double role, acting as a host which produces and consumes application traffic, and acting as a router that forwards the traffic of other nodes. The mobile node needs to be able to retrieve the QoS parameters from the application characteristics, triggers the check for QoS resources along the ad-hoc path, and checks the available resources in its wireless medium. It can also classify and mark the packets according to its class, ensure QoS differentiation, mark ECN bits and detect ECN marked packets in the case of congestion (See Fig. 2a).

The retrieval of the application QoS parameters is performed by the QoS Client; the Ad-hoc QoS Controller checks for QoS resources along the ad-hoc path and the available resources in its wireless medium. This is performed through the interaction with the MAC Measurement Module. The classification and marking of the packets is addressed in the Classification and (Re)-Marking module, and the QoS differentiation is realized by the hardware queues implemented in WLAN cards based on Atheros chipset (L2 differentiation) [9]. To address congestion situations in the ad-hoc network, the node has an ECN Marking module that obtains the congestion status information from the MAC Measurement module and marks the ECN bits to trigger the dynamic regulation of the flows.

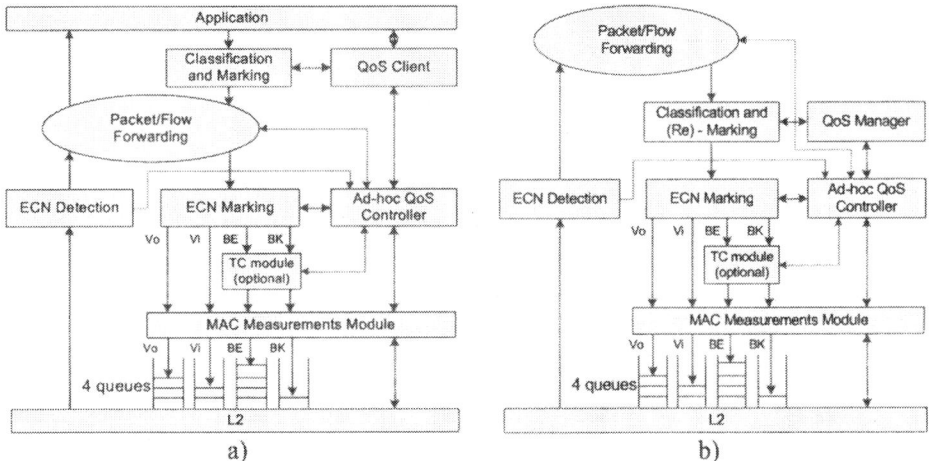

Fig. 2. Mobile (a) and gateway (b) node schemes.

The gateway is able to support the same functionalities of the mobile nodes, but does not have interaction with the application signalling (since it works only at the IP layer and below) (see Fig. 2b). Instead, it needs to perform interoperation between the QoS signalling in the ad-hoc and the infrastructure side.

2.2 Ad-hoc Network Technology Requirements

WLAN cards based on IEEE 802.11 a/b/g/e standards are considered within MANET of Daidalos II project [2], [3]. The IEEE 802.11e standard with MAC layer QoS support should be used to perform L2 service differentiation. The IEEE 802.11 a/b/g defines the number of physical layer types with maximum rates: 2 Mbps in 2.4 GHz band (IEEE 802.11), 11 Mbps in 2.4 GHz band (IEEE 802.11b), 54 Mbps in 2.4 GHz band (IEEE 802.11g), and 54 Mbps in 5 GHz band (IEEE 802.11a). The basic IEEE 802.11 standard was designed to operate in DCF (Distributed Coordination Function) and PCF (Point Coordination Function) modes. DCF is the fundamental access method used to support asynchronous data transmission on the best effort basis. IEEE 802.11b with DCF mode was used in Daidalos I project. As previously referred, MAC Measurement Module (MMM) is responsible for QoS measurements (L2 frames). These measurements are reported to L3 QoS modules to perform L3 service differentiation. MAC layer measurements are crucial to obtain L3 QoS.

IEEE 802.11e standard supports EDCA (Enhanced DCF Channel Access) which allows for L2 service differentiation. EDCA opens various parameters for service differentiation configuration, namely: CW_{min}, CW_{max}, AIFS, and TXOP. The EDCA is designed to provide differentiated, distributed channel accesses for frames with different priorities. The QoS support for EDCA can be achieved statistically by reducing the probability of medium access for lower priority traffic categories through the configuration of these parameters. Medium contention rules for EDCA are the same as in 802.11 DCF. Additionally, to achieve high medium utilization, the TXOP parameter was defined. TXOP is the time dedicated to the transmission of consecutive MAC frames of the same station. It is also recommended to always use RTS/CTS frames exchange before the data transmission to minimize the negative effect of hidden stations. L2 service differentiation can speed up the service differentiation process and allows to simplify L3 architecture.

3 SWAN Signaling

Similarly to Daidalos I, our enhanced SWAN concept has been chosen to build Daidalos II architecture due to the following features:
- it supports QoS negotiations and service differentiation for four different traffic classes, i.e., voice, video, best-effort, and background,
- it does not require any per-flow or aggregate state information in the intermediate nodes as it controls all traffic classes locally, with the use of MAC delay measurements,
- admission control is performed only at the source node with the use of the request/response probe which checks the available bandwidth on the path towards the destination.

When congestion/overload conditions occur (e.g., as a result of node mobility) the dynamic regulation of real-time sessions is possible with SWAN. In case a particular mobile node detects such a situation, it starts marking ECN bits in the IP header of real-time packets. When a destination node, monitoring the ECN bits, notices that they are marked, it sends a regulate message to the source so as to force it to re-establish the real-time session by sending a new probing request to the destination.

3.1 QoS Signalling and Dynamic Regulation

If the sender node is an ad-hoc mobile node, it sends an App_Sig Initiation message (message that corresponds to the signalling of the specific application start), and triggers a Probing Request message (Fig. 3). This request contains a Bottleneck Bandwidth (BB) field located in an IPv6 extension header that is updated in a hop-by-hop basis with the minimum available bandwidth of the corresponding class in the path, and the Requested Bandwidth (RB) for the flow (the mobile node includes a QoS client module that maps the application QoS parameters into network QoS parameters). The Probing Request message is updated by every intermediate node in the ad-hoc network with the BB of the corresponding class (i.e., minimum bandwidth in the path). Every mobile node has a L2 measurement module that measures the occupied bandwidth and the delay corresponding to each class in the wireless medium. After receiving this message, the gateway checks the BB and the RB (optionally, it can also check the delay values). If the BB is larger than the RB, this means that the ad-hoc network has sufficient available bandwidth. It replies to the probing request with a Probe Response message with indication of the available bandwidth in the ad-hoc path. Otherwise, an error message is sent to the sender. In the case of available resources in the ad-hoc path, the mobile node sends a NSIS_reserve message, and the gateway checks for the authorisation in the infrastructure network, issuing a QoS request to the ZQoSBr.

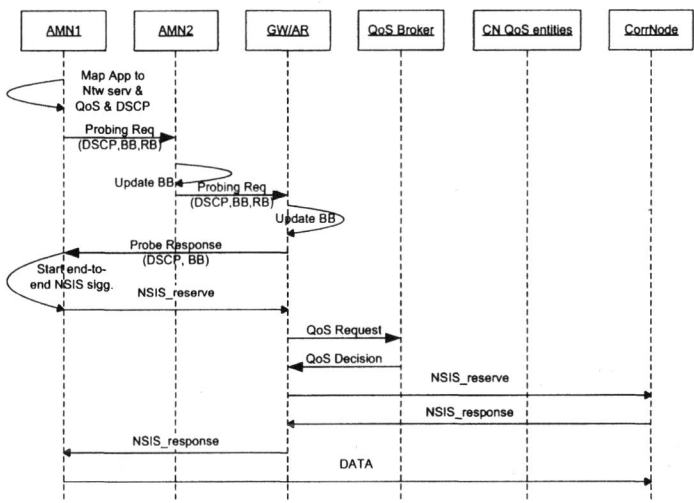

Fig. 3. Ad-hoc initiated session setup.

All out of path core signaling uses a Diameter protocol from the IETF DIME WG [11]. The gateway will request authorization from the ZQoSB that will make two verifications: first, if the user has permission for the service (communication with the A4C); second, if there are available resources for the request. In the case of a positive answer, it forwards the NSIS_reserve message to the receiver (correspondent node – CorrNode) through the core network (CN) entities (CN QoS entities). The correspondent node replies to the NSIS_reserve with an NSIS_response message and, if the session parameters are allowed in the two terminals, the setup process ends with an NSIS_response message to the sender node. If the sender is on the infrastructure side (Fig. 4), the GW recognizes the signalling message, asks the ZQoSBr for available resources in the infrastructure access network and sends a probe request message towards the receiver ad-hoc mobile node. If the probing process in the ad-hoc access network is successful the NSIS signalling is continued.

Fig. 5 depicts the case of dynamic regulation integrated in the infrastructure network. When an ad-hoc node detects an overload condition in a specific class (i.e., the target bandwidth for the class is exceeded), this node starts marking ECN bits in packets of the affected class. The gateway monitors the ECN bits, and upon their detection, notifies the sources by sending Regulate messages. When a source receives a Regulate message it should perform application adaptation, or else, should re-start the probing process.

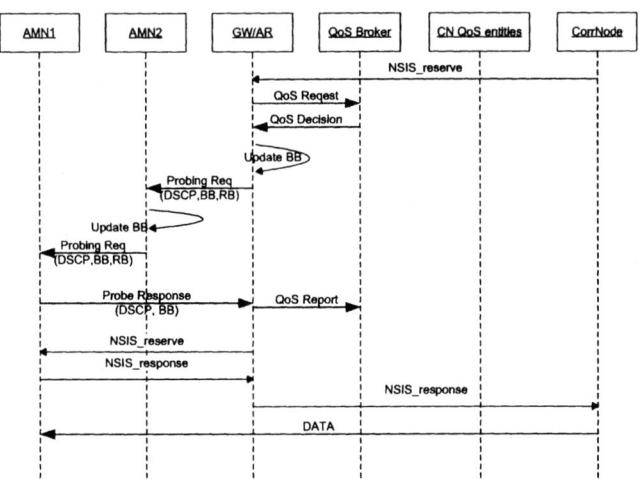

Fig. 4. Infrastructure initiated session setup.

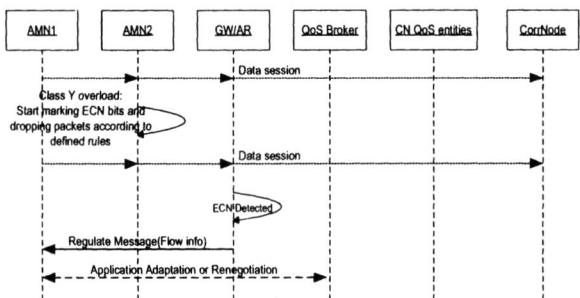

Fig. 5. Dynamic regulation.

3.2 QoS Differentiation

Service differentiation in the original SWAN concept assumes only two service classes, one specified for real-time UDP traffic and another one specified for best-effort TCP traffic. In Daidalos I the SWAN service level differentiation was expanded to four different traffic classes: voice, video, best effort, and background. Each of these classes has associated a set of service attributes. Request for bandwidth allocation for the desired traffic class is issued by a node that requests service at session setup time. The request can later be dynamically adjusted based on feedback from L2, especially for low priority classes. Bandwidth allocation for classes is controlled by rate limiters. It should be pointed that connection renegotiation can be required due to changing network conditions or excessive traffic passing through a mobile node in the ad-hoc network. The extended differentiation model in Daidalos I is composed by a classifier and by a cascade of priority schedulers, shapers and queues associated to each traffic class [7].

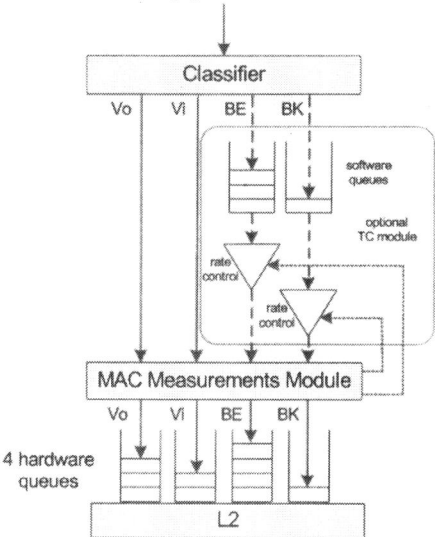

Fig. 6. The proposed Daidalos II ad-hoc service differentiation model.

The IEEE 802.11e WLAN cards possibilities (through hardware queues implementation) give an opportunity to realize L2 service differentiation. It should simplify L3 architecture and allows for faster L2 service differentiation. There is a plan to completely remove TC module (if L2 service differentiation will be sufficient) or use shapers only for two lower priority classes in new Daidalos II architecture. There is no sense to use shapers for conversational real-time services (voice) or streaming real-time services (video). It is better to renegotiate the new (lower) transmission rate (choose other voice or video codec type) for these real-time streams or discard new requests if we overload per class available bandwidth within these high priority classes than shape real-time streams. The shaping of real-time streams (voice or video) is unacceptable for most cases. The differentiation model needs to be complemented by per-class admission control of the two higher priority classes. An AIMD algorithm that has the MAC delay as feedback is proposed to control the shaping rate. The proposed service differentiation model is presented in Fig. 6.

3.3 MAC Layer Measurements

There is a need to obtain the MAC layer measurements information to assure a proper QoS level because fully distributed ad-hoc network is considered in this paper. Every mobile node has to perform some measurements in order to support admission control decisions and service differentiation functions (for two lowest priority classes only if L2 service differentiation will not be sufficient). The following parameters need to be measured: (1) Per class/overall delay – packet delay monitoring for four different classes/overall (from upper layer to the MAC layer and the time of the completion of RTS-CTS-DATA-ACK in EDCA; (2) Per-class/overall bandwidth utilization – achieved by sensing the media and constructing periodic statistics about overall and per class (DSCP code) bandwidth occupancy; (3) Transmission rate - current WLAN card transmission rate (in case of IEEE 802.11a/b/g the stations communicate using transmission rates from 1 to 54 Mbps); (4) Number of stations - the estimation of the number of active stations in the neighbourhood to determine the contention and to evaluate the available bandwidth using current rate information.

4 End-to-end Session Setup Signalling Strategies

To support most typical applications including the legacy and multimedia ones in our QoS architecture, we consider three general signalling strategies: explicit request using NSIS that is described in subsection 3.1; extracting QoS needs from a SIP [16] mediated session initiation, and network triggered. The first two strategies are straightforward: in the NSIS scenario the terminal requests QoS on behalf of the application; in the SIP scenario (similar to the proposed in IMS or TISPAN), the QoS needs are inferred from the session description object. In the third strategy, the network handles the QoS reservation by acting on every starting flow. This means that every packet is classified according to a user profile residing in the network.

All three strategies can coexist in the network but which will be used depends on the application selection and MN capabilities. The first scenario will cover also QoS unaware legacy applications, for which the user will have utility which associates, in real-time, QoS service level with a certain application or flow. This request then triggers NSIS signaling with predefined parameters. In the following subsections the second and third signaling strategies are described shortly; the first scenario is presented in section 3.

SIP Initiated End-to-end QoS Session Setup. The simplified QoS session setup for SIP with ad-hoc integration is presented in Fig. 7. The user application to reserve QoS resources signals its requirements to SIP-UA (*SIP User Agent*) module in MN. Then SIP-UA sends *QOS_request* message to QoSClient module which forwards the request to ad-hoc QoS Controller. Then, the ad-hoc entities check the resources on the path to the infrastructure network by the probing procedure. After positive response SIP-UA sends *SIP_invite* message to the MMSPP module in LMD. MMSPP authorizes the application request using signalling with A4C, and then signals for QoS requirements to the ZQoSBr. ZQoSBr performs QoS resources admission control and after successful procedure sets up pre-reservation state for the session, and then sends the response result to the MMSPP. After successful response MMSPP continues SIP session set-up forwarding *SIP invite* to the corresponding node. After successful QoS reservation in the correspondent node network the MMSPP confirms to ZQoSBr the SIP session establishment. Then ZQoSBr reserves the QoS resources for the session and MMSP signals to SIP-UA the successful session set-up.

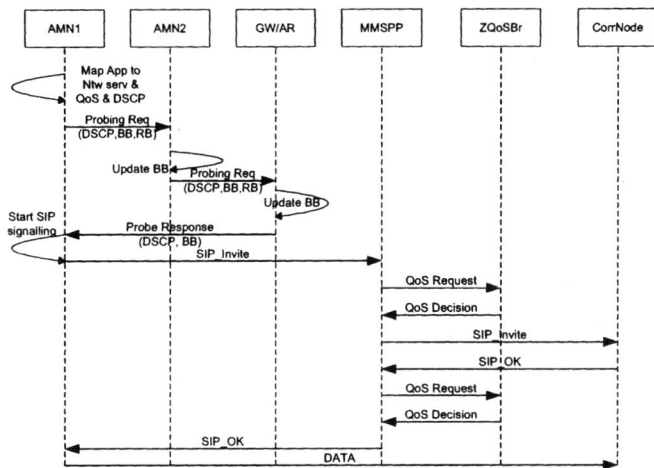

Fig. 7. SIP initiated session setup in ad-hoc.

Network Initiated QoS Reservations. A network initiated reservation is performed in the case when there are no QoS mechanisms implemented in Mobile Terminal. So this case does not support QoS in ad-hoc network due to assumption that the QoS mechanisms are not deployed in MNs. In that case, when AR detects a new flow it triggers QoS signalling with ZQoSBr to check the flow authorisation and QoS service level allocated to the flow in a user profile. To differentiate applications in the user service level agreement (SLA) has to be defined application flows filters and associated QoS service levels.

5 Conclusions

In the article the Daidalos II QoS architecture is described. The main focus is put on integration of ad-hoc networks with infrastructure ones to support end-to-end QoS for multimedia applications.

In the described architecture different schemes of signalling for session setup including SIP and NSIS protocols are supported. For DiffServ off-path signalling the Diameter with QoS extensions is proposed. The QoS mechanisms in ad-hoc networks, which make possible seamless integration with infrastructure, are described. The new IEEE 802.11e standard is used for L2 QoS differentiation. To the best authors knowledge the proposed QoS architecture for ad-hoc and infrastructure network integration exemplify one of the few complete solutions (the second after Daidalos I with four traffic classes support) for end-to-end QoS support in mobile ad-hoc networks.

This proposed architecture is being partially simulated in ns-2 and fully implemented in the Linux OS. Our future work concerns the validation of this QoS architecture through simulations and real experiments.

Acknowledgments. The authors wish to thank the partners of the Daidalos II Consortium, in particular partners of WP2.4 and WP3.3 for their collaborative work.

Disclaimer. The work described in this paper is based on results of IST FP6 Integrated Project DAIDALOS. DAIDALOS receives research funding from the European Community's Sixth Framework Programme. Apart from this, the European Commission has no responsibility for the content of this paper. The information in this document is provided as is and no guarantee or warranty is given that the information is fit for any particular purpose. The user thereof uses the information at its sole risk and liability.

References

1. Daidalos IST Project: "Designing Advanced Interfaces for the Delivery and Administration of Location independent Optimised personal Services". (FP6-2002-IST-1-506997)
2. IEEE 802.11 Standard for Wireless LAN: Medium Access Control (MAC) and Physical Layer (PHY) Specification, New York, IEEE Inc., 1999.
3. IEEE 802.11e: MAC Quality of Service Enhancements, New York, IEEE Inc., Nov. 2005
4. Gahng-Seop Ahn, A. T. Campbell, A. Veres, and Li-Hsiang Sun: Supporting Service Differentiation for Real-Time and Best-Effort Traffic in Stateless Wireless Ad-Hoc Networks (SWAN). In IEEE Trans. on Mobile Comp., vol. 1, no. 3, pp. 192-207, July 2002
5. Hannan Xiao, Winston K.G. Seah, Anthony Lo, and Kee Chaing Chua: A Flexible Quality of Service Model for Mobile Ad-Hoc Networks. In Proceedings of the IEEE Vehicular Technology Conference, Tokyo, Japan, pp. 445-449, May 2000
6. S. B. Lee et al., INSIGNIA: An IP-Based Quality of Service Framework for Mobile Ad-Hoc Networks J. Parallel and Distrib. Comp., Special issue on Wireless and Mobile Computing and Communications, vol. 60 n°4, pp. 374-406, Apr 2000
7. S. Crisóstomo, S. Sargento, M. Natkaniec, N. Vicari, A QoS Architecture Integrating Mobile Ad-Hoc and Infrastructure Networks - Workshop on Internet Compatible QoS in Ad-Hoc Wireless Networks, Egypt, 3-6 Jan 2005
8. A. Głowacz, M. Natkaniec, S. Sargento, S. Crisostomo: MAC Layer Measurements for Supporting QoS in IEEE 802.11 Ad-Hoc Networks – Workshop to QoS, held in conjunction with the Networking Conference, Coimbra, Portugal, 15-19 May 2006
9. MADWiFi – Multiband Atheros Driver for WiFi, http://madwifi.org
10. IETF, Next Steps in Signaling (nsis) charter: http://www.ietf.org/html.charters/nsis-charter.html
11. IETF, Diameter Maintenance and Extensions charter: http://www.ietf.org/html.charters/dime-charter.html
12. S. Blake (ed) et al., An Architecture for Diff. Services, IETF RFC 2475, Dec. 1998
13. Soliman et al., Hierarchical Mobile IPv6 Mobility Management (HMIPv6), RFC4140, IETF, august 2005
14. IETF, Network-based Localized Mobility Management charter: http://ww.ietf.org/html.charters/netlmm-charter.html
15. Draft Standard for Local and Metropolitan Area Networks: Media Independent Handovers Services (Draft .00). IEEE, March 2006
16. J. Rosenberg, et al, "SIP: Session Initiation Protocol", RFC 3261

17. M.C. Domingo, D. Remondo: Quality of Service Support in Wireless Ad Hoc Networks Connected to Fixed DiffServ Domains, Proc. IFIP TC6 9th International Conference on Personal Wireless Communications PWC 2004, Delft, The Netherlands, September 2004

Federating Personal Networks over Heterogeneous Ad-hoc Scenarios

Luis Sánchez, Jorge Lanza, Luis Muñoz

University of Cantabria, E.T.S. Ingenieros Industriales y de Telecomunicación, Avda. de Los Castros s/n, 39004, Santander, SPAIN

{lsanchez, jlanza, luis}@tlmat.unican.es

Corresponding author:
Luis Sánchez.
Departamento Ingeniería de Comunicaciones, E.T.S. Ingenieros Industriales y de Telecomunicación, Avda. de Los Castros s/n, 39004, Santander, SPAIN.
E-mail: lsanchez@tlmat.unican.es

Abstract. This paper will present the specification of the solutions and mechanisms that support the creation of a Federation between Personal Networks that are located in the same area and thus can establish a direct link between each other. The idea behind the Personal Network concept is that the user's personal devices organize themselves in a secure and private network independently of their geographical location or the access technologies used. Nevertheless, in order to fully exploit the benefits of this concept, it is necessary to complement it with the mechanisms that enable the interaction between several people with a shared objective (e.g. a hobby, common project, etc.). This concept of collaboration between several Personal Networks receives the name of Personal Network Federation. This collaboration can be established in multiple scenarios but in this paper the focus will be on the case in which clusters of nodes belonging to different users are located in the same area and they can establish links on peer-to-peer manner.

Keywords: *Personal Network, Federation, Secure Association, Heterogeneity*

1 Introduction

Personal communications have experienced an extraordinary boost in the recent years. One of the novel paradigms that have appeared is the Personal Network (PN), being an emerging concept which combines pervasive computing and strong user focus. PNs are a relatively new concept 1 that allows a user to transparently interconnect all his personal devices independently of their location (e.g. in the personal area, at home, at work or in his car). A PN is a virtual network where collocated personal devices organize themselves in *clusters* which are in turn interconnected over some interconnecting structure.

While Personal Networking is focused on the communication between personal devices only, many communication patterns need to extend the boundaries of the PN and involve the secure interaction of multiple people having common interests. Hence, personal communications cannot be restricted to the services provided by the devices the user owns, but the possibility to interact with other user's PN has to be enabled in order to support the user in his/hers private and professional activities. The concept of PN Federations (PN-F) is even a more challenging one since the relations between users have to be managed and the security has to be reinforced in order to not open security holes while allowing authorized users to cooperate with you.

Existing solutions such as virtual private networks 2 or peer-to-peer application overlays 3 can only offer a partial solution as they do not provide true self-organization and end-to-end security. Furthermore, they lack the notion of group trust and usually only focus on one specific software application 4. As it is discussed in 5 and 6 the current diversification of control planes requires a manual configuration of network interworking. The problem will increase in the future, with more dynamic topologies and integration of heterogeneous networks in a ubiquitous, reactive environment. Nevertheless, the solutions proposed in these cases, involved too generic assumptions and spam along a plethora of different kind of networks thus not providing practical solutions to the user centricity that is mandatory in personal networking. 7 represents a P2P Wireless Network Confederation (P2PWNC) model, in which a set of administrative domains is providing wireless Internet access to each other's users. The authors aim to replace the human administrator of roaming agreements by Domain Agents (DA), thus eliminating administrative overhead. While this research approach addresses many critical issues, it does not fully address all the emerging needs of future wireless and ubiquitous networks since it is too much focused on specific environment such as mobile ad-hoc networks.

The rest of the paper is structured as follows. In Section 2 the generic architecture of a federation between two PNs will be shown. The life cycle of the PN-F as well as the main functional entities of the PN-F architecture will be described. As already said, heterogeneity and security enforcement are key challenges for the federations' establishment. Thus, in Section 3 the proposed solutions to tackle them will be presented. Section 4 will present the specification of the mechanisms for create, form and use a PN-F over a heterogeneous Ad-hoc scenario. Finally, Section 5 will conclude the paper highlighting the main aspects of the work presented.

2 PN-F Architecture

A PN-F can be defined as a secure impromptu, situation-aware or beforehand agreed cooperation between Personal Networks of different people for the purpose of achieving a common goal or service by forming an efficient collaboration. More precisely, a secure overlay of the participating devices will be formed, that isolates a subset of the resources in the constituent devices. Within the federation, the devices can communicate with each other and allow each other access to specific services or the usage of specific resources for performing the common task.

The basic requirements are that the communication is secure, self-organized, confined within the subset of collaborating devices and that only the resources, applications and services needed to achieve the common goal are made accessible. The term federation describes the process where entities broker trust and exchange information across organizational boundaries. The WS-Federation specification 8 defines a federation as "a collection of realms that have established trust". For instance, in collaborative working, a PN-F could be formed between the relevant devices belonging to the different people working on a common project. Only the resources needed for the project (e.g. files, e-mails, project schedule, whiteboard, software, agenda...) are made available to the PN federation. Other resources (e.g. personal files...) are shielded from your colleagues or only available through other federations, for instance with family and friends. It is clear that this concept will rely heavily on the notion of group trust.

The possibilities of establishing a federation are wide and bring several ways of classifying the federations. Taking into account the duration, we can have *Short-lived* (conference network) versus *Long-term* (project network); from a triggering way standpoint we can have *Reactive* (emergency network) versus *Proactive* (family network); finally, depending on the scenario we can distinguish between *Ad-Hoc* (meeting room network) and *Infrastructure* (distant learning). While the first two ones falls under the administrative and context considerations, the last one affects highly the way the federation creation, formation and use processes can be tackled. While in the Infrastructure case, the mechanisms to be deployed can count on the existence of some entities on the Internet that provides support to all the procedures, in the Ad-Hoc case, no Internet access can be assumed and all the procedures have to be carried out among the nodes belonging to the federating PNs present at that particular moment and place. Additionally, while in the Infrastructure case the common ground can be assumed through the use of the IP protocol, in the Ad-Hoc case the connectivity level has to be solved first, meaning that the possible heterogeneity in terms of wireless technologies has to be tackled. In this paper we will focus on describing the architecture, mechanisms and solutions designed in order to allow a PN-F to be created, formed and used in an Ad-Hoc situation.

2.1 PN-F functional entities

Fig. 1 shows the generic architecture of a PN-F together with its main functional entities and the relations between them. These functional entities are:
- Federation Manager (FM): It is responsible for managing all the interactions between two PNs during the PN-F Participation phase, mainly focused on the exchange of PN-F profiles.
- Secure Context Management Framework (SCMF): It is a distributed framework that provides access to all the PN related context information. One of its responsibilities is to store the different profiles from the PN-Fs that the corresponding PN is involved in.
- MAGNET Service Management Platform (MSMP): It controls the service discovery and access. The FM will interact with it when the PN-F Participation profile is to be created. For the service discovery, the operation is centralized on

a Service Management Node (SMN), located on each of the PN clusters, while for the provision is fully distributed.
- Policy Engine (PE): Act as an interpreter and reasoner of the rules declared on the PN-F profiles to enable access control enforcement.
- Personal Network Directory Service (PNDS): Takes the role of a trusted third party. It can be used to verify the identity of the peer PN on the different phases of the PN-F and it can also host the publication of PN-Fs' and PNs' details.

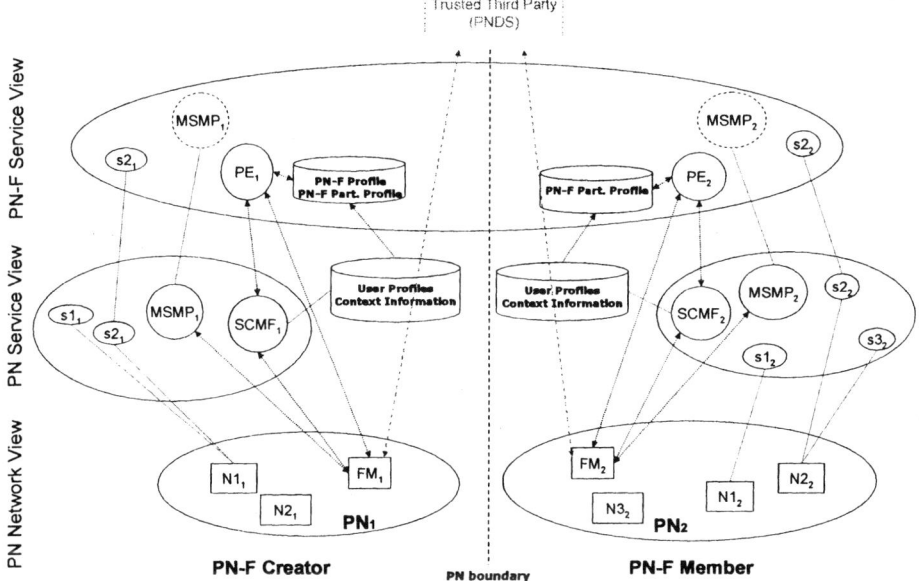

Fig. 1. PN-F Architecture and main functional entities

2.2 PN-F Life Cycle

We shortly introduce the three identified phases, as different policy rules have to be checked in each phase.

PN-F Participation: the objective of this phase is the agreement between the PN-F Creator (i.e. the PN that starts the PN-F) and one candidate PN-F Member (a PN aiming at participating on the PN-F). The basis for this agreement is the secure exchange of different PN-F profiles.

PN-F Formation: in this phase, the communication means between any two PN-F Members are deployed. The main issues solved during this phase are addressing, routing and security.

PN-F Use: This phase comprises both the discovery and provision of services offered to other PN Members. The authenticity provided by the networking solutions implemented below the PN-F service level allows the deployment of access policies for the shared resources.

3 Heterogeneity and Privacy Solutions

Up to know we have presented a generic architecture view of the PN-Fs. In the following sections we will present specific solutions for the case in which the PN-F is to be created, formed and used in an Ad-Hoc scenario.

3.1 Security association establishment

In 9 the baseline procedure for assuring privacy and security within a PN was described. This mechanism, so-called *imprinting*, basically consists on the establishment of a security association between a pair of personal nodes. Assuming that each pair of nodes within the PN had this security association, materialized into long-term bilateral shared secrets established under the supervision of the network owner, the authenticity, privacy and security in general is assured through leveraging these shared secrets whenever two personal nodes want to communicate with each other.

Following a similar approach, the communication between any two nodes belonging to different PNs can also be assured by leveraging a security association established in this case not on a node-node basis as it was the case of PNs, but on a PN-PN basis. The shared-secret will be used to protect the communication between any node of one of the PNs and another one from the other PN. Two main methods have been identified so far in order to accomplish the security association establishment. In any of the two cases, the result is that pair-wise secrets (so-called K_{PN}) are derived and associated to the peer PN identifier.

Basically, the enforcement of these keys assures the authenticity of the peer nodes (i.e. only the nodes belonging to a PN will know which K_{PN} is the correct one) and the privacy of the communication (i.e. K_{PN} is used to derive session keys that are subsequently used to encrypt the packets exchanged). Further authorization to make use of the services provided by the user PN should be based on this authentication. Besides, the solutions proposed enable node authentication but if really sensitive information might be disclosed, user authentication should be put on top.

3.2 Neighbor discovery and authentication

In the Ad-Hoc case the PN-F formation starts by discovering surrounding nodes belonging to peer PNs. A beaconing process has been implemented in order to be continuously aware of the immediate neighbors, both personal and foreign. Each node periodically sends one of these packets in order to advertise its presence not only to its personal neighbors but also to every other node belonging to other PNs. In case that the node belongs to a different PN, the receiving node acknowledges the reception and both nodes indicate to the Federation Manager the newly discovered PN indicating the PN ID.
- Mandatory payload fields:
  - Node ID: 20 bytes public identifier. Currently it is derived as a digest over the peer's public DH (Diffie-Hellman) key used during the imprinting procedure.

- PN ID: PN Name (or hash of the PN Name). Personal certificates that are used for the security association establishment are written for a certain PN Name. Uniqueness of the PN Name might therefore be guaranteed by the Authority that issues the certificate).

In case that a security association is already established with that PN, the node will be able to go into an authentication and session key derivation function in order to verify the identity the other node is claiming on its beacons.

The authentication is performed through a three way handshake (Request – Response – Success) in which the long-term shared key is used to verify the identity denoted by the identifier field in the beacon received.

The same procedure is used for neighbour authentication and for exchange of link session keys used at the Universal Convergence Layer (UCL – see Section 3.3) for privacy assurance through communications encryption.

The link layer session key is computed as HMAC SHA-256($LMSK_{1-2}$, N_1 | N_2) and is valid for T_2 seconds ($T_2 \leq T_1$). This procedure is run any time a new neighbour is discovered by a peer and whenever the derived session keys expire. $LMSK_{1-2}$ is calculated as HMAC_SHA_256 (K_{PN}, "MAC_1+MAC_2"). Use of the MAC addresses of the candidate radios in the derivation function ensures that different pairs of hardware adaptors share different link keys even for the same pair of devices.

3.3 Heterogeneity support and Security Association enforcement

The capability of working in a heterogeneous environment is a must for future personal networks. This heterogeneity will be mainly reflected in terms of the different air interfaces that will coexist in these scenarios requiring additional schemes to handle this heterogeneity.

The concept of isolating the upper-layers from underlying wireless technologies and thus providing real multi-mode can be achieved by introducing a Universal Convergence Layer 10. The UCL mainly will act as an enabler for backward and forward compatibility by defining a common interface towards the network layer while managing several different wireless access technologies independently of their PHY and MAC layers. In this sense, the solution adopted makes it possible for the nodes to have a single IP address independently of the number of air interfaces it has. This way the routing protocol placed in layer 3 will be able to settle routes embracing multiple radio domains in a complete seamlessly manner. The combination of these two techniques, UCL plus ad hoc routing protocol, enables the solution proposed to manage the heterogeneity that will appear in the Ad-Hoc PN-F environment.

From a security perspective, one of the most important design goals of UCL is to make sure that use of heterogeneous radio specific legacy security system does not cause any additional security vulnerabilities. In addition to making parallel use of different radio systems secure, presence of UCL also provides an opportunity to upgrade or even complement the legacy radio systems. Using the encryption capabilities provided by the UCL all the user data traffic sent is encrypted and signed to assure the integrity, authenticity and privacy of the information exchanged.

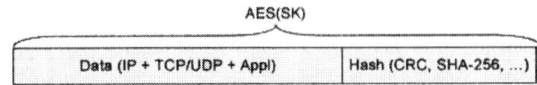

Fig. 2. Data PDU format

As shown in Fig. 2, a hash signature of the packet is applied to each packet. Additionally, the packet is encrypted using the previously exchanged session keys.

4 PN-F Specification

4.1 Scenario description

Fig. 3 presents the generic scenario over which Ad-Hoc PN-Fs might operate. As can be seen, clusters from different PNs communicates on a peer-to-peer manner where secure communications have to be assured starting from the connectivity level and that cannot rely on the support given by any third party located in the infrastructure. Additionally, many wireless technologies can be coexisting, thus at the connectivity level several radio domains can be identified forcing to establish a multihop communication path through combination of routing protocol and UCL techniques.

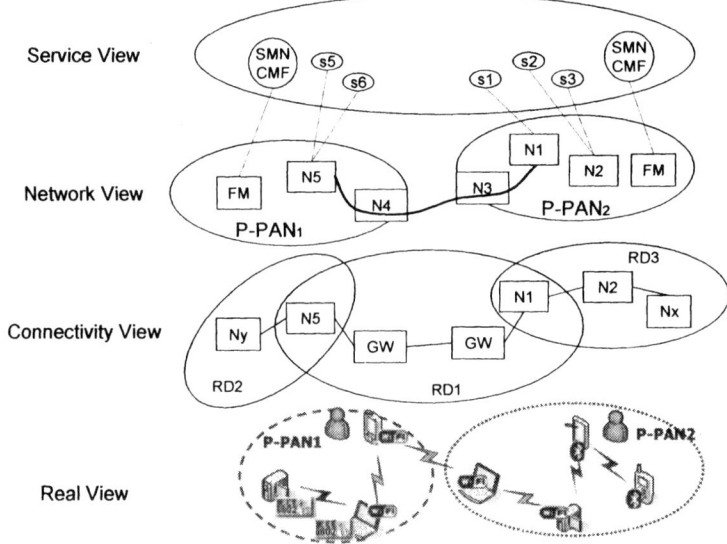

Fig. 3. High-level view of the Ad-Hoc PN-F scenario

4.2 PN-F Participation phase

The PN-F Participation phase comprises the procedures by which the information needed to join a PN-F is securely exchanged between the PN-F Creator and another PN aiming at being member of that PN-F.

Personal Wireless Communications   45

The event that triggers the participation phase can be flexible chosen from user action, context (time, location, presence) etc. Once the basic mechanism works, this type of scenario building can be done later on.

The PN-F Owner will first create/generate the PN-F profile. It will define the identity and the policies and rules within the PN-F. The minimum information (i.e. public part of the PN-F Profile – **PN-F Profile$_{PUBLIC}$**) that must be present is:

- PN-F Name/ID – A unique method for identifying the PN-F
- The goal or purpose of the PN-F
- Trigger conditions, activation/closing rules
- Creator Point of Contact (PoC) – The identity of the Creator and the address where negotiation to join the PN-F can begin

Optionally, the creator might also make public additional information such as:

- Participation rules: who else is or may be invited
- Minimum service list required

This phase starts when the Creator advertises the public part of the PN-F Profile. The Publication and Discovery step can have a great variety of possibilities. For example, an invitation could be issued to people already known to the Creator via e-mail, via a local broadcast to people in the vicinity of the Creator, or posted at a known 3rd Party location, where people visit to look for others to federate with. Once the Candidate is aware of the PN-F, and after a security association has been established between both PNs, it will edit its PN-F Participation Profile, mainly consisting on the resources that it makes available to this federation, and securely sends it to the Creator. The Creator then checks whether the Candidate fulfils the federation policies and if this is the case, the private part of the PN-F Profile, (**PN-F Profile$_{PRIVATE}$**) is sent to the new Member:

- Federation Broadcast Key
- List of current federation members
- List of currently available federated services

PN-F Participation: Publication and Discovery

In case that there is not access to a supporting infrastructure that provides a repository in which PN-Fs and potential members can advertise its existence, it is necessary to specify different procedures to fulfill this step in the PN-F Participation phase.

Fig. 4 shows the publication and discovery interactions as they would occur on an infrastructureless scenario.

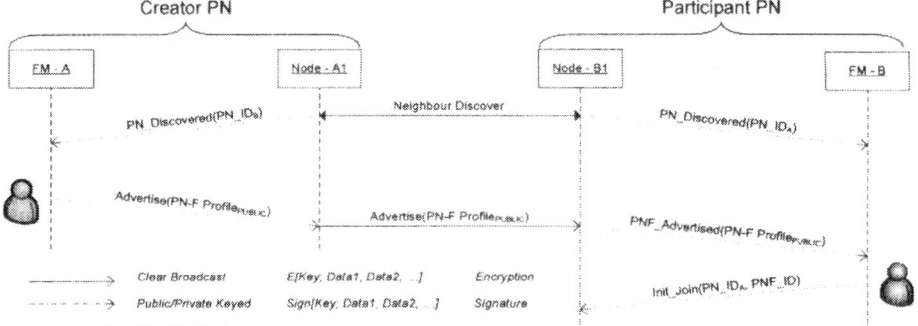

Fig. 4. Publication and discovery step on ad hoc scenarios

In this case, the PoC addresses advertised within the PN-F profile might not be usable. Typically, these addresses will be globally routable addresses for FM that is pointless in an infrastructureless environment. Hence, when the FM receives a *PNF_ADVERTISED* primitive it detects that this is an ad hoc scenario and issues an *INIT_JOIN* to the GW node from which the advertisement comes.

The *INIT_JOIN* primitive contains the following information:
- *PN_ID$_X$*: The identifier of the PN-F Creator so that the GW knows to whom address the forthcoming messages
- *PNF_ID*: The identifier of the PN-F to which the user is willing to join

This information will be used for feeding the Authentication step primitives as it will be shown in the next section.

PN-F Participation: Authentication and Security Association establishment

Up to this point in the PN-F Participation phase, the information must be taken as is with no guarantee of certainty either on the origin or the actual content.

In order to continue with the PN-F Participation phase, it is mandatory that both Creator and Candidate perform a mutual authentication and establish a security association between them so that the rest of the steps can be secured.

As stated in Section 3.1, two main authentication methods have been identified so far. The first one based on the typical PKI structure where certificates issued to PNs are used to verify its identity and to establish the bilateral security association. The second one is proposed in order to be infrastructure-independent and exploits PAC concept put forward for the imprinting of new personal nodes. In this section only a specification for the first type of authentication is presented, whereas further work will be done on the specification of an authentication method using this PAC.

It is important to note that if a previously established security association between the two PNs exists, this step can be omitted since they can use the shared secret resulting from that security association to assure the security in subsequent steps of the PN-F Participation phase. Hence, this step only makes sense for the first time the two PNs starts an interaction.

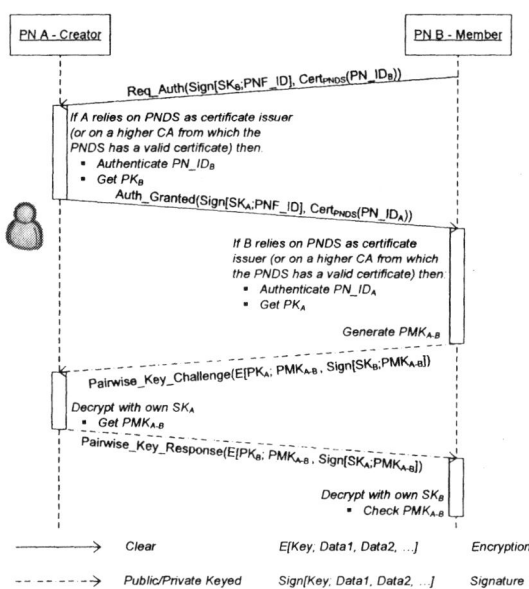

Fig. 5. Certificates based authentication

The common fields of the *REQ_AUTH* and *AUTH_GRANTED* primitives shown in Fig. 5 are:

- Sender Certificate: Issued by a commonly trusted (whether directly or by following a typical PKI hierarchy) third party, is used to both verify the sender's identity and to provide its Public Key (PK_X)
- Signature of PNF_ID: By encrypting the identifier of the PN-F subject of the interaction with its own private key (SK_X), the sender prevents using its certificate (it has to be remembered that this is public material) inadequately. This way it can be verified at the receiver that it is the actual certificate owner showing interest in starting the authentication.

Upon the correct verification of both *REQ_AUTH* and *AUTH_GRANTED* primitives, both PNs are mutually authenticated and they have each others PK.

Afterwards, the process of creating keys between the Creator and Member takes place. The Key Generation stage, used in the PN-F Formation, could be based on the Diffie-Hellman protocol, but since the Creator and Member already have a secure communication with the use of Public/Private keys then one can just generate a symmetrical key and inform the other about of it.

PN-F Participation: Join

Once this point in the PN-F Participation phase is reached the situation is as follows:

- Both PNs are mutually authenticated and share a pair-wise key that can be used to protect their communications providing privacy and origin authentication.
- The potential Member (potential since it has not yet provided its Participation Profile so the Creator can still prevent it to become full blown Member of the PN-F) has all the required information concerning the PN-F Profile and completed if required on an optional Additional PN-F info provision step.

Taking this into account, the potential Member may have already a participation profile for that type of federation, or may have none and would first prompt the user to edit it. When the user is ready and has decided to join, the *JOIN* message is sent to the creator (asynchronously to the invitation).

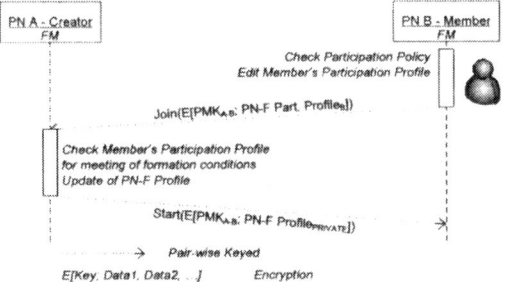

Fig. 6. Join step

Fig. 6 shows the primitives exchange during the Join step. The *JOIN* primitive contains information on the Member's PN-F Participation Profile as follows:
- Identifier of the PN-F for which this Participation Profile corresponds
- A Point of Contact for the Member containing the Identifier by which it can be recognized in the PN-F and an address where further negotiation concerning the PN-F can continue.
- Set of resources the participant intends to share with all other participants.
- Rules, policies for accessing these services (optional).

Once the Creator receives the *JOIN* message it will check if the Candidate fulfils the PN-F Participation rules. If they are, then the Creator can send the *START* primitive to the Member. However, it can occur that the general formation rules may not have been fulfilled yet, for example the federation use starts in two hours time, or certain participants have not joined yet, etc, so that the federation cannot start immediately. In this case, two options have been identified:
- Only when the federation can be started, the creator creates and sends the *START* primitive to all the registered Members
- The Creator sends the *START* primitive immediately and each of the Members is the responsible to enforce the general formation rules so that the PN-F is not used before.

In any of the cases, the *START* message will contain the **PN-F Profile$_{PRIVATE}$**.

4.3 PN-F Formation and Use phases

Once the new PN has fully joined the PN-F, it can proceed to the establishment of the PN-F with other surrounding members. In order to enable the communication between nodes belonging to different PNs, during the Formation phase, a network overlay is established among all the nodes of the PN-F members.

PN-F Addressing

Since a network overlay uses its own virtual addressing space, a PN-F addressing is defined (and this information is propagated as part of the PN-F profile) and every involved node will obtain a unique PN-F IP address within this addressing space. These addresses will be used for all communication that takes place within the federation.

This virtual address space is separated from the public IP addressing space and from the private PN addressing space of the participating PNs and guarantees that all PN-F communication is confined within the PN-F already at the network level. In addition, at the network level, specific primitives (e.g. PN-F wide multicasting) can be offered to the higher layers in order to facilitate for example service provisioning.

As it will be described in Section 4.3.3, part of the security checks implemented within the GW node at UCL level, assures that only those nodes belonging to PNs that are member of the PN-F can use this addresses on the IP header. As such, only members of the same PN-F will be able to become part of this overlay and all their communication will be shielded from the outside world, any other PN-F communication or PN communication.

Establishment of a network

In the ad hoc case using overlays, the PN-F formation takes place both at connectivity and at network level.

The main requirements are:
- Self-configuration (or minimal user intervention, if defined in the PN-F policies)
- Independence from third party entities not part of any of the involved clusters.
- Establishment of secure network overlay.
- Support for spontaneous PN-F creation.

The first step in the formation of the federation is the establishment of a secure connection. As described in Section 3.2, neighboring nodes are authenticated and link session keys have to be exchanged for securing traffic. Based on this cryptographic material, at UCL level the privacy and origin authentication of the communications at the connectivity level are assured being the basis for authorization and access control, which is the main aspect of federations, on upper levels.

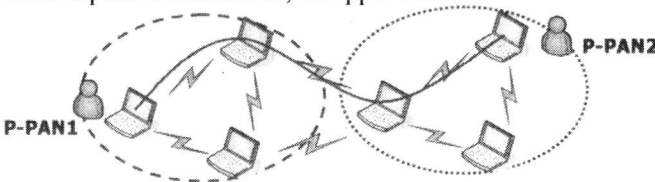

Fig. 7. Ad-hoc network establishment among cluster nodes

Once the connectivity level is assured, the network overlay can be established. As shown in Fig. 7, the routing protocol will provide end-to-end paths within the PN-F network. Different PN-Fs might require different routing alternatives. Nevertheless, reactive routing should be taken as the default choice since it fits better on the establishment of an on-demand network such as the PN-Fs would be in the majority of the cases. Route discovery process is authenticated since route response messages are unicast (protected by secure connectivity level).

Federation Use

The PN-F Use stage is triggered as soon as one of the nodes in any of the clusters involved in the PN-F request a service provided by some other node in a different person's cluster. In this sense, within this phase, both the service discovery and provision phases are comprised.

Using the encryption capabilities provided by the UCL all the user data traffic sent is encrypted and signed to assure the integrity, authenticity and privacy of the information exchanged. Fig. 8 shows the procedure followed on the transmission function in the UCL for selecting the appropriate key to secure the communication.

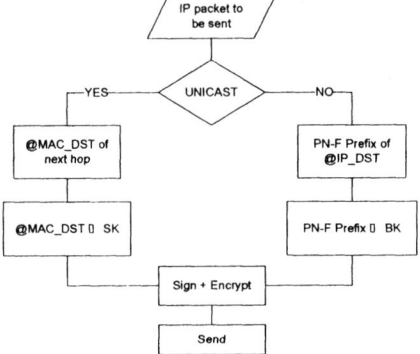

Fig. 8. Data transmission procedure.

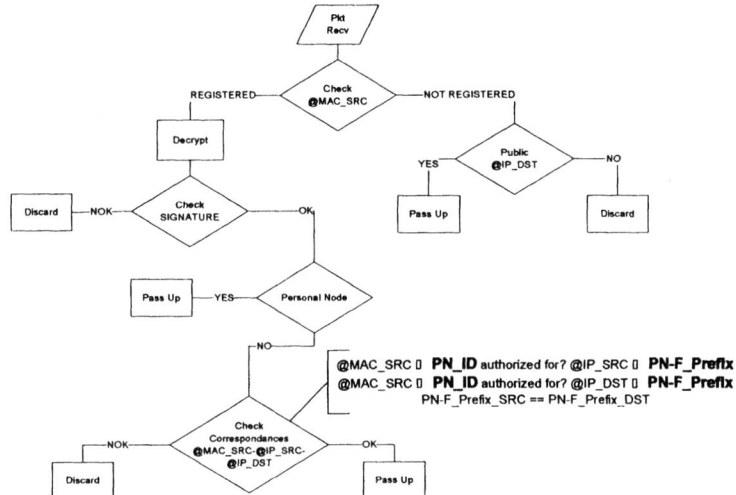

Fig. 9. Impersonation prevention procedure

Whenever a packet is received, the information on the datagram headers is checked to verify the authenticity of the information and to prevent impersonation as shown in Fig. 9. At the edges of the cluster, this conformance checks have to be reinforced in order to avoid misuse of the PN or PN-F resources. This allows upper layers to trust on the information contained on the packets and perform access control based on it.

In Fig. 10, a detailed description of the unicast and broadcast communications, both within the PN and under the auspices of a PN-F, is presented.

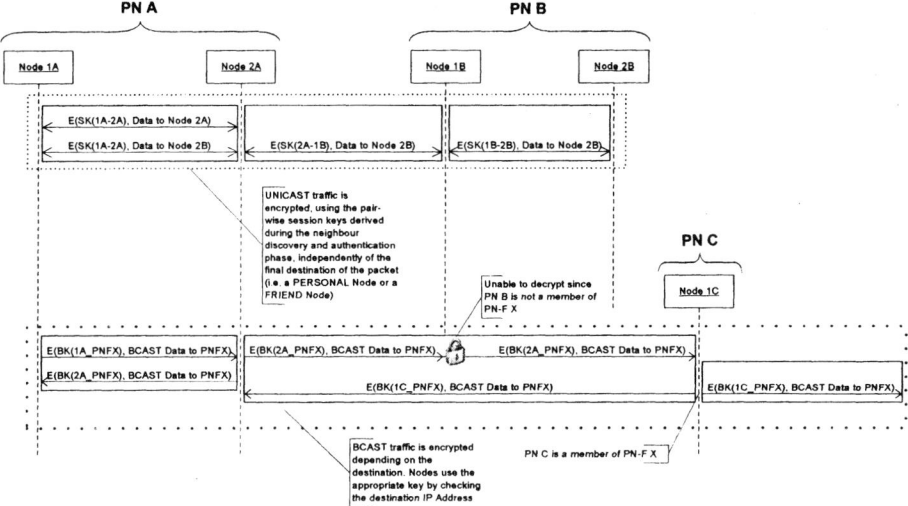

Fig. 10. Unicast and Broadcast communication encryption details

Finally, as shown in Fig. 11, the service discovery and provision is carried out.

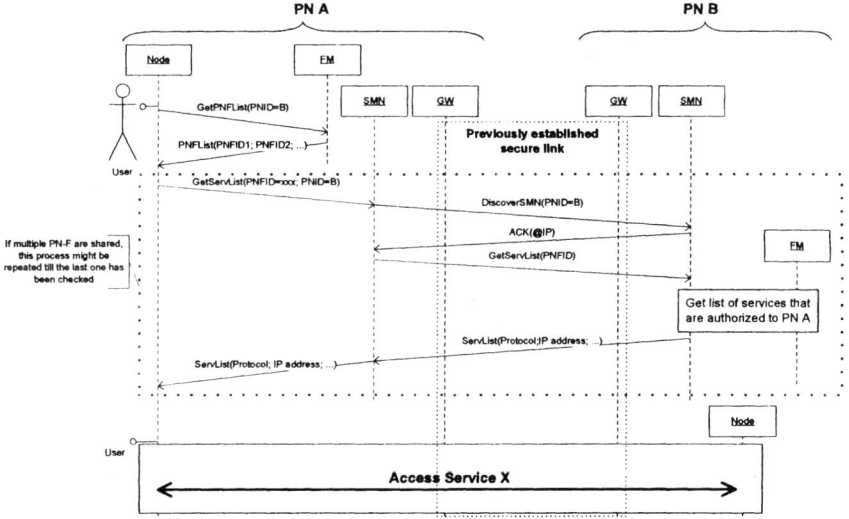

Fig. 11. Service discovery and access in ad hoc PN-F based overlays architecture

Whenever a node wants to know about the availability of services within the PN-F, it will direct its query to its cluster SMN. The SMN will first look for its counterpart in the other cluster and once it discovers the other SMN, redirects the query for available services. This query can try to discover all the authorized services (as shown in Fig. 11) or focus the search on specific ones. The SMN as the responsible for managing the services in the cluster is the one that has to check which of the complete list of available services the other person is authorized to access to. The information stored in the different PN-F Participation profiles is checked to see whether the available services are authorized or not. Once it has the response, it sends it back to the originating SMN who forwards the information to the requesting node.

After the discovery phase, the requesting node initiates directly with the server node the service provision phase. The authorization mechanisms used in the discovery phase, do not prevent the use of further mechanisms to authenticate and authorize the user against the service.

5 Conclusions

In this paper we have presented a detailed specification of the mechanisms that allows the creation, formation and use of Personal Networks Federations over a heterogeneous peer-to-peer wireless scenario.

The main requirements in terms of security and self-configuration imposed by the federation concept, and the scenario selected for its creation and use have been identified, and the corresponding procedures that cope with them described in detail. In this sense appropriate mutual authentication algorithms have been described and pair-wise secrets leveraged to provide a secure connectivity and network level so that upper layers authorization and access control can be easily supported.

Additionally, specific multi-standard convergence procedure has been implemented in order to tackle the heterogeneity in terms of access technologies such that network overlay establishment can be done transparently and independently of the underlying radio domains that compose the connectivity level.

This detailed specification is the basis for an implementation that is being carried out over real platforms consisting of laptops and PDAs that will raise true interest of industry and end-users as well as support the identification of future optimizations that could be achieved by enhancing the collaboration between the different components comprising the whole system.

References

1. I.G. Niemegeers and S. Heemstra de Groot, "From Personal Area Networks to Personal Networks: A user oriented approach", Journal on Wireless and Personal Communications 22 (2002), 175-186.
2. Charles Scott , Paul Wolfe , Mike Erwin, "Virtual private networks", O'Reilly & Associates, Inc., Sebastopol, CA, 1998
3. Sun Microsystems, "JXTA^{TM} Technology: Creating Connected Communities", January 2004
4. J. Hoebeke, G. Holderbeke, I. Moerman, Bart Dhoedt, P. Demeester "Virtual Private Ad Hoc Networking", Wireless Personal Communications, Volume 38, Issue 1, June 2006, pp. 125-141
5. C. Kappler, P. Mendes, C. Prehofer, P. Pöyhönen and D. Zhou, "A Framework for Self-organized Network Composition", WAC 2004 (IFIP Workshop on Autonomic Communication), Berlin, Germany, October 2004.

6. L. Fan, N. Akhtar, K. A. Chew, K. Moessner and R. Tafazolli, "Network Composition: A Step towards Pervasive Computing", 3G 2004, London, UK, October 2004.
7. Elias C. Efstathiou, George C. Polyzos: "A peer-to-peer approach to wireless LAN roaming", Proc. 1st ACM Int. workshop on Wireless mobile applications and services on WLAN hotspots, San Diego, CA, 2003.
8. Kaler, Chris, et al. "Web Services Federation Language (WS-Federation)." Online article, 8 July 2003. IBM, Microsoft, RSA Security Inc., and VeriSign. 22 March 2004. http://msdn.microsoft.com/library/en-us/dnglobspec/html/ws-federation.asp
9. IST-507102 MAGNET/WP4.3/UNIS/D4.3.2/R/PU/002/1.0, "Final version of the Network-Level Security", March 3, 2005.
10. L. Sanchez, J. Lanza, L. Muñoz, J. Perez Vila, "Enabling Secure Communications over Heterogeneous Air Interfaces: Building Private Personal Area Networks", 8th International Symposium on Wireless Personal Multimedia Communications - Aalborg, September 2005.

Analysis of Algorithms for Radial Basis Function Neural Network

Jiri Stastny[1], Vladislav Skorpil[2]

[1] Department of Automation and Computer Science,
[2] Department of Telekommunications,
Brno University of Technology,
Purkynova 118, 612 00 Brno, Czech Republic,
stastny@fme.vutbr.cz, skorpil@feec.vutbr.cz

Abstract. This paper describes the analysis of algorithms for the hidden layer construction of network and for learning of the Radial Basis Function neural Network (RBFN). We compared results obtained by using of learning algorithms LMS (Least Mean Square) and Gradient Algorithms (GA) and results are obtained by using of algorithms APC-III and K-means for hidden layer contruction of neural network. The principles and algorithms given below have been used in an application for object classification that was developed at Brno University of Technology. This solution is suitable for the research of personal wireless communications and similar systems.

Keywords: Radial basis function, Learning algorithm, Neuron, Hidden layer.

1 Introduction

The Radial Basis Function Network (RBFN) belongs to the most recent neural networks. It is a type of single-direction multilayer network and forward multi-layer network with counter-propagation of signal. This network has two layers with different types of neurons in each layer. Its advantage is mainly the speed of learning. The structure of this two-layer network is similar to that of the MLP (the Multi Layer Perceptron Neural Network) type of network but the function of output neurons must be linear and the transfer functions of hidden neurons are the so-called Radial Basis Functions – hence the name of the network. The characteristic feature of these functions is that they either decrease monotonically or increase in the direction from their centre point. Except for the input layer which only serves the purpose of handing over values an RBFN has a hidden layer (RBF) and an output layer is formed by perceptrons. This neural network can be used for wide scale of problems because it is able to approach arbitrary function and its training is quicker than for MLP neural network. Quicker learning is given by this, that RBFN has only two layers with weight and every layer may be determined sequential.

Please use the following format when citing this chapter:

Stastny, J., Skorpil, V., 2007, in IFIP International Federation for Information Processing, Volume 245, Personal Wireless Communications, eds. Simak, B., Bestak, R., Kozowska, E., (Boston: Springer), pp. 54-62.

2 Creation of Hidden Layer RBFN

Problems at creation of RBFN consist on determination of the number of neurons in hidden layer, on determination of the middles of these neurones and on determination of the neurones width. Powerful method for determination of the number and quality of neurons of hidden layer is the algorithm APC-III. This single-pass associating algorithm unlike other uses constant radial.

2.1 Algorithm APC-III

C: the number of neurons
c_j: the middle of j-th neuron
n_j: the number of samples in j-th neuron
d_{ij}:   the distance between x_i and j-th neuron

```
{
        C=1; c₁ ← x₁; n₁ = 1;
        for(i = 2; i =< P; i++)          // for every pattern from training set
        {
                for(i = 1; i =< C; i++)   // for every neuron
                {
                        calculate d_ij;
                        if(d_ij =< R₀)           // insert x_i into j-th neuron
                        {
                                c_j = (c_j n_j + x_i)/(n_j + 1);
                                n_j = n_j + 1;
                                break;
                        };
                };
                if(x_i is not in any neuron)     // create new neuron
                {
                        C = C + 1;
                        c_c ← x_i;
                        n_c = 1;
                };
        };
}
```

2.2 K-Means Algorithm

Initialize RBF neuron centres C in random.
Calculate $m()$ for all samples from the training set.
Calculate new centres C as the average of all samples that pertained to centre k by the pertinence function.
Terminate if $m()$ does not change, otherwise continue with point 2.

Specially for classification, the RBF network is simplified by refraining from searching for clusters, which have to be searched when approximating. The centres (prototypes) of neurons are set so that RBF neurons are represented by model clusters for the best.

2.3 Description of Implementation

Each neuron in the radial basis layer will give on the output the value that depends on how close the input vector is to each of the weight vectors of the given neurons. Thus the RBF neurons whose weight vectors are a bit different from input vector p have an output of about zero. On the contrary, the RBF neuron whose weight vector is close to the input vector will have a value of about 1. Individual neuron layers have the form of one-dimensional array. The weight matrix is in the form of two-dimensional array, where the index gives the number of neurons being connected. It is necessary to enter the number of RBF neurons n for one category. The principal computation methods are:

calculatePrototype() – by using the algorithm chosen (the K-means algorithm is used in the program) it will calculate the weight values of prototype C. On the output of RBF neurons the first $1...n$ neurons will calculate the output for the first category, $n+1...2n$ for the second category, etc.

calculate_sigma() – after calculating the weight values of prototypes, the size of the sphere of influence will be calculated for each RBF neuron.

calculate_h() – it will calculate the outputs of RBF neurons. The radial basis function reaches maximum of 1 when there is 0 on the input. The RBF neuron thus operates as an indicator that produces 1 always when the input vector is identical to its weight vector.

calculateOutputRBF() – on the output each RBF neuron contains the (in principle) percentage agreement with the input model. For the output neuron the value of the neuron which most agrees with the network input (maximum value) is chosen from n RBF neurons in the category given.

Algorithm APC-III is very powerful at creation of hidden layer of neural network. That is why it is enough to use this algorithm on input training set only once. That is great advantage against K-means algorithm. In addition APC-III is headed to creation of reasonable of the number of neurons and to determination of the radial of neurons on basis of training set distribution.

3 Determination of the Width of the Area (of Neuron)

For hidden layer it is necessary to determine the width of the single areas. The transient neurons function uses this width for calculation. In created testing program are used two variants of the areas width determination.
   a) all areas have the same width R_0

b) every neuron has its own width

3.1 Algorithm of Calculation of the Own Width of the Area (of Neuron)

Algorithm searches to every area the nearest one such, to this area belongs to the other pattern and at the same time it was the nearest to calculated area. By the help of distance of middles of these two areas is determined the width of the area like multiple of those distance. Like the multiplier is used the *distance coefficient*, which is adjusted in the programme and which specifies on how many per cent of those distance will be set-up the width of this area.

3.2 Pseudo-code of the Algorithm

Input: the distance coefficient of the neuron hidden layer
Output: the widths of neurons
rk: the distance coefficien*t*
C: the number of neurons
c_j: the middle of j-th neuron
v_j: the pattern allocated to j-th neuron
D_{ij}: the distance between i-th and j-th neurons
dst: the minimal distance between i-th and j-th neurons
σ_j: the width of j-th neuron

```
{
        for(i = 1; i =< C; i++)    // for every neuron
        {
                for(j = 1; j =< C; j++)    // pro every neuron
                {
                        if(i == j)
                        {        // the distance of neuron to itself is zero
                                Dij = -1;
                                continue;
                        };
                        if(vi == vj)
                        {        // the distance of neuron with the same patterns is
            omitted
                                Dij = -1;
                                continue;
                        };
                        $D_{ij} = \|c_i - c_j\|$;
                        Dji = Dij;
                };
        };
```

```
for(i = 1; i =< C; i++)   // for every neuron
{
            // location of minimal distance will be performed
            // it is necessary at calculation to ignore items, which have value –1
```
$$dst = \min_{j=1..c}(D_{ij});$$
            // we calculate the width of neuron
$$\sigma_i = dst\left(\frac{rk}{100}\right);$$
```
};
};
```

The distances between middles of areas are placed into the matrix of distances. This matrix must have on diagonal zero distances, that have be good for better filtering out on the minimization substituted by the value -1. Likewise the distances among the neurons with the same associated pattern are eliminated and in lieu of distance is into the matrix inserted the value -1. The distance coefficient is inserted in percents (in front of using is divided by 100).

Connection of hidden layer to output layer of neurons is the last step on network creation. These layers are connected by the system every with everyone from the other layer. The programme then randomly sets scales and thresholds of outgoing layer of the neural network. Thereby it is creation of the Radial Basis Function Network (RBFN) finished. The next phase is learning of neuron network.

4   Learning of RBFN

The learning process consist on precept of given network to answer correctly to engaged training set. As the hidden layer was in this network represented so-called areas and the middles of the areas are fast given to it, the learning process oversimplifies only to setting of scales and thresholds of the output layer. Gradient method and Last Mean Square (LMS) method were tested for learning of neuron network.

The gradient method (GA) uses relations derived for outgoing layer for algorithm Back-propagation (BP). On difference from the method BP this method optimizes only scales and thresholds of outgoing layer.

Method Least Mean Square (LMS) tries to find optimal scaled vector for general middle quadratic error of the network. This scaled vector is given by normal division:

$$w = (H^T H)^{-1} H^T y$$

where **w** is scale vector, H is suggestion matrix $H_{ij} = h_j(x_i)$ and **y** is vector of outgoing values. This method in contrast to of others ones uses like transient function of outgoing neurons layer in lieu of sigmoid the linear function.

Personal Wireless Communications 59

5 Programme Solution, Parameter Setting

At learning programme requires setting of following values:

a) The values common for all types of learning

- *The number of iteration on one etalon:* designates how many times will be submitted pattern set on learning of the network on the input. The optimal value for given to application was 5000.

- *The diagnosis accuracy:* designates, to what degree of accuracy the network will learn and subsequently will also diagnostic presented patterns. At setting of high value network will diagnostic high accurate, it is so errors in recognition will be almost zero, but the disadvantage will be inability of the network to generalize, which makes no-diagnosis of the damaged patterns. On the contrary at setting of low value the error count will grow at realising. The recommended value is 80%.

- *The minimal error:* past its over-fulfilment, i.e. if the error of learned network will be less than minimal error, reaches to stopping of calculation before reaching of given of the number of iterations. This election is recommended to use only while using of a large number of iterations.

- *The method of determination of the number of neurons:* here it is possible to choose the method of creation of hidden layer. The recommended setting is algorithm APC-III.

- *The coefficient α:* this coefficient is the main creation parameter for RBFN network. The coefficient is exploited to calculation of the radius R_0, by the help of it subsequently algorithm APC-III.

- *Option of the radius:* this option causes, so after creation of hidden layer oneself its own width calculates for every neuron in this layer.

- *The coefficient of the distance of the middles:* designates in percents how many from the distances of neurons will be the width of neuron.

b) The values of parameter for gradient method

- *The learning speed:* set the step size degree of learning. The recommended value is 0.5.

- *The adaptive step of learning:* it is possible to set adaptive (variable) step of learning

- *The maximum step:* designates the initial step of learning of the neural network. The recommended value is to the extend of 1 until 1.5.

- *The exponent of the curve of the learning step:* designates by the help of what curve will be the actual step of learning extracted. Whereby higher number, the speed of learning will be thereby moved towards finding of learning process, where it will decline very quickly.

The number of outgoing neurons forms dynamically, according to the number of learned patterns. The outgoing vector is after rounding binary and indicates the class of classified subject. For example at the number of classes 5 can the outgoing vector shapes $[1,0,0,0,0]$, which indicates first-class inside of order.

The values of all scales and thresholds on the outgoing layer are at the beginning randomly initialized to values from the interval $\langle -0.1; 0.1 \rangle$.

The current total mean square error and information on activities, that the programme just performs, displays along the learning.

6 Conclusion

The LMS learning method quickly navigates to aim. In contrast to the gradient method it needs only one iteration for all patterns together. LMS has also better characteristics on downsizing of a number of neurons in the hidden layer. The gradient method is not very suitable for learning of this network (in light of learning time). Learning of LMS is thanks to using algorithm APC-III reduced to learning of outgoing layer. The best results arranged the neuron network RBFN with the algorithm APC-III and with the learning method LMS (Least Mean Square).
Radial Basis Function networks can be designed very quickly. The time necessary for network learning was very little. The network was able to classify correctly 100% models and at the same time to recognize correctly even slightly damaged models. As the number of radial basis neurons is comparable the input space size and problem complexity RBF networks can be larger than MLP networks. Recognition with the aid of neural network is suitable where high-speed classification with randomly rotated objects is required and where we need to tolerate some differences between learned etalons and classified objects.

Acknowledgement. This research was supported by the grants:
MSM 0021630529 Intelligent systems in automation (Research design of Brno University of Technology)
MSM 6215648904/03 Research design of Mendel University of Agriculture and Forestry in Brno
No 102/07/1503 Advanced Optimizing the design of Communication Systems via Neural Networks. The Grant Agency of the Czech Republic (GACR)
Grant 1884/2007/F1/a "Innovation of computer networks participation in high-speed communication" (grant of the Czech Ministry of Education, Youth and Sports)
Grant 1889/2007/F1/a „ Repair of digital exchanges education in the course Access and Transport Networks" (grant of the Czech Ministry of Education, Youth and Sports)
No MSM 0021630513 Research design of Brno University of Technology " Electronic communication systems and new generation of technology (ELKOM)"
2E06034 Lifetime education and professional preparation in the field of telematics, teleinformatics and transport telematics (grant of the Czech Ministry of Education, Youth and Sports)

References

1. Miehie, D., Spiegelhalter, D. J., Taylor, C. C.: *Machine Learning, Neural and Statistical Classification.* Ellis Horwood, NY, 1994.

2. Lim, T., Loh, W. Y., Snih, Y.: *A comparison of of prediction accuracy, complexity and training time of thirty-three old and new classification algorithms.* 1999. http://www.stat.wisc.edu/~limt/mach1317.pdf

3. De Juan, Ch.: Contour Recognition Problem, [online]. 2001. Dostupné z: <www.cc.gatech.edu/classes/ay2000/cs7495_fall/participants/cnd/ps3/ps3.html

4. Goldberg, D. E.: *Genetic Algorithms in Search, Optimization and Machine Learning.* Addisson-Wesley, 1989.

5. Nilsson, N. J.: *Principles of Artificial Intelligence.* Springer Verlag, Berlin, 1982.

6. Ripley, B. D.: *Pattern Recognition and Neural Networks.* Cambridge University Press, Cambridge (United Kingdom), 1996.

7. Pavlidis, T.: *Algorithms for Graphics and Image Processing.* Bell Laboratories, Computer Science Press, 1982.

8. Wong, K. CH.: *A new diploid scheme and dominance change mechanism for non-stationary function optimization.* In Proceedings of the Sixth International Conference On Genetic algorithms, Pittsburgh, USA, 15. – 19. July 1995

9. Sarle, W. S.: *Neural Networks and Statistical Models.* Proceedings of the Nineteenth Annual SAS Users Group International Conference, Cary, NC: SAS Institute,1994, pp 1538-1550.

10. Šnorek, M., Jiřina, M.: *Neuronové sítě a neuropočítače.* ČVUT, Praha,1998. ISBN 80-01-01455-X.

11. Šťastný, J., Škorpil, V.: *Neural Networks Learning Methods Comparison.* International Journal WSEAS Transactions on on Circuits and Systems, Issue 4, Volume 4, April 2005, ISSN 1109-2734, pp. 325-330.

12. Vestenický,P. The Prediction Properties of Kalman Filter in Proceedings of TRANSCOM '95. UT&C, Zilina, pp. 243-248, 1995

13. Krbilová, I. and Vestenický,P.: Forgalomszabályozásés szolgáltatásminöség" Magyar távközlés 7, Vol.. 6/96, pp. 32-33, 1996

14. Vestenický, P. The Functions of ATM Interfaces in Proc. of DDECS '97 Conference Proceedings. VSB Technical University, Ostrava, pp. 186-191, 1997

15. Bubeníková, E. and Vestenický, P. Principles Of The Intranet Information System Creation in Proc. of ELEKTRO'99 Conference Proceedings, section Information & Safety Systems. University of Žilina, pp. 77-81, 1999

16. Vestenický, P. Optimization of Selected RFID System Parameters in Proc. of. AEEE 3, Vol. 2, pp. 113-114, 2004

17. Krbilová, I. and Vestenický, P. Use of Intelligent Network Services in Proc. of ITS. RTT , CTU Prague, 2004

18. Vestenický, M. and Vestenický, P. Evolutionary Algorithms in Design of Switched Capacitors Circuit in Proc. of International Workshop „Digital Technologies 2004". Slovak Electrical Society and University of Zilina, pp.34-37, 2004

Application of Path Duration Study in MultiHop Ad Hoc Networks

Alicia Triviño-Cabrera, Jorge García-de-la-Nava, Eduardo Casilari, and Francisco J. González-Cañete

Dpto. Tecnología Electrónica, Universidad de Málaga , Campus Teatinos,
29071 Málaga, Spain
{atc, ecasilari,fgc}@uma.es

Abstract. Some ad hoc protocols allow the discovery and/or storage of multiple routes to the same destination node. The selection of the path to utilize is commonly based on the criterion of the minimum number of hops. However, other strategies could be more appropriate to improve the network performance. In this sense, this paper proposes the criterion based on the estimation of the Mean Residual Path Lifetime. In order to compute this metric, a formal description of link duration in mobile ad hoc networks is presented. The model has been verified by means of simulations enclosing different mobility and transmission conditions. From the link duration model, an analytical expression for path duration and mean residual lifetime is then derived. Finally, the authors show that selected ad hoc routing paths live longer when the proposed criterion is employed.

Keywords: MANET, Path Duration, Route Selection.

1 Introduction

Mobile Ad Hoc NETworks (MANET) are formed by wireless devices that intercommunicate without the utilization of any centralized entity that manages the radio resources. As direct communication is only possible for neighbor nodes, the transmission of information between distant devices is supported by the cooperation of intermediate nodes that retransmit and route their packets. The sequence of intermediate nodes involved in the retransmission process constitutes a route or path. Due to the mobility of the nodes that are part of the established paths, routes present a finite lifetime (path duration). When the utilized route is down, the source node should initiate the procedure to discover a new path in order to continue the on-going communications. As this procedure is usually associated to a controlled packet flooding, the route discovery implies an increment in the network overhead and communication delay. Taking into account the energy consumption of the retransmissions as well as the wireless interferences associated to the flooding that may cause packet losses, the employ of route discoveries techniques should be minimized. In this sense, the criterion for selecting paths plays an important role. For two communicating nodes, there may exist multiple paths among which one route must be selected for the communication. Although the criterion of the minimum number of hops is commonly utilized, diverse metrics could be employed for the selection. Specifically, the consideration of the expected path duration could be advantageous so longer to live paths would be preferred to routes associated to shorter path durations.

Despite of the potential benefits that could be achieved if the employ of the expected path duration is considered, the a-priori computation of this parameter is not an easy issue as it requires the pre-determination of the future node movements. In order to overcome this limitation, other available information is utilized for the estimation of the residual path duration. For instance, some authors proposed that there is a direct relationship between the signal strength of the received packets along a path and its residual lifetime [2]. However, different propagation conditions may provoke that the previous assumption does not hold. From another point of view, the authors in [3] show that the inclusion of the mean link duration in the selection criterion implies the utilization of paths that live longer. This work is supported by approximating link and path duration to an exponential function, i.e. a memory-less model where the time that the route has been active since its creation is not considered. Although this approximation may become valid for those paths with a high number of hops [4], this condition is not satisfied in conventional ad hoc applications where paths usually contain 2 or 3 hops [5]. Under these circumstances, the present work proposes the lognormal fitting for the link duration in heterogeneous scenarios. From this model, the mean residual lifetime is computed. As the results show that the expected mean residual lifetime does depend on the survival time, this information is employed for the estimation of the path residual lifetime in the proposed criterion. The simulations show that the proposed criterion is associated to the selection of paths with greater duration.

The rest of the paper is structured as follows. Section 2 presents the related work. Section 3 shows the results of analyzing the link duration in heterogeneous scenarios. This analysis is utilized for the computation of the mean residual path lifetime presented in Section 4. With this parameter, a criterion for selecting paths is constructed as described in Section 5. The benefits associated to the employ of this criterion are checked by means of extensive simulations whose results are shown in Section 6. Finally, Section 7 draws the main conclusion of this work.

2 Related Work

In order to compute the mean residual path lifetime, that is, the time that an already established path is expected to live, a characterization of path duration is required.

Simulation has been the main method for analyzing the properties of path duration in ad hoc networks. One of the first studies concerning the analysis of path duration was due to Bai et al. [6]. Basing on experimental results obtained by simulations, these authors assume that the lifetime of a path with four or more hops can be approximated to an exponential distribution. However the study does not consider the fit of any other standard distribution. Moreover, the selection of an exponential distribution is not justified with any mathematical validation. To cope with this lack, Han et al., basing their work on Palm's theorem, state that, under some circumstances, the lifetime associated to those paths with a large number of hops converges to an exponential distribution [4]. The exponential fitting has also been utilized in [7] [8].

Another remarkable work related to the exponential fitting is described in [3]. In this study, each path is associated to a parameter (λ_{PATH}) of the exponential fitting of the route duration. The λ_{PATH} of the route is derived from the λ of the links (λ_{LINK}) that compose the route. With this purpose, the nodes estimate the λ_{LINK} as the inverse of the mean link duration in an interval of time (auto-regresive filters are suggested for this estimation). The path to employ is the one with a lowest λ_{PATH}, which is assumed to be related to longer path duration.

In spite of the popularity of exponential fitting for link duration in MANETs, this approximation presents a common disadvantage as they provide a solution for the analysis of paths which is valid only for routes with a large number of hops. Therefore, their study could not be fully applied to usual ad hoc networks and practical MANET applications where most paths normally only consist of 1 to 4 hops [5]. The importance of short paths is reinforced by the fact that the most relevant protocols utilize the minimum hop count as the metric to select the route in use in order to reduce the effects of the wireless medium on the performance of the network.

3 Link Duration

Although link duration is considered a fundamental parameter when evaluating the mobility in a MANET [9], few studies provide a formal description of this variable. Some authors have formally studied the link duration for specific mobility models. For instance, [10] have studied the mean value of link duration under a constant velocity model meanwhile [11] analyzed it for a deterministic, partially deterministic and Brownian mobility models. On the other hand, [12] formally describes the statistics associated to the link duration for a simplified random mobility model. In [13], a thorough analytical study of some significant parameters of link duration is presented. The main disadvantage is that it is only valid for the Constant Velocity Model.

In this section, we approach the fitting of the link duration in heterogonous scenarios characterized by different mobility models. The following standard distributions are considered: Normal, Gamma, Weibull, Rayleigh, Pareto, Exponential and Lognormal. The parametrization of the previous approximations are based on the correspondence of the two first moments of the data to the considered function moments. Similar results have been obtained if Maximum Likehood Estimators (MLE) are employed. Then, we evaluate the fittings by means of the Kolmogorov-Smirnof goodness-of-fit test (K-S test). The K-S test is an extended tool that captures the maximum deviation of the hypothesized Cumulative Distribution Function respect to the data to approximate.

The computation of link duration is obtained by the utilization of a developed module that was incorporated into the Matlab tool [14]. The basis of this module resides in the computation of the connectivity graph. A connectivity graph informs about the nodes that are neighbors, i.e., nodes that are directly connected because the distance that separates them is lower than the transmission range. Differences in the evolution of the connectivity graph imply the break or the creation of links.

The status of a wireless link depends on numerous system and environmental factors that impact both the transmitter and the receiver ranges. Although the transmission range of a node experiments spatial and temporal variations that concludes in a coverage area that is neither fixed nor symmetric, the exclusion of the modeling of these conditions simplifies the analysis of the network performance. Therefore the authors will consider the transmission range as a perfect circumference (in a two-dimensioned scenario).

In order to analyze link duration in the wide variety of scenarios where MANETs are expected to be deployed, two opposite mobility patterns are employed in order to conclude that the lognormal fitting is the best. Firstly, the Random WayPoint is utilized as it is quite extended in the studies of ad hoc network evaluation. We have varied the main parameters of this synthetic model. On the other hand, real movement traces are employed.

3.1 Random WayPoint

One of the most extended individual mobility models is the Random WayPoint (RWP). According to this pattern, the nodes of an ad hoc network move along a straight line between two destination points (waypoints) placed in a finite space. In the literature, this space is normally bi-dimensional and restricted to a rectangular area of dimensions x_{max} and y_{max}. Once a node reaches a destination point, a new one is uniformly selected from this area. The speed for a movement is also chosen from a uniform distribution in the interval $[v_{min}, v_{max}]$. Both speeds and waypoints are generated independently of all the previous destinations and speeds. In addition, the model allows nodes to pause between two consecutive trips for a certain period of time. This period (Pause Time) is habitually fixed to a constant value. By varying the values of the transmission range of the nodes, x_{max}, y_{max}, v_{min}, v_{max} and the pause time, it is possible to control the movement conditions of the simulated scenario.

The generic formulation of the Random WayPoint was proved to be unreliable by [15]. In order to suppress the instability in the results, authors in [15] propose the incorporation of a simple modification that consists of fixing a minimum not null speed in the nodes. This mobility pattern is called the Modified Random WayPoint. Due to its popularity as well as its capability of characterize multiple scenarios, the authors of this contribution will continue the analysis of path duration assuming nodes follow the Modified Random WayPoint Model.

Despite the Random WayPoint is widely utilized as a mobility pattern, at present, no analytical formulation has related the statistical properties of the link duration to the parameters of the modified RWP, the network dimensions, the node density or the transmission range. The references dealing with this issue are all based on simulations [16]. In general, the authors suggest the approximation to a standard distribution function without any mathematical support [6] [17].

In order to simulate the conditions on which multi-hop wireless networks can be deployed, we have evaluated the link duration in a wide set of experiments. The experiments estimates the link duration under different values of the area shape (1500x300 m^2, 1000x1000 m^2), the maximum speed (5,10,15 m/s), the pause time (0, 50 s), the node density (15, 50 nodes) and the transmission range (250, 100 m) in 10000-second simulations. Additionally, the distribution of link duration is also computed from traces obtained when the NS-2 Simulator tool is employed [18]. Because of space limitations, the results of the K-S tests are averaged and exposed in Figure 1 where the straight lines represent the standard deviation of the test statistics. As it can be observed, the best approximation is clearly obtained by a lognormal distribution function. It must be noted that for each of the performed simulations, the lognormal distribution is always the best fit according to the K-S test.

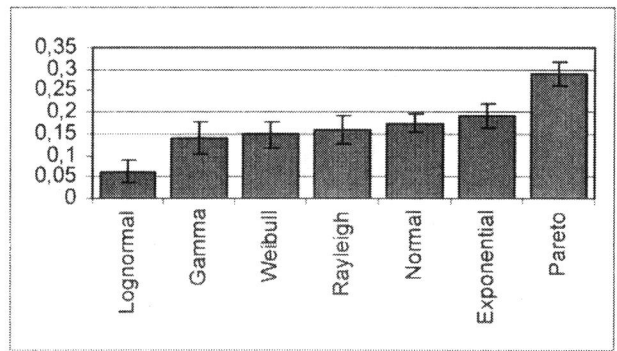

Fig. 1. Mean and deviation results of the K-S tests over 50 scenarios.

3.2 Real Mobility Traces

In order to study link duration in a VANET (Vehicular Ad Hoc Network) scenario, real position traces extracted from the public buses in Seattle are employed to compute link duration [19]. These traces are collected by the periodic transmission of the buses informing about their position along two days in 2001. Straight trajectories are assumed between consecutive position data. As conventional DSRC (Direct Short Range Communication), the transmission range was set to 1500 m. With these data, the connectivity graph is updated each second.

The fitting of the link duration is accomplished by means of the K-S test. The Fig. 2 shows that the best approximation corresponds to the lognormal function again.

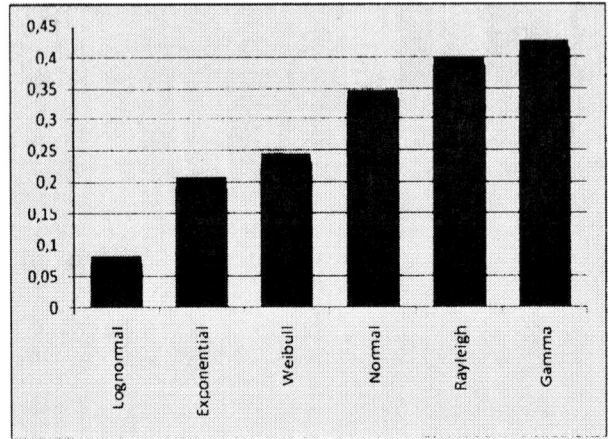

Fig. 2. Results of the K-S test for Link Duration Fitting in a VANET scenario.

3.3 Lognormal Fitting

From the previous sections, we can conclude that link duration could be accurately approximated by a lognormal function. This function is usually employed in the reliability research area as it is able to model the lifetime of electronic devices.

Formally, the probability that a link lifetime (L) is lower than t ($F(t)$) is computed by the Eq. 1 where μ y σ are the parameters of the lognormal function. These parameters are correlated to the characteristics of the scenario.

$$F(t) = \frac{1}{2} + \frac{1}{2} erf\left(\frac{\log(t/\mu)}{\sigma\sqrt{2}}\right) \qquad (1)$$

From this model, the path duration could be analytically described. The results are presented in [20].

4 Mean Residual Path Lifetime

In reliability theory, the mean residual life (MRL) is often used as a metric of the time that a component is expected to work given that the component has already lived a survival time t. In wireless multi-hop networks, this parameter can be employed as a measure of the stability associated to the routes of the nodes.

Formally, the MRL is defined as [21]:

$$MRL(t) = E[X - t / X > t] = \frac{\int_t^\infty \overline{F}(u)du}{\overline{F}(t)} = \frac{\int_t^\infty \left(\frac{1}{2} - \frac{1}{2} erf\left(\frac{\log(u/\mu)}{\sigma\sqrt{2}}\right)\right)du}{\frac{1}{2} - \frac{1}{2} erf\left(\frac{\log(t/\mu)}{\sigma\sqrt{2}}\right)} \qquad (2)$$

where $\bar{F}(t) = 1 - F(t)$ corresponds to the survavility function, i.e., the Complementary Cumulative Distribution Function or CCDF. In this case, the $F(t)$ corresponds to the distribution function of path duration which is analytically derived from link duration model in [20].

Figure 3 plots the MRL curves for different number of hops when all the links are characterized by the same μ and σ. It is shown that MRL values are strongly dependent on the number of hops. Additionally, they also vary on the survival time in a non-monotonic way as it corresponds to a lognormal-like MRL. The figure shows that this variation is smoothed when the number of hops increases. The limit when the number of hops is high makes the MRL curve tend to a constant value. Constant MRLs are related to exponential PDFs which is coherent with Palm's Theorem [4].

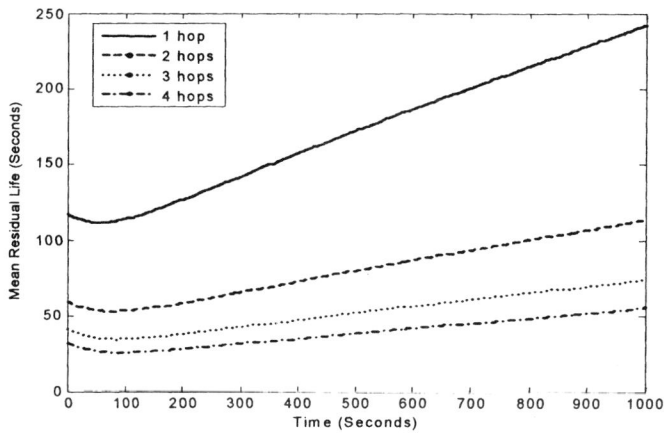

Fig. 3. MRL plots for 1,2,3 and 4 hops. Scenario parameters are: Uniform speed, $V_{max} = 2$ m/s, $V_{min} = 1$ m/s, Pause = 0 s, Tx range = 250 m, Nodes = 50, Area = 1500x300 m².

5 Proposed Criterion for Selecting Paths

Some ad hoc routing protocols allow the discovery of multiple routes to the same destination node. Under these circumstances, a criterion for selecting the route to employ is required. In this paper, a criterion based on the estimation of the path MRL is proposed. Specifically, among the paths with minimum number of hops, the route with a greater estimated MRL is selected. We consider that paths with minimum number of hops are preferred so that the number of wireless links can be minimized. The utilization along longer paths may increase the interferences as packets are retransmitted in each of the links that are part of the route.

To compute the Path MRL employing Eq. 2, the node that selects the path requires the parameters μ_i and σ_i of each i link within the path as well as the time when the link was established for a N-hop path ($1 \leq i \leq N$). Therefore, $3 \cdot N$ extra fields should be included in the ad hoc routing packets (Route Request or Route Reply). This is a clear limitation as it could noticeably increase the header length of packets. In order to overcome this restriction, each node estimates the MRL of the link that is going to be employed for the retransmission of the routing packet. From these data, the path MRL is approximated in a similar way to [3]. Formally, Eq. 3 shows the estimation of the Path MRL in a N-hop path:

$$\frac{1}{MRL\_PATH} = \sum_{i=1}^{N} \frac{1}{MRL\_Link_i} \quad (3)$$

Therefore, only one extra field (MRL_field) is incorporated into the header of the routing packets. This field accumulates the inverse of the MRLs of the links in the paths. As routes are learnt by the reception of Route Request (RREQ) and Route Reply (RREP) messages, the format of both packets should be modified. The Fig. 4 illustrates the process of retransmitting RREP packets and how the MRL_field is updated. In the example of the Figure 4, to select a route to Node C, Node A will select the path associated to a minimum value in the MRL_field as this field stores the inverse of the estimated path MRL.

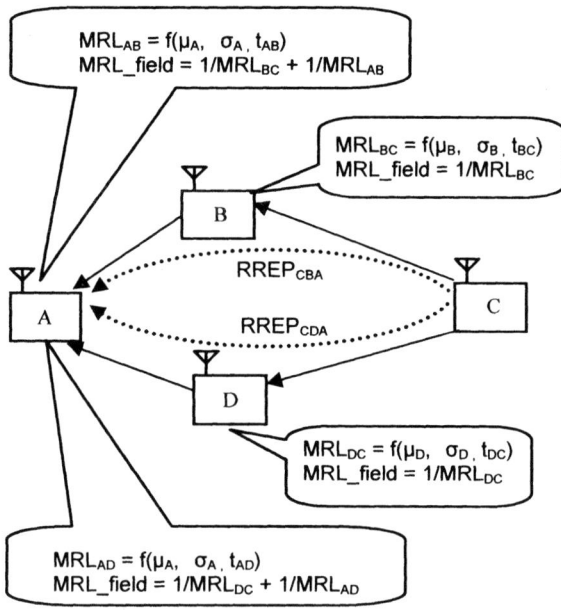

Fig. 4. Scheme for the retransmission of routing packets where the MRL_field is updated.

To compute the Link MRL, the i-th node utilizes the parameters μ_i, σ_i and the time when the link with the previous node j was established t_{ij}. The latter parameter could be easily stored in the neighbor list [22]. However, the μ_i and σ_i could only be estimated through the monitoring of the links. In [3], the λ parameter is derived from the observed mean link duration by means of an auto-regressive filter. In an analogous way, we estimate the μ_i and σ_i from the lifetimes associated to the links that are down in a given interval of time (T_{EWMA}). We compare this technique to the estimation of μ_i and σ_i from the lifetimes of the links that are at that moment up. The results show that there exist little differences when employing them. Therefore, we utilized the second method as its implementation is simpler.

6 Simulation Results

To evaluate the proposed criterion, a new module was developed in a Matlab framework [14]. In each 100000-second simulation, a source and a destination are randomly chosen. Then, four paths are selected accordingly to the different analyzed criteria. When one of the paths is broken, the path duration is stored in the data associated to the criterion by which the path was selected. The decision process is repeated once the four paths are down selecting again another set of four paths.

In each decision point, among the paths with a minimum number of hops, four of them are selected according to the following criteria:

- Maximum MRL Path. The path with the greatest estimation of the MRL is selected.
- Minimum λ Path. The path with a lowest λ is chosen. This criterion is included in order to compare the proposed criterion to the metric described in [3].
- Minimum MRL Path. The path with the lowest estimated MRL is selected. This criterion is included in order to establish a limit in the network performance.
- First Path. The first discovered path is employed. The route is randomly chosen among all the candidates with the minimum number of hops. This criterion does not consider any MRL estimation.

Five different scenarios characterized by a constant speed are employed for the evaluation. The simulation parameters are summarized in Table 1. The Fig. 5 draws the simulation results. As it is shown, the application of the Maximum MRL Path Criterion implies the selection of paths with longer lifetimes. These lifetimes exceed those obtained when the Minimum λ Path Criterion is employed.

Table 1. Parameters of the simulations.

Simulation Area	1500 m x 300 m
Mobile nodes	50
Mobility pattern	Constant Speed: [1 ,5] m/s. Pause Time : 0 seconds
Simulation Time	100000 s
Transmission Range	250 m

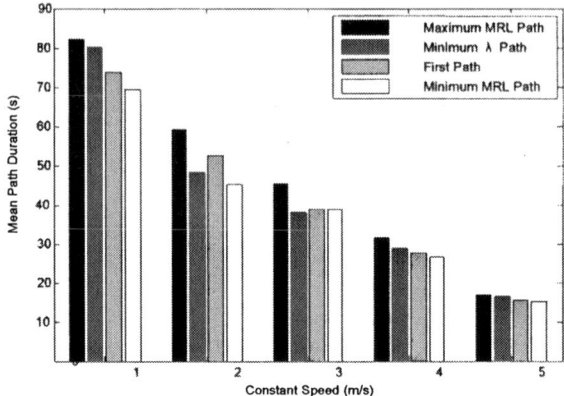

Fig. 5. Mean Path Duration associated to the application of 4 Path selection Criteria.

7 Conclusions

This paper presents two significant contributions. Firstly, the authors show that the link duration can be accurately approximated by a lognormal function in heterogeneous scenarios. Secondly and based on the previous result, the authors analytically compute path duration and path mean residual lifetime. From this study, it is possible to show that a memory-less model could only be employed when the number of hops that the route contains is high (greater than 4). For routes involved in conventional ad hoc applications (with 2 o 3 hops), the criterion for selecting routes in the network could be supported by the path survival time, i.e., the time elapsed since the route was established. The path mean residual lifetime is estimated by means of the link residual lifetime associated to the links that form the path. This data is included in the criterion for selecting routes in multihop ad hoc networks. The simulation results show that path live longer when this information is incorporated in the decision phase of path selection.

Acknowledgments. This work has been partially supported by the Public Spanish Project TEC2006-12211-C02-01.

References

1. S. Gwalani, E. M. Belding-Royer and C. E. Perkins, "AODV-PA: AODV with path accumulation," *Communications, 2003.ICC'03.IEEE International Conference on,* vol. 1, 2003.

2. O. Tickoo, S. Raghunath and S. Kalyanaraman, "Route fragility: a novel metric for route selection in mobile ad hoc networks," *Networks, 2003.ICON2003.the 11th IEEE International Conference on*, pp. 537-542,
3. Y. Han and R. J. La, "Maximizing Path Durations in Mobile Ad-Hoc Networks," *40th Annual Conference on Information Sciences and Systems, Princeton, NJ, March,* 2006.
4. Y. Han, R. J. La and A. M. Makowski, "Distribution of path durations in mobile ad-hoc networks--Palm's theorem at work," *16th ITC Specialist Seminar,* 2004.
5. J. Broch, D. A. Maltz, D. B. Johnson, Y. C. Hu and J. Jetcheva, "A performance comparison of multi-hop wireless ad hoc network routing protocols," *Proceedings of the 4th Annual ACM/IEEE International Conference on Mobile Computing and Networking,* pp. 85-97, 1998.
6. F. Bai, N. Sadagopan, B. Krishnamachari and A. Helmy, "Modeling path duration distributions in MANETs and their impact on reactive routing protocols," *Selected Areas in Communications, IEEE Journal on,* vol. 22, pp. 1357-1373, 2004.
7. S. Arbindi, K. Namuduri and R. Pendse, "Statistical estimation of route expiry times in on-demand ad hoc routing protocols," *Mobile Adhoc and Sensor Systems Conference, 2005.IEEE International Conference on,* pp. 16-23, 2005.
8. S. Jiang, D. He and J. Rao, "A prediction-based link availability estimation for mobile ad hocnetworks," *INFOCOM 2001.Twentieth Annual Joint Conference of the IEEE Computer and Communications Societies.Proceedings.IEEE,* vol. 3, 2001.
9. J. Boleng, W. Navidi and T. Camp, "Metrics to enable adaptive protocols for mobile ad hoc networks," *Proceedings of the International Conference on Wireless Networks (ICWN'02),* pp. 293 298, 2002.
10. S. Cho and J. P. Hayes, "Impact of Mobility on Connection Stability in Ad Hoc Networks," *Proc.of IEEE Communication Society, WCNC,* vol. 3, pp. 1650-1656, 2005.
11. D. Turgut, S. K. Das and M. Chatterjee, "Longevity of routes in mobile ad hoc networks," *Vehicular Technology Conference, 2001.VTC 2001 Spring.IEEE VTS 53rd,* vol. 4, 2001.
12. A. B. McDonald and T. Znati, "A path availability model for wireless ad-hoc networks," *Wireless Communications and Networking Conference, 1999.WCNC.1999 IEEE,* pp. 35-40, 1999.
13. P. Samar and S. B. Wicker, "On the behavior of communication links of a node in a multi-hop mobile environment," *Proceedings of the 5th ACM International Symposium on Mobile Ad Hoc Networking and Computing,* pp. 145-156, 2004.
14. http://www.mathworks.com
15. J. Yoon, M. Liu and B. Noble, "Random waypoint considered harmful," *INFOCOM 2003.Twenty-Second Annual Joint Conference of the IEEE Computer and Communications Societies.IEEE,* vol. 2,
16. M. Gerharz, C. de Waal, P. Martini and P. James, "Strategies for finding stable paths in mobile wireless ad hoc networks," *Local Computer Networks, 2003.LCN'03.Proceedings.28th Annual IEEE International Conference on,* pp. 130-139, 2003.
17. T. Dimitar, F. Sonja, C. Bekim and G. Aksenti, "Link Realiability analysis in ad hoc networks," *Proc.of XII Telekomunikacioni Forum TELFOR,* 2004.
18. K. Fall and K. Varadhan, "The ns Manual (formerly ns Notes and Documentation)," *The VINT Project,* vol. 1, 2002.
19. J. G. Jetcheva, Y. C. Hu, S. PalChaudhuri, A. K. Saha and D. B. Johnson, "Design and evaluation of a metropolitan area multitier wireless ad hoc network architecture," *Mobile Computing Systems and Applications, 2003.Proceedings.Fifth IEEE Workshop on,* pp. 32-43, 2003.
20. A. Triviño-Cabrera, J. Garcia-de-la-Nava, E. Casilari and F. J. González-Cañete, "An analytical model to estimate path duration in MANETs," *Proceedings of the 9th ACM International Symposium on Modeling Analysis and Simulation of Wireless and Mobile Systems,* pp. 183-186, 2006.

21. N. Ebrahimi, "Estimation of Two Ordered Mean Residual Lifetime Functions", Biometrics, Vol. 49, No. 2 (Jun., 1993), pp. 409-417
22. C. E. Perkins, E. Belding-Royer and S. Das, "Ad hoc on demand distance vector (AODV) routing. IETF RFC 3561, 2003.

WiMAX Throughput Evaluation of Conventional Relaying

Pavel Mach[1], Robert Bestak[1]

[1] Department of Telecommunications Engineering, Faculty of Electrical Engineering, Czech Technical University, Technicka 2,
166 27 Prague 6, Czech Republic
{machp2, bestar1}@fel.cvut.cz

Abstract. The paper analyses performance of IEEE 802.16 standard, also known as WiMAX, when relay stations are introduced. Thanks to this feature, coverage area and throughput of the system may be significantly enhanced. The article defines the increase of system capacity of the relay based cell deployment compared to a conventional single hop deployment of the same area size. Further, the parameters that have main impact on the system with relays and its capacity are determined.

Keywords: WiMAX, performance, throughput, Relay Station.

1 Introduction

In recent years, broadband wireless systems established themselves as one of the fastest growing and developing area in the field of telecommunications. The current trends and demands are to deliver multimedia services such as voice, video, high definition TV (HDTV) or interactive games with guaranteed Quality of Service (QoS). To support high quality multimedia services, a high data transmission are necessary. Since the most of the wireless systems operate in high frequency bands above 2 GHz [1], [2], the transmitted signal is highly attenuated in comparison with low bands. As a consequence, the cell size is smaller and more Base Stations (BSs) is needed. Another drawback in using of high frequencies is big power transmission requirement. An efficient way how to reduce power consumption and extend the range offers Mesh and Relay networks that are distinguished by multihop communication between individual nodes. Furthermore, the capacity of systems using Mesh and Relay topology can be enhanced when compared to conventional single hop deployment.

According to [3], three types of Relay Stations (RSs) are defined; fixed, nomadic and mobile. The relays are in most cases build in, owned and controlled by service provider. An RS is not directly connected to wire infrastructure and has the minimum functionality to support multihop communication. Multihop based network may also improve system performance when cooperation relay technique is put to use. This is accomplished by sending information simultaneously via multiple different paths and combining the received information at the side of the receiver [4].

Two concepts how to integrate multihop communication into IEEE 802.16 standard are presented in [5]. The first concept follows a centralized approach, where the BS has full control over the relay-enhanced cell and RS may be very simply. The second concept follows a semi-distributed approach, where RS coordinates the associated Subscriber Stations (SSs) itself. In the second case, the MAC protocol complexity of RS is comparable to BS.

The rest of the paper is organized as follows. Section 2 defines the simulation model, its parameters and basic assumption. Moreover, the scenarios that are taken into consideration are depicted. In section 3, the very process of simulation is described by path loss model and by method of throughput calculation. The following section provides simulation results in comparison of single and multihop deployment by means of system throughput. The last section gives conclusions of the paper and further contemplates the parameters that affect the systems performance.

2 Simulation model

The example of simulation scenario is illustrated in fig 1. The cell with one BS is enhanced by four symmetrically positioned fixed RSs. The simulation process is divided into 40 steps. In every step, one SS joins the network and after that throughput is computed. Thus, model includes 40 SSs which coordinates are pseudo randomly chosen so that 8 SSs are associated to the BS and other 32 are evenly distributed to each RS.

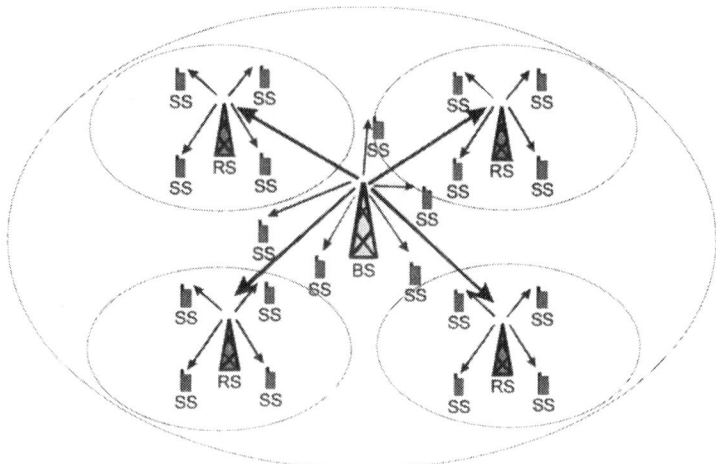

Fig. 1. Simulation scenario

The SSs establish link to the BS or RS according to received SNR. This means that SS try to join to station with the best signal. The BS and RS coverage depends especially on transmitted power, noise, height of antenna above the ground and finally on the path loss between the transmitter and the receiver. All necessary parameters used during the simulation process are summarized in Table 1.

Table 1. Simulation parameters

Parameter	Value	
BS range radius (m)	2600	
RS range radius (m)	1000	
BS-RS distance (m)	1470	
Frequency band (GHz)	5	
Number of BS	1	
Number of RS	4	
Number of SS	40	
Channel bandwidth (MHz)	3.5	20
Frame duration (ms)	20	
Symbol useful time t_b (μs)	64	11.64
t_g/t_b	1/16	1/4
CP time t_g (μs)	4	2.91
Symbol time t_s (μs)	68	15.55
Number of OFDM subcarriers	256	
Number of data subcarriers	192	
BS transmit power P_t (dBm)	30	
RS transmit power P_t (dBm)	26	
BS height (m)	30	
RS height (m)	25	
SS height (m)	2	
a	3.6	
b (m^{-1})	0.005	
c (m)	20	
Noise (dBm)	-108.54	-100.97

Table 2 demonstrates which type of modulation and coding rate is used in relation to received SNR [1]. Simultaneously, the number of SSs per each burst type is shown for both cases, with no RSs and with RSs. It is evident, that higher modulation orders are seldom utilized for its high SNR requirements. Consequently, the system throughput is largely degraded. But the unfavorable situation is largely improved by RSs utilization. Finally, Table 2 denotes how many bits can be allocated to one OFDM symbol. This information is vital to system capacity estimation.

Table 2. Type of modulation and coding rate according to SNR

Modulation and coding rate		Received SNR (dB)	Number of SSs		Bits/ symbol
			Without RSs	With RSs	
BPSK	1/2	3	4	0	96
QPSK	1/2	6	12	0	192
	3/4	8.5	8	9	288
16QAM	1/2	11.5	12	11	384
	3/4	15	4	16	576
64QAM	2/3	19	0	4	768
	3/4	21	0	0	864

During the simulation process, two basic scenarios were considered: i) without RSs and ii) with RSs (case 1). The first one reflects the situation when no RSs are introduced. This scenario may correspond to classical IEEE 802.16-2004 or IEEE 802.16e deployment. The second scenario introduces four RSs where every RS has dedicated own part of MAC frame for its 2nd hop transmission.

However, for a scenario with great number of relays is not profitable to reserve for every RSs transmissions individual part of the frame. The reason is that data transmitted on the 2nd or higher hops increase overhead. Thus, three additionally scenarios were investigated that study its reduction possibility: i) with RSs (case 2) – two RSs are allowed to transmit at once, ii) with RSs (case 3) – all active RSs are allowed to transmit at once and iii) with RSs (case 4) – all active RSs may transmit simultaneously and no interference is taken into account. This situation may occur when individual RSs are either in sufficient distance to each other or when some kind of obstacles are between them, e.g. trees, hill, etc.

Before the description of the simulator, it is important to report the assumptions under which the model was developed. These are following: i) all nodes are fixed ii) only DL transmissions are taken into account. But if we assume for the sake of simplicity, that transmit power of SS is the same as BS (RS) or its lower transmit power is compensated by higher BS (RS) received antenna gain, the results may also correspond to UL or combined UL/DL direction, iii) PMP network topology is assumed. Hence, all traffic always leads from the BS via direct path or via one intermediate RS (no communication between SSs is countenance), iv) TDD MAC frame is used.

3 Simulator description

All simulation and throughput estimation were generated in Matlab system. Fig. 2 describes the process of simulation step by step. First of all, the input simulation parameters are inserted (see Table 1) and derived parameters are computed. After association of the SS into the network, SNR calculation, number of data OFDM symbols and system throughput is estimated.

Personal Wireless Communications 79

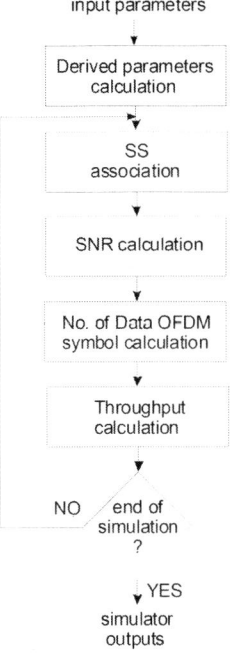

Fig 2. Description of simulation process

3.1 Path loss model characterization and received SNR calculation

Path loss estimation Channel model used in simulation was developed by Stanford University and was picked up by its suitability for fixed broadband wireless access [6]. The basic path loss equation with correction factors is calculated as,

$$PL = A + 10\gamma \log_{10}\left(\frac{d}{d_o}\right) + X_f + X_h \cdot \quad (1)$$

where, d is the distance between the transmitter and the receiver antennas in meters, $d_0 = 100$ m. The other parameters are defined as,

$$A = 20\log_{10}\left(\frac{4\pi d_0}{\lambda}\right) \cdot \quad (2)$$

$$\gamma = a - bh_b + c/h_b \cdot \quad (3)$$

where h_b is base station height above ground in meters and constant a, b and c are given in Table xx. The correction factors for the operating frequency and for SS height are given as,

$$X_f = 6.0 \log_{10}\left(\frac{f}{2000}\right). \qquad (4)$$

$$X_h = -20 \log_{10}\left(\frac{h_r}{2}\right). \qquad (5)$$

where f is the frequency in MHz and hr is the SS antenna height above ground in meters. Further, the received signal strength can by estimated by following equation,

$$P_r = P_t - PL. \qquad (6)$$

where P_t is the transmitted signal strength at the side of a receiver. The SNR is computed as,

$$SNR = P_r - Noise. \qquad (7)$$

where Noise represents either only thermal noise (see Table 1) when no interference is introduced, or summation of thermal noise and interference received by disturbers.

3.2 Calculation of number of OFDM symbols used for data transmission

Fig. 3 demonstrates how the throughput of the system is calculated. The overall number of OFDM symbols (variable N) in one MAC frame is derived from two parameters, i.e. length of OFDM symbol and frame MAC duration. According the Fig. 3, the number of OFDM symbols used only for data is calculated by following formula:

$$BS\_OFDM = N - (X_1 + X_2 + X_3 + X_4 + X_5). \qquad (8)$$

where:
- N is overall number of OFDM symbols in MAC frame.
- X_1 represents BS broadcast part include long preamble (2 OFDM symbols), frame control header (1 OFDM symbol) and finally DL/UL MAPs. Since the broadcast part of the frame is transmitted by the most robust modulation type and coding rate, only 92 bits/OFDM can be carried (see Table 2). The length of the maps and simultaneously the quantity of OFDM symbols depends on the number of IEs in MAPs.
- X_2 is a gap between DL and UL subframes depends mainly on the round trip delay (for this case 1 OFDM symbol is taken).
- X_3 are contention slots are composed from initial ranging slot and BW requests slots. The first one takes 5 OFDM symbols to overcome round trip delay and to

allow transmit ranging request message and long preamble. The BW request slot is 2 OFDM symbols long (short preamble and BW request message). The quantity of these slots is dependent on the number of users. It is supposed that more users need more slots, thus for every five SSs is dedicated one BW request slot.

- X_4 is before every UL burst, short preamble (1 OFDM symbol) is appended.
- X_5 is a gap between the UL and DK subframes (same as X_2).

Calculation of throughput for cases with RSs is slightly more complex (see Fig. 4). Compared with previous scenario, more OFDM symbols are taken for MAC overhead and that is by RS broadcast part of the frame (X_6). The quantity of OFDM symbols used for users data can be computed by following equation:

$$RS\_OFDM = N - (X_1 + X_2 + X_3 + X_4 + X_5 + X_6). \tag{9}$$

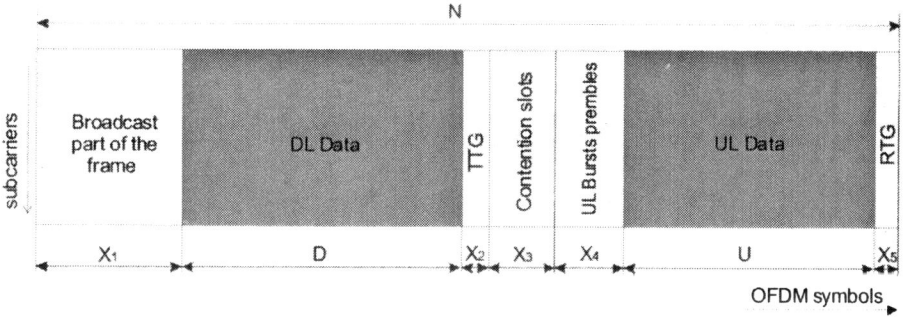

Fig 3. Calculation of OFDM symbol used for data (no RSs)

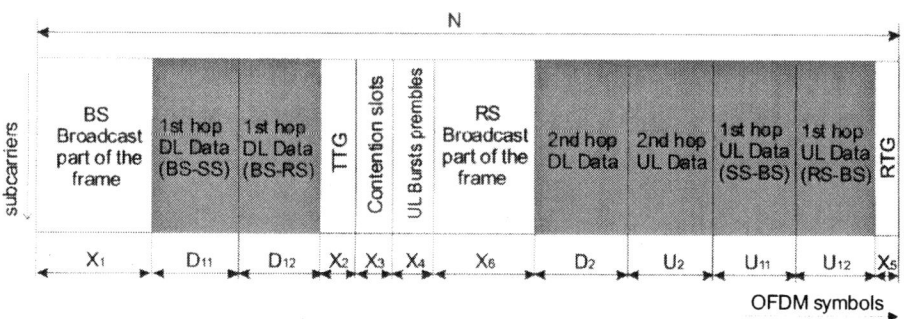

Fig 4. Calculation of OFDM symbol used for data (with RSs)

3.3 Throughput calculation

The rest of the OFDM symbols (BS_OFDM or RS_OFDM) are used for data transmission. The overhead introduced by other eventually MAC messages as well as by MAC PDU headers is neglected since the impact on overall throughput is meaningless. System capacity now depends on the modulation type and coding rate that can be applied to single SSs (according to received SNR). The WiMAX capacity to requested capacity can be estimated by equation 10 for cases without RSs and by equation 11 for cases with RSs:

$$WC\_to\_RC(BS) = \frac{BS\_OFDM}{R\_OFDM(BS)} \quad . \tag{10}$$

$$WC\_to\_RC(RS) = \frac{RS\_OFDM}{R\_OFDM(RS)} \quad . \tag{11}$$

where $R\_OFDM(BS)$ and $R\_OFDM(RS)$ represent how many OFDM symbol is needed to transfer given nominal bit rates for all active users. The WiMAX system is able to meet users' requirements as long as the WC_to_RC is higher than one. The $R\_OFDM(BS)$ can be estimated by the following formula,

$$R\_OFDM(BS) = BR * LoF * (\sum_{i=1}^{7} \frac{SSG1_i}{bps_i}) \quad . \tag{12}$$

where BR is the nominal bit rate in bits, LoF is frame duration in seconds, $SSG1_i$ represent the number of SSs that use individual modulation type and coding rate and finally bps_i express how many bits can be allocated to one OFDM symbol (see Table 2). The calculation of R_OFDM(RS) is computed by:

$$R\_OFDM(RS) = BR * LoF * (\sum_{i=1}^{7} \frac{SSG1_i}{bps_i} + \sum_{i=1}^{7} \frac{SSG2_i}{bps_i} + \frac{1}{bps} \sum_{i=1}^{7} SSG2_i) \quad . \tag{13}$$

where $SSG1_i$ has the same meaning like in previous case (number of SSs direct connected to the BS, the $SSG2_i$ corresponds to the quantity of SSs on the 2nd hop (between the RS and SS). The last term in the round bracket represents the transmission between the BS and RS and so bps express the quantity of bits that can be allocated to one OFDM symbol (dependent on the RS-BS link quality).

The throughput per users can be derived from equation 14 for "only BS" scenario and equation 15 for "with RSs" scenarios.

$$BR = \frac{R\_OFDM(BS)}{LoF * (\sum_{i=1}^{7} \frac{SSG1_i}{bps_i})} \quad . \tag{14}$$

$$BR = \frac{R\_OFDM(RS)}{LoF *(\sum_{i=1}^{7}\frac{SSG1_i}{bps_i} + \sum_{i=1}^{7}\frac{SSG2_i}{bps_i} + \frac{1}{bps}\sum_{i=1}^{7}SSG2_i)} \quad (15)$$

This process is repeated until the end of simulation when outputs of the simulation are obtained by means of results and graphs.

4 Performance evaluation

The overall system performance may be measured by its throughput. In Fig. 5 is shown how the throughput per user is decreased when number of SSs grows. Fig. 5a further indicates that for 3.5 MHz channel size the performance with or without RSs is nearly identical.

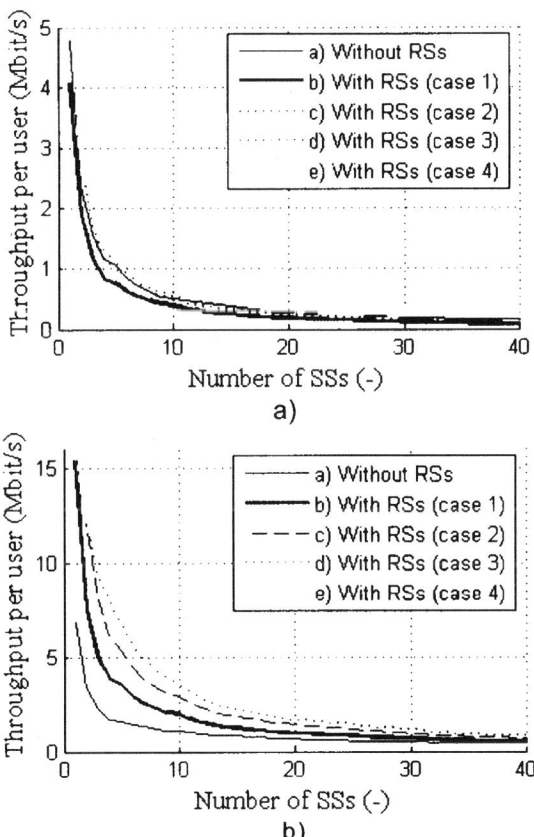

Fig. 5. System throughput per user for: a) 3.5 MHz channel size, b) 20 MHz channel size

This implies that for fully utilization of relay station merits, the use of wider channel is necessary. If this condition is satisfied, the system capacity may be significantly enhanced, e.g. for 10 active users, the bit rate per user is improved from 1.6 Mbit/s up to 4 Mbit/s depending on the case used (see Fig. 5b). In general, the improvement strongly rests on the distribution of SSs over the area and its number. Thus the positions of RS must be properly chosen.

Fig. 6 shows system behavior when nominal bit rates are set and the number of SSs is increasing. For reasons listed above, only 20 MHz channel size was taken into consideration. At y axis, the rate of the system to the requested capacity is computed. The capacity limit is denoted by the bold line. If the curve is above the bold line, the free resources are still available and more users can use given nominal bit rate. In opposite, some restrictions have to be applied to support more users. Expectably, the system performance with RSs is better, e.g. nominal bit rate 1 Mbit/s can be applied up to 35 users against 11 users for system without RSs, before congestion occurs.

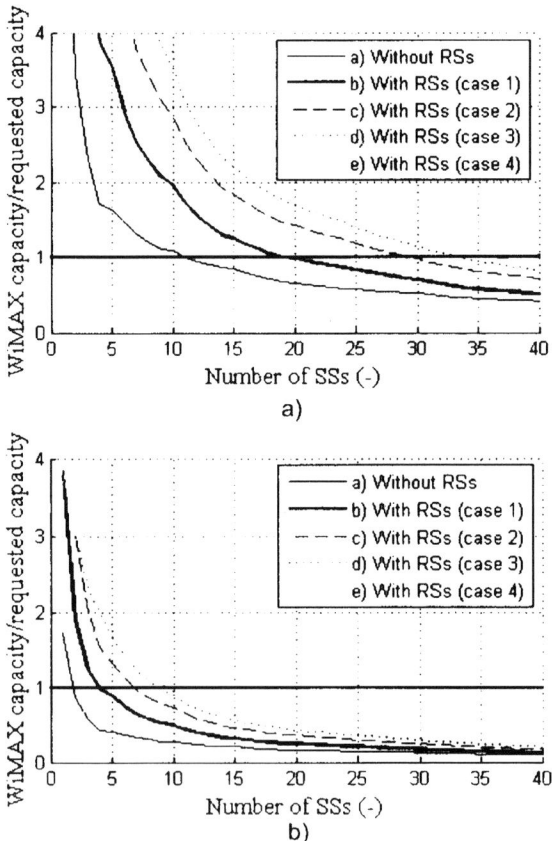

Fig. 6. System throughput for different nominal bit rate: a) 1 Mbit/s, b) 3 Mbit/s (20 MHz channel size)

Other important standpoints that must be taken into account is the height and the transmit power of RSs. The main reason is to avoid of co-channel interference and eventually also disturbance into adjacent channel. How the performance of the system is affected by adjustment of these parameters is shown in Fig. 7 and 8.

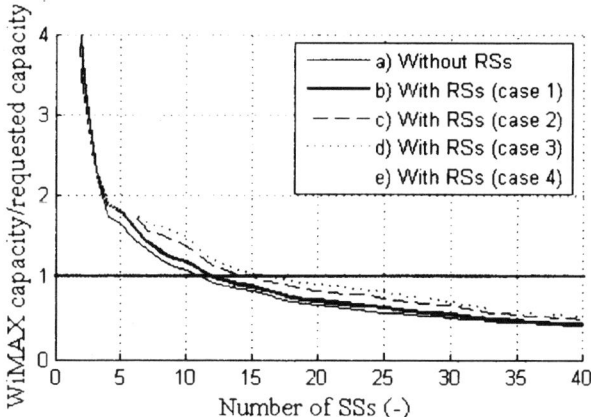

Fig. 7. System performance for RS height of 15 m (20 MHz channel size, nominal bit rate 1 Mbit/s)

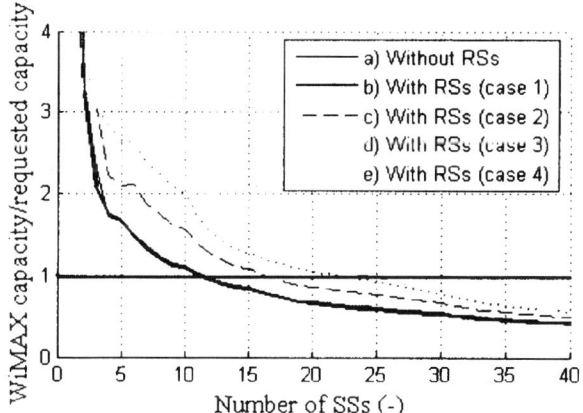

Fig. 8. System performance for RS transmit power 20 dBm (20 MHz channel size, nominal bit rate 1 Mbit/s)

It is plain that lowering of RS antenna height as well as reduction of its transmitted power has a great impact on the system performance. This is mainly caused by two facts namely i) RS area coverage is decreased and ii) the link between RS and BS is largely degraded. It could be observed that especially results with RS antenna height 15 m above the ground are almost comparable to scenario where no RSs were introduced.

5 Conclusions

The paper analyzed the performance of WiMAX system when Relay stations are employed. It was shown that for RSs deployment and their fully merits utilization, several aspects must be carefully considered for their effect on the system performance and that is i) height of RS antenna and its transmit power, ii) numbers of RSs transmit at once and iii) channel size.

The height of RS together with its transmit power have major impact at the system capacity by specifying of RS coverage area and link quality between the BS and RS. The BS-RS transmissions can be considered as overhead. The prospective possibility is to allow more than one RS transmission at the same time. The simulation results support this statement and show that throughput of the system is largely improved when more RSs transmit at once. So the time interval dedicated to the RSs can be significantly reduced. The improvement of the system capacity could be accomplished despite of interference rise.

At last, the results demonstrate that narrow channel size is not suitable for system with RSs. The main reason is that narrow channel size (e.g. 3.5 MHz) is distinguished by small number of OFDM symbols in MAC frame. Especially, this is true for shorter time MAC frame duration, e.g. 2.5 ms or 5 ms. In that case, big portion of frame is dedicated for BS-RS transmissions. As result, the overhead introduced by the RSs is much more significant. But this is not the case for the wider channel size (e.g. 20 MHz) where the percentage of OFDM symbols used for RS is insignificant.

Acknowledgments. This work has been performed in the framework of the FP6 project FIREWORKS IST-27675 STP, which is funded by the European Community. The Authors would like to acknowledge the contributions of their colleagues from FIREWORKS Consortium (http://fireworks.intranet.gr). Further, this research work was supported by Czech Technical University's grant No.CTU0715013.

References

1. IEEE Std 802.16e-2005, IEEE Standard for Local and metropolitan area networks, Part 16: Air Interface for Fixed and Mobile Broadband Wireless Access Systems, Amendment for Physical and Medium Access Layers for Combined Fixed and Mobile Operation in Licensed Bands, February 2006..
2. A. Molisch, "Wireless Communication", first edition, ISBN 0470848871, November 2005
3. J. Sydir, IEEE 802.16 Broadband Wireless Access Working Group – Harmonized Contribution on 802.16j (Mobile Multihop Relay) Usage Models, July 2006.
4. A. Nosratinia, T. E. Hunter, A. Hedayat, "Cooperative Communication in Wireless Networks", IEEE Communications Magazine, vol. 42, no. 10, October 2004, pp. 68-73.
5. Ch. Hoymann, K. Klagges, "MAC Frame concepts to Support Multihop Communication in IEEE 802.16 Networks", Wireless Word Research Forum, 2005
6. V.S. Abhayawardhana, et al., "Comparison of empirical proagation path loss models for fixed wireless access systems", IEEE Vehicular Technology Conference, 2005.

Study of the IEEE 802.16 Contention–based Request Mechanism *

Jesús Delicado, Francisco M. Delicado and Luis Orozco–Barbosa

Instituto de Investigación en Informática de Albacete (I^3A)
Universidad de Castilla–La Mancha (UCLM)
02071–Albacete, Spain
{jdelicado,franman,lorozco}@dsi.uclm.es

Abstract. Broadband wireless access systems offer a solution for broadband access and QoS–aware multimedia services through a wireless medium. The IEEE 802.16 standards specify the physical and medium access control layers for broadband wireless access systems as well as the various mechanisms to meet the quality of service (QoS) requirements of a wide variety of applications. Among the mechanisms defined by the standards, the bandwidth request and grant mechanisms play a central role on guaranteeing the QoS required by the subscriber stations.

In this paper, we undertake the study of the bandwidth request mechanisms defined in the IEEE 802.16 standards. We then propose an approach for reducing the overhead required by the signalling mechanisms. Our simulation results show that our approach outperforms the mechanisms proposed in the IEEE 802.16 standards.

Keywords: TDMA/FDD Resource Request, WiMAX.

1 Introduction

The design and adoption of broadband wireless access (BWA) systems is one of the most significant networking research and development activities nowadays. BWA systems provide fixed–wireless access to SS —Subscriber Station— (residential or business customers) to internet service provider (ISP) facilities. Their main advantages are: rapid deployment, scalability, low maintenance and upgrade costs. Their installation can be particular useful in (a) very crowded geographical areas such as urban areas or in (b) rural areas lacking broadband wired infrastructure. The IEEE 802.16 standards have been particularly designed for BWA systems. The IEEE 802.16 standards support a point–to–multipoint (PMP) architecture within the 10–66 GHz range of frequencies, achieving about

---

* This work has been jointly supported by the Spanish MEC and European Comission FEDER funds under grants "Consolider Ingenio–2010 CSD2006–00046" and "TIN2006–15516–004–02"; by JCCM under project PAI06–0106 and grant 05/112, and by UCLM under project TC20070084.

50 km of distance. Different products have been developed in the past few years [?].

The IEEE 802.16 medium access control (MAC) protocol is a centralized connection–oriented mechanism. Under this protocol, the BS is responsible of allocating the bandwidth required by the SSs. Furthermore, the BS has to classify, prioritize and schedule the SSs requests. Towards this end, the BS has to timely and efficiently manage the overall available bandwidth. The signalling protocol is therefore a central element allowing the SSs to place their requests and the BS to issue the grants to the SSs.

Regarding the IEEE 802.16 signalling protocols, the standard specifies that a SS has first to request the bandwidth according to the needs of each one of the individual connections associated to the SS. However, the requests can be granted using one of two modes: GPC (*Grants Per Connection*) or GPSS (*Grants Per Subscriber Station*). Under the GPC mode, grants are made to individual connections. The GPSS mode instead grants the bandwidth to the SS without specifying a particular connection. Under the latter mode, the responsibility of allocating the granted bandwidth is left to the SS.

In this paper, we develop and analyze a new algorithm implementing the request bandwidth primitives available to the SSs. Our proposal aims to improve the performance of the modes proposed by the standards in terms of three metrics, namely the overhead, throughput and delay.

The paper is organized as follows. Section 2 provides an overview of the IEEE 802.16 standard, which includes a brief description of the MAC protocol and request and grant bandwidth mechanisms. In Section 3, we analyze the relevant literature closely related to our proposal. Section 4 describes our proposal and in Section 5, we carry out a comparative performance study of our proposal with respect to the modes included in the standards. Finally, Section 6 concludes the paper.

2 IEEE 802.16

The IEEE 802.16 standards specify the physical (PHY) and MAC layer of the air interface of interoperable point–to–multipoint and optional Mesh topology BWA systems.

The range of frequencies supported by IEEE 802.16 is from 2 to 66 GHz, which includes the licensed and license–exempt bands. Line–of–sight (LOS) is sometimes necessary depending on the range of frequencies. Three types of modulation can be used: QPSK (Quadrature Phase–Shift Keying), 16–QAM (Quadrature Amplitude Modulation) and 64–QAM, but only QPSK is mandatory. The bit rate and robustness in the presence of errors depend on the type of modulation and frequency being used.

2.1 MAC layer

As already mentioned, the MAC layer is a centralized and connection–oriented mechanism, that is, the BS is responsible of allocating the resources and provisioning the system with QoS-aware mechanisms. To perform this task, the SSs are required to request to the BS their needs in a timely manner. That is to say, each SS has to request to the BS the required bandwidth on a frame by frame basis. In response to the resources required by all the SSs, the BS distributes the available bandwidth taking into account the requirements of all the outstanding SSs requests.

The communication between the BS and the SSs is realized by means of fixed-length frames, divided into two subframes: the downlink and uplink subframes, whose lengths are dynamically controlled by the BS. The mode of operation can be Frequency Division Duplexing (FDD) or Time Division Duplexing (TDD). The downlink and uplink communications are time multiplexed by means of a Time Division Multiple Access (TDMA) mechanism.

The downlink subframe starts with a Frame Start Preamble used by the PHY for synchronization and equalization. This is followed by the *frame control section*, which is composed of management messages. It contains the downlink and uplink maps stating the physical slots (PSs) at which bursts begin, that is to say, they comprise the bandwidth allocations for the data transmission in the downlink and uplink directions, DL-MAP and UL-MAP messages, respectively.

The DL-MAP contains the addresses of the first time slots being used to convey the data transmitted by the BS and its correspondent downlink burst profile (Downlink Interval Usage Code, DIUC).

The UL-MAP contains the specific data (Information Element, IE) which include the transmission opportunities, that is to say, the time slots at which each an every active SS can transmit. After receiving the UL-MAP message, the SSs transmit their data in predefined time slots as indicated in the IE. The BS scheduling module determines the transmission opportunities (IEs) using the bandwidth request (BW-REQ message) sent by the SSs to BS. The size of this message affects the remaining size of the downlink frame, that is to say, the longer this message, the less available bandwidth for the transmission of data over the downlink direction and vice versa. However, the UL-MAP size will increase as the number of grants in the uplink direction increases.

Following these maps, the DCD and UCD messages indicate the physical characteristics of the physical channels. Finally, it is introduced a TDM portion which carries data, organized into bursts depending on the negotiated burst profile between the BS and the SS, and therefore different levels of transmission robustness. Bursts are transmitted in order of decreasing robustness.

In the FDD case, the TDM portion may be followed by a TDMA portion used for transmitting data to any half-duplex SSs scheduled to transmit earlier than in the frame they are receiving data. Each burst begins with the Downlink TDMA Burst Preamble to regain synchronization. In this case, the DL-MAP message includes a map of, both, the TDM and TDMA bursts.

The structure of the uplink subframe is divided into three classes of bursts:

1. Contention opportunities reserved for Initial Maintenance. The RNG-REQ messages are transmitted by the SSs joining the network.

2. Contention opportunities defined by the Request Intervals. These are reserved to convey the replies to the multicast and broadcast polls, in which the transmitted BW-REQ messages are placed. These messages are issued by the SSs to indicate to the BS their needs.
3. Intervals defined by Data Grant IEs specifically allocated to individual SSs.

The bandwidth allocated for Initial Maintenance and Request contention opportunities may be grouped together and is always used with the uplink burst profiles. The remaining transmission slots are grouped by SS. The UL-MAP message, in the downlink subframe, grants bandwidth to specific SSs and indicates the uplink burst profile assigned to transmit. These data are used by each SS to transmit in its assigned allocation. These transmissions are separated by SS Transition Gaps in order to properly synchronize to the SS.

2.2 Requests

A request refers to the mechanism that SSs use to indicate to the BS that they need uplink bandwidth allocation. This request may come as a stand–alone bandwidth request header (BW-request) or as a piggyback request.

The IEEE 802.16 standard uses a random access mechanism in the uplink subframe for transmitting the opportunity requests to the SSs, during the request contention period defined into this subframe. The BS is responsible of establishing this reservation period at the beginning of each subframe. In this way, the SSs can place their reservations (BW-requests) for transmitting in the next subframe (or later, depending on the happening or not of collisions). The standard defines the *truncated binary exponential backoff* algorithm as the mechanism for resolving potential conflicts during this interval.

To limit the length of the contention resolution period is one of the main issues to be addressed. The longer the contention resolution period, the shorter the available capacity left for transmitting data.

According to the IEEE 802.16 standard, the requests have to be issued on an individual basis, that is, a SS has to issue a request for each connection associated to it. Each request has to be identified by a connection identifier denoted from now on as CID (Connection IDentifier).

The use of piggyback requests is out of the scope of this paper, because we are interested in improving the contention period in which it is only possible to make BW-requests using the backoff mechanism.

2.3 Grants

According to the IEEE 802.16 standards, the grants can be issued according to two schemes. The first version of IEEE 802.16 standard has defined a mechanism to grant bandwidth on a connection by connection basis, where each connection is associated to its corresponding SS. This mechanism has been named *Grants Per Connection* (GPC). The grant is associated to the connection by explicitly indicating on the grant the connection identifier, CID. In the second approach, the grants are issued on a station basis (*Grants Per Subscriber Station*, GPSS). Here, the bandwidth is granted to the SS and not explicitly to an individual CID. Each SS is then responsible of allocating the received opportunities among its different types of service flows (applications).

It is obvious that the GPC method requires more bandwidth to convey the grants messages, since it is necessary to individually indicate the bandwidth to each connection. In the GPSS method, only one grant per SS is sent to the concerned SS. There is therefore a trade–off to be considered between the amount of signalling traffic and the processing required for effectively distributing the bandwidth among the competing connections.

3 Related work

Many research efforts on the performance of MAC protocols have been carried out and have been reported in the literature. In [?], authors made an analysis of different parameters involved in the truncated binary exponential backoff algorithm to resolve collisions produced in the system. The optimal contention period has been studied in [?], according to the number of users under definition of a cost function where the channel throughput and delay of the system have been considered the two essential metrics. Finally, the authors have come to the conclusion that the optimal size of the contention period is $2M - 1$, where M is the number of SSs (users). In [?], the authors have proposed a new QoS architecture for IEEE 802.16 and have obtained the size of contention period accordingly to maximize the throughput. In this case, the result is a contention period equal to the number of competing SSs. In both papers, the results highly depend on the number of SSs, so the remaining uplink subframe to send data is smaller each time the number of SSs increases. An adaptive bandwidth request mechanism implemented at the SSs for real time traffic is proposed in [?]. In this approach, each SS predicts the arrival of real time packets prior to their arrival and requests bandwidth in advance. In [?], the authors have introduced a dynamic minislot allocation scheme. The number of required contention minislots in each frame period is based on an estimate of the maximum number of data packets that can be transmitted in a frame. The study formally analyzes the proposal, but it does not include numerical results. [?] introduces a new MAC protocol to prioritize traffic according to the waiting times required by the different types of traffic. The requests of the different types of traffic are assigned priorities. In [?], the authors introduce a new algorithm, called Multi–FS–ALOHA, which divides the contention period into two parts. The first is used by the SSs issuing bandwidth request for the first time while the second part is used by the SSs having previously attempted to transmit without success. These two parts are dynamically fixed on a frame by frame basis.

In all the studies having been reported, the authors make use of the mechanisms defined by the IEEE 802.16 standard to request and grant bandwidth. However, they do not make an in–depth analysis of the overhead. The overhead increases as the contention period increases. A simple modification can be used

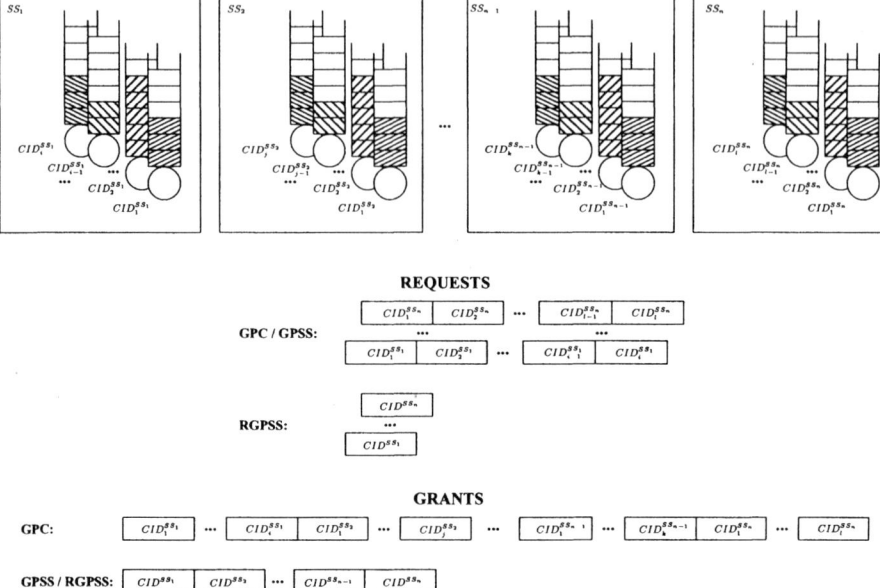

Fig. 1. Comparing GPC, GPSS & RGPSS modes regarding requests and grants.

to improve overhead in the system introduced by these mechanisms, especially requesting mechanism defined in the standard. This modification, which is explained below, will have a clear impact on the size of contention period.

4 Our proposal

As it has been described above, the IEEE 802.16 standard specifies a bandwidth request per connection made by SSs utilizing the corresponding CID. In this way, each SS has to make a request per active connection, requesting the amount of enqueued data associated to this connection. A drawback of this resource request scheme is that the internal collisions increase with the number of connections (an internal collision happens when two or more requests of different connections within a SS try to use the same contention slot). The other one is that to keep the probability of collision up when the number of connections increases, the contention period must increase too, reducing the size of uplink frame to be used for data transmission. This is derived from [?] and [?] when the number of connections per SS is bigger than one.

Regarding the grant bandwidth mechanisms, the standard describes two methods. The GPC, which grants bandwidth per connection, and the GPSS, which grants per subscriber station. Obviously the GPSS mode improves on the use of the frame, because the overhead in downlink is reduced, sending only a grant message in the DL-MAP per SS.

In the case of using the GPSS mode to grant resources, the BS does not need to know the individual request of each connection, since the grants are made by SS. In this case, the BS only needs to know the aggregated bandwidth request of all connections of each SS. To obtain this information, the BS can add up the all connection requests belonging to each SS. Obviously, this requires each connection to send its request, increasing the amount of request messages and therefore the overhead in the uplink.

In order to reduce the request messages in the uplink, we propose that each SS should send a single aggregated request containing the total amount of resources needed by all the active connections in this SS. This requesting scheme, which is called RGPSS (*Request and Grants Per Subscriber Station*), tries to minimize the number of requests to be sent, reducing the overhead in the uplink phase. A direct result of the reduction in the number of request messages is a decrease in the collision probability in the contention period, increasing the efficiency of the resource request mechanism based on a random access method. This operation mode reduces the length of the contention period since it only depends on the number of SSs.

The RGPSS scheme could be adapted to support classifications of connections in order to provide QoS. In this case, each SS will make a request per type of service flow, aggregating the resources that all the connections of a service flow need. So the BS will know bandwidth needs of each type of service flow of all SSs, and its scheduler could prioritize some service flows or SSs against another one.

5 Performance Evaluation

In this section, we carry out a performance analysis of our proposal. Throughout our study, all simulation are conducted using a model of IEEE 802.16 implemented in the OPNET Modeler v11.5 tool [?].

5.1 Scenarios

In our simulations, we consider an IEEE 802.16 wireless network consisting of several SSs and a BS describing a point–to–multipoint system. All nodes operate at 28 MHz, with a symbol rate of 22.4 MBaud. All transmissions are done using QPSK modulation with a bit rate of 44.8 Mbits/s. Accordingly to the standard a frame duration of 1 ms is used. The mode of operation is FDD. Ideal channel conditions are assumed, i.e., no packet corruption is due to the wireless channel. The system is assumed to operate in a steady–state, where the number of connections does not change over time. Each SS runs voice, video, background and best–effort applications, which are modelled as follows.

Constant bit–rate voice sources at a rate of 16 Kbits/s according to the G.728 standard are used. The voice packet size has been set to 384 bytes including RTP/UDP/IP headers, and all voice traffic are randomly activated within the interval [0,0.024] seconds, corresponding to a generating frequency of two consecutive packets. Each SS runs 37 voice connections.

For video applications, we assume H.264 variable bit–rate video traffic. This one is generated from the sequence *funny* encoded on CIF format at a frame rate of 30 frames/s are used. The average video transmission rate is around 1.1 Mbits/s with a packet size equal to 1,064 bytes (including RTP/UDP/IP headers). This type of traffic starts within the interval [0,0.5] seconds following a uniform random distribution. For each SS, we limit to one the number of video connection per SS.

For the background and best–effort traffics, we consider a *Pareto* distribution traffic model with an average bit rate for both types of traffic of 256 Kbits/s. The packet size has been set to a 552 bytes packet, including the TCP/IP headers. The sources are activated randomly within the interval [1,1.5] seconds. Associated to each SS, we assume six background connections and ten best–effort connections.

Since we are interested in studying the behaviour of the system with the new bandwidth request mechanism, all connections will request bandwidth utilizing the contention period. The scheduling algorithms used by the BS and SSs is FIFO (*First In, First Out*).

Throughout our study, we have simulated 10 seconds of operation of each particular scenario, collecting statistics after a warm–up period of 4 seconds. Each point in our plots is an average over 28 simulation runs, and error bars indicate 95% confidence interval. The study is carried out by varying the number of SSs (2, 5, 8 and 10) in order to increase the total offered traffic. The size of the contention period (#TOpp) is increased by 2 from 4 to 10 transmission opportunities. We evaluate the performance of the GPC, GPSS and RGPSS mechanisms allowing us to fairly compare them.

5.2 Results

In the case of the GPC mechanism, the collision probability is shown in Figure 2.(a). As expected, the collision probability decreases as the number of transmission opportunities increases. Moreover it increases as the network load increases. As a result, the request delay, defined as the time elapsed between the creation of a request in a SS and its arrival to the BS, is higher when the contention period size is shorter or when the traffic load offered increases. This is clearly depicted in Figure 2.(b). Since in GPC, the requests are sent per connections and the bandwidth grants are granted in the same manner; the grant delay (Figure 2.(c)), which is defined as the time elapsed between the first request of a connection and its associated grant, behaves in a similar manner to the request delay.

The results for the metrics described above in GPSS mode are shown in Figure 3. From the results, it is clear that the collision probability increases as a function of the traffic load offered. In the same way than in the GPC mode, if the contention period size decreases, the probability increases (Figure 3.(a)). The request delay is shown in Figure 3.(b), similar to the GPC mechanism,

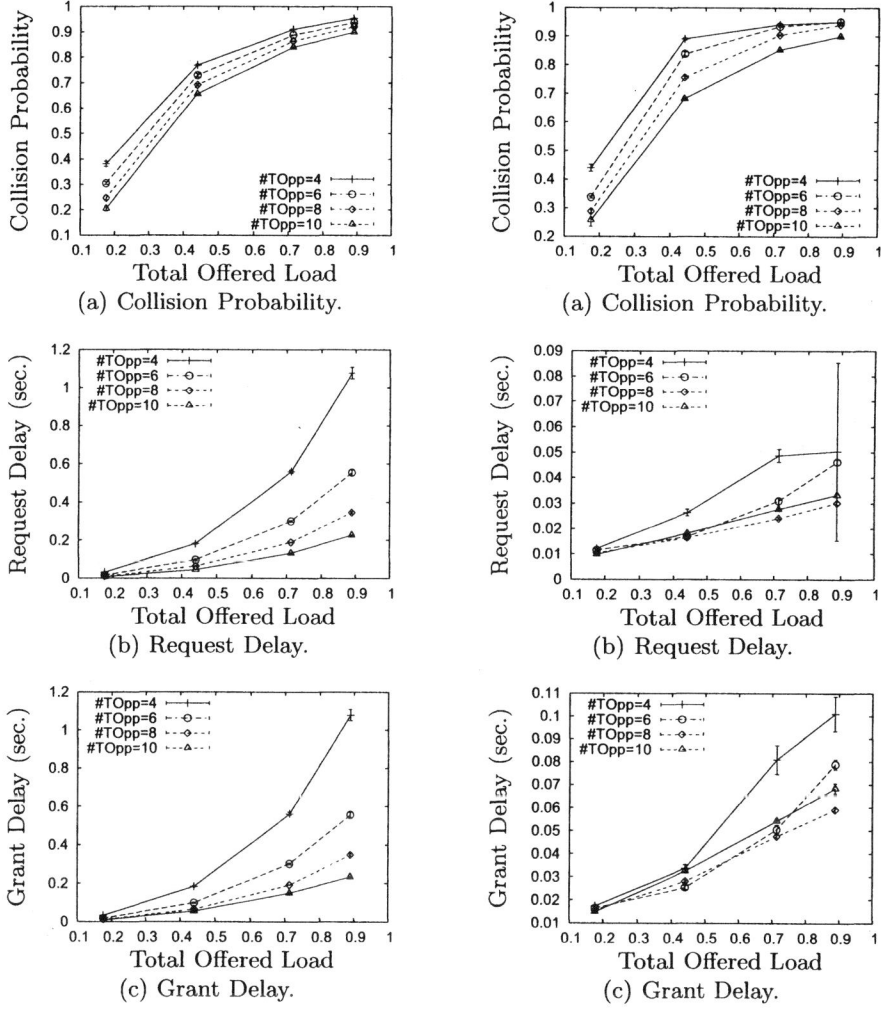

Fig. 2. Results in GPC mode.

Fig. 3. Results in GPSS mode.

it is smaller if the number of contention opportunities increases, except in the case of 8 and 10 transmission opportunities where this trend changes. This can be explained by the fact that for a longer contention period, more capacity is used by the resource request messages resulting in a reduction on the available capacity dedicated to data transmission. This means that the BS spends more time to serve a request. This can be clearly seen in Figure 3.(c). This increase on the time required to send a grant results on the expiration of the request timer. This event results in turn on the retransmission of the request. As a result of this retransmission, the request delay increases because upon receiving the request,

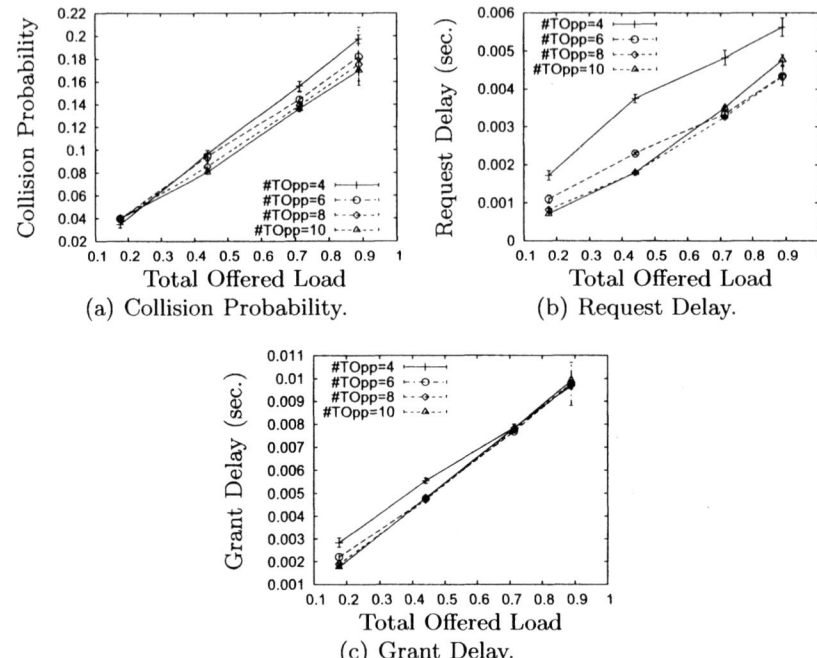

Fig. 4. Results in RGPSS mode.

the BS considers that it is the original request. Obviously, its creation time is the creation time of the original request.

In the RGPSS mechanism, the collision probability increases linearly as a function of the load (Figure 4.(a)), regardless of the contention period values. This is because in RGPSS only a request is sent per SS, aggregating all the connection bandwidth requests of the SS. Similar behavior is depicted in Figure 4.(b), where the request delay is represented. Not big difference could be appreciated in the case of 6, 8 and 10 contention opportunities, and only in the case of using 4 opportunities for contention, the delay is higher than 1 ms, which could be considered as acceptable since it matches the period of a MAC frame. The grant delay, shown in Figure 4.(c), presents a similar trend. It increases with the injected traffic and there are not big differences in the use of longer contention periods.

We now compare the performance of the three request and grant mechanisms. According to the previous results, the best results limiting the length of the contention period for each mechanism were: 10 transmission opportunities for the GPC; 8 for GPSS and 6 for RGPSS.

Figure 5.(a) depicts the collision probability. In it, the overall best result is provided by the RGPSS mechanism, proving effective the reduction of requests sent by the SSs.

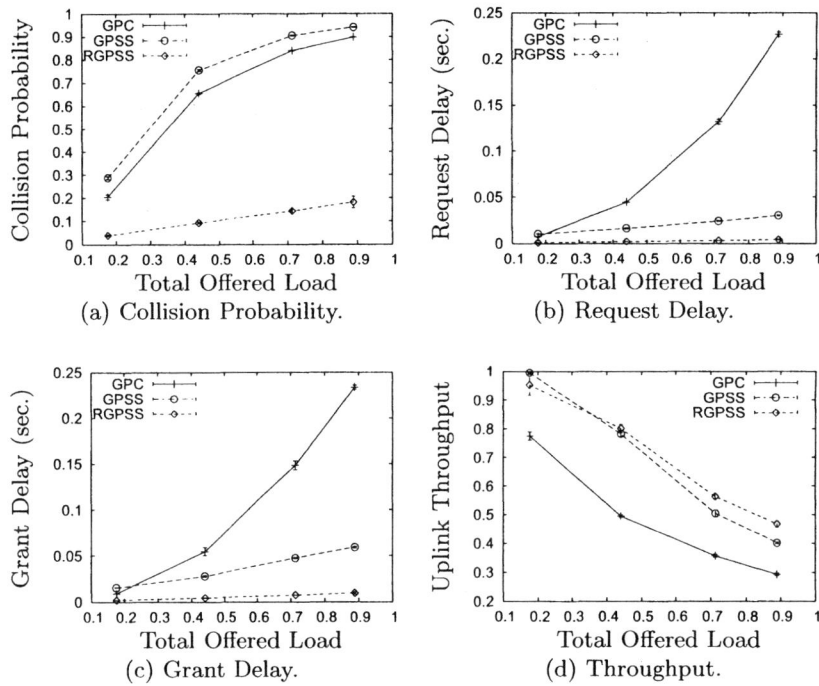

Fig. 5. Comparison between GPC, GPSS & RGPSS.

Regarding the request delay and grant delay (Figures 5.(b) and 5.(c), respectively), the difference between the mechanisms under study is clearly appreciated. This is particularly true as the network load is increased. The RGPSS exhibits the best behavior due to the limited number of requests.

As a result on the reduction in the grant delay in RGPSS, this mechanism presents a higher throughput in uplink direction, as seen in Figure 5.(d). This results in a lower collision probability, more requests arrive to the BS resulting on a shorter request delay. The BS is able to grant more connections in a shorter period of time. Other reason for this behavior is that contention period size in the RGPSS mechanism is shorter. In conclusion, more capacity is available to deliver actual data.

6 Conclusions

In this paper, a new mechanism to request in IEEE 802.16 network, called RGPSS (*Requests and Grants Per Subscriber Stations*), has been proposed. It sends only one request per SS, carrying the aggregated bandwidth requirements of all its associated connections.

It has been shown that the proposal exhibits a lower collision probability in the contention period and a higher throughput in the uplink direction than the schemes defined in the standard. These results demonstrate that with the RGPSS mechanism the network can support more traffic. It has also been shown that the contention period size used by RGPSS is shorter than the one required by the other two mechanisms.

This mechanism could be easily adapted to support different types of traffic (connections). It could be used as a means to provision a network with QoS support. In future works, we plan to evaluate the performance of such mechanisms and the impact of other parameters heavily impacting the length of the contention period.

Transmission Performance of Flexible Relay-based Networks on The Purpose of Extending Network Coverage

Ardian Ulvan[1] and Robert Bestak[1]

[1] Czech Technical University in Prague, Technicka 2 166 27 Praha 6, Czech Republic
{ulvana1, bestar1}@fel.cvut.cz

Abstract. Wireless mobile terminal may performs as relay for some other terminals which require relaying assistance. The deployment of flexible relay-based wireless networks is expected to increase the capacity of the network when placed within the base station. It also extends the coverage when placed at the border of base station coverage. In this paper, the performance evaluation of transmission between mobile nodes at the border coverage is carried out. The CDMA-based network with frequency hopped and direct sequence was applied. The Erlang capacity is used as the main parameter. The results show, the relay nodes can extend the coverage, however the transmission performance have been limited by blocking probability and signal to noise ratio (E_b/N_0). The results also show FH-CDMA has better performance than DS-CDMA.

Keywords: Mobile relays, transmission performance, transmission capacity

1 Introduction

A flexible relay-based wireless network is designed to extend the conventional point to multipoint communication. In addition, it also aimed to provide a broadband wireless access for the telecommunication users. The position of relays, where placed within the base station (BS) and at the border of BS's coverage, show different purposes. When relays are within the BS, they increase the capacity. Additionally, the relays extend the coverage when placed on the cell boundary [1].

In flexible relay-based system, beside a dedicated relay such as BS or access point (AP), the mobile wireless terminals in the network may perform as relays for some other terminals which require relaying assistance (mobile relays). The technology development to fulfill these characteristics is standardised by IEEE 802.16j which is still under progress.

In addition, since the family of IEEE 802.16 standards that based on OFDM is limited only to the air interface, it is also important to specify the network architecture [2]. The most common architecture is referred to Worldwide Interoperability for Microwave Access (WiMAX). In this paper we carried out some works in order to make the contribution for the implementations of flexible relay wireless OFDM-based networks that will improve the wireless broadband system of IEEE 802.16 families, yet in WiMAX network or others networks architecture.

The work is focused on transmission performance of mobile hosts that act as relay. In our scenario, the mobile hosts are placed at the border of BS coverage, with a relay capability they support the connections to other hosts which are out of BS range. In the scenario, the CDMA-based network is deployed as it provides all parameters needed. The spread spectrum techniques of Frequency Hopping (FH-CDMA) and Direct Sequence (DS-CDMA) are taking into account in order to analyze the characteristics and the transmission capacity. The transmission performance of both techniques are simulated and determined by spreading factor, blocking probability and signal to noise ratio (E_b/N_0).

The remainder of this paper is organised as follows. Section II gives brief description of flexible relay-based wireless networks scenario. Section III introduces our proposed scheme as well as all parameters that involved on the calculation. The performance analysis is given on section IV. As a final point, we present the conclusion in section V.

2 The Flexible Relay-based Network

The work of flexible relay-based wireless network which was to distribute the high capacity available in the area surrounding the BS has been carried out in [3]. In this work we performed the ability of flexible relay in the purpose of increasing the networks capacity. The flexible relay means it could be fixed or mobile. When mobile station (MS) is connected to the base station (BS) it may act as a relay to other MSs who have limited or cannot be served by BS or out of BS's range. When a MS does not have relay capability, it acts as conventional subscriber station. Every node with relay capability acts as router and takes place to determine and maintain the routing to other nodes within the network. The relaying and routing functionality is seems as ad-hoc network. The connection scenario is shown on the Figure 1.

The uses of ad-hoc network as extended networks have been showed on previous works. Authors in [3] had determined the position of flexible relay within the BS. It was found that the flexible relay could increase the transmission capacity. The others works concluded that the uses of ad-hoc network have been increased the throughput capacity and power efficiency as well [4] [5]. On the other works, the use of CDMA on ad-hoc network has been considered. The implementation of spread spectrum on ad-hoc network shown some benefits such as better security, interference resistance, simultaneous co-located transmission and large scalability [6][7][8].

The important parameter of spread spectrum system is spreading factor or processing gain, which is defined as the transmission to information bandwidth ratio.

In FH-CDMA, the system can tolerate all interferences as long as the interfering signals are under the processing gain limit. Total bandwidth, *W*, is divided by *M* orthogonal band frequencies *(W/M)*. In this case, the purpose of processing gain is to reduce the interfering signals. When two or more transmitters have used the same band frequency, then the collision occurs. However, the collision can be recovered by all time coding.

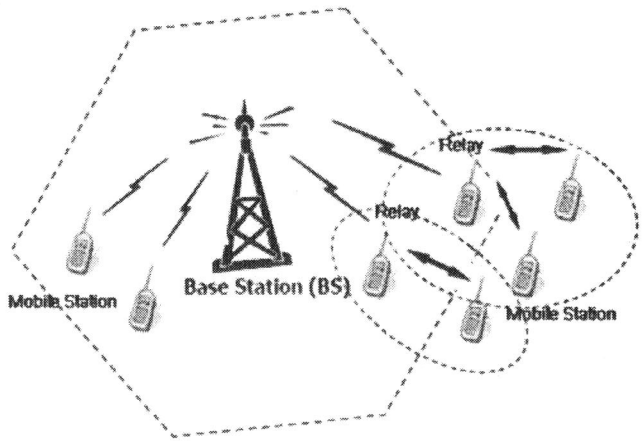

Fig. 1. Flexible relay-based wireless network. Some of the mobile hosts act as relay at cell boundary.

On the other hand, in DS-CDMA technique, the series of codes along the bandwidth, W, are multiplied by series of user data along the bandwidth, W/M. As a result is a series of signals which is transmitted along the bandwidth, W. The transmitted signal is spreading along the bandwidth which is M times higher than original signal. In DS-CDMA, M is known as spreading factor that analogous with processing gain in FH-CDMA. DS-CDMA uses spreading factor to reduce the minimum *signal to interference* noise *ratio* (SINR) at the receiver.

If the nominal requirement of SINR for FH-CDMA is β, therefore DS-CDMA reduces SINR to $\dfrac{\beta}{M}$ [7].

According to these characteristics, when they were deployed as a base system of flexible relay within the BS, FH-CDMA offers the interference avoidance. On the other hand, DS-CDMA is interference reduction [3]

In wide area wireless network with typical base station, the transmission intensity is defined by Erlang capacity. In CDMA system, the capacity is depend on the nominal value of interference. The blocking occurs when the interference power at reverse link reaches the given limit, which is assigned to maintain the quality of receiving signal [8]. Therefore, the users on CDMA system are allocated due to blocking probability. It is denoted as B_{CDMA}.

The call traffics in an individual time are random, as well as interference power. Due to these conditions, the blocking probability refers to call traffic distribution probability which is denoted as Erlang capacity.

3 System Scheme and Transmission Model

As describe in [7], based on analysis of user's interference power in reverse link, the blocking probability can be defined. Let says $P_{med} = e^{\gamma m_{dB}}$ with $\gamma = (\ln 10)/10$, thus the blocking probability for DS-CDMA can be calculated by following equation:

$$B_{DS-CDMA} = Q\left(\frac{\frac{W}{R_b}(1-\kappa_0) - \overline{M}\,\overline{v_r}P_{med}e^{\frac{1}{2}\gamma^2\sigma_{dB}^2}(1+\xi)}{\sqrt{\overline{M}\,\overline{v_r^2}P_{med}^2 e^{2\gamma^2\sigma_{dB}^2}(1+\xi')}}\right) \quad (1)$$

The Erlang capacity is denoted by $\overline{M}$. The interference parameter, k, is comparative to cell loading, x, where $k = 1 - x$. If k is changed into threshold value, k_0, then the loading threshold become $x_0 = 1 - k_0$. Substitute this into (1) gives as follow:

$$B_{DS-CDMA} = Q\left(\frac{\frac{W}{R_b}x_0 - \overline{M}\,\overline{v_r}P_{med}e^{\frac{1}{2}\gamma^2\sigma_{dB}^2}(1+\xi)}{\sqrt{\overline{M}\,\overline{v_r^2}P_{med}^2 e^{2\gamma^2\sigma_{dB}^2}(1+\xi')}}\right) \quad (2)$$

In case of FH-CDMA, the total interferences at BS occur only by users who have simultaneous transmitting on the same sub-channel, rather than all users. Therefore, the blocking probability is given as following:

$$B_{FH-CDMA} = Q\left(\frac{\frac{W}{R_b}\left(1-\frac{\kappa_0}{M}\right) - \overline{M}\,\overline{v_r}P_{med}e^{\frac{1}{2}\gamma^2\sigma_{dB}^2}(1+\xi)}{\sqrt{\overline{M}\,\overline{v_r^2}P_{med}^2 e^{2\gamma^2\sigma_{dB}^2}(1+\xi')}}\right) \quad (3)$$

Similar to DS-CDMA, where Erlang capacity is denoted by $\overline{M}$, the interference parameter is k, the cell loading is x and $k = 1 - x$. When k is changed into threshold value, k_0, then the loading threshold become $x_0 = 1 - k_0$. Then we substitute this into (3), the blocking probability can be calculated by:

$$B_{FH-CDMA} = Q\left(\frac{\dfrac{W}{R_b}\left(\dfrac{M-1+\chi_0}{M}\right) - \overline{M}\,\overline{v_r}P_{med}e^{\frac{1}{2}\gamma^2\sigma_{dB}^2}(1+\xi)}{\sqrt{\overline{M}\,\overline{v_r^2}P_{med}^2 e^{2\gamma^2\sigma_{dB}^2}(1+\xi')}}\right) \quad (4)$$

As we can see from equations (2) and (4), the blocking probability equations can be denoted as $Q_z(z)$, where $Q_z(z)$ is defined as *complementary cumulative probability distribution function* (CCDF) from zero-mean and variant unit of standard random variables. The values of Z with associated to some values of blocking probability are available on a table of CCDF for standard normal distribution. The table is very advantageous in order to determine the Erlang capacity for the transmission channels.

To analyse the performance of proposed scheme, we also assume some parameters and variables as shown on table 1.

Table 1. Parameters and variables

Parameters/Variables	Symbol	Assumption
Transmission Bandwidth	W	1.2288 MHz
Data rate	R_b	9.6 kbps
Signal to noise ratio (median)	$m_{dB} = \dfrac{E_b}{N_0}$	5 - 7 dB
Standard Deviation	σ_{dB}	2.5 dB
Cell loading	X_0	0.9
Numbers of band freq. (spreading factor)	M	6 - 128
Blocking Probability	B_{CDMA}	$10^{-3} - 10^0$

4 Performance Analysis and Results

The transmission performance of relay-based networks on the purpose of extending network coverage can be shown on Figure 2, Figure 3 and Figure 4. The calculation and simulation are done for FH-CDMA and DS-CDMA.

From Figure 2, it can be shown that the Erlang capacity is increase when the blocking probability is getting higher. It simply means when a number of users transmission are blocked, several numbers of idle channels are available, as indicate by Erlang capacity value. However, the system has a critical section where Erlang capacity reaches the maximum value. It approximately reaches when blocking probability is 0,5. Afterwards, the Erlang capacity decline gradually, even though the blocking probability is higher.

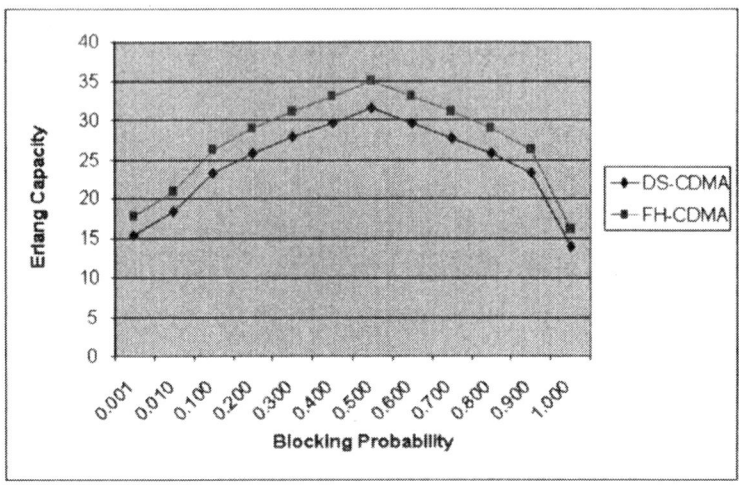

Fig. 2. The effect of blocking probability to Erlang capacity. FH-CDMA has better performance than DS-CDMA.

Furthermore as we can see from the graph, the values of FH-CDMA are higher than DS-CDMA. It can be explains that all interferences in DS-CDMA come from all active users. In contrast, all interferences in FH-CDMA come only from all active users using a same sub-band frequency.

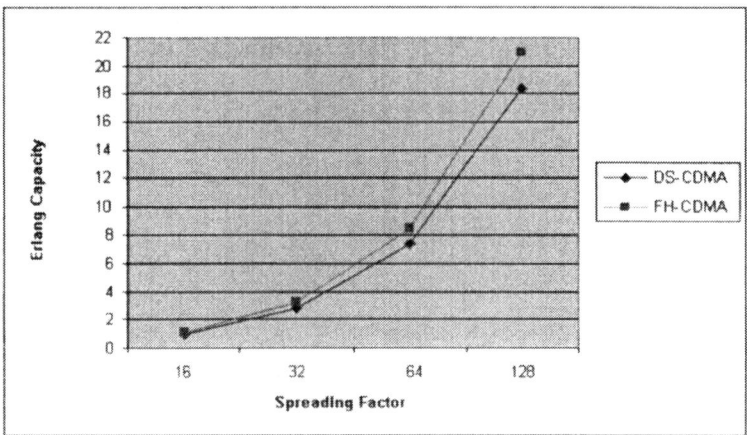

Fig. 3. The effect of spreading factor to Erlang capacity. FH-CDMA has better performance than DS-CDMA.

The second aspect which influences the Erlang capacity is spreading factor, as shown on figure 3. As can be seen on Figure 3, the Erlang capacity is incline gradually by the inclining of spreading factor. FH-CDMA shows a bit better performance than DS-CDMA. The Erlang capacity has tremendous inclination when the spreading factor is 128.

The last aspect that has been analysed to determine the transmission performance is the signal to noise ratio (E_b/N_0). Figure 4 shows the effect of signal to noise ration (E_b/N_0) to the Erlang capacity. As we can see from the graph, the Erlang capacity drop gradually while signal to noise ratio (E_b/N_0) is increased. This circumstance quite similar to the effect of spreading factor when applied and analysed from relays within the BS, as describe in [3].

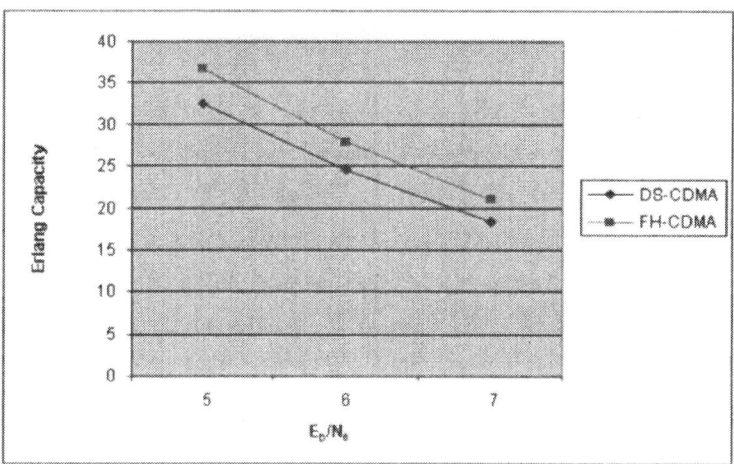

Fig. 4. The effect of E_b/N_0 to Erlang capacity. FH-CDMA has better performance than DS-CDMA.

The signal to noise ratio (E_b/N_0) expresses the signal's requirement power to get a successful transmission. In case of high interferences, the transmitter gains its signal power in order to reach better signal to noise ratio as well as to avoid the failed transmission. Unfortunately however, the gained power also increases significantly the probability of interference over the channel. It reduces the Erlang capacity. From the Figure 4, we can also see that in term of signal to noise ratio (E_b/N_0), the FH-CDMA has better performance than DS-CDMA.

5 Conclusions

In this paper we analysed the flexible relay CDMA-based transmission performance for the purpose on extending coverage. We have illustrated that the MS at the cell boundary of BS coverage can act as a relay. It generally extends the cell or networks coverage. According to the results, there are two aspects that limit the transmission capacity for the relay, those are blocking probability and signal to noise ratio (E_b/N_0). The blocking probability has certain value that match to maximum value of Erlang capacity. Based on simulation, the FH-CDMA shown better performance in all aspects rather than DS-CDMA.

Several aspects of this work deserve more analysis, especially the limitation use of band frequency on FH-CDMA. Another concern for future work is the implementation of 802.16d,e,j OFDM-based flexible relay on WiMAX network as wireless broadband access system.

Acknowledgments. This work has been performed in the framework of FP6 project FIREWORKS IST-27675 STP, which is funded by European Community. The authors would like to acknowledge the contributions of their colleagues from FIREWORKS Consortium (http://fireworks.intranet.gr).

References

1. C. Hoynmann, et.al, "Flexible Relay Wireless OFDM-based Networks". Available at http://fireworks.intranet.gr. (2006).
2. IEEE Standard P802.16e/D8: "Part 16: Air Interface for Fixed and Mobile Broadband Wireless Access System", (2005).
3. A. Ulvan and R. Bestak, "Transmission Performance of Flexible Relay-based Networks". 2006. International Workshop on Digital Technology, Slovak Republic. (2006)
4. P. Lungaro, "Coverage and Capacity in Hybrid Multi Hop Ad Hoc Cellular Access System". Wireless@KTH, Radio Communication System Laboratory. 2003.
5. K. Holger, "Using ad-hoc extensions to cellular networks for capacity and energy-efficiency improvements". Telecommunication Network Group. University of Berlin. Berlin. 2002.
6. W. Chunfeng, "An Introduction on Ad Hoc Network". Mita Lab, System Design Department, School of Science for Open and Environmental systems, Keio University. 2003
7. J. G.Andrews, X. Yang, A. Hasan, G. de Veciana, "The Flexibility of CDMA Mesh Networks". IEEE Wireless Communications Magazine. 2005.
8. P. Gupta and P.R. KUMAR,"The Capacity of Wireless Networks". IEEE Trans. Inform. Theory. vol. 46. No. 2, 2003.

Impact of Relay Stations Implementation on the Handover in WiMAX

Zdenek Becvar

Czech Technical University, Department of Telecommunication Engineering, Technicka 2,
166 27 Prague, Czech Republic
becvaz1@fel.cvut.cz

Abstract. Mobile WiMAX networks are generally based on the 802.16e standard. This version of standard supports a number of handover types and allows full mobility of users. However, the relay stations are not considered in this standard version. Relay stations are introduced in the originating 802.16j standard version. This paper deals with the impact of relay stations implementation from the handover point of view. The influences of fixed relay stations amount on the number of handover procedure initializations and on the diversity set size are investigated.

Keywords: WiMAX, Handover, Relay station, Diversity Set.

1 Introduction

Mobile WiMAX, based on 802.16e [1] standard, is a high-speed wireless networking communication system. WiMAX allows Non-Line-Of-Sight (NLOS) and Line-Of-Sight (LOS) communication in the frequency range 2-11 GHz and 10-66GHz respectively. The physical layer in 802.16e uses SOFDMA (Scalable Orthogonal Frequency Division Multiple Access) to achieve high transfer speed. WiMAX allows large coverage up to 50 km for NLOS [2]. Further, the QoS support and hard and soft handovers to support full mobility are implemented in WiMAX.
The handover according to 802.16e allows to users high speed (up to 160 km/h [3] mobility) and it can provide continuous data flow for all applications.
WiMAX, such as described in [1], do not consider the Relay Stations (RS). The implementation of these stations into WiMAX networks is the target of just originating standard IEEE 802.16j [4]. The RS can be used to the coverage extension or to the BS's capacity increase. In this paper, the RSs placed to allow increasing of the capacity will be assume.
The rest of paper is organized as follows. Next section describes the handover types and handover process. The third section gives an overview about the RS and outlines possibilities to the implementation of the RSs to the networks. In the further section, the simulation parameters and scenarios are described. The fifth section contains the simulation results and their discussion. Last section presents our conclusions and future work plans.

Please use the following format when citing this chapter:

Becvar, Z., 2007, in IFIP International Federation for Information Processing, Volume 245, Personal Wireless Communications, eds. Simak, B., Bestak, R., Kozowska, E., (Boston: Springer), pp. 107-114.

2 Handover procedure in WiMAX

The basic mean of the handover is to provide the continuous connection when a Mobile Station (MS) moves from an air-interface of one Base Station (BS) to an air-interface provided by another BS.

2.1 Handover types

IEEE 802.16e standard defines three basic types of handover [1], [5]: Hard Handover (HHO), Macro Diversity Handover (MDHO) and Fast Base Station Switching (FBSS). HHO is mandatory in WiMAX systems. Other two types of handover are optional. MDHO and FBSS can be called as the soft handovers.
Within hard handover, the MS communicates with just one BS in each time. All connections with old BS (called Serving BS) are broken before the connection to new BS (called Target BS) is established [5].
When the MDHO is supported by MSs and by BSs, the Diversity Set (in some publications noted as Active Set) is maintained by the MSs and by the BSs [5]. The Diversity Set is a list of the BSs, which are involved in the handover procedure. For downlink, one or more BSs from Diversity Set transmit data to the MS such that diversity combining can be performed at the MS. For uplink, the MS transmission is received by multiple BSs included in the Diversity Set and a selection diversity of received information is performed.
In the FBSS, the Diversity Set is maintained by the MS and by the BS similarly as in the MDHO. Opposite to the MDHO, the MS communicates only with an Anchor BS for all types of uplink and downlink traffic including management messages [5]. The Anchor BS is one of the BSs from Diversity Set in the MDHO. The MS is synchronized and registered to the Anchor BS; further the MS performs ranging and monitors the downlink channel for control information.

2.2 Handover procedure

The MS has to seek for suitable Neighbor BSs within normal operation mode to allow handover realization. Time dedicated for the searching for BSs in the neighborhood is assigned to the so-called scanning intervals. The MS determines the BS suitability to be a Target BS in these intervals. The scanning intervals are allocated via MAC management messages. After the MS finishes scanning of the Neighbor BSs, it sends the results to the Serving BS. There exist two types of results reporting. The information about used type is carried in the MOB_SCN-RSP and MOB_SCN-REP messages. In the first type, called Event trigger report, the MS reports based on the one of four defined trigger (CINR – Carrier to Noise plus Interface Ratio, RSSI – Received Signal Strength Indication, Relative delay, RTD – Round Trip Delay). Measurement report is sent to the Serving BS after each measurement in this case. In the second type, called Periodic report, the measurement reports are sent at periodic intervals. The spacing of the reporting intervals is indicated in number of frames. The length of each frame can be up to 20 ms. The number of frames between reports is contained in MOB_SCN-RSP message and the maximal number of frame is 256 (2^8). So the maximal value can be set up to 5.12 s. The handover should start immediately after the BS resolution based on the reported information.

3 Relay station

Relay stations are generally simplified kind of a BS. Two types of the RSs are defined: fixed and mobile. Fixed RS is permanently installed at the same place. Fixed and mobile RSs are connected to the network via radio interface and there is no connection to the wire backbone.

Wireless RS can be used to the coverage extension [6] or to increase of the capacity of the BSs [7]. The scenario with the RSs used for coverage extension is illustrated in Fig.1. The MS communicates with the BS via two RSs. The RS is placed close to the boarder of the previous RS (or BS). The communication frame has to be divided on two parts [8]. In the first part, the RS serving local users (it functions as a BS). In the second part, the RS relays the data from the BS (or RS) to another RS (or BS).

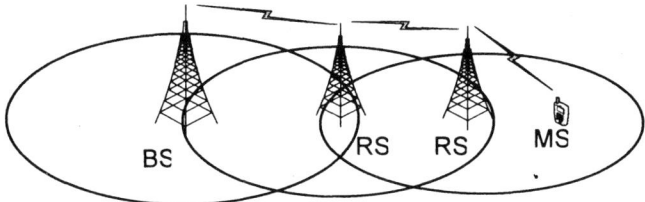

Fig. 1. Relay stations used to the enlarging of the coverage area.

Situation with the RS used to capacity increase is shown in Fig. 2. The frame has to be divided similarly as in previous case. The difference to the area enlarging is in the placement of the the RS. In this case, the RS is situated into the area covered by the BS so that whole RS's coverage area is inside of the BS coverage area.

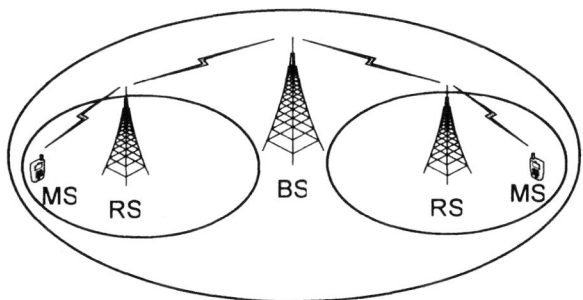

Fig. 2. Relay stations used to the increase of the capacity.

Two concepts of the RSs implementation into WiMAX are considered [8]. The first type is a centralized system. The functionality of the RS is very simple in the centralized system. The RS only forwards received messages in direction to the target station. The RS is fully controlled by the BS. Therefore, the RS needs no special intelligence. The second concept is a distributed system. Opposite to the centralized system, the RS can modify management messages according to the situation in neighborhood. The RS in the distributed system must have implemented some controlling and management logic on the similar level as it is in the BS.

Relating to the handover management messages, there are not needed changes in the centralized system, but the creation of new management messages is necessary in the distributed system because the RS has to obtain information about neighborhood. These messages and their exchange are described in [9].

4 Simulation scenario

We can assume both soft handover types (MDHO and FBSS) because there are no differences in the Diversity Set and the Anchor BS updating procedures. We assume the measurement reports sent periodically with the reporting period length equal to 1 s (20 ms frame length and reporting after each 50 frames). The mobile users are moving by 50 km/h speed. All systems operate in 5 GHZ frequency band. All signal strengths among the MS and the BS (RS) are evaluated based on the Okumura-Hata path loss model for small or medium city [10]. This model (1) takes into account a distance between a MS and a BS (or RS) (in equation noted as d), height of BSs and RSs antennas (h_t), transmitting frequency (f) and height of MS (h_r).

$$L = 69.55 - 13.82 * \log h_t - (1.1 * \log f - 0.7) * h_r + (1.56 * \log f - 0.8) \\ + 22.16 * f + (44.9 - 6.55 * \log h_t) * \log d \quad (1)$$

Creating and updating of the Diversity Set is based on the principles described in [1]. Firstly, the BS (or RS) with the highest signal level (Anchor BS) is selected by each MS. Next, the Diversity Set is created for each MS depended on the defined thresholds and the signal strengths in the MS's location. All BSs and RSs with the signal strength higher than Add_Threshold (2) are added to the Diversity Set. The BS and RS can be also deleted from the Diversity Set, in the further step, if the signal level is lower than Delete_Threshold (3). Add and Delete thresholds are set up to 4 dB and 3 dB respectively.

$$\text{BS signal level} > \text{Add\_Threshold} . \quad (2)$$

$$\text{BS signal level} < \text{Delete\_Threshold} . \quad (3)$$

Movement of the MSs is evaluated based on the Probabilistic Random Walk Mobility Model [11].

Four different scenarios were considered. The deployments of the MSs, BSs and RSs at the beginning of simulation for each scenario are shown in Fig. 1. Four BSs are placed on the same place near to the corners of simulated area in all scenarios. In the first scenario (*scenario A*), no RS is situated. In the second scenario (*scenario B*), only one RS in the middle of area is placed to improve signal conditions, because the signal level around the middle is week. In the third case (*scenario C*), 5 RSs are placed. One RS is placed in the middle, as in scenario B and each of the resting 4 RSs are situated between two Neighbor BSs. In the last case (*scenario D*), 9 RSs are placed. Five RSs is in the same location as in scenario C. Another 4 RSs are situated in the half way between BS and middle RS. The RSs in all scenarios are deployed to increase network throughput or number of connected user, but they do not enlarge the area. Fixed RSs are assumed. A movement of all MSs was same for all scenarios.

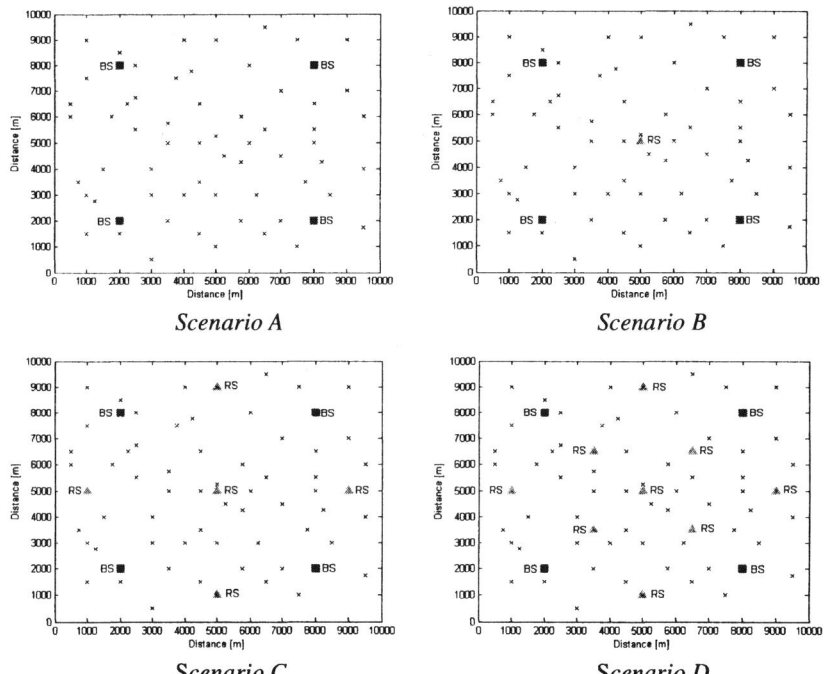

Scenario A   *Scenario B*

Scenario C   *Scenario D*

Fig. 3. Four cases of BSs (*rectangles*), fixed RSs (*triangles*) and MSs (*crosses*) deployment at the beginning of simulation. The *scenario A* is scenario without RS, 1 RS is in *scenario B*, 5 RSs are in *scenario C* and 9 RSs are in *scenario D*.

The MATLAB was used for the creation of simulation. Main parameters used in simulation are summarized in Table 1.

Table 1. Simulation parameters.

Parameter name	Value
Handover type	MDHO
Propagation model	Okumura-Hata suburban
Mobility Model	Probabilistic Random Walk
Size of simulated area	10 x 10 km
Number of MSs	60
Number of BSs	4
Number of RSs	0/1//5/9
BS transmitting power	170 dB
BS's range radius	4200 m
RS transmitting power	146 dB
RS's range radius	1000 m
Height of BSs and RSs	30 m
Height of MSs	2 m
Frequency	5 GHz
Speed of MSs	50 km/h
Delete threshold	3 dB
Add threshold	4 dB
Frame length	20 ms
Reporting period	50 frames
Length of simulation	900 s
Simulation step	1 s

4 Results

Because the part of data flow is processed in the RSs, the capacity or the number of users can be increased when the RSs are added to the network [7]. The MS can communicate with more BSs (RSs) in the case of the MDHO at one time. In this case, the MS communicates (including data traffic) with all BSs (RSs) from Diversity Set. In the uplink, the MS sends the data to all BSs (RSs) and the selection diversity is performed. In the downlink, all BSs (RSs) send the data to the MS and the diversity combining is performed. The changes in the average number of the BSs and the RSs communicate with one MS at one time are shown in Fig.4. If we assume each MS connected to at least one BS or RS (it means each MS is connected to the network), so the average number of connection per MS has to be equal or greater then 1; and the total number of connections in whole area at one time have to be equal or grater then a number of the MSs in the area. The independence of the average number of connections to the BSs (average number of BSs in the Diversity Set) on the count of RSs in area is caused by using of the absolute threshold (according to [1]). The BSs are added to the Diversity Set independent on the signal level of the other BSs or RSs. The increase of the average number of connections to the RSs (average number of the

RSs in the Diversity Set) is depending on the number of the RSs in the area. The size of the Diversity Set is equal to the average number of connections to the BSs plus the average number of connections to the RSs. Therefore the size of the Diversity Set is increasing with placing of the others RSs to the network.

In the FBSS, the situation is different. There, the MS exchanges user data with only one Anchor BS or RS at one time. Therefore, the total amount of BSs (or RSs) connected to the one MS is still same and is equal to the number of MSs in network.

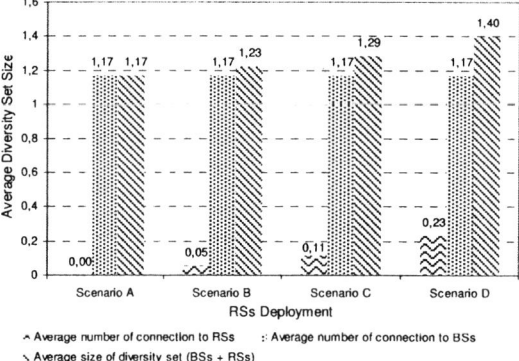

Fig. 4. The average number of connections at one time among MS and BSs (*bar filled with dot*), among MS and RSs (*bar filled with wavelet*) and among MS and BSs plus RSs (*bar filled with back slash lines*) per one MS.

Similar to this, with the increasing amount of the RSs in the area, the number of initialization of the handover procedures (in this case, handover procedures mean booth Diversity Set update and Anchor BS update) is rising too (see Fig.5). It is caused by increasing the area where the handover is initialized (the places where a signal level from BSs or RSs is higher than selected thresholds). Hence, users move across these areas more often. This results in an increasing number of handover messages exchanged in the network.

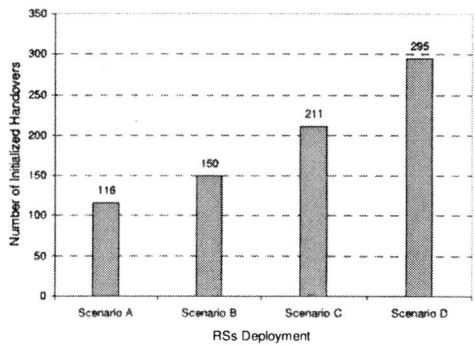

Fig. 5. Number of initialized handovers during simulation depending on RS deployment.

5 Conclusions and future work plans

We have investigated an impact of the Diversity Set size and an impact of amount of the handover procedure initialization on the number of RSs. The RSs location was selected based on the signal condition. It means that the RSs were placed into the areas with worst signal level, to increase throughput in this area.

The average number of BSs in the Diversity Set is independent on the count of the RSs in area if we assume absolute thresholds. Opposite to this, the average amount of the RSs in the Diversity Set is increasing with growing number of the RSs in the area. Therefore, average size of the Diversity Set (including BSs and RSs) is increasing to.

The increase of the Diversity Set size results to the raising of the number of initialization of the handover procedure.

Increase of the handover initializations and rising of the Diversity Set size brings more frequent management messages exchange and decrease the user's data throughput.

In the future, we will compare the Diversity Set size for absolute thresholds and relative thresholds and next, we will investigate an impact of the RSs placement on the ratio between user's and management data.

Acknowledgments. This research work was supported by Czech Technical University's grant No. CTU0715013.

References

1. IEEE P802.16e/D12: Air Interface for Fixed and Mobile Broadband Wireless Access Systems: Amendment for Physical and Medium Access Control Layers for Combined Fixed and Mobile Operation in Licensed Bands. New York (2005)
2. WiMAX Forum: WiMAX's technology for LOS and NLOS environments. (2004)
3. WiMAX Forum: Can WiMAX Address Your Applications?. (2005)
4. IEEE 802.16's Relay Task Group: IEEE 802.16j proposals. <http://www.ieee802.org/16/relay/>
5. Becvar, Z., Zelenka, J., : Implementation of Handover Delay Timer into WiMAX. 6th Conference on Telecommunication. Peniche, Portugal (2007)
6. Wei, Z. : Capacity Analysis for Multi-hop WiMAX Relay
7. Mach, P., Bestak, R.: Performance of IEEE 802.16 Enhanced by Relay Stations. 6th Conference on Telecommunication. Peniche, Portugal (2007)
8. Hoymann, Ch., Klagges K.: MAC frame concepts to support multihop communication in IEEE 802.16 Networks. Wireless World Research Forum, Shanghai, China (2007)
9. Lee, H., Wong, W. C., Jerry, S., Johnsson, K., Yang, S., Lee, M.: Overview of the proposal for MS MAC handover procedure in an MR Network. IEEE 802.16j proposal, No. S80216j-06_217r1.pdf (2007)
10. Korowajczuk, L., Souza, B., Xavier, A., Moreira, A., Filho, F., Ribeiro, L. Z., Korowajczuk, C., DaSilva, L. A.: Designing CDMA 2000 Systems, Wiley, Great Britain (2004)
11. T. Camp, J. Boleng, V. Davies, "A Survey of Mobility Models for Ad Hoc Network Research. (2002)

Optimization of Handover Mechanism in 802.16e using Fuzzy Logic

Ewa Kozlowska

Czech Technical University of Prague, Department of Telecommunication Engineering, Technicka 2, 166 27 Prague 6, Czech Republic

kozloe1@fel.cvut.cz

Abstract. This paper describes the handover mechanism in the standard 802.16e, which has implemented three handover methods – Hard Handover (HHO), Fast Base Station Switching (FBSS) and Macro Diversity Handover (MDHO) at this moment. Moreover, the handoff scheme is presented based on fuzzy logic. The mechanism takes under consideration three criteria: geographical data, received power level and used bandwidth of each base station. Fuzzy logic has been used to come with the decision about the target BS to be chosen such that the switching between BSs is more efficient.

Keywords: handoff, fuzzy logic

1 Introduction

Mobility is the most important feature of a wireless cellular communication system. Undoubtedly the future perspective of networking will demand to support mobile users traveling all over the world. In addition to that, the number of portable devices that need access to the Internet is exponentially increasing. Users on the other hand are no longer required to work in their company's home network while they may be moving from place to place. Usually, continuous service is achieved by supporting handover from one cell to another. Handover is the process of changing the channel (frequency, time slot, spreading code, or combination of them) associated with the current connection while a call is in progress. It is often initiated either by crossing a cell boundary or by deterioration in quality of the signal in the current channel. Handover is divided into two broad categories – hard and soft. They are also characterized by „break before make" and „make before break." In hard handover, current resources are released before new resources are used; in soft handover, both exist and new resources are used during the handover process. Poorly designed handover schemes tend to generate very heavy signaling traffic and, thereby, a dramatic decrease in quality of service (QoS). The reason why handovers are critical in cellular communication systems is that neighboring cells are always using a disjoint subset of frequency bands, so negotiations must take place between the mobile station (MS), the current serving base station (BS), and the next potential BS. Other related issues, such as decision making and priority strategies during overloading, might influence the overall performance [1].

Please use the following format when citing this chapter:

Kozlowska, E., 2007, in IFIP International Federation for Information Processing, Volume 245, Personal Wireless Communications, eds. Simak, B., Bestak, R., Kozowska, E., (Boston: Springer), pp. 115-122.

2 Overview of Handover Mechanism in 802.16e

Two crucial issues for mobile applications are battery life and handover. To enable power-efficient MS function Mobile WiMAX supports Sleep Mode and Idle Mode. 802.16e also supports seamless handover to allow the MS to switch from one base station to another at vehicular speeds without interrupting the connection. Handover schemes should be with latencies less than 50 milliseconds to ensure real-time applications such as VoIP perform without service degradation. Flexible key management schemes assure that security is maintained during handover [2].
There are three handover methods supported within the 802.16e standard: Hard Handover (HHO), Fast Base Station Switching (FBSS) and Macro Diversity Handover (MDHO). While the HHO is mandatory, FBSS and MDHO are two optional modes.

2.1 Hard Handover (HHO)

It begins with cell reselection proceeded by the MS. Information about adjoining BSs may be obtain from a decoded MOB_NBR-ADV message or a request to schedule scanning intervals or sleep-intervals can be sent to the serving BS. MOB_NBR-ADV message is broadcasted by the BS to identify the network and define the characteristics of the neighbour BS to the potential MS searching initial network entry or the handover. Intention of this stage is to check HHO possibilities. Procedure of searching available the neighbour BS does not involve termination of existing connection to a serving BS. Next phase is HO Decision and Initiation, which can be start either by the MS, by sending an MOB_MSHO-REQ message or the serving BS, by sending an MOB_BSHO-REQ message. In case of MS initiating handover with the BS, the serving BS replies with MOB_BSHO-RSP message containing instruction to the HHO with neighbour BSs list. After the handover preparation ends, execution follows. The target BS is selected by the MS and MOB_HO-IND message is sent; by sending this message the MS terminates service with the serving BS. As soon as the MS synchronize with the target BS, ranging process begins. It mainly aims to time synchronization and power adjustment. Then the MS negotiates basic capabilities, performs authentication and finally registers with the target BS. Whole this process can be shortened, if the target BS learns MS information from serving BS thru backbone network. Cancellation of the HHO is possible at any time till the expiration of Resource_Retain_Time interval after transmission of MOB_HO-IND message [3].

2.2 Macro Diversity Handover (MDHO)

Support of MDHO is optional for both the MS and the BS. The MS and the BS should hold a list of BSs involved in MDHO with the MS. The list name is Diversity Set. One of BSs in the Diversity Set is defined as an Anchor BS. Decision about performing handover in MDHO begins with an MS transmitting/receiving unicast message and traffic from multiple BSs at the same time interval. For downlink MDHO, two or more BSs provide synchronized transmission of MS downlink data such that diversity combining is performed at the MS. For uplink MDHO, the transmission from a MS is received by multiple BSs where selection diversity of the information received is performed. [3].

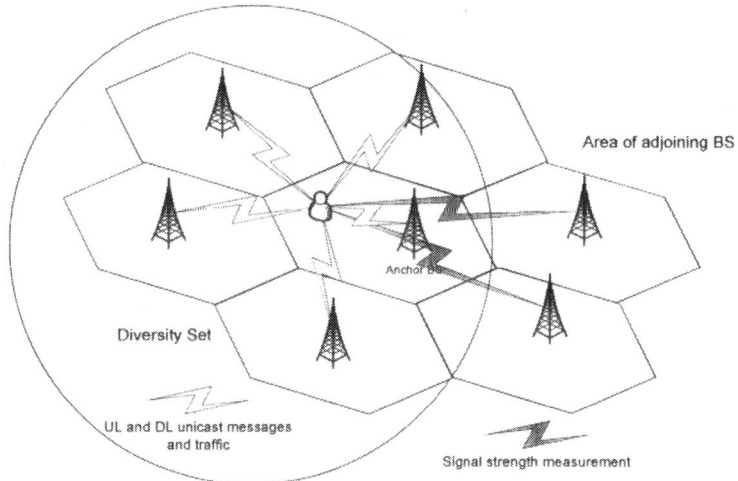

Fig. 1. Macro Diversity Handover.

2.3 Fast Base Station Switching (FBSS)

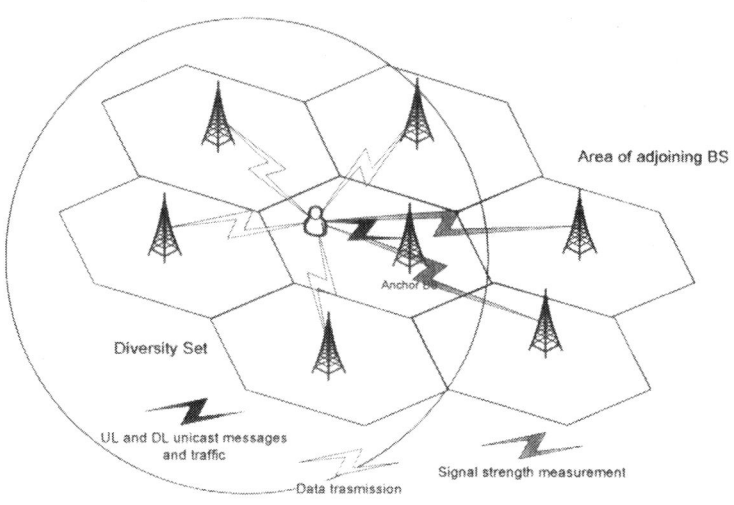

Fig. 2. Fast Base Station Switching.

Support of FBSS is optional for both the MS and the BS. Also in this mechanism the MS and the BS possess a list of BSs that are taking part in FBSS with the MS. The name of the list is exactly the same like in case of MDHO, Diversity Set and also the

Anchor BS is defined. In FBSS, the MS continuously monitors the base stations in the Diversity Set. FBSS begins with decision for an MS to receive/transmit data from/to the Anchor BS that may change within the Diversity Set. Switch from one Anchor BS to another is performed without proceeding HO signaling message. Scanning to search for suitable adjoining BS to update the set is made by the MS. By endless monitoring of the Diversity Set BSs signal strength, the MS is capable of choosing one as the Anchor BS. This selection should be reported by the MS to the network. An important requirement of FBSS is that the data is simultaneously transmitted to all members of an active set of BSs that are able to serve the MS [3].

2.4 Updating

Update of diversity se depends on the threshold contained in Downlink Channel Descriptor (DCD). Two thresholds are defined: H_Delete threshold and H_Add threshold. In case of dropping the serving BS from Diversity Set, the long-term CINR of serving BS has to be smaller than H_Delete Threshold. In case of adding a new BS to Diversity Set, the logn-term CINR of neighbour BS is bigger than H_Add Threshold. For both MDHO and FBSS the Diversity Set update is performed exactly the same.

Updating of Anchor BS is solved by two mechanisms. The first, called „Handover MAC Management Method", is based on exchanging five types of MAC management messages. The second method, noted as „Fast Anchor BS Selection Mechanism", uses Fast Feedback channel for exchanging anchor Bs selection information. A new anchor BS choice is made according to signal strength measurement reported by the MS. A new anchor BS should be included in the present Diversity Set [4].

3 Overview of Fuzzy Logic

Fuzzy Logic is a problem-solving control system methodology that lends itself to implementation in systems ranging from simple, small, embedded micro-controllers to large, networked, multi-channel PC or workstation-based data acquisition and control systems. It can be implemented in hardware, software, or a combination of both. FL provides a simple way to arrive at a definite conclusion based upon vague, ambiguous, imprecise, noisy, or missing input information. FL's approach to control problems mimics how a person would make decisions, only much faster.

FL requires some numerical parameters in order to operate such as what is considered significant error and significant rate-of-change-of-error, but exact values of these numbers are usually not critical unless very responsive performance is required in which case empirical tuning would determine them.

The fuzzy logic analysis and control method is, therefore:

1. Receiving of one, or a large number, of measurement or other assessment of conditions existing in some system we wish to analyze or control.

2. Processing all these inputs according to human based, fuzzy "If-Then" rules, which can be expressed in plain language words, in combination with traditional non-fuzzy processing.

3. Averaging and weighting the resulting outputs from all the individual rules into one single output decision or signal which decides what to do or tells a controlled system what to do. The output signal eventually arrived at is a precise appearing, defuzzified, "crisp" value.

The following is Fuzzy Logic Control/Analysis Method diagram:

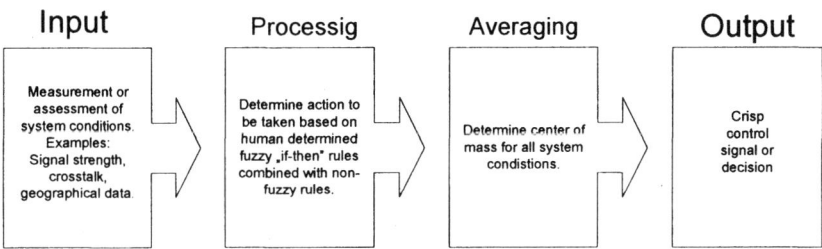

Fig. 3. The fuzzy logic Control-Analysis method.

Fuzzy Logic provides a completely different, unorthodox way to approach a control problem. This method focuses on what the system should do rather than trying to understand how it works. One can concentrate on solving the problem rather than trying to model the system mathematically, if that is even possible. This almost invariably leads to quicker, cheaper solutions. Once understood, this technology is not difficult to apply and the results are usually quite surprising and pleasing [5].

4 Proposed Optimization of Handover in 802.16e

In the fully mobile scenario, user expectations for connectivity are comparable to those experienced in 3G voice/data systems. Users may be moving while simultaneously engaging in a broadband data access or multimedia streaming session. The need to support low-latency and lowpacket-loss handovers of data streams as users transition from one BS to another is clearly a challenging task. For mobile data services, users will not easily adapt their service expectations because of environmental limitations that are technically challenging but not directly relevant to the user (such as being stationary or moving). For these reasons, the network and air interface must be designed upfront to anticipate these user expectations and deliver accordingly [6].

The one of idea in proposed mechanism is to predict movement of mobile user, by constructing an intelligent map of terrain with use of geographical data gathered from mobile users and base stations. Base stations position is fixed as there is no movement. An idea of using GPS receiver to provide the most accurate and robust source of time and frequency synchronization for WiMAX networks has been already given. Those receivers can be also use to determine geographical position of base station. As far as it goes for the mobile user, it is required to have implemented GPS receiver, which features we can use to create an intelligent map of terrain. Only three ways of movement we are interested in:

- by train, where destination point is known and the route is fixed,
- by car, where destination point is known and the route is predictable,
- on foot, where a velocity of mobile user is small, and by this predictable.

The main idea of presented handoff scheme is based on the fuzzy logic. The schemes consider not only received power level criterion, but also the bandwidth (data transfer rate in Kbps) and distance from the base station. Fuzzy logic here is used to reach a decision about the target BS to be chosen such that the handoff is performed as fast as possible with reasonable QoS levels.

I phase. First phase of fuzzy logic proceeding is to deliver input parameters of signal strength, distance from the base station and bandwidth use in the cell. Following parameters are fuzzified with use of pre-defined input membership functions, for example, membership function for distance shown on the figure Fig. 4:

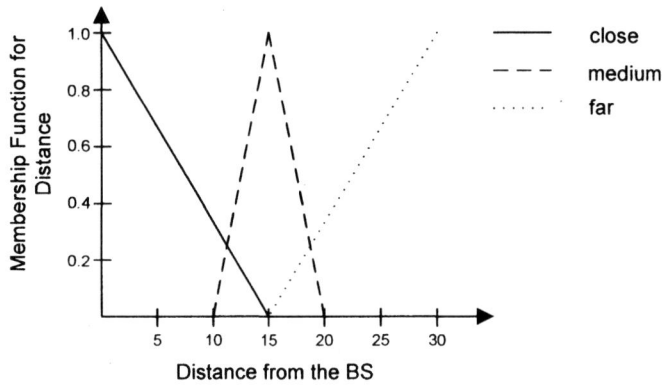

Fig. 4. Input membership function of distance from a BS.

As example triangular member function has been shown, but other shapes can also be chosen if their representation of real scenario in the 802.16e network is more suitable. The first step of the fuzzy logic Control-Analysis method is to transform the input value to names of member function and degrees of memberships in the function. For instance, given distance from BS=14, the following value can be read off the graph from Fig. 4.

Input	(Membership-Fn, Membership-Degree)	
Distance From BS (=14)	(C, 0.1)	(M, 0.75)

where, C – close, M – medium, F – far distance.

Second phase. Here the rule matrix is used, which has two input conditions, "error" and "error-dot" (rate-of-change-of-error), and one output response conclusion (at the intersection of each row and column). The rule matrix is a simple graphical tool for mapping the FL control system rules. It accommodates two input variables and expresses their logical product (AND) as one output response variable. The degree of membership for rule matrix output can take value of maximum, minimum or the average of the degree of previous of the rule [7]. It is often probable, that after evaluation of all the rules applicable to the input, we get more than one value for the degree of membership. In this case, the simulation has to take under consideration, all three possibilities, the minimum, maximum or average of the membership-degrees.

Third phase. It can be called defuzzification. The defuzzification of the data into a crisp output is accomplished by combining the results of the inference process and then computing the "fuzzy centroid" of the area. The weighted strengths of each output member function are multiplied by their respective output membership function center points and summed. Finally, this area is divided by the sum of the weighted member function strengths and the result is taken as the crisp output. One feature to note is that since the zero center is at zero, any zero strength will automatically compute to zero [7].

In fuzzy handoff decision mechanism, to the truth values have to have assigned a set of weights, represented by table of weight. According to these fuzzy handoff decision weight values and the degree of membership for fuzzy handoff decision output, the crisp value of output is determined by following formula:

$$fuzzy\_handoff\_decision = \frac{\sum M_i \times W_i}{\sum M_i} \quad (1)$$

where, M_i is the degree of membership in output singleton i, and W_i is the fuzzy handoff decision weight value for the output singleton i.

The crisp value of fuzzy logic decision is computed for each of the adjacent base stations. In result, the BS with highest value of decision is chosen as a target BS.

5 Conclusions

The theme presented in this paper is a part of the research work being done as the first phase of a study of handoff techniques optimization in standard 802.16e. We are currently working on the simulation model, to carry out simulation for all procedures, min-min, min-max, min-avg, max-min, max-max, max-avg, avg-min, avg-max, and avg-avg, in order to investigate the effect of the function choice on the result of the fuzzy logic process.

The handoff mechanism based on fuzzy logic presented in this paper hopefully will result in more efficient switching between BSs. Therefore, it can be used to provide sufficient QoS and better response time to handover request. Fuzzy logic handover scheme, by use of various input sets, is flexible and probably will give good results, even with wide set of input conditions. Proceeding based on fuzzy logic can be easily implemented through uncomplicated software procedures or dedicated fuzzy logic processing modules.

Acknowledgement. The article was supported by Czech Technical University's grant No. CTU0715013.

References

1. http://en.wikipedia.org/wiki/Handover [29.03.2007]
2. WiMAX Forum, *Mobile* WiMAX – Part I: A Technical Overview and Performance Evaluation, June, 2006
3. IEEE 802.16e-2005: Air Interface for Fixed and Mobile Broadband Wireless Access Systems, p. 244-257, February 2006
4. Bečvář, Z. and Zelenka, J.: Handovers in the Mobile WiMAX, In Research in Telecommunication Technology 2006 - Proceedings [CD-ROM]. Brno: Vysoké učení technické v Brně, 2006, vol. I, s. 147-150. ISBN 80-214-3243-8.
5. http://www.fuzzy-logic.com/Ch1.htm [04.03.2007]
6. Deepak Pareek: The business of WiMAX, John Wiley & Sons Ltd, 2006
7. http://www.seattlerobotics.org/encoder/mar98/fuz/flindex.html [04.03.2007]

Implementation of component based wireless communication protocol for SDR AT·

Junsik Kim[1], Sangchul Oh[1], Hongsoog Kim[1] Namhoon Park[1], and Nam Kim[2]

[1] Mobile Telecommunication Research Division
Electronics and Telecommunications Research Institute, Daejeon, Korea
{junsik, scoh, kimkk, nhpark}@etri.re.kr
[2] Dept. of Comput. & Commun. Eng., Chung-Buk Nat. Univ., Cheongju, Korea
namkim@chungbuk.ac.kr

Abstract. In this paper, we propose the SCARLET (SCA ReconfigurabLe middleware of ETRI) which is a middleware platform for the SDR (Software Defined Radio) AT (Access Terminal) and supports the SCA standard (Ver2.2) of JTRS. It guarantees the portability and reusability of the software, hardware, and also ensures the compatibility of mutual software component among products based on the SCA standard. The application component implementing the wireless communication protocol is arranged to the common terminal hardware based on this policy and requirement. Specifically, we propose and implement a design of the SDR AT middleware and the wireless protocol application components.

Keywords: SCA; SDR; Reconfiguration; Software Component; Middleware;

1 Introduction

It is expected that multiple radio access standards and systems will coexist in the same environment beyond 3G. SDR has made itself a key enabling technology in order to realize such a flexible and reconfigurable radio system. Therefore, it is considered as the core technology for positively coping with the change of this mobile communications market to one among the solutions working out the problem of not only the communications market but also the broadcasting market being faced. The SDR technology introduces the open architecture to the programmable digital radio technology. When it implements all kinds of the modeled devices, it could maintain an independency between each function module (the hardware, and the software). It provides the hardware reconfiguration ability, software programmability, flexibility, scalability, etc. It is consequently expected to form a new framework for the wireless telecommunications system implementation in the future.
In this paper, we propose a SDR AT developed on the basis of the SCA. The AT is built on the advanced SDR hardware platform on which provides an inherent

---

· This work was supported by the IT R&D program of MIC/IITA. [2006-S-012-01, Development of Middleware Platform Technology based on the SDR Mobile Station]

flexibility to support multiple air interface standards with different capacity ranges. Specifically, we propose a design of the SDR AT middleware and wireless protocol application components. The remainder of this paper is as follows. We describe an overview of the SCA which is a software platform for SDR and a flexible and reconfigurable environment in Section 2. In Section 3, we present the software, hardware architecture and functionalities of SDR AT, and section 4 includes application components and software platform architecture. In section 5, we describe the test environment and procedure of application component operation. Finally, conclusions are drawn.

2   SCA Overview

Ease of technology insertion, software reuse of waveform, which is radio access protocol and application, and software porting are feasible if the software architecture hides low-level system details[1]. The SCA hides a low level hardware and software details via;

- Encapsulation of hardware dependencies exclusively in SCA Device software which presents common interfaces defined in the Common Object Request Broker Architecture (CORBA) Interface Definition Language (IDL).
- The use of CORBA to hide details of the architecture, particularly the number and type of processors, their operating systems and communication mechanisms.
- A standard mechanism to describe the system and application configuration, the SCA Domain Profile[2]-[4].

The SCA defines an Operating Environment (OE) and specifies the services and interfaces that the applications use from the environment. The interfaces are defined by using CORBA IDL and graphical representations are made by using Unified Modeling Language (UML). The OE consists of a Core Framework (CF) that is the essential set for the open interface, a CORBA middleware and a POSIX-based OS.

The CF describes the interfaces, their purposes and operations [5]. It provides an abstraction of the underlying software and hardware layers for software application developers. The SCA compatible systems must implement these interfaces. The interfaces are grouped into Base Application Interfaces, Framework Control Interfaces and Framework Services Interfaces. The CF uses a Domain Profile to describe the components in the system. The Domain Profile is a set of XML files that describe the identity, capabilities, properties, inter-dependencies, and location of the hardware devices and software components that make up the system. The application components of the SCA are divided into two parts; CORBA components and Non-CORBA components. The communication between CORBA components and Non-CORBA components is made through a SCA adapter [6]-[8].

3 Platform for SDR AT

3.1 Hardware Platform

The functional modules on hardware platform are composed of mother board, processor board with GPP(General Purpose Processor), FPGA(Field Programmable Gate Array) for base band modem, IF board, RF module and reserved DSP board. For user interface ports, Ethernet port, USE2.0 port, UART port and JTAG for FPGA downloading/debugging are provided. The main control processor, PowerPC embedded in XC2VP30 operates at 150 MHz (Upto 300MHz is available according to Xilinx data sheet). 128 Mbyte FLASH is used for ROM and 128 Mbyte SDRAM is used for RAM. Two 1Mbit DPRAM(CY7C028) are provided for reserved data memory for MAC hardware. The both of RF and HSDPA modem are under construction, therefore, at the HSDPA mode, the lower transport layer of MAC is connected to LAN through a modem simulator and tested.

3.2 Software Platform

The main function elements of SCARLET are described in Fig. 1. They are comprised of the operating system, the CORBA middleware part, function components of the CF operated on CORBA that is the domain management unit, the device management unit, the XML configuration file and parsing unit, the application management unit, the component updating unit, and the external interface unit for the interface with Waveform Manager.

- **DomainManager** : The domain management unit performs the control and configuration of the file management and system domain.
- **DeviceManager** : The device management unit performs the function of managing the device component and interface with Device, loadableDevice, executableDevice, aggregateDevice of Framework Control Interface [1].
- **ApplicationManager** : The application management unit is comprised of compositional elements relating to the application management. It is the Application and the ApplicationFactory interface among the framework interface for the control of the system. The AssemblyController component does the resource controller part for the assembly of the components comprising one application. .
- **XML Parser** : The XML parsing part performs the function of the general XML parser. It converts configuration file into data type in which the program within SCARLET can use the Domain profile prepared to the XML format. For this, the XML parsing part provides the interface which the external program can use in the form of library. Moreover, it performs the function of detecting the error of the XML file, etc.
- **Component Update** : It is the function of renewing the application component of an error or the old version with the new application component. This function interfaces with the Core Framework part, and is not in the open interface of the

SCA CF standard but is the proprietary interface. The external operation from a user is received and started. After the operation of the application component is stopped and the corresponding component is removed, the component which it newly renews is inserted in the removed component column and it operates. An application has to be the CORBA component for the alternate interaction between the newly updated application, and existing application. And an interface has to be the open standard between the components.

> **External Interface** : This function manages an interface with the Waveform Manager in case the Waveform Manager is the Non-CORBA based system. The external interface part converts the message of the Non-CORBA based human-machine interface part into the interface of the CORBA based application component and delivers. It performs the message conversion function which is similar in the message of the opposite direction. The external interface part can omit in case the Waveform Manager is the CORBA based information network system.

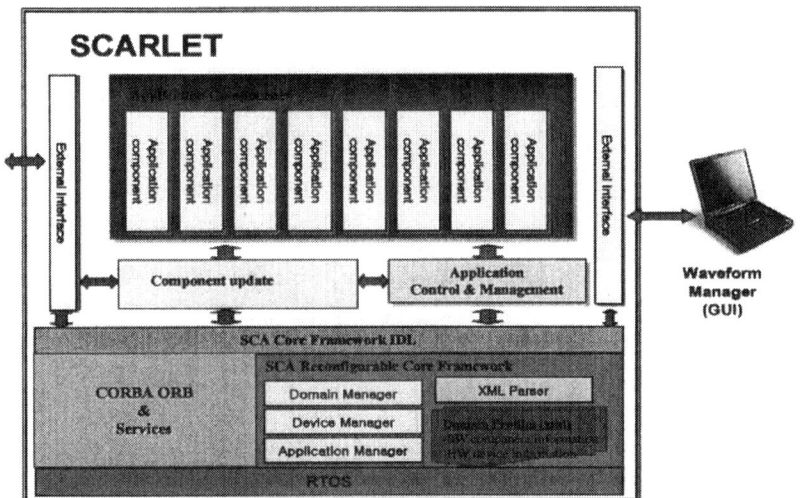

Fig. 1. SCARLET architecture

3.3 Waveform Manager

The waveform manager performs the management and the control function including the user interface for SCARLET middleware platform reconstructing the SCA base, and diagnostic function for the SDR AT. All kinds of manager software exist in the CF hierarchical layer for the reconfiguration of the component based application software. And it operates through the standard interface. SDR AT consists of TE(Terminal Equipment) and UE(User Equipment). The Waveform Manager is connected to the SDR TE, and it controls and manages the operation of the SCA application component software through an interface with the system. Moreover, it is developed in the GUI(Graphic User Interface) software environment(Fig 2.) of the

CORBA based and the system information is given and taken through the SCA application component and CORBA interface. And the event information is outputted. The SDR AT and Waveform Manager are connected with LAN and the CORBA based message delivery is made.

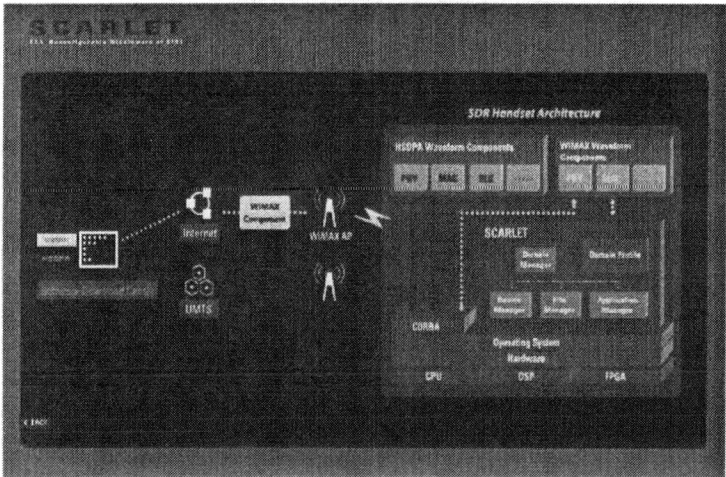

Fig. 2. Waveform manager GUI

4 Application component design

4.1 Application component architecture

Component software technology for maximizing software reuse is greatly helpful to overcome the extreme complexity of embedded software and reduce time-to-market. The component software technology is aimed at creating new software systems through the combination of deployable software, as opposed to ground-up development. We adopt the OMG's (Object Management Group) [9] definition of a component: a modular, deployable, and replaceable part of a system that encapsulates implementation and exposes a set of interfaces [10]. Components are typically in the form of shared library, depending on the deployment environment, and could be distributed as binaries, byte code, or even source files written in a scripting language.

The components for WiMAX and HSDPA were developed (Fig. 3), and performed under the SCARLET environment. The communication protocol of each mode utilized the used legacy protocol program which was developed before. Moreover, XML defining the connection structure between the application component, an attribute, etc. was built. It could be seen through a coupling with a middleware that complied with the standard of SCA in each application components.

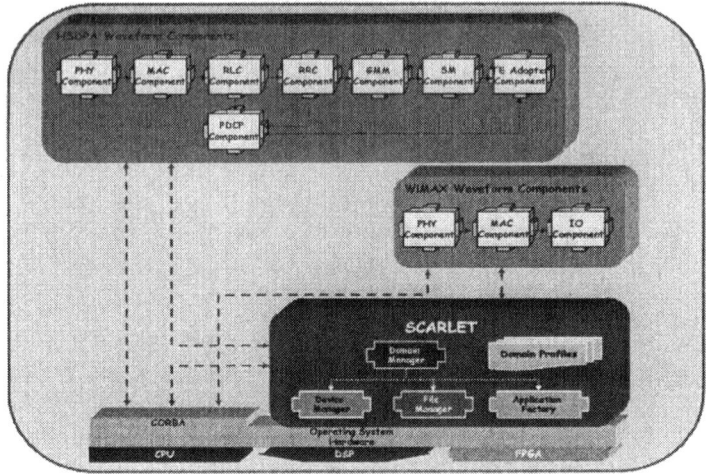

Fig. 3. Application component and middleware

4.2 Common function

The SCARLET system can be classified into the WF(Wave Form) subsystem and the CF subsystem. The software application components belonging to the WF subsystem operate with the CF subsystem through the interface as depicted in Fig. 4.

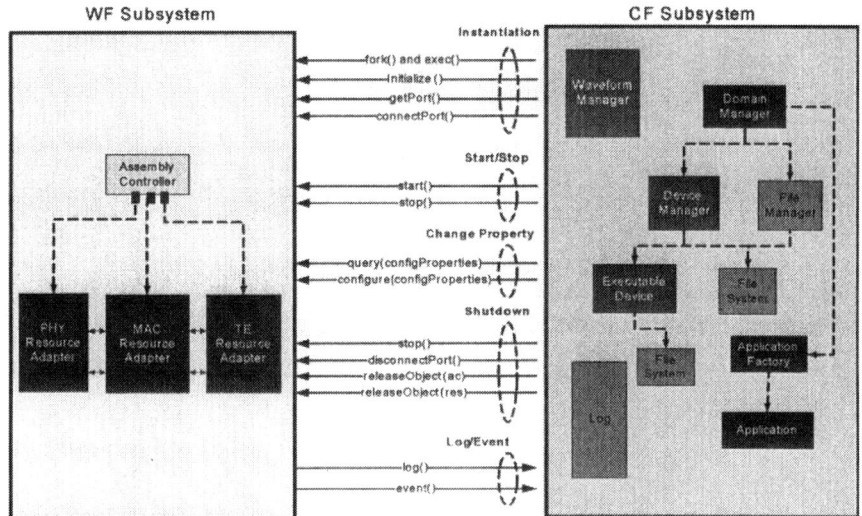

Fig. 4. Interface with CF

> Instantiation : The application components operate on the Executable Device with the fork() and exec() and are initiated by application with Initialize().

Moreover, the getPort() and connectPort() are used for the port information reference and attach of each component.
- ➤ Start/Stop : In Application, each application component is stopped through the stop(), and start() interface with a beginning the service.
- ➤ Change Property : In Application, the query() and configure() interface are used in order to inquire and change the property of the application component.
- ➤ Shutdown : In Application, the stop(), and the disconnectPort() and releaseObject() are used in order to make the application component shutdown.
- ➤ Log/Event : The application component delivers its own log information and event information to the log component and Application of the CF subsystem, through the log() and event() interface.

4.3 WiMAX Application component

As follows the Fig. 5, the WiMAX waveform application components are the WACB block which is assembly controller block, the WPAB block which is physical hierarchical layer adapter block, the WMAB which is MAC layer adapter block, and WIAB block which is the adapter block for the IO Resource and TE. The WACB block mutually interfaces with the WPAB, WMAB, and WIAB, which is the application resource of the WiMAX mode with WPAB, WMAB, and WIAB in order to perform the start / stop / property control (Fig. 5). WPAB, WMAB, and WIAB take charge of the interface function with the legacy software performed through the protocol layer of WiMAX and the SCA. Each one changes the existing legacy protocol software into the application component using adapter method complying with the standard of SCA and operates under the SCARLET environment.

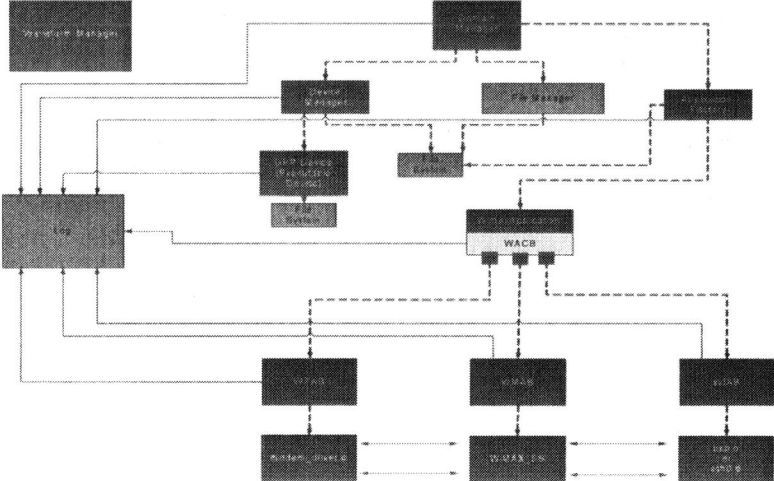

Fig. 5. WiMAX Architecture and Interface

4.4 HSDPA Application component

As follows the Fig. 6., the HSDPA application consists of 9 blocks. The HACB (HSDPA Assembly Controller Block) performs the start / stop / property control function of the HPHYB, HMACB, HRLCB, HRRCB, HPDCPB, HGMMB, HSMB, and HTEAB resources which is the components of the HSDPA waveform application. It is installed on the SCARLET domain. It registers itself at the Naming Service so that the other components can approach itself.

The HTEAB (HSDPA TE Adapter Resource Block) performs the role of interface between TE and UE. The HSMB (HSDPA SM Resource Block) performs the role of setting up the packet switched call and release. The HGMMB (HSDPA GMM Resource Block) performs the registration for the packet switched service and the authentication function. The HRRCB (HSDPA RRC Resource Block) performs the delivery of all parameters for using the radio resource between UE and UTRAN, and it controls each. In the HRLCB (HSDPA RLC Resource Block) performs trustworthy data transmission between higher layer and MAC, the HPDCPB (HSDPA PDCP Resource Block) performs the function it compresses IP packet header through the traffic path generated with RRC. The HMACB (HSDPA MAC Resource Block) classifies the performing function and the MAC protocol which transmits the data falling through the logical channel of the upper layer with the transport channel of the lower or transmits the data coming up to the transport channel of the lower with the upper layer through the logical channel of the upper layer according to MAC-d, MAC-c, and the MAC-hs entity and manages the respectively different transport channel and the MAC-hs protocol moreover handles transmitted data through HS-DSCH[11]. It manages the physical resources allocated for HSDSCH. In MAC, the HPHYB (HSDPA PHY Resource Block) transmits the data falling through the transport channel with the physical channel or performs the function of transmitting the received data with the physical channel through the transport channel to MAC.

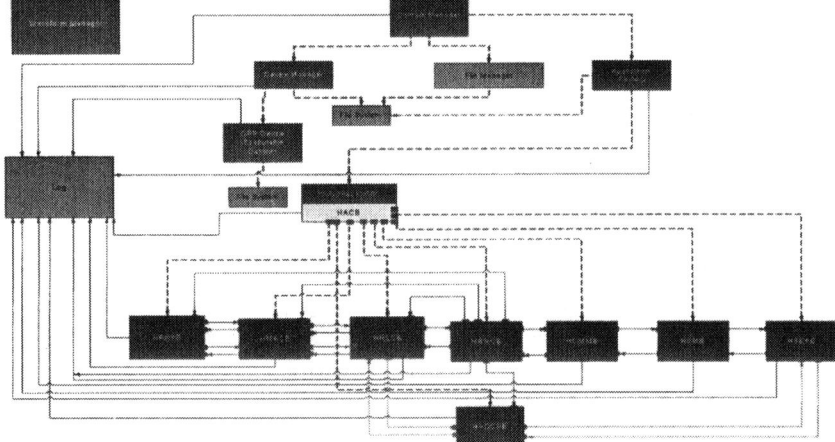

Fig. 6. HSDPA Architecture and Interface

5 TEST

The tests for verifying the SCARLET system consist of the application component test for the SCARLET, and middleware platform test. Fig. 7. shows the both test environment. The middleware platform and application component are verified on the test hardware platform. Moreover, the remote waveform Manager operates with the SCARLET middleware in terminal platform. After the HSDPA application components are set up on the test hardware in case a user desires the HSDPA service, the HSDPA service is provided. And after the WiMAX application components are set up on the test hardware platform in case a user desires the WiMAX service, the WiMAX service is provided.

Developed application components were smoothly performed under the common hardware platform and middleware. And it confirmed that the reconfigure operations including the loading / cancel/ release / perform / pause / mode switching, etc. were smoothly performed according to the direction of the CF. The applications like the web browser or the video player for the actual service, were performed in the TE.

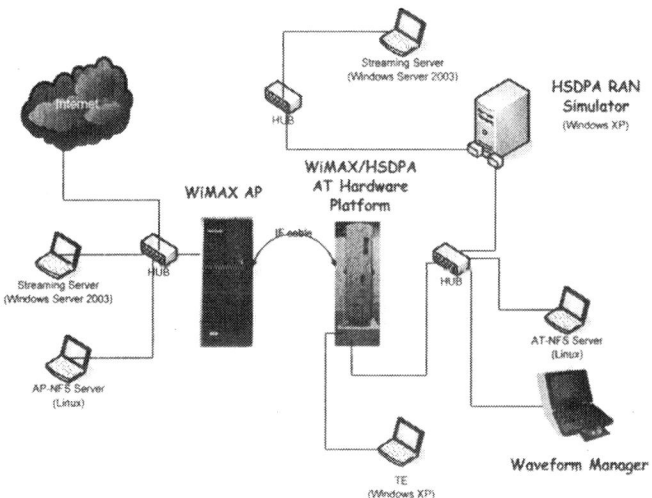

Fig. 7. Test environment

Using the Waveform Manager, the application components of the target mode are activated, and call processing is performed. The video on demand test which uses the internet service and Streaming Server are performed on TE. And, the WiMAX / HSDPA terminal platform and WiMAX AP were performed by using the NFS (Network File System) individual Server for the convenience of development.

Finally, by performing each service for each mode, right operation of the waveform application components and SCARLET were verified.

5 Conclusion

This paper represents a reconfigurable SDR AT developed on the basis of the SCA standard architecture and the WiMAX and the HSDPA service. We choose the bidirectional communication service combination of HSDPA and WiMAX in consideration of usability, performance and new wireless mobile communication technology. In order to support the standards of SCA for waveform application component, existing protocols were changed by using Adapter pattern. It currently adopted an interim architecture using a SCA adaptor. However, if there are more advances in the SCA technology as well as the SDR hardware, the architecture of the SDR AT will be upgraded for the system performance. Specifically, we designed SCA-based SDR AT software platform. The further research of reducing the reconfiguration time, FPGA componentization, and software modem improvement should be done.

References

1. Sofrware Communication Architecture(SCA) Specifications MSRC-5000SCAv2.2, 17 Nov. (2001)
2. Chia-Ching Lin, Hung-Lin Chou, Chin-Lien Chiu, Min-Chiao Wang and Shiao-Li Tsao: Design of a SDR Software Framework, SDR Forum document SDRF-02-I-0019-V0.00.
3. Shiao-Li Tsao, Chia-Ching Lin, Chin-Lien Chiu,S. Hung-Lin Chou and Min-Chiao Wang: Design and Implementation of Software Framework for Software Defined Radio System, Proceedings of the Vehicular Technology Conference 2002-Fall, Vol.4, September 24-28, (2002) 2395-2399
4. Saehwa Kim, Jamison Masse, Seongsoo Hong and Naehyuck Chang : SCA-based Component Framework for Software Defined Radio, Proceedings of the IEEE Workshop on Software Technologies for Future Embedded Systems 2003, May 15-16, (2003) 3-6
5. Joint Tactical Radio Systems SCA Developer's Guide, Contract No. DAAB15-00-3-0001, V1.1, June (2002)
6. JTRS website, http://www.jtrs.army.mil
7. http://www.sdrforum.org
8. Eun-Seon Cho, Chang-Ki Kim, Yeon-Seung Shin and Jin-Up Kim : SCA-based multi-LAN application development, Vehicular Technology Conference, 2004. VTC2004-Fall. 2004 IEEE 60th Volume 3, 26-29 Sept. (2004) 1978 – 1982
9. Object Management Group (OMG), http://www.omg.org.
10. Unified Modeling Language Specification Version 1.4 Appendix B - Glossary, Object Management Group, September (2001)
11. http://www.3gpp.org

A Hybrid WLAN-Bluetooth Access Network Solution for a More Efficient VoIP-Data & Video Traffic Management*

David Pérez[1], José Luis Valenzuela[1], Ángela Hernández[2], Antonio Valdovinos[2]

[1] Universitat Politècnica de Catalunya (UPC),
Avda.Canal Olimpic s/n, (08860) Castelldefels, SPAIN
dperez@tsc.upc.edu, valens@tsc.upc.edu
[2] Institute of Engineering in Aragón, I3A, University of Zaragoza,
María de Luna 1, (50018) Zaragoza, SPAIN
anhersol@unizar.es, toni@unizar.es

Abstract. In this paper we present a hybrid Physical and Medium Access Control protocol for a Wireless LAN which is designed to support both synchronous (voice) and asynchronous (data) traffic. The protocol is designed using a modified Bluetooth core and 802.11e radio access, where the voice service is supported by Bluetooth whereas data and video services are provided by 802.11e. The Bluetooth radio access is modified to operate using an orthogonal frequency hopping radio in bands of 20MHz coincident with 802.11e channels.

Keywords. Bluetooth, WLAN, MAC.

1 Introduction

The evolution of communication systems in last years has followed a clear tendency towards B3G systems (Beyond Third Generation) or 4G, that are intended to provide a clear support to the integration and coexistence of multiple and different Radio Access Technologies (RATs) in a unique and complex radio environment assuming the management of the whole system and coordinating the radio access with the Core Network. This core, the IP Multimedia Subsystem (IMS), provides the control functionalities and an ubiquitous and seamless wireless access system 2G (GSM), 3G (UMTS) and wide band (WLAN, WiMax), extended to auto organized network schemes with short range connectivity (Bluetooth) between intelligent terminals.

Wireless local area networks have been predominantly used to support data applications. However recent developments, particularly the ability of personal computers (PCs) to deal with real-time voice, have demonstrated the need for

---

* This work has been supported by CYCIT (Spanish National Science Council) under the grant TEC2006-09109, which is partially financed from the European Community through the FEDER program.

wireless LANs to efficiently support both voice and data traffic. Lately voice communications are heading to use data networks; usually IP networks. This kind of communications is known as VoIP. The requirements of voice and data traffic are however very different. In general voice networks can tolerate errors and packet loss (<1-5%) without degrading service, but have real-time constraints and fixed assignment wireless protocols are more appropriate such Time Division Multiple Access (TDMA). On the other hand data networks can tolerate packet delays, but cannot allow errors or packet losses, and random access protocols are more appropriate. Thus protocols designed for data networks typically use some form of CSMA/CA. Our goal is to be able to support high quality real-time voice conversations and, at the same time, provide a high data throughput. To overcome the limitations of current MAC technologies we have been investigating how to keep the advantages of both types of MAC protocols, without reducing the quality of service for each traffic type, and without increasing the overall complexity of the system. During system design a decision is taken to optimally support one or other traffic type and this leads to the adoption of a particular access mechanism.

In this work we propose a hybrid WLAN-Bluetooth Radio Access System, which allows taking profit of WLAN and Bluetooth technologies by defining an adaptive MAC implementation. The interaction in a multi system WLAN-Bluetooth network can facilitate, for example, coexistence of voice and data services. The first technology (WLAN), based on CSMA/CA is appropriate for data communications and the second one, based on TDMA, for voice communications. The main idea is to share radio resources among WLAN and Bluetooth, in a superframe time base of 20ms, divided in two parts. In the first part, a BT-TDMA access is employed by users with VoIP traffic demand. In the second one, a WLAN CSMA/CA access is employed by users with data traffic demands. The number of resources, assigned to WLAN and BT access, is not fixed but it depends on VoIP users. On the other hand, to provide high quality voice we have chosen to use an AMR voice codec which is identical to the one used by UMTS. This codec is widely used and enables us to offer good integration with the telephony network. Adaptive Multi-Rate (AMR) codecs are standardized by 3GPP for GSM [1] [2] the world's most widespread cellular technology, as well as for WCDMA and a payload format for IP-transport has also been standardized in IETF [3]. The paper is organized as follows. First, fundamentals of 802.11e/g and Bluetooth standard are briefly explained before introducing a detailed description of the proposal. In Section 3, the proposed hybrid Physical and Medium Access Control protocol is presented and simulation results obtained are discussed. Finally, conclusions are summarized in Section 5.

2 Fundamentals of Bluetooth and 802.11e/g

2.1 802.11

Since the appearance of the initial IEEE 802.11 standard [4], near ten years ago, wireless networks have experienced a vertiginous evolution. But, to be able to support voice and multimedia applications demanded by the market, just increasing transmission rates is not enough. It is also necessary a good management of these resources and to be able to offer to users what is called quality of service (QoS). Task group E of 802.11 IEEE working group, developed a standard [5] able to differentiate several traffic types based on the different QoS needs. This standard has evolved the original Distributed Coordination Function (DCF) and Point Coordination Function (PCF) to the new Enhanced Distributed Channel Access (EDCA) and Hybrid Coordination Function (HCF) Controlled Channel Access. The main difference is the establishment of four Access Categories (AC). Depending on the category to which a packet belongs, it has more or less priority. Fig. 1 shows the basic procedure.

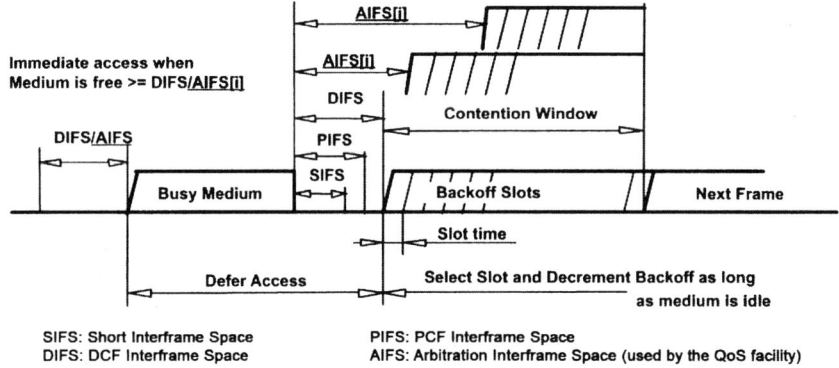

Fig. 1. 802.11e Back-off procedure

The 802.11g [6] standard proposes an extension of the transmission rates up to 54Mbps. The physical layer of 802.11 is modified and uses an OFDM modulation. At MAC layer level operation is the same as 802.11, so the medium access mechanism is CSMA/CA. The standard was designed to be backward compatible; so, the possibility of working in mixed mode is also considered (802.11b+802.11g). Nevertheless, we will use only the OFDM-PHY rates of 802.11g, that is, from 6 to 54Mbps. 802.11e, however, modifies the legacy 802.11 MAC. It establishes four AC: Voice, Video, Best effort and Background.

In each one of these access categories a set of parameters is defined that depends on the traffic priority. These parameters are also different for the Access Point (see Table 1). And they are:

- **AIFS[i]:** it indicates the minimum number of delay slots before the contention window for each access category, as it is possible to observe in Fig.1.
- **CWmin and CWmax:** the size of the contention window will depend on the AC assigned.
- **Transmission Opportunity (TXOP):** once won the access to the channel, a station is allowed to send packets as long as the duration of them is smaller

than the remaining TXOP. If TXOP is zero the station is only allowed to send one packet.

Table 1. Default 802.11e and 802.11g parameters

Parameter	AV-Voice(STA)	AC-Voice(AP)	AC-Best effort
AIFSN	2	1	3
CWmin	3	3	15
CWmax	7	7	1023
TXOP	1054μs	1054μs	0
PHY		ERP-OFDM	

The effective time to transmit a data packet and supposing that there are no collisions can be computed following (1) and Fig.2.

$$T_{ef}(\mu s) = T_{Access} + T_{PHY} + T_{DATA} + T_{ACK} \qquad (1)$$

where,

$$T_{access} = DIFS + \frac{CW_{min}}{2} \cdot SIFS$$
$$T_{PHY} = PLCP_{preamble} + Signal + Signal_{Ext}$$
$$Headers = MAC_h + IP_h + UDP_h + RTP_h$$
$$T_{DATA} = \left\lceil \frac{Service+Tail+8(Headers+MAC_{CRC}+DATA)}{N_{DBPS}} \right\rceil \cdot T_{sym}$$
$$T_{ACK} = T_{PHY} + \left\lceil \frac{MAC_{ACK} \cdot 8 + Service + Tail}{N_{DBPS}} \right\rceil \cdot T_{sym}$$

and N_{DBPS} is the number of bits per OFDM symbol and depends on the transmission rate. All the values are shown on Table II. And the operator ceiling ($\lceil\ \rceil$) returns the smallest integer greater than or equal the specified number.

Table 2. Considered parameters

Parameter	Value	Parameter	Value	PHY rate	N_{DBPS}
MAC header	24 bytes	Tail bits	6 bits	6 Mbps	24
MAC CRC	4 bytes	PLCP preamble	16 μs	9 Mbps	36
MAC ACK	14 bytes	Signal	4 μs	12 Mbps	48
IP header	20 bytes	Signal Extension	6 μs	18 Mbps	72
UDP header	8 bytes	Slot Time	9 μs	24 Mbps	96
RTP header	12 bytes	SIFS	10 μs	36 Mbps	144
AMR Data	31 bytes	AIFS	28 μs	48 Mbps	192
Service bits	16 bits	T_{sym}	4 μs	54 Mbps	216

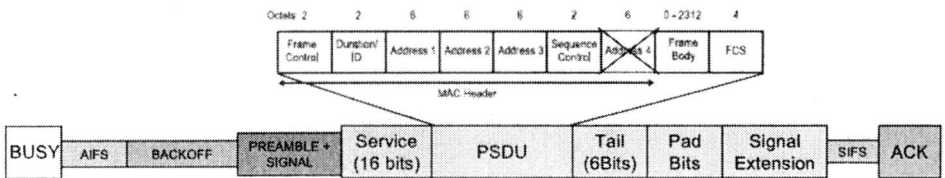

Fig. 2. 802.11 packet transmission structure

2.2 Bluetooth

Bluetooth [7] is a universal radio interface in the 2.4GHz ISM frequency band. Bluetooth is based on a centralized connection oriented approach. Bluetooth devices sharing a wireless channel form a piconet. One device in a piconet has the role of master and controls the channel access, while the others are slaves. There may be up to seven slaves in a piconet.

Bluetooth uses a Time-Division Duplex (TDD) scheme to divide the channel into 625ms time slots. Master and slave units transmit alternately. Each piconet is characterized by a particular fast frequency-hopping pattern; the frequency is uniquely determined by the master's address and is followed by all the devices participating in the piconet.

Two types of connections can be established in a piconet: the Synchronous Connection-Oriented (SCO) link, and the Asynchronous Connectionless (ACL) link. SCO links provide a circuit-oriented service with constant bandwidth based on a fixed and periodic allocation of slots. They require a pair of slots once every two, four or six slots, depending upon the SCO packet used. ACL connections, on the other hand, provide a packet-oriented service and span over one, three or five slots. For ACL links, Bluetooth uses a fast acknowledgment and retransmission scheme to ensure reliable transfer of data. The master controls traffic on ACL links by employing a polling scheme to divide the piconet bandwidth among the slaves. A slave is only allowed to transmit after the master has polled it.

SCO links have been designed to support voice services. Since these links require a periodic allocation of a pair of slots once every two, four or six slots. The SCO link is a symmetric, point-to-point link between the master and a single slave in the piconet. The SCO link involves reservation of slots and can therefore be considered as a circuit-switched connection between master and slave. The master can support up to three SCO links to the same slave or to different slaves. SCO packets have been designed to support 64kbps speech. The specifications define three pure SCO packets and one hybrid SCO packet, which carries an asynchronous data field in addition to a synchronous voice field.

The ACL link provides a packet-switched connection between the master and all active slaves in the piconet. A slave can send an ACL packet if it has been addressed by the master in the previous slot. To ensure data integrity, ACL packets are retransmitted. Only a single ACL link can exist between a master and a slave. The master schedules ACL packets in the slots not reserved for the SCO links. The specifications define several kinds of ACL packets, see Table 3.

Table 3. ACL Packets

Packet type	Time Slots	Modulation	Payload (user bytes)	Packet type	Time Slots	Modulation	Payload (user bytes)
DM1	1	GFSK	0-17	2-DH1	1	$\pi/4\ DQPSK$	0-54
DH1	1	GFSK	0-27	2-DH3	3	$\pi/4\ DQPSK$	0-367
DM3	3	GFSK	0-121	2-DH5	5	$\pi/4\ DQPSK$	0-679
DH3	3	GFSK	0-183	3-DH1	1	$8\ PSK$	0-83
DM5	5	GFSK	0-224	3-DH3	3	$8\ PSK$	0-552
DH5	5	GFSK	0-339	3-DH5	5	$8\ PSK$	0-1021
AUX	1	GFSK	0-29				

Bluetooth devices communicate with each other by using standard networking protocols to transport control and data packets over a network. Devices may use protocols such as TCP, IPv4 or IPv6. These protocols have dissimilar network packet formats. To provide seamless transmission of network packets over the L2CAP layer in the protocol stack, an intermediate protocol is required that encapsulates dissimilar network packet formats as a standard common format.

The Bluetooth Network Encapsulation Protocol (BNEP) provides this encapsulation by replacing the networking header, such as an Ethernet header, with a BNEP header. The L2CAP layer encapsulates the BNEP header and the network payload and sends it over the transport media. The Bluetooth Personal Area Networking profile describes how BNEP shall be used to provide networking capabilities for Bluetooth devices.

3 The hybrid physical-MAC protocol proposal

Once basic principles of WLAN and BT have been introduced, we describe the protocol of the physical-MAC level we propose. It is essentially a hybrid protocol (adaptive MAC) which combines both 802.11 and Bluetooth access mechanisms, Fig.3, in order to take profit of the advantages of both technologies.

Hybrid MAC protocols are not unique [8], [9] but the protocol described here incorporates features which ensure good performance under a wide range of conditions. The hybrid MAC protocol uses a superframe (whose duration is fixed to 20ms), which incorporates a contention free period (CFP) and a contention period. The access mechanism used during each CFP is TDMA (modified Bluetooth access), whereas the access mechanism used during the contention period is CSMA/CA. The first one will be used for VoIP communications (with codecs selected appropriately) and the second one, for data communications. The total amount of resources is distributed on an adaptive bandwidth partitioning strategy between Bluetooth and WLAN, depending on the number of VoIP users. The main idea is illustrated on Fig.4. However, before continuing describing

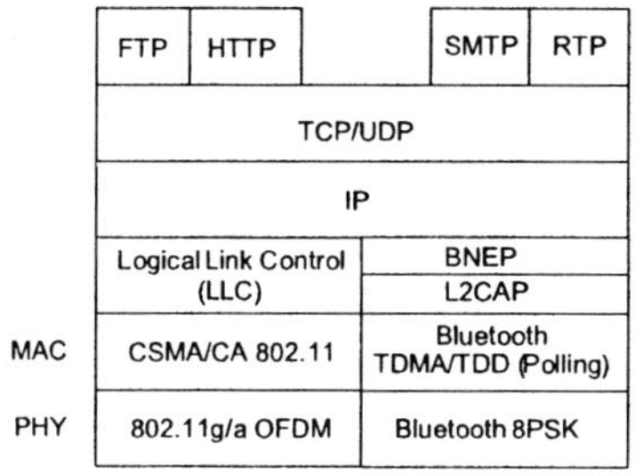

Fig. 3. Protocol stack

with more detail the MAC implementation and its associated parameter values, we review some aspects related to the required network architecture, the physical layer, the traffic sources and the associated framing structures.

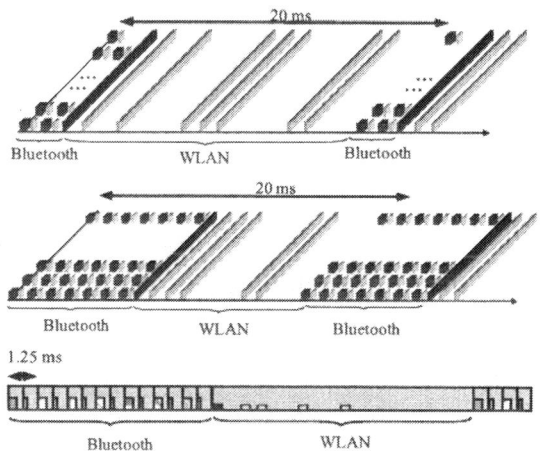

Fig. 4. Time-frequency proposed structure scheme

3.1 Network architecture

To support the proposed scheme, the network architecture consists of:

- Access Point (AP) that supports voice and data services.
- Voice Terminal that only uses the TDMA access mechanism to communicate with the Access Point.
- Data Node that uses the CSMA/CA access mechanism and can communicate with an Access Point and other data nodes.
- Voice and Data Node which can use both access mechanisms.

3.2 Physical layer

At the *physical level*, the system is designed to operate in the 2.4GHz ISM band, but Dynamic Frequency Selection (DFS) can be used to allow the possibility of using the 5GHz band. When working in WLAN mode, the behavior of the radio hardware is the same as an off the shelf 802.11g WLAN using the ERP-OFDM PHY layer that employs the whole 20MHz channel. On the other hand when working in Bluetooth mode, the specifications indicate that a Bluetooh receiver requires 5dB of Adjacent (1MHz) interference C/I_{1MHz} for 8PSK modulation and -25dB Adjacent (2MHz) interference C/I_{2MHz} [7]. However commercial Bluetooth modules achieve a C/I_{2MHz} up to -40dB [10]. So, with a band separation of 2MHz a right operation is guaranteed. Thus we propose that the radio hardware used works with ten new frequency hopping sequence sets in bands of 1MHz with a guard band of 1MHz. This guard band was considered instead of a continuous band in order to reduce adjacent interference when several frequencies have been used simultaneously. And moreover, the resultant spectrum is similar to the one of a WLAN.

3.3 Traffic and framing structure

We consider the VoIP transfer over Bluetooth using Bluetooth Network Encapsulation Protocol (BNEP). For that reason we looked for one of the voice codecs with smaller size, AMR codec [1]. The encoder outputs compressed speech data in octet aligned (by using bit stuffing) AMR-NB Interface Format 2, as defined in the 3GPP TS 26.201 [11]. Other codecs as can be *iLBC*, used in *Skype*, have similar characteristics.

The use of BNEP for transporting a voice ethernet packet is shown in Fig.5. BNEP removes and replaces the ethernet header with the BNEP header. Finally, both the BNEP header and the payload are encapsulated by L2CAP and are sent over the Bluetooth physical channel.

L2CAP Header	BNEP Header	IP/UDP/RTP	Payload
4 Bytes	4 Bytes	40 Bytes	13-31 Bytes

Fig. 5. Voice packet structure over L2CAP and BNEP

Bluetooth offers several types of packets with different user payloads and rates. We propose the use of 3DH1 packets to transport the encapsulated voice payload. For example, if we use the seventh mode of the AMR codec it has a bit rate of 12.2kbps. Thus, we generate a 31 bytes packet each 20ms, moreover we have to add to this 20 bytes of the IP protocol header plus eight of the UDP header, four of the BNEP and four of L2CAP. Looking at Table 3 we can see that it is possible to use the 3DH1 mode or even the 2DH1 mode for lower rate codecs.

With respect to data traffic, transmission over 802.11g uses the packet structure shown before in Fig.2, follows the access mechanism of Fig.1 and its default parameters were shown in Table 2.

3.4 The MAC Protocol

The start of the superframe is the point at which voice stations start to transmit. The duration of the superframe is fixed and equal to 20ms. This dwell period is fixed at 20ms to provide acceptable performance with respect to latency.

The length of the dwell period also means that each voice data message contains 20ms of AMR data. In addition each packet transmitted includes the necessary MAC and PHY headers. Each one of the contention free periods is divided into a number of pairs of fixed length slots (625ms), two per voice connection. The first slot in each pair is used to transmit voice data from the Access Point to a node (downlink) and the second is used to transmit voice data from a node to the Access Point (uplink) as shown in Fig.4.

Each Bluetooth station acting as Master can address up to seven clients simultaneously. As we have proposed the standard is extended to allow ten orthogonal hopping sequences. This implies we can handle up to ten voice users per each pair of slots that is a maximum of 70 voice users during the first 14 slots. These 14 slots have 8.75ms of duration which leave us 11.75ms remaining free of transmissions until the start of the next superframe. In a managed network a Beacon is transmitted immediately after the hop. This Beacon is used to maintain network synchronization, control the format of the superframe and manage when each node should transmit and receive data.

At the end of the first CFP in the superframe there is a space reserved for a service slot. The service slot is used by voice nodes to communicate with the Control Point. The time between the two CFPs, the contention period, is used for data transmissions using a CSMA/CA protocol similar to that specified in the 802.11 standard [4].

The MAC uses a slotted contention scheme, acknowledgement and retransmission of data messages and a fragmentation scheme to improve performance. If there is no voice connection active then the CSMA/CA period occupies the whole of the superframe, with the exception of the space required for the Beacon, maximizing data throughput. For example, 25 voice users would employ the first pair of slots for the first ten users, another pair for the next ten users and another pair for the remaining five users. That is a total of six slots (3.75ms) so the data transmission can employ 16.25ms.

3.5 Management

The primary function of the Beacon is to enable all nodes to synchronize to the timing of the network. The Beacon transmitted by the access point is also used to manage the network during the contention free periods. The beacon can include a list of active voice connections (and therefore frequency hopping and slot assignments), retransmission slot assignments for the current superframe, connection status information and paging information. Slot assignment and synchronization information does not change on a per frame basis, so if a node misses a Beacon it uses the information contained in the most recent valid beacon.

All connection and paging status requests and information are repeated until they are acknowledged by the target node.

4 Results

In this section some results are presented in order to establish the benefits of Hybrid MAC proposal opposite to results obtained with 802.11g/e. Just as an upper bound reference, we compute the effective time needed, for a VoIP transmission over 802.11g/e. Results are shown in Table 4. The number of admissible voice users, assuming perfect statistical multiplexing and no collisions is shown in the same table.

Table 4. Maximum capacity upper bound

PHY rate	Tx Time (μs)	Voice users
6	263.5	37.95
9	211.5	47.28
12	183.5	54.49
18	159.5	62.69
24	147.5	67.79
36	131.5	76.04
38	127.5	78.43
54	123.5	80.97

We compare these results with the ones following the hybrid scheme (showed in Fig.4) and those obtained with the real performance of the standard WLAN. Comparison is performed assuming a variable number of VoIP users multiplexed with an FTP user transmission. Combinations of physical transmissions of 6Mbps and 54Mbps for both VoIP and FTP traffic sources have been considered. Evaluation has been made in terms of Packet Loss Probability for audio traffic and Throughput (maximum possible transmission) for the FTP source.

Figures 6(a) and 6(b) show results for an FTP physical transmission at 6Mbps multiplexed with both 6Mbps and 54Mbps for VoIP sources. We can see that in a standard WLAN environment, 50 VoIP (54Mbps) users have unacceptable levels of 15% of losses and the FTP user works at 500kbps. In fact, maximum capacity under these conditions is 32 users (VoIP at 54Mbps) and the FTP user transmits at around 2Mbps. However with the hybrid proposed scheme we can handle 50 VoIP users (in addition, without losses due to congestion) and 3.33Mbps for the FTP, as was expected. Note that 50 VoIP users need $5 \cdot 1.25ms = 6.25ms$, so

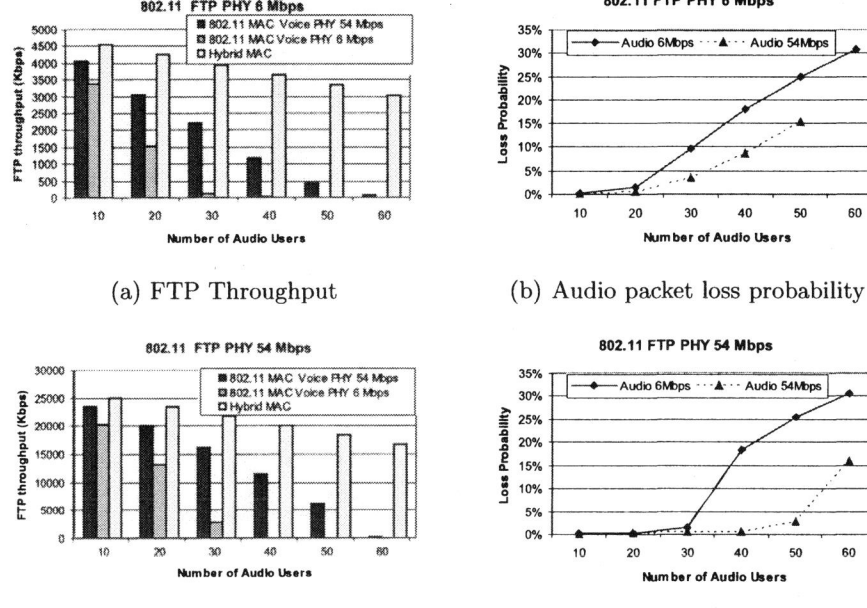

Fig. 6. Results

13.75ms can be employed for FTP. On the other hand, effective rate when all the radio resources are assigned to the FTP transmission is 4.8Mbps. So, if a fraction of 13.75/20 of radio resources is assigned, about 3.33Mbps can be achieved.

In figures 6(c) and 6(d), the same results are showed for FTP transmissions at 54Mbps. The number of simultaneous VoIP users supposing a 5% of losses at 54Mbps is near 60 users and transmitting at 6Mbps, just about 32. The FTP throughput associated is lower than 6.2Mbps and 3Mbps, respectively. However, in the hybrid case, we could even handle, without problems, 70 VoIP users and have enough remaining capacity to support a 13Mbps FTP transmission at 54Mpps.

In all the results, the maximum number of VoIP users in the hybrid protocol outperforms even the upper bound reference considered in Table 4.

Moreover commercial 802.11g/a products have sensitivities in a range between -90dBm(6Mbps) and -72dBm(54Mbps) On the other hand commercial Bluetooth products require just -85dBm which corresponds with a WLAN working at 9Mbps. This fact implies that the maximum number of users supported is reduced more than a half.

5 Conclusions

In this paper, a hybrid WLAN and modified Bluetooth based physical and MAC protocol layers have been proposed in order to improve system efficiency in heterogeneous traffic scenarios. Simulation results show that the new system outperforms system capacity over the WLAN-only solution when both VoIP and data transmissions are considered. Taking profit of the CSMA/CA access technique, used in WLAN, for data transmissions in addition to deterministic Bluetooth access for VoIP has a significant impact over the system capacity.

We have demonstrated along the paper that the theoretical capacity upper bound limit of a standard WLAN is overcame by the proposed solution.

The results show that not only the voice capacity of the system is improved, but also the CFP available for data transmissions, allowing a better throughput of data traffic.

References

[1] 3GPP TS 26.071, "AMR speech codec; general description"
[2] 3GPP TS 26.171, "Wideband AMR speech codec; general description"
[3] J Sjöberg et al., "RTP payload format and file storage format for the Adaptive Multi Rate (AMR) and Adaptive Multi-Rate Wideband (AMR-WB) audio codecs" 2002, IETF RFC 3267.
[4] ANSI/IEEE, 802.11-1999, "Part 11: Wireless LAN Medium Access Control (MAC) and Physical Layer (PHY) specifications", IEEE
[5] IEEE 802.11e. IEEE Standard for Information technology Telecommunications and information exchange between systems. Local and metropolitan area networks. Specific requirements Part 11: Wireless LAN Medium Access Control (MAC) and Physical Layer (PHY) specifications. Amendment 8: Medium Access Control (MAC) Quality of Service Enhancements
[6] IEEE 802.11g. IEEE Standard for Information technology. Telecommunications and information exchange between systems. Local and metropolitan area networks. Specific requirements Part 11: Wireless LAN Medium Access Control (MAC) and Physical Layer (PHY) specifications Amendment 4: Further Higher Data Rate Extension in the 2.4 GHz Band
[7] Specification of the Bluetooth System, Core Specifications, version 2.0+EDR, v1.2, v1.1 ; https://www.bluetooth.org/spec/
[8] K.S.Natarajan. "A hybrid medium access control protocol for wireless LAN". 1992 IEEE International Conference on Selected Topics in Wireless Communications. pp 134-7.
[9] B. A. Sharp, E. A. Grindrod, and D. A. Camm, "Hybrid TDMA/CSMA Protocol for Self Managing Packet Radio Networks" Proc. 1995 4th IEEE ICUPC, Tokyo, Japan, Nov. 6-10, 1995, pp. 929-33.
[10] BlueCore4-External Data Sheet. http://www.csrsupport.com/BC4Ext
[11] 3GPP TS 26.201. "Technical Specification Group Services and System Aspects; Speech codec speech processing functions; Adaptive Multi-Rate - Wideband (AMR-WB) speech codec; Frame structure. (Release 6)"

Interference Aware Bluetooth Scatternet (Re)configuration Algorithm IBLUEREA

Tomasz Klajbor[1], Jozef Wozniak[1]

[1] Gdansk University of Technology, Faculty of Electronics,
Telecommunications and Informatics, Gdansk, Poland
{klajbor, jowoz}@pg.gda.pl

Abstract. This paper presents a new algorithm IBLUEREA, which enables the reconfiguration of Bluetooth scatternet to reduce interference. IBLUEREA makes use of the complex model comparing ISM environment efficiency. The mechanism envisages the use of the assessment of the probability of successful (unsuccessful) frame transmission in order to take a decision concerning the co-existence of technologies which make use of the same ISM band (here Bluetooth and 802.11b).

Keywords: Bluetooth, IEEE 802.11b, interference, co-existence

1 Introduction

The number of various wireless technologies and network devices making use of ISM band (e.g. Bluetooth (BT) 2, IEEE 802.11b (Wi-Fi) 7 or IEEE 802.11g) is growing very fast. Due to this, it becomes more and more difficult to provide transmission parameters that can guarantee the quality of services required by co-existing networks. This specially refers to specific network devices operating in a close vicinity around other devices belonging to different independent networks, very often based on different technical and functional solutions.

In order to provide for a higher work efficiency of a number of technological solutions working within the same area, coexistence mechanisms have been worked out 1. Such mechanisms can be divided into two groups 1:

- Collaborative mechanisms, requiring information exchange between IEEE 802.11b and Bluetooth devices.
- Non-collaborative mechanisms, which can be adopted by 802.11b and/or Bluetooth devices without a direct collaborative system.

Apart from the mechanisms presented in 1, examples of different collaborative algorithms can be found in the literature. Isolated examples of solutions facilitating the co-existence of various technologies can be traced, which are based on predicting the propagation conditions variability. For example, in 4 Interference aware BLUEtooth Segmentation mechanism has been presented which is based upon a dynamic BT frame choice depending on the propagation conditions. This method relies upon the theoretical assessment of the probability of successful frames transmission and the queuing tasks analysis. Based on such information IBLUES "takes" decisions concerning the choice of a frame (from those defined in specification 2), through which the data will be transferred (e.g. DM1, DM3, DM5).

In this article a new coexistence mechanism has been presented which is correlated to the management of Bluetooth network topology. This mechanism has been named *Interference Aware BLUEtooth Scatternet (RE)configuration Algorithm* (IBLUEREA). IBLUEREA algorithm is based upon the idea of switching functions performed by those BT devices which more frequently use ISM band (i.e. masters) and are in close vicinity of receivers/transmitters of other technology solutions (e.g. 802.11b) and BT piconets. IBLUEREA involves operating as a master (in a given piconet) for a device which simultaneously causes and is susceptible to little interference (comparing to other BT piconet devices and networks using other technologies). For this analysis, IBLUEREA uses a new model comparing the efficiency of ISM complex environment.

Chapter 2 describes the above-mentioned model comparing the efficiency of ISM environment. Theoretical grounds of this model have been equipped with a simple example illustrating its mechanism.

Chapter 3 presents the idea behind the IBLUEREA algorithm. Chapter 4, in turn, presents the benefits of using this algorithm by giving scenarios, which has been the subject of simulations tests.

2 Model comparing the efficiency of ISM environment

While making comparisons regarding the efficiency of given ISM environments, among others, the number and function of various technology devices which are co-located need to be taken into consideration. In order to assess with accuracy the efficiency of a given ISM environment, the influence of all jamming devices on the receivers located within their range need to be accounted for.

In 6 the general principles of scatternet matrix description have been presented. The authors also suggested metrics (allowing for functions played by devices in given piconet and/or devices links), thanks to which the aggregated (and standardized) link capacity in scatternet assessment is possible. The metrics are of little significance while tackling the interference issue, which has only been mentioned in this article. Moreover, metrics do not enable the analyses of interference coming from other systems. Below, it has been presented the original methodology and key metrics necessary for the Bluetooth scatternet and IEEE 802.11b network co-effectiveness assessment.

Let us assume that a scatternet with the topology presented in Fig. 1 has been created. Devices 1 and 4 operate as masters. Piconet 1 includes slave devices with numbers 2 and 3. Whereas piconet 2 makes slave devices 3 and 5. Slave device no. 3 acts as a bridge between the two piconets.

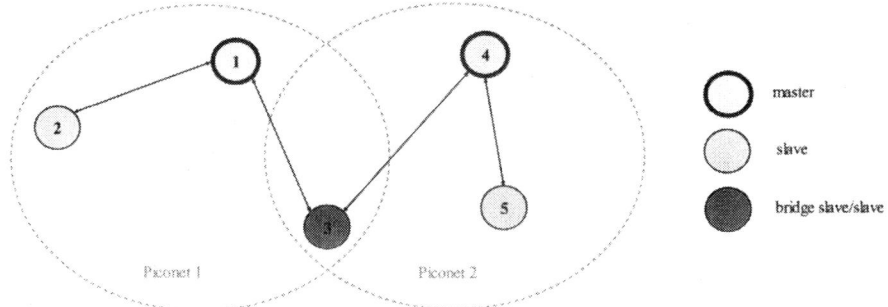

Fig. 1. An example of scatternet consisting of two BT piconets.

In order to illustrate co-existence mechanism, Fig. 2 presents also an example of ISM environment, with possible mutual interference areas of the Bluetooth piconet and the coexisting IEEE 802.11b network. It has been assumed that the mutual interference areas are those where given technology transmitters have a substantial negative impact on the receivers of other BT piconets or 802.11b network (for example, frame error rate can, in theory, exceed a given threshold value).

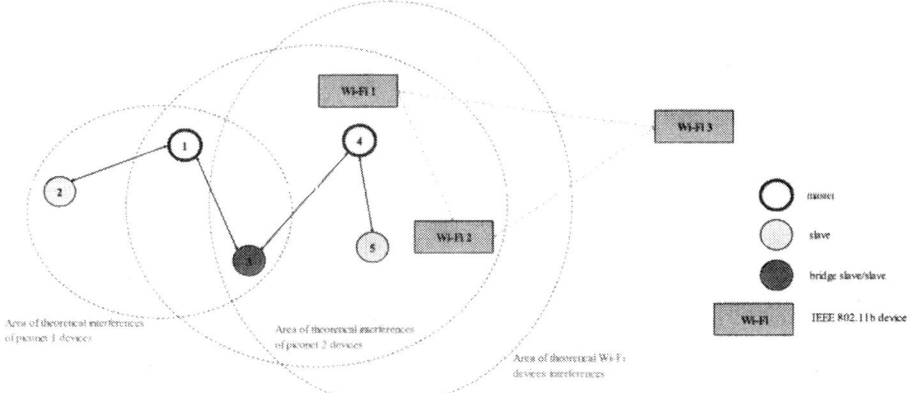

Fig. 2. Example of Bluetooth and Wi-Fi mutual interference areas

In accordance with the above illustration:
- Piconet 1 can potentially jam the transmission of device 3, in case when it operates as an element of 2^1 piconet,
- Piconet 2 jams the transmissions between 1 and 3 devices (piconet 1) and jams the receptions at IEEE 802.11b: Wi-Fi 1 and Wi-Fi 2,
- IEEE 802.11b network jams the transmissions of BT: 3, 4 i 5,
- Wi-Fi 3 is beyond interference over BT scatternet.

---

[1] To simplify the analyses, it has been assumed that if a device belongs to a given piconetwork, it should affect the interference under examination. In fact, the interference range of each individual device with another one should be examined.

Such an ISM environment, defined in such way, can be formally presented using the interference matrix $A_{(B+W)\times(B+W)}$:

$$A_{(B+W)\times(B+W)} = \begin{bmatrix} 0 & 0 & 1 & 0 & 0 & 0 & 0 & 0 \\ 0 & 0 & 1 & 0 & 0 & 0 & 0 & 0 \\ 1 & 0 & 0 & 1 & 1 & 1 & 1 & 0 \\ 1 & 0 & 1 & 0 & 0 & 1 & 1 & 0 \\ 1 & 0 & 1 & 0 & 0 & 1 & 1 & 0 \\ 0 & 0 & 1 & 1 & 1 & 0 & 0 & 0 \\ 0 & 0 & 1 & 1 & 1 & 0 & 0 & 0 \\ 0 & 0 & 0 & 0 & 0 & 0 & 0 & 0 \end{bmatrix} \qquad (1)$$

where B – the number of BT scatternet devices; W – the number of Wi-Fi network devices.

Columns of matrix **A** represent subsequent (as per numerical description): BT scatternet devices, and following IEEE 802.11b network devices. The elements of the matrix of $a_{i,j}$ are created in the following way: if a given i device potentially jams j device, then $a_{i,j}$ equals 1 (otherwise 0). Therefore, i lines of matrix **A** represent jamming devices, whereas columns j – the devices being jammed.

Matrix **A** presented in (1), informs which devices can cause interferences with other.

Let's mark additionally with: m – the master group, s – slave group (responding to given masters) and b – group of bridge devices for a given scatternet. For the network presented in Fig. 2, those sets are: $m=\{1;4\}$, $s=\{(2,3);(3,5)\}$ and $b=\{3\}$.

Let's mark by p_i a set of all devices of the Bluetooth piconet, within which the i device operates.

To estimate the impact of each i jamming device over j we will create matrix **X**. This matrix specifies the theoretical jamming frequency of j devices by i devices. Let us also assume that all devices have queued tasks (traffic load = 1). The mode of setting $x_{i,j}$ elemnets of **X** matrix has been presented below.

Each master device of $i \in m$ ($i \in p_i$) jamming a given device of $j \in p_j$, as per the above assumptions, manages p_i piconet creating every second frame, therefore:

$$\mathop{\forall}_{a_{i,j} \neq 0} \mathop{\forall}_{i \in m} \ j \in p_j \Rightarrow x_{i,j} = \frac{a_{i,j}}{2} = \frac{1}{2} \qquad (2)$$

Whereas the slave device of $i \in s$ ($i \in p_i$), interferes other devices at the frequency of calling up such device within p_i piconet, that is:

$$\mathop{\forall}_{a_{i,j} \neq 0} \mathop{\forall}_{i \in s} \ j \in p_j \Rightarrow x_{i,j} = \frac{1}{2 \cdot s_i} \qquad (3)$$

where s_i means the number of slave devices within p_i piconet.

For devices connecting Bluetooth networks $i \in b$ we assume formula (2), remembering that the bridge device can interfere all networks it links, and within which it is not currently performing transmission. For devices of other technologies (here 802.11b) we assume the maximum value under formula (2).

For the devices operating within 802.11b network we assume that the rival access to the medium requires the use of ISM band:

$$\underset{i,j \in mvs}{\forall} x_{i,j} = \frac{a_{i,j}}{L_{i,j}} \quad (4)$$

where: $L_{i,j}$ – number of IEEE802.11b devices in a given Wi-Fi network[2].

The above quoted formulas (2) - (3) relate directly to a BT scatternet, whereas (4) to IEEE 802.11b network. Using the relations form (2) to (4), matrix **A** can be modified, which now transforms to matrix **X**, presented as (5).

$$X_{(B+W) \times (B+W)} = \begin{bmatrix} 0 & 0 & 1/2 & 0 & 0 & 0 & 0 & 0 \\ 0 & 0 & 1/4 & 0 & 0 & 0 & 0 & 0 \\ 1/4 & 0 & 0 & 1/4 & 1/4 & 1/4 & 1/4 & 0 \\ 1/2 & 0 & 1/2 & 0 & 0 & 1/2 & 1/2 & 0 \\ 1/4 & 0 & 1/4 & 0 & 0 & 1/4 & 1/4 & 0 \\ 0 & 0 & 1/3 & 1/3 & 1/3 & 0 & 0 & 0 \\ 0 & 0 & 1/3 & 1/3 & 1/3 & 0 & 0 & 0 \\ 0 & 0 & 0 & 0 & 0 & 0 & 0 & 0 \end{bmatrix} \quad (5)$$

Each of the possible interferences of $a_{i,j}$ (or respectively $x_{i,j}$) features a given frame error rate, which can be specified for each situation under analysis. Therefore, it is possible to create the matrix $\mathbf{B}_{(B+W) \times (B+W)}$ – successful frame transmission probability matrix (at the entry of given *j* receiver). The elements of matrix **B** have been described in the following way:

$$b_{i,j} = x_{i,j} \cdot P_{S(i,j)} \quad (6)$$

where:

$P_{S(i,j)}$ – The probability of a successful frame reception by *j* device, allowing for a potential interference from *i* device (and other propagation conditions[3]).

---

[2] Just assumption for all Wi-Fi devices, in particular those whose transmission do not affect Bluetooth scatternet efficiency (that is located in a significant distance from BT devices), but using ISM band within a given BSS.
[3] Considering only a bit error rate would not allow for the dynamic ISM environment changes and the co-existence of various technologies (especially frame collision probability on the frequency level).

The probability of $P_{S(i,j)}$ can be in general written as follows:

$$P_{S(i,j)} = \sum_N P_S(P_E \mid n) \cdot P_C(n, N) \qquad (7)$$

where:

$P_C(n,N)$ – probability of given technology frame collision with n other technology frames (or the same technology for Bluetooth piconet) out of N possible collisions (frequency analysis),

$P_S(P_E \mid n)$ – respectively the probability of a successful reception of IEEE 802.11b frame(Bluetooth), which was subject, or was not, of a collision (time analysis).

Within $P_S(P_E \mid n)$ probability it is important to specify the bite error rate of P_E in case of a collision and in lack of collision. The relations of the bite error rates have been presented in 1.

For such created matrix **B** its metric β can be determined, which represents standard average efficiency measure of coexisting networks:

$$\beta = \frac{1}{2}\alpha_1 + \frac{1}{2}\alpha_2 \qquad (8)$$

where:
- α_1 – standardized BT network efficiency measure:

$$\alpha_1 = \frac{\sum_{i=1}^{B}\sum_{j=1}^{B} b_{i,j}}{\sum_{i=1}^{B}\sum_{j=1}^{B} a_{i,j}} \qquad (9)$$

- α_2 – standardized IEEE 802.11b network efficiency measure:

$$\alpha_2 = \frac{\sum_{i=B+1}^{B+W}\sum_{j=B+1}^{B+W} b_{i,j}}{\sum_{i=B+1}^{B+W}\sum_{j=B+1}^{B+W} a_{i,j}} \qquad (10)$$

Measure β is a numerical representation of the standardized sum of interference frequency affecting the devices within a given network (given technology solution), simultaneously allowing for the successful transmission probability (despite such interference) Having parameter β it is possible to specify the efficiency of the coexisting BT and IEEE 802.11b networks (and not only) as far as their mutual interferences are concerned. Metric β is this way a measure for comparing the efficiency of the complex ISM[4] environments. The lower the value of β ≥ 0, the lower

---

[4] A constraint of this method is the lack of allowing for the traffic analysis. In order to assess in with a greater accuracy this method can be extended by elements of e.g. the link average capacity within given network. For the purposes of the study of interference ratio, the analysis has been focused upon the model aspects irrespective of the traffic generated within given networks.

the interference ratio (mutual jamming) of a given ISM environment by various technologies. Optimal ISM environment is a such one which could be specified with a minimum measure β (criterion function).

Based upon the above described ISM environment efficiency comparison method, the reconfiguration (creation) of co-located networks topology is possible which would provide for the lowest mutual interference ratio. The algorithm enabling such function has been presented in the following chapter.

3 Nature of IBLUEREA algorithm

IBLUEREA algorithm, which is based upon the above described ISM environment efficiency method, is a suggests a reconstruction mechanism for Bluetooth scatternet. The idea behind IBLUEREA, relies upon the changes of functions fulfilled by a device (master, slave, bridge). IBLUEREA decisions are taken upon ISM environment features (the above described criterion function), which encompasses the analysis of potential interferences coming from different transmitters.

IBLUEREA algorithm encompasses the following operations cycle, preceded by the use of a mechanism generating scatternet of a minimal number of piconet[5]:

1. BT master devices control the traffic load of ISM[6] band and have information on the number of piconets within a given scatternet,
2. At spotting complex ISM[7] environment, BT master devices appoint randomly from them an IBLUEREA coordinator.
3. The coordinator triggers the procedure of establishing parameter β_1, that is a standardized efficiency of coexisting networks (for the existing topology), simultaneously assessing the network reconfiguration possibility to provide for the lowest possible interference ratio (β_2 for a new topology); information for the coordinator come from BT devices, which assess the frame successful transmission probability.
4. Upon collecting information from all BT master devices and the devices of other technologies (here Wi-Fi), the coordinator takes a decision on a possible network reconfiguration as per the new functions assigned.

IBLUEREA operating mechanism has been presented in Fig. 3 as a simple sequence of operations.

---

[5] e.g. *Law, Mehta and Siu* mechanism 5.
[6] The band control can be done e.g. every 30 time slot.
[7] Complex ISM environment is such which, for example, includes a number of piconets and the level of potential Wi-Fi interferences exceeds given threshold value.

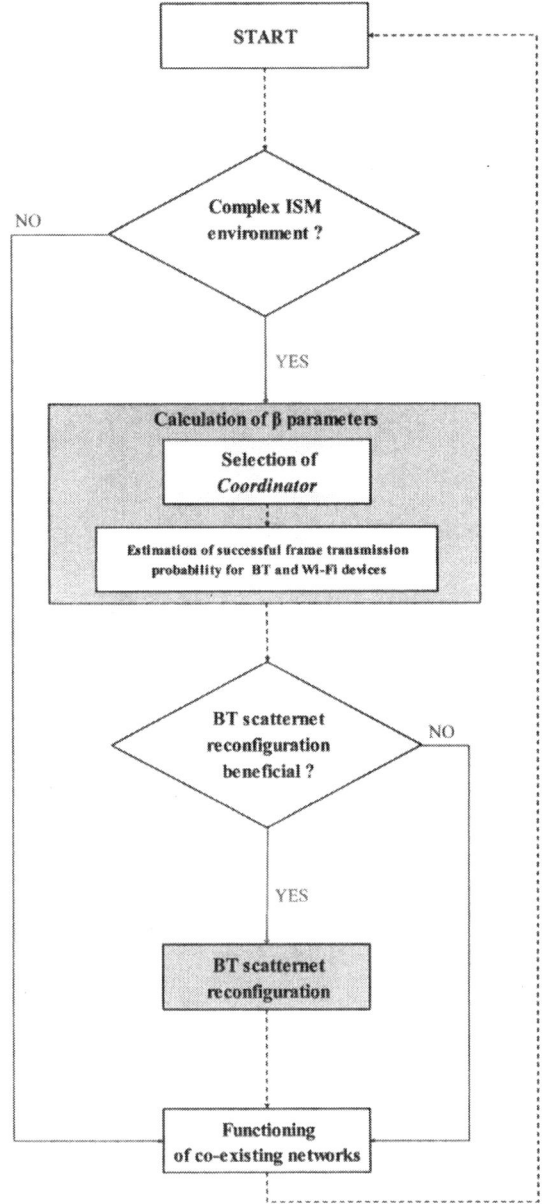

Fig. 3. IBLUEREA algorithm.

Each Bluetooth piconet is a subject of reconfiguration in order to provide for a function for such master device which is logically located as far as possible from the intererence source and/or can be in the least a source of interferation itself (here 802.11b network or other BT piconets). The coordinator appoints the new device upon a BT frame successful transmission probability (within a given piconet). The decision of the scatternet reconfiguration takes place when the minimal (lower[8]) parameter β is found (in relation to measure, β, ressulting from the current ISM environment).

To illustrate the algorithm mechanisms, examples of its functioning have presented further on (see Fig. 4).

4 Example of IBLUEREA functioning

In order to assess IBLUEREA usefulness, let us consider model operating scenarios of Bluetooth and 802.11b networks.

Fig. 4 gives examples of to theoretical topologies. The first is a result of a BT scatternet exemplary formulating. The letter case envisages the use of IBLUEREA algorithm for Bluetooth network reconfiguration. Fig. 4 includes some simplifed assmptions regarding network impact range and allowing for a better presentation of the mechanism itself.

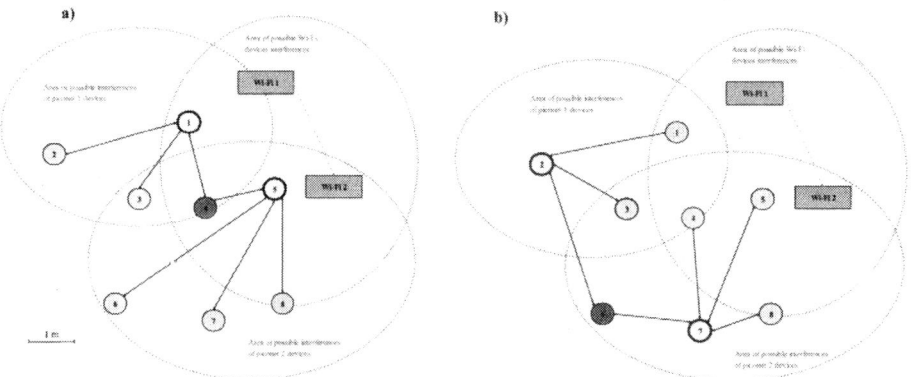

Fig. 4. Illustration of functioning of IBLUEREA algorithm: a) topology before reconfiguration, b) topology after IBLUEREA algorithm effecting changes.

Let us assume that the subject of our study are frames 1500B long for 802.11b and frames DH1 for Bluetooth generated by constant bit rate source. By establishing parameters β for the quoted scenarios w receive respectively: $β_a = 0,26 > β_b = 0,21$. Measure $β_b$ simultaneously specifies the minimum value of the parameter to be obtained in a such ISM environment (under the accepted comparison model).

The calculation of parameter β is necessary for taking the decision whether the suggested reconfiguration is likely to bring about expected benefits of ISM band efficiency enhancement through colocated network or not. It could be especially useful for highly complex ISM environments, in which in a relatively close vicinity coexits a nmber of various technologies and devices.

---

[8] Considerations concerning the search for only lower measures β can be a result of the implementation of IBLUEREA algorithm itself.

To illustrate the effect of IBLUEREA mechanism exemplary results of studies performed by way of simulations for a complex ISM environment (which allows for without limitations, the structure from fig. 4)[9] have been presented below.

Table 1 presents the results of simulation measurements for four cases (example 1). The coexistence of BT and 802.11b network was analysed which did not have coexistence mechanism triggered (FH – Frequency hopping BT) and using the AFH mechanism 1 (Adaptive Frequency Hopping) – see Fig. 4A. For the purposes of comparison a situation was analysed in which the topology was reconfigured as a result of IBLUEREA algorithm functioning (which corresponds with Fig. 4B).

Tab. 1. Average frame error rate of Bluetooth (class II devices) and 802.11b networks (example 1)

	Mechanism used			
	FH	AFH[10]	FH + IBLUEREA	AFH + IBLUEREA
Average BT FER [%]	57.0	3.9	56.8	2.4
Average 802.11 FER [%]	20.5	0.0	9.3	0.0

As presented in Tab. 1, making use of IBLUEREA mechanism (upon the analysed example) facilitates the efficiency of coexisting Bluetooth and 802.11b networks both for the scenario where no coexistence mechanism was used (FH), and where Bluetooh network triggered the AFH mechanism. As presented in the example under analysis, the band efficiency with triggered AFH was increased. It needs to be stressed out that in a situation where in a close viccinity function various technology solutions, using various transmission modes, often without one permanent band, the use of AFH algorithm can be hindered. Moreover, studies show that AFH algorithm is relatively ineffincent for dynamic ISM environment changes, in which the same frequence is used for a short period of time.

To illustrate the benefits of IBLUEREA algorithm also using AFH algorithm let us analyse a more complex ISM environment (example 2). Let us assume that the ISM environment in question consists of 3 independent 802.11b networks and 4 BT piconets (20 class II devices) scattered over the radius of 10m. We are examining two scenarios: in the first master devices are equally located within 4 m from the excentre and in the letter within 9m. Tab. 2 compares ISM environment efficiency using AFH mode.

Tab. 2. Average frame error rate of Bluetooth (class II devices) and 802.11b networks (example 2)

	Mechanism used	
	AFH	AFH + IBLUEREA
Average BT FER [%]	12.5	4.0
Average 802.11 FER [%]	0	0

---

9   It has been additionally assumed that 802.11b use two channels to enhance their throughput. It corresponds with the two independent Wi-Fi networks scenario.
10 For simplicity reasons, it has been assumed that only channels used by 802.11b network are avoided. Therefore mutual interferences only from BT piconets are possible.

In the scenarios from the second example, in fact, the issue relating the mutual BT piconet is being examined. As shown in Tab. 2, making use of IBLUEREA algorithm can have a positive impact on the enhancement of ISM band efficiency through the use of AFH devices. It can be particularily useful for sensor networks comprising of a large number of Bluetooth devices.

5 Summary

This paper presents a new IBLUERA control mechanism based upon the use of a new model comparing the efficiency of ISM environment. It has been proved that IBLUEREA can facilitate the coexistence mechanisms in use (including AFH mechanism 1).

An advantage of IBLUEREA algorithm is its potential to reduce mutual interferences of BT and 802.11b networks. It needs to be stressed out that IBLUEREA algorithm can be implemented togehter with other mechanisms (e.g. Law, Mehta and Siu algorithm 5), even at the BT scatternet formation stage. A disadvantage of such algorithm is the necessity of information exchange between coexisting technologis and additional LMP frames exchange (in Bluetooth). Moreover BT network needs to allow for the time slot to monitor the ISM band usage.

In future works, the authors plan to compare IBLUEREA mechanism while using chosen algorithms for Bluetooth network formation.

References

1. IEEE, IEEE Std 802.15.2-2003, *IEEE Recommended Practice for Information technology - Telecommunications and information exchange between systems - Local and metropolitan area networks - Specific requirements. Part 15.2: Coexistence of Wireless Personal Area Networks with Other Wireless Devices Operating in Unlicensed Frequency Bands*, August 2003.
2. Bluetooth SIG, Inc., *Specification of the Bluetooth System, Covered Core Package version: 2.0 + EDR*, 4 November 2004, http://www.bluetooth.org.
3. Klajbor, T.; Jaszcza, A.; *Effectiveness of IEEE802.11b system in neighbourhood of Bluetooth piconet* (in Polish), Proceedings of 3RD Conference on Information Technology, Gdańsk University of Technology, Faculty of Electronics, Telecommunications and Informatics, May 2005, Volume 7, pp. 369-376.
4. Cordeiro, C.D.M. Agrawal, D.P., *Employing dynamic segmentation for effective co-located coexistence between Bluetooth and IEEE 802.11 WLANs*, IEEE Global Telecommunications Conference, 2002. GLOBECOM '02., 17-21 Nov. 2002, Volume: 1, page(s): 195- 200.
5. C. Law, A. K. Mehta, and K.-Y. Siu, *A new Bluetooth scatternet formation protocol*, Mobile Networks and Applications, vol. 8, no. 5, Oct. 2003.
6. Cuomo, F.; Melodia, T.; *A general methodology and key metrics for scatternet formation in Bluetooth*, IEEE Global Telecommunications Conference, 2002. GLOBECOM '02. Volume 1, 17-21 Nov. 2002 Page(s): 941 - 945 vol.1.

7. IEEE, IEEE Std 802.11b-1999 (R2003), *Supplement to IEEE Standard for Information technology – Telecommunications and information exchange between systems – Local and metropolitan area networks – Specific requirements – Part 11: Wireless LAN Medium Access Control (MAC) and Physical Layer (PHY) specifications: Higher-Speed Physical Layer Extension in the 2.4 GHz Band*, New York, June 2003.
8. Golmie, N.; Rebala, O.; Chevrollier, N.; *Bluetooth adaptive frequency hopping and scheduling,* IEEE Military Communications Conference, 2003. MILCOM 2003., Volume 2, 13-16 Oct. 2003 pp.:1138 - 1142 Vol.2

Performance Studies of MPLS Based Integrated Architecture for 3G-WLAN Scenarios with QoS Provisioning

[1]Iti Saha Misra and [2]Chandi Pani

[1]Dept. of Electronics and Telecommunication Engg., Jadavpur University, Kolkata, India
[2]Dept. of Electronics and Communication Engineering
Meghnad Saha Institute of Technology, Kolkata-700150, India
[1]itimisra@cal.vsnl.net.in, [2]chandi_pani2309@yahoo.co.in

Abstract. Recently, the growing need of telecommunications such as Video-Conference, Voice over IP etc. and for the diversity of transported flows, Internet network does not meet the requirements of future integrated services satisfying Quality-of-Services (QoS). This is become much more complex for heterogeneous networks especially for wireless networks. In this paper, we propose an MPLS based integrated architecture between 3G(UMTS)/WLAN networks. We present the performance analysis of this MPLS enabled 3G-WLAN integrated framework for better QoS, throughput, less switching delay and less packet loss with the aim to fulfill the future demand of wireless communication. The mobility of the integrated framework is managed using hierarchical MIPv6 (HMIPv6). Integration point is considered at the MAP (Mobility Anchor Point) to restrict global update. Extensive simulation is done on mns-2.0 to evaluate the network performance in terms of packet forwarding, throughput and delay. For traffic engineering in the MPLS domain, CRLDP is used.

Keywords: 3G/WLAN integration, MPLS, UMTS, Traffic Engineering and QoS.

1 Introduction

In this paper end-to-end communications over an information transport consisting of heterogeneous communication platform mainly 3G/UMTS cellular networks and WLAN is provided. Mobile operators are transitioning towards third generation (3G) or beyond in order to access high-speed data rate in conjunction with IP. WLAN [1] provides high data rate with least installation cost but one of the major drawback of WLAN is limited area of coverage. Cellular systems [2] are one of the popular wireless access technologies in communication domain. It has moved through different generations like 1G, 2G (GSM, DAMPS), 2.5G (GPRS, EDGE) followed by 3G (UMTS) [2]. In 1992, International Telecommunication Union (ITU) has issued International Mobile Telecommunication for year 2000 (IMT-2000) which defines the basic characteristics of 3G. But due to the high deployment cost, 3G has not been accepted globally into the market. Advantages of cellular system are wide coverage and well known voice service, whereas limitations are low data rate and more costly compared with WLAN. The complementary characteristics of 3G cellular systems (slow, wide coverage) and WLAN (fast, limited coverage) make it attractive to integrate these two technologies to provide ubiquitous wireless access.

1.1 Focus of Integration

Considering the advantages and disadvantages of WLAN and Cellular systems, aim of integration is to develop a network, comprising the best features of these two heterogeneous technologies. The integration model will allow 3G/UMTS users to access the WLAN services from hotspot regions whenever they require high data rate and switch back in 3G/UMTS region when the service quality of the WLAN is not satisfactory or they are outside the WLAN region.

1.2 Related Work

There are numerous proposals regarding integration of these two technologies. [3] is the proposition of integration based on tight coupling (TCIA). But it has several disadvantages: TCIA inject the WLAN traffic directly towards the core network increasing the overall traffic in the core network. Secondly, the core network is completely exposed to WLAN network which violates the security and privacy of cellular network. Integration based on loose coupling (LCIA) [4] is introduced to overcome the disadvantage of TCIA. LCIA use mobile IP (MIP) [5] to provide seamless mobility. MIP has some disadvantage of triangular routing and seamless mobility in hand off intensive environment. So, MIP based integration introduces significant network overhead in terms of increased delay, packet loss and signaling when the user changes its point of attachment very quickly. To overcome these deficiencies hierarchical mobile IP (HMIP) based protocols [6-8] have been proposed which divides the network into domain and domains are further subdivided into subnets. HMIPv6 [9] is based on MIPv6 platform that introduces a new entity called mobility anchor point (MAP), the major idea is that once mobile node (MN) registers with MAP's CoA with home agent (HA), there is no requirement of further registration when MN moves locally i.e within the MAP. So this method provides low signaling overhead and less number of location update.

Multi protocol label switching (MPLS) [10] is a label-forwarding scheme provides a better solution for faster switching. MPLS is a versatile solution to address the problems faced by present-day networks-speed, scalability, quality-of-service (QoS) management, and traffic engineering. MPLS has emerged as an elegant solution to meet the bandwidth-management and service requirements for next-generation IP-based backbone networks. When it is used in conjunction with IP, conventional IP look up and forwarding within the network is replaced by faster label look up and switching, the IP header is analyzed only in entry and exit points of the network. There are so many advantages of MPLS systems [11], which motivate us to use MPLS in mobile wireless communication networks. Related works that support mobility in MPLS domain is mainly based on MIPv4 and HMIPv4 [12-14]. A quality of service (QoS) provisioning scheme is given in [15]. Integration of MPLS with hierarchical MIP based architecture and mobility management schemes are given in [16,17]. It is observed that the signaling cost for traffic transport is much reduced for MPLS and hierarchical mobile IP based integrated structure [16].

The current Internet Gateway Protocol (IGP) use shortest paths to forward traffic. It causes some of the links to be over utilized and some are under utilized. The purpose of traffic engineering (TE) [18, 19] is to enhance network utilization and to improve architecture of a network in a systematic way, so that the network becomes robust, adaptive and easy to operate. MPLS has extended routing capability that efficiently controls the network traffic by removing congestion and spreading the load over the different links. In [20-22] different route selection algorithms based on MPLS framework is provided. LDP (Label Distribution Protocol) is a new protocol that defines a set of procedures and messages by which one LSR (Label Switched Router) informs another of the label bindings it has made. CR-LDP [21] (constraint-based LDP) contains extensions for LDP to extend its capabilities. This allows extending the information used to setup paths beyond what is available for the routing protocol.

1.3 Motivation

The development and standardization process are currently underway for defining suitable efficient integrated architectures and are a challenging task hat needs a lot of research efforts. The growing worldwide deployment of public WLANs has a growing impact on what public wireless networks will look like and how public mobile services will be provided in the near future. WLAN provide significantly higher data rates than cellular networks that are expected to be available in the near future [23]. Public WLANs are the first wave of all-IP radio access networks making one step forward on their migrations to IP based wireless networks. It calls for new and innovative business models for public mobile services. 3G wireless networks allow mobile users to access Internet via standard IP with higher data rates. As the demand for multimedia data services increases, end users aim for enhanced performance through the greater coverage, higher data rate and lower overall cost. Whereas, mobile operators aims in generating large cost effective business to a wide range of applications such as mobile multimedia services (MMS). To meet this demand it requires the suitable integration of the existing networks maintaining the end-to-end QoS for data services. Moreover, the integration of QoS parameters will increase complexity in heterogeneous networks that include different types of networks (wired, wireless, mobile etc.).

Thus, in this paper we are exploring an integrated architecture between the 3G-WLAN paradigms for enhanced performance. We choose MPLS as the best choice for high switching rate and QoS maintenance. The proposed framework uses the enhanced type of MPLS router called label edge mobility agent (LEMA) that are placed at different hierarchical level of the integrated framework. LEMA takes an active part in mobility management. In the proposed model, MAP is placed at the integration point of the network that restricts the movement of the mobile node under MAP domain between the two heterogeneous networks. Use of MPLS in the proposed framework can satisfy the seamless mobility with fast packet forwarding, fast hand off, less number of packet loss during hand off and minimum delay that make the framework more scalable.

2 Proposed MPLS/HMIPv6 based Integrated Model and Mobility Management

Fig.1 shows the MPLS/HMIPv6 based integrated framework where two heterogeneous networks UMTS and WLAN are considered as two different domains. In the proposed architecture, MAP is placed at the integration point of two heterogeneous network to reduce signaling overhead and location update.

From the aspect of scalability, fast packet forwarding and QoS based Traffic engineering, MPLS domain is formed between MAP and the two heterogeneous networks where MAP is the ingress and 3G-AR, WGWR are the egress LER (Label edge router). So, all the edge routers have LEMA property and they are named as MAP LEMA, 3G-AR LEMA and WGWR LEMA whereas the intermediate routers are the LSR. Mobility of the network is handled by HMIPv6 protocol. Mobile IP initiates the MIP registration message to establish LSP. For signaling protocol, CR-LDP is used.

During registration, all the LEMAs are modified their label forwarding information base (LFIB) by placing the care of address (CoA) of MN in the forward equivalent class (FEC), also modify their label forwarding table (LFT) by placing the in-label and out-label value for next hop information and a new label switch path (LSP) is established afterward [17]. For seamless mobility to occur the user equipment (UE) must have dual mode terminal to interface 3G/UMTS and 802.11b.The 802.11b interface allow the user to operate in WLAN domain and UMTS driver allow the user to operate in UMTS domain. The different protocol stacks of integrated framework are shown in Fig.2. To describe mobility management, firstly it considers mobile node (MN) is under AR1 of UMTS domain. For intra domain mobility the movement of MN is restricted under same domain, 3G-AR LEMA handles the mobility issues, no requirement to inform MAP LEMA or HA. When MN moves from UMTS to WLAN domain, the mobility is inter domain mobility and managed by MAP LEMA. When MN moves outside the range of MAPLEMA, new registration up to HA be required henceforth, it is called the global mobility. After the registration process for the MN, CR-LDP is employed for establishment of traffic path. For dynamic route, CRLSP (Constrain Routing Label Switched Path) is configured to provide QoS and to follow automatic reconfiguration when a failure occur or the network state changes.

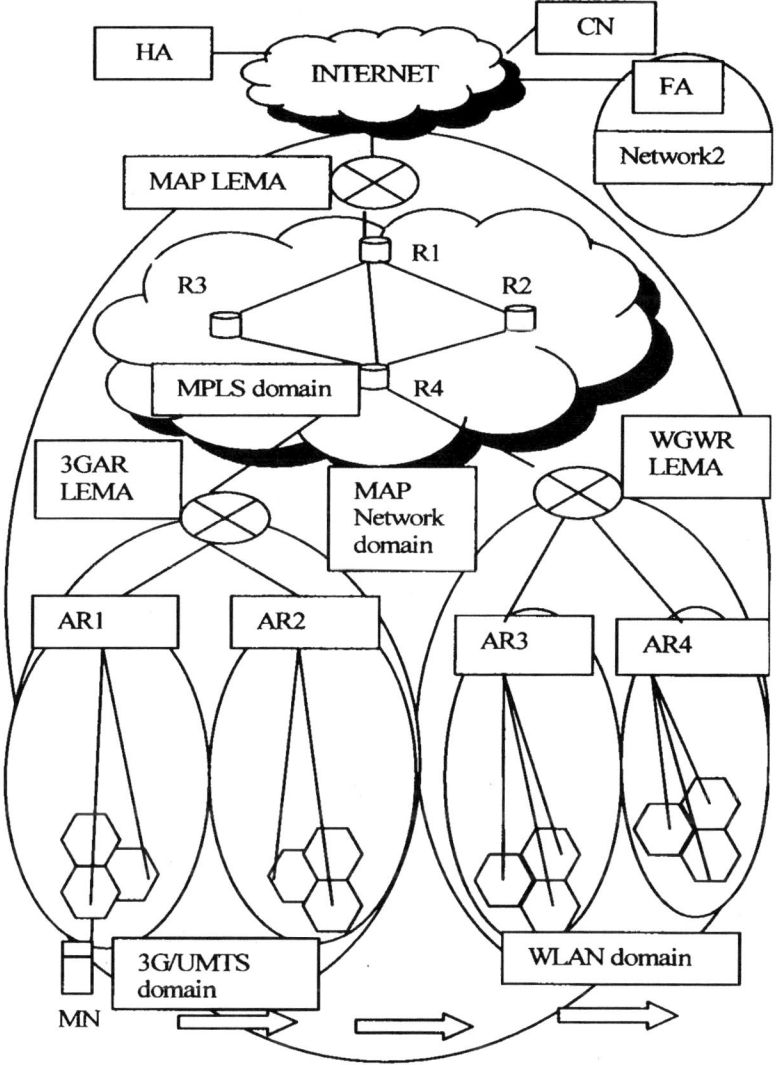

HA-home agent, CN- correspondent node, MAPLEMA-MPLS enabled MAP, 3GARLEMA-MPLS enabled 3G/UMTS gateway LEMA, WGWRLEMA-MPLS enabled WLAN gateway LEMA, MN-mobile node, AR- Access router, BSC- base station controller ⎕ MPLS Router ⟨⟩ Access point.

Fig.1 Proposed integrated framework

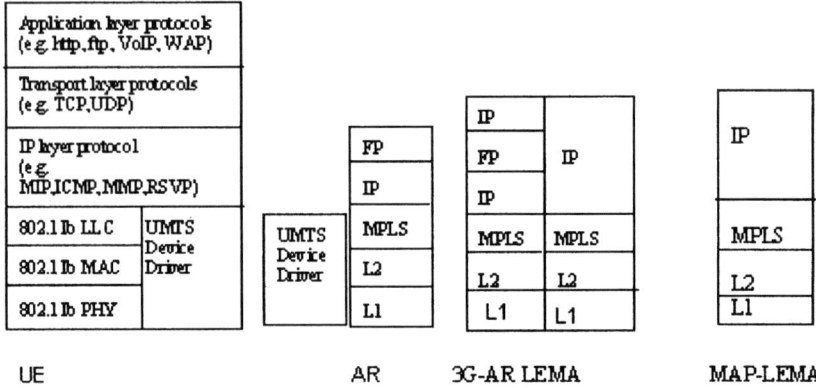

Figure 2. Protocol Stack for the integrated architecture

3 Network Performance Evaluation Through Simulation

We have performed series of simulations in order to evaluate the UMTS/WLAN/MPLS based architecture. The simulations are implemented using Network Simulator 2.26 (ns-2.26) [24].

The network topology as shown in Fig. 3 is used for simulation purpose. The entry node of the access network is referred to as MAP LEMA for integrated MPLS/HMIPv6 protocol. It will simply a MAP node when simulate only for HMIPv6 based mobility architecture. Within the MPLS domain, the internal nodes are LSR (Label Switched Router) otherwise they are IPv6 routers. The exit points of the domain are referred to as 3G LEMA or WGWR LEMA (for WLAN access router) for MPLS enabled architecture. For simulation, we focus on the traffic engineering and mobility management of the proposed architecture. In simulation scenario, we consider all the links have the capacity of 1Mb with 100ms delay. We assume that packets are arrived from CN to MN with exponentially varying packet size.

3.1 Simulation Results

This section discusses about the simulation results for both the architecture with and without the MPLS. To examine the TE performance, we use distance vector algorithm for route selection and Figure 4. LSP set up time for two cases with three exponential traffic of packet size 200, burst 2 sec, idle 1 sec and rate 100k, 200k, 300k respectively. Three different simulation scenarios are considered as given under.

Scenario 1

Case A: MN is registered with AR1 of Fig.3 under 3G domains. For traffic engineering, CR-LDP is used.

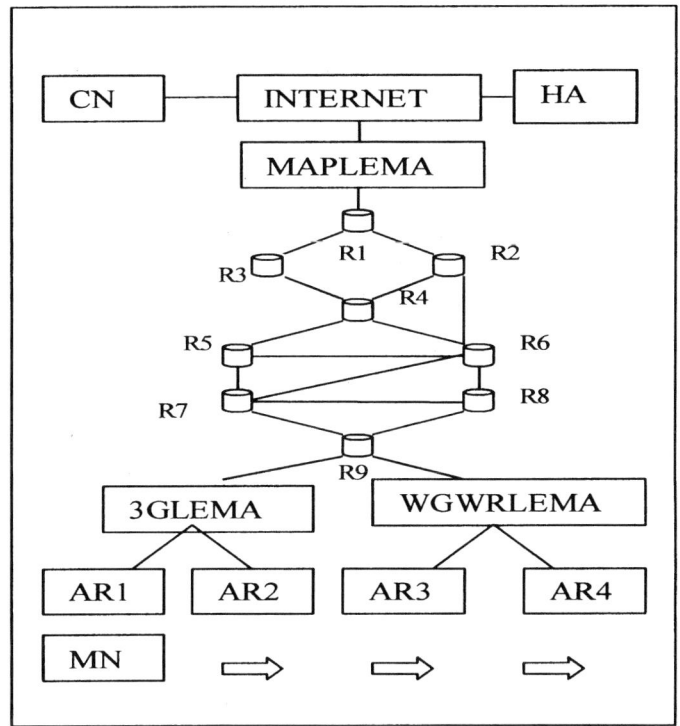

Figure 3. Simulated network

Figure 3. Network for simulation protocol is used to establish LSP in integrated MPLS domain and distant vector routing is used for HMIPv6 domain without MPLS. Fig.4 shows the path set up time from CN (corresponding node) to MN for two protocols (with MPLS, without MPLS). It is seen from Fig.4 that number of packet required to set up LSP is less in case of MPLS enabled framework. Also it requires less time for path setup.

Case B: Now three different packets having same size as mentioned earlier are sent from CN to MN under same scenario as in case A. Fig. 5 and Fig.6 show the bandwidth consumed by MN for different protocols with and without MPLS. It is seen that in Fig.5 only packet 3 is reached to MN over the simulation period whereas the other two packets are dropped. But in Figure 6, for HMIPv6 integrated network case of MPLS enabled framework all three packets are reached to MN over simulation period. Also the bandwidth consumption for packets forwarding is less in case of MPLS network.

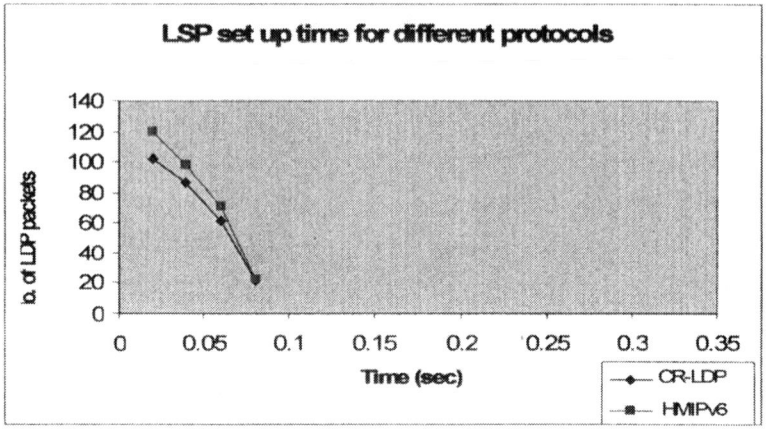

Figure 4. LSP setup time for two cases

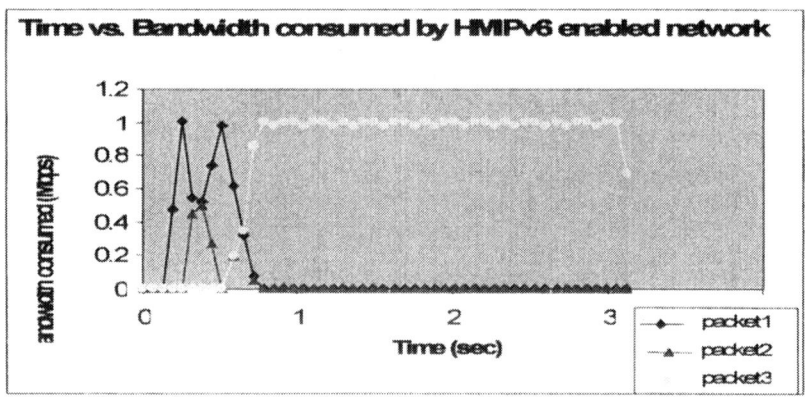

Figure. 5 Packet flow scenario for the simulation network for HMIPv6 integrated network

Scenario 2

Case A: MN is moved from 3G domains to WLAN domain (from AR2 to AR3) and same amount of packets are sent from CN to MN as considered in earlier cases. Fig.7 shows the amount of packet loss during hand off between these two domains
for two different protocols and approximately 68% gain is achieved in connection with dropped packet for MPLS network.

Case B: Then we increase the packet size in Mbytes and forward it from CN to MN for the flow of 3G domains to WLAN domain. Fig. 8 shows the average throughput in Mbps for two cases with achieved gain is approximately 28% for MPLS enabled integrated architecture.

From all the above results it can be said that for future generation network with increased network traffic load for mobile wireless communications, MPLS enabled mobility management network may fulfill the QoS requirement for bandwidth, throughput, delay and faster switching. Our proposed network model is one way of integrating heterogeneous networks. As HMIPv6 is the hierarchical mobility management protocol established to reduce signaling and delay overhead, the MPLS enabled model shows the improvement of the performance with respect to HMIPv6. This is because of the inherent advantages obtained from MPLS.

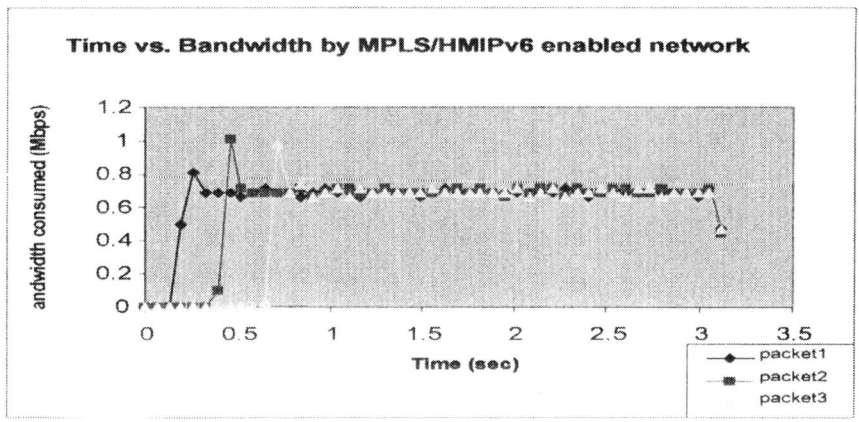

Figure 6. Packet flow scenario for the simulation network for MPLS/HMIPv6 integrated network

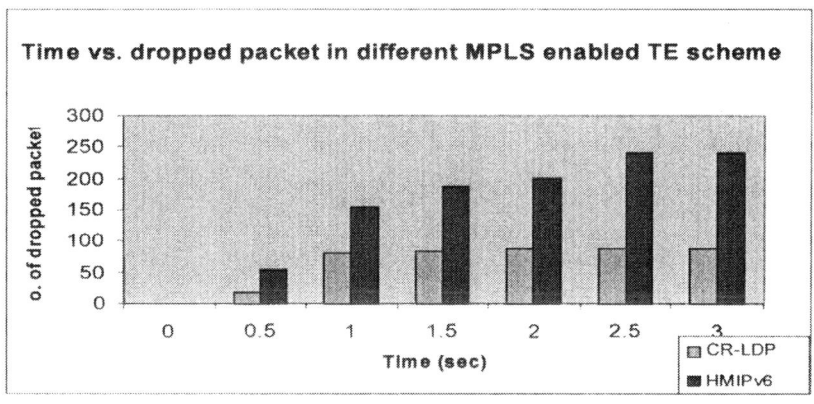

Figure 7. Packet Loss scenario

4   Conclusion

In this paper, we propose and evaluate an integrated architecture based on MPLS 3G-WLAN-integrated network. With the demand for multimedia services in wireless environment, faster packet forwarding is required along with enhanced throughput and reduced packet loss. MPLS integration with HMIPv6 for the proposed integration model satisfies the all criterion that is validated through the extensive simulation work. Use of MPLS with traffic engineering like CR-LDP may ensure QoS within the domain. Thus with increasing mobile users MPLS based integrated network model would provide seamless connectivity with less packet loss, less congestion due to faster switching and increased throughput compared to only HMIPv6 based model. Established HMIPv6 protocol reduces network signaling overhead and handoff delay. Use of MPLS further enhances the performance as clear from the simulated results.

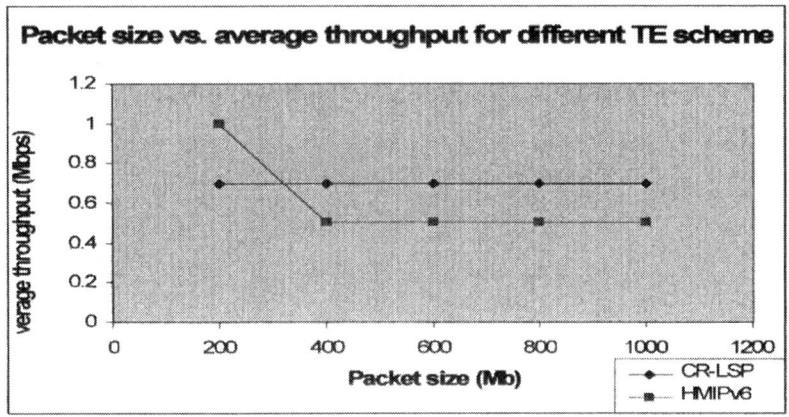

Figure 8. Throughput with increased packet size

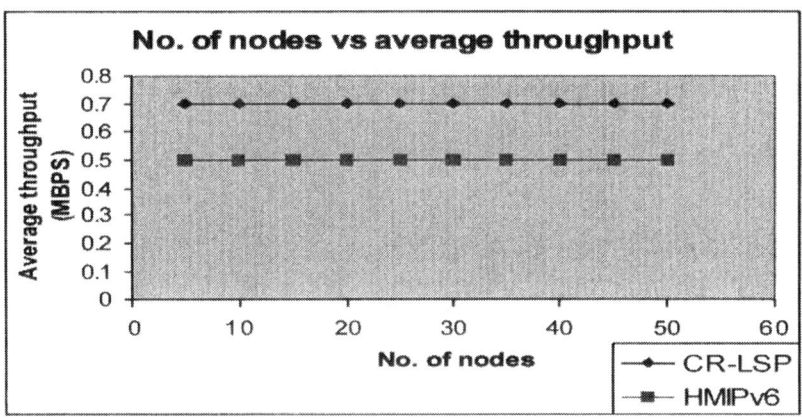

Figure 9. Throughput vs. number of nodes

Acknowledgement: Author Iti Saha Misra is thankful to AICTE for the financial support of this research work under AICTE, CAYT scheme, India.

References:

1. T.S Rappaport, "Wireless Communications- Principles and Practice"- Prentice Hall of India Private Ltd.
2. Jeffrey Bannister, Paul Mather and Stebastian Coope, "Convergence Technologies for 3G Networks – IP, UMTS, EGPRS and ATM", John Wiley and Sons, Ltd.
3. Muhammed Jassemuddin," Architecture for Integrating UMTS & 802.11 WLAN Network", IEEE Symposium on Computer Communications ISCC, Turkey, 2003, pp.1-8.
4. Milind M. Buddhikot, Girish Chandranmenon, Seungjae Han, Yui-Wah Lee, Scott.Miller, Luca Salgarellie, " Design and Implementation of a WLAN/CDMA 2000 Internetworking Architecture", IEEE communication Magazine, November 2003, pp.93-97.
5. C. Perkins, Ed., " IP mobility support", *IETF RFC 2002, Oct, 1996.*
6. Campbell et. al, "Design, implementation and evaluation of cellular IP", *IEEE Personal Communication, Aug . 2002*
7. R. Prasad et. al " IP Based Access Network Infrastructure for Next Generation Wireless Data Networks" *IEEE Personal Communications, Aug. 2000.*
8. S. Das et., al, "Tele MIP : Tele communication enhanced mobile IP architecture for fast intra domain mobility" , *IEEE Personal Communication, Aug. 2000.*
9. D.Johnson, C.Perkin & J.Arkho, " Mobility support in IPv6".
10. S. Jha and M. Hassan, Engineering Internet QoS, *Artech House, 2002.*
11. Fabino M. Chiussi, Denis A.Khotimsky, Santosh Krishnan, Lucent Technologies, "Mobility Management in 3G All IP Networks", IEEE Communication Magazine, September 2002.
12. Kaiduan Xie, Victor C.M Leung, "A MPLS Framework for Macro and Micro Mobility Management". *pp.549-555,Orlando,FL,USA,March2002.*
13. Tingzhou Yang, Yixin Dong, Bin Zhou, Dimitrios Makrakis, "Profile Based Mobile MPLS Protocol", Proc. of *IEEE Canadian Conference on Electrical and Computer Engineering, 2002.*
14. Tingzhou Yang, Yixin Dong, Bin Zhou, Dimitrios Makrakis, "Mobility Support for Differentiated Services in Next Generation MPLS Based Wireless Networks",*www.site.uottawa.ca/~tyang/paper/mobility_for_diffserv.pdf.*
15. Jesus Hamilton Ortiz, "Integration of Protocols HMIPv6 and M-RSVP over M-MPLS in order to Provide QoS in IP Network Mobility", University Polytechnic of Madrid, Spain, ETSI Telecommunication.
16. Iti Saha Misra, Sudipta Dey and Debashis Saha ," 4G All IP Integration Architecture for Next Generation Wireless Internet", ICWN'05.
17. Iti Saha Misra, Chandi Pani," MPLS/HMIPV6 based mobility management framework for GPRS/WLAN integration architecture," IEEE conference on Wireless and Optical Communication Network, April 2006.
18. White paper (Juniper Networks) "Multiprotocol Label Switching: Enhancing Routing in the new public Network"
19. Xipeng, Xiao, Alan Hannan, Brook Bailey, "Traffic Engineering with MPLS in the Internet", IEEE Network Magazine, Mar. 2000

20. Youngseok Lee, Yongho Seok, Yanghee Choi, Changhoon Kim, "A Constrained Multipath Traffic Engineering Scheme for MPLS Networks", Proc. of *IEEE Int. Conf. ICC2002. New York.*
21. Vasu Jolly, Shahram Latifi, "An Overview of MPLS and Constraint Based Routing", *www.Ee.unlv.edu/~venkim/opnet/AN%20OVERVIEW%20OF%20MPLS%20AND%20CONSTRAINT%20BASED%20ROUTING.pdf.*
22. Shyam Subramanian and Venkatesan Muthukumar, "Alternative Path Routing Algorithm for traffic Engineering",*www.ee.edu/~venkim/opnet/Alternative_Path_Routing.pdf.*
23. J.C. Chen and T. Zhang, IP Based Next Generation Wireless Networks- Wiley Publication, 2004
24. ns-2 home page, http://isi.edu/nsman/ns.

VoIP-PSTN Interoperability by Asterisk and SS7 Signalling

Jan Rudinsky

CESNET, z. s. p. o.
Zikova 4, 160 00 Praha 6, Czech Republic
rudinsky@cesnet.cz

Abstract. PSTN, the world's circuit-switched network, has employed Signalling System #7 as its protocol suite for international and national interconnection during past decades. VoIP networks however have developed different signalling protocols suitable for IP environment. Gateways interconnecting VoIP and PSTN networks are usually proprietary and expensive solutions. Today an open source software can perform this function.
As an example we have decided to test Asterisk PBX and two open source implementations of SS7, the SS7 channel driver and SS7 library. We have tested these solutions for interconnection to PSTN and run various tests to verify the implementation functionality.

1. Introduction

1.1. Signalling System #7

Common Channel Signalling System No.7 (also referred as SS7, CCS7 or C7) is a suit of control protocols used to provide instructions between elements within a public switched telephony network (PSTN). These instructions carry information about routing calls, requested services by subscribers, identities of participants and management information for signalling network.
Several organizations have written standards for SS7 networks. ITU-T standard is used on the international level of interconnection between signalling networks and it is the most common SS7 standard used on the national level in Europe. The national level of the network can generally use standard whatever exists within the country. An example is the ANSI standard used within the United States. We have tested the ITU variant of SS7 protocol suite as defined in ITU-T Q.7xx set of recommendations.

1.2. Asterisk PBX

Asterisk is a software private branch exchange (PBX). Unlike the traditional PBX being bound to vendor specific equipment, Asterisk can be run on almost any kind of hardware which uses Linux, BSD or Mac OS as their operating system. Asterisk supports functions such as traditional PBX services as well as set of additional functions including voice mail, interactive voice response or automatic call

Please use the following format when citing this chapter:

Rudinsky, J., 2007, in IFIP International Federation for Information Processing, Volume 245, Personal Wireless Communications, eds. Simak, B., Bestak, R., Kozowska, E., (Boston: Springer), pp. 169-173.

distribution.

The system can interface traditional circuit-switched systems such as PSTN network as well as packet based systems represented by Voice over IP (VoIP) networks. For interconnection with VoIP networks Asterisk doesn't need any additional hardware. However for interconnection with PSTN networks and almost all standards-based telephony equipment an additional hardware
is necessary. These hardware devices come in a form of PCI cards with analog or digital telephony interfaces. Related protocol support by Asterisk covers wide range of VoIP signalling protocols such as Session initiation protocol (SIP), Inter Asterisk Exchange protocol (IAX), Media gateway control protocol (MGCP), H.323 and Cisco Skinny protocol. The TDM protocol group includes most of the European and American standard signalling types, the analog E&M, FXO, FXS and multi frequency tones as well as the digital ISDN protocol group.

2. Configuration

2.1. System Configuration

Asterisk software PBX with SS7 support can be run on almost any kind of hardware. In this example we use Intel x386 family processor, standard network Ethernet adaptor and PCI bus controlled card with ISDN Primary Rate Interface (ISDN PRI) manufactured at Digium.

The system is based on GNU/Linux operating system, distribution Debian. The distribution includes device drivers handling the Ethernet card and other components. The PCI card specific driver has to be compiled from the source code given by the card manufacturer. Zaptel is a kernel interface device driver of Digium cards with analog or digital interface cards. The driver is then included in the system as a kernel loadable module.

Software requirements to install Asterisk application include following important packages:
- linux 2.6 kernel headers
- bison and bison-dev
- ncurses and ncurses-dev
- zlib1g and zlib1g-dev
- libssl and libssl-dev
- libnewt-dev
- initrd-tools, cvs and procps

2.2. Asterisk SS7 Installation and Configuration

The Asterisk source code is available at the developers web page [1]. Important packages include "Asterisk" and "libpri", an ISDN PRI library for E1 interface. The installation is done by standard make - make install procedure.

Each signalling protocol is implemented in Asterisk as a channel, that represents incoming and outgoing protocol sessions. First signalling solution, the SS7 channel driver is implementated in a form of SS7 channel that process SS7 messages. SS7 channel driver includes implementations of MTP2 layer, bare essentials of MTP3 and a large subset of ISUP functions. It can be downloaded from the developers web page [2] and compiled into the Asterisk loadable module.

Second solution, the SS7 library makes use of standard Zaptel channel. This channel is a part of the Asterisk source code and is used to manage sessions via Zaptel compatible hardware cards with E1 interface.

Configuration of lowest protocol layers is stored in "zaptel.conf" and it is similar for both SS7 solutions. This file is read by zaptel driver, which accordingly sets the link encoding, timing source for synchronisation, type of error detection code and mapping of signalling and traffic channels.

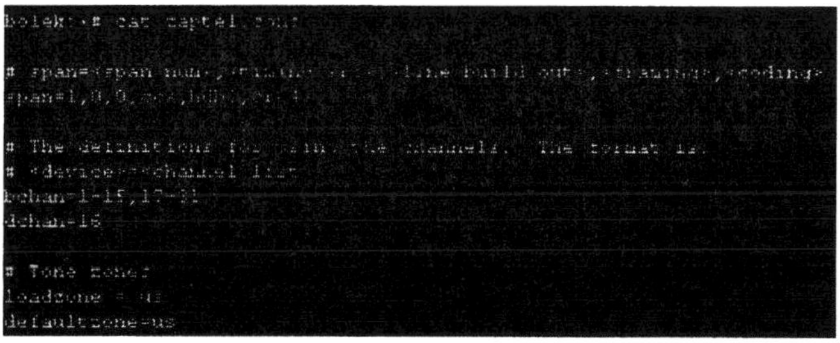

Fig. 1. Zaptel driver configuration

Higher layer configuration of SS7 message transfer part and user part is different for each solution. SS7 channel driver configuration is stored in file "ss7.conf". The content of the file defines SS7 signalling linkset properties, parameters of signalling links including signalling and traffic channel allocation and point code assignments. SS7 library configuration file zapata.conf contains similar parameters definition only in a different format.

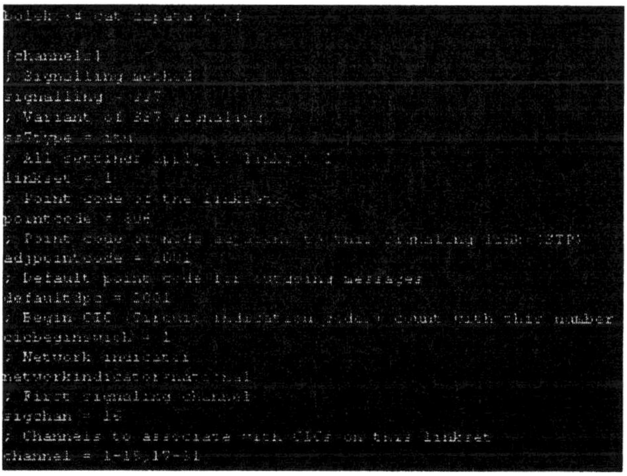

Fig. 2. SS7 higher layer configuration

To define the Asterisk behaviour we have to set the Asterisk dialplan. Configuration file extensions.conf contains necessary information to control the execution of call flow and related operations.

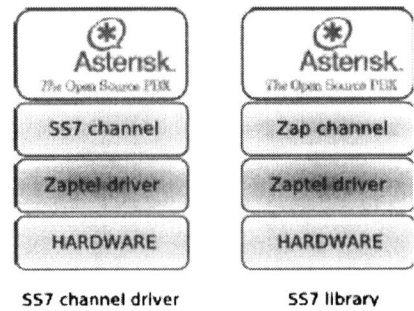

Fig. 3. Asterisk SS7 stack

3. Signalling and Voice Traffic Tests

We have verified the functionality of Asterisk PBX with SS7 support in various tests.
In the first scenario we have used Tektronix K1297 protocol tester to test the SS7 message transfer part (MTP) layer implementation of SS7 channel driver. The tests ended with positive result leaving SS7 link "In service" and voice channels in "Idle" mode.
Second scenario was represented by two Asterisk boxes, each with one SS7 solution installed. The servers were interconnected via E1 crossover cable plugged into the Digium TE110P interface cards at each end. After the initial configuration and Asterisk startup we have manually loaded the SS7 channel and Zap channel drivers on both servers. The initial procedure of MTP alignment and a group reset of E1 circuits ran successfully and ISUP controlled calls were exchanged in both directions.
The next phase was focused on Asterisk to PSTN interconnection. The telephone network was represented by Ericsson AXE platform mobile switching center (MSC). We have used standard E1 interface with 30 traffic channels and a single SS7 signalling channel. We have tested subsequently both SS7 solutions. After solving initial problems with SPCs (Ericsson and SS7 library using decimal, while SS7 channel driver hexadecimal numbers) and Asterisk exception of counting Circuit Identification Codes from 0, we have successfully interconnected the exchanges. Initialization process of the SS7 connection was identical to the previous case and successful outcome was confirmed by Asterisk on the console. We have passed calls in both directions.
Besides the basic call setup we have tested the SS7 channel driver also for several supplementary services. The implementation of calling line identity services - presentation and restriction corresponds to the recommendations, while the connected line presentation and restriction seem not to be supported. The call waiting and call hold supplementary services were tested with successful outcome, while call forwarding was not supported.

4. Conclusions

Signalling System #7 is today's most common signalling protocol in the PSTN worldwide and a share of PSTN on the telecommunication market is large. However the market represented by VoIP networks is still growing and will eventually prevail PSTN networks. Therefore there is a need for interconnection between these types of network.

Current solutions are proprietary and usually expensive systems. As an option we can use Asterisk PBX as a gateway between VoIP and PSTN using Signalling System #7. It represents an open source solution with wide range of supported protocols (SIP, MGCP, H.323) among which the SS7 has been added. The system is running on GNU/Linux operating system ensuring credible amount of reliability.

As the test have proven, both SS7 solutions, the SS7 channel driver and the SS7 library, can perform large subset of SS7 functionality including routing and call setup and a basic subset of supplementary service. Though some functions remain to be implemented.

Additionally Asterisk can perform as a media gateway by supporting number of codecs (G.711, G.723.1, G.726, G.729, GSM, iLBC, LPC-10 or Speex) and translation between them.

Acknowledgement. This project is carried out as a part of research of IP telephony group at CESNET z.s.p.o. CESNET is a Czech academic network operator with the mission of research in advanced network technologies and applications. The major framework for our research is the official research plan titled *Optical National Research Network and its New Applications* which is implemented since 2004 until 2010.

References

1. http://www.asterisk.org
2. http://www.sifira.dk/chan-ss7
3. T. Russell, "Signalling System #7", McGraw-Hill (2002), ISBN 0-07-138772-2

Experimental NGN Lab Testbed for Education and Research in Next Generation Network Technologies

Eugen Mikoczy[1], Pavol Podhradsky[1], Ivan Kotuliak[1], Juraj Matejka[1]

[1] The Faculty of Electrical Engineering and Information Technology, Slovak University of Technology in Bratislava, Ilkovicova 3, 812 19 Bratislava, Slovakia
{ mikoczy, podhrad, ikotul, Matejka }@ktl.elf.stuba.sk

Abstract. The main evolution trends of NGN architecture towards unified service control based on IMS principles are presented in the article. The actual implementation of NGN testbed platform including extensions and integration of application is described. We also provide an overview of the ongoing incorporation and integration of IMS core elements within the actual architecture. The existing NGN solution includes communication, collaborative community and e-learning applications and highlights the benefits of our experience in the integration of those applications and the usability in education process. We explain also the main NGN technology issues and topics for possible future research and education activities planed for the NGN Lab testbed.

Keywords: NGN, converged network, IP Multimedia Subsystem, mobility, education

1 Introduction

The IP Multimedia Subsystem (IMS) as apart of the Next Generation Network (NGN) architecture is major core network architecture that enables to provide multimedia services in both, wireline and wireless network environments (standardized by all major standardization bodies, namely ITU-T, ETSI, 3GPP). We shortly take a look into the network convergence trends and also compare functionalities in two major evolutionary steps in NGN (softswitch based and IMS based NGN architecture).

We focus mainly on the service control and application layer because in the NGN concept they are independent and aware of underlying transport technologies (concept of unified service control). Therefore future converged networks architecture will probably be based on similar concept as we have realized in our testbed platform. This enhanced NGN concept is based on IMS (evolved from SIP/softswitch architecture) and more details are provided in Section 2. The description of NGN platform and integration work already done is also given there. Specific enhancements to the implementation of IMS are described in Section 3 together with an overview about interoperability and testing issues researched at Slovak University of Technology (STU). Final part of the article focuses on employing NGN platform in educational process. Final remark and future plans are discussed in Conclusion.

1.1 Network convergence and NGN standardization

The trends of the convergence touch several levels which the process of convergence can take place within the communication networks, services or used media. The NGN technology has been evolving for several years. Let us give you at least the most frequently presented reasons substantiating the need for converged network architecture represented usually by the next generation network platform:

- Several specialized networks for a certain type of services, some of them being ineffective to be developed; it is necessary, however, to ensure their tasks.
- Each network platform has, more or less, its own architecture and specifics, though it does not cover all communication needs.
- Duplicity of resources, vertical architecture, and therefore cost is less effective.
- More complicated securing of Network Management System – NMS and operation as well.
- Reduction in costs of infrastructure and more flexible development of network and services as well.
- The need to respond more flexibly to the advancement within the ICT technologies development.

Actual standardization of ITU-T [1] and ETSI TISPAN [2] address most of those requirements in NGN are based on IP Multimedia Subsystem (IMS), as new fixed mobile converged architectural framework within first releases of NGN standards. IP multimedia subsystem [3] originally resulting in standardization process within 3GPP during the standardization of the third generation of mobile networks [4]. Later, it was extended by ETSI TISPAN also for fixed access networks. Several other standardization or technical organizations try to employ also IMS in their conceptual architecture (e.g. CableLabs for cable infrastructure [5]). Actual standardization activities of several standardization bodies incorporate more independency of service control from transport networks or heterogeneous access networks (including wireless access) as well as address also the requirements needed for providing IPTV services.

1.2 Evaluation of IMS based NGN technologies

The evolution of converged networks can be easily illustrated on the evolution from the pure VoIP architecture based on SIP protocol (SIP servers, SIP application servers) to the NGN architecture based on softswitch technologies and towards currently most preferred IMS based architecture. Basic differences of the two evolutionary steps in NGN concept (softswitch based vs. IMS based architecture) can be seen in their simplified characteristics (shown in Table 1). Comparing both approaches explains main advantages and drivers also for our selection of IMS principles for implementation in our NGN Laboratory. Every operator willing to deploy IMS (most often on top of existing NGN softswitch based infrastructure) needs to take into account several principal differences between these two NGN architectures. Some of them are analyzed more deeply in Section 3

Table 1. Evaluation of main differences in IMS based vs. softswitch based NGN architectures [6]

General characteristics	Softswitch based NGN architecture	IMS based NGN architecture
Standardization	just vendors & industry driven specifications	NGN standards published by 3GPP, TISPAN, ITU-T
Modularity and Open protocols	low, alternative and proprietary protocols	high, standardized open protocols (e.g. SIP, Diameter,...)
Important reference points	usually not accessible	accessible via standardized interfaces
Media delivery and service control separation	separated, but control functions are highly integrated	separated, control functions can be distributed
Control functions	call control oriented	session control oriented
Transport control functions	missing specialized elements, lack of end-to-end QoS	specified in the architecture for providing end-to-end QoS control (e.g. over NASS, RACS)
Network convergence	PSTN and IP networks, more fixed network oriented	unified service and control layer independent from fixed, wireless, mobile access
Mobility	nomadic mobility	seamless mobility, roaming user, device and inter-domain mobility
Databases and profiles	usually data stored separately for each service	centralize databases with user and service profiles
Registration and user identities	unique ID based on per service principle	centralized and service independent
Security	network and service security out of softswiching concept	specialized border and security functions incorporate in standards
Inter-working between vendors	usually only the same vendor's products could be guaranty for inter-working	various vendor's products should inter-work based on interfaces and standards
Services	More-less limited to voice, however, set of services may be extended by adding application servers	Multimedia platform: voice, data and video across heterogeneous platforms and network domains
Applications	Dominantly integrated applications	IMS interworks with more type of AS: SIP, OSA/Parlay, IN - 3rd party service creation
Service Capabilities and Enablers	Limited for each service on application sever capability	Shareable service enablers to support a number of more complex applications
Service integration	Limited to each service platform	Possible across service layer

2 NGN testbed platform

The NGN Lab was developed to support the research and development activities in the area of NGN, as well as the educational activities based on e-learning. This platform has distributed architecture and services provided by pilot platform available for academic and student community over any internet access.

2.1 NGN platforms at STU Bratislava

One of the key outputs of the research and development project "Convergence of the ICT Networks and Services in the Slovak Communication Infrastructure" acting within the State Research and Development Programme "Building of Information Society" the pilot NGN platform has been build up. This platform has distributed architecture across several Slovak universities. The operational and management platform is located at Telecommunication Department of the Slovak University of Technology in Bratislava. The partial segments of this NGN pilot platform are dislocated also at TU Kosice and University of Zilina interconnected via 2 Gbit IP core network of Slovak Academic Network (SANET) across Slovakia.

The segments of the testbed NGN platform at the Telecommunication Department of the Slovak University of Technology in Bratislava consist of two parts:
1. Pilot NGN platform with operational and management segment to support the pilot deployment and also used for national research and development activities within distributed national academic testbed.
2. NGN Lab with the basic NGN conception in local department laboratory environment supporting our research and development activities, as well as educational activities.

2.2 NGN Laboratory architecture

We have tracked and implemented those technology trends in the area of NGN described previously and have systematically built testbed platform in our NGN Lab laboratory. Our main effort was to analyze the possibility of NGN implementation using mainly open source applications (more information is provided on [7]). We recognize significant benefits to use this kind of technology in terms of the ability to provide additional extensions and modification for required integration purposes. The configuration of the NGN Lab was designed to enable future experimentation of students and researchers, to support them in the research and development activities in real environment of NGN testbed with architecture, protocols and services provided by real NGN platform accessible from wireline or wireless access networks.

NGN testlab environment integrates several NGN core component using major open protocols on all interfaces. Application layer is providing several applications and services (Video, Voice, Web, and Collaborative projects, e-learning). We have also developed a unified user interface to those applications. We are introducing developmental interfaces to enable later development of new converged services and manage services, users profiles and all laboratory components. NGN Laboratory should be therefore well exploited for educational activities based on experiments, practical exercises, simulation, modelling, measuring and testing of NGN principles. The conception of the NGN Lab platform and activities supporting by the NGN Lab are described below in more details.

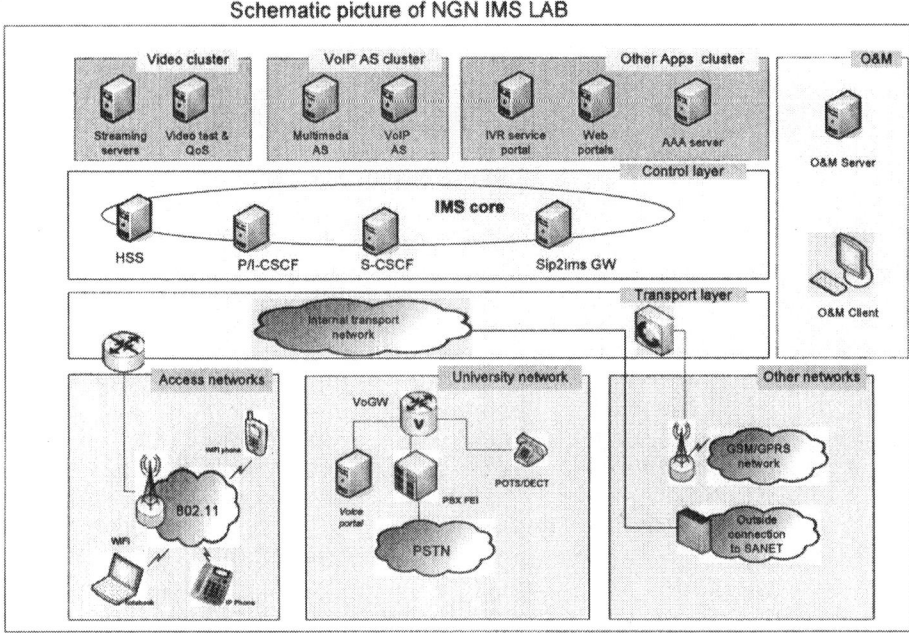

Fig. 1. Schematic picture of NGN Laboratory

2.3 Integration of NGN application

Within the NGN Laboratory a technologic environment was developed to enable integration of various applications and services and providing them over any IP networks independently from access technologies. Among main services we consider multimedia communication, collaborative tools, video streaming services, enhanced messaging and also plenty of different web services.

The most important part of NGN Lab capabilities is providing in application layer (application server farm) several types of different communication and collaboration tools for users of the laboratory and project team members. The NGN Lab offers several types of services we can split into several major groups:
- web/portal based informational services,
- communication services,
- communities and collaborative work services,
- multimedia streaming services,
- e-learning services.

2.4 Integration of LMS to NGN Laboratory

Learning management system (LMS) was based on managed e-learning infrastructure. We can look on it as a solution for planning, delivering, and managing learning events. These events include online, virtual classroom, and instructor-led courses. The evolution of information technologies (ICT) allowed LMS to advance in deployment of multimedia materials (text, animations, voice and video, interactive & collaborative tasks) or to integrate real network/service simulations. IMS platform proposed new solutions in this area. LMS server became one type of application server with integrated user management with NGN network. In our testbed platform, we have partially integrated LMS server. We have started with user databases integration and self-care portal. Another significant enhancement was the interactive communication (multimedia telephony and messaging services), which is used in the classroom among students and for communication between lector and students.

2.5 NGN Laboratory used for communication, education and practical experience with NGN technologies

The integration of LMS application into NGN technology has improved the e-learning system deployment in several main areas. The main area is the user management and integration with other applications and collaborative tools. We have already used integrated solution for several courses. In the same time, research activities in the area of convergence are ongoing. We have performed extensive testing of QoS impact on the application performance. This study has been based on NIST-NET emulator. For the study, the real-time traffic measurements software probes have been deployed and the impact of the delay and PLR has been analyzed.

Similarly, testing of converged e-learning application via wireless access on mobile terminal (e.g. PDA) was in question. But perception within the research testing group was not fully positive. The inconvenience was the price and size of the equipment needed and relatively uncomfortable work and studying of materials.

3 Additional IMS integration to NGN Lab platform and full interoperability testbed

We have already installed, configured and tested the principles of NGN architecture based on IMS core by introducing main IMS components like P-CSCF, S-CSCF, I-CSCF (Proxy-/Serving-/Interrogating- Call Session Control Function), HSS (Home Subscriber Server) using open source software components based on Open IMS Core [8] as given in Fig. 1. But there are several aspects necessary to be changed from softswitch approach to IMS architecture [6]. For example, in softswitch NGN there is each user registered separately in each service (application servers). With IMS the operator has to change its implementation of services with the introduction of CSCF and HSS functions leading to flexibility for the new user oriented services (Fig. 2.).

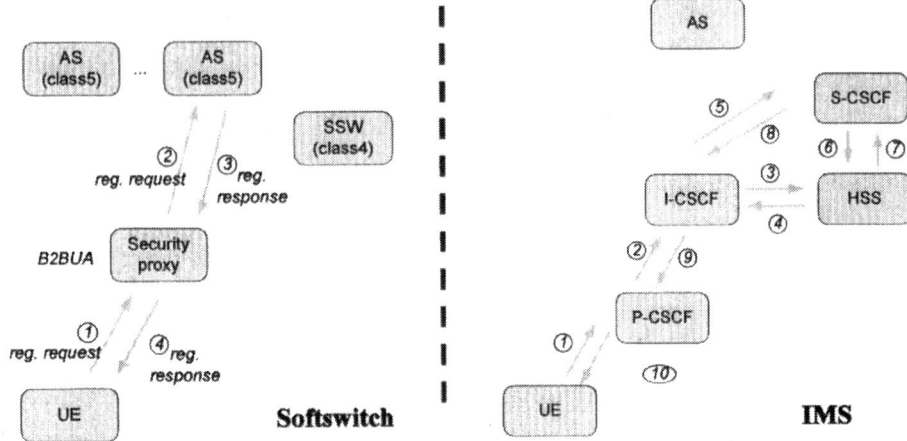

Fig. 2. Centralized registration approach within IMS core

Next important difference is the user data storage (Fig. 3.). In softswitch based NGN architecture user's data are stored in each application server; separately for each service this means difficulties to integrate them and also other limitations. IMS brings HSS central data storage for all subscribers and for all services in operator's NGN network. Diameter is AAA protocol determined to use at Cx, Sh and Si interfaces.

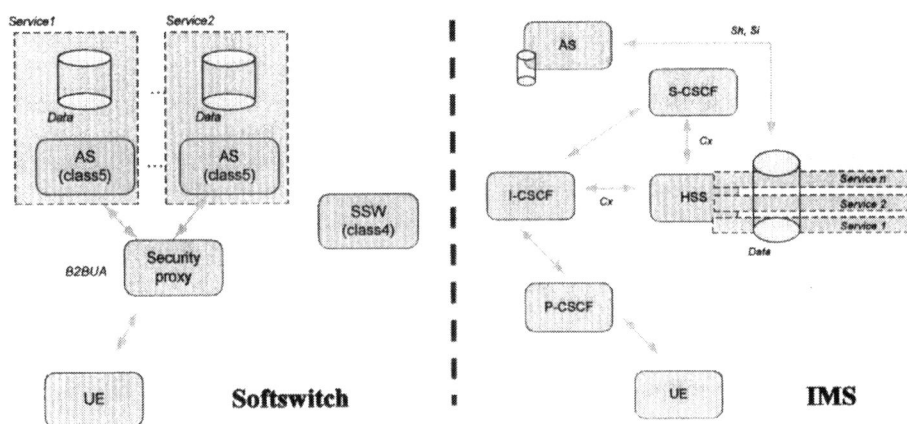

Fig. 3. User data storage approach within IMS core

Of course, there are more differences between softswitch and IMS based NGN which have to be taken into account in the process of migration and IMS integration. This new architecture principle includes the introduction of a new charging concept, public and private identities, and way of integration of heterogeneous networks including wireless access networks (e.g. WMAN, WLAN, and WPAN). The NGN Lab testbed need to provide secure interface to interconnect with other similar testbeds or external networks (operators) via SCSF elements and border gateway functions.

3.1. IMS interoperability and integration of SIP based service

Main issues are related to the complexity of architecture, availability of different IMS client implementation and last but not least the application migration from pure SIP implementation to IMS based application servers. We are currently focusing on the integration and testing of different components towards IMS core and also including really important area of research according to the IMS interoperability and standard compliance. Although there are common goals in IMS based NGN architecture (ETSI TISPAN, 3GPP IMS, CableLabs 2.0, ITU-T, ATIS, etc.) behind the common approach, existing status-quo, and the focus on some specific priorities naturally influencing the scope and ways to reach fully multimedia NGN networks in each standardization body. Additionally, various ways of IMS implementations by vendors, maximization of reuse of their current technology portfolio and their different roadmaps towards fully supported standards lead to the operator's wish to deploy this technology and services to uncertain position. With focus on wireless problems, there are at least several areas with the requirement of further study:

- Interconnection of NGN networks with aspect on roaming scenarios with focus on cooperation between different service components.
- Security aspects of vertical handovers.
- QoS establishment and guarantying between heterogeneous access networks.
- Service interworking among different domains and different vendors.
- Standardization of implementation NGN service in NGN networks in terms of support handover/mobility techniques.

4 Testlab experiments and applications used in education

The main purposes of NGN Laboratory is to provide the facilities for experimenting in real laboratory environment and evaluate real measurements in order to investigate and approve some of the NGN principles or compare them with expected simulation or modelling results.

The education activities focus on several areas and issues:

- NGN architecture, components, interfaces of the NGN platform (e.g. integrate additional access networks like wireless, mobile),
- NGN protocols and procedures (call control protocols, media gateway control protocols, transport protocols, protocols for AAA, QoS and support) including core IMS protocols and others (SIP, Diameter, etc.),
- Test, enhance or integrate NGN services and applications implemented in the NGN Lab environment,

The laboratory environment enable during educational process involve students also to learn also by practical experience by case scenarios where students need to setup, measure and evaluate real network characteristics and their relation to services and impact to user subjective quality perception. A simple example is given in next subsection.

4.1 Subjective quality evaluation for video streaming services according to different wireless access and encoding parameters

Measuring of video quality was processed in two types of network environment [9]. Transmission was measured in fixed LAN and wireless WiFi network (WiFi signal strength -66 dBm and -76 dBm). As VOD server was used VLC (ver. 0.8.2). Video files used for streaming was transcoded by VLC [10] with MPEG-2 codec in three bitrates 512, 1024 and 2048 kbps (Fig.3).

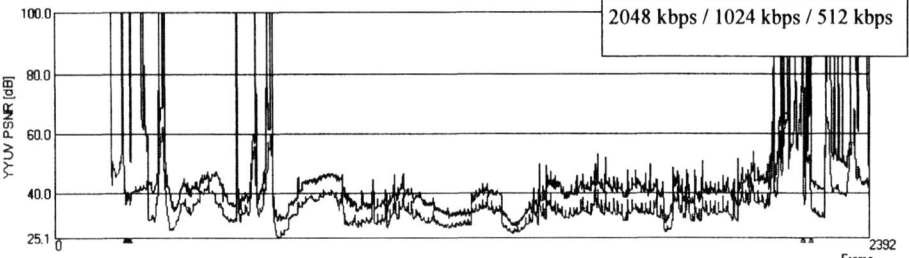

Fig. 4. PSNR comparison of reference video 2048 kbps with 1024 / 512 kbps samples (Average PSNR = 38,57 dB for 1024 kbps, Average PSNR = 33,12 dB for 512 kbps)

The streamed files were separately captured on VLC client (v.0.8.6a). Measuring and comparison on video files were processed on MSU Video Quality Measuring Tool (v.1.4).

Table 2. Dependency of network environment and video bitrates with relation to video quality

Video Sample bitrate	LAN	WiFi [-66dBm]	WiFi [-76dBm]
512 [kbps]	1.25	1	0.75
1024 [kbps]	3.25	2.25	0.5
2048 [kbps]	3.75	2.5	0.25

Evaluation of simple experiment was based on subjective rating of 20 subjects with MOS-like [11] uses the Mean Opinion Score (MOS) like rating in telecommunications industry for the voice quality subjective metrics. The values of the MOS are: 1= bad; 2=poor; 3=fair; 4=good; 5=excellent. The difference in video quality between 2048 and 1024 kbps bitrate is from subjective rating in the LAN environment relatively small. Negative influence on quality is presented only for 512 kbps where the very high level of compression reduces the quality with macro-blocking effect even though there are no artefacts as dropouts during the streaming.

The rating given in Table 2 show that users in wireless network accept small damage of video stream in case there is used video stream with higher bitrate. Bitrate compensation of video quality is not applicable where signal strength of WiFi is very low. In this case, users are very sensitive to accrued artefacts as dropouts and interlacing. Video streaming in this high error environment is more preferable with lower bitrates even the artefacts as macro-blocking are presented in captured video. Finally, in some mobile devices is low bitrates also more feasible because of computational power of hardware in such devices (PDA, Mobile Phones, etc.)

5 Conclusion

The article deals with some aspects and trends of the network convergence towards IMS based NGN. The unified service control concept and flexible application layer allows to provide any services to any type of terminal across heterogeneous access networks including wireline, wireless, mobile access technologies. The conception of the NGN testbed platform implemented at the STU Bratislava is also described in this paper. Additional value of presented NGN testbed was not just about building and running such a NGN platform, but mainly to provide technology playground allowing the setup for several research projects dealing with different NGN issues (architecture, protocols, mobility, QoS, service, network simulation, interoperability & compliance testing) including areas of future studies.

Acknowledgments

This NGN platform and further outputs resulted from the research and development activities realized within the following projects:
- State Research and Development Program "Building of the Information Society" granted by Slovak Ministry of Education
- National project - basic research "VEGA No. 1/3094/06 and 1/4084/0", granted by Slovak Ministry of Education.
- International educational project "Leonardo ICoTeL", granted by EU Leonardo da Vinci Programme

References

1. ITU-T Recommendation Y.2011: General principles and general reference model for next generation networks. International Telecommunication Union, Geneva, Switzerland. (2004)
2. ETSI ES 282 007: TISPAN NGN Release 1; Core IMS architecture". European Telecommunications Standards Institute, (2006)
3. ETSI TS 123 002: Digital cellular telecommunications system (Phase 2+); Universal Mobile Telecommunications System (UMTS); Network architecture (3GPP TS 23.002). European Telecommunications Standards Institute, (2005)
4. ETSI TS 123 228: Digital cellular telecommunications system (Phase 2+); Universal Mobile Telecommunications System (UMTS); IP Multimedia Subsystem (IMS); Stage 2 (3GPP TS 23.228). European Telecommunications Standards Institute, (2005)
5. PacketCable™ 2.0, IP Multimedia (IM) session handling; IM call model; Stage 2 Specification 3GPP TS 23.218, PKT-SP-23.218-I02-061013, CableLabs, (2006)
6. Mikoczy, E.: Unified service control based on IMS. In 3rd Annual Congress IMS, Implementation and migration strategies to an all-IP network, Berlin, Germany, (2007)
7. NGN Lab web page, http://ngnlab.ktl.elf.stuba.sk, (2007)
8. Open IMS core project web page, http://www.openimscore.org/, (2007)
9. Tomek, R.: Multimedia Services in the NGN environment. Diploma thesis, FEI STU, Bratislava, Slovakia, (2007)
10. VLC – VideoLan software webpage, http://www.videolan.org/vlc/ (2007)
11. ITU-T Recommendation P.800, Methods for subjective determination of transmission quality. International Telecommunication Union, Geneva, Switzerland. (1996)

Enhancing IEEE 802.11n WLANs using group-orthogonal code-division multiplex

Guillem Femenias and Felip Riera-Palou

Mobile Communications Group - Dept. of Mathematics and Informatics
University of the Balearic Islands
07122 Palma, Mallorca (Illes Balears), Spain.
{guillem.femenias,felip.riera}@uib.es

Abstract. The definition of the next generation of wireless networks is well under way within the IEEE 802.11 High Throughput Task Group committee. The resulting standard, to be called IEEE 802.11n, is expected to be a backward-compatible evolution of the successful IEEE 802.11a/g systems also based on multicarrier techniques. It can be anticipated that 802.11n systems will outperform its predecessors in terms of transmission rate and/or performance, mainly, due to the use of multiple antennae technology for transmission and reception. In this paper we propose to incorporate group-orthogonal (GO) code division multiplex (CDM) into the IEEE 802.11n specifications to further enhance its performance. It is shown how GO-CDM can take full advantage of the diversity offered by the multiple antennae and multicarrier transmission by using an iterative maximum likelihood (ML) joint detector. Furthermore, the use of GO-CDM does not compromise the backward compatibility with legacy systems.

Keywords: IEEE 802.11n, multicarrier, code-division multiplex, maximum likelihood detection, turbo receivers.

1 Introduction

The last decade has seen an explosive growth in the deployment of wireless local area networks (WLANs) which has made the concept of nomadic computing a reality. Most of these networks are based on one of the flavours of the IEEE 802.11 family of standards. The most recent and powerful set of specifications currently deployed, IEEE 802.11g, operates on a bandwidth of 20 MHz and its physical layer is based on a particular form of multicarrier transmission, namely, orthogonal frequency division multiplexing (OFDM) enabling these systems to achieve transmission rates up to 54 Mbps. At present, the standardisation of what should be the next step, named IEEE 802.11n, is being pursued by the IEEE 802.11 High Throughput Task Group committee. The new standard will support much higher transmission rates thanks to the use of multiple antenna technology and other enhancements such as the possibility of operating on a

40 MHz bandwidth (employing more subcarriers) and transmission modes using a reduced guard interval. In its fastest mode, 802.11n is expected to surpass a transmission rate of 700 Mbps. Despite all the enhancements introduced, it is mandatory for the new standard to remain compatible with multicarrier legacy systems (802.11a/g) and therefore, 802.11n-compliant devices should have means to fall back to older 802.11 specifications.

Like its predecessors, 802.11a and 802.11g, the new 802.11n standard will also be based on OFDM allowing the use of other enhancements which have recently been proposed for this type of physical layers. A powerful improvement over conventional OFDM was the introduction of multicarrier code division multiplex (MC-CDM) OFDM by Kaiser in [1]. In MC-CDM, rather than transmitting a single symbol on each subcarrier as in conventional OFDM, groups of symbols are multiplexed together by means of orthogonal spreading codes and simultaneously transmitted on a group of subcarriers. This technique resembles very much the principle behind multicarrier code-division multiple access (MC-CDMA) [2] where different users share a group of subcarriers by using each of them a different spreading code. More recently, group-orthogonal MC-CDMA (GO-MC-CDMA) [3,4] has been introduced as a particular flavour of MC-CDMA whereby users are split in groups and each group exclusively uses a (small) subset of all the available subcarriers. The subcarriers forming a group are chosen to be as separate as possible in the available bandwidth in order to maximise the frequency diversity gain. A GO-MC-CDMA setup can be seen as many independent MC-CDMA systems of lower dimension operating in parallel. This reduced dimension allows the use of optimum receivers for each group based on maximum likelihood detection at a reasonable computational cost. Group-orthogonality has also been proposed for (uncoded) MC-CDM systems in [5] where results are given for group dimensioning and spreading code selection. Nevertheless, as shown in [6], the benefits of CDM-OFDM are rather limited when measuring the coded performance in typical operating scenarios conforming to IEEE 802.11a specifications, especially, when using decoders based on soft/iterative procedures. This is due to the large subcarrier correlation found in many wireless environments which severely limits the achievable frequency diversity. This issue is largely solved with the introduction of multiple transmit and receive antennae as then subcarrier correlation within a group can be greatly reduced by exploiting the spatial dimension. A comparison between two of the possible techniques to exploit the spatial diversity, namely, space-time block coding (STBC) and cyclic delay diversity (CDD), within the context of IEEE 802.11n has recently appeared in [7] but in there, the receiver structures proposed are based on linear techniques (minimum mean square error detectors) and hence are not suitable for iterative processing.

This paper, after reviewing the physical layer of the current IEEE 802.11n draft, proposes the use of GO-MC-CDM to further improve its performance. It will be shown how using an iterative reception scheme based on a maximum likelihood multisymbol detector (ML-MSD) with soft decoding allows the exploitation of both, frequency and spatial diversity. Simulation results are presented for two possible spatial configurations, one uniquely based on cyclic delay diversity (CDD) and another one based on the combination of spatial division multiplexing (SDM) and CDD.

We conclude this introduction with a brief notational remark: throughout this paper vectors and matrices are denoted with lower and upper cases bold characters, respectively. Scalars are represented with non-bold characters (either lower or upper case). We use the notation $\mathcal{D}(\mathbf{x})$ to denote a diagonal matrix with vector $\mathbf{x}$ at its main diagonal. Vectors are assumed to be column-oriented and the notation $(\cdot)^T$ is used to denote the transposition of a vector or matrix. Finally, $\mathbf{x}_{[k]}$ represents a vector $\mathbf{x}$ with its kth entry removed.

2 System model for current IEEE 802.11n proposal

The current draft for the physical layer of IEEE 802.11n can be found in [8]. As mentioned in the introduction, a qualitative difference of IEEE 802.11n with respect to previous standards is the introduction of multiple antenna technology. Moreover, the standard is defined in such a way that allows the use of different methods to exploit the spatial dimension, namely, spatial division multiplexing (SDM), cyclic delay diversity (CDD) and space-time block coding (STBC). In this work we have focused on the design of an enhanced physical layer based on the application of GO-CDM when using CDD and/or SDM[1]. The transmission architecture is depicted in Fig. 1. Note that this figure already includes the optional GO-CDM extension (in dashed lines) which will be covered in detail in the next section.

We focus on the IEEE 802.11n specification operating in a bandwidth of B=20 MHz utilized by means of $N_c = 64$ orthogonal subcarriers of which $N_c^d = 52$ subcarriers are used to transmit user data while the rest, $N_c^p = 12$, correspond to pilot and guard subcarriers. Transmitter and receiver are assumed to have N_t and N_r antennas, respectively, with $1 \leq N_t, N_r \leq 4$.

The user data to be transmitted is generated and segmented into frames of N_b bits. It is assumed that frames are independent of one another, and therefore, to simplify notation, there is no need to specify the segment index. Each generated frame is subsequently fed to a rate R_c punctured convolutional encoder (RCPCC box in Fig. 1) to generate N_b/R_c coded bits. The coded bits are then separated into $N_s \in \{1, 2, 3, 4\}$ spatial streams as specified by the spatial parsing equation in [8]. It should be pointed out that for the cases of employing CDD and/or SDM, it should hold that $N_s \leq N_t$. The bits on each stream are then interleaved (blocks Π_1 in Fig. 1) and subsequently mapped to symbols from an m-point constellation (BPSK, m-QAM) with $m = 2^M$ where M represents the number of bits per symbol, yielding the set of signal vectors to be transmitted $\mathbf{s}^w = \begin{pmatrix} s_1^w & s_2^w & \ldots & s_{N_{\text{qam}}}^w \end{pmatrix}^T$ where $1 \leq w \leq N_s$ represents the spatial stream index and $N_{\text{qam}} = N_b/(R_c N_s M)$ denotes the number of symbols per frame to be

---
[1] The application of GO-CDM in combination with STBC in the context of IEEE 802.11n is a topic of current research.

transmitted on each spatial stream. The symbols on the different spatial streams are at this point properly segmented into N_{OFDM} OFDM symbols, each made of N_d (QAM) symbols holding that $N_{\text{qam}} = N_c^d\, N_{OFDM}$. We will denote by $\mathbf{s}_l^w$, $1 \leq l \leq N_{OFDM}$, the lth OFDM symbol of the wth spatial stream in the current frame. Skipping for the moment, the GO-CDM processing, the resulting OFDM symbols from all streams are linearly mixed according to

$$(\mathbf{x}_l^u)^T = \mathbf{W} \left(\mathbf{s}_l^1\ \mathbf{s}_l^2\ \ldots\ \mathbf{s}_l^{N_s}\right)^T \tag{1}$$

where $1 \leq u \leq N_t$ and $\mathbf{W}$ represents an $N_t \times N_s$ spatial spreading matrix. The role of the matrix $\mathbf{W}$ is twofold: on one hand, for the cases where $N_s < N_t$, the spatial spreading serves to distribute the incoming streams among all transmit antennae thus making full use of the available spatial diversity. On the other hand, the specific selection of $\mathbf{W}$, in combination with a suitable choice of the cyclic delays $\delta_1, \ldots, \delta_{N_t}$ (see Fig. 1), provides the system with either CDD, SDM or a combination of the two. Examples are later provided showing how this matrix and cyclic delays can be chosen to exploit the spatial diversity in a prescribed manner. Finishing the transmission procedure, and as shown in Fig. 1, the different spatially spread OFDM symbols, $\mathbf{x}_l^u$, are then IFFT-converted, expanded with a cyclic prefix and processed by the RF transmission chains.

Taking for granted that transmit and receive antennas are sufficiently apart, the $N_t \times N_r$ propagation channels between transmitter and receiver are safely assumed to be independent and derived from a common and scenario-dependent power delay profile

$$S(\tau) = \sum_{p=0}^{P-1} \phi(p)\delta(\tau - \tau_p) \tag{2}$$

where P denotes the number of independent paths of the profile and $\phi(p)$ and τ_p denote the power and delay of each path, respectively. A single realization of the channel impulse response between transmit antenna a_t and receive antenna a_r at time instant t will then have the form

$$h^{a_t,a_r}(t;\tau) = \sum_{p=0}^{P-1} h_p^{a_t,a_r}(t)\delta(\tau - \tau_p) \tag{3}$$

where it will hold that $E\left[|\,h_p^{a_t,a_r}(t)\,|^2\right] = \phi(p)$. The corresponding frequency response will be given by

$$\bar{h}^{a_t,a_r}(t;f) = \sum_{p=0}^{P-1} h_p^{a_t,a_r}(t)\exp(-j2\pi f\tau_p) \tag{4}$$

which, evaluated at the N_c equispaced subcarrier frequencies, $f_0, f_1, \ldots, f_{Nc-1}$, yields the $N_c \times 1$ vector $\bar{\mathbf{h}}^{a_t,a_r}(t) = \left(\bar{h}^{a_t,a_r}(t;f_0)\ \ldots\ \bar{h}^{a_t,a_r}(t;f_{N_c-1})\right)^T$. Assuming without loss of generality that the channel is static over the duration of a frame and independent from other frames, to simplify the notation, we will

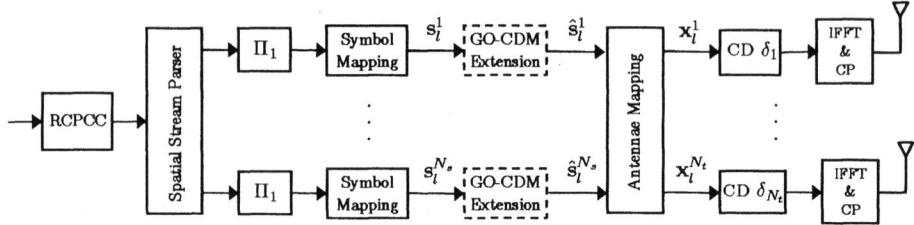

Fig. 1. IEEE 802.11n transmitter proposal supporting SDM and/or CDD.

express the frequency response of the data subcarriers (i.e. pilots and guard subcarriers excluded) simply as $\bar{\mathbf{h}}^{a_t,a_r} = \left(\bar{h}_0^{a_t,a_r} \; \bar{h}_1^{a_t,a_r} \ldots \bar{h}_{N_c^d-1}^{a_t,a_r} \right)^T$.

At this point it is necessary to distinguish which form of spatial diversity is employed, which is basically determined by the choice of $\mathbf{W}$ and cyclic delays. In setups based exclusively on SDM, it will hold that $N_s = N_t$ and $\delta_{a_t} = 0$ for $1 \leq a_t \leq N_t$. This implies that each antenna is in charge of transmitting one spatial stream. In contrast, in configurations where transmit antennae are configured to exploit CDD, it will hold that $N_s = N_t/\Lambda$, where Λ denotes the number of antennae involved in the CDD processing (CDD factor) of each spatial stream. That is, when using CDD, the available antennae are split in N_s groups of Λ antennae and each group is used for the transmission of one spatial stream. In CDD, all the antennae in a CDD group transmit the same information but each antenna applies a different cyclic delay [9]. It is easy to show [9] that this amounts to transmit a given spatial stream information from a single antenna over the composite channel reaching receive antenna a_r given by

$$\bar{\mathbf{h}}_w^{a_r} = \sum_{a_t^w=0}^{\Lambda-1} \bar{\mathbf{h}}^{a_t^w,a_r} \exp\left(-j \frac{2\pi \delta_{a_t^w}}{N_c}\right). \tag{5}$$

where $\bar{\mathbf{h}}^{a_t^w,a_r}$ denotes the different channel frequency responses between the antennae belonging to CDD group w and receiver antenna a_r and $\delta_{a_t^w}$ represents the specific cyclic delay applied by each antenna in the CDD group. Taking into account that the available transmit antennae will be split into parallel CDD systems, it can be derived from [9] that an advantageous delay choice will be given by

$$\delta_{a_t} = \begin{cases} 0 & , \text{ if } mod(a_t-1,\Lambda) = 0 \\ N_c/(N_t/\Lambda) + \delta_{a_t-1} & , \text{ otherwise} \end{cases} \tag{6}$$

where $mod(a,b)$ denotes the modulus of a/b. It is important to stress that, to all effects, the composite CDD channel $\bar{\mathbf{h}}_w^{a_r}$ can be considered as a conventional channel with the distinctive feature of an increased frequency selectivity with respect to the original power profile defined by $S(\tau)$.

Having defined the transmission scheme and channel, it is now possible to specify

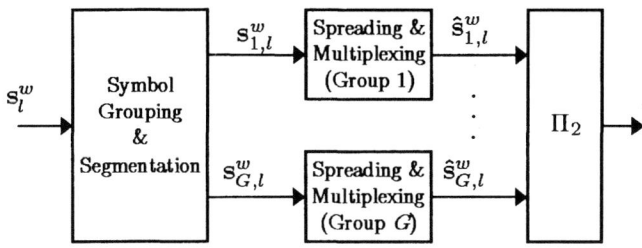

Fig. 2. GO-CDM extension.

the reception equation for the N_s OFDM-symbols which have been transmitted in parallel at a given instant as

$$\begin{pmatrix} \mathbf{r}_l^1 \\ \vdots \\ \mathbf{r}_l^{N_r} \end{pmatrix} = \begin{pmatrix} \mathbf{H}^{1,1} & \cdots & \mathbf{H}^{1,N_s} \\ \vdots & \ddots & \vdots \\ \mathbf{H}^{N_r,1} & \cdots & \mathbf{H}^{1,N_s} \end{pmatrix} \begin{pmatrix} \mathbf{x}_l^1 \\ \vdots \\ \mathbf{x}_l^{N_s} \end{pmatrix} + \begin{pmatrix} \mathbf{v}_l^1 \\ \vdots \\ \mathbf{v}_l^{N_r} \end{pmatrix}, \qquad (7)$$

where $\mathbf{H}^{i,j}$ will be of the form $\mathcal{D}\left(\bar{\mathbf{h}}^{a_r,a_t}\right)$ for the case of SDM ($N_t = N_s$) or the form $\mathcal{D}\left(\bar{\mathbf{h}}_w^{a_r}\right)$ for the case where CDD is used ($N_s = N_t/\Lambda$). The $N_c^d \times 1$ vectors $\mathbf{v}_{g,l}^1, \ldots, \mathbf{v}_{g,l}^{N_r}$ represent the additive white Gaussian noise samples on each receiving antenna. Every scalar noise sample, is assumed to be distributed according to a zero-mean Gaussian pdf with variance σ_n^2. The original data symbols to be transmitted s_i^w, are suitably scaled to have power $E\left\{|s_i^w|^2\right\} = 1$ in the case of SDM transmission and $E\left\{|s_i^w|^2\right\} = 1/\Lambda$ if the CDD component is present. In this latter case, this factor represents the distribution of energy available per symbol among the various transmit antennas used for the transmission of the wth spatial stream. For both cases, SDM and CDD (and their combinations), this power normalisation allows the operating signal to noise ratio to be written as $E_s/N_0 = 1/\sigma_n^2$.

3 Group-orthogonal CDM for IEEE 802.11n

The introduction of multiple transmit and receive antenna brings along the possibility of further exploiting the potential frequency selectivity, specially, whenever a CDD component is present. To this end, GO-CDM can be used to distribute the source symbols energy among multiple subcarriers. Figure 2 shows a detailed diagram of the GO-CDM mechanism which would correspond to the dashed boxes in Fig. 1. The GO-CDM processing is performed on each spatial stream and consists of splitting the incoming OFDM symbols, $\mathbf{s}_l^w$, into G groups of $Q = N_c^d/G$ QAM symbols each. Each group, $\mathbf{s}_{g,l}^w$, is then assigned Q subcarriers which will be used to jointly transmit the Q source symbols in the group. The distribution of the group symbols among the Q subcarriers is carried out by means of a

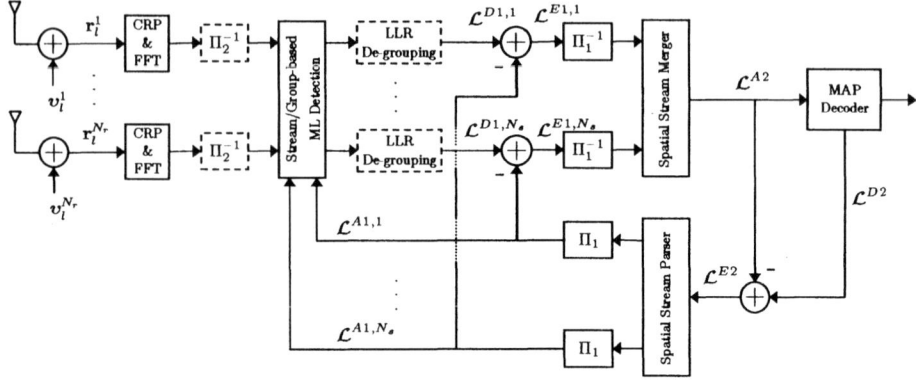

Fig. 3. ML-based turbo receiver for GO-CDM enhanced IEEE 802.11n.

spread and multiplex operation defined by

$$\hat{\mathbf{s}}_{g,l}^w = \mathbf{C}\,\mathbf{s}_{g,l}^w \tag{8}$$

with $1 \leq g \leq G$ and with $\mathbf{C} = \begin{pmatrix} \mathbf{c}^0 & \mathbf{c}^1 & \ldots & \mathbf{c}^{Q-1} \end{pmatrix}$ representing the $Q \times Q$ spreading matrix with each $\mathbf{c}^q$ denoting the $Q \times 1$ spreading code associated with the qth symbol in the gth group. A typical choice of spreading matrix is the Walsh-Hadamard matrix due to its column-wise orthogonality and low complexity implementation. We note that the spreading matrix is common to all groups since each group utilizes separate (and orthogonal) sets of subcarriers and therefore, there is no need to provide extra separation. The resulting spread and multiplexed symbols from all groups are then time interleaved (block Π_2 in Fig. 2) to increase the resilience against noise. The idea is to transmit (QAM spread) symbols from different OFDM-symbols in each group so that, upon deinterleaving at the receiver, each symbol in the group has experienced uncorrelated noise samples. To avoid confusion with the spatial spreading performed by the linear transformation given by $\boldsymbol{W}$, from this point onwards, we will refer to the GO-CDM processing as temporal spreading. As can be inferred from Fig. 1, if GO-CDM processing is present, the spatial spreading in (1) operates on the temporally spread symbols resulting from (8) rather than on $\mathbf{s}_l^1, \ldots, \mathbf{s}_l^{N_s}$. We conclude this section by noting that backward compatibility with IEEE 802.11a/g can easily be insured by simply setting the group size to $Q=1$.

4 ML-based turbo reception

As in conventional OFDM systems, the reception process on each receive branch (illustrated in Fig. 3) begins by removing the cyclic prefix (CPR) and performing the FFT on each of the OFDM-symbols forming the transmitted frame. In case GO-CDM has been employed at the transmitter, as we assume through the rest of this section, the corresponding deinterleaving (Π_2^{-1}) is then applied resulting in N_r vectors $\mathbf{r}_l^{a_r}$ of dimension $N_c^d \times 1$ (pilots and guard subcarriers are not taken into account). These vectors are split into groups, $\mathbf{r}_{g,l}$, in concordance with the grouping performed at transmission. It should be noted now that, as inferred from (7), each received group vector $\mathbf{r}_{g,l}$ contains information from the groups of all transmit spatial streams $\mathbf{s}_{g,l}^w$, making mandatory some form of interference cancelling device to reduce the intersymbol interference in order to estimate the different symbol groups from the different spatial streams. The main reason to split the symbols in groups when spreading and multiplexing (i.e. applying group-orthogonality) is to make ML-based detection computationally feasible within each single group. In line with this, we have based our detection mechanism on a bank of list sphere detectors (LSD) [10], each targeting a single-group. The LSD detector is based on the sphere detection procedure introduced in [11] which is an efficient method of performing an exhaustive search among a set of candidates (i.e. ML detection). The main feature of LSD is that, apart from symbol estimates, it is able to produce soft information regarding the bits of the estimated symbol in the form of likelihood ratios (LLRs).

Since the LSD detector works on a group basis, it is useful at this point to write the reception equation for a single group as

$$\begin{pmatrix} \mathbf{r}_{g,l}^1 \\ \vdots \\ \mathbf{r}_{g,l}^{N_r} \end{pmatrix} = \begin{pmatrix} \mathbf{H}_g^{1,1}\mathbf{WC} & \ldots & \mathbf{H}_g^{1,N_s}\mathbf{WC} \\ \vdots & \ddots & \vdots \\ \mathbf{H}_g^{N_r,1}\mathbf{WC} & \ldots & \mathbf{H}_g^{1,N_s}\mathbf{WC} \end{pmatrix} \begin{pmatrix} \mathbf{s}_{g,l}^1 \\ \vdots \\ \mathbf{s}_{g,l}^{N_s} \end{pmatrix} + \begin{pmatrix} \mathbf{v}_{g,l}^1 \\ \vdots \\ \mathbf{v}_{g,l}^{N_r} \end{pmatrix} \quad (9)$$

or, more compactly,

$$\tilde{\mathbf{r}}_{g,l} = \mathbf{A}_g \, \tilde{\mathbf{s}}_{g,l} + \tilde{\mathbf{v}}_{g,l} \quad (10)$$

where $\tilde{\mathbf{r}}_{g,l} = \begin{pmatrix} \mathbf{r}_{g,l}^1 & \ldots & \mathbf{r}_{g,l}^{N_r} \end{pmatrix}^T$ represents the N_r reception vectors evaluated on the frequencies assigned to the gth group, $\tilde{\mathbf{s}}_{g,l} = \begin{pmatrix} \mathbf{s}_{g,l}^1 & \ldots & \mathbf{s}_{g,l}^{N_s} \end{pmatrix}^T$ are the symbols from all transmitted spatial streams in group g, $\mathbf{A}_g$ corresponds to the matrix in (9) jointly modelling the channel, temporal spreading and spatial spreading effects, and $\tilde{\mathbf{v}}_{g,l} = \begin{pmatrix} \mathbf{v}_{g,l}^1 & \ldots & \mathbf{v}_{g,l}^{N_r} \end{pmatrix}^T$ are the noise samples added on the group subcarriers. We define now the mapping $\tilde{\mathbf{s}}_{g,l} = \mathcal{M}(\mathbf{b})$ as the modulation mapping to arrive to symbol vector $\tilde{\mathbf{s}}_{g,l}$ comprising all the symbols belonging to group g of the lth OFDM symbol from the N_s transmitted spatial streams from the corresponding group bits $\mathbf{b} = \begin{pmatrix} b_1 & b_2 & \ldots & b_{(N_s MQ)} \end{pmatrix}^T$ (to simplify notation, we skip the group and OFDM-symbol indices when referring to the bits). Making use of the Max-log approximation, the LLR for a given bit, b_i, can then be approximated by [10]

$$\mathcal{L}_{g,l}^{D1}(b_i) \approx \frac{1}{2} \max_{\mathbf{b} \in \mathcal{B}_{i,+1}} \left\{ -\frac{1}{\sigma_n^2} \|\tilde{\mathbf{r}}_{g,l} - \mathbf{A}_g \mathcal{M}(\mathbf{b})\|^2 + \mathbf{b}_{[i]}^T (\mathcal{L}_{g,l}^{A1})_{[i]} \right\}$$
$$- \frac{1}{2} \max_{\mathbf{b} \in \mathcal{B}_{i,-1}} \left\{ -\frac{1}{\sigma_n^2} \|\tilde{\mathbf{r}}_{g,l} - \mathbf{A}_g \mathcal{M}(\mathbf{b})\|^2 + \mathbf{b}_{[i]}^T (\mathcal{L}_{g,l}^{A1})_{[i]} \right\} \quad (11)$$

where the symbols $\mathcal{B}_{i,+1}$ and $\mathcal{B}_{i,-1}$ represent the sets of 2^{N_sMQ-1} bit vectors whose i^{th} position is a '+1' or '-1', respectively. The $(N_sMQ - 1) \times 1$ vector $(\mathcal{L}_{g,l}^{A1})_{[i]}$ contains the a-priori LLR for each bit in **b** except for the ith bit. All a-priori LLRs are assumed to be zero for the first iteration. Since moderate values of M and/or Q make the sets $\mathcal{B}_{i,+1}$ and $\mathcal{B}_{i,-1}$ extremely large and therefore the search in (11) computationally unfeasible, LSD limits the search to the sets $\hat{\mathcal{B}}_{i,+1} = \mathcal{B}_{i,+1} \cap \mathcal{C}$ and $\hat{\mathcal{B}}_{i,-1} = \mathcal{B}_{i,-1} \cap \mathcal{C}$ where $\mathcal{C}$ is the set containing the bit vectors corresponding to the N_{cand} group candidates closer, in an Euclidean sense, to the received group vector, i.e., $\mathcal{C} = \{\mathbf{b}^1, \ldots, \mathbf{b}^{N_{cand}}\}$ where $\mathbf{b}^j = \mathcal{M}^{-1}(\tilde{\mathbf{s}}_{g,l}^j)$ with $\{\tilde{\mathbf{s}}_{g,l}^1, \ldots, \tilde{\mathbf{s}}_{g,l}^{N_{cand}}\}$ being the N_{cand} group candidates for which

$$\|\tilde{\mathbf{r}}_{g,l} - \mathbf{A}_g \tilde{\mathbf{s}}_{g,l}\|^2 \tag{12}$$

is smallest. Notice that when computing the list of group candidates using (12), the a-priori information from each bit is not taken into account. However, when N_{cand} is chosen sufficiently large most often the list of candidates minimizing (12) contains with high probability the maximizer (ML solution) of (11). Two points are to be taken into account when applying the LSD. Firstly, it is possible that either $\hat{\mathcal{B}}_{i,+1}$ or $\hat{\mathcal{B}}_{i,-1}$ is empty for a given bit. Such a situation arises when for a given bit position, all the candidates turn out to have a '+1' (or '-1') in that position. In these cases, that bit position is assigned an extreme (positive or negative) value [10]. Secondly, the limitation of the space search to a limited pool of candidates may cause the a-priori information $\{\mathcal{L}^{A1,1}, \ldots, \mathcal{L}^{A1,N_s}\}$ to be of limited value or even deleterious whenever the candidate pool does not contain the solution hinted by the a-priori information. This problem can be alleviated using the constrained LSD detector proposed in [12].

The computed LLRs for the bits in the G groups of the N_{OFDM} OFDM-symbols for the N_s spatial streams, $\mathcal{L}_{g,l}^{D1} = \left(\mathcal{L}_{g,l}^{D1}(1) \ldots \mathcal{L}_{g,l}^{D1}(N_sMQ)\right)$ are then de-grouped and de-segmented to arrange the LLRs in frame format for the N_s transmitted streams $\{\mathcal{L}^{D1,1}, \ldots, \mathcal{L}^{D1,N_s}\}$. The a-priori information $\mathcal{L}^{A1}$ is subtracted from the computed LLRs and the resulting extrinsic information $\{\mathcal{L}^{E1,1}, \ldots, \mathcal{L}^{E1,N_s}\}$ is then temporally de-interleaved and spatially merged (e.g. inverse spatial parsing) resulting in the signal labeled as $\mathcal{L}^{A2}$ in Fig. 3, which is fed to a soft-in/soft-out maximum a-posteriori (MAP) decoder. The MAP decoder output can be supplied to a bit slicer to obtain bit estimates or fed back to the LSD module for further LLR refinement. In this latter case, see Fig. 3, only extrinsic information ($\mathcal{L}^{E2}$) is sent back to the detector.

5 Numerical results

Numerical simulations have been conducted to evaluate the performance of the GO-CDM extension in a IEEE 802.11n setup. The system has been configured to operate on a 20 MHz bandwidth with a total of N_c=64 subcarriers of which

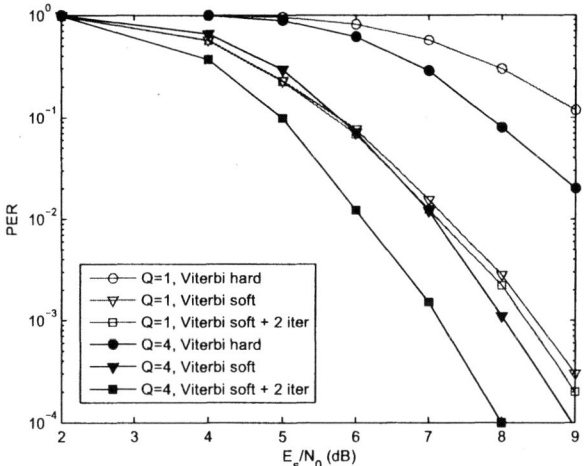

Fig. 4. SDM-CDD transmission. $N_t=4$, $N_r=2$, $N_s=2$. MCS 9.

$N_c^d=52$ are used to carry data. Without loss of generality, information bits are generated in packets of $N_b=416$ bits. The transmitter uses a rate terminated punctured convolutional coder with generator polynomials [133 171]. Perfect channel knowledge and subcarrier synchronisation are assumed at the receiver. It has been also presupposed that the duration of the cyclic prefix appended to each OFDM symbol exceeds the root mean squared channel (rms) delay spread, thus eliminating any interference among consecutive OFDM symbols. As in [7], we have configured the system to have $N_t=4$ and $N_r=2$ transmit and receive antennae, respectively. The simulations have been conducted using channel model E-NLOS as defined in [13]. This channel profile corresponds to a large office environment and it is made of 38 independent paths distributed among 4 clusters and it has an rms delay spread of 100 ns. Two different spatial configurations have been tested by considering different numbers of spatial streams and particular selections of the spatial spreading matrix. In all simulations, an ML-based turbo reception scheme (with 2 iterations) as the one introduced in the previous section was employed with the number of considered candidates set to $N_{cand}=64$. For comparison, the results obtained with hard and soft non-iterative Viterbi decoding will also be shown. In the first spatial configuration, the four antennas are configured as two CDD systems (CDD factor $\Lambda=2$) operating in parallel. It is easy to see that this setup can be achieved with a choice of spatial spreading matrix given by:

$$W = \begin{pmatrix} 1 & 0 \\ 1 & 0 \\ 0 & 1 \\ 0 & 1 \end{pmatrix} \tag{13}$$

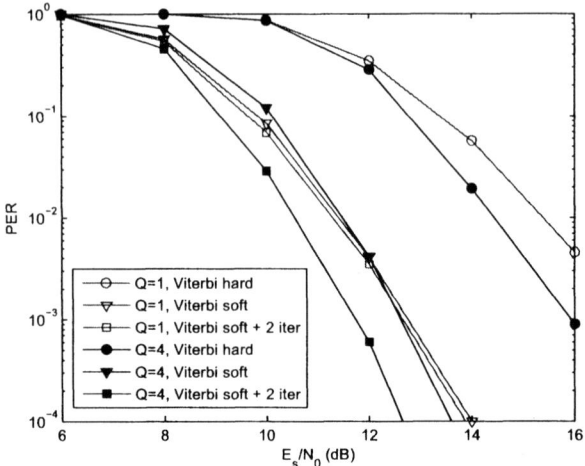

Fig. 5. CDD transmission. $N_t=4$, $N_r=2$, $N_s=1$. MCS 3.

and cyclic delays defined by $\delta_1 = \delta_3 = 0$ and $\delta_2 = \delta_4 = 32$. This system is an example of a combination of CDD and SDM, as two independent spatial streams are simultaneously transmitted over different antennae and each stream employs CDD. The system is configured to use QPSK modulation which, in combination with the chosen number of spatial streams and bandwidth, would correspond to the modulation coding scheme (MCS) number 9 in [8] achieving a transmission rate of 26 Mbps. Figure 4 shows the attained packet error rate (PER) when employing GO-CDM processing with $Q=4$ (black lines with solid markers) in comparison with the standard system without GO-CDM ($Q=1$, red lines with hollow markers). It can be seen that the configuration with GO-CDM clearly outperforms the conventional system when both employ iterative detection. In particular, gains between 1 and 2 dB are observed across the range of PER values of practical interest. Note that in the standard configuration, the iterative detection hardly provides any PER reduction over soft Viterbi decoding.

In the second tested configuration only one spatial stream is assumed to be present ($N_s=1$) and the four transmit antennae are used to perform CDD ($\Lambda=4$). Choosing $W = (1\ 1\ 1\ 1)^T$ and setting the cyclic delays to $\delta_1 = 0$, $\delta_2 = 16$, $\delta_3 = 32$ and $\delta_4 = 48$, configures the system as a 4-branch CDD setup. For this scenario, the modulation format has been set to 16-QAM (MCS number 3 in [8]) resulting again in a transmission bit rate of 26 Mbps. The results with ($Q=4$) and without ($Q=1$) GO-CDM are shown in Fig. 5. As in the previous spatial configuration, the simulations outcome leaves no doubt that the system with GO-CDM has the potential, using iterative detection, to significantly outperform the current standard.

6 Conclusions

This paper has proposed the introduction of group-orthogonal code-division multiplex within the context of the IEEE 802.11n (draft) standard. A reception scheme has been developed using an iterative ML-based detection procedure. Simulations results have shown that the combination of GO-CDM with iterative detection brings along important performance gains in terms of PER reduction irrespective of the spatial configuration used. Moreover, the GO-CDM extension is simple to incorporate to the current standard draft and it can easily be configured to be backward compatible with IEEE 802.11a/g.

Acknowledgments

This work has been supported in part by the MEC and FEDER under project MARIMBA (TEC2005-0997), Govern de les Illes Balears under project XISPES and grant PCTIB-2005GC1-09, and a Ramon y Cajal fellowship, Spain.

References

1. S. Kaiser, "OFDM code-division multiplexing in fading channels," *IEEE Trans. Commun.*, vol. 50, pp. 1266–1273, Aug. 2002.
2. N. Yee, J.-P. Linnartz, and G. Fettweis, "Multi-carrier CDMA in indoor wireless radio networks," in *IEEE PIMRC*, Yokohama (Japan), Sept. 1993, pp. 109–113.
3. X. Cai, S. Zhou, and G. Giannakis, "Group-orthogonal multicarrier CDMA," *IEEE Trans. Commun.*, vol. 52, no. 1, pp. 90–99, Jan. 2004.
4. F. Riera-Palou, G. Femenias, and J. Ramis, "On the design of group-orthogonal MC-CDMA systems," in *Proc. IEEE SPAWC*, Cannes (France), July 2006.
5. ——, "Downlink performance of group-orthogonal multicarrier systems," in *Proc. IFIP PWC*, Albacete (Spain), Sept. 2006, pp. 389–400.
6. F. Riera-Palou and G. Femenias, "Combining multicarrier code-division multiplex with cyclic delay diversity for future WLANs," in *Proc. IEEE SPAWC*, Helsinki (Finland), June 2007, (Accepted for publication).
7. W. Yan, S. Sun, Y. Li, and Y. Liang, "Transmit diversity schemes for MIMO-OFDM based wireless LAN systems," in *Proc. IEEE PIMRC*, Helsinki (Finland), Sept. 2006, pp. 1–5.
8. S. A. Mujtaba, "TGn Sync Proposal Technical Specification. doc.:IEEE 802.11-04/0889r7," Draft proposal, July 2005.
9. G. Bauch and J. Malik, "Cyclic delay diversity with bit-interleaved coded modulation in orthogonal frequency division multiple access," *IEEE Trans. Wireless Commun.*, vol. 8, pp. 2092–2100, Aug. 2006.
10. B. Hochwald and S. ten Brink, "Achieving near-capacity on a multiple-antenna channel," *IEEE Trans. Commun.*, vol. 51, pp. 389–399, March 2003.
11. U. Fincke and M. Pohst, "Improved methods for calculating vectors of short length in a lattice, including a complexity analysis," *Math. Comp.*, vol. 44, pp. 463–471, Apr. 1985.
12. J. Liu and J. Li, "Turbo processing for an OFDM-based MIMO system," *IEEE Trans. Wireless Commun.*, vol. 4, pp. 1988–1993, Sept. 2005.
13. V. Erceg, "Indoor MIMO WLAN Channel Models. doc.: IEEE 802.11-03/871r0," Draft proposal, Nov. 2003.

Saturation Throughput Analysis of IEEE 802.11g (ERP-OFDM) Networks

Krzysztof Szczypiorski, Jozef Lubacz

Warsaw University of Technology, Institute of Telecommunications,
ul. Nowowiejska 15/19, 00-665 Warsaw, Poland
{ksz, jl}@tele.pw.edu.pl

Abstract. This paper presents the saturation throughput analysis of IEEE 802.11g (ERP-OFDM) networks. The presented work is based on the Markov model previously introduced and validated by the authors in [7]. In the present paper the saturation throughput is evaluated in different channel conditions as a function of frame length.

Keywords: WLAN, IEEE 802.11, CSMA/CA, modeling

1 Introduction

The paper presents saturation throughput analysis of IEEE 802.11g (ERP-OFDM) networks. We use our general Markov-based model introduced and validated in [7]. The model presented in [7] is in line with the extensions of the basic Bianchi's model [1] which were proposed in [8] and [6]. The essential difference of the presented model with respect to the latter two is that it takes into account the effect of freezing of the stations' backoff timer along with the limitation of the number of retransmissions, maximum size of the contention window and the impact of transmission errors. The results presented in [7] showed that our model has good accuracy both in the case of error-free and error-prone channels. For both error-free and error-prone cases the proposed model shows better accuracy than the literature models with which it was compared (including: [1], [6] and [8]), especially for large number of stations.

The paper is organized as follows: Section 2 contains a summary of the model presented in [7]. Section 3 presents the saturation throughput analysis based on the given model. Finally, Section 4 presents essential conclusions.

2 The Model

We consider saturated conditions: stations have no empty queues and there is always a frame to be sent. n stations compete for medium access (for $n=1$ only one station sends frames to other station which may only reply with ACK frame). Errors in the transmission medium are randomly distributed; this is the worst case for the *frame error rate* – FER. All stations have the same *bit error rate* (BER). All stations are in transmission range and there are no hidden terminals. Stations communicate in ad hoc mode (BSS – *Basic Service Set*) with basic access method. All stations use the same physical layer (PHY). The transmission data rate R is the same and constant for all stations. All frames are of constant length L. Only data frames and ACK frames are exchanged. Collided frames are discarded – the capture effect ([5]) is not considered.

The saturation throughput S is defined as in [1]:

$$S = \frac{E[DATA]}{E[T]} \quad (1)$$

where $E[DATA]$ is the mean value of the successfully transmitted payload, and $E[T]$ is the mean value of the duration of the following *channel states* ([7]):

T_I – idle slot,
T_S – successful transmission,
T_C – transmission with collision,
T_{E_DATA} – unsuccessful transmission with data frame error,
T_{E_ACK} – unsuccessful transmission with ACK error.

Above channel states depend on:

T_{PHYhdr} – duration of a PLCP (*PHY Layer Convergence Procedure*) preamble and a PLCP header,
T_{DATA} – duration to transmit a data frame,
T_{ACK} – duration to transmit an ACK frame,
T_{SIFS} – duration of SIFS (*Short InterFrame Space*),
T_{DIFS} – duration of DIFS (*DCF InterFrame Space*),
T_{EIFS} – duration of EIFS (*Extended InterFrame Space*).

The relation of the saturation throughput to physical channel characteristics is calculated similarly as in [6]:

$$\begin{cases} T_I = \sigma \\ T_S = 2T_{PHYhdr} + T_{DATA} + 2\delta + T_{SIFS} + T_{ACK} + T_{DIFS} \\ T_C = T_{PHYhdr} + T_{DATA} + \delta + T_{EIFS} \\ T_{E\_DATA} = T_{PHYhdr} + \delta + T_{DATA} + T_{EIFS} \\ T_{E\_ACK} = T_S \end{cases} \quad (2)$$

where σ is the duration of the idle slot (*aSlotTime* [2]) and δ is the propagation delay.

For OFDM (*Orthogonal Frequency Division Multiplexing*) PHY, i.e. 802.11a [3] and 802.11g [4]:

$$T_{ACK} = T_{symbol} \left\lceil \frac{L_{SER} + L_{TAIL} + L_{ACK}}{N_{BpS}} \right\rceil \quad (3)$$

$$T_{DATA} = T_{symbol} \left\lceil \frac{L_{SER} + L_{TAIL} + L_{DATA}}{N_{BpS}} \right\rceil \quad (4)$$

where:

T_{symbol} – duration of a transmission symbol,
L_{SER} – OFDM PHY layer SERVICE field size,
L_{TAIL} – OFDM PHY layer TAIL fields size,

N_{BpS} – number of encoded bits per one symbol,
L_{ACK} – size of an ACK frame,
L_{DATA} – size of a data frame.

Values of σ, T_{PHYhdr}, T_{SIFS}, T_{DIFS}, T_{EIFS}, T_{symbol}, N_{BpS}, L_{SER}, and L_{TAIL} are defined in accordance with the 802.11 standard ([2], [3], or [4]).

Probabilities corresponding to states of the channel are denoted as follows:
P_I – probability of idle slot,
P_S – probability of successful transmission,
P_C – probability of collision,
P_{E_DATA} – probability of unsuccessful transmission due to data frame error,
P_{E_ACK} – probability of unsuccessful transmission due to ACK error.

Let τ be the probability of frame transmission, p_{e_data} the probability of data frame error and p_{e_ACK} the probability of ACK error. These are related to channel state probabilities as follows:

$$\begin{cases} P_I = (1-\tau)^n \\ P_S = n\tau(1-\tau)^{n-1}(1-p_{e\_data})(1-p_{e\_ACK}) \\ P_C = 1-(1-\tau)^n - n\tau(1-\tau)^{n-1} \\ P_{E\_DATA} = n\tau(1-\tau)^{n-1} p_{e\_data} \\ P_{E\_ACK} = n\tau(1-\tau)^{n-1}(1-p_{e\_data})p_{e\_ACK} \end{cases} \quad (5)$$

The saturation throughput S equals

$$S = \frac{P_S L_{pld}}{T_I P_I + T_S P_S + T_C P_C + T_{E\_DATA} P_{E\_DATA} + T_{E\_ACK} P_{E\_ACK}} \quad (6)$$

where L_{pld} is MAC (*Medium Access Control*) payload size and $L_{pld} = L - L_{MAChdr}$, where L_{MAChdr} is the size of the MAC header plus the size of FCS (*Frame Checksum Sequence*).

S can be normalized to data rate R (called normalized S):

$$\overline{S} = \frac{S}{R} \quad (7)$$

where

$$R = \frac{N_{BpS}}{T_{symbol}} \quad (8)$$

Let $s(t)$ be a random variable describing DCF backoff stage at time t, with values from a set $\{0, 1, 2,...,m\}$. Let $b(t)$ be a random variable describing the value of the backoff timer at time t, with values from a set $\{0, 1, 2,..., W_i-1\}$. These random variables are dependent because the maximum value of the backoff timer depends on backoff stage:

$$W_i = \begin{cases} 2^i W_0, & i \leq m' \\ 2^{m'} W_0 = W_m, & i > m' \end{cases} \quad (9)$$

where W_0 is an initial size of contention window and m' is a maximum number by which the contention window can be doubled; m' can be both greater and smaller than m and also equal to m. W_0 and $W_{m'}$ depend on CW_{min} and CW_{max} [2]:

$$W_0 = CW_{min} + 1 \tag{10}$$

$$W_{m'} = CW_{max} + 1 = 2^{m'} W_0 \tag{11}$$

The two-dimensional process $(s(t), b(t))$ will be analyzed with an embedded Markov chain (in steady state) at time instants at which the channel state changes. Let (i,k) denote the state of this process. The one-step conditional state transition probabilities will be denoted by $P = (\cdot, \cdot | \cdot, \cdot)$.

Let p_f be the probability of transmission failure and p_{coll} the probability of collision. The non-null transition probabilities are determined as follows:

$$\begin{cases} P(i,k \mid i,k+1) = 1 - p_{coll}, & 0 \le i \le m, 0 \le k \le W_i - 2 \\ P(i,k \mid i,k) = p_{coll}, & 0 \le i \le m, 1 \le k \le W_i - 1 \\ P(0,k \mid i,0) = (1 - p_f)/W_0, & 0 \le i \le m - 1, 0 \le k \le W_0 - 1 \\ P(i,k \mid i-1,0) = p_f/W_i, & 1 \le i \le m, 0 \le k \le W_i - 1 \\ P(0,k \mid m,0) = 1/W_0, & 0 \le k \le W_0 - 1 \end{cases} \tag{12}$$

Let $b_{i,k}$ be the probability of state (i,k). It can be shown that:

$$b_{i,0} = p_f \cdot b_{i-1,0} \tag{13}$$

$$b_{i,0} = p_f^{\,i} \cdot b_{0,0} \tag{14}$$

and

$$b_{i,k} = \begin{cases} \dfrac{W_i - k}{W_i(1 - p_{coll})} p_f^{\,i} \cdot b_{0,0}, & 0 < k \le W_i - 1 \\ p_f^{\,i} \cdot b_{0,0}, & k = 0 \end{cases} \tag{15}$$

From:

$$\sum_{i=0}^{m} \sum_{k=0}^{W_i - 1} b_{i,k} = 1 \tag{16}$$

And

$$\sum_{i=0}^{m} b_{i,0} = b_{0,0} \frac{1 - p_f^{\,m+1}}{1 - p_f} \tag{17}$$

we get

$$b_{0,0}^{-1} = \begin{cases} \dfrac{(1-p_f)W_0(1-(2p_f)^{m+1}) - (1-2p_f)(1-p_f^{\,m+1})}{2(1-2p_f)(1-p_f)(1-p_{coll})} + \dfrac{1-p_f^{\,m+1}}{1-p_f}, & m \le m' \\ \dfrac{\Psi}{2(1-2p_f)(1-p_f)(1-p_{coll})} + \dfrac{1-p_f^{\,m+1}}{1-p_f}, & m > m' \end{cases} \tag{18}$$

where

$$\Psi = (1-p_f)W_0(1-(2p_f)^{m'+1}) - (1-2p_f)(1-p_f^{\,m'+1}) + W_0 2^{m'} p_f^{\,m'+1}(1-2p_f)(1-p_f^{\,m-m'}) \tag{19}$$

The probability of frame transmission τ is equal to:

$$\tau = \sum_{i=0}^{m} b_{i,0} =$$

$$= \begin{cases} \left(\dfrac{(1-p_f)W_0(1-(2p_f)^{m+1})-(1-2p_f)(1-p_f^{m+1})}{2(1-2p_f)(1-p_f)(1-p_{coll})} + \dfrac{1-p_f^{m+1}}{1-p_f} \right)^{-1} \dfrac{1-p_f^{m+1}}{1-p_f}, & m \leq m' \\ \left(\dfrac{\Psi}{2(1-2p_f)(1-p_f)(1-p_{coll})} + \dfrac{1-p_f^{m+1}}{1-p_f} \right)^{-1} \dfrac{1-p_f^{m+1}}{1-p_f}, & m > m' \end{cases} \quad (20)$$

For $p_{coll}=0$ the above solution is the same as presented in [6].
The probability of transmission failure

$$p_f = 1-(1-p_{coll})(1-p_e) \quad (21)$$

where p_e is the frame error probability:

$$p_e = 1-(1-p_{e\_data})(1-p_{e\_ACK}) \quad (22)$$

where p_{e_data} is FER for data frames and p_{e_ACK} is FER for ACK frames. p_{e_data} and p_{e_ACK} can be calculated from bit error probability (i.e. BER) p_b:

$$p_{e\_data} = 1-(1-p_b)^{L_{data}} \quad (23)$$

$$p_{e\_ACK} = 1-(1-p_b)^{L_{ACK}} \quad (24)$$

The probability of collision

$$p_{coll} = 1-(1-\tau)^{n-1} \quad (25)$$

Finally

$$p_f = 1-(1-p_{coll})(1-p_e) = 1-(1-\tau)^{n-1}(1-p_e) \quad (26)$$

Equations (20) and (26) form a non-linear system with two unknown variables τ and p_f which may be solved numerically.

3 The Analysis

All diagrams presented in this section show values of the normalized saturation throughput. All calculations were made for $n \in \{1, 2, 3, 4, 5, 10, 15, 20, 30, 40\}$. For $L=1000$ bytes frame the following values of *BER* were used $\{10^{-4}, 5 \cdot 10^{-5}, 10^{-5}, 5 \cdot 10^{-6}, 10^{-6}, 0\}$. For $L \in \{100, 250, 500, 1000, 1500, 2000\}$ bytes $BER \in \{0, 10^{-5}, 10^{-4}\}$. We considered IEEE 802.11g – ERP-OFDM i.e. "g" only mode and data rate $R=54$ Mbps (with the exception of the last diagram, which consists evaluation of R impact on S).

Fig. 1 presents normalized S as a function of n for $L=1000$ bytes frame and different values of *BER* (Table 1). Along with increasing value of BER saturation throughput S is reduced. Also maximum of S is shifted from $n=2$ (for two smallest BER 0 and 10^{-6}), through $n=3$ for BER $5 \cdot 10^{-6}$ and 10^{-5}, $n=5$ and $5 \cdot 10^{-5}$, into $n=10$ for $BER=10^{-4}$. Along with increasing value of BER presented curves are flattened. For a given *BER* reduction of *S* with increase of *n* is related to increasing number of collision in medium. Reduction of *S* between *BER*=0 and *BER*=10^{-6} is very small.

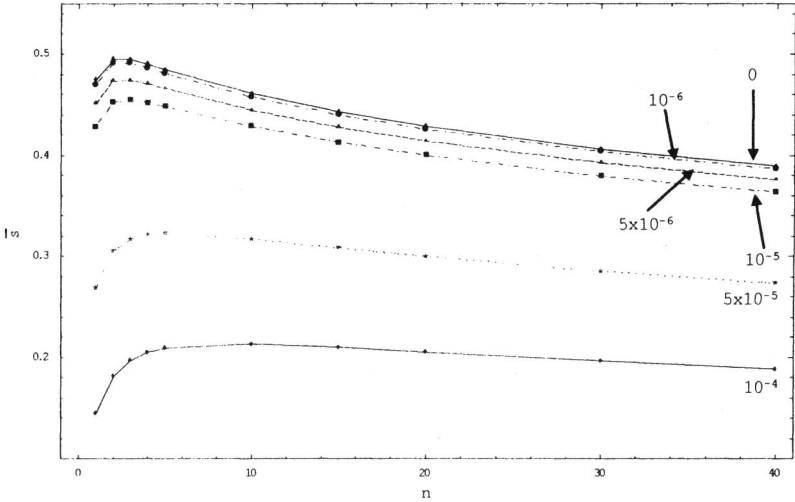

Fig. 1. Normalized *S* as a function of *n* – for *L*=1000 bytes and different values of *BER*.

Fig. 2 presents normalized *S* as a function of *n* for different values of frame length and *BER*=0 (Table 2). For a given *n* along with increase of frame length the value of *S* increases. Maximum value of *S* depends on *n* and falls into [2;5].

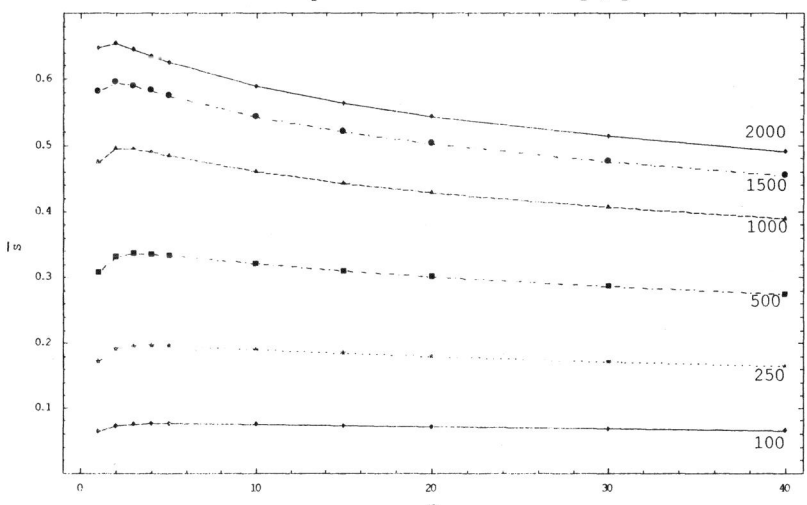

Fig. 2. Normalized *S* as a function of *n* – for different values of frame length and *BER*=0.

Fig. 3 shows normalized *S* as a function of *n* for different values of frame length and *BER*=10^{-5} (Table 3). For a given *n* along with increase of frame length the value of S increases. Maximum value of S depends on n and falls into [2;5]. In comparison to $BER=0$ S is reduced because of channel errors.

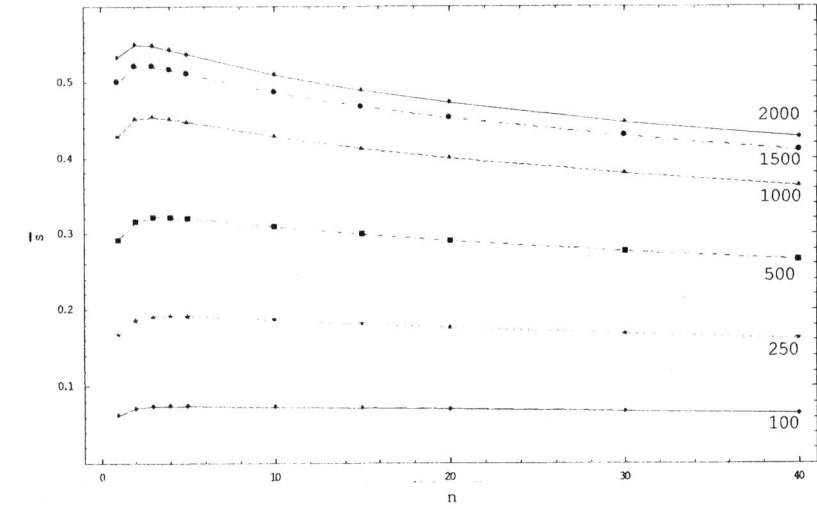

Fig. 3. Normalized S as a function of n – for different values of frame length and $BER=10^{-5}$.

Fig. 4 presents normalized S as a function of n for different values of frame length and $BER=10^{-4}$ (Table 4). For a given n along with increase of frame length the value of S increases but only to the limited level; S decreases for frames greater than 500 bytes and for $n\geq 3$. This is an influence of increasing FER with length of frame.

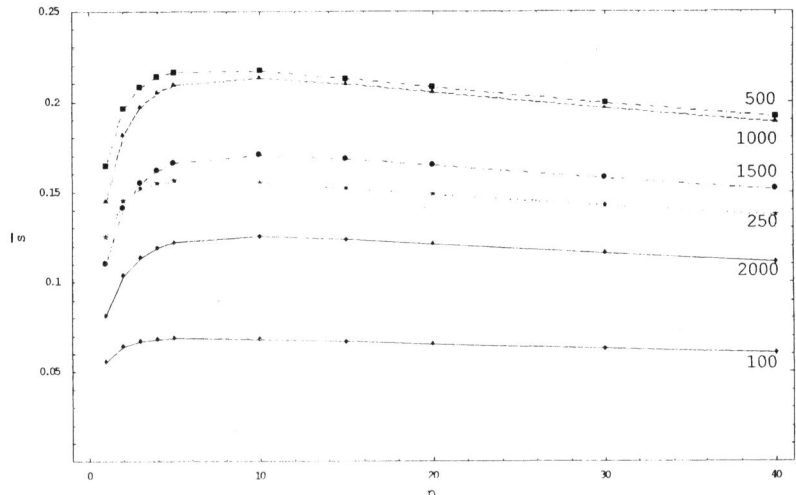

Fig. 4. Normalized S as a function of n – for different values of frame length and $BER=10^{-4}$.

Finally, we evaluate normalized S as a function of n for different IEEE 802.11g data rates $R \in \{6, 9, 12, 18, 24, 36, 48, 54\}$ Mbps. Results show (Fig. 5 and Table 5) that the channel usage for lower rates is better than for upper; for 6 Mbps and $n=1$ is close to 85%, while for 54 Mbps and $n=1$ is close to 47%.

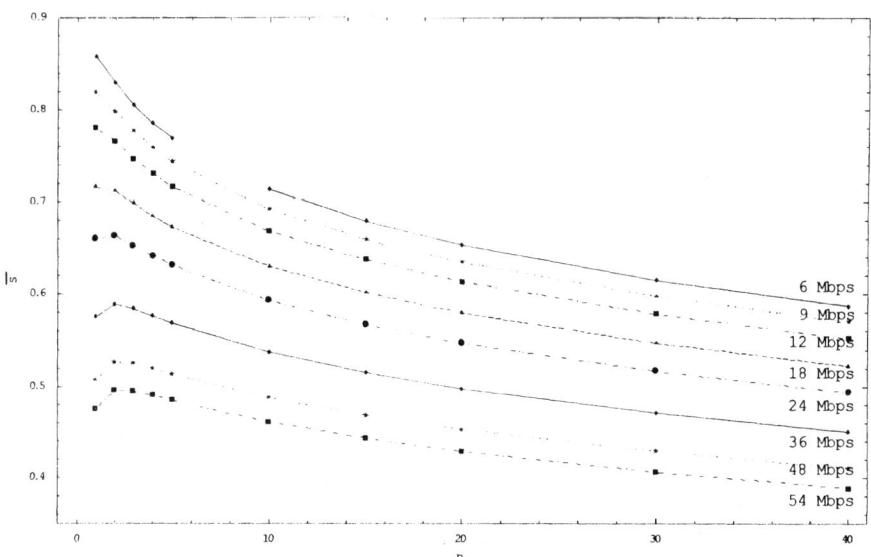

Fig. 5. Normalized S as a function of n – for $L=1000$ bytes, $BER=0$, and different values of R.

Table 1. Normalized S as a function of n – for $L=1000$ bytes and different values of BER.

BER n	10^{-4}	$5 \cdot 10^{-5}$	10^{-5}	$5 \cdot 10^{-6}$	10^{-6}	0
1	0.1446	0.2688	0.4281	0.4510	0.4697	0.4745
2	0.1816	0.3054	0.4521	0.4734	0.4909	0.4953
3	0.1971	0.3170	0.4542	0.4740	0.4904	0.4945
4	0.2050	0.3213	0.4517	0.4705	0.4860	0.4899
5	0.2092	0.3226	0.4479	0.4660	0.4808	0.4845
10	0.2131	0.3171	0.4285	0.4443	0.4574	0.4607
15	0.2097	0.3082	0.4126	0.4275	0.4396	0.4427
20	0.2052	0.2998	0.3997	0.4138	0.4254	0.4284
30	0.1963	0.2853	0.3792	0.3925	0.4034	0.4061
40	0.1885	0.2735	0.3631	0.3758	0.3863	0.3889

Table 2. Normalized S as a function of n – for different values of frame length and $BER=0$.

L n	100	250	500	1000	1500	2000
1	0.0637	0.1717	0.3074	0.4745	0.5808	0.6471
2	0.0723	0.1904	0.3319	0.4953	0.5949	0.6541
3	0.0748	0.1949	0.3360	0.4945	0.5896	0.6450
4	0.0756	0.1958	0.3354	0.4899	0.5817	0.6346
5	0.0758	0.1954	0.3334	0.4845	0.5738	0.6249
10	0.0744	0.1899	0.3208	0.4607	0.5422	0.5880

L\n	100	250	500	1000	1500	2000
15	0.0725	0.1842	0.3098	0.4427	0.5197	0.5626
20	0.0708	0.1793	0.3008	0.4284	0.5021	0.5430
30	0.0678	0.1712	0.2862	0.4061	0.4751	0.5132
40	0.0653	0.1646	0.2747	0.3889	0.4544	0.4905

Table 3. Normalized S as a function of n – for different values of frame length and $BER=10^{-5}$.

L\n	100	250	500	1000	1500	2000
1	0.0629	0.1668	0.2909	0.4281	0.5004	0.5330
2	0.0715	0.1855	0.3162	0.4521	0.5211	0.5502
3	0.0740	0.1902	0.3212	0.4542	0.5208	0.5484
4	0.0749	0.1914	0.3213	0.4517	0.5166	0.5431
5	0.0751	0.1912	0.3199	0.4479	0.5114	0.5372
10	0.0738	0.1862	0.3089	0.4285	0.4872	0.5105
15	0.0719	0.1808	0.2988	0.4126	0.4682	0.4900
20	0.0702	0.1760	0.2902	0.3997	0.4529	0.4736
30	0.0673	0.1681	0.2763	0.3792	0.4289	0.4480
40	0.0648	0.1617	0.2652	0.3631	0.4103	0.4281

Table 4. Normalized S as a function of n – for different values of frame length and $BER=10^{-4}$.

L\n	100	250	500	1000	1500	2000
1	0.0556	0.1251	0.1643	0.1446	0.1103	0.0812
2	0.0643	0.1450	0.1960	0.1816	0.1412	0.1038
3	0.0671	0.1519	0.2081	0.1971	0.1548	0.1137
4	0.0682	0.1549	0.2137	0.2050	0.1619	0.1190
5	0.0687	0.1561	0.2165	0.2092	0.1659	0.1219
10	0.0682	0.1553	0.2172	0.2131	0.1705	0.1254
15	0.0667	0.1520	0.2129	0.2097	0.1682	0.1237
20	0.0652	0.1485	0.2081	0.2052	0.1647	0.1212
30	0.0626	0.1423	0.1992	0.1963	0.1576	0.1159
40	0.0604	0.1370	0.1916	0.1885	0.1512	0.1111

Table 5. Normalized S as a function of n – for $L=1000$ bytes, $BER=0$, and different values of R.

R\n	6	9	12	18	24	36	48	56
1	0.8574	0.8186	0.7793	0.7158	0.6592	0.5752	0.5070	0.4745
2	0.8290	0.7979	0.7651	0.7120	0.6629	0.5892	0.5267	0.4953
3	0.8043	0.7762	0.7463	0.6978	0.6524	0.5840	0.5248	0.4945
4	0.7844	0.7583	0.7301	0.6844	0.6412	0.5762	0.5193	0.4899
5	0.7681	0.7432	0.7162	0.6725	0.6309	0.5684	0.5133	0.4845
10	0.7136	0.6920	0.6682	0.6297	0.5928	0.5370	0.4872	0.4607
15	0.6791	0.6591	0.6370	0.6013	0.5668	0.5147	0.4679	0.4427
20	0.6534	0.6345	0.6135	0.5796	0.5468	0.4973	0.4526	0.4284
30	0.6152	0.5978	0.5784	0.5470	0.5165	0.4706	0.4288	0.4061
40	0.5867	0.5703	0.5520	0.5224	0.4936	0.4501	0.4105	0.3889

4 Conclusions

The presented analysis shows that the saturation throughput essentially depends on bit error rate – for a given number of stations and the length of frame the lower *BER*, the greater is *S*. This is an influence of channel error on the effective volume of transmitted data. Increasing the number of stations implies collisions and finally reduces the value of saturation throughput. The saturation throughput depends on frame error rate because *FER* is a function of *BER* and *L*.

References

1. Bianchi, G.: Performance Analysis of the IEEE 802.11 Distributed Coordination Function. IEEE Journal on Selected Areas in Communications, Vol. 18, No. 3 (2000) 535-547
2. IEEE 802.11, 1999 Edition (ISO/IEC 8802-11: 1999) IEEE Standards for Information Technology – Telecommunications and Information Exchange between Systems – Local and Metropolitan Area Network – Specific Requirements – Part 11: Wireless LAN Medium Access Control (MAC) and Physical Layer (PHY) Specifications (1999)
3. IEEE 802.11a-1999 (8802-11:1999/Amd 1:2000(E)), IEEE Standard for Information technology – Telecommunications and information exchange between systems – Local and metropolitan area networks – Specific requirements – Part 11: Wireless LAN Medium Access Control (MAC) and Physical Layer (PHY) specifications – Amendment 1: High-speed Physical Layer in the 5 GHz band (1999)
4. IEEE 802.11g-2003 IEEE Standard for Information technology – Telecommunications and information exchange between systems – Local and metropolitan area networks – Specific requirements – Part 11: Wireless LAN Medium Access Control (MAC) and Physical Layer (PHY) specifications Amendment 4: Further Higher-Speed Physical Layer Extension in the 2.4 GHz Band (2003)
5. Kochut, A., Vasan, A., Shankar, A., Agrawala, A.: Sniffing Out the Correct Physical Layer Capture Model in 802.11b. In: 12th IEEE International Conference on Network Protocols (ICNP 2004), Berlin (2004)
6. Ni, Q., Li, T., Turletti, T., Xiao, Y.: Saturation Throughput Analysis of Error-Prone 802.11 Wireless Networks. Wiley Journal of Wireless Communications and Mobile Computing (JWCMC), Vol. 5, Issue 8 (2005) 945-956
7. Szczypiorski, K., Lubacz, J.: Performance Evaluation of IEEE 802.11 DCF Networks. In: 20th International Teletraffic Congress (ITC-20), Ottawa, Canada, June 17-21, 2007; Lecture Notes in Computer Science (LNCS) 4516, Springer-Verlag Berlin Heidelberg (2007) 1082-1093
8. Wu, H., Peng, Y., Long, K., Cheng, S., Ma, J.: Performance of Reliable Transport Protocol over IEEE 802.11 Wireless LAN: Analysis and Enhancement. In: IEEE INFOCOM'02 (2002)

A Tiered Mobility Management Solution for Cellular/WLAN Integrated Networks with Low Handoff Delay

[1]Iti Saha Misra, [2]Sibaram Khara, and [3]Debashis Saha

[1]Dept. of Electronics and Telecommunication Engg, Jadavpur University, Kolkata, India
[2]Dept. of ECE, College of Engineering and Management, Kolaghat, Midnapur, India
and [3]MIS Group, IIM Calcutta, India,
[1]itimisra@cal.vsnl.net.in [2]sianba@rediffmail.com, and
[3]ds@iimcal.ac.in

Abstract. Demand for the Internet services through GPRS/3G networks is proliferating day-by-day. But low bit rate puts constraints on multimedia or mobile business services. In contrast, Wireless Local Area Networks (WLANs) provide high bit rate with low deployment cost. Therefore, integration of these two complementary systems would be the focus of next generation wireless networks. However mobility and handoff management become the key issues for such integration. If a cellular operator deploys as Operator's IP network (OIN), it will be connected to GPRS at Gi interface and to the Internet through firewall. Alternatively, the Internet can be connected directly at Gi interface and OIN can be connected to the Internet. These connection mechanisms restrict on scalability due to large packet forwarding delay. In this paper, we propose a new way to integrate GPRS/UMTS, OIN and the Internet for better scalability and smaller packet delay. We deploy one gateway for users of cellular (GPRS/UMTS) and WLAN host. The gateway is in OIN and is connected directly with GGSN at Gi interface. The home agent functionalities are distributed in two level hierarchies within the OIN to reduce the packet and handoff delay. The proposed architecture would be useful for large network size.

Keywords: 3G/WLAN integration, GPRS, handoff, mobility management, MIP, UMTS, and WLAN.

1 Introduction

During last couple of years an increasing number of Internet users, in conjunction with IP based services e.g. e-commerce and applications (e.g., WWW and email), has created a huge demand for wideband access to the Internet. In near future, it is expected that next generation Internet will be the combination of different wireless access technologies, each having different data rates, different coverage areas and different architecture. The interest in cellular/WLAN integration is proliferating for such next generation networks. Design of multimode terminal and mobility management are the key issues of inter-working architectures [4]. The protocol stacks of cellular and WLAN systems can be converged in single dual mode terminal equipment [1, 2, 3, 5, 6]. A software-defined radio (SDR) can dynamically configure itself depending upon best available access network. Tight coupling and loose coupling are two basic methods of cellular/WLAN integration. Although faster handoff is achieved in tight coupling architecture, it incurs design complexity and congestion in cellular network [5, 7, 8]. Other hand, loose coupling architecture will be more promising for next generation all-IP based heterogeneous network [9, 10].

General Packet Radio Services (GPRS) operator can earn huge revenue deploying its own WLAN as operator's IP network. For example, in [1, 2, 3], an operator's IP Network (OIN) is connected at Gi interface with a GPRS network. The Internet is connected to the OIN through firewall. Therefore, the Internet connection for pure GPRS host takes place through OIN. In such an architecture, the hop distance between GPRS host and correspondent node (CN) increases. Obviously, this increases the packet delay compared to the delay when Internet is directly connected to Gi interface. All traffic travel through core of OIN and may suffer extra delay due to congestion in OIN. The traffic related to cross connection between two OIN would usually be affected by the Internet.

When mobile station (MS) is in GPRS or UMTS network, all incoming packets are routed to GGSN by external IP network. If MS moves to WLAN of OIN, GGSN is the responsible node to forward the packets to IP network. This mandates the deployment of Home Agent (HA) functionality at GGSN so that it can forward IP packet using mobile IP (MIP) based tunneling through OIN. In such case GGSN becomes the unique gateway for GPRS as well as OIN. Therefore, when MS is in WLAN, all incoming packets come via GGSN. GGSN also handles both incoming and outgoing traffic of GPRS service. This architecture increases both packet and handoff delay. To reduce this delay, HA functionality is deployed in OIN. All incoming packets are routed to HA by external IP network. All traffic for GPRS MS is also routed to HA. If HA does not find any MIP binding for destination MS, it may give packets to GGSN. The subscribers having only GPRS connection will also received packets through HA.

We take up the scenario when MS moves within the WLANs under OIN. If MS changes the access router, the MIP based mobility management is always performed with HA. MIP based handoff delay increases with traffic load in OIN.

From the review of related works we find that a better inter-working architecture is very much in demand such that traffic of OIN and Internet do not affect each other. We identify a few more important points in view of *user's requirement* and *operator's interest* in cellular/WLAN integration. A cellular user will prefer specific options for WLAN service at hot spots. One, **user may prefer for WLAN service with same subscription** without changing terminal equipment. Two, **user may opt for selective WLAN services of multiple networks**. Three, a user with same terminal can have **separate subscription for WLAN** service of other network.

Again, a cellular operator may wish to have following options. One, it can provide **its own WLAN service** for its cellular users. Two, the operator may have **agreement with other networks** so that it can provide their WLAN service to their subscriber.

We address this space and propose an integrated network that fulfills the **requirements of users and operators**. Moreover, the problem of Mobile IP (MIP) [10] is also taken care of and then we propose tiered mobility agent based solution for fast and seamless handoff intensive environments.

To validate the proposed architecture, extensive simulation results are provided based on ns-2.26 simulation environment. We emphasize on the design of network model to meet the desired goal for seamless connectivity of large networks with varied mobility patterns of the users and for increasing number of mobile users. Also a performance comparison of the proposed protocol is made with Mobile IP.

2 Proposed Network Architecture

Figure 1 shows the inter-working architecture between GPRS, UMTS, Internet and OIN. We deploy a global gateway router for each operator. This router is the gateway for OIN. It is also connected to GPRS or UMTS at Gi interface. This node also provides HA functionality for entire operator's network. We say it a global HA (GHA). GPRS/UMTS operator's IP network and Internet are also connected to it through separate IP level port. GHA provides HA functionality for GPRS MS while moving into IP networks. GHA of GPRS operator and that of UMTS operator can be interconnected for inter-working between GPRS and UMTS operator's IP network. The inter-working networks would be expanded interconnecting other GHA(s) too.

Figure 1 also shows the different roaming scenario for cellular MS. A GPRS MS can move from cellular to own OIN, to Internet or to UMTS OIN. The MS can also move from GPRS OIN to Internet and from there to UMTS OIN. Similarly, UMTS MS can move to its own OIN, to Internet or to GPRS OIN. It can also move from UMTS OIN to Internet and from there to GPRS OIN. Concerned GHA provides HA functionality for mobility management in each of the roaming scenario.

Figure 2 shows the architectural details of OIN. This network provides WLAN service to its cellular users at hot spots. Each OIN has hierarchical structure and connected to cellular network through GHA at Gi interface. We deploy another HA functionality in OIN. These HAs are implemented in lower hierarchy of GHA. We say it as middle HA (MHA). In hierarchical architecture they are deployed in between foreign agent (FA) and GHA. FAs are essentially the access router controlled by MHA. The Operator's WLAN IP network is divided into number of domains. Within a domain, MHA provides the HA functionality. Domains are again divided into sub-domains under the control of a FA or AR. An access point (AP) is placed at small WLAN area of each sub-domain under the control of AR. FA and MHA are IPv4 routers with some additional functionality. They can act as proxy HA for their respective service areas.

Four types of mobility scenario are considered for the proposed cellular/WLAN integrated network as given below.

Case 1: Local mobility - across the APs, MS moves from one WLAN area to other

Case 2: Micro mobility - across the ARs, MS moves from one sub-domain to other

Case 3: Macro Mobility - across the MHAs, MS moves from one domain to other, and

Case 4: Global Mobility – across the GHAs, MS moves from one network to other.

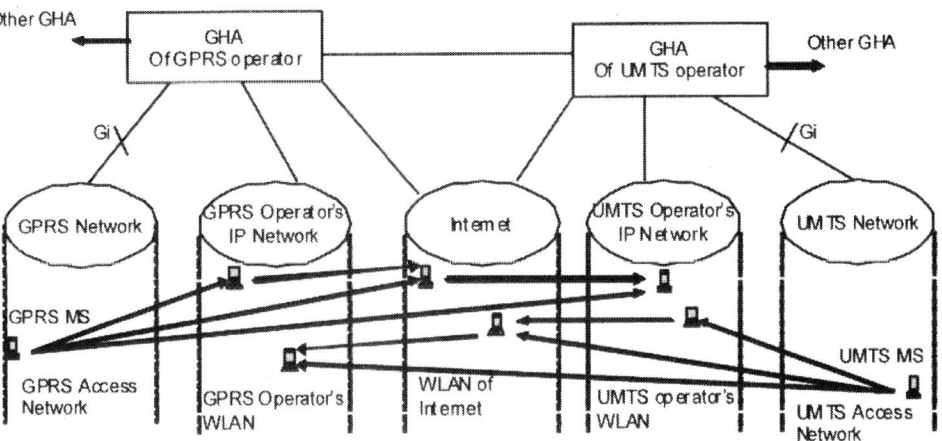

Fig 1: Proposed architecture of cellular (GPRS and UMTS) and WLAN interworking network.

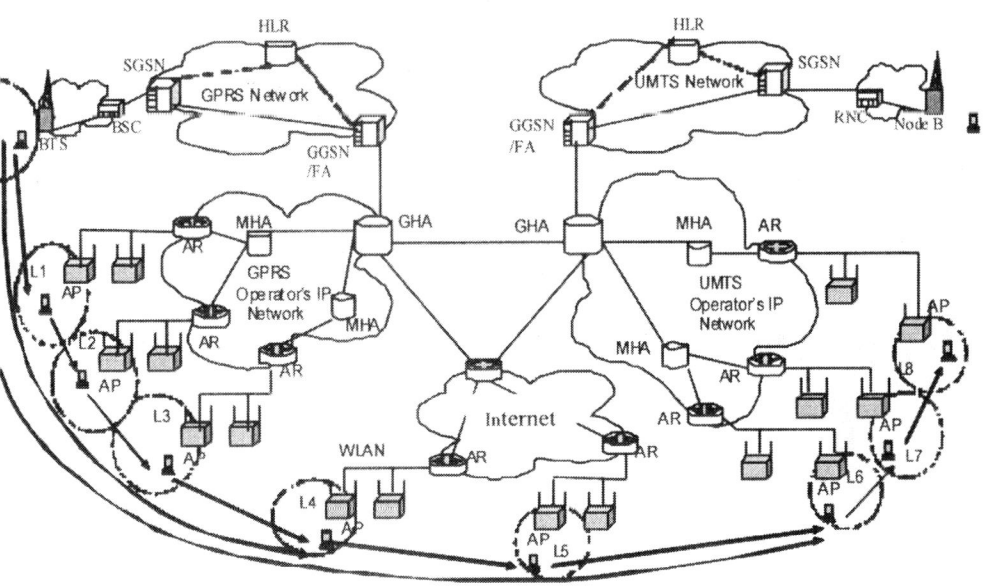

Fig 2: A complete network architecture using three level distributed agents in operator's IP network.

In global roaming scenario MS performs mobile IP based mobility signaling with HA at GHA level (example, when MS moves from location L0 to L1). During MIP based handoff signaling, MHA0 is updated with the authentication and service profile of the mobile terminal. Subsequently MHA0 can provide HA functionality if MS moves within the domain of MHA0. This distributed structure would reduce the registration update latency providing scalability to the network.

3 Mobility Management of the Proposed Architecture

The HA functionalities are imposed in GHAs and MHAs. GHA is deployed in gateway router at Gi interface. MHA can be configured in any IPv4 router in operator's IP network. Any cellular MS can have additional subscription for WLAN service at hot spots. HA contains the AAA (Authentication, Authorization and Accounting) profiles of WLAN service of cellular MS. GHA acts as gateway router for GPRS/UMTS and operator's IP network. The incoming packet destined for any cellular MS such as GPRS MS is routed to GHA by other network. It checks whether any mobility binding exists for destination MS. If it does not exist GHA assumes that MS might be under cellular coverage. It then sends the packet to GGSN through Gi interface. GGSN handles the IP packet usual way. The mobility entity MHA intercepts each IP packet; both control and data packets. On receipt of MIP registration request packet it checks whether MIP binding already exists in router. If not, it forwards the packet further up in hierarchical network. For example, when first time MS moves from location Lo to location L1, the MHA0 and GHA0 in operator IP network do not have any location information of such MS. Therefore MHA0 cannot provide HA functionality. It forwards the MIP registration request packet to GHA0. GHA0 updates the MIP binding records to map between home address and concerned MHA0. The MIP registration response packet is sent to concern MHA0 providing all AAA information. The MHA0 creates necessary binding information to correlate between home address and Foreign Agent (FA).

Each incoming data packet is intercepted by GHA before it reaches to GGSN. The destination address of each received data is checked by GHA. The destination address of data packet is the home address of the cellular MS. If binding exists for this home address, the GHA retrieves the address of responsible MHA. This data packet is tunneled to MHA using MIP. MHA retrieves the address of responsible FA and tunnel the data packet to it. FA sends the IP packet to destination usual way after decapsulation of the packets.

The implementation of GHA and MHA and their actual position in OIN are the prime task in proposed architecture. The implementation mechanism requires some changes in MIP specifications. The MIP registration request packet remains unchanged. But MIP registration **response packet must be modified** such that it carries the AAA context. Therefore, a new type of information is to be specified for response packet. The type field can be followed by length and the AAA context. The unused bit pattern can be assigned to designate the new type of information. In a particular hierarchical path, FA must be pre-configured with the address of its MHA. MHA is pre-configured with the address of GHA. FA always forwards the MIP registration request packet to its MHA. If MHA cannot provide HA functionality it forwards MIP registration packet to concern GHA. MS has always to send to MIP registration request to FA. MS should acquire the foreign agent care-of address. The MIP registration response packet is to be sent to the agent from which the request was received.

4 Simulation Results and Discussions

Creating MIPv4 platform in ns-2.26 simulates the proposed architecture. We develop GPRS tunneling protocol (GTP) to provide the bearer service of GPRS and UMTS for IP packets in cellular network. HA functionality is imposed to provide services needed by GHA and MHA locally. Initially, for MIP based simulation, GHA is replaced by HA and the MHA and AR act as normal IP routers. Then, for the proposed distributed architecture, GHA, MHA and AR are deployed with their respective functionalities. To evaluate the network performance, handoff delay is computed under different roaming scenarios. Typical parameter values used for simulation are provided in Table 1.

First, handoff delay is computed for the global and macro mobility scenario. In both scenario MS performs handoff signaling with GHA. A global mobility scenario occurs when MS moves from position L0 of GPRS to position L1 of OIN as shown in Figure 2. Macro mobility occurs when MS moves from one domain to another domain in OIN. For example such mobility happens when MS moves from position L2 to position L3 in OIN. For such mobility scenarios we compute the handoff delay in pure MIP based system and in GHA/MHA based system. It is seen that the handoff delay in GHA/MHA based system is 8.73% more than pure MIP based handoff delay. This is because of the user's service; subscription and authentication profiles are imposed in MIP response packet in GHA/MHA based procedure. Thus the size of MIP registration response packet from GHA is more than usual MIP response message. The extra handoff delay is tolerated for first time when global or macro mobility occurs. In such handoff process the MHA gets ready to provide HA functionality for subsequent movement of MS within same sub-domain. Users may frequently perform MIP based handoff signaling due to changing WLAN area. In such cases GHA/MHA based handoff procedure will be more fruitful for a long-term service.

Figure 3 shows the comparison between handoff delays in micro mobility scenario with varying packet size. In MIP based system the HA is implemented at GHA node position. In contrast to micro mobility for GHA/MHA based system; here always-macro mobility management is initiated, whenever MS moves within the IP networks. The handoff delay increases by 32ms when packet size is varied from 100 to 1000 bytes as compared to 28 ms in GHA/MHA based case under same mobility pattern. This effect will be more prominent with the increase of hop distance. Similar results are observed when UMTS MS moves under same environment. In case of a large IP network, the entire IP network domain is divided into sub-domains, which again will be divided into areas. Therefore, the micro and macro mobility scenarios may occur frequently. In such situation benefit of handoff delay may be achieved than the usual MIP based technique making the network scalable one.

We compute the handoff delay **under varying traffic load** in inter-working network. We increase the number of connection between GPRS and Internet hosts.

Table I: Typical values used for simulation

Quque type	DropTail
Interface Quque length	400
Link Layer Overhead	25μs
Link Layer delay	50μs
Adhoc routing	DSDV
WLAN Host bit rate	4 Mbps
GPRS host bit rate	144Kbps
UMTS host bit rate	2Mbps
Transmitting Power	0.5818w
Physical Channel frequency	2.4 Ghz
TCP Packet size in bytes	500
BS Omni Antenna height	1.5m
High speed channel between Two wired node	Duplex channel, 40Mbps

We assume that GPRS MSs at location L0 sends traffic to Internet hosts at location L5. All packets pass through GHA to Internet. We compute the handoff delay in micro mobility scenario, when MS moves from location L1 to L2. First we compute the handoff delay in usual MIP based system. Here GHA provides HA functionality. We observe that when GPRS connection is varied from 1 to 10, the handoff delay increases from 9.44% to 109% as shown in Figure 4. Under same mobility condition and connection environment, we compute the handoff delay in GHA/MHA based system. It is observed that handoff delay remains almost constant. The reason is that the MHA provides the HA functionality. The handoff signals are not dealt by GHA node. The GHA/MHA based handoff signals avoid the queuing and congestion effect of node GHA. The signals also travel shorter hop distance.

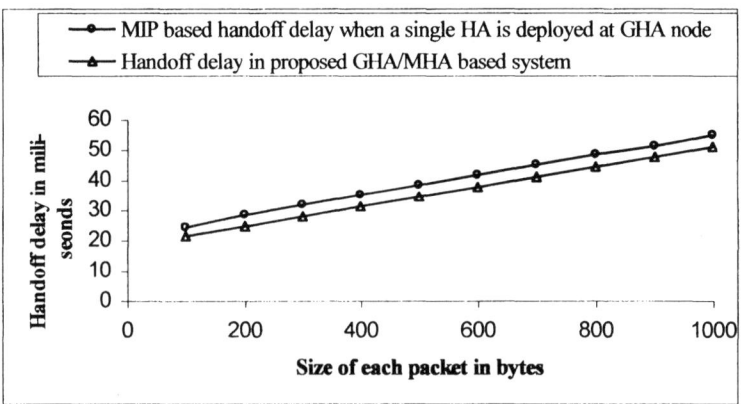

Fig. 3. Handoff vs. Packet size in micro-mobility scenario

Under same traffic environment the throughput of the system is also computed. We deploy a transmitting MS in location L1 and corresponding receiving MS at location L3. The receiving MS is employed mobility and enters at location L2. If receiving MS performs pure MIP based handoff with HA at GHA, then all incoming packets for receiving MS will come via GHA node. But if receiving node performs a GHA/MHA based handoff procedure, then all incoming packets will come via MHA. Because after handoff, MHA behaves as HA for receiving MS and it tunnels all packets to concern FA destined for receiving MS. We compute the throughput at receiving end under varying GPRS connection. It is seen that the throughput reduces by 35% when number of connections increases by 10 in MIP based system [Fig 5]. In GHA/MHA based system the throughput remains almost constant

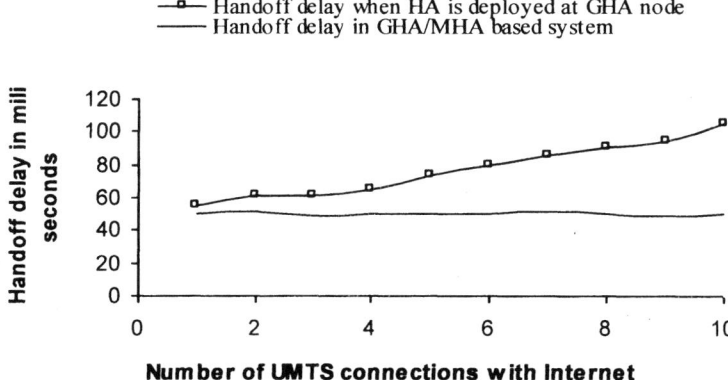

Fig. 4. Handoff Vs. Number of UMTS connections with Internet

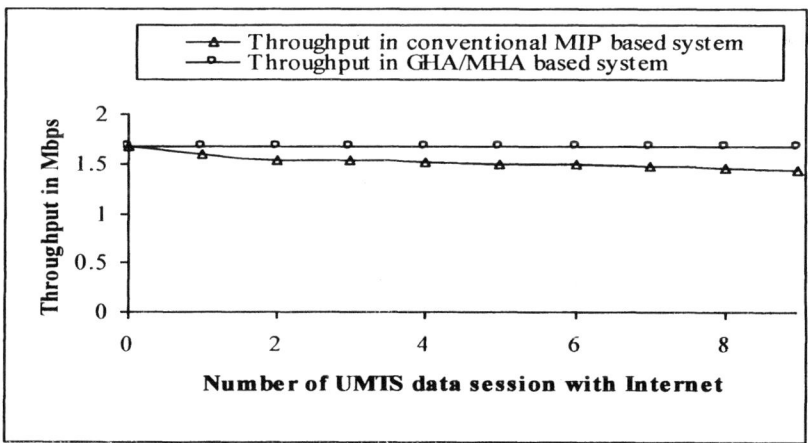

Fig. 5. Throughput vs. number of GPRS connection with Internet in MIP and GMA/HMA based architecture

The jittering of packets at receiving MS is also observed. The inter-arrival delay of received packets at location L2 for MIP based as well as GHA/MHA based system is computed. It is seen that this delay in pure MIP based system is more fluctuating than that in GHA/MHA based system [Fig 6 and 7]. When receiving MS is in location L2, the incoming packets are affected by the GPRS/Internet traffic at node GHA in MIP based system. This makes uneven packet delay. When receiving MS is in location L2, incoming packets do not suffer the effect of GPRS-Internet traffic.

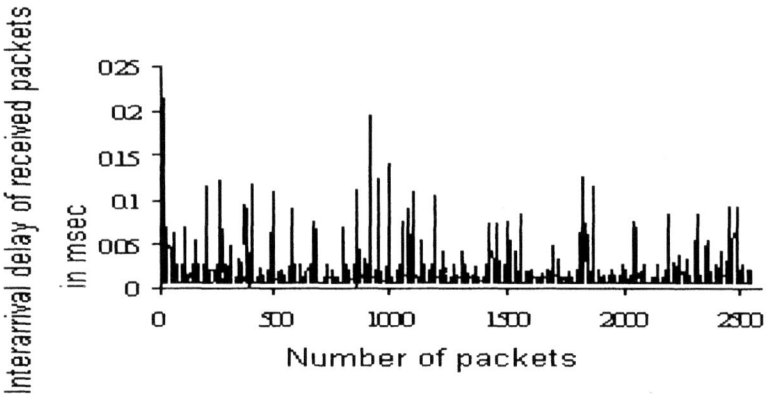

Fig. 6. Inter arrival delay of received packets in MIP based system when transmitting MS is in L1 and receiving MS is in L2

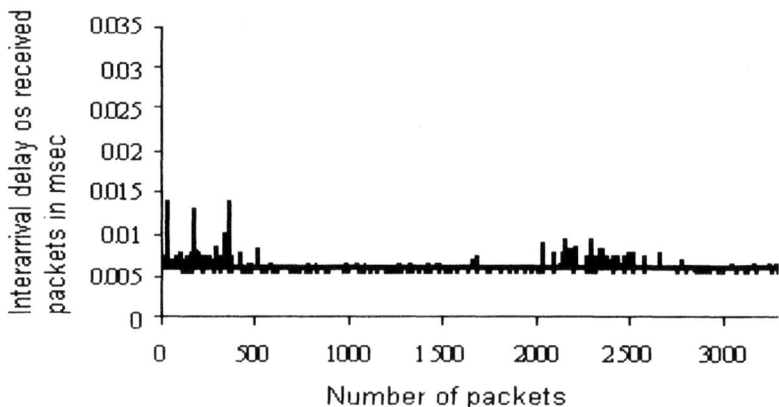

Fig. 7. Inter arrival delay of received packets in GHA/MHA based system when transmitting MS is in L1 and receiving MS is in L2.

5 Conclusion

In proposed cellular/WLAN integrated network, any other operator of GPRS or UMTS technology can join through separate GHA node. Thus architecture allows easy deployment of Operator's IP Networks. Operators can also provide separate corporate networks using other GHA node. In such integrated network, users can have subscription for sole GPRS/UMTS service or for sole WLAN service or for both services. Although as the initial vertical handoff time from cellular to WLAN increases by 5%, for subsequent movements at micro level, the handoff delay reduces substantially. In GHA/MHA based systems, if source and destination exist in the same locality, the traffic does not affect the performance of network outside the locality. Moreover, jitter in GHA/MHA based systems is smoother than that in MIP based systems. Thus the proposed integrated network architecture may be useful for future wireless networks with increased size and traffic.

Acknowledgement: Author Iti Saha Misra is thankful to AICTE for the financial support of this research work under AICTE, CAYT scheme, India.

References:

1. J. Ala-Laurila, J. Mikkonen, and J. Rinnemaa, "Wireless LAN Access Network Architecture for Mobile Operators", IEEE Communications Magazine, Novmber 2001, pp-82-89.
2. K. Salkintzis, C. Fors and R. Pazhyannur, "WLAN-GPRS Integration for Next-Generation Mobile Data Networks", IEEE Wireless Commun, Oct 2002, pp-112-124.
3. H. Luo, Z. Jiang, B. Kim, N. K. Shankaranarayan and P. Henry, "Integrating Wireless LAN and Cellular Data for the Enterprise" IEEE Comp Soc, March-April 2003, pp-25-33.
4. P. Mahonen, J. Riihijarvi, M. Petrova, and Z. Shelby, "Hop-by-Hop toward Future Mobile Broadband IP", IEEE Commun Magazine, Mar 2004, pp. 138-146.J. Clerk Maxwell, A Treatise on Electricity and Magnetism, 3rd ed., vol. 2. Oxford: Clarendon, 1892, pp.68–73.
5. S. Khara, I. S. Mishra and D. Saha, "An Anternative Architecture of WLAN/GPRS Integration" 63rd Vehicular Technology Conference (VTC 05), organized by IEEE , Melbourne, Australia, May 7-10, 0-7803-9392-9/06/$20.00 © 2006 IEEE.
6. H.-W. Lin, J-C. Chen, M-C. Hiang, C-Y. Huang, "Integration of GPRS and Wireless LANs with Multimedia Applications", Proceedings of IEEE Pacific Rim Conference on Multimedia 2002, pp. 704-711.
7. M. Jaseemuddin, "An Architecture for Integrating UMTS and 802.11 WLAN Networks", Proceedings of IEEE Symp on Comp. and Commun.(ISSC-2003), pp. 716-723.
8. M. B. R. Murthy and F. A. Phiri, "Performance Analysis of Downward latency in a WLAN/GPRS interworking System." Journal of Computer Science 1(1), Science Publications, 2005, pp 24-27.
9. K. Salkintzis, "Interworking techniques and Architectures for WLAN/3G integration toward 4G Mobile Data Networks" IEEE Wireless Commun, June 2004, pp-50-61.

10. F. Akyildiz, J. Xie and S. Mohanty "A Survey of Mobility Management in Next-Generation All-IP-Based Wireless Systems", IEEE Wiress Communications, Aug 2004, pp 16-28.
11. M. Shi, X.(S) Shen and J.W. Mark, "IEEE802.11 Roaming and Authentication in Wireless LAN/Cellular Mobile networks." IEEE Wireless Communication, Aug 2004, pp 66-75.
12. C. Perkins, "IP mobility support " IETF RFC 3220 Jan 2002.

The Positioning of Base Station in Wireless Communication with Genetic Approach

Yong Seouk Choi[1], Kyung Soo Kim[1], Nam Kim[2]

[1]ETRI, 161 Gajeong-Dong, Yuseong-Gu, Daejeon, Korea
choiys@etri.re.kr
[2]Chungbuk National University, 12 Gaeshin-Dong, Heungduk-Gu, ChungJu, Korea

Abstract. This paper addresses the displacement of a base station with optimization approach. A genetic algorithm is used as optimization approach. A new representation that describes base station placement, transmitted power with real numbers and new genetic operators is proposed and introduced. In addition, this new representation can describe the number of base stations. For the positioning of the base station, both coverage and economy efficiency factors were considered. Using the weighted objective function, it is possible to specify the location of the base station, the cell coverage and its economy efficiency. The economy efficiency indicates a reduction if the number of base stations for cost effectiveness. To test the proposed algorithm, the proposed algorithm was applied to homogeneous traffic environment. Following this, the proposed algorithm was applied to an inhomogeneous traffic density environment in order to test it in actual conditions. The simulation results show that the algorithm enables the finding of a near optimal solution of base station placement and it determines the efficient number of base stations. Moreover, it can offer a proper solution by adjusting the weighted objective function.

1. Introduction

Base station placement is a highly important issue in achieving high cell planning efficiency. It is expected that third generation wireless systems will provide a great variety of services. Thus, cell planning should be carried out considering inhomogeneous traffic. The placement of base stations depends on the traffic density, channel conditions, interference scenario, the number of base stations, and the other network planning parameters; as a result, it is a very complex issue. A genetic algorithm is useful for solving this type of complex problem. This method represents feasible solutions in terms of individuals with genomes, and determines which individuals could survive in a certain criterion formulated to maximize (or minimize) a given objective function. In several studies, a genetic approach has been used to find the best possible base station placement [1, 2]. Binary string representation is applied in [1], and a hierarchical approach is considered in [2]. However, those approaches have a representation limit, and a lot of trials can not guarantee an optimum result, as the possible base station positions are discrete.

Please use the following format when citing this chapter:

Choi, Y. S., Kim, K. S., Kim, N., 2007, in IFIP International Federation for Information Processing, Volume 245, Personal Wireless Communications, eds. Simak, B., Bestak, R., Kozowska, E., (Boston: Springer), pp. 217-229.

In this paper, a new representation describing base station placement is suggested, and is one which uses a real number and introduces new genetic operators. The proposed representation can determine not only the locations of the base stations but also the number of base stations.

In addition, the transmitted power of base station is considered as a factor of the propposed algorithm. To consider both coverage and the economy efficiency, an objective function with a weighted factor is established. The proposed algorithm is verified by applying it to homogeneous traffic density case as an obvious optimization problem. In addition, the approach is tested in an inhomogeneous traffic density environment.

2. Overview of Genetic Algorithm

Like other computational systems inspired by natural systems, genetic algorithms have been used in two ways: as techniques for solving technology problems, and as simplified scientific models that can answer questions about nature [3]. Genetic algorithms (GA) are evolutionary optimization approaches which are an alternative to traditional optimization methods. GA approaches are most appropriate for complex non-linear models where location of the global optimum is a difficult task. It may be possible to use GA techniques to consider problems which may not be modeled as accurately using other approaches. Therefore, GA appears to be a potentially useful approach. GA performance will depend very much on details such as the method for encoding candidate solutions, the operator, the parameter setting, and the particular criterion for success. As for any search, the way in which candidate solutions are encoded is very important. Many genetic algorithm applications use fixed-length, fixed-order bit strings to encode candidate solution. However the algorithm proposed in this paper uses real-valued encoding schema to represent solutions. In GA, feasible solutions are modeled as individuals described by genomes. A genome is an arrangement of several chromosomes, which symbolize characteristics of the individual. Population is the total amount of individuals. Some of them can survive and others will die in the next generation by their own fitness and a given selection rule. Fitness is evaluated by a given objective function. Genetic operations such as crossover and mutation are performed to produce new individuals in subsequent generations. The crossover operator defines the procedure of generating a child from its parent's genomes. The mutation is carried out chromosome by chromosome, and its exploration and exploitation helps the algorithm to avoid local optimum. If the current population accepts the given termination condition, new generation is no longer produced. Otherwise, dominant individuals are selected and genetic operators reproduce new individuals from them. The best individual of each generation is transferred over to the next generation if elitism is adopted.

The theoretical basis of GA relies on the concept of schema. A schema is defined as the similarity of templates describing a subset of genomes with similarities in certain chromosomes. Schemata are available to measure the similarity of individuals. John Holland's schema theorem and building-block hypothesis [4] have often been used to explain how the GA works. According to the schema theorem, short, low-order, and above-average schemata receive exponentially increasing trials in subsequent generations. This proves that the individuals with high fitness will have a high survival probability when a suitable representation is applied. The building-block hypothesis suggests that the GA will perform well when it is able to identify above-average-fitness and low-order schemata and recombine them to produce higher-order schemata of higher fitness. In sum, individuals with similar characteristics must be represented by a similar genotype.

3. Proposed Algorithm for Base Station Placement

The processing of the proposed algorithm is implemented in a two-dimensional map; therefore representation in binary form is difficult to present for the genome which describes the number of base stations and the location of the base stations. For a good approximation, it is necessary to have a longer genotype. A real value representation is more efficient than the representation of a binary genome in this case. Consequently, in this paper the genotypes that have real value representations for the optimization algorithm were chosen. Given the allowable transmitted power of a cell site in a traffic map, this chapter introduces GA that optimizes the cell site location, the number of cell sites and the transmitted power. A GA that works well in terms of the base station placement problem is proposed. The main characteristics considered for the development of the proposed algorithm are:

<1> The genome must represent all of the base station locations, and the genotype can describe the number of base stations as well as the position of the base station.
<2> A chromosome expresses one base station position.
<3> The number of possible base station locations must be unlimited; therefore, there are infinite candidates of base station locations.
<4> Similar genotypes represent the genomes of the closely located base stations.

An algorithm satisfying the above factors is consistent with the building-block hypothesis and schema theorem.
The three things that must be defined in order to solve a problem through genetic algorithms are as follows:

- Define a representation
- Define the genetic operators
- Define the objective function

How one defines a representation, genetic operators, and objective function determines the algorithm. It is essential to design the genetic algorithm by considering <1>~<4>. The following sections explain the proposed algorithm in detail.

3.1 Representation

Fig. 1 illustrates the representation of the genomes. A genome is denoted as a vector $g = (c_1, \cdots, c_K)$ where $c_k = (x_k, y_k, pwr_k)$ is the chromosome for the k-th base station position. This method fulfills <1> and <2>. K is the maximum number of base stations, and all of these can be located in the x-range $[-X_{max}, X_{max}]$ and y-range $[-Y_{max}, Y_{max}]$ with origin $(0,0)$.

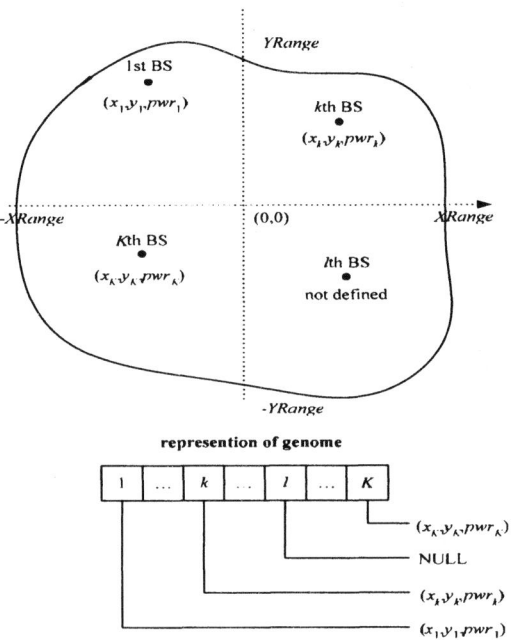

Figure 1. Representation of the genome for the placement of the base station

If the position of a base station is not defined, it is expressed as NULL. This method applies for a case in which there are fewer base stations than in K, so that it fulfills <1>. $n(g)$ is defined as the number of EXISTENCE in g. In order to satisfy <3> and <4>, x_k, y_k and pwr_k must be real numbers. M is assumed as population size.

3.2 Genetic Operators (Crossover and Mutation)

It is necessary to design an initialization and a termination method, a crossover and mutation operator, and a selection strategy in order to define the reproduction procedure.
A proper initial population can provide a fast convergence to the optimum point. It is desirable for a user to define initial positions of base stations intuitively. The first

individual, $c_{1k} = (x_{1k}, y_{1k})$ for $k = 1, \cdots, K$, is determined by a user and other individuals (for $m = 2, \cdots, M$) are determined by the following rule: If c_{1k} = NULL, then c_{mk} = NULL with probability P_n^I or $c_{mk} = (\upsilon_1, \upsilon_2)$ with probability $1 - P_n^I$, where $\upsilon_1 = U(-X_{max}, X_{max})$ and $\upsilon_2 = U(-Y_{max}, Y_{max})$. If c_{1k} is defined ($c_{1k} \neq$ NULL), then c_{mk} = NULL with probability $1 - P_v^I$ or $c_{mk} = (x_{1k} + \xi_1, y_{1k} + \xi_{2k})$ with probability P_v^I, where $\xi_1, \xi_2 = N(0, \sigma_S^2)$. $U(a,b)$ is a uniformly distributed random variable between a and b. $N(\bar{x}, \sigma^2)$ denotes a Gaussian distributed random variable with mean $\bar{x}$ and variance σ^2. P_n^I and P_v^I indicate the probability of producing NULL from NULL and that of producing EXISTENCE from EXISTENCE, respectively. However, it may require further trials in order to determine the global optimum if the initial value, as user defined, is close to the local optimum. When the user does not define any initial positions, it is decided that c_{mk} = NULL with $\widetilde{P}_n^I$ or $c_{mk} = (\upsilon_1, \upsilon_2)$ with probability $1 - \widetilde{P}_n^I$ for $m = 1, \cdots, M$, where $\widetilde{P}_n^I$ denotes the probability of producing NULL.

A termination criterion is used to determine whether or not a GA is finished. Generation, convergence, or population convergence can terminate the procedure of genetic algorithm. The easiest scheme is termination upon generation. When the number of current generations is larger than the specified number of generations, the algorithm is finished. Termination upon convergence compares the previous best-of-generation to the current best-of-generation. If the current convergence is less than the requested convergence, the reproduction procedure is ceased. Termination upon population convergence compares the population average to the score of the best individual in the population.

In the proposed application, one child crossover operator is used. A single child c_k^{child} is born from its father and mother, c_k^{dad} and c_k^{mom}. Fig. 2 shows the procedure of one child crossover operation in the proposed algorithm. If one of the parents is NULL, the child receives the other parent's attributes. Otherwise, the child is generated by (1), where σ_C is the parameter of the crossover operation. $|x_k^{dad} - x_k^{mom}|$ and $|y_k^{dad} - y_k^{mom}|$ can be used as a measure of closeness. This method is based on the fact that if the attributions of both parents are similar, the child's attributions are also similar to its parents.

Mutation is performed chromosome by chromosome with probability P_{mut}. Fig. 3 shows the procedure of the mutation operation in the proposed algorithm. The mutation is very close to the initialization scheme with the user-defined base station position. If c_{mk} = NULL, redefine c_{mk} = NULL with probability P_n or $c_{mk} = (\upsilon_1, \upsilon_2)$ with probability $1 - P_n$. If $c_{mk} \neq$ NULL, redefine $c_{mk} = (x_{mk} + \chi_1, y_{mk} + \chi_2)$ with probability P_v or c_{mk} = NULL with probability $1 - P_v$, where χ_1 and χ_2 are Gaussian distributed random variables with zero mean and variance σ_m^2. P_{mut} and σ_m^2 are the parameters of the mutation operation.

A roulette wheel method is applied for the selection scheme. This selection method chooses an individual based on the magnitude of the fitness score relative to the rest of the population. The higher the score, the more selective an individual will be. Any individual has a probability p of the choice where p is equal to the fitness of the individual divided by the sum of the fitness of each individual in the population. Therefore, the individual with a high fitness level can survive with high probability.

$$x_k^{child} = \frac{x_k^{dad} + x_k^{mom}}{2} + \varsigma_1, \; \varsigma_1 = N\left(0, \left(\frac{(x_k^{dad} - x_k^{mom})\sigma_C}{2}\right)^2\right)$$

$$y_k^{child} = \frac{y_k^{dad} + y_k^{mom}}{2} + \varsigma_2, \; \varsigma_2 = N\left(0, \left(\frac{(y_k^{dad} - y_k^{mom})\sigma_C}{2}\right)^2\right). \quad (1)$$

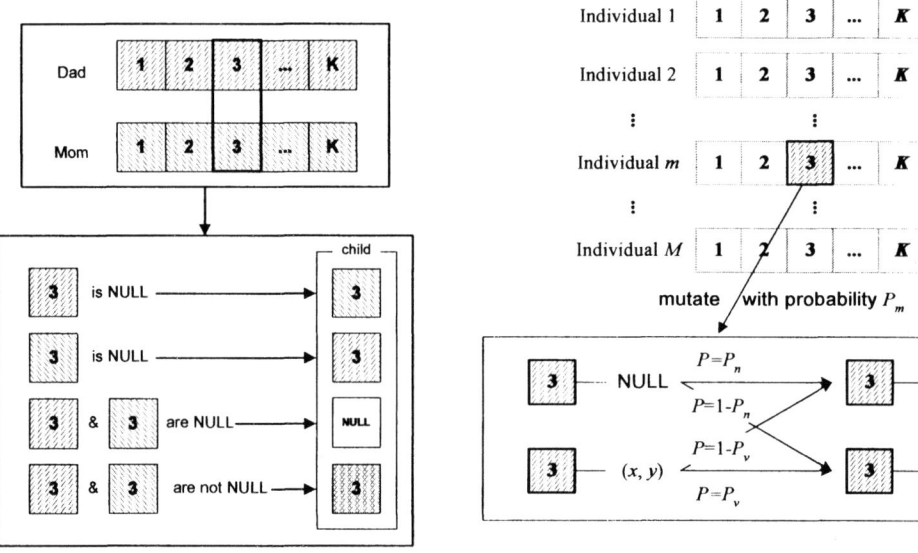

Figure 2. One child crossover operation

Figure 3. Mutation operation

3.3 Fitness Evaluation

Fig. 4 illustrates the fitness evaluation procedure composed of an evaluator and an objective function. The evaluator calculates the covered traffic by using a propagation model, traffic map, and map for a path loss prediction. Cell area covered by the base stations is evaluated, and the covered traffic is then obtained. Considering coverage, power and economy efficiency, the objective function is defined as

$$f(G) = \omega_t \cdot f_t(G) + \omega_p \cdot f_p(G) + \omega_e \cdot f_e(G)$$

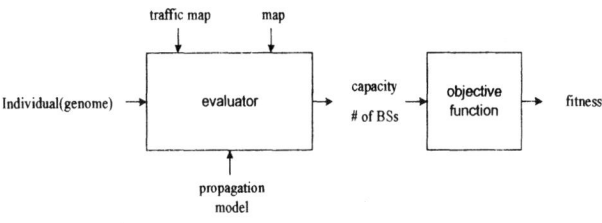

Figure 4. Fitness evaluation

where f_t, f_p and f_e are the objective functions for coverage, power and economy respectively, and these are defined as:

$$f_t(G) = \text{traffic coverage rate} = \frac{\text{covered traffic}}{\text{total traffic}}$$

$$f_p(G) = \text{BS power fitness} = \frac{\text{AvailableMaximumBS power} - \text{Used BS power}}{\text{AvailableMaximumBS power}}$$

$$f_e(G) = \text{economic fitness} = \frac{\text{Available Maximum BSs} - \text{Used BSs}}{\text{Available Maximum BSs}}$$

As the covered traffic area widens corresponding to the given propagation model, $f_t(G)$ increases. Conversely, $f_e(G)$ increases when fewer base stations are placed. Total fitness is calculated with w_t, w_p and w_e subject to $w_t + w_p + w_e = 1$. The weights are determined by the user's preference. If coverage is more important, then one may choose a large w_t. Otherwise, a large w_e may be chosen to be more desirable using fewer base stations. Therefore, the purpose of optimization in this paper is to determine the maximum traffic coverage with the minimum number of base stations and minimum amount of power.

This paper uses Hata's model to obtain the coverage of the base station.

It is possible for each individual can have K(the maximum number of base stations). To achieve the cell coverage, it is necessary to compute the path loss K times. If the population is large, the computing power required becomes very large. In this paper, to reduce processing time, Hata's model was used, which is fast for computing the path loss with height information.

4. Testify Algorithm

To test the proposed algorithm, a one-tiered hexagonal cellular environment is considered, where traffic is distributed uniformly in each hexagonal cell whose radius is 2.5 km. In this case, the optimum position of the base station is in the center of hexagon, and the optimum number of base stations is obviously seven. A path loss prediction is carried out using the equation $L = L_0 \times (d/d_0)^{-4}$, where $L_0 = 140$ dB and $d_0 = 2.5$ km. As the generation increases, the base stations tend to be placed where they are optimum, and the number of base stations is also converged automatically. After the 1000th generation, a base station placement that guarantees 99.78% coverage can be determined.

Figure 5. Homogenous traffic density for verification

The maximum number of base stations depends on the width of the target area. The wider the target area, the more likely a greater amount of computing time for convergence is needed. Population size is the solution set. If the population size is large, the convergence of the solution can be quickest. However, in this case the total computing time is larger, as a processing of the propagation model will be needed for each individual in the population. As the individuals with low fitness values are removed, the initial values of base station's maximum number and location are not related to the entire performance. Therefore, a null-to-null probability and Pos-to-Pos probability is loosely coupled with the fitness relationship, and the mutation probability in a real-value representation is the main factor in speeding the convergence.

Fitness with various mutation probabilities in each generation is shown in Fig. 6. The higher the mutation probability, the better the fitness. However, too high a mutation probability has a tendency to downgrade the performance, as it has a frequently changing possible solutions set. In the given homogenous traffic in Fig. 5, it is known that the best performance is shown when the mutation probability is 0.1. (Fig. 6)

Fig. 7 shows that a high deviation of mutation will be good for performance.

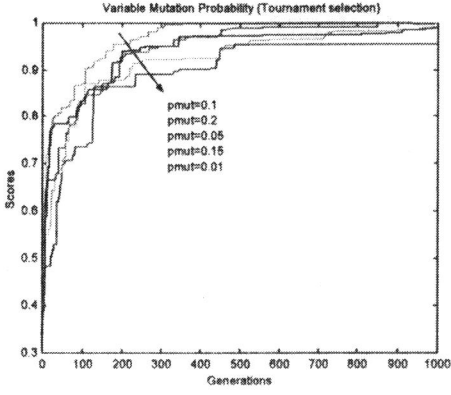
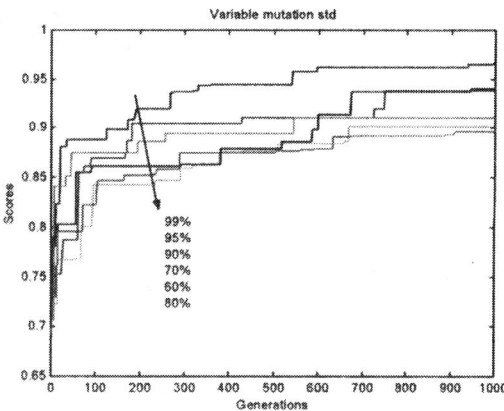

Figure 6. Fitness in various mutation probabilities

Figure 7. Fitness in various mutation deviations

Figs. 8 to 12 show the optimization processing of base station displacements. Fig. 8 shows the initial random location of the base stations, and in this case five base stations have covered 69% of the target area. In Fig. 9, seven base stations have covered 92% of target area with uniform selection, but it is still not optimized. Fig. 10 is the result of a Roulette wheel selection, and this is an improvement over the uniform selection. It covers 93.85% of the target area. The Rank selection covers 97.90%; this is a very good result. The Tournament selection offers 99.78% coverage. This is approximately at the optimization level. As fitness is sensitive in terms of selection schemes, optimization processing needs appropriate selection schemes.

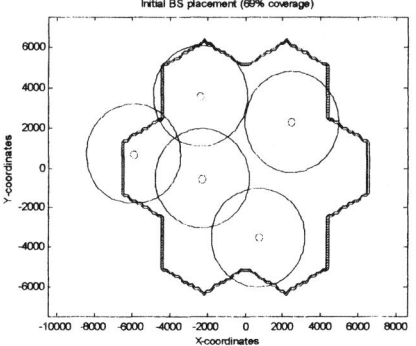
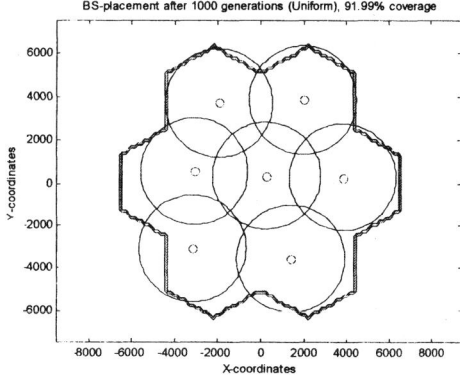

Figure 8. Initial base station location

Figure 9. After the 1000th generation, base station location with uniform selection

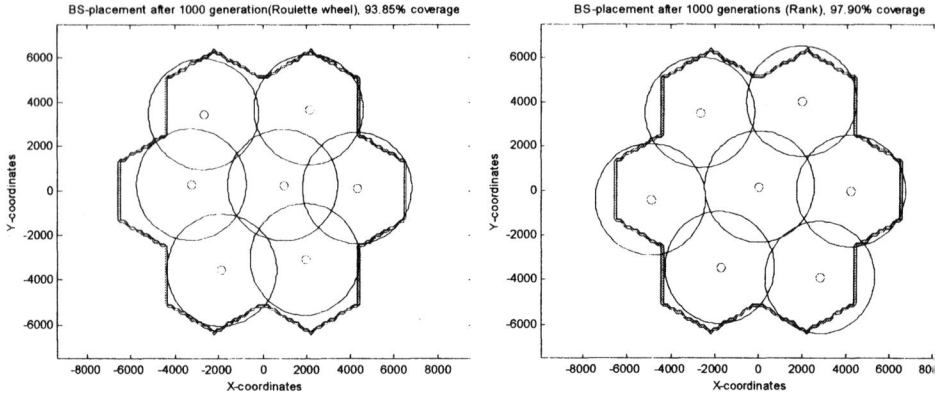

Figure 10. After the 1000th generation, base station location with Roulette wheel selection

Figure 11. After the 1000th generation, base station location with Rank selection

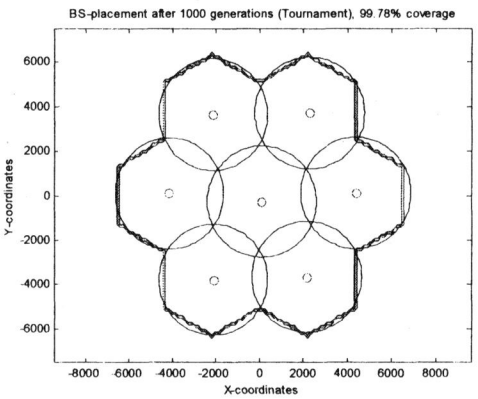

Figure 12. After the 1000th generation, base station location with Tournament selection

5. Simulation Results

To demonstrate if the proposed algorithm determines which positions match optimum location, a simulation was conducted on areas similar to that in Figs. 13 and 14 (inhomogeneous traffic). The actual valued representations in this paper, as mentioned above, consist of the candidate location of the base station's transmit power. Fig. 13 shows the altitude map of the target areas, and Fig. 14 shows the traffic density map. The traffic density is inhomogeneous and the target area for simulation is an urban pattern. The width of the area for simulation is 12Km x 12Km and the size of the bin is 120m x 120m. Therefore, the total number of bins is 10,000.

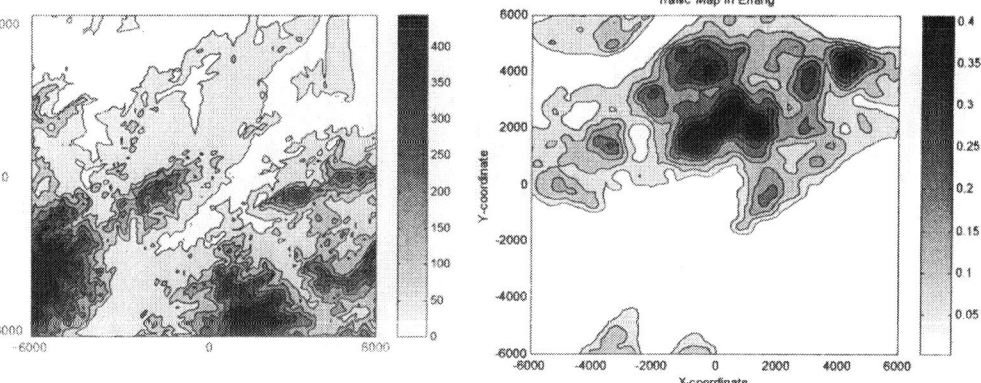

Figure 13. Altitude Map

Figure 14. Traffic density map

Figs. 15 and 16 show the location of the base station from one generation to 500 generations, when the weighting condition of their object function is $(\omega_t, \omega_p, \omega_e) = (0.9, 0.0, 0.1)$. The assigned transmit power range of each base station is from 22.63 dBm to 39.36dBm, and its mean value is 33.84 dBm. In this case, the coverage rate is 82.62% and the fitness value is 0.74258.

In the case where the condition of object function is $(\omega_t, \omega_p, \omega_e) = (0.8, 0.1, 0.1)$, the results are shown in Figs. 17 and 18. The coverage rate is 77.47%, and the fitness value is 0.663181. The assigned transmit power range of each base station is from 21.1752 dBm to 38.57794dBm, and its mean value is 32.3230 dBm. As the traffic capacity is limited, the cell boundaries of the high traffic density are less than those of the low traffic density. The coverage rate is decreased according to the changing weight of the traffic factor, from 0.9 to 0.8. As the weight of the power factor increases, the actual assigned transmit power value decreases. In the results shown in Fig. 17, the overlapped base station is clearly shown. The cause of this is the decrease of the weighted economy factor. The traffic map that was used for the simulation consisted of high traffic density areas and very low traffic density areas such as mountains and rivers. Therefore, traffic is scattered in all directions on the map; consequently, the search space becomes larger. To obtain a better coverage rate, the population size can be enlarged or the mutation probability can be increased. Additionally, it is necessary to process more generations.

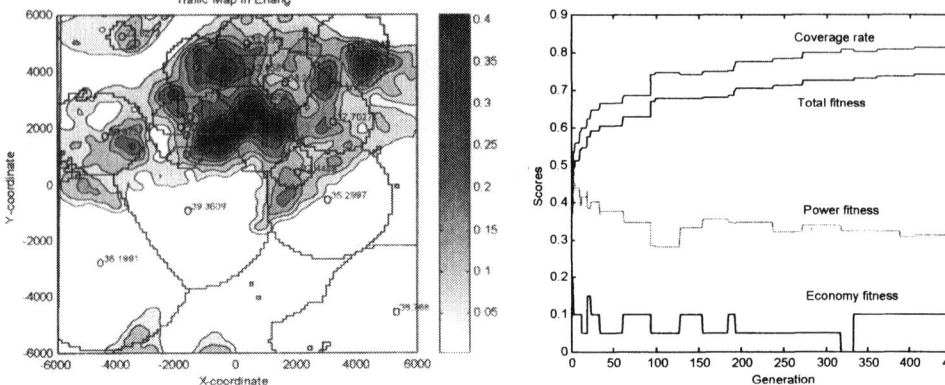

Figure 15. After 500 generations, the location of the base stations, $(\omega_t, \omega_p, \omega_e) = (0.9, 0.0, 0.1)$

Figure 16. Fitness Value, $(\omega_t, \omega_p, \omega_e) = (0.9, 0.0, 0.1)$

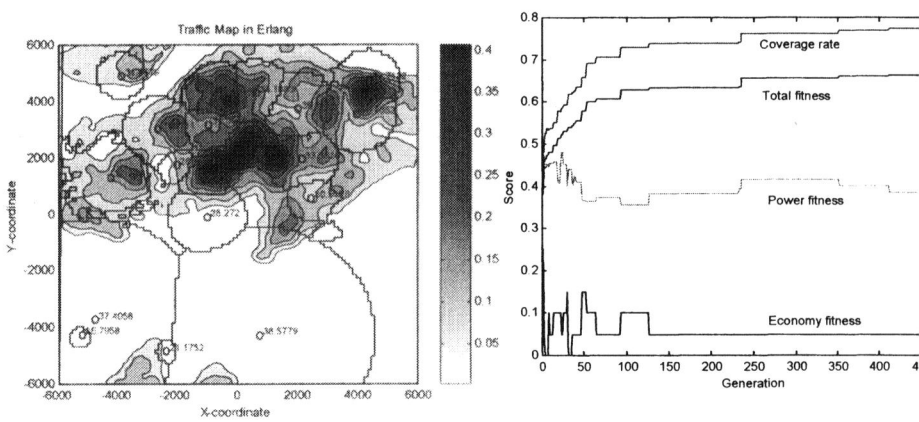

Figure 17. After 500 generations, the location of the base stations, $(\omega_t, \omega_p, \omega_e) = (0.8, 0.1, 0.1)$

Figure 18. Fitness values, $(\omega_t, \omega_p, \omega_e) = (0.8, 0.1, 0.1)$

6. Conclusion

In this paper, given inhomogeneous traffic information and the map for the propagation model, a new algorithm was proposed that enables the optimization of the locations and transmitted power of a base station. In addition, this algorithm includes an economic factor (the number of base stations). Good use was made of the genetic algorithm and it was excellent for obtaining a solution of complex problems. Genetic operators using the real valued representation are also suggested, and the objective function is defined in consideration of the coverage, the transmitted power of base station and the economy efficiency through an adjustment of crossover and mutation. The selection, input parameters and scaling are shown to be tightly coupled with the algorithm performance. Therefore, there is a need for these to be harmonized. From a simulation, the proposed algorithm was verified.

References

[1] X. Huang, U. Nehr, and W. Wiesbeck, "Automatic Base Station Placement and Dimensioning for Mobile Network Planning," Proc. IEEE VTC 2000 Fall, vol. 4, pp. 1544-1549, 2000.
[2] Melanie Mitchell, An Introduction to Genetic Algorithms, The MIT Press, 1996.
[3] Holland J.H., Adaptation in Natural and Artificial Systems, University of Michigan Press, Ann Arbor, 1975
[4] Thomas Back, Frank Hoffmeister, Hans-Paul Schwefel, "A Survey of Evolution Strategies", Proceedings of the 4th International Conference on Genetic Algorithms, pp.2-9, San Diego, CA, July, 1991
[5] Xuemin Huang, Ulrich Behr, Werner Wiesbeck, "Automatic Base Station Placement and Dimensioning for Mobile Network Planning", IEEE Vehicular Technology Conference, October, 2000
[6] A. H. Wright, Genetic Algorithms for Real Parameter Optimization, in Foundations of Genetic Algorithms (Ed. J. E. Rawlins), Morgan Kaufmann, 1991
[7] Hata. M, "Empirical Formula for Propagation Loss in Land Mobile Radio Services", IEEE Transactions on Vehicular Technology, Vol.VT-29, No.3, pp.317-325, August 1980
[8] C.M. Fonseca and P.J. Fleming, "An Overview of Evolutionary Algorithms in Multiobjective Optimization," Evolutionary Computation 3(1), Massachusetts, MA: MIT-Press, 1995
[9] Aracena,J, Lamine.SB, Mernet,MA, Cohen,O and Demongeot,J "Mathematical modeling in genetic networks: relationships between the genetic expression and both chromosomic breakage and positive" IEEE Transactions on Systems, Man and Cybernetics, 2003
[10] Hirasawa.K, Okubo.M, Katagiri.H, Hu.J, Murata.J, "Comparison between Genetic Network Programming (GNP) and Genetic Programming (GP)", Proceedings of the 2001 Congress on Evolutionary Computation, 2001.
[11] Nuaymi,L., Godlewski,P., "Association of uplink power control and base station assignment in cellular CDMA systems" Proceedings ISCC 2000 Fifth IEEE Symposium on Computers and Communications, 2000.
[12] Amaldi,E., Capone,A., Malucelli,F., "Optimizing UMTS radio coverage via base station configuration", The 13th IEEE International Symposium on Personal, Indoor and Mobile Radio Communications, 2002.
[13] Lee,C.Y., Kang,H.G., "Cell planning with capacity expansion in mobile communications: a tabu search approach", IEEE Transactions on Vehicular Technology 2000.
[14] Santiago,R.C., Lyandres,V., "A sequential algorithm for optimal base stations location in a mobile radio network" 15th IEEE International Symposium on Personal, Indoor and Mobile Radio Communications, PIMRC 2004.
[15] Yufei Wu, Pierre,S., "Base station positioning in third generation mobile networks", IEEE CCECE 2003.
[16] Hurley,S., "Automatic base station selection and configuration in mobile networks", IEEE VTS-Fall VTC 2000.

A Door Access Control System with Mobile Phones

Tomomi Yamasaki[1], Toru Nakamura[1], Kensuke Baba[2,3], and Hiroto Yasuura[2,3]

[1]Graduate School of Information Science and Electrical Engineering
[2]Faculty of Information Science and Electrical Engineering
[3]System LSI Research Center
Kyushu University
Motooka 744, Nishi-ku, Fukuoka, 819-0395, Japan
{yamasaki,toru,baba,yasuura}@c.csce.kyushu-u.ac.jp

Abstract. This paper proposes a door access control system with mobile phones which allows off-line delegations of an access. A model of door access control with mobile phones is introduced, and then the delegation is formalized as a copy of a door-key. On the previous model, secure copy by off-line is realized using the essential idea of the proxy signature. Moreover, the proposed system is implemented on mobile phones, and then the execution time of a copy and a verification are estimated. As a result, it is shown that the proposed system is feasible.

Keywords: Access control, door-key management, mobile phone, off-line delegation

1 Introduction

Service providing systems using portable devices, for example, a door access control system with smart cards, are being popular in our daily life. In most of such systems, the scheme to control authorities to receive a service is based on entity authentications (or identifications) by communications with electronic data, and therefore the portable device stores secret information for the identification. In this sense, the portable device (such as a smart card, a PDA, a mobile phone, and so on) is called a "token".

In some practical systems providing a service, a delegation of the authority for a service can be a very useful function. For example, in a door access control system, copying a door-key may be the most common requirement. In an authority management system with token based identifications, passing the token is the naive scheme to realize a delegation, however it is not practical if the token is for multiple services. Therefore, the scheme based on cryptographic technologies with passing only electronic data is necessary to meet the requirement. In this approach, a delegation is straightforwardly realized if we allow a communication with the entity who manages the authorities. However, it is not clear how to realize an off-line delegation, that is, a delegation without any communication with the third party.

We consider the situation that an entity who has an authority (hereinafter called an "owner") wants to delegate the authority to other entity (hereinafter called a "deputy"). The idea of proxy signature [4, 3] can be a method to realize a off-line delegation of an authority. The entity creates the signature for the public-key of the deputy, and the owner sends the digital signature to the deputy as a warrant. The deputy sends the warrant to the verifier with the deputy's signature. Similar methods can be found in systems of electronic cash [6, 1]. The idea of "transferability" corresponds to the off-line delegation. A user issues the license from the bank. The user issues electronic cash from the bank by using the license, and use cash for the settlement in the retail store. When an owner delegate it to a deputy, the owner sends a certification which denotes the delegation. Although there are many of related work about the protocol of the authority delegation, applying the ideas to a practical system needs more discussion. This difficulty depends on at least the following two factors:

- the difficulty of modeling practical systems which manages authorities,
- the gap between the ideal environment required from the theoretical schemes and the current technologies.

This paper focuses a door access control system and proposes a technique for applying the essential idea of the proxy signature to the authority delegation. First, we introduce a model of door access control to make clear the requirements of practical systems. If we consider the delegation of a door-key, we can ignore a double-use of the authority in most case. This is a critical difficulty for the transferability of electronic cash. Using mobile phones solves the second factor of the previous difficulty, since the input devices and the screen which displays the result of the data exchange are provided. Moreover, some kinds of mobile phones have the infrared rays communication function and the integrated circuit chip [5], while the intercommunication function is not provided in smart card. The mobile phone possesses the input function and the screen, therefore it is suitable for exchanging data. Then, we have the following assumptions on our model:

- the delegation allows a double use of the authority,
- an owner and a deputy can have secure communication without any third party.

Even on the strong assumptions, it is not trivial to realize the off-line delegation. Our scheme is a simple application of the idea of an encryption or an digital signature technology. Moreover, we implement our scheme on practical mobile phones and evaluate the feasibility.

This paper is constructed as follows: in Section 2, we describe some requirements of door access control systems with mobile phone from a practical viewpoint; in Section 3, we introduce a model of door access control and interpret the requirements to the model, and then propose a scheme to control authorities which allows off-line delegations; in Section 4, we show the environment and experimental results of the implementation of the proposed scheme.

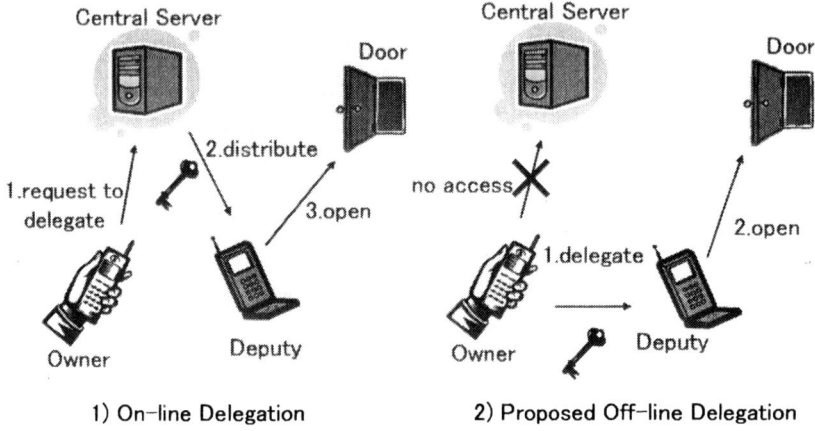

Fig. 1. On-line and off-line delegations

2 Requirements of Door Access Control Systems

In this section, we show effective application examples of a door access control system we are considering, and describe some requirements from two viewpoints. Fig. 1 shows the outline of our system.

2.1 Examples of Practical Use

Temporal Delegation When we consider a door-key, for example, of our private room, our office room, a public conference room, and so on, it often happens the case that we want to lent the door-key to our friend, colleagues, students, and so on. In the case of a physical door-key, we must lent our master key itself or a copied one. It is easy to imagine insecure or uneconomic situations. These problems can be solved by electronic keys with detailed information about the key such as a restriction of doors or time for use. If we use a portable device such as a smart card for the electronic key, the authority control is important especially for a multi-services system. Do you want to lent your smart card with a credit service to let your student use a conference room?

Load Distribution An off-line delegation key management system can provide better performance than on-line, from the viewpoint of reducing their burdens to change key management databases on an central server. In on-line system, the central server must change the database every time when a user asks the server to delegate his/her key to others. On the other hand, in off-line system, a user can delegate his/her key without access to the central server. Additionally an off-line delegation realizes a hierarchical manage of user's keys. For example, in the case where a boss tries to manage 1,000 keys for 100 company members, the boss just has to delegate 100 keys to 10 general managers of him and change his database for only the part with respect to the managers.

2.2 Information as a Door-key

The identification between the server and a user is well studied and there exist schemes which are secure in a practical sense for the situation that the communicating entities have computing resources. Especially, the identification of the server by any user is straightforward in the case that the server is regarded as an entity connected to a door.

If we consider the delegation of a door-key, we can ignore a double-use of the authority, since in most case the service for the authority does not disappear by a use. Therefore, we can consider as a delegation of an authority a copy rather than the delegation in the strict sense.

In some application of a door access control system, the authority is delegated for free as the examples in this section. In such cases, it is not necessary to confirm that a delegated door-key is the correct one. Therefore, we can use a simple scheme of a delegation and a verification of an authority. If the confirmation is necessary, an idea of the digital signature is used instead of a private-key encryption.

2.3 Communication between Mobile Phones

Although the intercommunication function is not provided between smart cards, there exist some kinds of mobile phones which have an infrared communication function [5]. Therefore, we can allow as processes of a door-key not only the communication between a server and a user but also two users. This is the essential technology to realize an off-line communication.

Some mobile phones have also enough resource for computations of encryption, and therefore we can assume an ideal identification between users. Moreover, the following two considerations show an appropriateness of the previous assumption: 1) we usually face to the other entity when we use a mobile phone to send data, hence we can identify the entity in the sense of a real communication rather than electronic process; 2) we carry our mobile phone all the time, hence the correspondence between a human and his/her mobile phone is guaranteed.

3 Formalization

In this section, we introduce a model of a door access control system with mobile phones and interpret the requirements in the previous section to the model. Then, we propose a scheme of a verification and a delegation of the authority to open a door.

3.1 Door Access Control System

A *door access control system* is constructed by *users* who want to use their authority to open a door and a *server* who examines whether a user has the authority. We denote by $u_1, u_2, \ldots \in U$ the users and by s the server. Any $u_i \in U$ and s can communicate another entity along with a given protocol. By the argument in Subsection 2.2, we formalize this protocol that a $u_i \in U$ submit a string to s to open a door, and s outputs whether the u_i has the authority. We call this process a *verification* by s of u_i with respect to a door. In the rest of this paper, we consider a door access control system with a single server in the situation that the server controls a single door, and therefore we regard the server and the door as a single entity.

For the previous model, we allow some process between two users. Any $u_i \in U$ can operate interactive processes with another $u_j \in U$ along a given protocol, which is reasonable for systems with mobile phones by the argument in Subsection 2.3. Additionally, by the argument, we assume an ideal identification between any two entities in the users and the server, which enable any user to confirm the identifier, called the *ID*, of the other user.

Assumption 1 *Any pair of entities in $U \cup \{s\}$ has a scheme of an identification.*

By the previous assumption, any entity can know the ID of the other entity. The ID of a user u_i is denoted by n_i. In some protocols in the rest of this paper, the identification process is not described explicitly. Note that the ideal identification is not realized by the ID, but the ID is confirmed as the result of the identification. It is not in the scope of this paper how to realize a secure identification.

As mentioned in the argument of the previous assumption, we assume any pair of communicating entities can use a suitable cryptographic technology, since the system is realized by a server computer and mobile phones which have enough resource for computations. We consider the following two conditions.

Assumption 2 *Each user in U and s have a private-key encryption scheme, respectively.*

By the previous assumption, u_i has a function ϕ_{u_i} for an encryption and s can know (the result of an application of) $\phi_{u_i}^{-1}$. The condition also yields that any user has a scheme of a message authentication code to s, that is, s can verify the data integrity of the message from the user [2].

Assumption 3 *Any entities in $U \cup \{s\}$ has a public-key encryption scheme.*

By the previous assumption, any entity in $U \cup \{s\}$ has a scheme of a digital signature [2]. In addition to the situation by Assumption 2, this condition yields that any entity can verify the integrity of a message with the digital signature from any user.

3.2 Copy of an Authority

We introduce an idea of "trust" on a door access control system. The ideal assignment of the authority to open a door to the users is expressed by a function $F: U \to \{0,1\}$ such that $F(u_i)$ is 1 if u_i should have the authority, and 0 otherwise. Now we consider a door access control system Σ in which any verification by s terminates for any user in U. Then, the result of practical verifications in Σ is also expressed by a function $G_\Sigma : U \to \{0,1\}$ such that $G_\Sigma(u_i)$ is 1 if the verification outputs that u_i has the authority, and 0 otherwise. A door access control system Σ is *trustful* if, for any u_i, $G_\Sigma(u_i) = 1$ if and only if $F(u_i) = 1$.

Let Σ be a door access control system such that $G_\Sigma(u_i) = 1$ and $G_\Sigma(u_j) = 0$ for $u_i, u_j \in U$. Then, a *copy of the authority* is a process to change Σ into Σ' such that $G_{\Sigma'}(u_j) = 1$ and $G_{\Sigma'}(u_k) = G_\Sigma(u_k)$ for the other $u_k \in U$. Therefore, the ideal assignment $F' : U \to \{0,1\}$ after the copy to u_j is

- $F'(u_k) = 1$ if $F(u_k) = 1$ or $u_k = u_j$,
- $F'(u_k) = 0$ otherwise.

We denote the ideal assignment F' of the authority after a copy from u_i to u_j by $F \cup f_{u_i \to u_j} : U \to \{0,1\}$. The previous notation can be extended straightforwardly to the ideal assignment after plural copies with a suitable definition of $\cup$. Then, in the followed subsection, we show that the system after a copy operated by our scheme is still trustful.

Note: By allowing the copy of the authority, a part of the access control is entrusted to the users who has the authority. If we express by $F_s : U \to \{0,1\}$ the assignment of the authority by s on an initial condition, and by $F_{u_i} : U \to \{0,1\}$ u_i's will whether he/she wish to copy his/her authority to each user, then the ideal assignment $F_s \cup F_{u_i} : U \to \{0,1\}$ can be defined as

- $F_s \cup F_{u_i}(u_k) = 1$ if $F(u_k) = 1$ or there exists $u_i \in U$ such that $F_s(u_i) = F_{u_i}(u_k) = 1$,
- $F_s \cup F_{u_i}(u_k) = 0$ otherwise.

Since a mapping from a set to $\{0,1\}$ is defining a subset of the set, it is easy to extend our idea with this expression to more complex controls, for example, authorization with recommendations by plural authorized users.

3.3 Procedures

As we mentioned in this section, the basic protocol of the verification of the authority is a submission of a string by a user to the server after the identification between the two entities. The string submitted in the protocol is called a *door-key* and denoted by K. Then, we show the procedures to verify or copy a door-key.

By Assumption 1, if we consider a system in which any copy of the authority is not allowed, then it is trivial that a trustful system can be realized by the ideal identification of $u_i \in U$ by s. s has only to prepare the function of the ideal

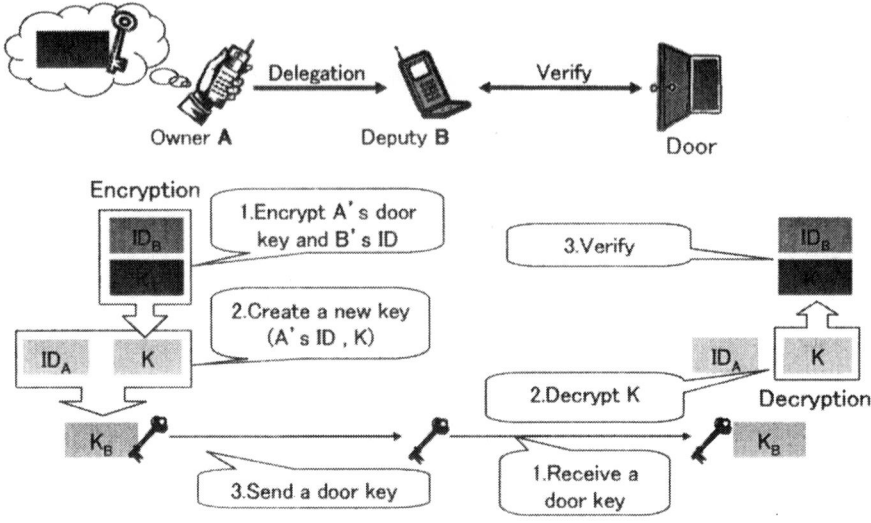

Fig. 2. Overview of the proposed delegation scheme.

assignment of the authority to the users. It is also trivial that an on-line copy, that is, a copy by modifying the function for every copy yields a trustful system. We consider a system with off-line copies. The outline of the proposed system is illustrated in Fig. 2.

Consider a copy of an authority from u_i to u_j. The following is the procedure of u_i after the identification between u_i and u_j, where ϕ is a function for encryption.

Procedure 1 (copy) u_i *submits* $K = (n_i, \phi_{u_i}(n_j))$ *to* u_j.

Then, the following is the procedure of the verification by s of u_j. Before the procedure, s and u_j operate the identification and u_j submit K which is copied from u_i along the previous procedure. Let $K[i]$ be the ith element of K.

Procedure 2 (verify) s *outputs*

- 1 *if* $F(u_j) = 1$,
- 0 *if* $F(u_j) = 0$ *and* $F(u_i) = 0$,
- 1 *if* $F(u_i) = 1$ *and* $\phi_{u_i}^{-1}(K[2]) = n_j$,
- 0 *otherwise*.

In the previous procedure, the first step is same as the procedure for a system does not allow any copy. The second and third step correspond to the ideal assignment after the copy from u_i to u_j. Therefore, a criterion of trust of the system depends on ϕ, in other words, this system is trustful if ϕ is ideal. In the system with Assumption 2, the entity who receive the door-key by a copy cannot

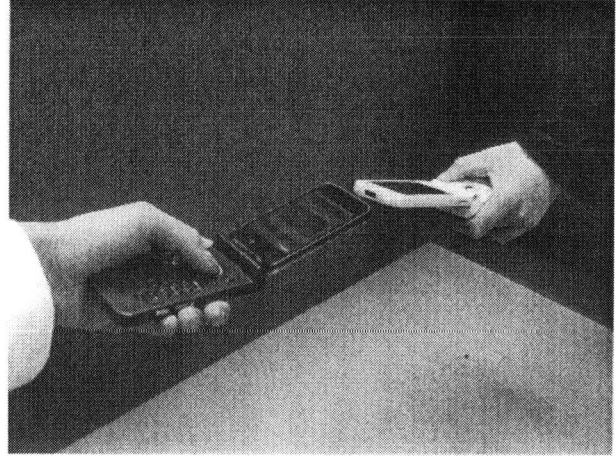

Fig. 3. An infrared communication between mobile phones

confirm the door-key is correct. Therefore, the entity who copy the door-key can success an attack to get a compensation for the door-key. In a practical sense, this situation is not fatal in some cases, as we mentioned in Subsection 2.2. In the system with Assumption 3, the attack is prevented since any user can confirm the door-key. In this case, some messages which describes what K is are submitted with K.

Note: We can easily extend the system to allow recursive copies by applying ϕ to K in Procedure 1. Then, Procedure 2 is modified to repeat the second and third step till the ID of the communicating entity appears. In this case, the length of K depends on the depth of the "sublease", and hence a restriction of the number of copies or a synchronization with the server for a period is required.

4 Implementation

In this section, firstly, we show the implementation environment. Secondly, we show the experimental result of the proposed system in this environment. Finally, we discuss the feasibility and usefulness of the proposed system.

4.1 Environment

We implement the idea in the previous section into a mobile phone which has a infrared communication function (see Fig. 3) and a contactless IC chip [5]. Table 1 shows the execution environments of the mobile phone and the server in this implementation.

At the present time, in mobile phones which is usualy avairable, there is no cryptographic co-processor which can be used freely. Moreover, any significant

Table 1. The environment of our implementation

Device		
	Model	NTT DoCoMo F902i and F902iS
	OS	Symbian OS
	CPU	SH-Mobile
	Bandwidth of infrared communication	max 100 Kilo-Bytes
	IC chip	FeliCa chip (64 Bytes memory)
R/W	Model	RC-S440C
Server	OS	Microsoft Windows XP
	CPU	Pentium4 3.20GHz
	RAM	2.00GB

cryptographic algorithm implemented as a software can not execute in a practical time due to the poor resource of a mobile phone. Therefore, we estimate the execution time for the cryptographic algorithm from the result of another experiment on an JAVA card. In the implementation for mobile phones, we use an exclusive-or operation as a dummy cryptographic function with the user's identifier and secret information of length 4 bytes, respectively.

We consider the situation that a user u_i copies his/her door-key K to another user u_j, and the server s verifies u_j' authority. Let σ_i be the u_i's secret and rw the Felica chip Reader/Writer. Then, the protocols of a copy and verification of a door-key are as follows, where $\oplus$ denotes the bitwise exclusive-or.

Copy-Protocol:

STEP1: u_i calculates $K = (n_i, n_j \oplus \sigma_i)$;
STEP2: u_i submits K to u_j by using the infrared communication function;
STEP3: u_i writes K to the access area of u_j's IC chip.

Verify-Protocol:

STEP1: rw waits for accession u_j with polling;
STEP2: rw reads K from u_j's access area;
STEP3: rw sends K to s;
STEP4: s verifies K and σ_j.

4.2 Evaluation

We measured the execution time of the previous protocols on mobile phones and a server. Each execution time is conducted as the average of 10 trials. In Copy-Protocol, the process consists of three parts: generating K, an infrared communication, and writing K on an IC chip. In Verify-Protocol, the process consists of three parts: reading K from an IC chip, a radio communication (including a PIN authentication), and a verifying K. The experimental results of the execution time are shown in Table 2.

Table 2. The experimental results of the execution times

Protocol	Operation	Execution time (sec)
Copy	Generating K	0.002
	Infrared communication	3.302
	Writing K	0.690
	Total	3.994
Verify	PIN authentication	0.112
	Reading K	0.036
	Verifying K	0.002
	Total	0.150

4.3 Discussion

According to the result with respect to Copy-Protocol, the execution time of the infrared communication occupies a large portion of the total time. As we mentioned, in the step of generating K, an exclusive-or operation is used in place of a complex cryptographic function. Therefore, we measure the execution time of the cryptographic part on a smart card and estimate the time on a mobile phone.

Some smart cards have a cryptographic co-processor which can be used freely. Mobile phones will have the same cryptographic co-processor as smart cards since mobile phones with an IC chip can use for the usage to a smart card. Therefore, as another experiment, we measured the execution times of DES and RSA on JAVA Card to estimate the performance of mobile phones. In the experiment by JAVA card, the execution time of DES and RSA are 0.05 seconds and 0.21 seconds, respectively. The times are short compared with the execution times of an infrared communication and writing K on an IC chip. Thus, the gap between the execution times on a mobile for DES/RSA and the exclusive-or operation seems not to lead fatal inconsistency on our implementation.

As to the result for Verify-Protocol, the execution time of the PIN authentication occupies a large portion of the total time. Supposing that we adopt an advanced cryptographic function, the time for the verifying K will be longer. We can expect that the influence for the total execution time is small, since server has a lot of resources.

From the previous discussions, we can show that mobile phones can unite practicable operation time and safety in the future as adopted our proposed system.

5 Conclusion

We proposed a door access control system with mobile phones which allows off-line copies of a door-key. A model of door access control with mobile phones was introduced, and then a copy of a door-key and a criterion of security were formalized. On this model, a secure copy by off-line was realized using the essential idea of the proxy signature. Moreover, the proposed system was implemented on mobile phones.

Acknowledgment

This work has been supported partially by the Grant-in-Aid for Scientific Research No. 17700020 of the Ministry of Education, Culture, Sports, Science and Technology (MEXT) from 2005 to 2007.

References

1. T. Eng and T. Okamoto. Single-term divisible electronic coins. In *Proceedings of EUROCRYPT' 94*, volume 950, 1994.
2. O. Goldreich. *Foundations of Cryptography*. Cambridge, 2001.
3. M. Mambo, K. Usuda, and E. Okamoto. Proxy signatures: Delegation of the power to sign messages. *IEICE TRANSACTIONS on Fundamentals of Electronics, Communications and Computer Sciences*, E79-A(9):1338–1354, 9 1996.
4. B. Neuman. Proxy-based authorization and accounting for distributed systems. In *Distributed Computing Systems, 1993., Proceedings the 13th International Conference on*, 1993.
5. NTT DoCoMo Inc. http://www.nttdocomo.com/.
6. T. Okamoto and K. Ohta. Universal electronic cash. In *Proceedings of the 11th Annual International Cryptology Conference on Advances in Cryptology*, volume 576. Springer-Verlag, 1991.

UWB Radar: Vision through a wall

Ondrej Sisma[1], Alain Gaugue[1], Christophe Liebe[1], Jean-Marc Ogier[1]

[1] Laboratoire Informatique, Image et Interaction (L3i), Université de La Rochelle
Pôle Science et Technologie, 17042 La Rochelle Cedex 1
sismao1@fel.cvut.cz, christophe.liebe@univ-lr.fr

Abstract. There exist a lot of methods for vision through an opaque medium. At present UWB (Ultra-Wideband) technology is used more and more because it is suitable for localization and detection of a human body behind a wall. First of all this paper describes known methods for vision through walls, which can be divided into two general groups - imaging and non-imaging systems. Secondly it describes the state of the art UWB radar for this application and its specifics. Finally it depicts our UWB radar system (centre frequency 4,7GHz) and our practical procedures relevant to the detection of a human body presented behind a concrete wall.

Keywords: UWB, radar, through-the-wall surveillance, imaging system, non-imaging system, bistatic radar.

1 Introduction

Detection of persons and things behind an opaque medium is a very interesting theme for a wide range of industries and in following years this issue will constitute a strategic point of applicability in biomedical engineering, security service, agricultural industry, etc. For example, a through-the-wall surveillance (TWS) could be utilized in searching for people in rubble and in buildings on fire. The military could use this technology for bomb-disposal, neutralization of aggressors, hostage rescue, etc.
Different technologies are applicable for through-the-wall surveillance: radio frequency (RF) technologies (from the UHF radar to the submillimeter wave imager), acoustic, X-ray scanner. These all appear to offer a partial solution. The desire to see through a wall creates a technical conflict between the possibility of successfully penetrating the wall, which implies lower frequencies, and the possibility of obtaining maximum resolution for the image, which implies higher frequencies. Perhaps one exception to this is the X-ray system which provides both good resolution and good penetration, but is limited to very short ranges and has safety risks.
The text below elaborates on vision through walls system based on radio frequency. These can be classed in two categories: low frequency non-imaging systems that have good wall penetration capabilities but low spatial resolution, and high frequency imaging systems with limited wall penetration capabilities.

2 Imaging systems

The through-the-wall imaging systems (TWIS) allow an image to be obtained from behind a wall (only the shape of the person or object, not an optical image). Two categories of system exist: active millimetre wave cameras (AMMW), a system functioning like a microwave radar and passive millimetre wave cameras (PMMW), a system similar to a radiometer. The principal characteristics of these systems described below are given in Table 1.

2.1 AMMW camera

In the middle of 1990, Millimetrix with the consortium MIRTAC developed a 94GHz imaging radar for TWS. The latest version was developed by Millivision. This system consists of a frequency modulated continuous wave (FMCW) transmitter (allowing 256 range bins over a 25m range to be achieved) with a broad-beam antenna and a 256 element receiver array, 16 azimuth elements by 16 elevation elements (plus acquisition system, display, etc.). The cover area represents 11° by 11° by 25m of coverage. To cover larger areas, antennas can scan to obtain a coverage at 33° by 33° by 25m, but this process requires up to 3 minutes to collect a single image. The output of the radar can be displayed in either 2-D or 3-D formats. Software was developed for interpreting the images.

In 1997, Hugues Adavanced Electromagnetic Technologies developed a millimetre wave radar that provides 2-D or 3-D images of scenes through interior walls. The radar operates at 50GHz with a 33cm aperture and is mechanically scanned over an angular sector of up to +/- 30° to obtain an image.

In 2004, the Swedish Defence Research Agency (FOI) developed a millimetre wave imaging radar. This radar system is a 94GHz imaging pulse radar system comprising a 32 element receiver array, 4 azimuth elements by 8 elevation elements. The resulting field of view is approximately 3° by 5°. Typical indoor wall materials (plasterboard and chipboard) were tested: the distance to the person was 20m, and a wall was placed 5m from the radar between the radar and the person.

2.2 PMMW camera

In 1996, Thermotrex developed an original concept of passive millimeter wave camera for security applications. The camera combines a phase array radio receiver with an acousto-optic processor to create a real time millimetre-wave to visible-light converter. Then the visible-light is detected using a standard video camera. The millimeter camera captures the incoming millimetre radiation through a 32 channel, $0.9m^2$ array antenna. The antenna signal is downconverted from 94GHz to 9GHz for modulating the acousto-optic converter. For TWS applications, they tested their camera indoors. However to see something through a wall, the scene needs to be "backlit" by a metal mirror used to reflect external light. Since then, this architecture has been abandoned, because it was too expensive.

Table 1. State-of-the-art imaging system of TWS (through-the-wall surveillance technology)
Notice: ? = Unknown data

Authors Company System	Freq. (GHz)	Input type/ size	Spatial Resolution	Field of View	Output Dim/ Type	Type (power)
N. Currie Hugues AETC [1]	50	Lens 300mm	6cm	+/- 30°	3 D Screen monitor	AMWI Mechanic. scanned.
Huguenin Millivision [2]	94	Lens 300mm	12cm	33°x33°	2 D or 3 D Screen monitor	FMCW AMWI FPA 16x16 Mechanic. scanned.
J. Svedin SDRA (FOI) [3]	94	Lens 300mm	4cm	3°x5°	2 D Screen monitor	Impulse AMWI FPA 4x8
Lovberg Thermotrex [4]	94	Array 0.9x0.9m^2	4cm	6°x8°	2 D Screen monitor	PMMW 32 channel
Huguenin Millitech MAPS [5]	95	Lens 300mm	7cm	?	2 D Screen monitor	PMMW

Millitech Corporation has also developed a PMMW imaging system for security applications. A demonstration of the system's ability to perform through-the-wall imaging was performed with an interior partition mounted on an external platform. A person behind this partition was well seen, and the good image quality showed that he was carrying a gun. Such a system can operate, but only with a high radiometric temperature difference between the person and the wall.

3 Non-imaging systems

These systems allow users to detect the presence of an individual through an intervening door or wall (any non-metallic wall) using a radio frequency sensing technique. Usually these systems use RF radar technologies at low frequencies, from UHF to microwaves.

3.1 FMCW radar

In 1992, Raytheon developed a monitoring system through-the-wall, called the Motion Detection Radar (MDR). This system consists of a UHF radar using a sensitive continuous wave (CW) phase detector (centred at the frequency of 915MHz). This radar detects motion and creates an audio tone which varies in pitch proportional to the motion that is being detected. These tones fluctuate (from 60Hz to 300Hz) giving a relative indication of the distance from the radar and the speed at which the person is moving. The detected motion is between 0.1m/s and 1.7m/s. For the MDR, the range in the open is approximately 30m. Through 15cm of steel reinforced concrete this drops down to 10m. Through 1m of concrete block wall it drops to about 6m.

In 1996, a new capability was added: measuring the distance to moving targets. The system provides not only an indication of movement of the target and its distance, but can also indicate that several individuals are moving. This new 2D-TWS radar system, known as the "2D Concrete Penetration Radar (2D-CPR)" was also developed by Raytheon. This radar is a frequency modulated continuous wave system (FMCW) with a 950MHz centre frequency and a sweep frequency band of 700 to 1200MHz. For the 2D-CPR, the range in the open is approximately 30m. Through a 20cm thick concrete-steel reinforcing, a target can be detected at more than 25m from the inside wall, with a radar set up at a distance of 6m from an outside wall. The principal characteristics of these two radars are given in Table 2. Finally, in 1999 a lightweight field portable version, called the Motion And Ranging Sensor (MARS), was developed by Raytheon. The key limitation of these radars is that if the target is not moving, an object can not be differentiated from a person.

In 1997, the Georgia Tech Research Institute developed a microwave radar, called the RADAR Flashlight. This system was designed to detect the respiration of a motionless human behind a wall. The laboratory unit is a homodyne FMCW radar which operates on a frequency near 10.525GHz (different frequencies had been tested up to 35GHz). The radar output signal displays rises and falls of the rhythmic respiration response. The RADAR Flashlight can detect the respiration of a human standing up to 5m away behind a 20cm hollow-core concrete block wall.

At the same time, SRI International developed a laboratory UHF TWS system built with a network analyser, that is operated as a FMCW radar. The FMCW signal is transmitted from a centre antenna and received by two separate spatial receiver antennas. This particular set-up allows an ultra-wideband differential radar to be

Table 2. State-of-the-art non imaging systems - FMCW.
    *: Frequency is either the centre frequency or the frequency band

Authors Company System	Freq.* (GHz)	Input type / size	Spatial Resolution	Field of View	Output Dim / Type	Type (power)
L. Frazier [6] Hugues MDR-1A	0.915	2 Antennas •omnidir • 9dB	-	+/- 45 °	0 D Audio tone	CW RADAR (10mW)
L. Frazier [7] Hugues 2D-CPR	0.95 0.7-1.2	2 Antennas • array (9dB)	15cm	+/- 45 °	2 D Screen monitor	FMCW RADAR (40mW)
E. Greneker Georgia Tech Radar [8-9] Flashlight	10.52	Antenna • Parabolic (40dB)	-	+/- 16 °	0 D Bar-graph	FMCW RADAR (30mW)
D. Falconer SRI [10] International	0.2– 0.45	3 Antennas • omnidir.	1m	Large	2 D Screen monitor	FMCW diff. RADAR (10mW)

obtained using a sweep-frequency and trilateration technique. The frequency sweep provides the downrange measurement and the trilateration provides the cross range coordinate. The location accuracy of the system is determined by the transmitted bandwidth and the distance between the receiver antennas. The radar output signal is displayed on a screen monitor. The nominal range in the open is approximately 25m. Through 15cm of steel reinforced concrete this drops down to 12m.

3.2 UWB radar

UWB radar uses very narrow or short duration pulses that result in very large, or wideband, transmission bandwidths. The Federal Communication Commission (FCC) adopted a report on using this technology with imaging systems, but with certain frequency and power limitations: systems must be operated below 960MHz or within the frequency band 1.99-10.6GHz.

There are less than ten laboratories throughout the world, which work on the UWB radar application "detection/vision through the walls". Company Time Domain (USA) is first producer of this system. Lawrence Livermoore National Laboratory (LLNL) with radar MIR (Micropower Impulse Radar) is able to detect a man through a concrete wall of 40cm. The laboratory Cambridge Consultants (UK) evaluates their first prototype, PRISM 200. The Moscow Aviation Institute (Russia) realized some UWB radar prototypes that operate in different frequency (from 800MHz to 1,5GHz). Technical specifics are mentioned at table 3.

Other companies or laboratories also work on the subject, but they separate the communication relating to the projects. Thus, information about their prototypes is

Table 3. State-of-the-art UWB technology

Authors, Laboratory/ Company	Prototype, Centre Frequency (GHz)	Bandwidth/ spatial resolution	Range	Field of view/ Power
L. Fullerton Time Domain [11]	RadarVision 3,85	3,5GHz 5cm	10m	Hor. : +/- 60° Ver. : +/- 45° 50µW
Mc Ewan LLNL [12]	MIR 2,5	1GHz 15cm	50m	-
Cambridge consultants [13]	PRISM 200 1,7 – 2,2	0,5GHz 30cm	20m	Hor. : +/- 70° Ver. : +/- 60°
I. Immoreev MAI [14]	1	0,8GHz 0,5m	3m	240µW
J. Tatoian [15] Eureka aero	ImpSAR 2	3,5GHz 5cm	100m	-

unavailable. For example, there can be mentioned the company Camero (Israel) and its radar system Xaver 800, then companies Satimo (France), Akela (USA), University of Rome (Italy) and Defense R&D Canada.

4 UWB radar technology

4.1 Description of UWB radar

Mostly, clear picture of the watching scene is not necessary, because the relevant information: number of people, positions, speed of movement, etc is sufficient. UWB radar is a very good solution for obtaining this information, because it presents a lot of advantages compared to other systems [16]:
- A good capability to penetrate wall and floor materials.
- A good picture spatial resolution (approximately ten centimeters). The wider the bandwidth radar uses, the finer the resolution it obtains.

$$\Delta R = \frac{V}{2n\Delta F} \quad (1)$$

where ΔR is the spatial resolution, V is propagation speed of impulse, n is frequency step and ΔF is bandwidth.
- A possibility to identify targets due to the great numbers of emitting frequencies.
- A noise robustness due to large bandwidth.

UWB radar is based on emitting ultra short impulses from dedicated wide bandwidth. As its name indicates, UWB radar is a system allowing the transmission and then reception of a signal, which is reflected by a target. In this case, the signals are represented by impulses, which have the property to traverse walls and to be reflected back by human skin.

The radar core is based on Gaussian impulse transmitter/receiver modules which allow the evaluation of the features of UWB technology. These two modules permit variable configuration. Firstly the module can be configured as the impulse transmitter and secondly as the impulse receiver. The centre emitting frequency is 4,7GHz and its bandwidth is 3,2GHz.

These modules (initially designed for communication in UWB) have reconfigured hardware (antennas) and software, in order to be able to use these as the radar UWB. Our prototype is the bistatic radar type. An omnidirectional antenna is mounted on the output of a Gaussian impulses generator (transmitter) and a directional antenna on the receiver input. Then it follows ADC (analog-to-digital converters), correlators and the directional antenna (gain 7 dB, field of view +/- 45°) [17]. This disposition allows a spatial resolution of 5cm (formula 1; n=1 and V=$3 \cdot 10^8$m/s)

4.2 Results and discussion

First of all, the deformation and the delay of UWB impulse as it goes through the wall were studied.

UWB impulses were emitted in free space to verify if correct propagation speed in environment was obtained (figure 1). Then the deformation and the delay of UWB impulses in this scene configuration were measured: distance between transmitter and receiver was 3 metres. Just in front of the receiver a 22cm thick wall was situated.

Figure 1 shows the observed propagation delay 1,5ns which is caused by the wall. It is difficult to compare this experimental value with theoretical value, because the wall thickness is not homogenous; the structure of the wall is breeze block. In practice, if UWB impulse is going through the 22cm thick wall then the UWB impulse is propagated through a thickness of 11cm of concrete. This corresponds to a theoretical propagation delay: $\tau = (d \times \sqrt{\varepsilon_r})/c = 1{,}1\text{ns}$, where d is the thickness of the concrete itself, ε_r is dielectric constant of medium ($\varepsilon_r = 10$), c is the speed of light in free space (air). The difference between the calculated and experimental value is due to simplification of wall thickness. The deformation of impulse which was propagated through the wall (see figure 1) is caused partially by the dispersion during propagation inside the wall but also by the imperfect impedance adaptation of antenna situated just near the wall.

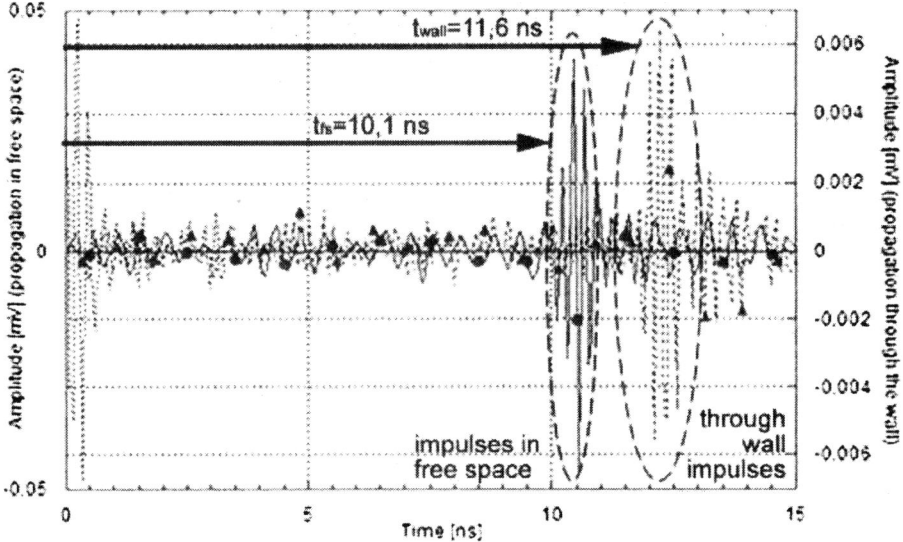

Fig. 1: Influence of the 22cm thick wall on the UWB impulse propagation. t_{fs} corresponds to impulse propagation time in free space, t_{wall} corresponds to impulse propagation time going through the wall. The difference between propagation time t_{fs} and t_{wall} is equal 1,5ns, which indicates propagation delay caused by the wall.

Secondly, an experiment was carried out for the validation of our UWB radar with an observation scene through a 15cm thick concrete wall. The scene configuration, the parameters of the measured scene such as the presence (or not) of a human body and the open (or closed) door, are represented in figure 2. Our disposition actually permits observation of the scene in one direction only. The field of view is associated with an antenna pattern.

Figures 3 to 6 present the different echoes observed through the wall. The scene configuration is described in figure 2. In figure 3 and 4 respectively the observed scene is shown with an open and closed door. In figure 5 and 6 respectively the observed scene is shown with the absence of a human body and presence of a human body in the corridor.

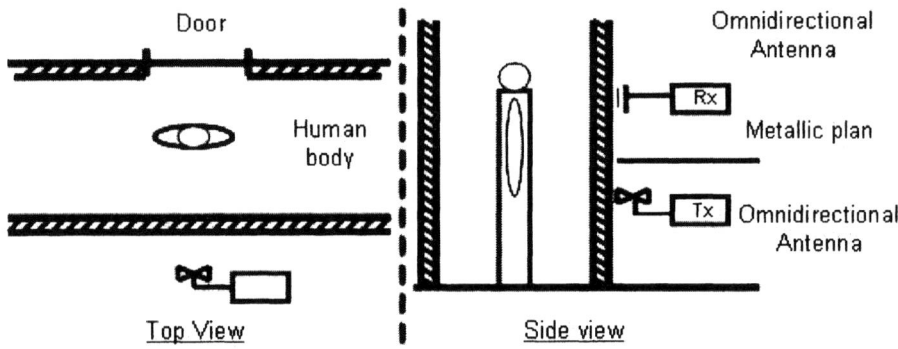

Fig. 2. Measured scene. The thickness of the concrete wall is 15cm.

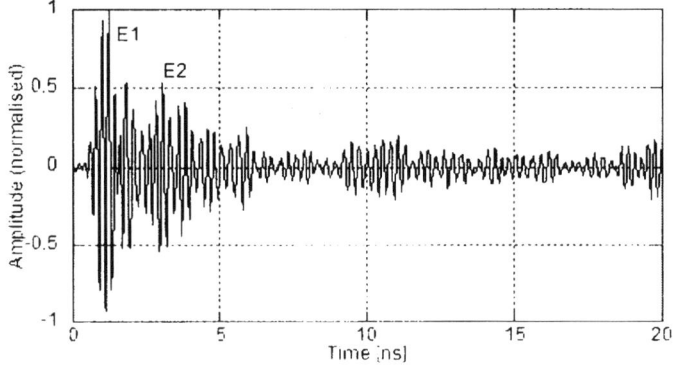

Fig. 3. Radar echoes received through a 15cm thick concrete wall. Measured scene with open door. The echo "E1" corresponds to the direct field received by the receiver, the echo "E2" is due to the reflection on the wall.

Pulses marked "E1" correspond to the direct field received by the receiver, the echoes "E2" are due to the reflection on the wall. The presence of the echo "E3" in figure 6, and its absence in figure 5, highlights the detection of the human body located in the corridor. The echoes "E4" correspond to a wave reflected by the door when it is closed. These signals show the possibilities to recover complementary information, other than the presence or absence of a human body located behind a wall.

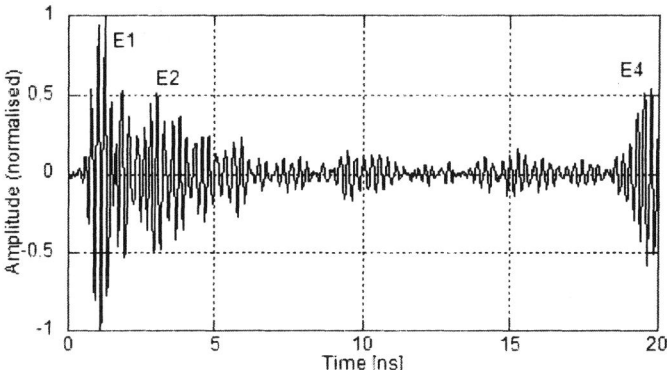

Fig. 4. Radar echoes received through a 15cm thick concrete wall. Measured scene with closed door. The echo "E1" is due to direct field received by the receiver, the echo "E2" is due to the reflection on the wall. The echo "E4" is due to a wave reflected by the closed door.

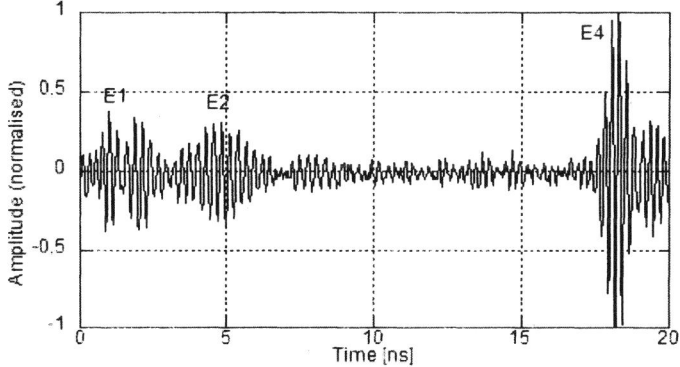

Fig. 5. Radar echoes received through a 15cm thick concrete wall. Measured scene with absence of a human body. The echo "E1" corresponds to direct field received by the receiver, the echo "E2" is due to the reflection on the wall. The echo "E4" is due to a wave reflected by the closed door.

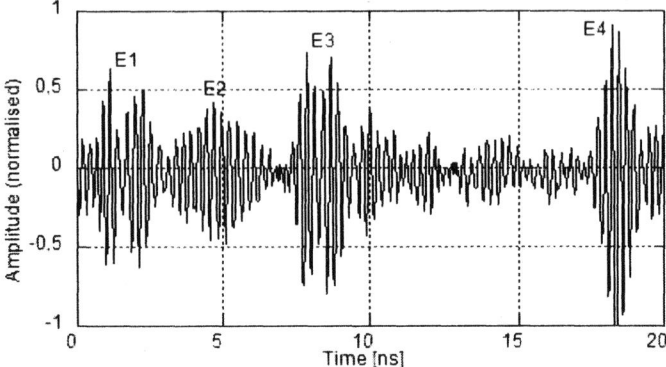

Fig. 6. Radar echoes received through a 15cm thick concrete wall. Measured scene with presence of a human body. The echo "E1" corresponds to direct field received by the receiver, the echo "E2" is due to the reflection on the wall. The echo "E3" corresponds to presence of the human body. The echo "E4" is due to a wave reflected by the closed door.

5 Conclusion and future implication

The currently developed UWB radar system can detect a human body located behind a wall and some parameters of the scene. But this system is limited by scene observation in one direction. We are developing a mechanic sweep system to permit the acquisition of a scene in 2D (typically a space of $10 \times 10 m^2$). This radar will be able to give us the number of people in a space and their positions.

Acknowledgements. This research work was supported by Czech Technical University's grant No. CTU0715013. Authors would like to thank Ms J. Coates for help with English expression.

References

1. N. Currie, D. Ferris, R. McMillan and M. Wicks, "New law enforcement applications of millimeter wave radar", Proceedings of SPIE, Vol. 3066, pp. 10-23, (1997).
2. G. Huguenin, "millimeter-wave video rate imagers", Proceedings of SPIE, Vol. 3064, pp. 34-45, (1997).
3. J. Svedin and L-G. Huss, "A new staring 94GHz focal plane array", Proceedings of SPIE, Vol. 5619, pp. (2005).
4. R. Olsen, J. Lovberg, R. Chou, C. Martin and J. Galliano, "Passive millimeter-wave imaging using a sparse phased-array antenna", Proceedings of SPIE, Vol. 3064, pp. 63-70, (1997).
5. B. Blume, J. Wood and F. Downs, "Naval special warfare PMMW data collection results", Proceedings of SPIE, Vol. 3378, pp. 86-94, (1998).

6. L. Frazier, "Surveillance through walls and other opaque materials", IEEE AES Systems Magazine, pp. 6-9, (Oct 1996).
7. D. Ferris and N. Currie, "A survey of current technologies for through-the-wall surveillance (TWS)", Proceedings of SPIE, Vol. 3577, pp. 62-72, (1998).
8. E. Greneker, "Radar flashlight for through-the-wall detection of humans", Proceedings SPIE 3375 pp. 280-285 (1998).
9. E. Greneker, J. Geisheimer, D. Adreasen, O. Asbel, B. Stevens. B. Mitchell "Development of inexpensive RADAR Flashlight for Law enforcement and corrections applications", Final Technical report on contract NIJ 98-DT-CX-K003, (Apr 2000).
10. Falconer, K. Steadman and D. Watters, "Through-the-wall differential Radar", Proceedings of SPIE, Vol. 2938, pp. 147-151, (1997).
11. K. Siwiak Time Domain Corp. "An introduction to Ultra-Wideband wireless technology", IEEE VTC (Vehicular Technology Conference), May 2001.
12. F. Nekoogar, F. Dowla, A. Spiridon "Rapid synchronisation of Ultra wide band transmitted reference receivers" Wireless, July 2004. Calgary Canada.
13. Gordon Oswald Cambridge consultants "UWB radar Applications" Ultra-WideBand conference, 1999. Downtown Washington DC.
14. Igor Y. Immoreev, P.G. Sergey, V. Samkov, The-Ho Tao "Short – distance Ultra-Wideband Radars. Theory and designing", Radar 2004. International Conference on Radar Systems, october 2004, pp. 211 – 213. Toulouse. France.
15. http://eurekaaerospace.com/impsar.php
16. Ultra-Wideband Radar Technology, J. D. Taylor, 2001, CRC Press.
17. A.C. Lepage, X. Begaud, G. Le Ray and A. Sharaiha « F-probe fed Broadband Triangular Patch Antennas Mounted on a Finite Ground Plane », IEEE Antennas and Propagation Symposium, Monterey, USA, June 2004.

SDU Discard Function of RLC Protocol in UMTS

Vit Krichl[1]

[1] Czech Technical University in Prague, Faculty of Electrical Engineering, Department of Telecommunication Engineering, Technicka 2, 16627 Prague, Czech Republic
krichv1@fel.cvut.cz

Abstract. This paper deals with Radio Link Control protocol (RLC) in UMTS mobile network. It describes in brief the RLC protocol, its functions, properties, parameters, data structure and functional modes. In detail this study reports the acknowledged mode SDU discard procedure and its effect on the protocol performance. For the purpose of monitoring behavior of two RLC entities and performance evaluation of the RLC protocol with different values of RLC parameters a simulation program was developed. Graphic simulation outputs reporting the effect on performance are attached in the paper.

Keywords: RLC, UMTS, performance, Radio Link Control, SDU discard

1 Introduction

UMTS (Universal Mobile Telecommunication System) is a standard of the 3rd generation of mobile telecommunication systems (3G) under specification of 3GPP organization (3rd Generation Partnership Project). WCDMA (Wideband Code Division Multiple Access) on the radio link interface, spread spectrum and high data rates (above 1 Mbps) are the significant features of this technology.

1.1 UMTS Architecture

The architecture of UMTS network is very similar to the 2G (GSM) network architecture. The network elements can be divided in two groups. Radio Access Network - RAN (often called UTRAN - UMTS Terrestrial RAN) provides all the radio service and the access to the core network for the mobile user via radio interface. Analogy in GSM is the BSS (Base Station Subsystem). Core Network CN is responsible for switching and routing functions of voice and data calls and cooperation with external circuit switched (CS) and packet switched (PS) networks. Analogy in GSM is NSS (Network Switching Subsystem). The last part of the UMTS network are the User Equipments (UE) such as mobile phones or laptops. In the UMTS architecture many interfaces (Iu, Uu ...) are also defined and standardized.

Personal Wireless Communications   253

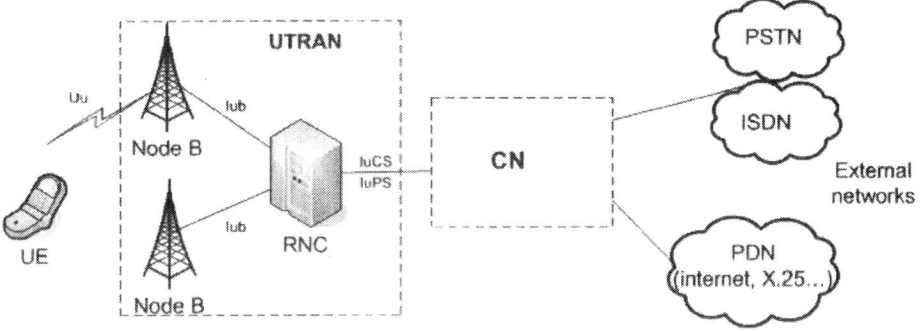

Fig. 1. General UMTS architecture ([1]).

1.2 UTRAN Protocol Architecture

Fig. 2 shows the UTRA interface protocol architecture and relation to the OSI reference model. The data link layer (L2) is divided in the control plane into MAC (Medium Access Control) and RLC (Radio Link Control) protocol layer. In the user plane there are 2 more service-dependent protocols: PDCP (Packet Data Convergence Protocol) for header compression and BMC (Broadcast/ Multicast Control Protocol). In the control plane of network layer (L3) is RRC (Radio Resource Control).

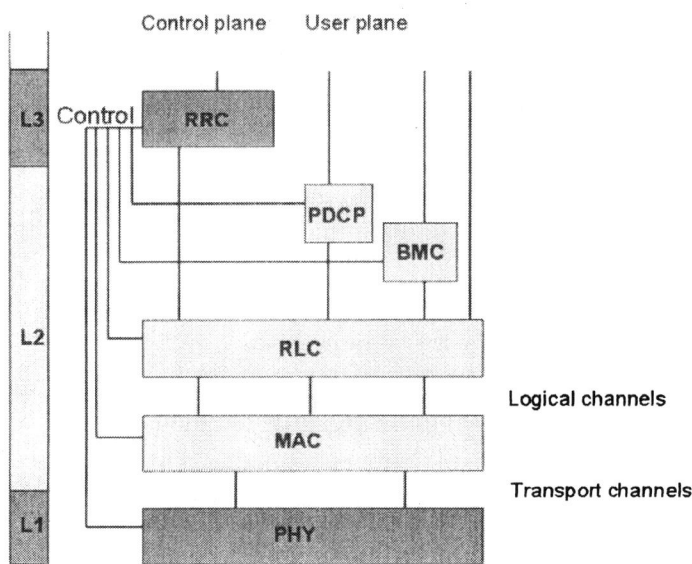

Fig. 2. UTRA protocol architecture ([2]).

2 Radio Link Control Protocol

RLC (Radio Link Control) protocol provides the functionality of the data link layer on the WCDMA radio interface in the UE (user end) and in the RNC (network end). It is responsible for the correct data blocks delivery. It involves the configuration of the radio link parameters and reconfiguration in case of condition change. It works simultaneously in the user and control plane and provides the services for the CS and PS links. RLC protocol provides services for user even for control data. Segmentation, reassembly, concatenation of data blocks, padding, error correction, in-sequence delivery of upper-layer data blocks, duplicate detection, flow control, sequence number check, ciphering is example of protocol layer functions.

2.1. RLC Data Structure

RLC protocol receives upper layer data packets called RLC SDU (RLC Service Data Units). For ensure the requested functions and cooperation with the lower layer (MAC), RLC protocol uses its own data structure by changing format, length and header of the SDU. These protocol specific blocks are called RLC PDU (Protocol Data Units). There are two types of RLC PDUs. Data PDU is used for upper-layer user data (RLC Service Data Unit, RLC SDU) transport and is defined for every functional mode. Control PDUs are used only in Acknowledged Mode for indication of correctly received/ corrupted/ missing PDUs and changing values of protocol parameters.

2.2 RLC Modes

The upper-layer protocol RRC (Radio Resource Control) chooses one of three possible functional RLC modes.

Transparent Mode (TM) adds to the upper-layer data no header. The Receiver simply deletes or marks the erroneous data blocks. TM can be used for voice services, video-telephony or multimedia applications with stream data transfer.

Unacknowledged Mode (UM) uses the sequence numbering of data blocks (PDU) but has no implemented mechanism for their retransmission that's why the data blocks delivery can not be guaranteed. Unacknowledged mode is suitable for Voice over IP (VoIP) and cell broadcast services.

Acknowledged Mode (AM) is using sequence numbering of data blocks and sophisticated retransmission mechanism (requires backwards channel for acknowledgements) for correct user data delivery. The RLC AM is usually used in packet services such as FTP data traffic, or web browsing. Only acknowledged mode network entities will be assumed considered in the next parts of this paper.

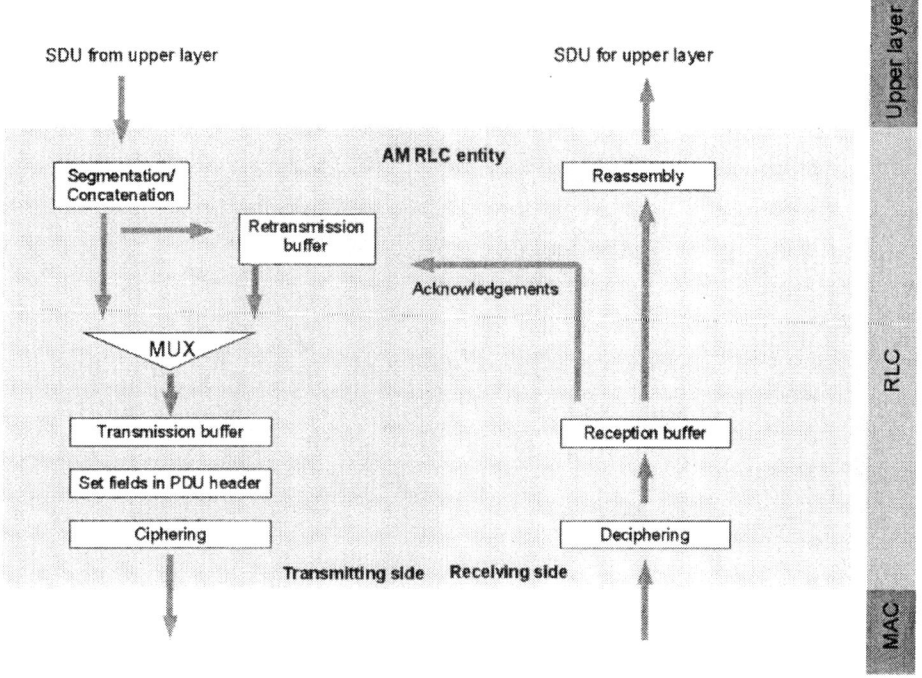

Fig. 3. Acknowledged mode RLC entity model ([3]).

Error correction in the AM is provided by the selective ARQ (Automatic Repeat Request) mechanism. The receiver acknowledges the correctly received PDUs. If erroneous PDU is detected, the receiver requires repeated transmission of this PDU. For effective data transfer a transmitting window is used, it is not necessary to acknowledge every single PDU. On the other side it is not effective to repeat the PDU transmission many times. Many attempts of transfer 1 PDU can delay other data blocks. That's why a SDU discard function is used.

3 SDU Discard Function

The SDU discard function is used by the Sender to discharge RLC PDUs from the RLC PDU buffer. If the transmission of single PDU with a given sequence number does not succeed in defined time or in defined number of transmissions, this PDU and all other PDUs belonging to the same SDU are deleted and no more indicated for transmission. SDU discard function allows avoid buffer overflow. There are several alternative operation modes of the RLC SDU discard function. For an acknowledged mode RLC entity, one of these modes is always configured by upper layers control:
- SDU discard after limited number of transmissions
- Timer based discard with explicit signaling
- No discard after limited number of transmissions

3.1 SDU Discard after Limited Number of Transmissions

Every PDU scheduled for transmissions increases its counter in the Sender. If the counter reaches a given threshold, the relevant SDU will be discarded. The Receiver is notified about discarded SDU by the Status PDU message directly from the Sender. This form of SDU discard function is dependent on the channel rate and its changes and it can not guarantee the maximum delay value. On the other hand it strives to keep the SDU loss rate constant for the connection.

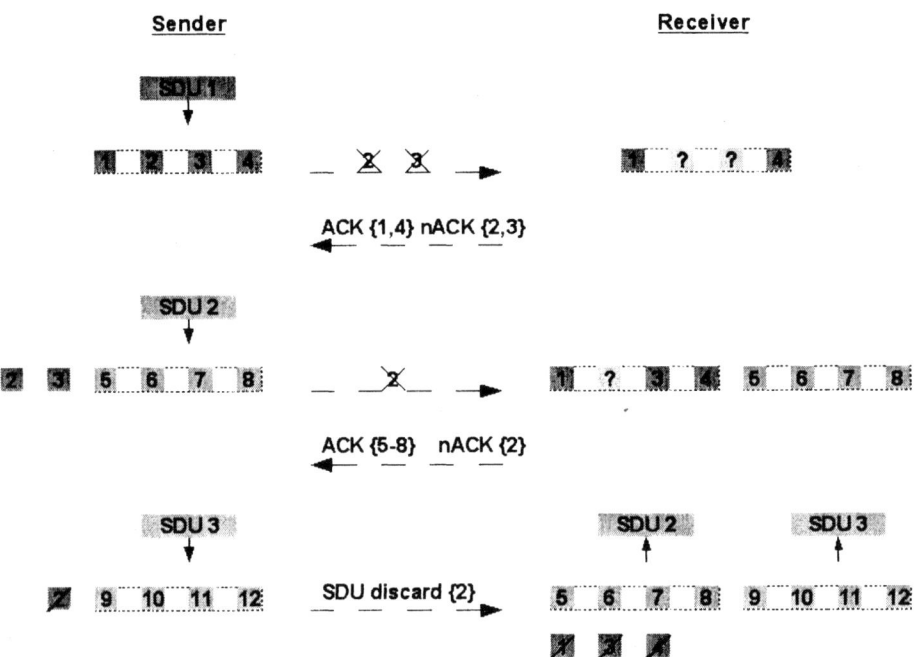

Fig. 4. Simplified SDU discard procedure with 1 retransmission allowed.

For an example of SDU discard procedure (configured with one possible retransmission) see Fig. 5. The first incoming data block SDU 1 is segmented in the RLC sublayer into 4 RLC PDUs. During the transmission of these data blocks PDU 2 and PDU 3 were corrupted that is reported by the Receiver in Status PDU. The Sender will transmit these specified PDUs again together with other (new) segmented PDUs. Because of In-sequence delivery configuration, the Receiver can not submit the reassembled SDU 2 to the upper layer before reception of all PDUs of SDU 1. After reception of the second Status PDU, the Sender does not submit PDU 3 for transmission, because of reached limited number of retransmissions. All fragments of SDU 1 are deleted from (re)transmission buffer and Status PDU is generated to inform the Receiver (move the reception window).

3.2 Timer Based SDU Discard with Explicit Signaling

This version uses a timer based triggering SDU discard. For every SDU received from upper layers, the Sender starts a discard timer. When the discard timer of a SDU expires, the Sender will discard the appropriate SDU and inform the Receiver by a Status PDU (including Move reception window command). This SDU discard mode is insensitive to variations in the channel rate and provides means for exact definition of maximum delay. However, the SDU loss rate of the connection is increased as SDUs are discarded. Possible values of discard timer parameter are specified in [4].

3.3 No Discard after Limited Number of Transmissions

In this alternative mode, the RLC reset procedure is initiated by the Sender after reaching the limited number of PDU transmissions. This version of SDU discard function is not further more described in this paper.

4 Simulation

The created model evaluates behavior of two RLC entities (UE and RNC), their data PDU and control PDU exchange. It provides mainly RLC functions such as segmentation and reassembly of upper layer data blocks, padding of PDU, data PDU and control PDU transfer, ARQ mechanism for error correction and in-sequence delivery of upper-layer data blocks. For better understanding of the RLC parameters influence, the backward channel (for acknowledgments) is considered error-free. In this part of simulation the input data rate was considered constant-size SDU stream with constant time intervals between SDUs. This concept could be simplified model of FTP traffic. For simulating the radio channel conditions - error rates, the Gilbert model was chosen for its easy implementation (often used in wireless networks models). Assigning the probability values was based on data frame trace analysis of GSM radio link with CSD (Circuit Switched Data) service [5]. Modifying these values, we can simulate the radio channel conditions with different BLER.

Table 1. Simulation parameters default values.

Simulation parameter	Parameter value
Transmission time interval	10 ms
SDU size	1280 bytes
SDU data rate	512 kbps
PDU size	80 +2 header bytes
Status timer period	100 ms
PDU per TTI	max 16
PDU one-way delay	45 ms
Window Size	256 PDUs

Table 1 shows the default simulation parameters values. Every incoming SDU of 1280 bytes is first segmented. During one transmission time interval 10 ms up to 16 PDUs each of 82 bytes can be sent. The Receiver generates every 100 ms a Status PDU to acknowledge the correct received PDUs and to report the corrupted PDUs. The average time interval between submitting PDU for the lower layer in RNC (network side) and PDU processing in UE (user side) is fixed 45 ms. This time interval includes one TTI, network propagation time and delays caused by a queue of other traffic PDUs in Node B. For performance evaluation and parameter values effect estimation we simulate 5000 ms of network traffic.

5 Simulation Results

5.1 Limited Number of Retransmission Effect

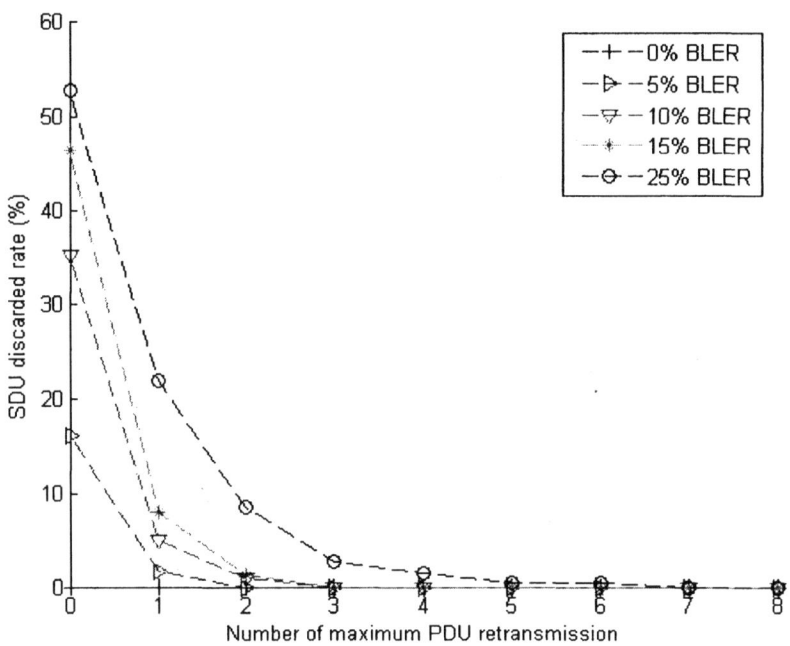

Fig. 5. Effect on SDU discarded rate with different BLER.

The Fig. 4 shows that the portion of SDU discarded during data transmission rapidly increase with more possible attempts for transmission. With 2 retransmissions the share of discarded SDU is lower than 10% in all simulated error rates. Radio channel with BLER <15% has this share under 10% with only one retransmission. Configuring even higher number of retransmission causes only small decrease of the SDU discarded share. High number of possible retransmissions causes rapid increase of RLC SDU's delay due to requested in-sequence delivery of reassembled RLC SDU to upper layer, transmission buffer queuing delay. The SDU delay is also influenced by the status PDUs interval (time period elapsed between 2 control PDUs reporting the corrupted PDUs). Marginally, the transmitting window can be blocked by the PDUs for retransmission.

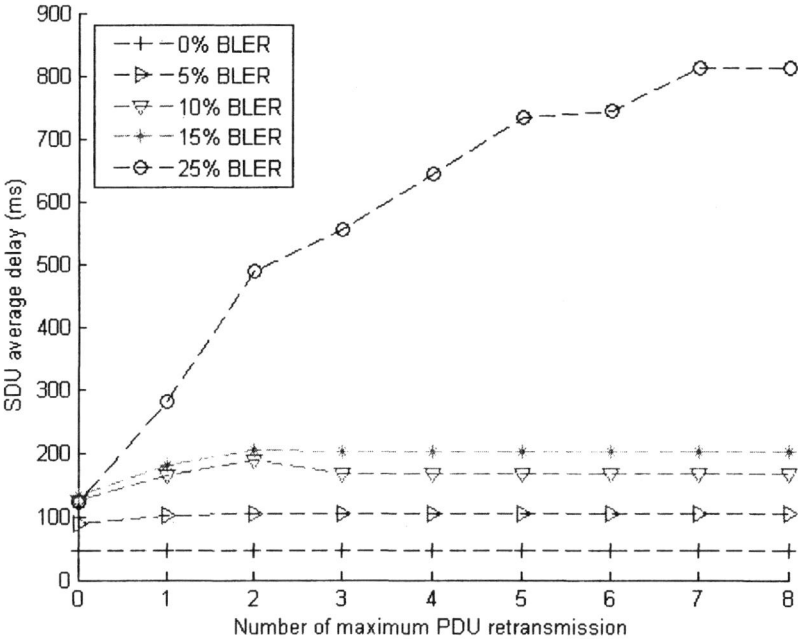

Fig. 6. Effect on SDU average delay with different BLER.

Because of increasing SDU delay it is not suitable to configure the RLC link (with configuration given above) with more than 4 attempts at RLC SDU delivery. The RLC link on the radio channel with low level error rate can be even configured with only 1 or 2 retransmission attempts. In case of RLC PDU transport failure even with reaching the limit of retransmission, an additional error correction of this RLC SDU transfer can be performed by higher layer protocols such as TCP (Transmission Control Protocol) in transport layer.

5.2 Discard Timer Value Effect

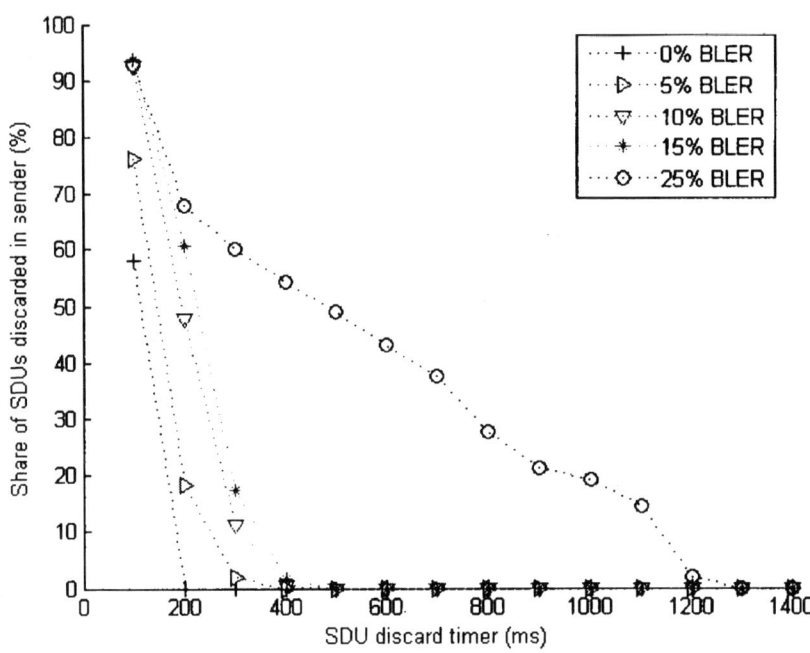

Fig. 7. SDU discard ratio in Sender.

Even with error-free radio channel, with 100 ms of discard timer, there is non-zero SDU discard share because the PDUs are deleted in Sender before Status PDU acknowledges a correct PDU delivery. On the radio channel with block error-rate less than 20 %, it is not necessary to configure the RLC link with SDU discard timer 300 ms or more – leads only to higher SDU delay, SDU discard share is lower than 20%. In conditions of higher error rates, this function provides transmission buffer overflow protection, a significant part of the buffer memory (more than 30%) is periodically cleaned. The number of discarded SDU is also influenced by the time interval between Status PDU reports. Mainly with lower values of discard timer, the portion of SDU discarded in Sender is really high, but it does not mean necessarily SDU loss. As we can see (Fig. 8) SDU discard ratio in Receiver is lower. SDU discard is triggered by the elapsed discard timer in Sender, all PDUs belonging to the SDU are deleted from (re)transmission buffer and a Status PDU report is generated. Before this Status PDU (reporting SDU discard) reaches the Receiver, all relative PDUs sent in previous TTIs could be completed and SDU could be reassembled. Reception of Status PDU in Receiver then only move reception window (no deletion of PDUs from receiving buffer).

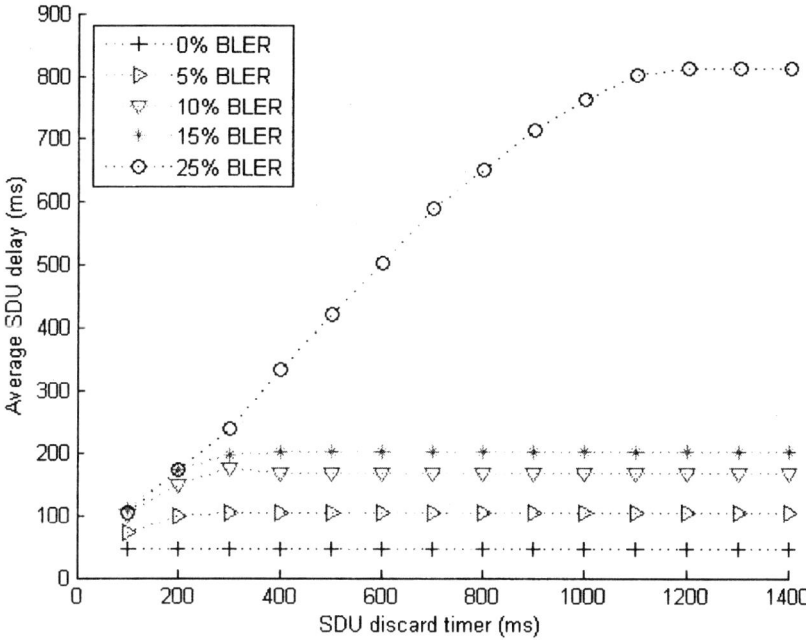

Fig. 8. SDU delay with variable discard timer value.

As a consequence of bigger discard timer value and in-sequence delivery configured in the Receiver, the delay of SDUs is increasing (Fig. 8.). First, the delay increases as a linear function of the discard timer value, the second part of the curve is constant delay line. Thanks to discard timer it can not rise more, the delay is saturated. This is typical feature of this discard function version, which can be used for Quality of Service (maximum delay) guarantee. In theory, maximum possible delay can be determined as a sum of discard timer value and PDU one-way delay, in case of blocked transmission window (very high error rate, too small transmission window).

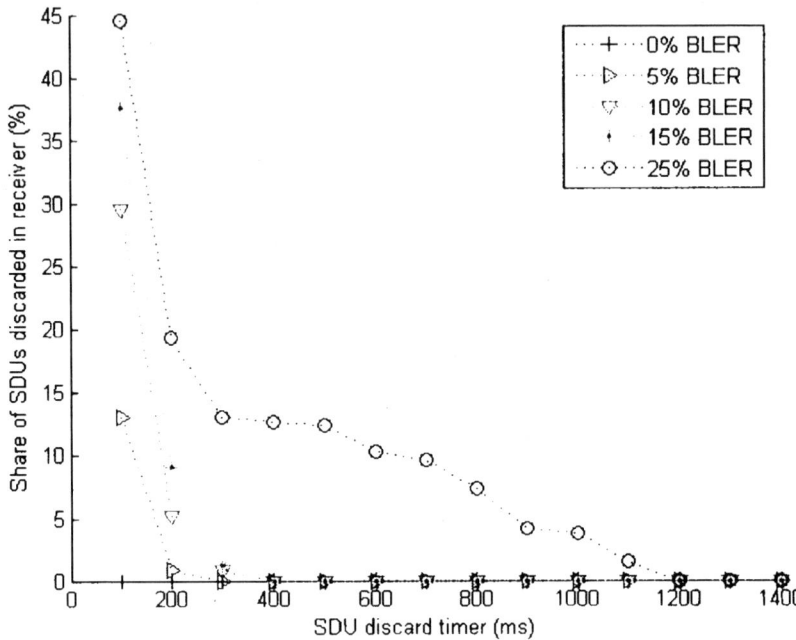

Fig. 9. SDU discard ratio in Receiver.

6 Conclusion

RLC protocol is effective mechanism supporting reliable connection and user data transfers on radio interface in UMTS. Some performance evaluation of RLC protocol concerning mainly the polling mechanism and Status PDU report interval was published in the past [6]. Effect of SDU discard function can be found in our study, which has only evaluated an effect of SDU discard triggers on the RLC performance (considered independently of the other PDU retransmissions triggers, but there are more triggers that can invoke transmission of a status report). The future studies can focus on the optimization of two or more triggers which are simultaneously active. The model developed in Matlab environment is easily to extend or improve.

Our simulation results indicate that the RLC protocol with number of retransmissions triggered SDU discard function can have the optimal performance with given configuration when the number of retransmission is set to 1 or 2. With higher number of possible PDU retransmission, the SDU discarded rate is low, but the SDU delay is extremely high and for many required services unacceptable. Timer triggered SDU discard function provide means of guarantee QoS, data blocks delay in detail. The more intensive data rate, the more SDU discarded ratio because of higher probability of transmission window blocked and heavy retransmission. When no PDU retransmission is possible (because of discard timer, PDU transmission counter) PDUs would be discarded and the higher-layer protocol (i.e. TCP) would have to provide correct delivery of these data blocks.

Acknowledgments. This research work was supported by Czech Technical University's grant No. CTU0715013.

References

1. Kaaranen, H.: UMTS networks: architecture, mobility and services, 2nd edition, Chichester: Wiley (2005) ISBN 0-470-01103-3
2. Holma, H., Toskala, A.: WCDMA for UMTS: Radio Access for Third Generation Mobile Communication, 3rd edition. Chichester : Wiley (2004) ISBN 0-470-87096-6.
3. Radio Link Control (RLC) protocol specification (Release 7), 3GPP TS 25.322 V7.2.0 (2006-10). http://www.3gpp.org/ftp/Specs/html-info/25322.htm
4. Radio Resource Control (RRC) specification (Release 7): 3GPP TS 25.331 V7.2.0 (2006-09) http://www.3gpp.org/ftp/Specs/html-info/25331.htm
5. Konrad, A., Zhao, B.: A Markov-Based Channel Model Algorithm for Wireless Network, ACM Wireless Networks (2003) http://bnrg.cs.berkeley.edu/~adj/publications/paper-files/
6. Li, J., Montuno, D., Wang, J.: Performance Evaluation of the Radio Link Control Protocol in 3G UMTS, MITACS (2004) http://www.scs.carleton.ca/~canccom/Publications/LiPerfEvalRLC04.pdf

Scheduling algorithms for 4G wireless networks

Jaume Ramis, Loren Carrasco, Guillem Femenias and Felip Riera-Palou

Mobile Communications Group - Dept. of Mathematics and Informatics
University of the Balearic Islands
07122 Palma, Mallorca (Illes Balears), Spain.
{jaume.ramis,loren.carrasco,guillem.femenias,felip.riera}@uib.es

Abstract. Scheduling algorithms are fundamental components in the process of resource management in mobile communication networks with heterogeneous QoS requirements such as delay, delay jitter, packet loss rate or throughput. The random characteristics of the propagation environment and the use of complex physical layers in order to combat this random behavior further complicates the design of simple, efficient, scalable and fair scheduling algorithms. This paper presents the main criteria used in the design of scheduling algorithms for 3G/4G mobile communications networks and provides a survey of scheduling mechanisms proposed for use in TDMA and CDMA based systems.

Keywords: scheduling, 4G, GPS, utility functions

1 Introduction

One of the most challenging issues for next-generation wireless networks is the provision of Quality-of-Service (QoS) guarantees when dealing with the high number of emergent multimedia applications. This necessitates the development of high-performance physical-layer technologies, as well as powerful resource management strategies to provide high throughput and efficient use of resources. Among these strategies, scheduling algorithms, which distribute the available 'capacity' among existent connections, have been recognized as key components of these QoS aware wireless systems. In order to support the provision of QoS in wireless networks a large number of traffic scheduling algorithms have been proposed in the literature. The knowhow in wireline schedulers has been the basis for the development of these scheduling strategies; however, the service heterogeneity, the scarcity of resources, and the hostility and variability of mobile radio channels, have rendered unavoidable the adaptation of wireline proposals to the more challenging wireless scenario. Scheduling algorithms for wireless networks can be classified into two categories: centralized and distributed algorithms. Distributed proposals are mainly applied in adhoc or uplink operation, where users contend for channel access; these strategies do not achieve the efficiency, fairness and fulfilment of QoS requirements that can be reached with centralized algorithms. In this paper we propose an exhaustive overview of centralized wireless scheduling techniques, which are evaluated in accordance with a set of relevant performance criteria for next generation wireless networks scenarios.

2 Scheduling criteria for 4G wireless networks

The main criteria used to evaluate wireless schedulers in their application to 3G/4G wireless networks are:

- **Efficiency:** In highly loaded 4G scenarios, efficiency (measured in terms of total achieved *throughput*) is one of the most significant performance criteria. In TDMA the maximum efficiency is achieved when at each time instant the user with the highest available throughput is selected for transmission (multiuser diversity). If the physical layer includes a CDMA component the *soft-capacity* phenomena should be considered in the scheduling process in order to increase the efficiency. That is, the scheduler should not only take into account the capacity variations due to changes in the existent interference level, but should also consider the capacity variations provoked when selecting one or another particular combination of services to be transmitted simultaneously.
- **Applicability:** Using this concept we group the issues of algorithm complexity, amount of signalling involved in the scheduling process, parameter settings (i.e. GPS weights determination) and the considered channel model. With respect to this last point, a scheduler for 4G wireless systems should be capable of managing an adaptive channel model with multiple possible states corresponding to the different transmission modes available in 3G-4G physical interfaces. This implies that the scheduler design strongly depends on the physical layer characteristics and that the use of a cross-layer design of both layers is highly desirable.
- **QoS support:** The existing service heterogeneity in new multimedia networks is directly translated into multiple and distinct QoS traffic requirements that should be jointly managed and guaranteed by the scheduler. Three levels of QoS support are considered: best-effort traffic, data traffic with throughput and delay guarantees, and real-time traffic.
- **Fairness:** A fair distribution of resources between connections of the same type is recommended. In wireless scenarios different kinds of fairness could be implemented: fairness in terms of the data rate assigned to each connection or in terms of resource consumption.

3 Classification of scheduling algorithms

Wireless networks schedulers proposed in the literature could be grouped in different families. Figure 1 shows a non exhaustive classification that nevertheless includes the main existing scheduling families.

3.1 GPS-based scheduling algorithms

GPS (*Generalized Processor Sharing*) [4], also known as FFQ (*Fluid-flow Fair Queueing*) [5], is a fair, work-conserving, flexible and efficient algorithm originally devised for error-free wireline networks. The fundamental concept in GPS-based algorithms is that the amount of service session i receives from the switch

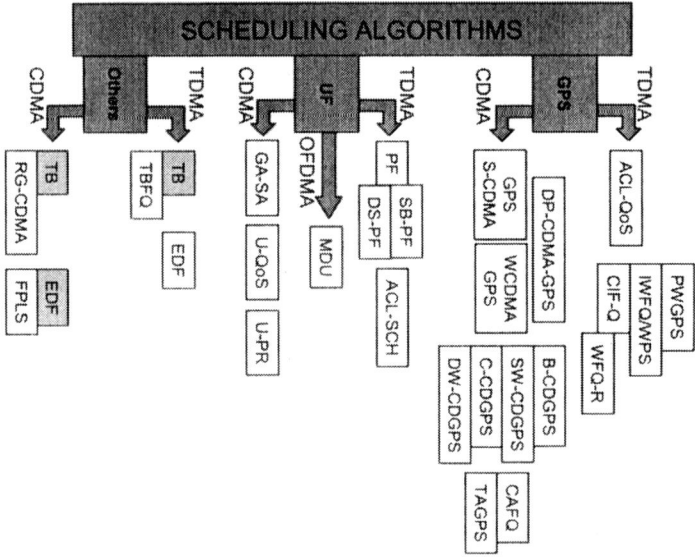

Fig. 1. Wireless schedulers classification.

(in terms of transmitted packets) is proportional to a positive weight also called relative service share ϕ_i, provided that the session is continuously backlogged in this interval. Defining $S_i(t,\tau]$ as the amount of service received by session i in the time interval $(t,\tau]$, then it holds that $\frac{S_i(t,\tau]}{S_j(t,\tau]} = \frac{\phi_i}{\phi_j}$ for all sessions j that have received service in this time interval. It follows that in the worst case, the minimum guaranteed rate r_i given to session i is $r_i = r\phi_i / \sum_{j=1}^{M} \phi_j$, where M is the maximum number of sessions that could be active in the system and r is the available total throughput. Therefore, from a user perspective, GPS guarantees that network resources are allocated to the sessions irrespectively of the behavior of other sessions (isolation property), and that the distribution is performed with perfect fairness, that is, whenever a session i generates traffic at a rate less that r_i, the 'spare' bandwidth is allocated to other sessions proportionally to their respective weights. In addition, if session i is (σ_i, ρ_i)-leaky bucket constrained, where ρ_i is the token generation rate and σ_i the size of the token bucket, and the minimum guaranteed rate is such that $r_i \geq \rho_i$, then the maximum delay is bounded by $D_i^{max} \leq \sigma_i/\rho_i$. The main consequence of this property is the possibility of using GPS for delay constrained traffic. Therefore, session rate and delay guarantees depend on an adequate choice of the service share ϕ, but the determination of sessions weights could be a challenging task. When the sessions have a long term average rate, the ϕ_i allocation appears to be straightforward. The ϕ_i determination is more complex when the traffic is bursty or self-similar. To fix its value, minimum session rate and maximum accepted delay have to be taken into account.

GPS considers traffic as a fluid, therefore ideal GPS cannot be implemented in practical schedulers, since it requires that the scheduler serves multiple flows simultaneously and that the traffic is infinitely divisible. There are many packet-level algorithmic implementations of this model. Essentially, the goal of these algorithms is to serve packets in an order that approximates GPS as closely as possible. The first GPS packet adaptation was the PGPS (*Packet-by-packet GPS*) [4] also called WFQ (*Weighted Fair Queueing*) [5]. This algorithm is based on the 'system virtual time' definition. The virtual time concept is used to track the progress of the system under the GPS discipline and will lead to a practical implementation of PGPS. When a new packet arrives to the PGPS server it is stamped with its 'virtual starting time' t_{vs} and its 'virtual finishing time' t_{vf} defined as the time instant at witch this packet would start/ finish service under the fluid GPS respectively. Packets are served according to virtual finishing time order. Both times could be calculated on the packet arrival as long as the set of active sessions at that moment is known. Obviously, this adaptation of the fluid GPS system to the packetized transmission implies a significant increase of complexity due to the fact that the scheduler has to tag each packet and has to maintain the system virtual time permanently updated. There are several proposals derived from PGPS that either approximate more accurately the GPS behavior [6], [7], [8] or decrease the PGPS complexity [9], [10], [11].

When adapting wireline fair queuing algorithms to wireless channels it may happen that some flows can transmit but other flows cannot due to location dependent channel errors; therefore only a subset of flows can be scheduled on the channel at a given time instant, and this subset is dynamically changing as time evolves. All the wireline algorithms cited above assume that the channel is error-free, or at least that either all flows can be scheduled or none of them can. In order to succeed in maintaining the long term fairness, most wireless GPS-based algorithms introduce compensation models (flows that at a certain time interval receive less channel resources than in an error-free environment will be compensated by receiving extra resources when their channel conditions improve).

GPS-based schedulers for TDMA networks incorporate compensation models to cope with the channel variability, e.g. IWFQ (*Idealized Wireless Fair Queueing*) and WPS (*Wireless Packet Scheduler*) [2]. The compensation model is based on several concepts: the *error free service* of a flow is the service that it would have received at the same time instant if all channels had been error free under identical offered load. A flow is said to be *leading* if it has received an allocation (*lead*) in excess of its error free service, in contrast, a flow is said to be *lagging* if it has received an allocation (*lag*) less than its error free service. The compensation for the service loss suffered by *lagging* sessions will be performed when these connections have a radio channel that can ensure the meeting of their QoS requirements. Obviously the extra service assigned to *lagging* sessions during compensation will be subtracted from *leading* sessions. In IWFQ and WPS strategies, the packet from the subset of flows with a good channel and the lowest *virtual finishing time* is selected to be transmitted. Since the virtual time of a session increases only when it receives service, this may result in a large difference between the virtual time of an error session i and a error-free session. If session i exits from error it will then have the smallest virtual time among all sessions. The server will select session i exclusively for service until its virtual time catches up with those of other sessions. By artificially bounding the *lag* and *lead*, it can be established a trade off between long-term fairness and the starvation of error-free sessions. To solve these problems the CIF-Q (*Channel Condition Independent Fair Queueing*) [12] algorithm defines an additional parameter called *lag* that keeps track of the difference between the service a session should receive in the error-free reference system and the service it has actually received. The packet selected to be transmitted is the one with the lowest starting virtual time in the error-free system, but if the selected packet corresponds to a leading flow that has already received its guaranteed share of service (according to its weight) the slot will be assigned to the *lagging* flows proportionally to their weights. This mechanism guarantees the delay bound and throughput of error-free sessions and provides long-term and short-term fairness.

Another algorithm that tries to solve the problem of starvation is the PWGPS (*Packet Wireless GPS*) [13]. Packets are served in increasing order of a simplified virtual finishing time. In this case the compensation is achieved increasing the GPS weights ϕ of the error sessions until they have obtained their corresponding service share. This scheduler does not require the maintenance of an error-free system thanks to a redefinition of the virtual finishing time. However, the changes in the weights imply that those virtual times can not be calculated at the packet arrival and should be recalculated at each scheduling interval.

All the above algorithms assume a perfect channel prediction before transmission, but this is a very difficult task and in real networks link layer mechanism as ARQ retransmissions are required. The algorithm WFQ-R (*WFQ with link level Retransmission*) [14] distributes the scarce wireless resources among all flows according to their weights, but considering also the resource consumption of the retransmissions. It combines the CIF-Q algorithm with a compensation system where the share used for retransmissions is regarded as a debt of the retransmitted flow to the others. The compensation could be charged to the retransmitted flow only (an error-prone flow should take responsibility for its own channel condition) or the compensation is distributed over all flows proportionally to their weights. The algorithm ACL-QoS (*Adaptive Cross-Layer scheduler with prescribed QoS guarantees*) [15] is a recent proposal of a TDMA-scheduler with a cross layer design. Instead of an on-off physical layer model it assumes a system with AMC (*Adaptive Modulation and Coding*). The channel is modeled as a finite Markov chain with multiple states, i.e. depending on the channel state a certain modulation and coding scheme will be selected. The scheduler distinguishes between two traffic types: QoS-guaranteed traffic and best-effort traffic. The resource distribution for QoS-guaranteed traffic does not follow a GPS discipline but takes into account the amount of data in the transmission buffer and the channel state, that is, the scheduler uses information from physical and data link layers. The resources left after this process are distributed among best-effort connections using GPS.

As we already mentioned, the fluid model of the GPS discipline assumes that multiple sessions with different assigned rates can be served simultaneously. The adaptation of this fluid model to TDMA requires the maintenance of a system virtual time and therefore a significant increase in system complexity. Otherwise CDMA offers the possibility of simultaneous transmissions, and the resource distribution could be adjusted by using the spreading factor, the amount of assigned power, etc.

GPS-based schedulers for CDMA networks are derived from the original GPS algorithm, and they consider slotted-CDMA physical layers, i.e. the distribution of resources is performed frame by frame.

B-CDGPS (*Basic CDGPS*) [16] is a simple scheduler that considers an ideal channel with a fixed capacity estimated as a lower bound of the system soft capacity. This implies a severe loss of efficiency and, therefore, the rest of considered algorithms take the variability of the CDMA capacity into account. For instance, GPS S-CDMA (*GPS in Slotted CDMA*) [17], SW-CDGPS (*Static Weighted CDGPS*) [16], C-CDGPS (*Credit based CDGPS*) [16] and the WCDMA GPS [18], consider the capacity variations in an ideal channel. The S-CDMA [17] scheme, derived from PGPS, considers a unicellular system in which the system resources (power and bandwidth) are distributed proportionally to the connections weights ϕ_i. BER and delay requirements of each connection are used to adjust the transmission powers and GPS weights. The algorithm considers that any transmission rate is possible. The SW-CDGPS (*Static Weighted CDGPS*) [16] algorithm introduces the concept of *nominal capacity*, which corresponds to the amount of available resources in a cell, and this can be determined if the intercell interference level is known. The set of SIR values and transmission rates corresponding to all active users in the cell can be derived from the users capacity requests and the nominal cell capacity. The calculation process maximizes the aggregate throughput and satisfies the GPS fairness property. To increase SW-CDGPS efficiency, the C-CDGPS (*Credit based CDGPS*) [16] algorithm sacrifices the short-term fairness of non real time traffic. This is implemented by using a credit counter to track the difference between the received service share of a session and the fair GPS service that would correspond to that session. This counter is bounded in order to achieve a long-term fairness. The WCDMA GPS algorithm [18] also makes use of the *nominal capacity* concept, but instead of following the basic GPS strategy, resources are distributed using the PGPS discipline and considering a discrete set of possible transmission rates.

As we have already mentioned, the described schedulers consider an ideal channel. The introduction of a non-ideal channel implies, as well as in TDMA proposals, the addition of compensation mechanisms to guarantee long-term fairness to those sessions with a 'bad' channel. Some examples of this kind of schemes are DW-CDGPS (*Dynamic Weighted CDGPS*) [16], CAFQ (*Channel Adaptive Fair Queueing*) [19] and TAGPS (*Traffic Aided GPS*) [3]. The proposal of [16] firstly schedules real time transmissions using SW-CDGPS, and secondly distributes the remaining capacity among the non real time connections according to the DW-CDGPS algorithm: the GPS weights of those connections will be adjusted dynamically every scheduling period depending on their channel conditions in order to increase the data throughput. In this compensation mechanism there is a trade-off between efficiency and short-term fairness. Similarly to many GPS-based TDMA algorithms, the CAFQ [19] algorithm has to keep an error-free system as a reference, in which the SFQ discipline [11] is used to schedule transmissions. In the real system, sessions are arranged in increasing order of *lagging* credits (the bigger the difference of service received in the real system in comparison to the service received in the reference system, the more *lagging* credits), and error-free sessions are chosen to be served. There is an exception: if the reference system selects a lagging session with an error-free channel to be transmitted, the real system will maintain this selection. TAGPS [3] uses a similar mechanism. Similarly to other proposals [17] system resources are represented by the power index g_i. It distinguishes two user types: voice with a constant transmission rate and data with variable bitrate. The distribution of resources consists in determining the g_i values of data users. If a 'lagging' user perceives a 'good' channel, its GPS-weight, ϕ_i (and consequently its power index, g_i) will be increased in order to compensate its lack of service.

The DP-CDMA-GPS (*Dynamic Programming CDMA-GPS*) [20] algorithm tries to avoid delay and bandwidth coupling. To that end it uses variable weights, which are determined by using a cost function that minimizes the queuing delay experienced by active connections. This cost function also takes into account the QoS requirements and the radio channel conditions of the different connections.

3.2 Utility-based scheduling algorithms

This theory has its origins in economics, where the utility functions are used to quantify the benefit of using certain resources. Similarly, utility theory can be used in communications networks to evaluate the degree to which a network satisfies service requirements of user's applications, rather than in terms of system centric quantities like throughput, outage probability, packet drop rate and power [21]. The utility function maps the network resources a user utilizes into a real number that tries to reflect the level of user satisfaction derived from using these resources. Different applications have different utility function curves or even different parameters. For instance, the utility functions of best-effort applications consider throughput, whereas those of delay sensitive traffic applications take into account delay. Finding the most adequate utility functions for the different types of applications is one of the key elements to ensure a correct performance of this kind of techniques. Examples of different utility functions can be found in [22–24].

Once the utility functions of all applications in the system have been defined, the objective is to achieve an optimum scheduler that maximizes the aggregate utility in the system subject to the capacity limit determined by the physical layer techniques. Usually, utility functions are non linear and the problem can be regarded as a non linear combinatorial optimization of a cost function (the aggregated system utility) with a large dimension. This optimization is a complex process and the algorithms designed for this purpose constitute the other key element of the utility-based scheduling architecture. The optimization algorithms should be efficient (in terms of computational effort) and they have to work with a discrete set of possible solutions (the physical layer offers a discrete set of possible rates and/or delays). For instance, in [25] a set of algorithms for OFDM wireless networks are proposed, and some of them could easily be adapted to CDMA networks.

Proportional fair schedulers. The PF (*Basic Proportional Fair*) [26] is a downlink algorithm well suited for best-effort traffic. The scheduler tries to take advantage of the independent channel variations experienced by different users by scheduling transmissions that perceive strong signal levels. The scheduler sends data to the mobile with the highest DRC_i/R_i ratio, where DRC_i represents the highest feasible rate out of all possible rates under those channel conditions in a given slot, and R_i represents the average rate received by the mobile over a window of appropriate size. This way, each user is served in periods where its requested rate is closer to the peak compared to its recent requests. This channel-aware scheduling can substantially improve network performance through *multiuser diversity*. It can be shown that this system amounts to use a logarithmic utility function for all users $U_i(R_i) = \ln(R_i)$. In this case, the aggregate system utility optimization can be performed using a utility-based gradient scheduling (assigning the resources to the connection with a higher DRC_i/R_i in each scheduling period). PF is an example of an utility-based solution whose aggregate cost function optimization is very simple thanks to the chosen utility function. This algorithm is used in current systems (i.e. HDR channel in the CDMA2000 system).

The main drawback of the PF scheduling stems from the use of a rate-dependent-only utility function that makes the system unable to ensure traffic delay requirements. Several proposals try to solve this limitation [27, 28] maintaining the PF utility function for best-effort traffic and defining new utility functions for delay sensitive traffic flows. In the DS-PF scheduler (*Delay Sensitivity PF*) [27] real-time sessions increase their priority when a certain delay value is trespassed, while in the SB-PF (*Sender Buffer PF*) [28] proposal the objective is to avoid playout buffer starvation. Unfortunately, in these mixed cases with different utility functions, the aggregate system utility optimization is much more complex. To preserve PF simplicity, both Barriac in [27] and Koto in [28] propose sub-optimum optimization algorithms.

Multimedia utility-based schedulers define different utility functions for different traffic types and use optimization algorithms to maximize the aggregate system utility (GA-SA [22], MDU [23], U-PR [24], U-QoS [29], ACL-SCH [30]). Most of these algorithms are designed for CDMA networks [22, 24, 29], but there are also TDMA [30] and OFDMA [23] proposals.

3.3 Other scheduling algorithms

Token Bucket-based schedulers include a data bucket and a token bucket per connection. Each unit of traffic (usually a packet) to be transmitted consumes a token of the token bucket. Tokens are generated at a constant rate derived from the contracted data rate of the connection. The maximum number of tokens in the token bucket limits the allowed peak rate for that connection. The TBFQ (*Token Bucket Fair Queueing*) [31] algorithm was designed for TDMA systems. In real time multimedia applications it is very difficult to predict traffic profiles and out-of-profile degradations may be detrimental to the overall QoS experienced by the end user. The TBFQ tries to adapt to this unpredictable workload by accepting traffic profile violations when excess bandwidth is available, provided the session does not exceed its bandwidth allocation in the long term. To this end, each connection has a sufficiently large input data buffer, and a token bucket holding tokens for one packet transmission only. In addition there is a common token bank and connections can borrow tokens from the bank when its token pool is depleted and there are still packets to be served, or give tokens to the bank during periods when the incoming traffic rate is less than its token generation rate. Each connection has an associated counter keeping track of the number of tokens borrowed from or given to the token bank. The connection with the highest counter value has the highest priority in borrowing tokens. There is also a debt limit to preserve fairness, below which the connection can no longer borrow from the bank.

The RG CDMA (*Rate Guarantee in CDMA*) [32] proposal is adapted to the downlink of a W-CDMA system and it allows connections with variable data rate by using an OVSF (Orthogonal Variable Spreading Factor) code tree. It assigns a guaranteed data rate to each traffic flow and associates a credit counter to it. Every scheduling period the credit counter increases proportionally to the guaranteed data rate and decreases with each transmitted packet. In this way, the credit counter value represents the difference between the guaranteed and the actually executed transmissions. Connections are ordered according to the number of credits and higher rate OVSF codes are assigned to the connections with a higher credit counter value. It should be pointed out that this algorithm does not consider the CDMA soft capacity phenomena and assumes an ideal channel.

EDF-based algorithms. The EDF (*Earliest Deadline First*) [33] is a proposal for TDMA networks. In EDF each packet is tagged with its deadline. This deadline is calculated according to the delay requirements of that traffic flow. Packets are served in the order derived from their deadlines. In mixed TD-CDMA networks the FPLS (*Fair Packet Loss Sharing*) [34] algorithm schedules multimedia packets transmission ensuring that packet losses are distributed among all connections according to their QoS requirements. For this purpose packets are ordered according to their timeout. If the available system capacity is not enough to transmit all backlogged packets, the packets to be discarded (packets reaching their timeout) are selected among packets of all connections taking into account each connection's BER (Bit Error Rate). The main drawback of this algorithm is the assumption of an ideal channel without intercell interference.

4 Conclusions

An obvious conclusion of this study is the impossibility of designing a wireless scheduler that is simultaneously: fair, simple, efficient and able to ensure real-time delay guarantees. Focusing on the GPS-based algorithms, several aspects have to be highlighted:

- The perfect fairness provided by the GPS discipline is not possible due to channel errors affecting a varying subset of connections. Therefore it is necessary to incorporate compensation mechanisms to maintain the long-term fairness. Currently these mechanisms use too simplistic channel models (usually on-off models) to represent de behavior of adaptive 3G-4G physical layers and they affect the isolation property of GPS resulting in an increased complexity.
- If the traffic is leaky-bucket constrained, the maximum packet delay can be bounded. However, since in GPS-based schedulers rate and delay guarantees are controlled by a single parameter, the session weight ϕ_i, there is a coupling effect between rate and delay requirements. This implies that to ensure low delays in a certain traffic flow it is necessary to assign a high portion of the available bandwidth to that connection (i.e assign a high weight). This behavior suffers from a non efficient use of system resources mainly in multimedia environments. As a consequence, GPS-based schedulers are not suitable to manage heterogeneous multimedia traffic environments.
- One of the most significant drawbacks of GPS-based algorithms is the connection's service share ϕ_i determination, as a correct operation of the algorithm depends on a suitable selection of the sessions weight values. Surprisingly many existent proposals do not treat this problem.
- In addition, the packetized adaptations of the GPS fluid discipline require the introduction of the virtual time concept. The calculation of this virtual time significantly increases the GPS original complexity. The GPS-based proposals for TDMA networks will always require the maintenance of a system virtual time because the simultaneous transmissions are not possible. Some of these proposals include a simplified definition of virtual time although these simplifications compromise the fairness requirements.
- In CDMA systems a discrete set of transmission rates is possible. This fact is not taken into account by many proposals that are solely based on GPS fluid model, where transmission rates can take any value.

Therefore if we apply the criteria defined in section (2) we can conclude that in GPS-based techniques efficiency is sacrificed in favor of fairness and that the applicability of GPS algorithms to 3G/4G wireless networks is quite limited and in any case restricted to best-effort traffic and rate-delay guaranteed data.

The basic idea behind utility-based schedulers is the mapping of the resources use (bandwidth, power, etc.) or performance criteria (data rate, delay, etc.) into the corresponding utility function and its optimization, instead of directly measuring the network performance parameters. For instance, if a small increase in the transmission rate allocated to a multimedia application makes it able to start its transmission, the benefit of this small rate increment is highly valuable, whereas if the same increase of the data rate is allocated to a best-effort application that has already started transmission at a high bit rate it will not imply a significative utility increase. The key is the definition of the most adequate utility function for each of the different application classes. The main drawback of utility-based schedulers is that the required aggregate system utility optimization usually requires the resolution of a nonlinear combinatorial optimization problem of a large dimension. The computational cost of this optimization process can be minimized by using efficient optimization algorithms. Some proposals, as for instante the PF algorithm, use simple utility functions and, therefore, optimization can be performed using a very simple technique. It is worth mentioning that physical layer information can be included by means of restrictions to the aggregate system utility [23]. The inclusion of this information in the definition of the cost function makes it necessary to solve the optimization problem every time the channel state changes [30]. Utility-based techniques are a promising scheduling discipline that is able to cope with heterogeneous QoS requirements, can achieve high efficiency and a certain degree of fairness between connections of the same type. The applicability of these techniques in 3G/4G systems is conditioned by the existence of simple and efficient optimization algorithms.

Acknowledgments

This work has been supported in part by the MEC and FEDER under project MARIMBA (TEC2005-0997), Govern de les Illes Balears under project XISPES and grant PCTIB-2005GC1-09, and a Ramon y Cajal fellowship, Spain.

References

1. H. Zhang, "Service disciplines for guaranteed performance service in packet switching networks," *Proc. IEEE*, vol. 83, no. 10, pp. 1374-1396, 1995.
2. S. Lu et al., "Fair Scheduling in Wireless Packet Networks," *IEEE/ACM Trans. Networking*, vol. 7, no. 4, pp. 473-489, 1999.
3. D. Liao and L. Li, "Traffic aided fair scheduling using compensation scheme in CDMA cellular networks," *ICC'05*, vol. 1, pp. 363-367.
4. A. Parekh, "A Generalized Processor Sharing Approach to Flow Control in Integrated Services Networks: The Single-Node Case," *IEEE/ACM Trans. Networking*, vol. 1, no. 3, pp. 344-357, 1993.
5. A. Demers et al., "Analysis and Simulation of a Fair Queueing Algorithm," *SIGCOMM'89*, pp. 1-12, 1989.
6. J. Bennet and H. Zhang, "WF2Q: Worst-case Fair Weighted Fair Queueing," *IEEE INFOCOM 1996*, vol. 1, pp. 120-128, March 1996.

7. J. Lee et al., "WF2Q-M : a worst-case fair weighted fair queueing with maximum rate control," *IEEE GLOBECOM 2002*, vol. 2, pp. 1576-1580, Nov. 2002.
8. J. Gallardo and D. Makrakis, "Dynamic predictive weighted fair queueing for differentiated services," *IEEE ICC 2001*, vol. 8, pp. 2380-2384, June 2001.
9. J. Bennett and H. Zhang, "Hierarchical packet fair queueing algorithms," *IEEE/ACM Trans. Networking*, vol. 5, no. 5, pp. 675-689, 1997.
10. S. Golestani, "A self-clocked fair queueing scheme for broadband applications," *IEEE INFOCOM 1994*, vol. 2, pp. 636-646, June 1994.
11. P. Goyal et al., "Start-time fair queueing: a scheduling algorithm for integrated services packet switching networks," *IEEE/ACM Trans.Networking*,vol.5,no.5, pp.690-704, 1997.
12. T. Ng et al., "Packet fair queueing algorithms for wireless networks with location-dependent errors," *IEEE INFOCOM 1998*, vol. 3, pp. 1103-1111, March-April 1998.
13. M. Jeong et al., "Wireless packet scheduler for fair service allocation," *IEEE APCC/OECC 1999*, vol. 1, pp. 794-797, Oct. 1999.
14. N. Kim and H. Yoon, "Packet fair queueing algorithms for wireless networks with link level retransmission," *IEEE CCNC 2004*, pp. 122-127, Jan. 2004.
15. Q. Liu et al., "Cross-layer scheduling with prescribed QoS guarantees in adaptive wireless networks," *IEEE JSAC*, vol. 23, no. 5, pp. 1056-1066, 2005.
16. L. Xu et al., "Dynamic Fair Scheduling With QoS Constraints in Multimedia Wideband CDMA Cellular Networks," *IEEE Trans. Wireless Com.*,vol.3,no.1,pp.60-73,2004.
17. M. Arad and A. Leon-Garcia, "A generalized processor sharing approach time to scheduling in hybrid CDMA/TDMA," *IEEE INFOCOM 1998*, vol. 3, pp.1164-1171.
18. X. Wang, "An FDD Wideband CDMA MAC Protocol with Minimum-Power Allocation and GPS-Scheduling for Wireless Wide Area Multimedia Networks," *IEEE Trans. Mobile Comp.*, vol. 4, no. 1, pp. 16-28, 2005.
19. L. Wang et al., "Channel Adaptive Fair Queueing for Scheduling Integrated Voice and Data Services in Multicode CDMA Systems," *IEEE WCNC 2003*, vol. 3, pp. 1651-1656, March 2003.
20. A. Stamoulis et al.,"Time-varying fair queueing scheduling for multicode CDMA based on dynamic programming," *IEEE Trans.Wireless Com.*,vol.3,no.2, pp.512-523,2004.
21. G. Song and Y. Li, "Cross-layer optimization for OFDM wireless networks. Part I: Theoretical framework," *IEEE Trans. Wireless Com.*, vol.4, no.2, pp. 614-624, 2005.
22. W. Zhao and M. Lu, "CDMA downlink rate allocation for heterogenous traffic based on utility function: GA-SA approach," *CNSR'04*, pp. 156-162, May 2004.
23. G. Song, "Utility-based resource allocation and scheduling inOFDM-based wireless broadband networks," *IEEE Commun. Magazine*, pp. 127-134, 2005.
24. X. Duan et al., "A dynamic power and rate joint allocation algorithm for mobile multimedia DS-CDMA networks based on utility functions," *PIMRC'02*, vol. 3, pp. 1107-1111, Sept. 2002.
25. G. Song and Y. Li, "Cross-layer optimization for OFDM wireless networks. Part II: algorithm development," *IEEE Trans. Wireless Com.*, vol.4, no.2, pp.625-634,2005.
26. A. Jalali et al., "Data throughput of CDMA-HDR a high efficiency-high data rate personal communication wireless system," *IEEE VTC'00*, vol. 3, pp. 1854-1858, May.
27. G. Barriac and J. Holtzman, "Introducing Delay Sensitivity into the Proportional Fair Algorithm for CDMA Downlink Scheduling," *IEEE ISSSTA'02*,vol.3,pp.652-656.
28. H. Koto et al., "Scheduling Algorithm based on Sender Backlog for Real-Time Application in Mobile Packet Networks," *IEEE WCNC 2005*, vol. 1, pp. 151-157.

29. S. Shen and C. Chang, "A utility-based scheduling algorithm with differentiated QoS provisioning for multimedia CDMA cellular networks," in *VTC'04 Spring*,vol.3,pp. 1421-1425.
30. K. Johnsson and D. Cox, "An adaptive cross-layer scheduler for improved QoS support of multiclass data services on wireless systems," *IEEE JSAC*, vol. 23, no. 2, pp. 334-343, Feb. 2005.
31. W. Wong et al., "Soft QoS provisioning using the token bank fair queuing scheduling algorithm," *IEEE Wireless Commun.*, vol. 10, no. 3, pp. 8-16, 2003.
32. A. Kam et al., "Supporting Rate Guarantee and Fair Access for Bursty Data Traffic in W-CDMA," *IEEE JSAC*, vol. 19, no. 11, pp. 2121-2130, 2001.
33. Q. Pang et al., "Service scheduling for general packet radio service classes," *IEEE WCNC 1999*, vol. 3, pp. 1229-1233, Sept. 1999.
34. V. Huang et al., "QoS-Oriented Packet Scheduling for Wireless Multimedia CDMA Communications," *IEEE Trans. Mobile Comp.*, vol. 3, no. 1, pp. 73-85, 2004.

Precise Time Synchronization over GPRS and EDGE

Petr Kovar, Karol Molnar

Brno University of Technology, Dept. of Telecommunications, Purkynova 118,
612 00 Brno, Czech Republic
kovapetr@phd.feec.vutbr.cz, molnar@feec.vutbr.cz

Abstract. Modern communication services are more and more focused towards real time mobile multimedia applications like mobile video and Voice over IP. To assure sufficient Quality of Service control, appropriate measurements are needed. One of the most difficult tasks is to measure the exact value of one way delay, because very precise time synchronization between host and peer is needed. Most multimedia applications use JavaME platform and GPRS or EDGE technologies without any additional services or hardware. This leads to the very challenging task: To assure precise time synchronization using application layer and high latency networks.

Keywords: Time offset, time synchronization, EDGE, GPRS, JavaME

1 Introduction

Precise time synchronization over data networks is not a new problem. In 1985, the first version of Network Time Protocol was released [1]. The Network Time Protocol is marginally used on low latency networks or at least on networks with constant traffic delay. Using GPRS or EDGE services is much more problematic, because traffic delay is very variable and is depending on actual network load.
Other problem arises when JavaME platform is used. JavaME is platform designed for usage on resource constrained devices ant it uses sandbox model for assuring better compatibility end security. Lower network layers cannot be accessed directly, this can be made only with Java API.

2 Theoretical expectations

2.1 Available directions

Most publications focused on the problem are counting with high, bud invariable delay [2][3]. The solution of this problem should be, at first, to propose whole new standard. The second possibility is to analyze Network Time Protocol and its adaptation, if it is possible, to application layer and networks with very variable latency. For the following work, the second possibility was chosen. The very important condition is that accuracy in ones or few tens of milliseconds is sufficient for the job.

2.2 Network Time Protocol

As were said, the Network Time Protocol specification was proposed in 1985 [1]. We have the third version of NTP specification at the moment [4].

For exact definition of the clock offset between host and peer, the NTP protocol specifies the clock offset as following function:

$$\theta = \frac{(T_{i-2} - T_{i-3}) + (T_{i-1} - T_i)}{2}, \qquad (1)$$

where T_i, T_{i-1}, T_{i-2} and T_{i-3} are timestamps described by Fig. 1.
Roundtrip delay is defined as

$$\delta = (T_i - T_{i-3}) - (T_{i-1} - T_{i-2}). \qquad (2)$$

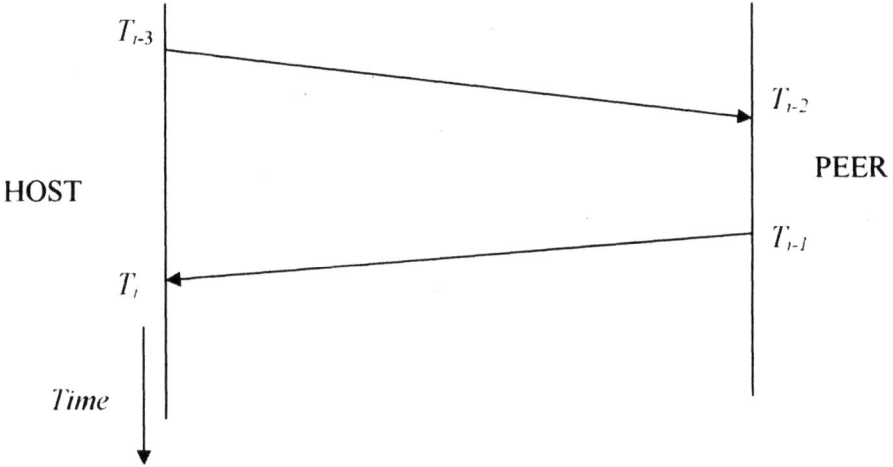

Fig. 1. Network Time Protocol principle

T_{i-3} is system clock of host in which he sends request to peer. T_{i-2} is system clock of peer when the request arrives. T_{i-1} is system clock of peer when the answer is send and T_i is the time of its arrival at host system. NTP is implemented on network layer and thanks for that, it is able to recognize exact time when the packet was received or sent.

2.3 Network Time Protocol application layer adaptation

Software designated for JavaME platform, which is ideal for developing universal and highly compatible mobile applications, is not able to access network layer directly. Network communication is possible only using JavaME API libraries. Because of that, it is necessary to adapt NTP to work on application layer. In this case, probably the biggest problem is that it is not possible to detect exact moment in which is packet sent or received. For this case, much more corresponding traffic scheme is shown on Figure 2.

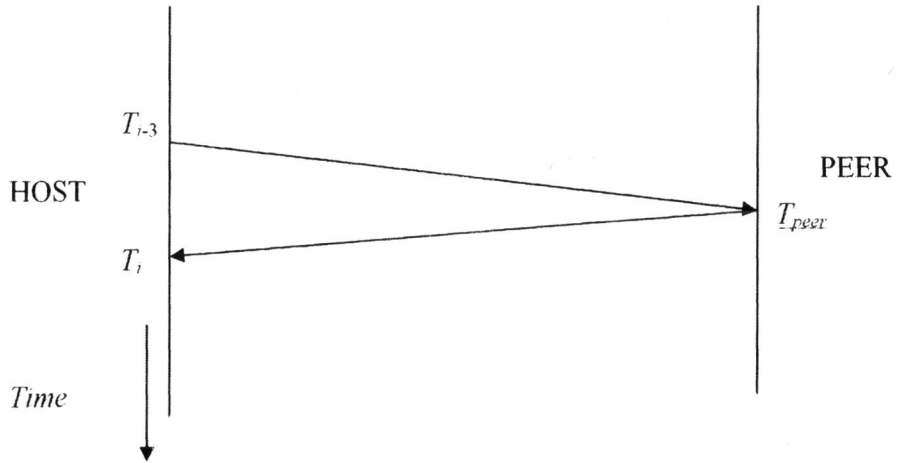

Fig. 2. Network Time Protocol adapted to application layer

When it is not possible to detect exact moment of packet receiving or sending, the time variables T_{i-2} and T_{i-1} are merged to only one variable, T_{peer}, because when application is optimized to count with application layer limitations, it is not useful to try to differentiate the moments when the packet was received and sent. For this case, clock offset and roundtrip delay are defined as

$$\theta_1 = \frac{(T_{peer} - T_{i-3}) + (T_{peer} - T_i)}{2} = T_{peer} - \frac{T_{i-3} + T_i}{2}, \quad (3)$$

$$\delta_1 = (T_i - T_{i-3}) - (T_{peer} - T_{peer}) = T_i - T_{i-3}. \quad (4)$$

2.4 High latency networks - two-way algorithm

Time synchronization in high latency networks with very high jitter, as is in GPRS or EDGE, is very challenging task. The roundtrip delay for GPRS/EDGE usually varies between 400 and 1200 ms. The delay in the host to peer direction can also vary from peer to host direction very heavily. If this happens, it is absolutely necessary to perform measurements not only in one direction, but in both directions. Also, it is necessary to perform the second way measurement as earliest as possible, because network load and other conditions influencing traffic delay can change in any second. With regard to this fact, the two way algorithm is proposed (Fig. 3).

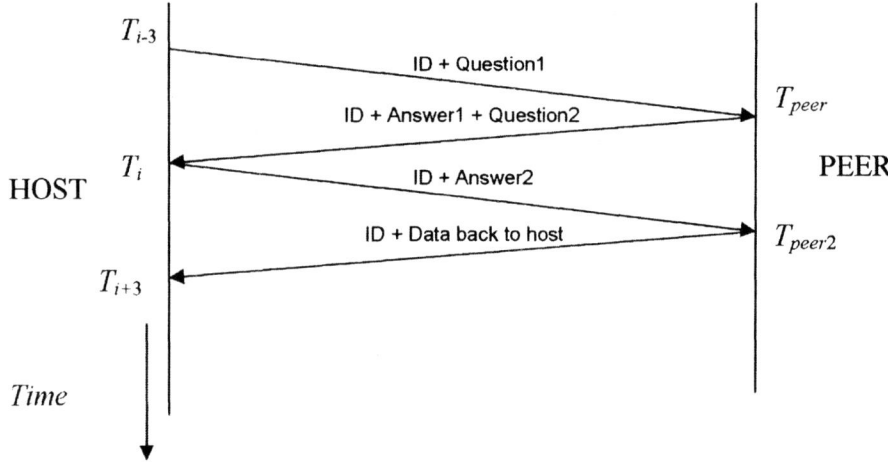

Fig. 3. Network Time Protocol adapted to application layer with two way measurements

Using this algorithm, it is possible to obtain two measurements at almost one time, the first for host – peer – host direction and the second for peer – host – peer direction. The first clock offset remains as (3) and the second clock offset can be enumerated as

$$\theta_2 = \frac{(T_i - T_{peer}) + (T_i - T_{peer2})}{2} = T_i - \frac{T_{peer} + T_{peer2}}{2}. \quad (5)$$

The second roundtrip delay is equal to

$$\delta_2 = (T_{peer2} - T_{peer}) - (T_i - T_i) = T_{peer2} - T_{peer}. \quad (6)$$

Using this two way algorithm, we believe that it is possible to eliminate traffic delay variation and obtain clock offset value in precision of tens. Also, two way algorithm helps us to determine values that are encumbered with errors and should be disqualified.

3 Obtained results, discussion

3.1 Java application

For the first analysis, both the server and the client side of the Java based test application using the modified NTP system was implemented on JavaSE platform. After successful tests and proper algorithm investigation, the JavaME version of the client will also be developed and tested.

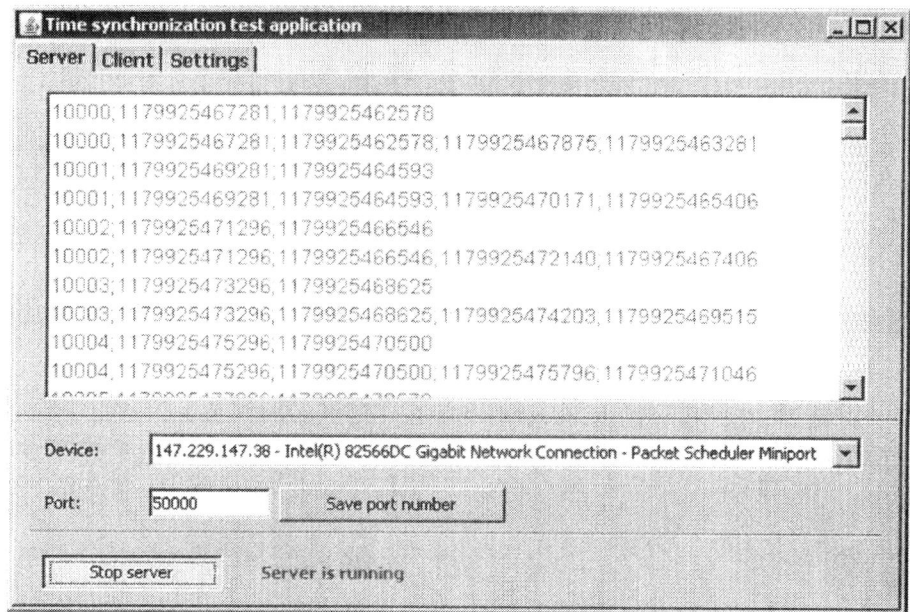

Fig. 4. Time synchronization test application, server side

The UDP connection was used, because the TCP error correction methods are not acceptable here. Even if all packets don't carry the same amount of information, we decided to set all packets to same size, filling spare place with blank data.

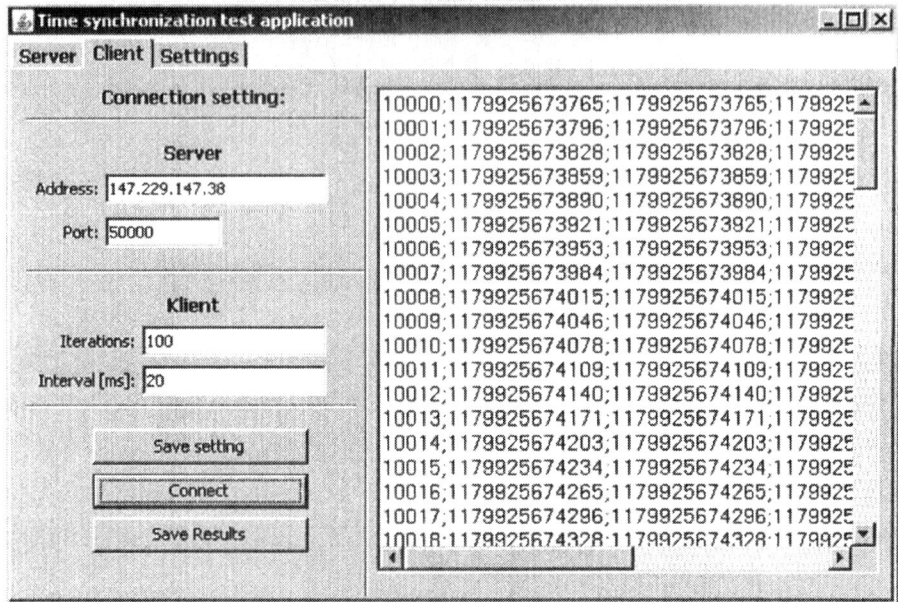

Fig. 5. Time synchronization test application, client side

3.1 Measurements

The server – client application used to collect samples was successfully developed and tested. The next part is to analyze the results and to propose evaluation methods that should lead to proper results.
In Figure 7, there is a typical distribution of the results achieved by a 100-cycle measurement. The right time offset value is 3816ms and is marked by line. Graph shows us, that some of the results are very inaccurate and whole measurement is encumbered with random errors, which mainly pushes the time offset to lower values than it actually is. Also the graph shows us, that systematic faults are also very probably involved, pushing the peak of distribution to wrong position. The most difficult task is to analyze the results in way, which will separate only inadequate samples and obviates affects of systematic fault.
After deep analysis of the problem, we believe that most time offset measurement errors arise from situation, when traffic delay in forward direction is different from traffic delay in backward direction. With increasing difference, the error size is increasing, too. With two-way measurement algorithm we can partially eliminate this problem, because some of the error encumbered samples can be identified and disqualified.
Many tests were carried out and proved that, for precision in ones, only about 100 iterations are needed. This exceeded out theoretical expectations. With low network load, only about 15 - 25% samples have to be disqualified. Tests also proved that it is

not effective to start one iteration when another was not finished yet. This can partially overload the network and lead to unusable results. The sufficient inter-iteration interval is about 2000 ms.

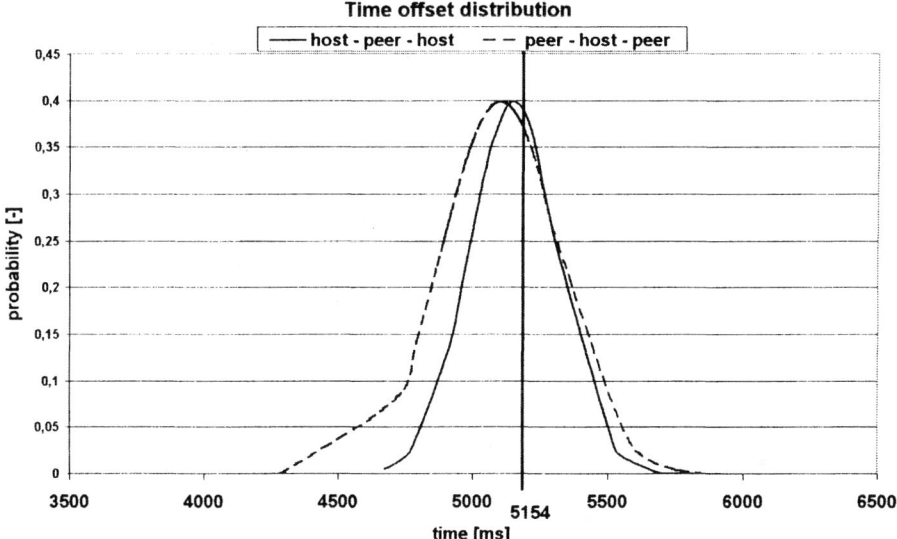

Fig. 6. Two-way measurements algorithm results distribution for 100 iterations.

To achieve still sufficient precision in few tens of milliseconds, 20-iteration measurement is needed, but this still needs about 40 seconds.

4 Conclusion and following work

Time synchronization over high latency network with huge delay variation is very challenging, but not impossible task. Using proposed methods, the success rate partially depends on network load, when the biggest influence has an additional traffic generated by GPRS-connected side of communication. Because of this, we recommend that time synchronization should be only task performed on GPRS-connected terminal in one time.

The following work on this topic will be focused on proposing other optimalizations, which will reduce the time needed for measurement with sufficient precision. The target time is about 15 seconds. When finished, algorithm will be used to determine characteristics of communication channel and will help with setting optimal traffic parameters before its start.

Acknowledgments: The paper was written with support of the MŠMT 1K04116, GAČR 102/06/1569, MŠMT MSM0021630513 and FRVŠ 1835/G1 project.

References

1. Mills, D.L.: Network Time Protocol, Sterling, 1985 Internet Engineering Task Force [online] <http://www.ietf.org/rfc/rfc0958.txt>
2. Syed, A.A., Heidemann, J: Time Synchronization for High Latency Acoustic Networks, INFOCOM 2006. 25th IEEE International Conference on Computer Communications. Proceedings, Barcelona, page 1 – 12, ISSN: 0743-166X, ISBN 1-4244-0221-2
3. Elson, J., Estrin, D.: Time synchronization for wireless sensor networks, Parallel and Distributed Processing Symposium., Proceedings 15th International Volume , Issue , Apr 2001, page1965 – 1970, ISSN: 1530-2075, ISBN: 0-7695-0990-8
4. Mills, D.L.: Network Time Protocol (Version 3) - Specification, Implementation and Analysis, Sterling, 1992 Internet Engineering Task Force [online] <ftp://ftp.rfc-editor.org/in-notes/rfc1305.pdf>

GSM Base Station Subsystem Management Application

Vit Novotny, Pavel Svoboda

Faculty of Electrical Engineering and Communication, Dpt. of Telecommunication,
Brno University of Technology
{novotnyv@feec.vutbr.cz, xsvobo63@stud.feec.vutbr.cz}

Abstract. The Faculty of Electrical Engineering and Communication of the Brno University of Technology offers several subjects focused on mobile communication networks. For practical exercises the experimental mobile network has been building at our faculty. As we do not have OMC or NMC in our network and as the low level network management is quite complex we designed the software that is useful tool for configuration and management of the network on the higher level. The paper deals with our experimental network and with software we have developed.

Keywords: GSM, BSS, BSS database, BSS administration

1 Introduction

Mobile networks and mobile communication have become very common in many parts of our lives. Even if usage of mobile services becomes simpler and simpler, the mobile network and also its management are more and more complicated. This is due to the more complex management procedures of the network resources allocation that is required by the operators to make network resources allocation much more effective so that the network infrastructure costs are lower and the operator can offer lower prices for its services and to attract more subscribers and to earn more money as the result. Of course also reliability of the network infrastructure and quality of the provided services are very important so that the operator puts quite large investments into the operation and maintenance centres (OMCs) and network management centres (NMC).

2 Experimental mobile network

Operators require more and more educated engineers in the area of mobile networks. Therefore also at the Faculty of Electrical Engineering and Communication of Brno University of Technology several subjects specialized on mobile networks are taught. To enable students to get practical experience from the area of mobile networks we have been building experimental mobile network. It is complicated and quite expensive but we hope that we will succeed at the end.

What we maybe will not manage at all are the OMC and NMC elements. So that when we want to configure the access network and core network elements we need to do it at the low level, e.g. over terminal connection with the control processor of given part of mobile network. Such low level configuration is quite complicated operation and the management staff has to be skilled.

Therefore we decided to design the communication software that will simplify the management process of our experimental network.

Structure of the mobile experimental network at the Faculty of Electrical Engineering and Computer Science is shown in Fig. 1. It consists of two main parts – base station subsystem BSS and network switching system NSS. It can be seen that it is incomplete. Nevertheless the GSM part, i.e. circuit switched has been already put into operation. The main element of network subsystem, i.e. MSC is not physically equipped in the network, but it is simulated by the Tekelec simulator which also simulates functionality of Home Location Register (HLR), Visitor Location Register (VLR) and Authentication Centre (AuC).

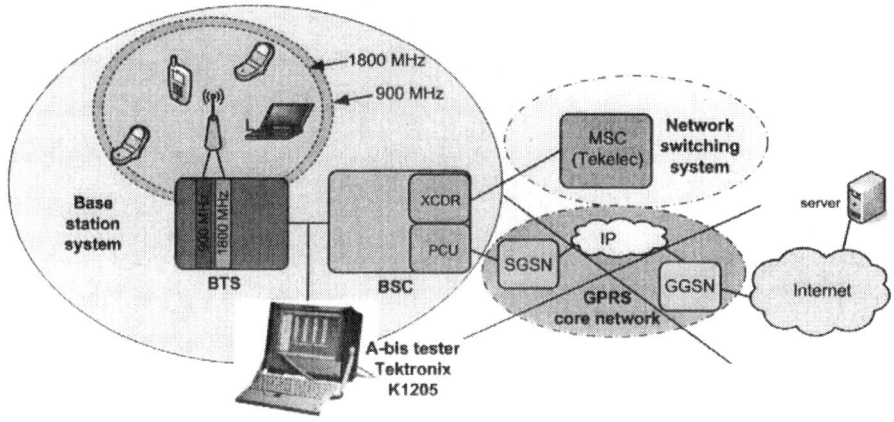

Fig. 1. Experimental mobile network

Building of GPRS network is at the beginning because only Packet Control Unit PCU is equipped. We suppose to equip other GPRS components during next two years and put it into full operation.

Our experimental network is equipped mainly by components manufactured by Motorola company, see Fig. 2. More precisely the BTS, BSC with transcoder and the PCU components are the products of Motorola company ([2]).

Personal Wireless Communications 287

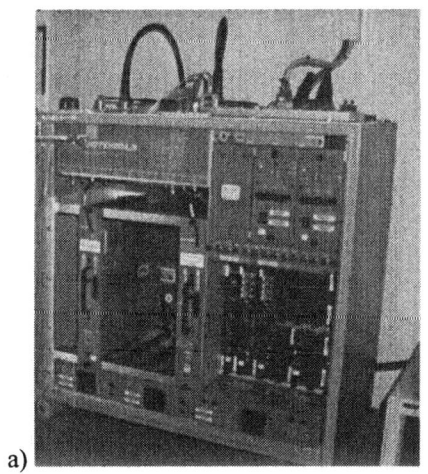 

Fig. 2. a) BTS Motorola Horizon*macro*™, b) BSSC2 cage equipped by BSC and XCDR functionalities

3 BSS database

BSS database is a code object which specifies the hardware equipment of the BSS, hardware functionality and the configuration parameters of the hardware equipment, so that it specifies the overall behaviour of the BSS ([1]). The database can be configured and modified via configuration script. Complete configuration script has to have certain structure so that it contains several configuration steps which are displayed in Fig. 3.

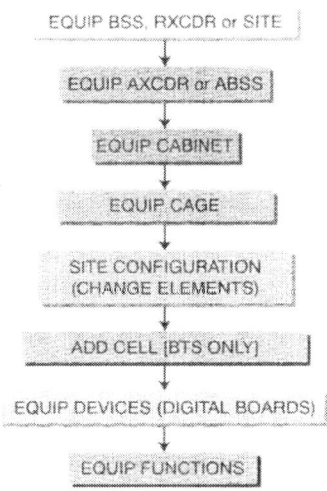

Fig. 3. Database programming sequence

The database can be changed via **MMI** (Man-Machine Interface) interface either from the OMC or from the terminal physically connected to the M-GPROC (master) in the BSC. Database changes are controlled by the **CM** (Configuration Management) process and by **CA** (Central Authority) process that proves or refuses the requested changes. To realize the command appropriate security level is required. There are three security levels protected by passwords. When the command is accepted all changes have to be distributed to all control equipment of the BSS to put them in operation. The commands are necessary to enter in certain order because majority of them depend on the status of equipage of the BSS by previous commands. Therefore the BSS system units should be equipped in order that is shown in **Fig. 4**.

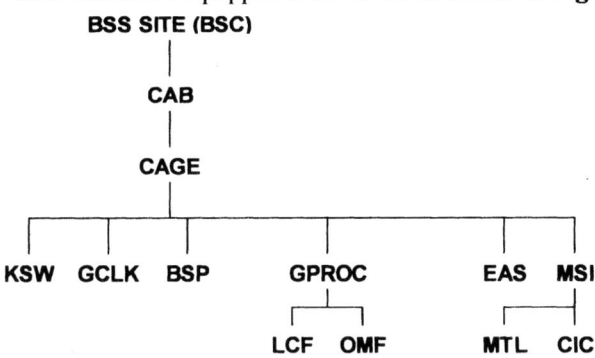

Fig. 4. Hierarchical system for the order of equipping different parts of BSS system by command *equip*

Two examples of commands follow. First one shows the equipage of the BSC by Kiloport Switch (TDMA switch) and the second one shows the equipage of the BSC by processor card GPROC.

equip bsc KSW	it specifies Kiloport Switch module in the BSC cage
0	0- 3 – TDM position identifier for KSW
0	0 or 1 – KSW module position, 0 = slot 27, 1 = slot 1
0	0-15 – unit number that contains KSW module
equip bsc GPROC	GPROC module specification
1	1-111 – GPROC identifier
0	0-13 – unit identifier where the GPROC is located
19	0-15 – slot number in the unit

4 Management application

The primary goal of the application is to simplify management of the base subsystem of the GSM mobile network. The application provides BSS administrator help in the form of the syntax and set of valid parameters for required command. Therefore the BSS administrator does not need very expensive OMC centre or to remember syntax of all commands that the administrator otherwise must know to administer the BSS from command line over terminal connection. Therefore the administration is much simpler and faster. The application was written in C++ in integrated development environment Borland® C++ Builder™ version 6.0. The application is 32-bit and it was designed for Windows® operating systems.

4.1 Libraries and components

Standard libraries and components of C++ Builder™ integrated development environment were used. BSS commands, their parameters and help texts are stored in the Paradox 7 format and the access to it is realized over standard database components. Either SQL interface is used to store or to get information or thick client is used direct access.

Application communicates with the base subsystem over serial interface RS232-C. ComPort Library v.3.0 by Dejan Carnilla is used to serve the communication.

4.2 Program structure

The application consists of several form and dialog windows and of one data module. The main interface window is shown in Fig. 5. Majority of operations are realized via this window. The other forms windows are used to access to the datbase tables containing commands and their parameters. The other window is displayed when the help contains a lot of text and it is valuable to present it in own window. Remaining windows are used for configuration of the program parameters and to display some supplementary information.

When the program starts the main window is displayed. Data module fills ComboBox components by the appropriate data, i.e. by commands, their parameters and other auxiliary values. At the bottom of the main window the most frequently used parameters are displayed to easily revoke them whe necessary.

The parameters for communication via serial line are already preset (COM2, 9600 bps, 8bit words, 1 stop bit and no parity). Of course the parameters can be changed when needed.

The commands are edited in the top edit box of the main window. The command can be inserted manually or selected from the ComboBox bellow the edit box. When the latter way is used the brief command description appears in the gray frame.

The communication between the application and the BSS is stored in the buffer and displayed in the bottom part of the main window. If convinient the commands are responses exchanged between the program and BSS can be shown in standalone window. When necessary they can be exported in text format to the text file and displayed in any text editor when needed.

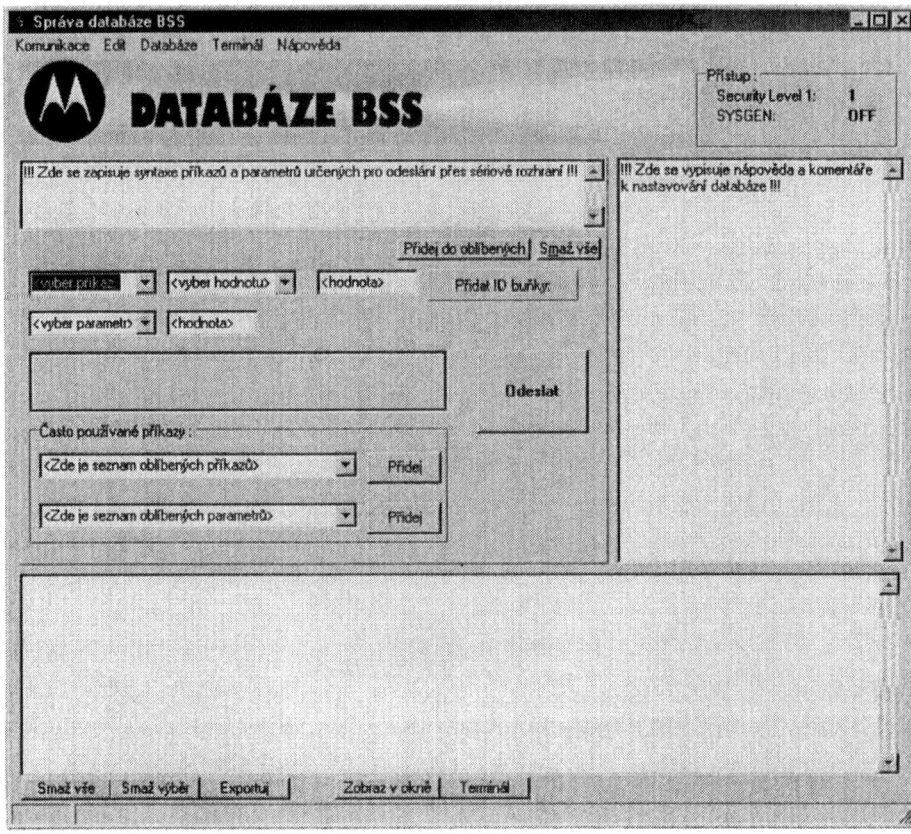

Fig. 5. The main application window

In the case of commands that require the cell ID specification auxiliary form window can used, see Fig. 6.

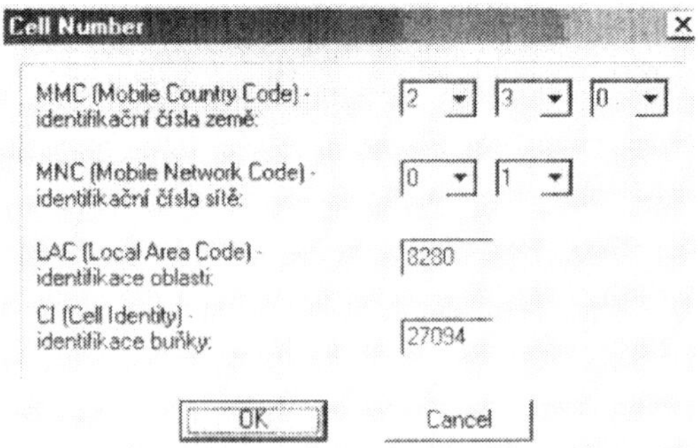

Fig. 6. The form window cell ID specification

Some commands can be realized only when higher security level is active. There are three security levels and the lowest level is default level. Only small set of commands that change the BSS settings can be run. The user needs to know appropriate password to enter appropriate level. Also some commands can be executed only in Sysgen ON mode. The main window allows the change both of the security level and of the SYSGEN mode, see Fig. 5.

4.3 Database of command and parameters

It is possible to manage the database of commands and their parameters. This gives the application flexibility so that it can be quite easily adapted to the other equipment to be configured.

To open the database we select item "Databáze" from the main menu. Then we select the item *"Databáze příkazů a parametrů"* and the window shown in Fig. 7. Here the user can modify the database of commands, their arguments and parameters.

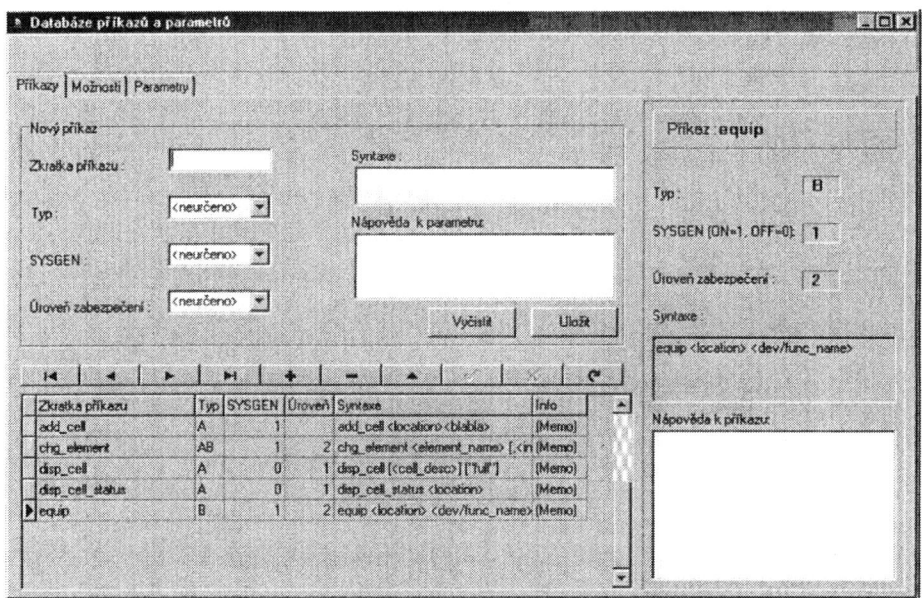

Fig. 7. Window for access to the database

As it can be seen from Fig. 7 also other important information can be added. Mainly the syntax of the command, then help text and also the SYSGEN mode and the security level needed to proceed the command.

To store frequently used commands not only for current session duration but for later use is also very helpful function. This operation is realized via pressing button *"Přidej do oblíbených"* in the main window of the application (Fig. 5). Selected command is analysed and SYSGEN mode and security level is determined. Then the command is stored. New window displayed as shown in Fig. 8. Table is sorted in alphabet order of the command as default.

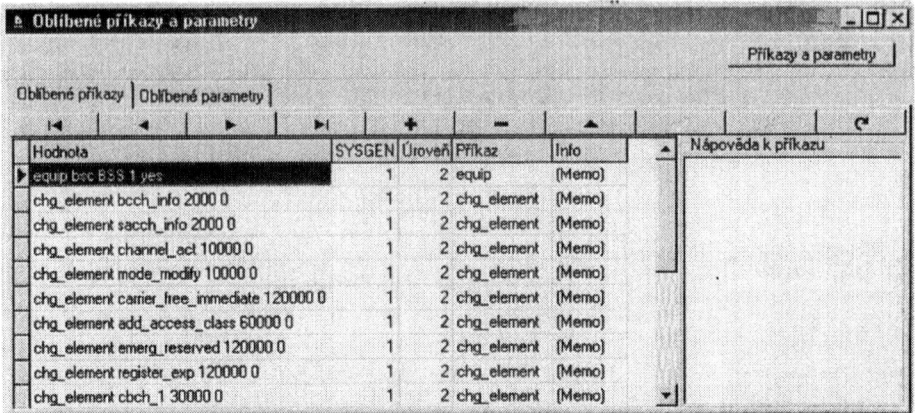

Fig. 8. Window for access to the database of frequently used commands

4.4 Command exchange

After the communication parameters are set the program automatically sends "Carriage return" character and the system should display prompt, e.g. "**[29/04/07 17:08:26] MMI-RAM 0115 ->**" as the response. Then the user can enter the appropriate command or set of commands and when button "Odeslat" is pressed the command is sent to the BSS and providing the syntax is correct, appropriate security level is set and SYSGEN mode is at the required state the command is processed.

5 Conclusion

Because of the communication network management is more and more complicated we designed the software to simplify the administration of the BSS of our experimental network. The software works well and it helps us to manage BSS in faster way. The application has been designed in flexible way so that it would be not a big problem to adapt it for administration of other equipment.

Acknowledgement. The paper has been worked out under the support of the project granted by Grant Agency of the Czech Republic No 102/06/1569

References

1. MOTOROLA Inc. *Maintenance of BSS Database GSR7*. Motorola, documentation, Swindon, 2003,
2. MOTOROLA Inc. *BSS Operational Theory*. Motorola, documentation, Swindon, 2002

3. SOLDANI,D., MAN,L., CUNY,R. *QoS and QoE Management in UMTS Cellular Systems*. John Wiley and Sons, ISBN 0-470-01639-6, England, 2006,
4. BATES, R., J., BATES, R., J., B. *GPRS: General Packet Radio Service* (Professional Telecom), ISBN 0071381880, McGraw-Hill Professional, 2001.
5. CASTRO, J.P. *All IP in 3G CDMA Networks: The UMTS Infrastructure and Service Platforms for Future Mobile Systems*. John Wiley & Sons, ISBN 0470853220, USA, 2004.
6. BOSSE,J.G. *Signalling in Telecommunication Networks*. John Wiley & Sons, Inc., 1998, ISBN 0-471-57377-9, New York 1998

Transport resources reservation in IMS frameworks: Terminal vs. PDF driven

Antonio Cuevas[1], Jose I. Moreno[1,2], Hans Einsiedler[2]

[1] Universidad Carlos III de Madrid, Dpto. Ing. Telemática
Avda. Universidad 30,
28911 Leganés, Madrid, Spain
{antonio.cuevas, joseignacio.moreno}@uc3m.es
Deutsche Telekom Laboratories
Ernst Reuter Platz 7
D-10587 Berlin, Germany
{joseignacio.moreno, hans.einsiedler}@telekom.de

Abstract. IMS is a good candidate to become the service platform for next generation networks (NGN). Among one of its key characteristics is the ability to keep the Internet paradigm of application and transport separation while designing interfaces between the two layers. In IMS there is an interaction between the CSCFs (call session control function), SIP proxies managing the application setup, and the PDF (Policy Decision Function), controlling the transport network. Still, the terminals have to perform the transport resource allocation (activating the PDP context). In future networks, a similar behavior is possible but also another approach can be followed: the PDF is in charge of allocating the transport resources between the terminals, allowing to reinforce the network control. In this paper we analyze and compare both approaches, including by resorting to simulation.

Keywords: IMS, PDP, RSVP, ns2, simulations, resource reservation, transport.

1 Introduction

It is word of mouth that telecommunications are suffering a major revolution and that, like all major changes, lots of uncertainties exist. However some trends can be identified: technologically, a migration to a universal IP network and, in the commercial field, the raise of new business models and relationships between the parties.

Migrating to a universal IP network gives birth to many opportunities: the use of any access technology, wired or wireless, of any device, any application and under all circumstances (fixed, mobile) is a promise of these new networks, that are being designed to be user centric. In fact, most of the telecom operators and service providers have internal roadmaps to migrate to IP technology completely within the next 5 years trying to reduce OPEX (OPerational EXpenditures) and CAPEX (CAPital EXpenditures). This can be done by migrating all their networks (e.g. SDH,

ATM, FR, X.25) to a universal IP network [6]. The technological challenges are also important, struggling mainly in integrating different functionalities and solutions, for instance, IP mobility and enabling QoS in data transport

In the business arena, tensions emerge and the directions to be taken by the telecommunications world are not so clearly identified. However, the raise of new business models, like the semi-walled garden one [18], is identified along with the necessity for all the possible models to coexist. The semi-walled garden model takes advantage of the openness of the Internet to build services but, still, lets the network operators keep a central role in the business value chain, making them "service brokers".

Service platforms are a key part in next generation networks. They enable new business models and commercial relationships between the parties. NTTs' i-mode [18], and 3GPP Open Service Access (OSA) and IMS systems are known service platforms based on the semi-walled business model [8].

The technological challenges implied the need of supporting and coordinating several business models, entities and services, requires large research developments. It is thus no surprise that aspects that have "long ago" attracted attention and reached satisfactory solutions are again "hot topics" striving to cope with these challenges, bringing technology and business concepts together. One of these topics is QoS (Quality of Service). And one of the biggest concerns is how to integrate and manage the QoS-enabled data transport in next generation networks (NGN) and service platforms.

IMS service platform directly targets the operator's traditional business of multimedia communications (phone calls). IMS "core" applications demand strong QoS requirements and, thus, the QoS interfaces to the network are an important part of the IMS design. While i-mode or, even OSA, still adopt "traditional cellular networks" protocols, IMS is based on IPv6 and other open IETF protocols. IMS respects the Internet paradigm of transport and application layers separation, yet building the interfaces between them. This paper embraces IMS' approach in managing QoS-enabled data transport in next generation networks and service platforms.

IMS key element in dealing with QoS is the PDF (policy decision function) that is like a Broker between the QoS defined at the application level and its actual enforcement at the network level. Still, in IMS, the terminals must perform the transport resources reservation. This paper explores the possibility of the PDF performing the transport reservation and not the terminals. It will analyse and compare both approaches in an IMS-like framework.

To perform our analysis the paper is structured as follows: in the next section we provide the reader with the basic IMS knowledge so that it can understand the discussion. Section 3 is the core of this paper: it analyses and compares two strategies to perform transport resource allocation in IMS-like frameworks. An important part of this analysis is done by resorting to simulation and this is dealt with in section 4. The conclusion gathers the main results, tries to answer which strategy is better and opens for future works.

2 IMS overview

This section will present what we consider will be the core of NGN service platforms, the IMS.

The IMS goal is not to deliver any service; rather IMS is designed to assist peers in establishing, managing and tearing down their (multimedia) sessions. Roughly, we can say that IMS is a SIP proxies infrastructure –termed CSCFs-. CSCF are ready to be used by costumers and their devices' SIP User Agents (SIP-UA). Figure 1 presents the IMS architecture. Some of the main IMS features are:

- o IMS uses IPv6 as network layer and its interfaces employ open IETF standardized protocols.
- o Like any SIP based infrastructure, it assists peers in setting up, controlling and tearing down their sessions (although IMS is not a session participant).
- o It interacts with the network operator infrastructure to fit the QoS given to transport the session flows.
- o IMS informs the network operator about the sessions so that it can do service bundling and the "semi-walled garden" business model is enabled.
- o IMS respects the Internet paradigm of transport and application separation.

In the following discussion, for ease of understanding, we will consider the IMS infrastructure as an entity independent of the network operator. This is not really so, indeed, IMS is owned by the network provider, but this "ownership" does not reduce the validity of our exposition.

The IMS platform works at the service level. IMS follows the Internet paradigm and, as such, it is completely independent of the data transport level which is handled by the UMTS network. However, a fundamental IMS feature is to achieve negotiation with the network operator, on a per session basis, of the transport-level parameters so that they fit the application-layer requirements. The IMS' PDF is the key entity: It is contacted using Diameter protocol [3] by the IMS's CSCFs nodes. Using another IETF standardized protocol, COPS [9], the PDF contacts the network routers (or the GGSN in UMTS networks). The GGSN (gateway GPRS support node) connects the UMST network to other networks, such as IP based ones like the Internet. Instructed by the IMS's PDF, the GGSN will enforce the QoS at network level.

Following the semi walled garden business model, the user's profile, his Authentication, the Authorization to consume resources and the accompanying Accounting (AAA) and posterior billing are handled by the network operator. IMS depends on the network operator's AAA services to control their users. Users trust and pay the network operator who, in its turn, trusts and pays (retaining, for example, 10 percent of the total amount) the IMS.

When a user gains access to UMTS networks, an authentication process is performed with the network operator's home subscriber server (HSS), which holds user's data and credentials. When IMS (which does not hold the user credentials) needs to authenticate a user, it delegates and depends on the HSS and the previous authentication of the user to this system. The IMS CSCF interfaces with the HSS following the IETF Diameter protocol.

Concerning authorization, this is done by the IMS/CSCF itself but is based on the user profile and context obtained, also using the Diameter protocol, from the HSS.

Accounting (and charging) follow the same philosophy as QoS: they are done at two separate levels (application and data transport) and an entity, this time part of the UMTS network infrastructure, processes, correlates and consolidates the relevant parts of the information and generates call detail records for the UMTS billing system. Thus, richer charging schemes (and tariffs) than the current ones are possible. The central entities are the charging data function (CDF) and the charging gateway function (CGF). IMS entities (namely CSCFs) communicate with the CCF following the Diameter base protocol. CSCFs can instruct the CCF about the type of session (e.g. audio or video call), its duration, or the number of participants. The interaction between the GGSN and the CGF is also defined to receive session information in the network plane (e.g., number of bytes sent and received).

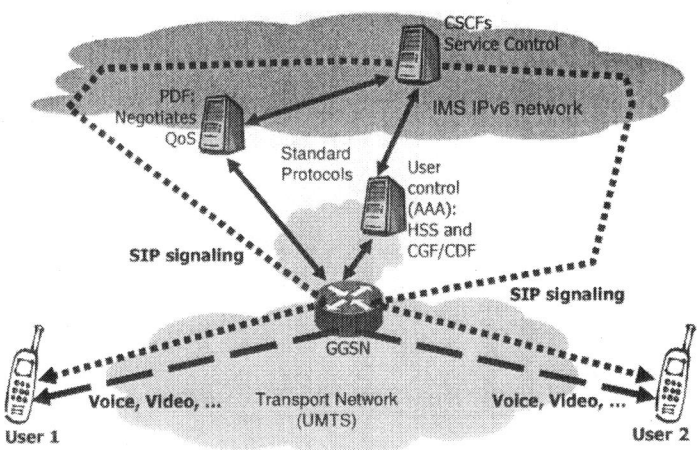

Figure 1 IMS architecture

A short usage scenario will help the reader to better understand IMS. A user powers on his mobile. A registration is done with the network, following a pure 3G, non-IMS-related process. The central node in this process is the HSS. Latter on, this user wants to setup a call using the IMS infrastructure. He sends SIP signaling to the callee. This SIP signaling traverses IMS CSCFs. CSCFs, before letting the call setup proceed, consult the HSS. The HSS "knows" the user (who registered before) and gives a positive answer to CSCFs. CSCFs then let the signaling proceed. CSCFs inform the PDF that network resources will be needed to accommodate the voice transport between the users. Once this is done, the users allocate the transport resources in the GGSN. The GGSN, before allowing this allocation consults the PDF. The PDF configures the GGSN to allocate the needed resources for the voice call. CSCFs may account aspects such as quality of the voice call and its duration. GGSN, on its side, can account for bytes sent and received. All the accounting information is gathered and correlated by the CGF/CDF. Depending the network policies the user will (or will not) be billed both by the call service and by network resources consumed. The use will finally receive one bill from his network operator with all the services he enjoyed, including this voice call.

IMS has other features, for example, content adaptation capabilities. Besides, some details are still to be completely solved like the integration with Mobile IPv6. For the purposes of our discussion, we do not need to describe them further; the interested reader can check, for instance, [16] or [4]. The reader now understands the main IMS characteristics and we are ready to analyze and compare the IMS way of allocating transport resources (letting the terminal allocating the PDP context) to the approach where the PDF is in charge of such task.

3 Terminal vs. PDF/SIP Proxy driven resource reservation

The IMS service platform presented in the previous section provides a good solution for managing and coordinating the different aspects of service provision, including the various issues affecting QoS, for instance, matching the codecs chosen and the transport resources allocated. In IMS, the terminal is responsible for the transport resource reservation. It does so by activating a PDP context, a well-known 3G networks procedure. In NGN, only IP mechanisms, not tied to any network technology, should be employed and the PDP context activation should be replaced by protocols like RSVP [2]. We point out that, for scalability reasons, the most accepted approach ([10], [14]) is to design IP networks with RSVP and IntServ [1] capabilities only in the edges; the core network is endowed with less accurate, more scalable QoS capabilities. This discussion is completely out of this paper's objectives. Regaining our main thread, we consider two models for resource reservations:
   o Terminal driven, in which terminals allocate transport resources using RSVP. A PDF "policing" this resource allocation is a possible scenario (see [11]).
   o Network driven, in which the PDF is responsible for allocation of transport resources between the communication endpoints. This resource allocation will be done following the "guidelines" that the PDF is given by the CSCFs (the IMS' SIP proxies) to police the resource allocation (see section 2). COPS [9] and its COPS-PR [5] extension are the protocols that the PDF could employ to configure the routers (or GGSN in UMTS) to accommodate the needed resources.

Let's recall that the main goal of this paper is comparing these two approaches: the terminal vs. the PDF doing the transport resource allocation in IMS-like frameworks.

Salsano and others ([15], [17]) have already considered and compared these two possibilities in a SIP proxy infrastructure (CSCFs in IMS) interacting with bandwidth brokers. Bandwidth brokers (or QoS brokers) are nodes comparable to the PDF in IMS, taking policy decisions affecting the transport resources but also managing the whole IP network. Salsano et al. do not consider other elements present in IMS and NGN service platforms like the ones performing accounting (CDF/CGF) or user control (HSS). Thus, their analysis and comparison of the two transport reservation techniques has to be broaden to IMS like frameworks; this is the goal of this paper.

Let's note that Salsano's proposed interaction follows this sequence: SIP proxy, routers, bandwidth broker and routers again, while we profit from the IMS defined PDF to CSCF interaction to follow a simpler sequence: CSCF, PDF, routers (or GGSN). For reader's comfort we recall that SIP proxy is assimilated to IMS' CSCFs and bandwidth brokers to IMS PDF.

In the PDF driven transport resource reservation, terminals will need to support only basic SIP, they do not need to implement any resource reservation mechanism like RSVP, since the appropriate inter-working of CSCF, PDF and routers (or GGSNs) is able to "deduce" from the SIP message exchange (e.g. looking at the codecs) the required transport resources and allocate them in the transport network. Terminals, thus, can be simpler. Besides, heterogeneous networks may employ different resource allocation mechanisms, provide that the SIP signaling and the correct interpretation of the transport resources needed by the codecs is kept end-to-end.

Next generation networks and service platforms will enable complex "ecosystems" where different parties interact making profit and joining their efforts to offer richer customized services to the users. QoS enabled data transport will not be an exception. It will evolve (keeping the "user interface" simple) from current flat rates and best-effort characteristics to differentiated services with various performances and will become a tool to develop different products and tariffs for different market segments, boosting the competition (as stated e.g. in [19]). Pushing the resource reservation process from the terminals to networks nodes like the PDF (besides letting those nodes take policy decisions) will improve the integration and management of the transport service into the overall service provisioning chain. The network could provide, for instance, multicast capabilities with content adaptation for the branches of the multicast tree directed to terminals with little performance and slow connections.

Orchestration of all the nodes in these rich ecosystems will be a key aspect in next generation networks and service platforms. Each node will have different interfaces. For instance, IMS' CSCFs interact with many more nodes (e.g. PDF, HSS, CDF). To avoid overloading the nodes and making them potential bottlenecks, the functionalities they assume must be chosen carefully. And this includes letting the CSCFs and PDF doing the transport resources allocations instead of assigning this task to the terminals. This may specially impact the SIP proxies (in IMS actually the P-CSCFs) since they would need to keep state of the transport reservation and of the call (SIP session) associated to it. Note that, in IMS P-CSCFs are not stateful, this is let to the S-CSCFs. We will focus on the processing costs for the different nodes of the two proposed resources reservation strategies (transport resource reservation driven by the terminals or by the PDF/CSCFs), by resorting to simulations as we will discuss and describe in section 4 of this paper.

Mobility also impacts on QoS model, choosing one the two transport reservation strategies. In large networks, it is sensible to have several SIP proxies scattered along the network (like the IMS' P-CSCFs). The scenario that is relevant to us is when the terminal hands over between parts of the network assigned to different P-CSCFs. Mobility in NGN seems to be supported by two mechanisms: make-before-break and context transfer [13]. Performing a context transfer in the handover process implies gathering state information from all stateful entities involved in a session. In case of transport QoS configurations, the context transfer must be immediate. The parts of the context to be transferred immediately during a handover should be kept to the minimum. From this point of view, it is interesting to place a large fraction of the context in the common element during handover: the terminal. Due to this reason it is better to let the terminals doing the QoS-enabled transport reservation rather than the PDF instructed by the CSCFs.

Another point to consider is that, if the terminal needs to handle the transport resource reservation process, more messages need to be sent over the access link, which may have scarce bandwidth. Thus, the transmission time of these messages may be significant and may slow the session setup process.

4 Simulation based analysis and comparison

Previous section analyzed the pros and cons of letting the terminal or the PDF - instructed by the CSCFs- doing the transport resource reservation. We saw the processing load of the different nodes was an aspect to be analyzed. There are essentially three ways to do so: creating and analyzing a mathematical model, the implementation of a prototype where a number of tests are executed and resorting to a simulation model. Due to the many nodes and protocols involved in next generation networks and service platforms, the development of a mathematical model that is simultaneously faithful and tractable is difficult. Same happens with deploying a test-bed. Besides, in lab prototypes it is not possible to perform some kind of tests like overloading the network with thousands of calls. Simulation is, therefore, a valuable tool in obtaining indicators early in the engineering process and refining the system as it is being implemented. Simulation is our choice for comparing the processing load of the two resource reservation strategies studied in this paper.

We employ ns2 [12] to build our simulation model. We extended ns2 to support the protocols present in an IMS like scenario [7]. We developed a "two level" SIP proxy infrastructure emulating the IMS' P-CSCF and S-CSCF+I-CSCF SIP proxies. The P-CSCFs can interact with a PDF and accounting nodes (the CDF in IMS). The S-CSCF + I-CSCFs can interact with an authentication and authorization infrastructure (like the HSS in IMS). Roaming was supported. Our main goal was to evaluate the processing delays in the nodes. Each message, before exiting a node, is put in a FIFO queue where it has to wait for the other messages to be processed and an own processing time. User terminals do not have any processing delay. This reflects the fact that terminals have to handle few messages while "network nodes" have to process several hundreds of them, corresponding to the terminals they serve. To our knowledge, the novel "IMS like" model we developed can be a valuable tool to asses in the engineering and deployment of the complex IMS infrastructures.

The developed IMS model will serve us to evaluate and discuss the impact of the different transport resource strategies. We used a scenario (Figure 2) with 4 domains, each domain divided in one "Core Network" with a S+I CSCF and a HSS and 8 "Access Networks" each with one P-CSCF, a PDF and the accounting infrastructure (CDF). The "networks" are connected by ERs (edge routers). The processing delays of the nodes were chosen proportionally to the number of messages they had to process. For the P-CSCFs the calculation was done for the case they do not have to handle the transport reservation messages.

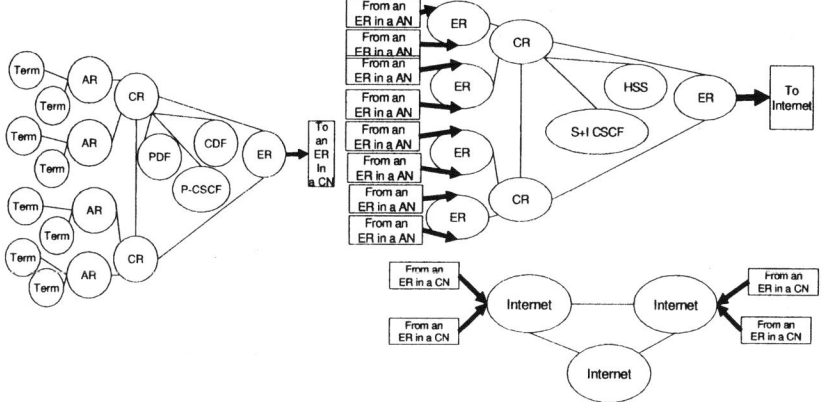

Figure 2 Simulation scenario: Terminals, an "access network" (AN), a "core network" (CN) and the connection between the four domains

Under a low load of sessions (calls) setup attempts to be handled by the IMS infrastructure, the difference in session setup time between the two strategies (terminal vs. PDF/P-CSCF driven transport reservation) matches the extra processing time needed to be done by the P-CSCF in the second strategy. But when the network load increases this time difference increases (Figure 3).

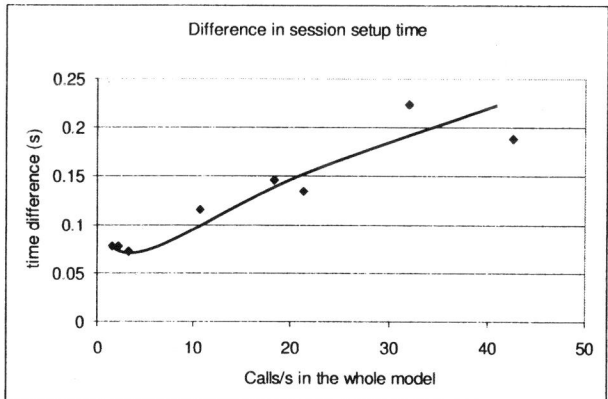

Figure 3 Difference in average session setup between the two strategies for reserving transport resources. This difference increases when the IMS infrastructure starts to be highly loaded. Points represent the difference in average session setup time for the PDF driven and the terminal driven strategies. Measures are done for different call loads. Each point is a measure.

To try to explain this difference we analysed individually the behaviour of the different nodes (HSS, PDF, CDF and the two kinds of CSCFs). We saw that none of the nodes reached an overloaded state even for the highest call load. By overloaded sate we mean that the rate of input messages into a node is constantly higher than the rate of messages exiting this node, thus, due to the FIFO characteristic of the message processing queues, the processing time of the messages increases steadily. We stress that the overloaded state is not reached for any of the nodes and that the processing time of the messages has a similar behaviour in both strategies. But, for the P-CSCF/PDF transport resource driven reservation strategy, the processing time in the P-CSCF suffers big fluctuations and so do the other nodes. Since in this strategy this node has to process more messages, it is more exposed to randomly distributed instants of higher processing load. This node interacts with all the others (but with the HSS) thus this phenomenon affects the whole system. This results in a higher session setup time in this mentioned strategy.

An important conclusion here is that in complex scenarios, like the IMS, with many nodes interacting, the whole system performance can be influenced by the behaviour of one central node and this node is the P-CSCF.

We performed another set of tests, also for both strategies. We exposed the system to a constant and low load, introduced a short peak of calls and, then, injected again the same low and constant load. In the transport resource allocation driven by the terminal strategy, during this peak of calls, the session setup time increases fewer than in the P-CSCF/PDF driven strategy. Besides, the lapse to "recover" the "normal" session setup time in stable conditions is shorter in the P-CSCF/PDF driven strategy. The tests were repeated but increasing the processing power of the P-CSCFs in P-CSCF/PDF driven strategy, proportionally to the extra number of messages it had to manage respect to the terminal driven strategy. Results were similar. This stresses the conclusions we presented before: the strong influence in complex systems in the behaviour of a node (and of the whole system) of the many other nodes inter-working with it.

4 Conclusion

We analyzed two possible strategies for performing transport resource allocation in IMS and next generation network (NGN) scenarios. Those were letting the terminals perform this process (as it is done in IMS) or assigning this task to the PDF (or bandwidth broker in NGN). As usual, both possibilities have pros and cons. Making the PDF deal with this task presents many benefits allowing simpler terminals and making easier integrating the QoS enabled transport into the whole service delivery chain. However, as we saw thanks to simulation, if this strategy is chosen, the SIP proxies interacting with the PDF need to be carefully dimensioned since they become a central point interfacing with many nodes and may be potential bottlenecks. Mobility and context transfer issues are also an important part to analyze in future works.

Acknowledgments. This work was partly funded by Spanish Minister of Education (MEC) under CASERTEL-NGN project (TSI2005-07306-C02-02) and Madrid National Research Program e-Magerit (S-0505/TIC/000251).

References

1. Braden, R. et al. RFC 1663, "Integrated Services in the Internet Architecture: an Overview", June 1994
2. Braden R. et al. RFC 2205, "Resource ReSerVation Protocol (RSVP) -- Version 1 Functional Specification" September 1997
3. Calhoun P. et al. RFC 3588, "Diameter Base Protocol" September 2003
4. Camarillo, G. and García Martín, M.A.. "The 3G IP Multimedia Subsystem (IMS): Merging the Internet and the Cellular Worlds", Wiley, 2004.
5. Chan, K. et al., RFC 3084 "COPS Usage for Policy Provisioning (COPS-PR)". March 2001
6. Cherry, S. "Nothing but Net: Britain switches its entire phone network to the Internet Protocol." IEEE Spectrum. January 2007.
7. Cuevas. A. "Contribution to Design Suitable Session Setup Solutions in 4G Networks". Ph.D. Thesis. Universidad Carlos III de Madrid. 2006
8. Cuevas, A. et al., "The IMS Service Platform, the Key for Next- Generation Network Operators to Be More than Bit Pipes", IEEE Comm. Mag., vol. 44, no. 8, Aug. 2006
9. Dirham, D. et al., RFC 2748 "The COPS (Common Open Policy Service) Protocol" January 2000
10. García G. et al. "Soporte de QoS en Redes de 4° Generación" Revista IEEE America Latina, Volume 4, Issue 1, March 2006
11. Herzog, S., RFC 2749 "COPS usage for RSVP", January 2000
12. Information Sciences Institute, University of Southern California, Network Simulator 2 http://www.isi.edu/nsnam/ns/
13. Jähnert, J. et al. "The pure-IP Moby Dick 4G architecture" Computer Communications Vol 28/9 pp 1014-1027
14. Marques, V. et al. "An IP-based QoS architecture for 4G operator scenarios" IEEE Wireless Communications, June 2003
15. Papalilo, D., Salsano, S., Veltri, L. "Extending SIP for QoS support", Joint Planet-IP NEBULA workshop, Courmayeur, 2002.
16. Poikselkä, M. "The IMS : IP multimedia concepts and services in the mobile domain" Wiley & Sons ISBN: 047087113X (2004)
17. Salsano, S. Veltri, L. "QoS Control by Means of COPS to Support SIP-Based Applications" IEEE Network, March/April 2002
18. Scott-Joynt, J. "The secret of NTT's i-mode success" BBC news available at http://news.bbc.co.uk/1/hi/business/1835821.stm
19. WirtschaftsWoche, "Google & Co. müssen zahlen" WirtschaftsWoche, 23rd february 2006, Germany

Enhancing access control for mobile devices with an agnostic trust negotiation decision engine*

Daniel Díaz-Sánchez, Andrés Marín, Florina Almenárez

Telematic Engineering Department, Carlos III University of Madrid
Avda. Universidad, 30, 28911 Leganés (Madrid), Spain
{dds, amarin, florina}@it.uc3m.es

Abstract. Dynamic open environments demand trust negotiation systems for unknown entities willing to communicate. A security context has to be negotiated gradually in a fair peer to peer basis depending on the security level demanded by the application. Trust negotiation engines are driven by decision engines that lack of flexibility: depend on the implementation, policies languages or credentials types to be used. In this paper we present an agnostic engine able to combine all that information despite its origin or language allowing to select policies or requirements, credentials and resources to disclose, according to user preferences and context using iterative weighted Multidimensional Scaling to assist a mobile device during a trust negotiation.

Key words: trust negotiation, access control, flexible

1 Introduction

"Access Control" requires to determinate if an entity is entitled to use a service or not. Moreover, it should decide other parameters as which quality of service should be granted to an entity. Determining wether a user or entity can access or not to a resource, can be simplified in finding an answer to the question: "Can entity E perform action A on resource R?".

The answer can be found in different ways but in general comprises authentication, authorization and policy enforcement. Moreover, the requirements for access control depend on the context of usage, application and the sensitivity of the resources to be accessed and the credentials to be disclosed.

Different Access Control systems have been proposed: Mandatory Access Control (MAC), Access Control Lists (ACL), Role Based Access Control (RBAC) and some recent XML based efforts like eXtensible Access Control Markup Language [1]. Besides, different credentials are used: key centric bindings as KeyNote described in RFC 2704, SPKI in RFC 2693; and unique name binding credentials as Public Key Infrastructure (PKI) in ITU Recommendation X.509 or RFC

---

* This work has been partially supported by the Spanish Ministry of Science and Education throught the ITACA project (TSI2006-13409-C02-01).

3280, Privilege Management Infrastructure (PMI) in RFC 3281 or XML based like Security Assertion Markup Language (SAML) [2].

Modern distributed authorization models employ the Role Based Access Control (RBAC) that uses roles instead of identities. Roles can be expressed with PKCs [3], ACs ITU Recommendation X.509 that suffers from name-binding. SAML and other schemas [4] can be used also but suffer from limitations of key-centric approaches, or do not provide desirable definitions as separation of duties or authorization management as explained in [5].

This work presents a solution for assisting users to select policies or requirements, credentials and resources to disclose, according to their preferences and context, during a P2P trust negotiation. The work described here uses a human-mimicking decision engine able to simplify problems and to graphically present those problems to the user in a way that allows he/she to understand what is occurring despite his/her technical training.

1.1 Trust negotiation

Section 1 describes some authentication and authorization mechanisms that assume that every involved entity is known and trusted to the system in advance, for instance, a Certification Authority, the entity that vouches for an identity in PKI, should be trusted or the role/group should be accepted by both parties.

Promising efforts on trust negotiation as [6][7] allow **strangers to negotiate trust**, disclosing credentials, and even properties-based credentials (based on properties of the user rather than identities and capabilities). Thus, a stranger can be authenticated and authorized. Trust negotiation systems are based on the fact that any resource is protected by a policy that sets which credentials should be disclosed to obtain access to it. In [8] some requirements that a trust negotiation system must satisfy are described. Policies play an important role in trust negotiation, [6] recognizes that policies should be disclosed gradually according the level of trust reached since contain sensible information. Besides, rogue peers might build policies that force other entities to disclose more credentials and information than the necessary, so the credential disclosure should be done gradually in a peer to peer basis: providing only the necessary credentials to the peer holding the resource and asking the resource holder for credentials if more information than the necessary is required. Policy and credential disclosure should be driven by a decision engine to ensure the fairness of the process.

Systems supporting different credentials together with new trust negotiation systems are on the road to success since even strangers can be authenticated and authorized. This is the cornerstone for a real peer to peer secure interaction.

1.2 Multidimensional Scaling

Multidimensional Scaling [9], MDS, is a set of techniques widely used in behavioral, psychological and econometric sciences to analyze similarities of entities. From a pairwise dissimilarities matrix, usually m-dimensional Euclidean distances [10], MDS can be used to represent the data relations faithfully providing a geometrical representation of these relations. MDS is used to reduce the dimensionality of a problem to a small value. MDS techniques have been used for several problems with good results: to determine the distance among elements of sensor networks [10], to classify music, browse it and generate playlists [11] or to derive an *interaction distance* measure for network selection [12].

MDS can consider not only Euclidean distances but also any other evaluation of dissimilarities: qualitative or quantitative. The dissimilarities from attributes of data can be weighted (weighted MDS), thus, assigning a different weight to each attribute allows to obtain more particular results depending on the problem. So, a complex m-dimensional problem can be simplified preserving the essential information using MDS.

In classical scaling the proximities are treated as distances, however, any (di)similarity can be derived from data attributes in order to obtain a metric: in case of ordinal data, another procedure has to be followed than the use of singular value decomposition since we want to recover the order of the proximities and not the proximities or a linear transformation of the proximities. A solution to this problem was given by Shepard [13] and refined by Kruskal [14]. These solution iteratively minimize a fit measure called *Stress* by an iterative algorithm, which is suitable for processing.

We have used an algorithm called ALSCAL [15], which uses alternate least-squares, combined with weighted (di)similarities, can combine both metric and nonmetric analysis and can also deal with spare proximity matrixes so it is suitable in the absence of some data.

1.3 Article organization

Trough this section we have described the related work, section 2 introduces the *Agnostic trust negotiation decision engine* starting with a set of definitions in section 2.1 and the architecture in section 2.2. Moreover, sections 2.3 and 2.4 describe how to classify, extract and combine security and context information and also how to derive (di)similarities. Section 3 shows results for an example and finally we summarize the goals achieved in 4.

2 Agnostic trust negotiation decision engine

Along this section we will introduce algorithms to combine data that can be used to assist the user during a trust negotiation. We will use an example to make the algorithms easier to understand.

2.1 Definitions

A negotiation process involves information disclosure between peers according to a strategy that warranties that the process is fair for all parts. The strategy avoids rogue peers asking for unneeded credentials to reveal sensible information.

To help the reader to understand the article, a set of definitions are provided here:

Policies are pieces of data issued by a resource manager, a domain administrator or a provider. A **resource** can be protected by more than one policy. Those policies should be combined to obtain requirements. A **requirement** represents the information to be disclosed in order to satisfy part of a policy or combined set of policies. So from a policy or set of policies a requirement or set of requirements can be derived. A **policy item** is a formal definition for a requirement that can be used by other peers to find out which credential should be disclosed in order to satisfy a requirement. A **credential** is a piece of information to be disclosed to satisfy a requirement. A **resource** is any information, service or mechanism which its disclosure implies a risk. Credentials are also considered resources and should be protected by policies (so some requirements should be satisfied to be accessed).

2.2 Architecture

To take access control decisions the information available at a given instant of time and the context, should be taken into account. The context defines for instance, the connection speed of the available network interfaces, the location, the level of battery... Thus, given a context, a resource is described by its own properties or attributes and the constraints that the context imposes.

Consider a mobile device governed by a set of policies. The policies are written by the user, the domain administrator of the user's company and the UMTS provider. Those policies are controlled by different access control engines (ACEs). Every ACE processes the policy or set of policies it understands and extracts **requirements** and **policy items**. Then they register requirements and policy items to the decision engine. Figure 1 shows the architecture.

2.3 Merging policies, context and constraints

In this section we show how different requirements, extracted from different policies in different languages, can be combined using a single decision engine. Policies might be written in any language and might be general enough to cover all the possible types of users over a domain (mobile network operator or company policies). (ACEs) process the policies and extract the requirements. We require the Access Control Engines (ACEs) (possibly with the aid of the user) to be able to extract the part of each policy which applies to the device during setup (first time used).

We will use an example in which the ACEs extract four pieces (policy items) of the policies (P_n), and four requirements Rq_n. The policy items can be sent to the other part so the other peer can find out what credentials should disclose to satisfy a requirement. Policy items should be also protected. Disclosing a policy implies some risk, for instance, a policy item asking for a credential issued from a bank that asserts user's account balance might be disclosed only if other requirements has been previously satisfied. In the example, resources are named

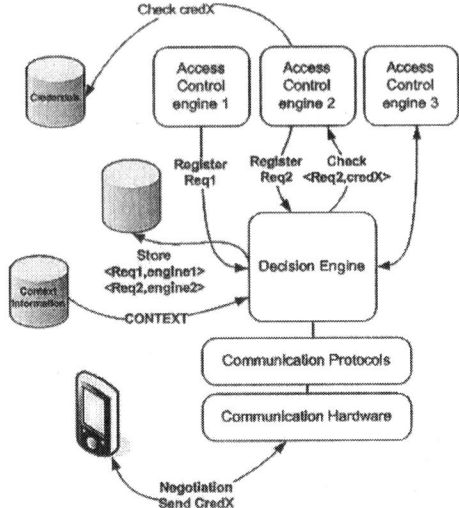

Fig. 1. Access control structure

C_n with n ranging from A to E for credentials and R_n with n ranging from A to H for other resources (not credentials). The table 1 illustrates the example.

Besides security attributes, properties of resources are also considered. In the example (table 1), **P/W** distinguishes among resources that can be accessed for personal (0) or work (1) usage. **RT** classifies resources as web-services (0), ftp (1), file sharing (2), and agenda resources (3). **Loc** specifies the resource location: 1 at mobile device and 0 outside mobile device (for situations where the device acts as a proxy). **CE** express battery consumption (0 for no consumption or connected to power line).

2.4 Computing dissimilarities

As can be seen in Table 1 we define a new item in that table to represent the negotiation (**Neg**). The attributes of **Neg** vary during a negotiation, thus, deriving dissimilarities including the negotiation makes possible to determine which resources are near the **Neg** point. After that calculation, the closer a resource is to the negotiation, i.e. its disclosure implies less risks.

(Di)Similarities between pairs of elements (rows in the table) can be derived as follows:

$$\delta_{i,j,\alpha} = \frac{|u_{i,\alpha} - u_{j,\alpha}|}{max(u_\alpha) - min(u_\alpha)}, \text{ for quantitative data} \quad (1)$$

$$\delta_{i,j,\alpha} = \frac{|rank(u_{i,\alpha}) - rank(u_{j,\alpha})|}{max(rank(u_\alpha)) - 1}, \text{ for ordinal data} \quad (2)$$

$$\delta_{i,j,\alpha} = \begin{cases} 0 : u_{i,\alpha} = u_{j,\alpha} \\ 1 : otherwise \end{cases}, \text{ for membership data} \quad (3)$$

	Reqs.	Loc.	P/W	RT	EC
Neg	Variable	1	V	V	V
P_1	none	1	U	U	U
P_2	Rq_1	1	1	U	U
P_3	Rq_2	1	1	U	U
P_4	$Rq_2 \& Rq_3$	1	0	U	U
C_A	Rq_1	1	0	U	U
C_B	$Rq_1\|Rq_2$	1	U	U	U
C_C	Rq_2	1	1	U	U
C_D	Rq_1	1	0	U	U
C_E	$Rq_1\|Rq_2\|Rq_4$	1	0	U	U
R_A	$Rq_1 \& Rq_2$	1	1	0	0
R_B	Rq_1	1	0	3	0
R_C	$Rq_1\|Rq_4$	1	1	3	0
R_D	$Rq_1 \& Rq_3 \& Rq_4$	1	1	0	0
R_E	$Rq_1 \& Rq_3 \& Rq_4$	1	0	2	0
R_F	$(Rq_1\|Rq_2)\&(Rq_3\|Rq_4)$	1	1	2	0
R_G	$Rq_1\|Rq_3$	1	1	2	0
R_H	$Rq_1 \& Rq_2 \& Rq_3$	1	0	0	0

Table 1. Attribute values in a possible decision scenario. U:Unspecified, V:Variable

where $u_{i,\alpha}$ is the αth attribute value of element i (policy item, credential or resource). We consider different types of data: quantitative (for trust relations [16] and distances [10]); ordinal (for QoS classes, and service differentiation); membership (to distinguish credential types).

Table 1 shows logical operators that are used to combine requirements. Sequences of requisites combined with logic operators (as *and, or*) cannot be compared using equations 3,4 and 5. For that reason we define the equation below to compare expressions:

$$\delta_{i,j,Rq_\alpha} = \begin{cases} 0: if\ f_{i,Rq_n} \neq f(Rq_\alpha) \\ 0: if\ f_{j,Rq_n} \neq f(Rq_\alpha) \\ eval(f_{i,Rq_n})\ \&\ eval(f_{j,Rq_n}): otherwise \end{cases} \quad (4)$$

where f_{i,Rq_n} is the logical function that combines the requirements satisfied during the negotiation for element i. $f_{i,Rq_n} \neq f(Rq_\alpha)$ means that the logical function that combines the requirements for element i does not depend on requisite Rq_α. $eval(f_{i,Rq_n})$ returns the result of evaluating the logical function using as parameters the requirements satisfied during the negotiation.

Once the (di)similarities are calculated, they are weighted (according to user preferences or context) in order to obtain an unique weighted (di)similarities matrix (weighted MDS, see 1.2). These weighted (di)similarities are defined for a set of n objects with q attributes as follows:

$$\delta_{i,j} = \left(\frac{\sum_{\alpha=1}^{q} w_{i,j,\alpha} w_\alpha \delta_{i,j,\alpha}^\lambda}{\sum_{\alpha=1}^{q} w_{i,j,\alpha} w_\alpha} \right)^{\frac{1}{\lambda}} \quad (5)$$

where $w_{i,j,\alpha}$ takes value 0 if objects i and j can not be compared on the αth attribute and 1 otherwise, w_α is the weight given to attribute α and $\delta_{i,j,\alpha}$ is the (di)similarity between objects i and j on the αth attribute.

The first element represents the negotiation in a time t and will be used to measure the *distance* from the negotiation to the resources: the nearer a resource is from the negotiation, the less risk the user experience. Even more, it is not possible to assign values to every attribute of every element during a negotiation. For example, the *type of service* attribute of a credential is **undefined**. MDS is suitable here due to the fact that is able to work in absence of some data.

Another key element in our model is the weights vector. The following equation shows the weights calculations. Table 2 gives the weights for the example at different instants of time:

$$w_\alpha = \begin{cases} \alpha \in Rq_n : \begin{cases} 0 : if\ \alpha \in \mathbf{Neg} \\ \frac{k}{ua} : if\ \alpha \notin \mathbf{Neg} \end{cases} \\ \alpha \notin Rq_n : \begin{cases} 0 : if\ \alpha \notin \mathbf{Neg} \\ \frac{k}{ua} : if\ \alpha \in \mathbf{Neg} \end{cases} \end{cases} \quad (6)$$

w_α is weight for α^{th} attribute. A weight value allows to express how important is a given attribute compared to the others. During the first round we use the same value for the unspecified attributes. The reader should note that a requirement cannot be unspecified: if not fulfilled a requirement is equal to 0 (false) and it is considered unspecified for weight calculation.

We maintain constant the sum of weights, so as long as the other peer provides credentials (fulfilling requirements) and the negotiation item becomes more specified, we uniformly spread the value among the attributes: if the attribute represents a requirement and it is already unspecified, we cannot grant access to any resource protected by that requirement, so we give to that attribute a weight of $\frac{k}{ua}$. k is a constant and ua the number of unspecified attributes.

Moreover, if the attribute represents a requirement but it has been fulfilled, the weight given to that attribute is 0. Otherwise, for attributes that represents properties of resources, we give them the value of 0 if unspecified and $\frac{k}{ua}$ if specified.

At this point we had the necessary data to run the MDS algorithm. We solved for two dimension and set $\lambda = 2$ to handle attributes as euclidean distances.

2.5 Computing risk limits

We use MDS to reduce the complexity of the problem to a visible number of dimensions, so we are able to graphically present to the user the decision space. Thus, the user is aware of the risk that involves a given interaction. Typically, users do not spend many time checking, for instance, website's certificates or other credentials, furthermore, when a user is prompt to accept or not a given credential, usually he/she accepts without wondering about the risks. A graphical

t	Rq_1	Rq_2	Rq_3	Rq_4	Loc	P/W	RT	EC
$t=0$	0	0	0	0	1	1	Un.	0
w[ ]	1.16	1.16	1.16	1.16	1.16	1.16	0	0
$t=1$	1	0	0	0	1	1	Un.	0
w[ ]	0	1.4	1.4	1.4	1.4	1.4	0	0
$t=2$	1	1	0	0	1	1	Un.	0
w[ ]	0	0	1.75	1.75	1.75	1.75	0	0
$t=3$	1	1	1	0s	1	1	Un.	0
w[ ]	0	0	0	2.33	2.33	2.33	0	0

Table 2. weight calculation during the negotiation. Requirements not fulfilled have 0 value. $K(t=0) = 7$.

presentation of the problem can be useful to be aware of the services that are been exposed to outside: the user can see which resources are similar, in terms of requirements, so he/she becomes aware of the resources affected by a decision.

Once a two dimensional presentation of resources has been obtained, we consider necessary to define a limit for the accepted risk: the risk can be displayed as a circle whose center is the **Neg** point and the radius depends on the context. Resources inside that circle can be disclosed since the risk is inside the accepted limits. But, how can we define that limit? The most restrictive approach should defend that resources can be disclosed only if their distance to the **Neg** point is 0.0. However, being some of the attributes unspecified, an exact match is likely difficult to achieve. We consider that the risk should vary depending on the context in the following fashion: the less defined a negotiation is, the more attributes are undefined and the biggest are the dissimilarities, so the bigger are the distances between the **Neg** and the resources, thus the more risk can be assumed since the less points will be inside the circle. To derive a value for the risk circle radius we propose the following equation:

$$radius = \begin{cases} 0 \text{ if } ua < uaMin \\ (\frac{ua}{attrNum}) * (\frac{maxDist}{attrNum}) \text{ otherwise} \end{cases} \quad (7)$$

where $maxDist$ is the maximum distance, $attrNum$ the number of attributes, ua the number of unspecified attributes and $uaMin$ the minimum number of attributes that should be specified to derive a radius different from 0.

3 Proof of concept

In this section we present the results of a negotiation. The resources to be analyzed have been already shown in Table 1. The negotiation advances in the following fashion: peer A, the one that holds the resources, discloses policy items, that express requirements, and peer B discloses credentials to fulfill those requirements. The negotiation evolves as displayed in table 2: includes both the requirements fulfilled at a given instant of time and the evolution of dissimilarities weights.

At $t = 0$ peer B tries to access a resource held by peer A. Peer A's decision engines has registered several requirements that affect that resource: the resource is disclosable only to company's employees, so it turns the attribute **P/W** to 1 and *Loc* to 1 (located at the mobile device). Since battery is full charged and there is no restriction about the type of resource **RT** and **EC** remains undefined. No requirement has been already fulfilled, so the values for Rq_n are 0. The mobile device compute the weights (see table 2) for $t = 0$. Then the mobile device compute dissimilarities and simplify the problem to a two dimensional problem. Figure 2 shows the decision space and table 3 shows the distances.

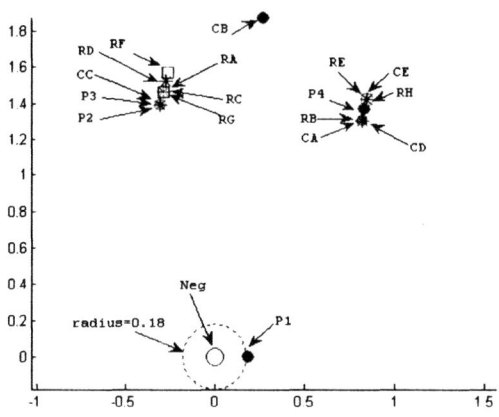

Fig. 2. Negotiation space at $t = 0$

As can be seen in figure 2, P_1 is the only resource that fits into the risk circle, so it can be disclosed. The rest of the resources form two groups, the first with the center located approximately at $[0.4, 1.5]$ is composed by the resources that can be disclosed for work purposes. The other group is composed by the resources that can be disclosed only for personal purposes.

Figure 3 shows the decision space at $t = 1$, just when the other peer discloses credentials to fulfill Rq_1. Under this condition, peer A can disclose P_2, R_G and R_C since those resources can be disclosed for work and requires Rq_1 to be satisfied. Resources C_E, C_A, C_D and R_B require also Rq_1 to be satisfied, but are separated enough from the **Neg** since they can be disclosed only for personal issues. Resources R_A, P_3 and C_C are grouped together since depends on Rq_2 to be fulfilled and should be disclosed only for work purposes. Despite resources R_H, R_E and P_4 can be disclosed only for personal issues as C_E, C_A, C_D and R_B, they form a different group, separated even more from **Neg**, since they have more complex security requirements. Moreover, resources R_D and R_F are located closer to the group of P_3 since they depend on Rq_2 also.

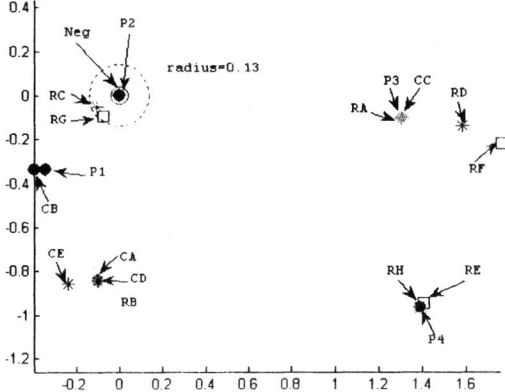

Fig. 3. Negotiation space at $t = 1$

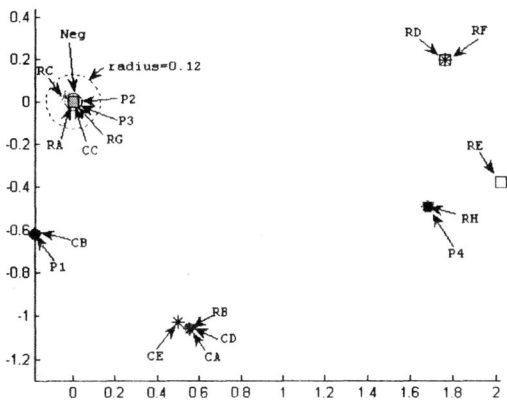

Fig. 4. Negotiation space at $t = 2$

Decision space at $t = 2$ is represented in figure 4. Rq_2 has been disclosed: the group of resources R_A, P_3 and C_C is now inside the circle, so can now be disclosed. The rest of resources remains far from the **Neg** element due to either their requirements or their personal nature.

In Figure 5 it can be seen that once peer B fulfill Rq_3, the resource R_F becomes available, and the rest remain far from the center. P_4 can not be disclosed since it is personal and, for that reason, many resources that depend on Rq_4 are not disclosed.

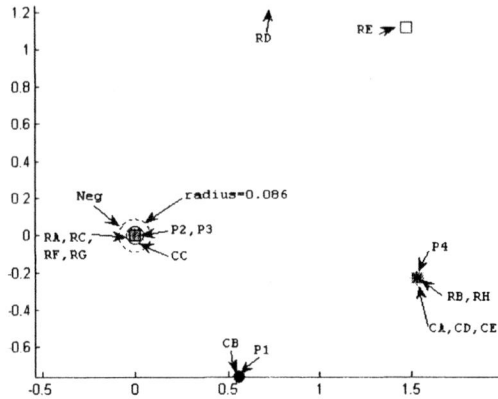

Fig. 5. Negotiation space at $t = 3$

t	Neg	P_1	P_2	P_3	P_4	C_A	C_B	C_C	C_D
0	0	0.18	1.42	1.42	1.60	1.53	1.89	1.42	1.53
1	0	0.48	0.00	1.30	1.69	0.84	0.51	1.30	0.84
2	0	0.64	0.00	0.00	1.75	1.19	0.64	0.00	1.19
3	0	0.94	0.00	0.00	1.54	1.54	0.94	0.00	1.54
t	C_E	R_A	R_B	R_C	R_D	R_E	R_F	R_G	R_H
0	1.65	1.49	1.53	1.49	1.54	1.65	1.59	1.49	1.65
1	0.89	1.30	0.84	0.11	1.58	1.69	1.77	0.11	1.69
2	1.14	0.00	1.19	0.04	1.76	2.05	1.76	0.00	1.75
3	1.54	0.00	1.54	0.00	1.44	1.84	0.00	0.00	1.54

Table 3. Distances

4 Conclusions

Through this paper we have demonstrated how simple, easy to understand by the user and strong is our decision engine. It is simple since we use just a distance to find the resources that can be disclosed, we treat resources, credentials and policies in the same way (agnostic). Thus, if during the negotiation, B changes the role and starts asking for credentials to peer A, to protect itself against rogue peers, peer A can use the results of the engine to determinate whether a credential can be disclosed to B or not.

The decision engine is easy to understand by the user and minimize the risks since the user is now aware of the consequences: the user can see which group of resources is affected by a decision so he/she can consider a larger picture just having a look to the resource clustering. Our decision engine is strong: no resource is disclosed unless every requirement is fulfilled.

We have also demonstrated that there is no need to find a common model to represent every element involved in a negotiation: despite its language, origin, and attributes, we provide a mechanism able to work even with unspecified data.

We propose also some rules to adopt logic operators and to derive a risk limit that depends also on the context. Furthermore, access control engines that, instead of returning true or false when verifying a credential, returns a continuous value can be used also just computing those values as shown in equation 3.

References

1. OASIS: eXtensible Access Control Markup Language (XACML) (2003) http://www.oasis-open.org/apps/org/workgroup/xacml/.
2. Mishra, P.: Saml v2.0 oasis standard specification. Technical Report SAML v2.0, OASIS Security Services TC (2005)
3. Herzberg, A., Mass, Y., Michaeli, J., Ravid, Y., Naor, D.: Access control meets public key infrastructure, or: Assigning roles to strangers. In: SP '00: Proceedings of the 2000 IEEE Symposium on Security and Privacy, Washington, DC, USA, IEEE Computer Society (2000) 2
4. Chadwick, D., Otenko, A.: The PERMIS X.509 role based privilege management infrastructure. Future Generation Computer Systems **19** (2003) 277–289
5. Bhatti, R., Bertino, E., Ghafoor, A.: An integrated approach to federated identity and privilege management in open systems. Commun. ACM **50** (2007) 81–87
6. Squicciarini, A.C.: Trust negotiation systems. In: EDBT Workshops. (2004) 90–99
7. Bertino, E., Ferrari, E., Squicciarini, A.: X -tnl: An xml-based language for trust negotiations. policy **00** (2003) 81
8. Bertino, E., Khan, L.R., Sandhu, R., Thuraisingham, B.: Secure knowledge management: confidentiality, trust, and privacy. Systems, Man and Cybernetics, Part A, IEEE Transactions on **36** (2006) 429–438
9. Borg, I., Groenen, P.: Modern multidimensional scaling, theory and applications. In: IEEE SECON 2004, New York, NY, USA, Springer-Verlag (1997)
10. Shang, Y., Ruml, W., Zhang, Y., Fromherz, M.P.J.: Localization from mere connectivity. In: MobiHoc '03: Proceedings of the 4th ACM international symposium on Mobile ad hoc networking & computing, New York, NY, USA, ACM Press (2003) 201–212
11. Platt, J.C.: Fast embedding of sparse music similarity. In: Advances in Neural Information Processing Systems vol. 16. (2004)
12. Díaz, D., Marín, A., Alménarez, F., García-Rubio, C., Campo, C.: Context awareness in network selection for dynamic environments. In: 11th IFIP International Conference on Personal Wireless Communications "PWC'06", Springer (2006)
13. Shepard, R.N.: The analysis of proximities: multidimensional scaling with unknown distance function part i. In: Psychometrika 27. (1962)
14. Kruskal, J.B.: Multidimensional scaling by optimizing goodness of fit to a nonmetric hypothesis. In: Psychometrika 29. (1964)
15. Takane, Y., Young, F.W., de Leeuw, J.: Nonmetric individual differences multidimensional scaling: an alternating least squares method with optimal scaling features. In: Psychometrika 42. (1977)
16. Alménarez, F., Díaz, D., Marín, A.: Secure Ad-hoc mBusiness: Enhancing WindowsCE security. In: 1st Conference on Trust Digital Business (TrustBus'04). (2004)

Wireless Technology in Medicine Applications

Otto Dostal and Karel Slavicek

Institute of Computer Science
Botanicka 68a, 60200 Brno, Czech Republic
{otto,karel}@ics.muni.cz
http://www.ics.muni.cz

Abstract. Institute of Computer Science of Masaryk University is working on the field of supporting medicine multimedia data transport archiving and processing more than ten years. Since first steps like transport of ultrasound and CT images across private fibre optics network these activities have grown up to regional PACS archive. Today more and more hospitals are participating on this project know under the name MediMed. For deployment of MediMed to some remote locations and especially to residence users (medicine specialists) it was necessary to utilise wireless and microwave technologies. Experiences with usage of wireless technology for delivery of medicine picture data are described in this paper.

Keywords: PACS, DICOM, Medicine Picture Data

1 Introduction

Institute of Computer Science of Masaryk University operates shared regional PACS (Picture Archiving and Communication System) since 2002. That solution was from its start designed to serve as a reliable and accessible communication node and also as an educational and research centre available to any hospital or other healthcare institution, including medical faculties, interested in participating.

The aim of our project was to establish a collaborative platform supporting daily routine in radiology community, to develop a communication channel supporting the exchange of information and special consultations among various medical institutions and also to support medical training for practicing radiologists and medical students. We enable the users from outside the hospitals to have the same access and functionality allowing them to have almost the same working conditions as in the radiology departments of their hospital.

One of the goals of our solution is also to establish an open, collaborative environment to support coordinated research and education among cooperating health care institutions and faculties of medicine by exploiting the large potential of databases of medical image information being processed in hospitals today.

2 PACS

The PACS (Picture Archiving and Communications System) is a currently used procedure and methodology for processing medical multimedia data obtained from picture acquisition machines like computer tomography, ultrasound, x-ray etc. Multimedia medicine data obtained from these machines (in PACS terminology called modalities) are stored in central PACS server. The PACS server then provides these multimedia data to viewing stations. Viewing stations serve to radiologists for analysing the multimedia data. This approach offers much more capabilities than former film medium. Viewing stations allow image transformation, combination of images from more modalities etc. National Electrical Manufacturers Association (NEMA) has developed a standard for communications between modalities, PACS servers and viewing stations. This standards is DICOM. Currently DICOM version 3.0 is used in mostly all modalities and PACS servers. The structure of PACS is cleanly presented on the figure 1.

Fig. 1. Common structure of PACS system. Modalities serve for acquisition of medicine multimedia data. These data are stored in PACS server and examined and analysed in viewing stations.

The system deals with transmitting, archiving, and sharing medical image data originating from various medical modalities (computer tomography, magnetic resonance, ultrasound, mammography, etc.) from hospitals. The central PACS serves as a metropolitan communications node as well as a long term archive of patients' image studies [7]. The technological solution is described in [4] and [6]. Deep theoretical background of PACS principle and int basic protocol DICOM is discussed in [1] and [2].

3 MediMed

Today most hospitals are using local PACS system serving only to one hospital. The goal of the MediMed project [5] is to start a collaboration among hospitals with respect to archiving and use of medical multimedia data and to provide the necessary technological infrastructure for the System.

The realisation of the project facilitates fast communication among individual hospitals, allows decision consultations, and brings various other advantages due to direct connections via optic networks. In general the MediMed project is clearly designed to support society-wide healthcare programmes in the Czech Republic as well as programmes implemented by other countries. The system is also supposed to serve as a learning tool for medical students of the Masaryk University as well as physicians in hospitals.

The gradual development of the joint system for processing and archiving image information is a natural step towards an increasing health care standard in the city of Brno and the whole region. Information on a patients treatment in his own healthcare centre as well as in other centres would be available. Consultations by more specialists will be enabled over the patent's picture, in case that a required specialist is not available in the centre in question. Image information evaluation can be carried out in another place, general practitioners in the country will be able to consult specialists in hospitals, etc. Examination results will be available for the doctors in much shorter time than before.

Outsourcing of the hospitals' archiving and communications technology permits cooperation among hospitals and usage of existing patient multimedia data. The Shared Regional PACS is more than just a set of computer network applications. Gradually, it changes the thinking of medical specialists and gets them to cooperate and share data about patients in electronic form. It builds a network of medical specialists. The impact of this work is not only in patient care but also in the education of medical specialists. The implementation of the project has increased the speed of communication among individual hospitals, allowed decision consultations, and brought various other advantages due to dedicated network connections.

4 Technology background

The main idea of private network serving for regional PACS archive is the following: We put our firewall in front of a hospital's router or firewall. Our firewall is connected to PACS server's firewall via any secure way. In the case of hospital reachable via private fiber optics network of Masaryk university a dedicated fiber optics pair is used. For remote hospitals the most common way is to use IPSEC tunnel on top of any public data network - in most cases the Internet.

This approach allows us, as administrators of the application, to control the access to central resources and allows the administrators of the hospital's network to control the access to their network. That way everybody can control access to the part of the network that he is responsible for. The situation is easy to see from figure 2.

Fig. 2. PACS connectivity for hospitals reachable via university private fiber optics network

There is a large fibre optic cable network owned and operated by the universities in the city of Brno. The development of this network started in 1993. It is a private network of all Brno universities connected to the National Research and Education Network (NREN), operated by CESNET association, through the Masaryk University computer centre. The network connects all universities in the city and their faculties spread around the city, various departments of the Academy of Science, most hospitals and some other institutions. Currently the network consists of about 100km of optic cables and more than 100 nodes are using it. Since the Brno Universities own the fibre optic network, they can ensure that there is sufficient fibre optic cable available to implement new applications and to support new initiatives. The ownership provides the freedom to establish private connections dedicated to these applications.

For hospitals beyond the scope of university's optocable network we need to use public data network. The usage of the public data network forces us to use strong cryptography to secure the tranported data. For this purpose we use IPSEC with AES or in case of equipment with pure CPU performance 3DES encryption algorithm. The logical view of intercity connections to central PACS archive is in figure 3.

Fig. 3. Remote hospitals interconnection.

Since Regional PACS system is used on a regular production basis it should provide users with reliable and safe services. Because we are dealing with very sensitive information we strongly relay on data storage and transport security. Regional PACS archive is running on dedicated network infrastructure with mostly no interaction with another data networks. In case that it is necessary to use public data networks for servicing remote hospitals we are using strong cryptography for securing medicine data transport.

As the MediMed project was growing up it was necessary to duplicate critical components into remote location to provide required reliability and availability of medical data. The regional PACS system is situated in two locations. The distance between them is about 3 km. All crutical data are stored in both locations. there are used PACS servers from different vendors to have backup for possible (however not probable) breakdown of server system vendor and especially services supported by that vendor. The overall situation is illustrated on picture 4.

Fig. 4. For reliability improvement there exist backup servers in other location. Backup servers are about 3km away from primary ones.

Medicine multimedia data processing is based on DICOM standard. In the case of a single hospital, the problems of user authentication and authorisation are easily solved. In the case of the network with several hospitals, authorisation can be for instance based on the IP address of the requesting computer station. This approach is acceptable in the case of medical modalities and computer stations used by a single person. With greater use of the metropolitan grade of PACS there is the problem of several users sharing the same PACS viewing station. This problem is solved by the use of IPSEC. All users sharing the one PACS viewing station use IPSEC for communication with the central PACS archive. After successful authentication, the IPSEC server gives each user a unique private IP address to the endpoint of the IPSEC tunnel. This allows us to use IP based authorisation for access to medical data. Because authenticated users of the PACS archive need to change locations several times a day, it is necessary to provide a fast and easy method of authentication while maintaining a high level of security. For these reasons we use the Public Key Infrastructure (PKI) authentication with the private key stored on a USB key. The keys are stored in the certification authority (CA) operated by Masaryk University.

The amount of various equipments used in this project is increasing during its development and deployment of new functionality. Recently the number of used devices enforced development of centralised management system. This system provides all the necessary supplemental services like collecting of traffic statistics, backup of configurations, time synchronisation, authentication of network managers, etc.

Two Linux based servers are used as central management stations. These stations form redundant solution for all the goals listed above. The primary management station is located at the Institute of Computer Science of Masaryk University and the second one is located at the Faculty of Medicine. Each management station provides full set of services and both stations work independently.

The open source solution - Nagios - was selected for network status monitoring. This software is used for monitoring of network component status, CPU utilisation number of active users, number of running processes, local disk usage of servers and also monitoring of disk usage of central storage system. The whole monitoring system is accessible via web interface and provides the current status as well as the history of all equipments and services availability. Critical alarms are propagated via SMS messages distributed via SMS terminals Siemens 35i connected via RS232 interface directly to both management servers.

5 Wireless technology

As already mentioned, the backbone system uses optical wires as a transport medium. Nevertheless, only the hospitals in the city and several of the others in the republic are connected by optical wires.

Another transport medium which is being used is the radio connection. It may be utilized for the main connection of the locality, however in that case the bigger hospitals require at least 20-30 Mbit/s speed. We are speaking mainly about sending and storing pictures from MR, CT, and such where there is a high demand for transport capacity.

Radio connection is also being used as the so called last mile. We have optical connections between cities, but the radio connection is needed to connect the hospitals inside the city. This concerns establishing traffic in the paid band. Another large area where we employ radio connection is to provide a backup connection. In that case, the requirements for data capacity are lower than for the main connection, however, they still must be sufficient for the operation of the medical facilities.

One of the biggest groups of the users of the wireless networks are the radiologists. They very frequently use the opportunity to create the descriptions of the pictures at home. That way, they may react to urgent cases immediately. There is no need to go to hospital and start working on a pressing case after a significant delay. Simultaneously, they save their time, because the work on the picture may often take significantly less of it, than the voyage. The map of wifi coverage of city is on figure 5.

There are some issues with the usage of the wireless systems. The performance may drop during bad weather such as heavy snowing and there are sometimes problems in the cities when a new building (or an extension of an old one) emerges in the way of the original signal. Still, it can be said, that their employment accelerates the exploitation of systems for archiving, exchange and processing of data as a whole.

Fig. 5. WIFI coverage of the city of Brno. The black triangles are home offices connected via university's private wifi network.

The way of usage of these ways to transfer data is different in different medical facilities. One of the possibilities is, that the doctors have radio connections to their houses, so they may describe pictures at home, consult over medical information with a specialist in another medical facility in another city (or even state), etc. Also, the cable television connections are being used for this purpose if the doctors have them.

As for the doctors working outside their workplace, there is a frequent request for a transportable notebook with mobile telephone for a connection to the central system. However, the limited speed available limits the range of modalities to work with.

Another technology, which we use for the MediMed project, is the satellite. Within the scope of HEALTHWARE project (which is a 6th EU framework program project), there are being installed terminal satellite devices to the places, where any proper connections are not existent. Some facilities, such as medical institutions for patients with tuberculosis, may be found in woodlands, areas without industrial burden. Than, the usage of a satellite system is one of the few ways, we may use for transfer of a medical information. It is therefore used despite its limited data capacity, that is so needed in the cases of urgent demands for transfer and processing of medical image data.

6 International collaboration

We cooperate it the area of processing of medical image information also in international scale. For example the Healthware (Standard and interoperable satellite solution to deploy health care services over wide areas) project within the sixth framework programme of EU covers many telemedicine activities.

The goal of this project is developing of healthcare services over the satellite network to increase quality and comfort in European medical practices. The aim is to bridge the medical digital divide in Europe by designing, integrating and validating interoperable telecoms and services platforms to provide existing and future health care services. The satellite based platforms can interact with mobile and terrestrial technologies to supply effective and reliable end-to-end healthcare services and boost the deployment of large-scale satellite communications telemedicine services.

Additionally, Healthware will have a beneficial effect on training and education as far as 7 Universities and Research Centres are concerned. For undergraduate, post-graduate and PhD students, the participation in such programs is a unique opportunity to be exposed to team work with regular reporting and evaluation by the partners. The research performed is usually of very high quality due to the number experts involved in the group and the concentration of financial resources. It is also the occasion to be exposed to a multicultural environment and to establish international relationships that are very useful to build and strengthen the European Research Area.

7 Activities supporting medical training

Teaching has always been one of the most important parts of radiology. Nurture an excellent radiologists in this technological age involves more resources, new methodologies, reorganization of radiological training.

The core of our solution is tailored PACS. That PACS can be used as a "PACS trainer" for students and young radiologists but also forms the basis for additional educational and research applications such as for example the Case Studies describing treatment of real patients. The Case Study is an integrated hypertext document forming didactic unit and consists of short texts, structured clinical data, radiological images of various kinds, images from nuclear medicine modalities, macroscopic and microscopic pathology images or demonstration of the video movies recorded during surgeries.

Images appropriate for teaching and research purposes are made anonymous (i.e. the personal data of the patient and other information that may disclose the identity of the patient is replaced with fictitious information or modified in such a way so as not to lose any relevant information but so as to prevent disclosure of the patients identity) when sending into Educational and Research PACS.

One of the basic principles when sending images into the Educational and Research PACS is the coordinated assignment of fictitious patient identity, so it can offer a more complex view of the evolution of the patients health in situations where the patient is being treated in different healthcare facilities. Therefore, the legal barrier preventing access to sensitive and confidential patient data is removed.

The Case Study can be accessible via standard web browser and if the users have DICOM diagnostic workstation installed on their computers, then the referenced image study can be manipulated and processed in all ways supported by the particular workstation. It means that medical students can access large amounts of systematized medical cases related to their subject. The labs equipped with appropriate software can also serve as training simulators for those training to be radiologists. The students can learn more practical lessons instead of wasting their time in the library.

8 Conclusion

The efficiency and rationalization of technological and human resources must be considered in connection with improvements of the quality of healthcare. Teleradiology in the Czech Republic is crucial because medical experts and specialists can be available in urgent cases permitting qualified external medical experts to be involved in the diagnosis. Regional-level procurements of systems and services are preferred to achieve major savings and improve the quality of healthcare including the learning processes.

The evolution of educational and research services provided by this solution is also influenced by emerging wireless communication technology. This technology permits the appropriate services to be accessible also through the satellite network covering the Europe. The satellite based platforms can interact with mobile and terrestrial technologies.

References

1. K.J.Dreyer- D.S.Hirschorn-J.H.Thrall- A. Mehta , PACS A Guide to the Digital Revolution, 2006, Springer Science +Business Media, Inc., USA, ISBN 978-0387-26010-5
2. H. K. Huang, DSc, PACS and Imaging Informatics: Basic Principles and Applications, Hoboken, NJ: Wiley, 2004 ISBN 0-471-25123-2.
3. M. Petrenko - P. Ventruba - O. Dostal, First application of Picture Archiving and Communication System (PACS) in Gynecologic Endoscopy, ISGE - 6th Regional Meeting, Bangkog. ISGE 2002, p.53-56.
4. O. Dostal - M. Filka - M. Petrenko, University computer network and its application for multimedia transmissin in medicine, WSEAS Int. Conf. on Information Security, Harware/Software Codesign, ECommerce and Computer Networks, Rio de Janeiro, Brazil; WSEAS 2002, 1961-1964
5. O. Dostal - M. Javornik - K. Slavicek, MEDIMED-Regional Centre for Archiving and Interhospital Exchange of Medicine Multimedia Data, Proceedings of the Second IASTED International Conferee on Communications, Internet and Information Technology. Scottsdale, Arizona, USA : International Association of Science and Technology for Development- IASTED, 2003. ISBN 0-88986-398-9, p.609-614.
6. O. Dostal - M. Javornik - K. Slavicek - M. Petrenko - P.Andres, Development of Regional Centre for Medical Multimedia Data Processing, Proceedings of the Third IASTED International Conferee on Communications, Internet and Information Technology. St.Thomas, US Virgin Islands: International Association of Science and Technology for Development- IASTED, 2004. ISBN 0-88986-445-4, p.632-636.

7. M. Schmidt - O. Dostal - M. Javornik, MEDIMED - Regional PACS Centre in Brno, Czech Republic, Proceedings of the 22th International Conference of EuroPACS & MIR (Managenment in Radiology) Conference, 16 - 18 September, Trieste, Italy
8. Dostal O. - Javornik M.: Regional Educational and Research Centre for Processing of Medical Image Information, CARS 2005 Computer Assisted Radiology and Surgery, June 22.-25. 2005, Berlin, Germany, p.911-915. ISBN 0-444-51872-X, ISSN 0531-5131.

Wireless Military Communications - NNEC Enabler

Miroslav Hopjan, Zuzana Vranova
University of Defence, Department of Communication and Information Systems,
Kounicova 65, 60200 Brno, Czech Republic
Miroslav.Hopjan@unob.cz, Zuzana.Vranova@unob.cz

Abstract. The paper discusses the NATO Network Enabled Capability concept, mainly from the communication point of view. The changes involve complete new requirements on the role of command and control to increase flexibility and effectiveness. Integration of Modeling and Simulation with Command, Control, and Information Systems increases the number of risks but it promises to leverage the projected capability and interoperability.

Keywords: NNEC, simulation, tactical communication, CCIS.

1 Introduction

This paper does not introduce new technological solution in wireless communications, the point of view is closer to the customer side of the house – how to use the emerging communication technologies in an optimal way supporting the concept of NATO Network Enabled Capability (NNEC) which emphasizes the role of information superiority in modern warfare. The aim is to evaluate number of aspects when implementing this technology in the Czech Army. What role is adequate for contemporary microwave devices, why the implementation is delayed, what risks must be outweighed by benefits of these solutions. Communications networking is the clearly visible part of the solution, and suitability of selected, mostly wireless communication approach, is discussed.

Although it is not the core functionality for Command and Control Information Systems (CCIS) it is apparent that NNEC encompasses also Modeling and Simulation (M&S) capability. These two domains have developed different architectures, standards but further progress of one system is related to the other. Modern wireless communication means promise to fill one gap between these two worlds.

This article tackles two oncoming trends:
- raising demand of (wireless) communication means to individual combatants to share the knowledge (operational picture)
- integration of M&S capability into CCIS

2 The Need of Flexibility and Agility

The processes of changing doctrine, force structure, inevitably accompanied by change in procedures and equipment, called "transformation" is symptomatic for all armies recently balancing the East-West military powers trained and prepared for another World War. While not completely abandoning the imperative that armed forces of particular country must be prepared to defend the state territory from massive attack it became clear that building and maintenance of military forces according to Cold War era would be prohibitive simply for economic reasons.

So, instead of traditional, fixed unit organization that is both costly to move, accommodate, supply, protect, and slow to maneuver we are trying to enable any possible task force, autarchic enough for months-long deployment, and tuned for so called Effect Based Operations according to current task, adversary, and conditions (Mission-Enemy-Task-Time-Civilians). It is apparent that in case of technological superiority the enemy will use more concealed, guerrilla-like ways of combat. This does not exclude using the same contemporary technology by the enemy whenever it is effective. Building such modular, well trained forces, capable of deployment anywhere in the world as well as acting as an element of territory rescue system, or doing police job in unstable regions, is difficult from a number of points of view. At the same time, implementation of NNEC principles enables getting over limits flowing from traditionally strict military organization; improved access to information both concentrated in databases and gathered by front-line sensors and units together with increased decision autonomy (Loosely Coupled Management Process [2]) enables creation of informal networks. Using Complexity Theory [1] information entropy of such system is lower, system with non-centralized decision making and information sharing allows all parts to learn and adapt, see Fig. 1.

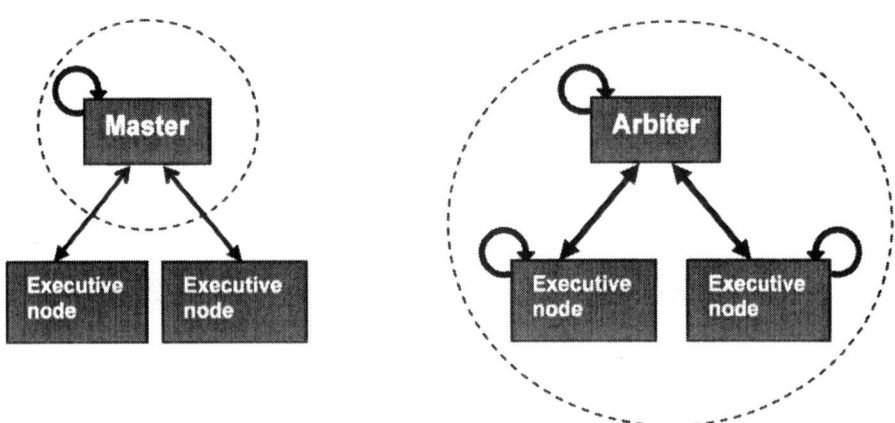

Fig. 1. Loosely Coupled Management Process
The left hand side depicts the traditional organization where subordinate units (generally executive nodes) perform tasks, higher commander (master) is the only learning and adapting one (dashed circle). Sharing the information and decision power is illustrated on the right side.

Obviously, the world is not only black and white; the level of autonomy varies depending on current situation. The technology level that enables extensive information sharing makes not only the decision making process at lower level more effective and optimal, it enables the transfer of information in opposite direction, too.

The feasibility of getting detailed and timely information from front-end units may be tempting for high-level commanders to skip the chain of command and apply direct control.

2.1 The NEC Concept

The simplified view how the network concept can contribute to enhanced operational capabilities is depicted in Fig. 2. New capability in such complex environment cannot ensue only from extended technology level as this creates new challenges in related domains. For example, extended range and higher precision of weapon system is undoubtedly operational advantage. But if any delay occurs in communication, the IFF (Identification Friend or Foe) systems are not fully interoperable, or different procedures of decision making process do not support the timely mission plans alignment we have a problem with euphemistic name "friendly fire". If components in all layers are not in accord then an individual improvement can be disturbing, even risk increasing factor.

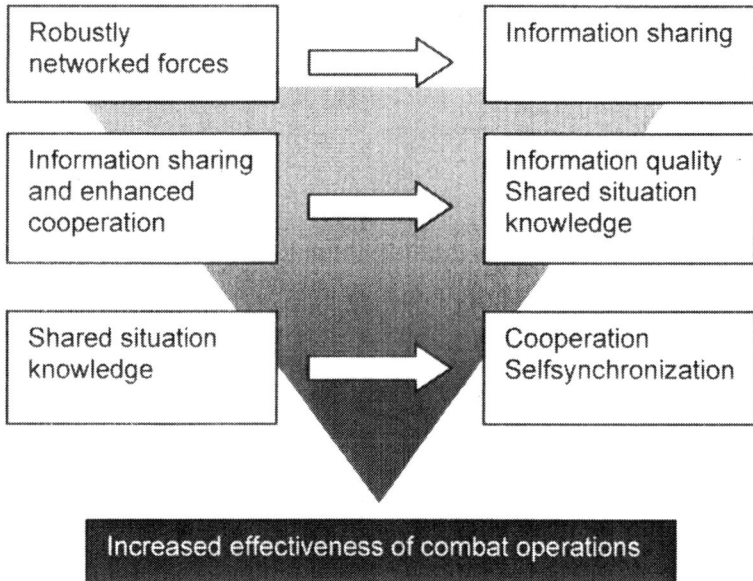

Fig. 2. The NEC concept

The NEC does not mean simple network layer extension vertically and horizontally (vertical connection of Ground Forces and Air Force communication and information systems, horizontally - extending connectivity to smallest units); according to [3] there is an impact on all domains:
- Physical (tactical) – sensors, surveillance systems, data acquisition
- Information (operational) – information analysis, tools and models supporting decision making
- Cognition (military strategic) – deep understanding, highest decision level

- Social (affecting all individuals)

The changes will affect not only technical infrastructure but also the organization structure, decision making and planning procedures as well as the education and training.

2.2 Bringing Modeling and Simulation closer

Both CCS and military M&S systems have their own long history. Information technology serving in commander decision process must not increase notably the weight, complexity, power consumption, and maintenance staff as it might easily become ballast increasing vulnerability, requiring special treatment and long time to process raw data. This was absolutely true for high-resolution virtual simulators as well as for computation intensive staff simulators requiring special hardware, months or weeks to create terrain database and exercise scenario, and experienced personnel to run the simulation. When low-cost PCs connected in a network became appropriate the hardware limits could be overcome.

There are many problem areas where we still have not reached final solution; standards should be developed to allow seamless integration of M&S systems within overarching NNEC environment.

The ACR technology lag is quite apparent in the communications domain. Well, it is difficult to get any satellite communication except the contracted one but the NNEC concept was implemented in information systems and communication domain rather with respect to legacy systems than with broader, future vision.

Such a goal has not been pronounced by the ACR representatives yet but thee future development should continue in diminishing the difference between the combat and training by embedding the M&S into equipment. Despite this raises many new questions how to eliminate related problems it is clear that modern wireless communication systems will be the key enabler of new capabilities.

3 Aspects to Consider

Let us make closer look at problems accompanying the NNEC concept realization in the Czech Army. Based on experience from previous projects we should identify risks, they are mutually related:
- Political risks – despite there is limited chance to avoid wrong political decisions their impact on technology can be eminent (further development costs, interoperability, strong bond to certain supplier, etc.),
- Economical risks – life cycle cost of a system is extremely difficult to plan, also poor user specification and changing requirements often produce collateral damage when resource shortage stops or puts behind linked projects; this comes to unsatisfactory functionality, creating new demands on manpower or growing system obsolete
- Technical risks – when implementing new technology it must be thoroughly analyzed whether this technology is mature enough to avoid

life cycle shortening due to reaching its technical limits; is it likely that this technology will be standardized?

3.1 Bandwidth

Building the communication infrastructure using traditional SW tactical radios may become fatal barrier when involving sensor grid such as UAVs or integrating simulation assets. There is backbone network using modern commercial technology capable of transferring satellite imagery and databases in no time. How can we deliver and process data from numerous ground sensors? ZigBee rather than WiFi could be the right solution, although there is not capacity reserved in today's communication infrastructure. The ACR wireless tactical intranet is capable of sharing situation reports, short messages, and orders, nothing else.

3.2 Interoperability

Before any new technology is considered for military implementation an analysis of its impact on acknowledged standards or possible need to create new one must be carried out. The standardization processes are NATO is somewhat cumbersome in areas where vital functionality is not affected. Whenever possible, international commercial standards are adopted. Communication parameters belong to critical areas with little freedom to use different coding, data model and formats, frequency bands, modulation, etc.

On the other hand, the M&S area was for many years "rear-echelon" activity. High intensity warfare was not favorable environment for building deployable training facilities, and technology limits did not allow truly portable systems. Then, ground and air systems were isolated; the same was true for combat and logistics. Tactical and operational simulators were useful for early training phases, rather than for mission planning or mission rehearsal, which becomes an important feature today.

When networked simulations started to play important role in NATO activities standards emerged to allow even global and secure multinational exercising.

Strong effort is being spent now to find an appropriate architecture, data models, and other aspects that would enable seamless CCIS and SIM integration.

3.3 Robustness

The first most important requirement on the military communication channels is high probability of successful message delivery. In terms of the network topology and management there must be redundant or backup infrastructure, and network condition monitoring to keep the reliability high. Depending on classification level and unlikely in the public Internet, mostly dedicated, and therefore expensive, channels e.g. satellite or radio-relay links are necessary.

3.4 Security

Information security should minimize access to knowledge to unauthorized system user while not affecting the overall capability (Need to Know versus Need to Share). Today's closed, individual service owned CIS do not support multi-layered information security concept, and there are mutually incompatible cryptographic products in use. Secure Communications Interoperability Protocol (SCIP) will be necessary to interconnect system elements. Wireless data network solutions such as WiFi, WIMAX, or ZigBee are becoming respected players after proving their suitability for classified networks.

Security is a big issue; when connecting voice and data networks together plus linking intelligence databases further increases the security urgency. Attacks against military networks are still mostly proof-of-concept but this does not mean that such a weakness will not be exploited in critical situation when the advantage can be maximized.

We pay considerable attention to data channel encryption, still, the security is not just about unbreakable cipher – it is alarming that stolen computer discs from military installations, sometimes with data extremely useful for the adversary, can be bought on markets in Afghanistan.

3.5 EW Insusceptibility

Electronic warfare has changed the view of military operations due to its potential to gain a superiority using smart, high-tech means. Sometimes the cutting edge technology is of little help against enemy that uses low-tech means or an effective and inexpensive countermeasure appears soon after spending fortune on technologically advanced system. Defence research and development projects have been drivers for commercially successful products, the opposite direction of valuable solutions propagation is analyzed eagerly too. Though, the cost concerns cannot be overlooked.

The intense wireless computer network expansion left the military decision makers untouched till the security concerns prevailed. Still, military implementation must look farther to balance the good and bad. Overcoming the security issues of wireless communication there is still problem of traceability of the transmitter; accuracy of which increases in higher frequency bands. Another example is the painful Improvised Explosive Devices (IED) problem. To disable remote control of IEDs powerful broadband jamming must be used.

Satellite communication becomes inevitable when forces are deployed abroad, not only as a home country link but also in the theatre. Today's doctrine did not count on vast distances covered by small, tactical units; their radio reach is insufficient, especially when operating in mountainous terrain. But, what risk we take if contracted SAT connection can be tracked by an enemy connected to the Internet?

3.6 Small Footprint

The measure of effectiveness of any military campaign compares the effect of armed units with total costs. While lack of important material can slow down or even disrupt operation plans, and therefore certain surplus is preferred, opposite case increases the need of transport capacity, affects the mobility, requires more people and time to move, maintain, and protect the logistics tail. Supporting units are more vulnerable in low intensity conflicts when fewer personnel are operating in large territory. Generally, the relationship between new technology and its footprint is nonlinear, depending on the technology maturity; users often perceive in the beginning that new functionality is well balanced by the additional weight, shorter battery life, and high complexity affecting reliability and reparability in harsh environment, ease of use when under stress. The increasing networking and computer control trend is faster than progress in lightweight, high-capacity accumulator or fuel cell technologies. Data communication is taking over the traditional voice communication channels with the aim of unified, simple, and new possibilities offering infrastructure.

4 Conclusion

The Army of the Czech Republic is facing new challenges related to doctrine change, coalition bonds, global security risks, and latest technology achievements. The traditional territory defence has evolved into highly mobile expeditional units. The NNEC concept striving for information superiority allowing higher combat effectiveness places new requirements on communication infrastructure (not only that, the impact will be much more complex) in terms of connectivity, capacity, and security. The short term goal is seamless connection of link and wireless encrypted communication channels. The system is scheduled for security certification but we cannot include multiple security levels as requested. Current solution of wireless "tactical intranet" would not handle the sensor grid data volume. Integration of M&S and CCIS increases the demand on the network bandwidth while enabling training and mission rehearsal in combat zone, even different options evaluation during the decision making process will be possible.

References

1. Moffat, J.: Complexity Theory and Network Centric Warfare. Washington DC, USA: CCRP Publication Series, 2003.
2. Atkinson, S.R., Moffat, J.: The Agile Organization. Washington DC, USA: CCRP Publication Series, 2006.
3. Smith, E.A.: Complexity, Networking, & Effect-Based Approaches to Operations Theory and Network Centric Warfare. Washington DC, USA: CCRP Publication Series, 2006.

Wired Core Network for Local and Premises Wireless Networks

Jiri Vodrazka[1], Tomas Hubeny[1]

[1] Czech Technical University of Prague, Faculty of Electrical Engineering,
Technicka 2, 166 27 Prague 6, Czech Republic
{vodrazka, hubenyt}@fel.cvut.cz

Abstract. The local and premises wireless networks have been expanding into many areas. The core network access point or basic station infrastructure can be made on wireless systems, on fixed cable systems with metallic twisted pairs, and, prospectively with optic fibers. The VDSL2 line systems with VDMT modulation and line bounding will be used for a data speed of hundreds Mb/s and distance between active nodes can be measured in hundreds meters. A suggestion of active access network and simulation of data bit rate is presented.

Keywords: Core network, Active access network, VDSL2, VDMT modulation

1 Introduction

The backbone or core network access point or basic station infrastructure can be made on wireless systems, or on fixed cable systems with metallic twisted pairs and prospectively with optic fibers. Optical fibers are gradually replacing the metallic lines, and data bit rates have increased from tens Mb/s to hundreds Mb/s or Gb/s in LANs and in the access network (first miles of digital subscriber lines). However, the information capacity of metallic lines can be increased if the systems use a VDSL2 [8], physical layer with vectored DMT modulation and line bounding concept. The methods for partial crosstalk cancellation are discussed and simulation results are presented for the middle-range used for connection of access points. The data bit rates 100 Mb/s and 1 Gb/s are currently used for interfaces and connection to backbone network.

2 Conception of core network

If it is necessary to cover some area by high-speed wireless technology [6], the connection of access points to the central point of the network gateway has to be set. For connection of the access points sufficient throughput is necessary and hence the reason for building an efficient core network arises.

Wireless point-to-point or fixed point-to-multipoint connections are used for attachment of access points. This solution is not suitable under all circumstances. There can be problems with low throughput, interferences and insufficient bandwidth.

For extreme throughput of the core network a fiber optic FTTB (Fiber to the Building), or FTTN (Fiber to the Node) FTTN conception can be used (Node is a wireless access point AP - see fig.1). Here the length of the cables is restricted for common low-coast interfaces to 10 - 20 km, which is more than satisfactory and also the transmission speed 1Gb/s, eventually 10 Gb/s, offers a large reserve.

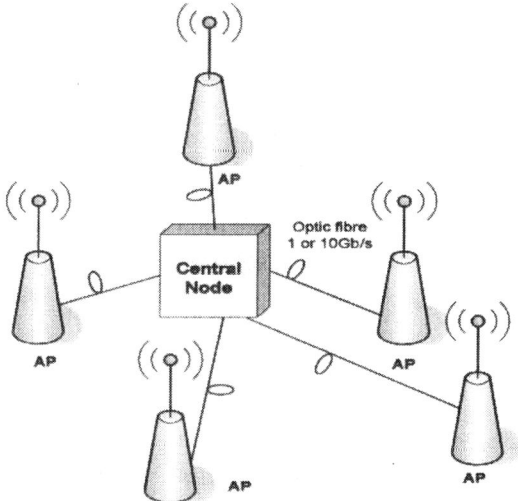

Fig. 1. The conception of hybrid optic-wireless FTTN network for core of wireless network.

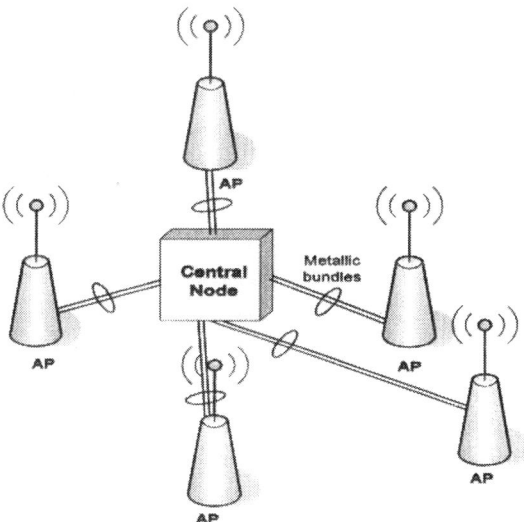

Fig. 2. The conception of wired core network for connection of wireless access points.

The installation of new optic fiber infrastructure is not profitable anytime. On the other hand, classic metal cabling in the LAN networks has a restricted range of 100 m. The solution is to use a VDSL2 physical layer with multi-pair bounding (see fig. 2) taking existing lines (e.g. telephone lines) into account. The metallic line provides speeds up to 100 Mb/s. Using Multi-pair bounding which is working on the Inverse Multiplexing Principle will result in multiplying the transfer speed, hence it is possible to achieve a transmission speed to 1 Gb/s.

A problem in multi-pair cables is crosstalk between pairs, which is the main source of noise and affects the transfer speed. This can be solved by echo cancellation (section 2.1 and 2.2). Nevertheless, the transfer speed is only 1Gb/s for a distance of 500 m (for 10 pair bounding – section 3.3). In this case it is possible to move from a star to a tree topology and use a Hierarchic active core network concept (see fig. 3) on Hierarchic Active Access Network (HAAN). Total coverage for 1Gb/s will increase to a radius, only the lengths of the tree sections are restricted (approximately 500 m for local cable with diameter 0.4mm).

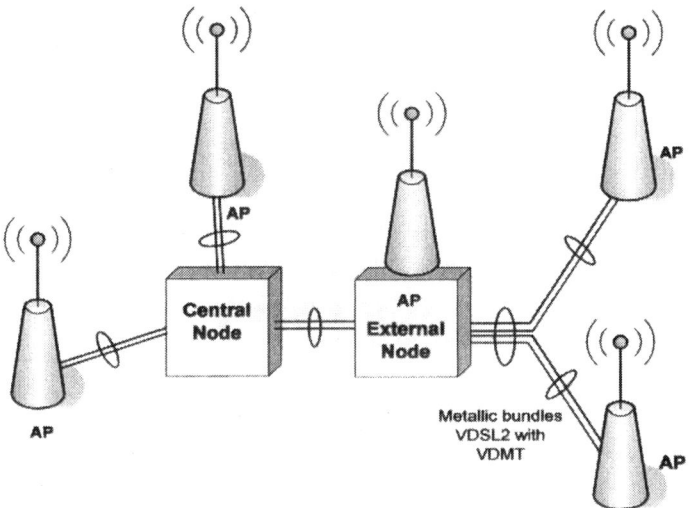

Fig. 3. The conception of wired Hierarchic active core network with VDSL2 (VDMT) for connection of wireless access points.

2.1 Crosstalk cancellation

A method of crosstalk cancellation is used in the GbE interface for cables of the 5E category for distances up to 100 m. For further distance it is suitable to dispose of crosstalk at the near end using FDD frequency duplex and then deal with cancellation at the far end (see fig. 4). DMT very effectively cancels crosstalk, where cancellation between sub-channels is used. This modulation is then called vectored DMT (VDMT) and can be classed into a group of MIMO systems. (Multiple Input – Multiple Output) [10], [11].

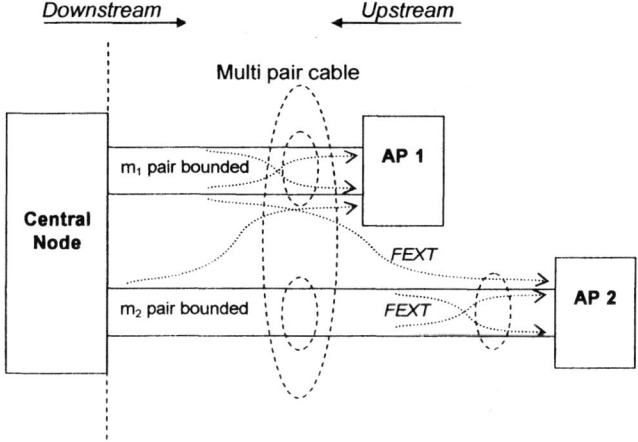

Fig. 4. The FEXT crosstalk for downstream direction in multi pair cable with pair bounding. The crosstalk exists between bounded pairs and between pairs from other bounding groups too.

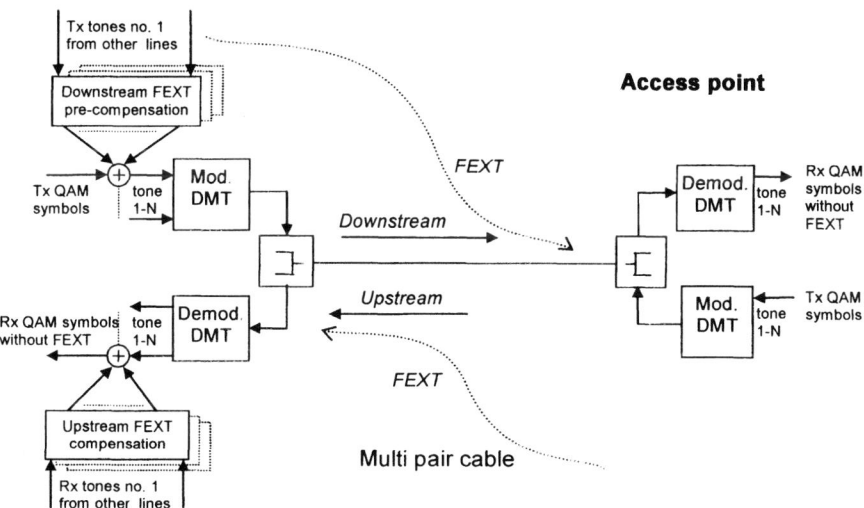

Fig. 5. The principle of VDMT for multi pair cable – FEXT compensation on receiver side (Rx) in frequency domain for upstream and pre-compensation on transceiver side (Tx) in frequency domain for downstream.

The advantage of VDMT is a possibility to create multipoint networks, where the cancellation for both directions is done in a central network element. For the upstream direction there is compensation at the receiver side and for downstream there is pre-compensation at the transmitter side (fig. 5).

For correct operation of VDMT it is necessary to know the transmission path parameters, along with the nature of crosstalk from surrounding (neighboring) lines, placed in identical metallic cables. Transmission path parameters [9] are identified during the process of establishing connection between access points and central node. To compensate for crosstalk it is required to have signals sent from all lines. They are present in central mode, but not in each access point. Therefore, compensation has to be performed in both direction of transmission in central node, or more precisely for direction upstream crosstalk compensation on receiving side and for direction downstream signal's pre-compensation on transmitter side.

2.2 Hierarchical active core network with VDMT

Conception of a hierarchic active access network solves the problem of markedly restricted distance when the extreme transmission speeds are required. The increased reach is possible using active elements in cable tree structure analogous to a hybrid access network, FTTC. The only difference is that the active element is not connected to the access point via an optic fiber but via metallic wire bundles using inverse multiplex principle (IM)

For maximization of the transfer speed usage of crosstalk cancellation in a bundle of 10 pairs (5 quads) is suitable. In local branching there are access points as well as active elements of core network. In active network elements the multiplexing of data from users is accomplished.

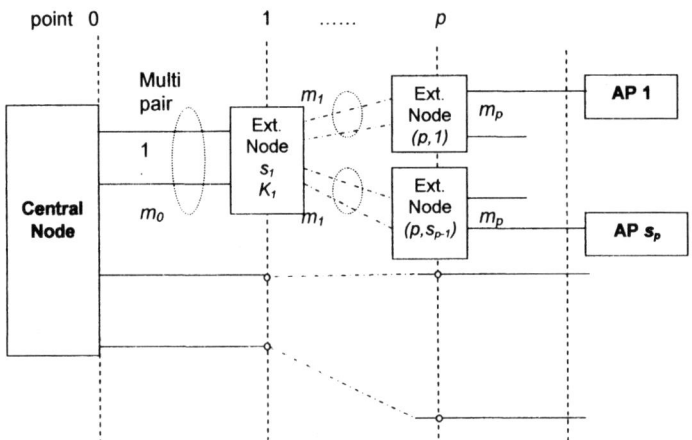

Fig. 6. The structure of hierarchic active core network for connection access points (AP) to central node trough external nodes (can include access points too) with p-levels.

In general, layout of the access network is figured out with a structure (see fig. 6). The central node is connected via bundle of m_0 pairs to the external node. Transmission in the bundle is ensured by VDSL2 subscriber lines with VDMT modulation. Supposing identical transfer speeds of all lines which operate at distance of d_0 in the bundle $R(d_0)$ the resulting transfer speed of this section is $R_0 = m_0 \cdot R(d_0)$. With this speed the external node is connected to the central node.

Users can be connected via external DSLAM or another hierarchic level of external DSLAMS can follow. Supposing the number of external DSLAM's s_1, connected with the number of pairs m_1 at distance d_1 the speed is $R_1 = m_1 \cdot R(d_1)$. The traffic concentration is used, hence $R_0 \leq s_1.R_1$. The K_1 concentration factor is established, so it can be expressed:

$$R_0 = R_1 \frac{s_1}{K_1} \quad \text{and generally for } p \text{ levels of hierarchy} \quad R_0 = R_p \prod_{i=1}^{p} \frac{s_i}{K_i} \qquad (1)$$

Followed by:

$$m_0 R(d_0) = m_p R(d_p) \cdot \prod_{i=1}^{p} \frac{s_i}{K_i} \qquad (2)$$

Assuming the speed at the distance of p:

$$R_p = m_0 R(d_0) \cdot \prod_{i=1}^{p} \frac{K_i}{s_i} \qquad (3)$$

The resulting speed for the user will be limited by throughput of all sections:

$$R_s = \min\left\{ m_p R_p(d_p); m_{p-1} R_{p-1}(d_{p-1}) \cdot \frac{K_p}{s_p}; \ldots; m_1 R_1(d_1) \cdot \prod_{i=2}^{p} \frac{K_i}{s_i}; m_0 R_0(d_0) \cdot \prod_{i=1}^{p} \frac{K_i}{s_i} \right\} \qquad (4)$$

For the practical applications it is not suitable to create more than two hierarchy levels. The concentration rate can be chosen, in ranges according to types of provided services, e.g. $1:K_1$, where K_1 is from interval 2-20 at one level hierarchy, eventually $1:K_1:K_2$ where K_1 is from interval 2-5 and K_2 is from interval 2-8 at two level hierarchy.

3 Simulation results

3.1 Simulation of transmission environment

Metallic lines in local networks begin at the main distribution frame, then they run as multi-pair cables to line distribution frames, from which they branch into smaller groups of subscriber lines or to individual lines leading to the subscribers' premises. In this network there are various types of cables in common use with copper core (mostly 0.4mm in diameter) and various numbers of pairs, basically arranged in quads. The principal factors limiting the transmission of high-speed signals are attenuation of the line, and crosstalk between the individual pairs [3], [4].

Since there is a need to estimate the available transmission speed, we have used a MATLAB Web Server to design a simulator of xDSL lines that is available at our web pages [1]. The program accepts many input parameters for the calculation and provides results in various forms. The most important simulation result is the estimated transmission speed for upstream and especially for downstream direction.

3.2 Crosstalk modeling

The described method for summarizing of contributive crosstalk (NEXT and FEXT) has been recommended by the FSAN (Full Service Access Network) consortium. The simplification of crosstalk computation consists in crosstalk parameters. These parameters are averaged over the total length of the subscriber line, not considering the real cascade structure. In addition, the position (in the same group, in different groups) of disturbing and disturbed pairs is ignored. The values of crosstalk constants are preventively calculated for the worst-case disturbance environment scenario. However, this worst case results in a more pessimistic noise level.

The typical attribute of crosstalk is very high variance of values. It is clear that if we interleave the minimum attenuation values, we obtain the worst-case disturbance scenario, which is approximately 10dB worse than average attenuation values. The maximum attenuation values are 12 or 15dB (or more) higher then the worst-case scenario [5].

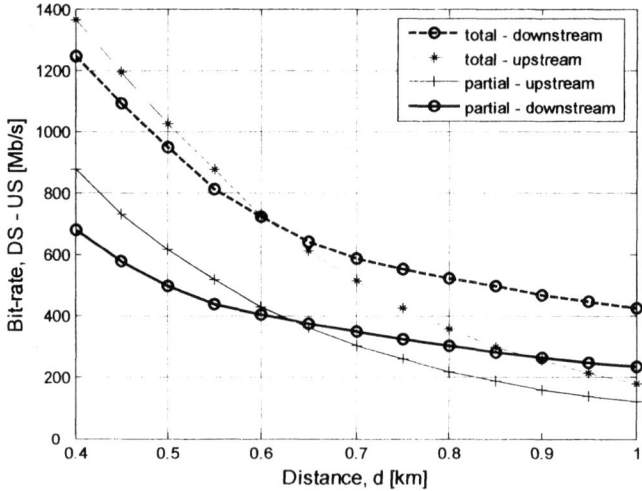

Fig. 7. Total transmission speeds of a bundle for 997 plan with full (dash line) and partial (full line) VDMT coordination.

The frequency dependence of crosstalk transmission function has a random character. With respect to the construction of standard cables used in access networks, it is possible to divide the symmetric pairs into several groups. Each group shall contain pairs with the same position. The knowledge of crosstalk transfer function between all pairs will be necessary for transmission using Vectored DMT modulation.

3.3 Results for hierarchic active core network

For estimation of reachable transfer speed let's assume that crosstalk cancellation is used in a bundle with 10 pairs (5 quads). Hence let's focus on estimation of transfer speed reachable in the bundle of ten lines. Supposing total FEXT cancellation is used, a residual noise AWGN will take effect only (-130dBm/Hz). Then the downstream and upstream speeds will occur – as shown in fig. 7 (frequency plan 997) and fig. 8 (frequency plan 998).

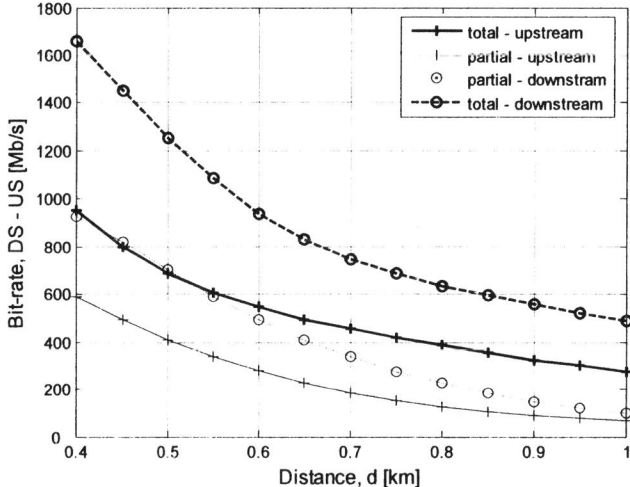

Fig. 8. Total transmission speeds of a bundle for 998 plan with full (dash line) and partial (full line) VDMT coordination.

Practically, in a cable other subscriber's lines will be present – ADSL (assumed 20% from the cable profile), ADSL2+ (assumed 10% from the cable profile) and SHDSL (assumed 6% from the cable profile). Hence slower transfer speeds will occur – in figures described by the full line.

4 Conclusion

Existing metallic cables which were originally designated for low-frequency telephone transmissions are used for digital transmission with speeds up to hundreds Mb/s. In the paper, the application of these systems for core network of access points was shown. Using VDMT modulation brings a significant increase in transfer speed in cables which are already filled with other subscriber's line in comparison to common way of transmission. This places considerable demand on the computation capacity of signal processors. The solution is to use a partial crosstalk cancellation with a selection of the most interfering DMT tones. According to our experiments this method does not have the results as we expected and hoped.

Nevertheless for transmission within medium distances with a bundle of lines, speeds up to 1Gb/s can be reach using a physical layer VDSL2 (plan 997). Therefore transmission among network elements in core network for wireless access points can be performed thus.

The active core network on a basic Hierarchic Active Access Network (HAAN) was performed. Global coverage for speeds of 1 Gb/s will differ according to the number of hierarchical levels. The lengths of the cables are restricted to 500 m for cables with wire diameter of 0.4 mm, as follows from the results of simulations. For extreme transfer speeds the metallic infrastructure has to be replaced by an optic fiber.

Acknowledgments: This paper has originated thanks to the support of the Grant Agency of the Czech Republic within project GA102/07/1503 and Czech Technical University's grant No. CTU0715013.

References:

1. Vodrazka, J. – Jares, P. – Hubeny, T.: xDSL simulator. Matlab Server on-line. http://matlab.feld.cvut.cz/en/
2. Vodrazka, J.: Downstream Power-Back-Off Used for ADSL. EC-SIP-M 2005. Bratislava: Slovak University of Technology, 2005, pp. 349–353. ISBN 80-227-2257-X
3. Vodrazka, J. – Simak, B.: Theoretical Limits of DSL Lines. Communication and Information Technologies – Conference Proceedings. Liptovský Mikuláš, 2005, pp. 243–246. ISBN 80-8040-269-8
4. Simak, B. – Vodrazka, J.: Limits for Broadband Transmission on the Twisted Pairs and Other System Co-existence. CTU Workshop. Prague: CTU 2006
5. Jares, P.: Throughput Modelling and Spectral Compatibility of Digital Subscriber Lines. EC-SIP-M 2005. Bratislava: The Faculty of Electrical Engineering and Information Technology of the Slovak University, 2005, s. 228-233. ISBN 80-227-2257-X.
6. Konhäuser, W.: Broadband Wireless Access Solutions - Progressive Challenges and Potential Valueof Next Generation Mobile Networks. Wireless Personal Communications: An International Journal archive. Kluwer Academic Publishers Hingham, MA, USA. Volume 37, Issue 3-4 (May 2006). Pages: 243 - 259 ISSN:0929-6212
7. Tardy, I. - Bråten, L.E. - Bichot, G. - Settembre, M.....: Hybrid architecture to achieve true broadband access in rural areas. BroadBand Europe Brugge, BELGIUM 08-10 December 2004. Session 07 – Paper 07-02. Page 1 of 6.
8. Very-high-speed Digital Subscriber Line (VDSL2). ITU-T Recommendation G.993.2
9. Rauschmayer, D. J.: ADSL/VDSL Principles: A Practical and Precise Study of Asymmetric Digital Subscriber Lines and Very High Speed Digital Subscriber Lines. Indianapolis, USA: Macmillan Technical Publishing, 1999
10. Cendrillona, R. – Ginisb, G. – Moonena, M. – Acker, K.: Partial Crosstalk Precompensation in Downstream VDSL. Signal Processing 84. Elsevier (2004), pp. 2005–2019
11. Brady, M. H. – Cioffi, J. M.: The Worst-Case Interference in DSL Systems Employing Dynamic Spectrum Management. Hindawi Publishing Corporation EURASIP Journal on Applied Signal Processing. Volume 2006, Article ID 78524, pp. 1–11.

Classification of Digital Modulations Mainly Used in Mobile Radio Networks by means of Spectrogram Analysis

Anna Shklyaeva[1], Petr Kovar[1] and David Kubanek[1]

[1] Department of Telecommunications,
Faculty of Electrical Engineering and Communication
Brno University of Technology, Purkynova 118,
612 00 Brno, Czech Republic
xshkly00@stud.feec.vutbr.cz, kovarpetr@phd.feec.vutbr.cz, kubanek@feec.vutbr.cz

Abstract. In this paper a new method of modulation classification is proposed. For the analysis, modulation signals and their spectrograms were obtained in the Matlab program. The classification method is based on spectrogram image recognition and it can discriminate between various digital signal modulations, such as FSK, BPSK, MSK, QPSK and QAM. The new method was tested using modulated signals corrupted by Gaussian noise, and it is well usable with signal-to-noise ratios as low as 10 dB.

Keywords: Modulation, Classification, Spectrogram.

1 Introduction

In connection with the requirement for faster and more reliable communication, the present digital processing methods and digital communications are mainly used. Together with the rapid growth of cellular technologies, PCS (Personal Communication Services) and WLAN services in the last decade, a number of different wireless communication standards were proposed and employed, and each of them has its own unique modulation type, access technique, etc. To realize a seamless inter-communication between these different systems, a multiband, multimode smart radio system, such as software radio, is becoming the focus of commercial and research interests. The automatic modulation classification technique, which is indispensable for the automatic choice of the appropriate demodulator, plays an important role in such a multimode communication system [1]. Automatic identification of the type of digital modulation has found application in many areas, including electronic warfare, surveillance, and threat analysis [2].

ASK (Amplitude Shift Keying), BPSK (Binary Phase Shift Keying), QPSK (Quadrature Phase Shift Keying), FSK (Frequency Shift Keying), QAM (Quadrature Amplitude Modulation), MSK (Minimum Shift Keying) belong to the best-known digital modulations. These modulation types are used in modern radio telecommunication systems (GSM, WiFi, WiMAX, etc.).

In recent years, various methods of the modulation classification were developed. However, most of them are based on the knowledge of some parameters of received signal and the other methods are computationally very intensive.

This paper describes a new method of modulation classification, which is based on spectrogram image recognition.

2 Obtaining Signal Spectrograms

Spectrograms for long signal intervals were analyzed. Simulations were performed in the Matlab simulation software, where the examined signal was obtained from modulator models. The signals with FSK, MSK, BPSK, QPSK and QAM16 modulation types were submitted to spectral analysis. For simplification, the same settings were used for all modulators, i.e. the modulation signal with the same data-signaling rate was used for all modulators. All the modulators used the same carrier frequency. For the FSK modulation, frequencies f_1 and f_2 were set so that the medium frequency was equal to the carrier frequency of the other modulations. The sampling rate for all types of modulation was chosen identical.

It is necessary to take into account some requirements while determining the segment size for spectrum calculation. The first one is the requirement that solely signal elements with the same value in the segment should appear. The simplest solution of this problem is to choose the segment size equal to the signal element size. The second requirement, which must be satisfied, is sufficient discrimination ability along the frequency axis. It is necessary for the discrimination of nearby frequencies that are present in MSK modulation. Then the obtained module and phase spectrograms (Figs. 2.1. – 2.6.) were analyzed by means of the recently proposed analysis.

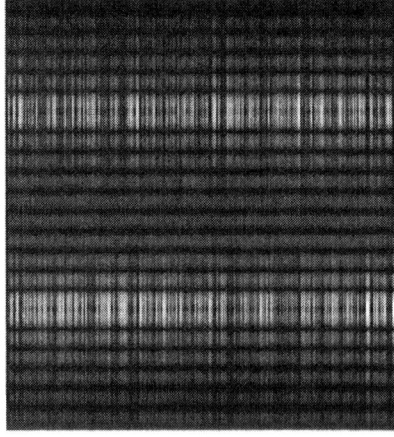

Fig. 2.1. Module spectrogram of BPSK, QPSK signals

Fig. 2.2. Module spectrogram of FSK signal

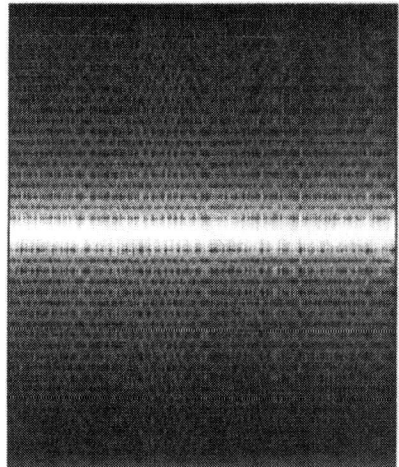
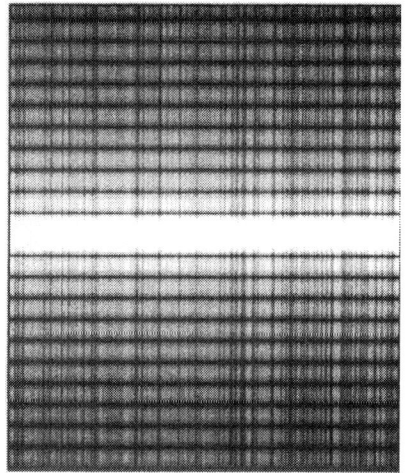

Fig. 2.3. Module spectrogram of MSK signal

Fig. 2.4. Module spectrogram of QAM-16 signal

Fig. 2.5. Phase spectrogram of BPSK signal

Fig. 2.6. Phase spectrogram of QPSK signal

For the purpose of classifying noisy signals, the broadband white noise was added to the modulated signals so that the signal-to-noise ratio decreased to a value of 10 dB. The spectrograms of noisy signals were obtained in the same way.

3 Detection of Signal Element Size

As previously said, the analysis results are very dependent on the segment size used for spectrum calculation. The characteristic properties of modulation types are apparent only in spectrograms that were obtained with the segment size equal to the signal element size. If the segment size is equal to several lengths of signal element, the modulation recognition method based on the spectrogram image analysis is unusable. Therefore it is necessary to find the signal element length prior to obtaining a spectrogram. For this purpose we suggested three methods: wavelet transform, cepstrum analysis, and autocorrelation function.

4 Analysis of Signal Spectrograms

4.1 Analysis of Module Spectrograms

For the estimation of spectrogram features, it is advantageous to use histograms of their images. The digital grey-scale image can be presented as a matrix of numbers $\mathbf{A}(i, j)$, where i in the range [0, M-1] and j in the range [0, N-1] are indexes of rows or columns of the image (matrix).

For the analysis of module spectrograms it is suitable to observe the count of maximum brightness levels (amplitudes) in separate rows (on separate frequencies). The analysis results are shown in the following figures.

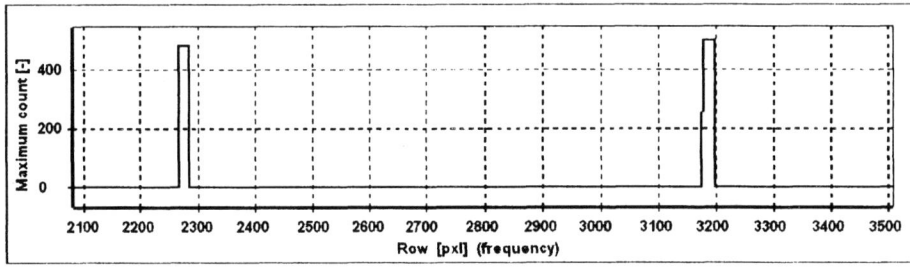

Fig. 4.1. Occurrence count of maxima in separate rows of FSK spectrogram

The FSK modulation type is easiest for recognition, because two distant carrier frequencies occur in the module spectrogram. Therefore the method of finding the maximum occurrence in separate rows easily detects two characteristic maxima for this modulation (Fig. 4.1).

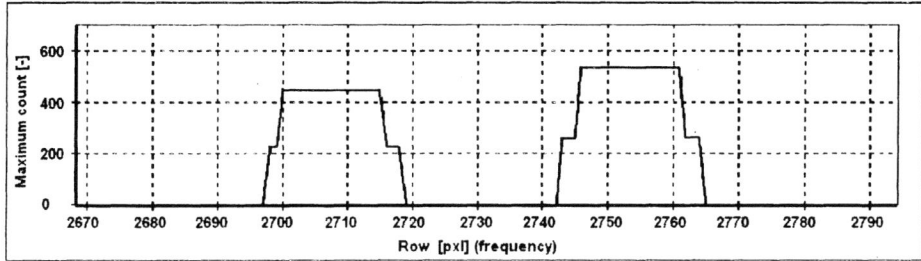

Fig. 4.2. Occurrence count of maxima in separate rows of MSK spectrogram

Similar to FSK modulation, two carrier frequencies can be found in the module spectrogram of the MSK modulation. However, it is not easy to recognize them from the spectrogram. Therefore a detailed cut of the spectrogram part was made in order to find more easily the areas of maximum brightness values. These areas allow the recognition of two carrier frequencies close to each other, as apparent in Fig. 4.2.

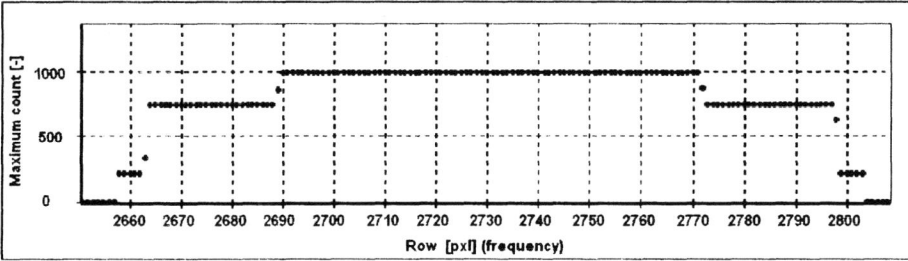

Fig. 4.3. Occurrence count of maxima in separate rows of QAM-16 spectrogram

The QAM-16 modulation has one carrier frequency with varying amplitude. It can be found from the constellation diagram of QAM-16 modulation that three different values of amplitude occur in the QAM-16 module spectrogram. In Fig. 4.3, the three amplitude levels are shown in the form of three main values of maximum count (in the form of three symmetrical steps in this discrete function).

4.2 Analysis of Phase Spectrograms

The analysis of module spectrograms described in chapter 4.1 can detect three different modulations with varying frequency or amplitude. For the remaining two modulation types (BPSK and QPSK) it is necessary to analyze the phase spectrogram. The graph of maximum occurrence count in separate rows does not provide any usable properties.

From the detailed view of phase spectrograms it is apparent that several different values of brightness (phases) occur in the area around the carrier frequency. The number of these values corresponds to the number of phase positions used in the modulation. Thus the analysis must evaluate how many phase values occur at the carrier frequency. If the carrier frequency is not known, it can be easily found from the module spectrogram. The counts of phase values at the carrier frequency of BPSK, QPSK and QAM16 spectrograms are shown in the following three figures.

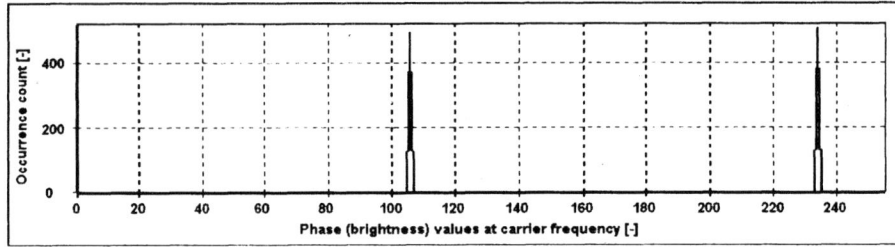

Fig. 4.4. BPSK: Occurrence count of phase (brightness) values at carrier frequency

The two maxima in the graph in Fig. 4.4 correspond with the theoretical expectation that BPSK modulation has two phase positions.

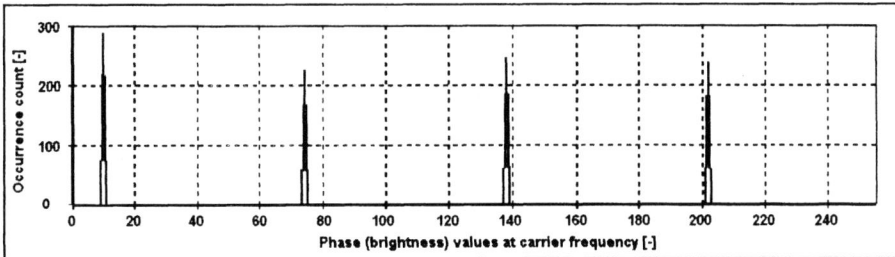

Fig. 4.5. QPSK: Occurrence count of phase (brightness) values at carrier frequency

From the analysis results, which are shown in Fig. 4.5 is apparent that the modulation has four possible phase positions. This corresponds again with the theoretical expectations for the QPSK modulation.

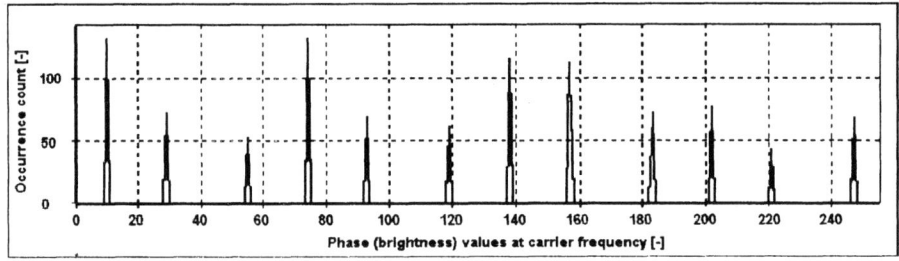

Fig. 4.6. QAM-16: Occurrence count of phase (brightness) values at carrier frequency

In addition to the analysis of the amplitude spectrogram of the QAM-16 modulation described in chapter 4.1, we can analyse also the phase positions. A constellation diagram of the QAM-16 modulation shows 12 various phase positions. Fig. 4.6 confirms this fact because 12 maxima are apparent there. This additional phase analysis is a suitable for the reliable determination of the modulation type because the discriminability of amplitude values markedly decreases with increasing noise.

4.3 Analysis of noisy signals

The module and phase spectrograms of noisy signals were analyzed in the same way. The signal-to-noise ratio of 10 dB was used in all cases.

The added noise in all the spectrogram analyses caused smoothing of acute transitions in the resulting graphs. This fact is insignificant for the FSK and MSK modulations and the characteristic properties are still clearly visible in the graphs. For the QAM-16 modulation, the smallest of the three characteristic symmetrical steps of discrete function (see Fig. 4.3) was smoothed due to the noise. However, the characteristic steps are still perceptible and the QAM modulation is distinguishable from the other modulation types. Due to the added noise, the determination of possible amplitude values is more complicated for signals with low amplitude variation. Therefore, the analysis of phase spectrograms is advisable for additional evaluation. The influence of noise is not significant in the analysis of phase spectrograms.

5 Conclusion

We designed a new method for the classification of digital modulations by means of spectrogram image analysis. Spectrograms, in which the characteristic properties of modulations are apparent, were obtained with the segment size equal to the signal element length. Thus it was necessary to find the signal element length prior to the spectrogram computation. Analysis of histograms of spectrogram images was used for the survey of modulation properties. By means of this analysis, the typical parameters of each modulation were found (carrier frequency, number of amplitude levels, number of phase positions). Spectrograms of noisy signals were also obtained and analysed. It has been proved that the recently designed method is also suitable for signals disturbed by noise.

Acknowledgments. This work was supported by Ministry of Education project No. 1835/G1 and Ministry of Education project No. 1446/G1.

References

1. Dai, W., Wang, Y., Wang, J.: Joint power estimation and modulation classification using second- and higher statistics. WCNC 2002 - IEEE Wireless Communications and Networking Conference, no. 1 (2002) 767 – 770
2. Hong, L., Ho, K. C.: Identification of digital modulation types using the wavelet transform, MILCOM 1999 - IEEE Military Communications Conference, no. 1 (1999) 427 – 431
3. Shklyaeva, A., Riha, K., Sysel, P., Ciz, R., Rajmic, P., Vondra, M. Rozpoznavani digitalnich modulaci pomoci analyzy obrazu spektrogramu. (Research report in Czech)
4. Xiong, F. Digital Modulation Techniques. London: ARTECH HOUSE, INC., 2000. ISBN 0-89006-970-0

5. Yu, Z., Shi, Y. Q., Su, W. M-ary frequency shift keying signal classification based-on discrete Fourier transform. MILCOM 2003 - IEEE Military Communications Conference, 2003, p. 1167 – 1172
6. Hsue, S. Z., Soliman, S. S. Automatic Modulation Recognition of Digitally Modulated Signals. MILCOM 1989 - IEEE Military Communications Conference, 1989, p. 0645-0649.

Security Issues of Roaming in Wireless Networks

Jaroslav Kadlec[1], Radek Kuchta[1], Radimir Vrba[1]

[1] Dept. of Microelectronics, Faculty of Electrical Engineering and Communication
Brno University of Technology, Udolni 53
CZ-62100 Brno, Czech Republic
{kadlecja, kuchtar, vrbar}@feec.vutbr.cz

Abstract. This paper is focused on the secure roaming problematic in wireless automation applications. Description of a roaming procedure and security issues is presented along with the newest techniques for fast roaming according to prepared network roaming standard IEEE802.11r and security trends defined in IEEE802.11i. Methods for minimizing handoff delay for wireless automation applications and requirements on the sufficient security level are also presented in this paper.

Keywords: Wireless networks, roaming, handoff, security

1 Introduction

Wireless digital communication starts to increase its prominence for the industrial automation domain. Wireless LANs based on IEEE 802.11 and other wireless concepts based on 802.15 (Bluetooth and ZigBee) have been introduced and still more and more producers of automation systems try to offer complete wireless solution for some specific automation applications but the harsh noise environment, and the multiple propagation behavior limit the use of many technologies and require further research and development. To design remote mechanisms (tele-supervisory, tele-operation, tele-service) using wireless communication, an increasing number of communication technologies are available. Using of these technologies is limited by strict quality of service requirements.

We can divide typical automation applications into several scenarios. The first one is permanent wireless connection of automation device. The second is temporary connected wireless automation device, which moves in the range of one wireless network, and the third one is temporary connected wireless automation device which moves across a several wireless networks. For both of temporary connected scenarios it is necessary to guarantee uninterrupted communications with fixed QoS and this strongly limits the current use of these systems within the automation domain. Basic principles and possible solutions for roaming between wireless networks describes IEEE802.11X standard.

2 IEEE 802.11 roaming scenario

The IEEE 802.11 [4] MAC specification defines two basic modes of operation. The first mode is ad-hoc which allows peer-to-peer communications between two or more wireless devices. The second mode of operation is infrastructure mode which defines connections between access point (AP) and wireless mobile stations. In this mode AP provides wireless network connectivity to connected wireless mobile stations. Network connectivity services in one AP create Basic Service Set (BSS). A number of access points could be used in one wireless network. Collection of BSS from access points is extended with services (ESS – Extended Service Set) for solving moving problems, access points communications to authorization server and other network devices (routers etc). Definition of services for moving issues within one wireless network is described as a handoff. Mobile wireless station can also move not only in one wireless network, but across several wireless networks.

Handoff mechanism is composed from sequence of message between AP and wireless mobile station. Important part of handoff mechanism is definition of wireless mobile station identifying exchange across access points in one wireless network. Detailed explanation of handoff process is shown in Fig. 1.

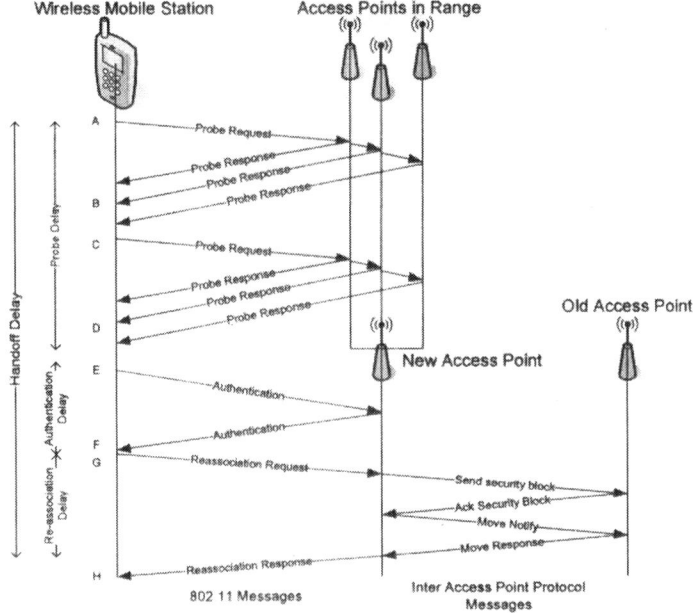

Fig. 1. Handoff process according to 802.11 standard

Preconditions of handoff process in Fig. 1 are valid connections of access points to authorizing server and no association of mobile station to access point. In the first step of handoff process, the mobile station scans for available access points by active scanning through sending probe message or by passive listening of broadcasted beacon messages from access points. Steps A to D in the figure show active mode scanning of access points in communication range. Mobile station selects the new AP with the best signal strength and data rates after finished scanning. Probe Delay is the time which needs mobile station to select the new AP. After the probe delay, the STA and new AP start Authentication process according to 802.11. Authentication delay is time which is necessary for successful authentication of mobile station to AP. After authentication, the STA sends re-association request to the AP (message G) and receives re-association response from the AP (message H) which completes the handoff process. Re-association process is done by the Inter Access Point Protocol (IAPP).

2.1 Inter Access Point Protocol

The IEEE 802.11f standard specifies two types of information exchange [3]. The first set of interaction is between access points during a handoff process. The second type is between AP and authorization server. IAPP provides secure communication link during a handoff for mobile station information exchange for reducing time cost in re-association delay. When a station first associates to an AP, the AP broadcasts an Add-Notify message notifying all other access points in the network association of the new mobile station. After access points receive an Add-Notify message all old associations for the new mobile station are cleared. For securing of the mobile station information, International Association of Privacy Professionals (IAPP) recommends the use of a RADIUS server (shared keys encryption) to secure the communication between access points [1].This IAPP re-association mechanism ensures a unique association for the mobile station in the network structure.

3 IEEE 802.11r fast roaming scenarios

The IEEE802.11r standard [5] should provide solution for fast roaming applications with high level of security. Unfortunately, securing of wireless communication goes against the fast roaming process. The IEEE 802.11i standard includes several new mechanisms for speed up authentication process as pair wise master key caching and pre-authentication, but application of the newest definition of security standards from 802.1i with temporally key integrity protocol (TKIP) to wireless network can results into handoff delay in hundreds of millisecond.

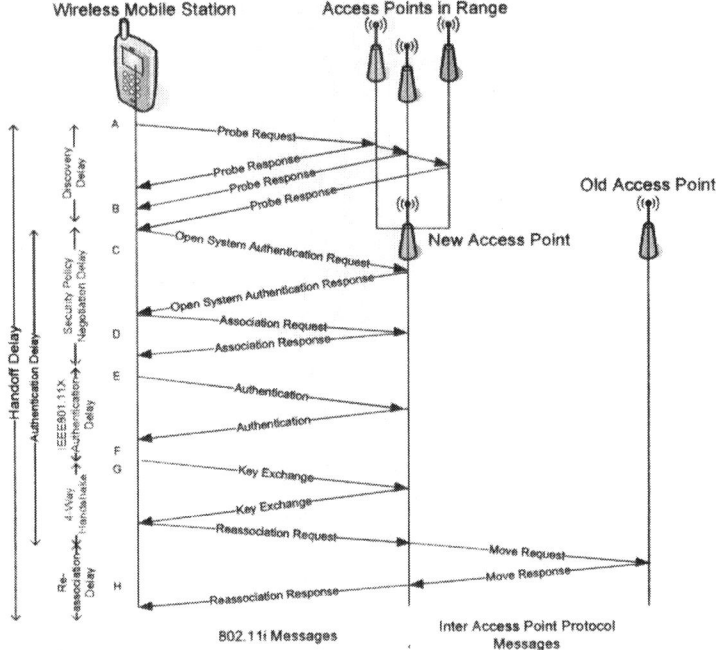

Fig. 2. Handoff process according to 802.11i standard

Fast roaming solution based on the 802.11r standard predicts handoff delay only about 50 ms. Necessity of fast roaming process with sufficient level of security results in a compromise solution based on the nature of application.

Fig. 3. Association of Wireless Mobile Station to Access Points in roaming process according to 802.11i standard

IEEE 802.11r ensures that the authentication processes and encryption keys are established before roaming request.

Main principle consists in roaming across the network with cryptographic information derived from the first authentication into the network. This pre-authentication reduces load on the authentication server and decreases handoff delay.

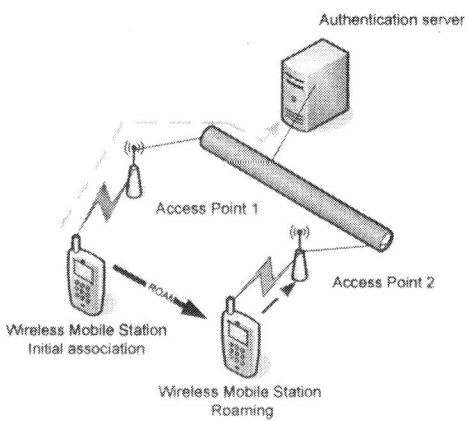

Fig. 4. Association of Wireless Mobile Station to Access Points in roaming process according to 802.11r standard

The 802.11r standard includes a new hierarchy of key-management for secure key caching and distributing. In this hierarchical key-management, the top level key holder has access to original cryptographic information and has to derive keys for holders in lowest level. Deriving of keys is based on the one-way hash algorithm for preventing of decryption original cryptographic information from lower level keys [2]. The main difference between IEEE802.11i standard (Fig. 3) authentication process and IEEE802.11r standard (Fig. 4) consists in requirements on the performing a full authentication with the authorization server for each re-association of wireless mobile station and AP in the IEEE802.11i standard case and initial association and redistributing information about new AP through whole network according to IEEE802.11r standard.

4 Conclusion

We proposed a set of problems of routing scheme for wireless automation applications in this paper. We focused on the routing and security technologies definition for wireless mobile platform and on the selection of available technologies from wireless network applications. Possibilities of integration security mechanisms to the roaming by IEEE802.11r standard were described in this paper, too. Otherwise IEEE802.11r standard hasn't been officially released we can derived main principles of roaming techniques from it. The main gap between the used security level and a handoff delay by IEEE802.11r standard will be rapidly decreased. Based on those findings it should be possible to develop wireless automation system with well defined parameters of wireless communication link. We presented basic scenarios for typical wireless automation application and our future work will be application of roaming functionality into real wireless automation system.

Acknowledgments. The research has been supported by Czech Ministry of Education in the frame of Research Program MSM1850032 *MIKROSYN* and by the European Commission in the 6th Framework Program under the project IST-016969 *VAN - Virtual Automation Networks.*

References

1. Hill, Joshua. An Analysis of the RADIUS Authentication Protocol. *Untruth Networks* . http://www.untruth.org/~josh/security/radius/radius-auth.html. [Online] 10 24, 2001. [Cited: 5 1, 2007.]
2. Alexander Wiesmaier, Marcus Lippert, Vangelis Karatsiolis. *The Key Authority – Secure Key Management in Hierarchical Public Key Infrastructures.* Department of Computer Science. Darmstadt, Germany : Proc. of the International Conference on Security and Management (SAM 2004), 2004. p. 5
3. IEEE. *Draft 4 Recommended Practice for Multi-Vendor Access Point Interoperability via an Inter-Access Point Protocol Across Distribution Systems Supporting IEEE 802.11Operation.* s.l. : IEEE, 2002. Draft 802.1f/D4
4. IEEE. *Part 11: Wireless LAN Medium Access Control(MAC) and Physical Layer (PHY) Specifications.* 1999. IEEE Standard 802.11
5. Molta, Dave. 802.11r: Wireless LAN Fast Roaming. *Network Computing.* [Online] 4 16, 2007. [Cited: 5 1, 2007.] http://www.networkcomputing.com/channels/wireless/showArticle.jhtml?articleID=198900107.

Secure Networking with NAT Traversal for Enhanced Mobility

Lubomir Cvrk[1], Vit Vrba[1]

[1] Brno University of Technology, Dept. of Telecommunications, Purkynova 118, 61200 Brno, Czech Republic
{cvrk, vrba}@westcom.cz

Abstract. When a peer in a public network opens a connection to another one being behind a network address translator, it encounters the network address translation problem. So called "UDP hole punching" approach allows to open a public-to-private or private-to-private network connection. This article deals with this approach to propose new security architecture for IPv4 communication introducing so called "implicit security" concept. Main contributions are ability to interconnect to any host behind NAT using just a host's domain name, enhanced mobility, and encryption and authentication of all data transmitted through this connection right from a packet sender to a local receiver. Secure channel is established on-demand automatically and is independent on any application. No additional modification of current NAT, IPv4 or DNS is required.

1 Introduction

Communication in IPv4 networks encounters the network address translation (NAT) problem. When a peer needs to initiate communication with another one which is behind a NAT box, it is basically impossible.

But there are several approaches, how to obviate this problem. One of them uses so called TCP/UDP hole punching, introduced by Daniel Kegel in [3], fairly well described in [4]. This approach uses NAT characteristic behavior when a connection from a private to public network is established. A NAT box creates a "hole" – a rule that allows packets from the host in the public network pass into a private network destined to a host which has opened the connection. In case of the TCP protocol this rule exists until TCP connection is closed. In case of the UDP protocol the situation is a little bit different, because the UDP does not use the institute of a two sided connection. Packets sent by the UDP across a NAT box out to the public network force it to create appropriate address translation rule. But there is no sequence for closing this communication (it is not connected) and therefore there is no object or flag usable for the control of the translation rule existence. Hence NAT implementations use time-outs for incoming packets. The time-out is as short as 20 seconds for a reply, but this is not standardized constant, just experimental results (see [4]).

Application can cryptographically secure its data transmissions with several approaches. The first one, and probably mostly used, is security built into an application. This approach requires a good crypto library and developers with cryptographic skills. A user of a software system is then in mercy of its developer, whether she builds her system in such a way that it provides cryptographic protection of a transmitted data, or not. He has to look for the information about encryption to know how confident data could he enter.

Better approaches how to protect communication operate at lower layers than application level security providers. The SSL virtual private networking (VPN, [7]) or IPSec ([5]) are well known.

Both these VPN implementations encounter several disadvantages: VPN server must exist in the public network or at least NAT box must be configured to route VPN traffic to a VPN server in a private network.

After correct authentication to the VPN server all data transmitted over the Internet between the user's computer and the VPN server are encrypted, messages are reply-protected and authenticated. But when a user starts to communicate with some other node in the Internet, everything is as bad as without VPN protection. It is needed to remark that even if in the VPN, the communication with another local network host behind the VPN server (on the local network) is also insecure.

2 Implicit security model

Implicit security is inspired by the opportunistic encryption – the idea introduced by John Gilmore in the FreeS/WAN project [9]. Implicit security, the concept we introduce, establishes a secured channel whenever it is possible. When a destination host is behind NAT, no VPN is needed and the channel is decrypted by that host so the end-to-end communication security problem is solved. Nobody but the sender and the receiver reads the messages.

3 Architecture components

Architecture of the system works with IPv4 and requires three main components: 1) Connector server in the public network tightly cooperates with appropriate DNS server; 2) Connector client establishes connections using connector server (initiates the UDP hole punching process); 3) Packet processor is installed on every participating host.

3.1 Packet processor

The network packet processor is bumped in the protocol stack below the network level. It captures outgoing packets, asks the connector client to create a secure channel and passes packets into this channel.

The packet's IP payload is packed in the UDP protocol with appropriate port and destination setup. Incoming packet comes back to the specified UDP port, so packet processor detects it and unpacks. The restored original packet is then passed up to the user-space. This simple technique allows the channel to be very easily traversed across NAT boxes or firewalls, because the data of all protocols are packed and transmitted over one single UDP port.

When the packet processor detects outgoing packets, it automatically tries to establish secure connection. For this service it asks the connector client.

3.2 Connector client

The connector client logs in a connector server when it starts (usually with the operating system, it runs as a service), and waits for the packet processor's request to connect to another host.

3.3 Connector server

The connector server must be connected in the public network, so connector clients are able to connect from private networks.

The server handles the UDP hole punching process. For details of the UDP hole punching, see [3], [4].

The server also handles client's login process. This requires some cooperation with appropriate DNS server.

The server may keep a database with public keys and may be configured in the strict mode to accept only signed login requests.

4 Architecture overview

In the following figures these terms are used: "Detec" – we named this system "Detec" [ditek], the abbreviation of "Decentralized end-to-end communicator"; "Detec peer" or "Caller" is a host which requests communication with another host – "called host", which may be behind NAT.

Below is described the architecture from the caller's point of view.

4.1 Login to connector server

When a client starts the connector client, it tries to log in the connector server. IP address or domain name is known from the configuration of the client. This example is configured in such a way that client (Detec peer) is behind NAT in a private network.

The fig. 1 shows the messages sent between client and server:

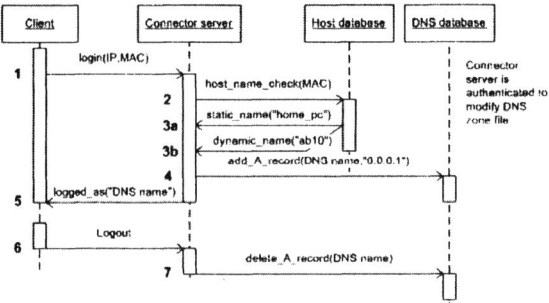

Fig. 1. Login process

First, the client sends the server a message with a login request, and as a parameter it passes the MAC address of its network interface and its local IP address (1).

The server checks the static host database (2) whether this client should be assigned a static (3a) DNS name or dynamic (3b). When a client is to receive a static DNS name, the login request message must be therefore signed and the server must know client's public key. In this case the server also authenticates itself to the client using public key cryptography.

The server then saves appropriate DNS name to the DNS server's database (4) and assigns host's DNS name to IP address 0.0.0.1. This IP address is not routable and is from the reserved address space 0.0.0.0/8 meaning "this network". Below is explained why.

The logout process is very similar to login and the server deletes client's A record from the DNS server database.

The DNS server should have a special zone (zones) reserved for the purpose of this system, i.e. third (fourth etc., depends on administrator) level domain and must be primary. In real life this is the DNS server handling public-IP-to-name mapping of the local network gateway's public interface. The connector server and the DNS may be installed on the same host (better for security).

4.2 Opening a connection

When the client is logged in the connector server it can open a connection. Every connection starts with the DNS query, represented by the function "gethostby-name". The result of this function is IP address of the host.

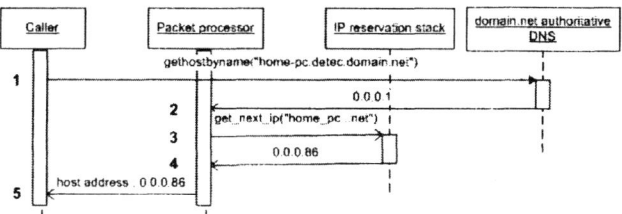

Fig. 2. Finding host's IP address

The fig. 2 shows the example how to get home_pc's address. First, the gethostbyname socket API function is called. This function generates DNS query to resolve the domain name "home-pc.detec.domain.net". The authoritative name server answers "0.0.0.1". The query was notified by the packet processor. When the response is received, the packet processor detects that IP is 0.0.0.1. This tells to the packet processor that some application from an upper layer needs to connect to a host being behind NAT. The packet processor accepts the DNS UDP packet but does not pass it to upper layers. Instead it takes a look in its "Detec IP reservation stack", where DNS name and appropriate assigned IP is stored. The stack answers 0.0.0.86. This means, that the stack saved the pair (home-pc.detec.domain.net, 0.0.0.86). Note that the stack stores global IP addresses too, where no modification is required. The packet processor modifies then the resolved DNS data and replaces "0.0.0.1" with "0.0.0.86" and passes the packet up. The global 0.0.0.1 IP address assigned to every Detec's host is on the local machine transformed to the first available Detec IP address. This address space provides more than 16 millions of single IP addresses. It is a very low probability that this space would ever exceed, because one host usually does not communicate with so many others roughly in the same time. Every established connection is automatically closed after the network inactivity time is out.

4.3 Data transmission and packet processing

Once the application receives the destination's IP address (0.0.0.86) it tries to open a communication channel (i.e. TCP). On the fig. 3 it is represented by the tcp_SYN (1). This packet is captured by the packet processor, because its destination IP (0.0.0.86) is in the IP reservations stack. Because the IP is from the network 0.0.0.0/8 the processor knows that a destination (called host) is behind NAT. That is why a UDP holes on both NAT boxes (caller's and called host's) must be opened.

4.3.1 Tunnel creation
Opening these holes is done this way: The packet processor asks the connector server for the called host's public IP address and a UDP port where a secured tunnel will be established (2a). It opens a UDP connection to the server (2b). Connector server sends a request to open a UDP connection (3a) to the called host and it opens it (3b). Server sends the message 3a through a TCP connection that the connector client has opened

to the connector server at login process. The connector server is the called host's one, not the caller's. Its IP is translated from detec.domain.net.

Once both clients opened the UDP connection to the server it saves the UDP source port values (received from NAT boxes). Then it sends both sides the public IPs and the UDP ports to use (4a, 4b). When they receive this data they start sending UDP packets and the UDP hole is created on the both sides (5, 6). In the event of received packet from the opposite host (7) they both send udp_hole_punching_done messages to each other (8,9). If any of them does not receive this message in a configured time-out (in seconds), then the process state resets and begins again. If three attempts fail, then the application packet is dropped. In future versions this may need some interaction with a user to decide what to do.

The results of the UDP hole punching (NAT public address and the UDP port) are associated with appropriate 0.0.0.0/8 address in the internal connection database.

4.3.2 Tunneling

The original packet is processed and encapsulated into the UDP secure tunnel (11). On the called host's side it is processed by the packet processor and sent to appropriate application. The response is packed by the called host and sent to the caller. After message number processing, source IP verification and message authentication the packet is decrypted and passed to the upper layers, otherwise it is dropped.

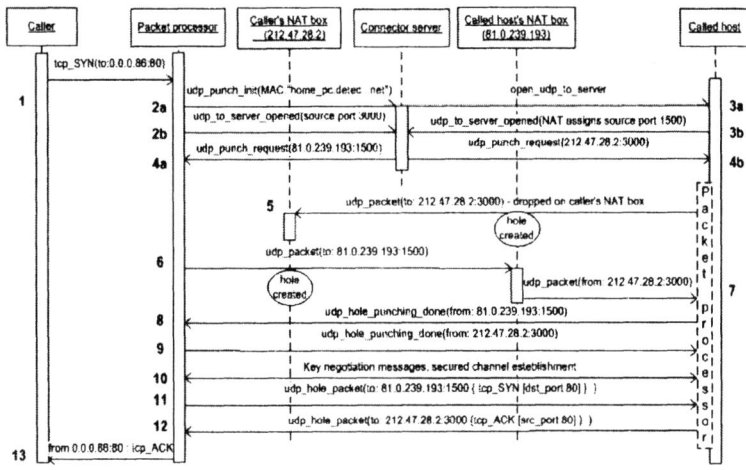

Fig. 3. Data transmission

5 Packet processing

The packet processor encrypts and authenticates the outgoing packets and decrypts and verifies incoming packets. Outgoing packets are immediately sent over the UDP channel and incoming are received from the UDP channel.

This requires a protocol that supports encryption and authentication with reply protection. That is why we have designed the protocol SPUT (Secure protocol for UDP tunneling).

5.1 Secure protocol for UDP tunneling

This protocol defines some overhead fields required for successful decryption and reply protection.

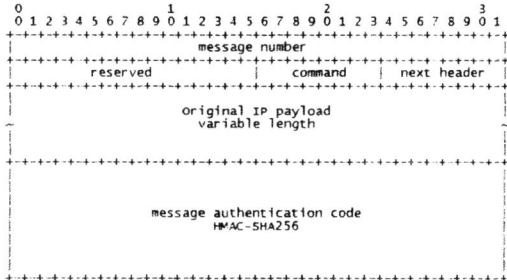

Fig. 4. SPUT protocol definition

The field *message number* indicates the number of currently processed packet. When it exceeds 32 bits, the key negotiation process is started to negotiate new session key. The system keeps track of received messages and once it receives a message it can not accept a message with the same message number. A 32 bit value is used internally where each bit is a flag informing the packet processor whether the message with current message number was already received or not. This value provides a window for 32 messages, because they may come out of order. When a message is received bits in this flag are rotated left by appropriate number of bits and a flag is set to 1. Messages with the message number smaller than current message number decreased by 32 are dropped as well as messages already received. The field *command* is used to control the encryption / decryption process. Nowadays it uses only the value 1, which closes the connection. In this case an empty packet is sent. The *next header* field indicates the protocol number encapsulated in SPUT. The *original IP payload* is the field of variable length used for TCP or UDP (or any other) data from the original packet. It does not contain original IP header. The last field of the protocol is *MAC*. HMAC-SHA-256 is used for authentication computation, exactly as described in [8].

The fig. 5 shows encrypted and authenticated fields of the SPUT protocol. SPUT is designed to use encrypt-first-authenticate-then method of MAC computation.

```
|messageNumber{32}|reserved{16}|command{8}|nextHeader{8}|payload{?}|MAC{256}|
                 +---------------------------------------------------+
                          E n c r y p t e d       AES-CTR-128
+------------------------------------------------------------------------+
                    A u t h e n t i c a t e d     HMAC-SHA-256
```

Fig. 5. Encrypted and authenticated fields

The SPUT protocol's overhead is 320 bits. Data are transported as the UDP payload whose overhead is 64 bits. That is why the system requires decrement of the maximum transmission unit by 384 bits = 48 bytes.

In case of multiple hosts behind NAT talking to one host (H) the tunnel and the payload destination ports must be remapped in such a way that H stores the original packet's source ports as future destination ports and substitutes the IP payload's port (if payload is TCP or UDP) for another port assigned by H. This allows H to identify more hosts hidden behind the public network interface of one NAT box.

5.2 Encryption process

Encryption process uses the AES-128 in CTR mode. The whole process requires computation of a key stream using the `message number` (and a secret key of course) as an input.

The key stream k is generated as follows:

$$k_j = E(K, j \| i \| s); j = 0..l/128$$

j is a number of current 128 bits long block of data to encrypt (8 bits). i is a SPUT message number (32 bits). s is a constant made of first (least significant) 88 bits of the secret session key. l is a number of bits of the data to encrypt. These values are concatenated to a 128 bits input block for AES encryption. k_j is 128 bits key-stream segment.

6 Experiments and results

We have tested several parts of the architecture. The most important information we had to find out was whether the packet processing overhead is acceptable. That is why the overall performance has been tested and compared to the implementation of IPSec.

6.1 Packet processor overall performance

We have implemented this architecture in C++ for the Microsoft Windows platform. This platform requires an NDIS kernel driver that hooks over the network interface and captures all traffic and passes it to the user space.

There is one disadvantage in out current implementation. It transfers captured packets from the kernel-mode to user-mode so we expected this to impact performance especially on the high speed lines. The measurements have proven this (see fig. 6).

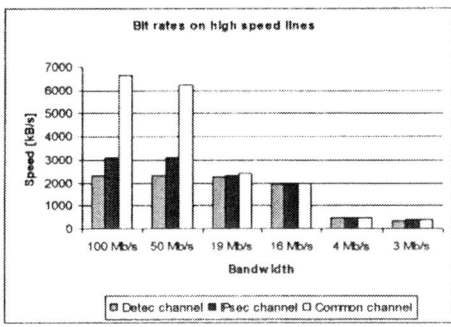

Fig. 6. Bit rates on high speed lines

The speed limit of this kind of implementation is 2293 kB/s – about 17.9Mb/s. This is acceptable for most use-cases because the system is designed for end-users not routers or gateways which must not use it.

In comparison with Microsoft's IPsec implementation in Microsoft Windows XP Professional our implementation is slower but not significantly. The speed limit of IPsec is around 3000 kB/s. IPsec is probably implemented in kernel mode and its code is optimized. Our code is just an initial version built with debug information and with no source code or compiler optimizations.

6.2 Performance on low bandwidths

On low bandwidths (1.5Mb/s and lower) the packet processor is as fast as IPsec and common (insecure) channel (see fig. 7). The fact of encryption and authentication is not perceptible since the bandwidth is 19 Mb/s or slower (see fig. 6).

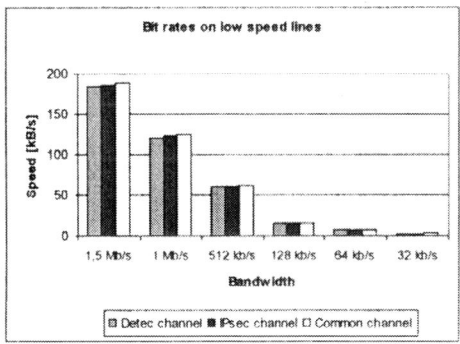

Fig. 7. Bit rates on low speed lines

6.3 Performance impact ratio on low speeds

We have compared the performance impact ratio of IPsec and Detec to the common (insecure) communication channel of the IP protocol - fig. 8.

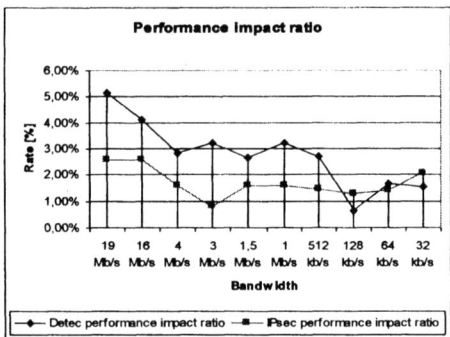

Fig. 8. Performance impact ratio

The rate means how many percents is particular encrypted connection slower than insecure one.

From the graph can be seen that on very low speeds (128kb/s and less) IPSec and Detec produce almost the same performance. This is due to the overhead of ESP protocol which is greater than that of SPUT protocol.

6.4 Packet analysis

For the reason of encryption checking we have analyzed the data transmitted through the network. The fig. 9 shows sequence of plain ICMP echo-request messages.

```
0000  00 11 d8 b8 46 31 00 0c  76 40 89 cc 08 00 45 00   ....F1.. v@....E.
0010  00 3c 0e 1d 00 00 80 01  a7 38 c0 a8 02 10 c0 a8   .<...... .8......
0020  02 0b 08 00 42 5c 02 00  09 00 61 62 63 64 65 66   ....B\.. ..abcdef
0030  67 68 69 6a 6b 6c 6d 6e  6f 70 71 72 73 74 75 76   ghijklmn opqrstuv
0040  77 61 62 63 64 65 66 67  68 69                     wabcdefg hi

0000  00 11 d8 b8 46 31 00 0c  76 40 89 cc 08 00 45 00   ....F1.. v@....E.
0010  00 3c 0e 1e 00 00 80 01  a7 37 c0 a8 02 10 c0 a8   .<...... .7......
0020  02 0b 08 00 41 5c 02 00  0a 00 61 62 63 64 65 66   ....A\.. ..abcdef
0030  67 68 69 6a 6b 6c 6d 6e  6f 70 71 72 73 74 75 76   ghijklmn opqrstuv
0040  77 61 62 63 64 65 66 67  68 69                     wabcdefg hi
```

Fig. 9. Sequence of ICMP messages

The fig. 10 shows these messages transported by the Detec system. They are transported within the UDP protocol port 32589, both encrypted by AES-128-CTR and authenticated using HMAC-SHA-256.

```
0000  00 0c 76 40 89 cc 00 11  d8 b8 46 31 08 00 45 00    ..v@.... ..F1..E.
0010  00 6c 11 76 00 00 80 11  a3 9f c0 a8 02 0b c0 a8    .l.v.... ........
0020  02 10 7f 4d 7f 4d 00 58  c4 d7 01 00 00 00 25 05    ...M.M.X ......%.
0030  4d c6 63 80 af 3e 33 21  f5 86 c5 2a 4c b9 57 6e    M.c..>3! ...*L.Wn
0040  76 f6 38 61 87 07 58 f8  96 e9 71 0e 15 42 81 ce    v.8a..X. ..q..B..
0050  4d d3 dc 95 1f 96 7d 37  a9 19 8d ce 29 9a df a1    M.....}7 ....)...
0060  90 42 c3 ce 9b 7c d0 16  ab 8a f7 74 85 88 d7 3e    .B...|.. ...t...>
0070  2e 91 b7 23 c5 6a e8 a9  14 ea                      ...#.j.. ..

0000  00 0c 76 40 89 cc 00 11  d8 b8 46 31 08 00 45 00    ..v@.... ..F1..E.
0010  00 6c 11 77 00 00 80 11  a3 9e c0 a8 02 0b c0 a8    .l.w.... ........
0020  02 10 7f 4d 7f 4d 00 58  9e 4b 02 00 00 00 a8 5b    ...M.M.X .K.....[
0030  13 66 59 f7 2e 07 95 5d  56 d9 8f 22 62 d8 f8 3a    .fY...]  V.."b..:
0040  9d 4b d0 c6 3e ad 45 34  19 f6 5e 38 f8 55 b0 14    .K..>.E4 ..^8.U..
0050  04 c3 1e 17 0a a6 de 37  94 fb 1c 58 fd 66 4e 5d    .......7 ...X.fN]
0060  a8 04 2f 3b 55 f7 ed 04  c0 0b 1c f5 59 69 2d cd    ../;U... ....Yi-.
0070  cd d4 8b 89 22 8e 62 29  48 d8                      ....".b) H.
```

Fig. 10. Encrypted ICMP sequence

The same plaintext from the source messages is encrypted to two different cryptograms in encrypted messages.

7 Conclusion

This article has introduced the new secure communication architecture for IPv4 with NAT traversal. Key benefits it brings are ability to interconnect to any host behind NAT using just a host's domain name, and encryption and authentication of all data transmitted in this connection. Secure channel is established on-demand automatically and is independent on any application. Applications do not need to care of encryption of the data they transmit over the network.

The architecture is based on a packet processor which captures packets to encrypt, authenticates and sends over the UDP to the destination. The DNS database is used for the global naming service of clients behind NAT. It assigns these clients a special IP address so the packet processor can recognize that the host it needs to connect to is behind NAT and therefore the process requires the UDP hole punching to establish a LAN-LAN channel.

Performance of the packet processor has been measured and the results are showing it to be as fast as IPsec implementation in Microsoft Windows XP Professional. The future releases are expected to be faster than IPsec because of kernel-level implementation, lower protocol overhead and optimization of the source code.

Every user of this system is reachable by a single DNS name from any public or private network in the world which strongly simplifies his mobility.

Acknowledgement: *This article has been sponsored by the Ministry of Industry and Trade of the Czech Republic, project no. FT-TA3/011.*

8 References

[1] Fergusson, N., Schneier, B., *Practical Cryptography*, Wiley Publishing, Inc., Indianopolis USA, 2003

[2] A. J. Menzes, P. C. van Oorschot, S. A. Vanstone, *Handbook of applied cryptography*, CRC Press LLC, Florida, USA, 1997.

[3] D. Kegel, "NAT and Peer-to-peer networking", *Web page*, http://alumnus.caltech.edu/~dank/peer-nat.html. 1999.

[4] B. Ford, P. Srisuresh, and D. Kegel, "Peer-to-Peer Communication Across Network Address Translators", *Web page*, http://www.brynosaurus.com/pub/net/p2pnat/ 2005.

[5] S. Kent, R. Atkinson, "Security Architecture for the Internet Protocol", *RFC 2401*, 1998.

[6] T. Dierks, C. Allen, "The TLS Protocol Version 1.0", *RFC 2246*, 1999.

[7] OpenVPN project, http://openvpn.sourceforge.net

[8] H. Krawczyk, M. Bellare, and R. Canetti, "HMAC: Keyed-Hashing for Message Authentication", *RFC 2104*, 1997.

[9] FreeS/WAN project, http://www.freeswan.org

[10] S. Kent, R. Atkinson, "IP Encapsulating Security Payload (ESP)", *RFC 2406*, 1998.

[11] L. Cvrk, V. Zeman, D. Komosny, "H.323 Client-Independent Security Approach". *Lecture Notes in Computer Science*, 2005.

[12] S. Kent, and R. Atkinson, "IP Encapsulating Security Payload (ESP)", *RFC 2406*, 1998.

Simulation model of a user-manageable quality of service control method

Karol Molnar

Dept. of Telecommunications, FEEC
Brno University of Technology, Purkynova 118, Brno, Czech Republic
molnar@feec.vutbr.cz

Abstract. Diffserv, the most often used Quality of Service support technology, offers fast and efficient data flow processing. On the other hand these advantages are achieved at the cost of static, network-driven data-flow metering, classification, marking and queue management, which does not support any interaction with the end-user services. The following work offers a solution how to allow the end-user or end-system to cooperate with the edge router of a DiffServ domain and further increase the efficiency of the available QoS-support system.

Keywords: QoS, DiffServ, MIB, SNMP

1 Introduction

Modern packet-switched network technologies are designed to support different network services and, under a moderate network load, these services can offer an appropriate level of service quality. In the network nodes the incoming data flow is usually classified into a relatively small number of service classes and each class is assigned a specific queue. The queues are then served by a deterministic packet scheduler, very often implementing the Priority Queuing (PQ) or the Weighted Fair Queuing (WFQ) mechanism. This solution offers stateless and so far very fast traffic differentiation and differentiated packet processing.

The above principles are also implemented in the technology of Differentiated Services (DiffServ), which is nowadays the most extended solution for Quality of Service (QoS) support. Because of its relative simplicity and stateless packet processing, DiffServ offers high scalability and can efficiently operate in large data networks too. This is the reason why DiffServ is practically the only world-wide extended QoS support technology in use.

A serious limitation of the DiffServ technology is its network-oriented data-flow processing. All packet-processing rules, including service classification, traffic metering, packet marking, queuing and packet scheduling, are configured in network elements, usually in the edge and the core routers, and operate totally independently of the user stations.

The missing interaction between DiffServ network components and the user terminal represents a severe gap, which greatly conduces to the slow growth of QoS–support technology implementations. By allowing an application to define its requirements on network resources and specify the desired service class, this situation could be changed. However, this improvement requires the accomplishment of two main requirements. First, the networking application must be able to define its requirements on the network resources. Since the QoS support is required mainly in multimedia applications, with both conversional and streaming character of traffic, the prediction of the required resources is directly related to the type and parameters of the codecs used and usually can be specified quite precisely. In the second step, the requirements specified must be communicated to the network, more precisely to the QoS support technology used in this network. The objective of our work is to offer a solution for this communication.

Since the DiffServ specification does not reckon with direct control mechanisms between the edge router and the end end-station, there is no dedicated communication protocol designed for this purpose. In addition, the edge router is not designed to offer extensive control functions to the end-stations. In order not to burden the edge router, such an extensive communication should be avoided.

The solution presented in this document tries to affect the classical DiffServ specification as little as possible. Instead of trying to control the edge router we focus on how to retrieve DiffServ-related configuration data from this router and transform this information into a form easily understandable to the end-user or an application. Based on this information the user or the application can automatically select the service class which best suits its requirements.

We divided the experimental implementation of the mechanism presented into several phases. First, the suggested method is evaluated in the OPNET Modeler simulation environment. Since the OPNET Modeler uses a C-based programming language to describe the models on the process level, the majority of the algorithms implemented in the simulation model can also be used for practical evaluation in a laboratory environment. This document presents the results of the theoretical preparation work.

2 The communication model

The key component of the user-manageable QoS control method mentioned above is an information exchange between the end-station and the edge router. As a result of this process the end-station can obtain a limited set of DiffServ configuration parameters from the edge routers. It would be possible to design a proprietary protocol for this purpose, but using the standardized Simple Network Management Protocol (SNMP) [4] offers the same results with much less effort. In addition, the SNMP manager, through the built-in SNMP agent, has direct access to the majority of (if not all) configuration parameters of the edge router, and no further tool is required to obtain these data. Using the SNMP can greatly simplify the situation and allow us to concentrate on the basic functionalities of the suggested extension.

For security reasons, normal users are usually not allowed to access directly the SNMP agent. On the other hand, our solution requires just read-only access and only to a clearly specified part of the Management Information Base (MIB) [5]. The amount of data, that should be accessible, can thus be largely reduced.

The communication process used in the suggested user-manageable quality of service control method is shown in Fig.1.

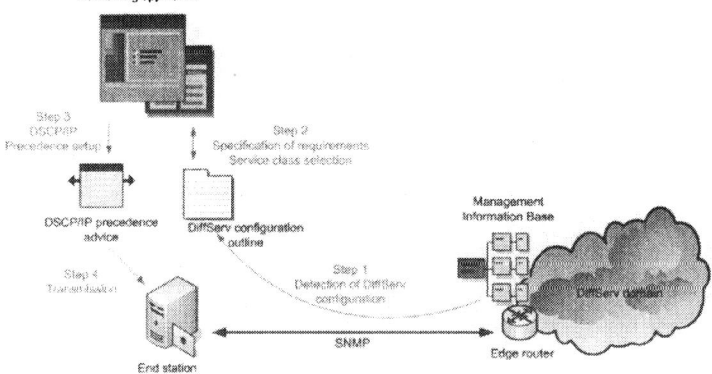

Fig. 1. Concept of the user-manageable quality-of-service control method

First, the end-station must detect the current DiffServ configuration by retrieving the related data from the edge router's MIB. The method suggested uses the SNMP to collect this information from the edge routers. The configuration is usually accessible from a special MIB controlled by the SNMP agent implemented in the router, so the end station with sufficient access rights could retrieve this information.

In the next step, the acquired information must be processed and transformed into a form that the user-application can simply evaluate. Based on the information about the available service classes and their parameters, the application automatically, or in cooperation with the end-user, can select the most suitable service class. The desired service class can be indicated by setting the corresponding DiffServ Code Point (DSCP) value in the header of the transmitted IP packets. These packets with the desired DSCP value are then sent to the edge router. Packet processing in the edge router will be realized in a standard way, meaning that the traffic is first classified and then metered. If there are enough network resources and the incoming traffic flow does not violate the Service Level Agreement (SLA), the edge router can keep the demanded service class for the data flow. If there is any conflict, the packet marking will be based on the results of the measurement process and the desired DSCP will be overwritten with a value determined by the edge router. Using this solution the final decision will be made by the edge router so that the DiffServ domain will not be compromised by occasionally or intentionally misclassified traffic.

2.1 Simplified SNMP manager

We have defined that the SNMP protocol will be used between the end-station and the edge router to obtain the DiffServ configuration. Of course, our solution does not require a complex SNMP manager full of features. Rather a small and fast manager application should be used which is able to connect to one specific device, the edge router, and can acquire a part of the implemented MIBs and can do this task repeatedly with a relatively long repetition interval. According to our requirements it is enough to support only three SNMP operations, in particular GetRequest, GetNextRequest and GetResponse, in this manager application. For the purpose of simulation it is not really important, but for practical realization we want to assign this functionality to a service or daemon running directly under the operating system.

2.2 DiffServ-related MIBs

The most crucial part of the designed system is represented by the Management Information Base containing the required DiffServ configuration parameters. The complications come from the fact that there is not a single universal DiffServ-MIB used in all network elements. The Internet Engineering Task Force published a Request For Comment 3289 [1] containing a concept for DiffServ-MIB, but practically it is not in use. The manufacturers usually define their own proprietary solutions. The next chapters shortly describe the DiffServ MIB designed by IETF and the Class- based QoS MIB used by Cisco Systems Inc.

DiffServ Informal Management Model

In the year 2002, IETF published An Informal Management Model for DiffServ Routers in RFC 3290 [2]. This document introduces a management model which could be used for modelling management and configuration tasks in DiffServ routers. The management model includes all elementary components of the DiffServ technology, called DiffServ data path elements in the document. The model of each data path element contains the corresponding configuration parameters and a way of linking them with other elements. The cascade of these elements builds up a Traffic Conditioning Block (TCB). These modelled elementary components are:
- Classifiers (e.g. Behaviour Aggregate and Multi-Field Classifier),
- Meters (e.g. Average Rate, Exponential Weighted Moving Average, and several Token Bucket Meters),
- Action elements (e.g. DSCP Marker, Absolute Dropper, Multiplexor or Counter),
- Queuing elements (e.g. FIFO Queue, scheduler, algorithmic Dropper, etc.).

The Traffic Conditioning Block thus represents a set of data path elements providing specific traffic treatment. There is no fixed position for the data path elements in the TCB. They can be linked together arbitrarily according to the desired traffic policy. An example of a TCB [2] is shown in Fig.2.

In spite of the complexity of the TCB shown in the Figure its functionality is quite evident from the blocks used. It represents a exemplary way of processing data flow, starting with packet classification and followed by the measurement of different traffic classes, marking them accordingly, adding selective discard to prevent router form congestions, storing the processed packets in a queue, and scheduling packet transmission from the queues.

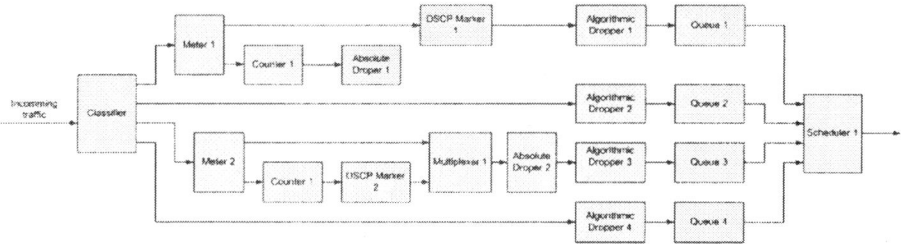

Fig. 2. Example of a Traffic Conditioning Block

IETF DiffServ MIB

Based on the DiffServ Informal Management Model, IETF defined the DiffServ MIB designed for configuration and management purposes. The DiffServ MIB is designed to allow both remote monitoring and configuration of DiffServ routers. The Object Identifier of the MIB is 1.3.6.1.2.1.97 and is located under the iso.org.dod.internet.mgmt.mib-2.diffServMIB tree. There are three basic branches defined in the MIB: diffServMIBObjects, diffServMIBConformance and diffServMIBAdmin. The diffServMIBObjects branch contains the management structure for each data path element. The diffServMIBConformance branch defines the MIB statements for full and for read-only compliance implementations and defines the MIB groups which should be implemented. The third branch, diffServMIBAdmin, defines a management model for concrete implementations of token-bucket meters and schedulers.

The management models of data path elements are stored in tables and the related table entries are linked together using pointers called RowPointer. The data path elements are divided into the following tables:

Data path table (diffServDataPathTable) defines the starting point of the data path identifying the interface, its direction and assigning a pointer pointing to the first functional data path element.

Classifier table (diffServClassifier) stores information about classifier elements, which are used to identify a specific part of the network traffic. In RFC 3289 [2] only the Multifield Classifier is implemented but the pointer-based structure is flexible enough to define additional types. The traffic, according to the filtering criteria, is detached and then sent to the next data path element.

Meter table (diffServMeter) contains the DiffServ traffic meters used in the router. Meters have two outputs, one for the traffic conforming to the metering parameters and one for the traffic exceeding these parameters. The next data path element for both directions is identified by a separate RowPointer.

Token-bucket parameter table (diffServTBParam) contains the list of token-bucket meters which can be implemented in the metering elements.

Action table (diffServAction) contains the data path elements which provide different actions applied to the traffic. These actions are DSCP marking or packet counting. Packet counting is used mainly for statistics and measurement. After providing the required action, packets are forwarded the next data path element identified by a pointer.

Algorithmic dropper table (diffServAlgDrop) contains controllable droppers used to prevent queues from being overloaded. RFC 3289 [2] defines a management model for several dropping algorithms, like tail drop, head drop or parametric random drop algorithms.

Queue table (diffServQueue) stores information about the queues implemented in the router. Each queue is modelled by a FIFO queue, but using a scheduler element the FIFO queues can be organized into a complex queue system.

Scheduler Table (diffServScheduler) contains the list of scheduler elements which are responsible for packet scheduling and are managing the related queues according to the configured relationships between them.

The IETF's DiffServ MIB offers very flexible monitoring and configuration of DiffServ elements, but on the other hand it is quite complex. The biggest disadvantage is that it is not widely used by the manufacturers of active network elements. As an alternative encountered in practice we can mention the Cisco Class based QoS MIB.

Cisco Class-based QoS MIB

The Cisco Class-based QoS MIB is a proprietary Management Information Base for Cisco devices supporting the Modular QoS Command-line Interface. In contrast to the DiffServ MIB, the main purpose of the Class-based QoS MIB is to provide read access to QoS configuration in the DiffServ router and allow the network manager to collect statistical information about the traffic processed.

The Class-based QoS MIB uses twenty tables to describe configurations and statistics for the implemented traffic policies, class mappings, classifiers and queues, for packet marking and for the Random Early Detection algorithm. One of the tables, called QoS objects, is used to collect all the implemented components of the DiffServ model. The QoS objects table consists of ClassMaps, Match Statements, PolicyMaps and Feature Actions.

- The **Match** statement specifies match criteria which are used to identify packets for classification purposes.
- The **ClassMap** object represents a user-defined traffic class that contains one or more match statements used to classify packets into several categories.
- The **Feature Action** represents the way the selected part of traffic is processed. Features include policing, traffic-shaping, queuing, random dropping using the RED method, and packet marking.
- The **PolicyMap** table contains the associations between the QoS action and the traffic class defined in the ClassMap table.

The process of reading information form the MIB starts by learning the cbQosServicePolicyTable and cbQosObjectsTable MIB tables. The corresponding indexes are cbQosPolicyIndex and cbQosObjectsIndex. cbQosPolicyIndex is designed to identify the service policies attached to logical interfaces and the cbQosObjectsIndex is designed to identify each QoS feature on a specified device. The DiffServ related configuration parameters are stored in the system corresponding to the structure of command-line commands.

2.3 The simulation model

The simulation model for the user-manageable quality of service control method proposed in this document is developed in the OPNET Modeler environment. The OPNET Modeler is an up-to-date simulation environment capable of simulating the behaviour of network processes (communication protocols), network components (servers, workstations, switches, routers, etc.), applications (http, ftp, email, VoIP, database, etc.) and their extended combinations (subnetworks, fixed and wireless networks, etc.). It also supports Differentiated Services with the configuration process quite similar to the configuration of real systems.

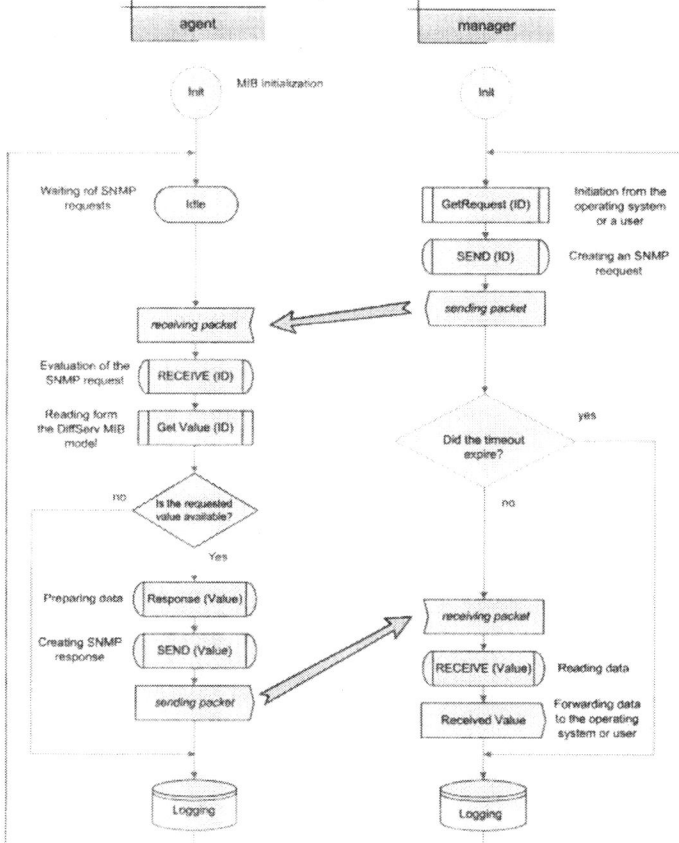

Fig. 3. The state-machine of the manager and agent

Since, the structure of the Cisco Class-based QoS MIB is simpler and it is more suitable for practical implementation, we decided to use IETF's DiffServ MIB in our model. Since we need to simulate only a limited part of the DiffServ model, we could simplify enough the DiffServ MIB to implement it into our simulation model. Our system requires information only about the available service classes metering parameters and the corresponding queuing system. We do not need to deal with packet dropping, processing of packets not conforming to the metering criteria and we do not require statistics in our model. Using these simplifications a relatively compact model of DiffServ MIB can be defined.

On the other hand, to implement the simulation model we had to redefine the standard application part available in the process model of a workstation and server. Since the SNMP is not supported in the current version of the OPNET Modeler (version 12.0 Patch Level 5) we had to define a new SNMP application and link it with the UDP transport protocol. For this purpose we used the Interface Control Information (ICI) [6] functions, which are able to build virtual connections between different network entities. The state machine of the implemented agent and simplified manager are shown in Fig. 3.

At the time of writing this document the functional evaluation is in progress. The data exchange between the models of workstation and server is evaluated together with the algorithms working with the simplified MIB.

3 Conclusion

The technology of Differentiated Services can significantly improve the efficiency of network transmissions if it is correctly configured. On the other hand, DiffServ is not able to directly interact with the end-user applications and that is why it is not able to support end-to-end QoS. In this document a user-manageable QoS control method is introduced which partially overcomes this limitation. For the communication between the edge router and the end-station a standardized SNMP protocol has been chosen. As a result of this communication the end-station is able to obtain the current DiffServ configuration and the application can suggest for the data to be sent the service class which fits its requirements the most.

In the current state of realization a theoretical preparation has been finished and the building of a simulation model on OPNET Modeler is in progress. By now the standard process models of a workstation and a server have been extended with a simplified version of an SNMP agent and manager, respectively. The communication between these components is solved using the Interface Control Information API. Functions working with the MIB are also implemented. Some complications were caused by the fact that there is no general DiffServ-related MIB defined and each manufacturer uses their own proprietary solution. After several evaluations we chose for our simulation model the DiffServ MIB defined by the Internet Engineering Task Force.

The future work will be concentrated on finishing the simplified DiffServ MIB implementation and optimizing the communication between the end-station and the edge router. After this task is successfully finished, we will start to work on the evaluation in laboratory environment.

Acknowledgement: This paper has been supported by the Grant Agency of the Czech Republic (grants No. 102/06/1569 and No. 102/05/P585), Ministry of Industry and Trade of the Czech Republic (FT-TA3/011) and the Ministry of Education of the Czech Republic (grants No. 1K04116 and No. MSM0021630513).

References

1. Bernet, Y., Blake, S., Grossman, D., Smith, A.: An Informal Management Model for Diffserv Routers, RFC 3290, Internet Engineering Task Force (2002)
2. Baker, F., Chan, K., Smith, A.: Management Information Base for the Differentiated Services Architecture, RFC 3289, Internet Engineering Task Force (2002)
3. Fung, A.: Cisco Class-Based QoS MIB, Cisco Systems Inc., http://tools.cisco.com/Support/SNMP/do/BrowseMIB.do?local=en&mibName=CISCO-CLASS-BASED-QOS-MIB-CAPABILITY (2001)
4. Case, J., McCloghrie, K., Rose, M., Waldbusser, S.: Structure of Management Information for version 2 of the Simple Network Management Protocol (SNMPv2) RFC 1442, Internet Engineering Task Force (1993)
5. Case, J., McCloghrie, K., Rose, M., Waldbusser, S.: Management Information Base for version 2 of the Simple Network Management Protocol (SNMPv2) RFC 1450, , Internet Engineering Task Force (1993)
6. Opnet Technologies, Inc.: OPNET Modeler Product Documentation Release 12.0, OPNET Modeler, 2006

Voice Quality Planning for NGN including Mobile Networks

Ivan Pravda[1], Jiri Vodrazka[1],

[1] Czech Technical University of Prague, Faculty of Electrical Engineering,
Technicka 2, 166 27 Prague 6, Czech Republic
pravdai@fel.cvut.cz, vodrazka@fel.cvut.cz

Abstract. The plan of transmission parameters or the network plan was created for Czech Republic in 2005 and this plan has had key effect on traffic in the telecommunication networks providing public telephone service, including mobile networks, NGN networks based on IP protocol and private networks connected to public telephone network. This paper describes practical effects of aforesaid plan on some typical situations in the network. Related research projects of the Department of Telecommunications Engineering, Faculty of Electrical Engineering, CTU in Prague are presented here, such as tandem arrangement of codecs and non-intrusive monitoring of voice quality.

Keywords: Voice quality, Transmission plan, E-model, R-factor

1 Introduction

The global telephone network transmission environment has rapidly changed. The IP based Next generation networks are build, the networks providers are interconnected and the private networks are connected to the public networks. Today, dominant part of telephone traffic uses the mobile and other wireless networks. The traditional transmission planning methodologies and documents are no longer flexible enough to account for all these new factors. For modern transmission planning, the following issues have to be considered:

- Multinational networks (especially private networks) become common and require planning which takes into account regional differences in loss plan requirements and inter-network transmission plans.
- Due to the liberalization of the telecommunication markets there are no longer laid down ranges of values for transmission parameters by regulation.
- The changing scenario in the public network operator domain is impacting transmission performance.
- The network plan should be applicable to the use of new technology within the networks under consideration, including cordless or mobile sections and transmission of packet voice (VoIP).

New public telephone network plan of the transmission parameters for Czech Republic was published by regulator – CTU – in 2005 [1].

Please use the following format when citing this chapter:

Pravda, I., Vodrazka, J., 2007, in IFIP International Federation for Information Processing, Volume 245, Personal Wireless Communications, eds. Simak, B., Bestak, R., Kozowska, E., (Boston: Springer), pp. 376-383.

The plan was created in special workgroup constituted at The Association of Public Telecommunication Network Operators in Czech Republic (APVTS) with both authors of this article actively participating.

2 Basic Parameters of Network Plan for the Czech Republic

The plan of transmission parameters defines a reference configuration with two interconnected networks A and B, or with third possible network C connecting both networks A and B (see fig. 1). The parameters are defined between acoustic interfaces of subscribers, eventually between access points of these networks. As terminal equipment can be used a classic analog or ISDN telephone, a wireless telephone (DECT), a mobile terminal (GSM, UMTS), an IP telephone, a Wi-Fi phone, etc. The plan includes possible private automatic branch exchanges (PABX) and private networks (PN) of companies or institutions.

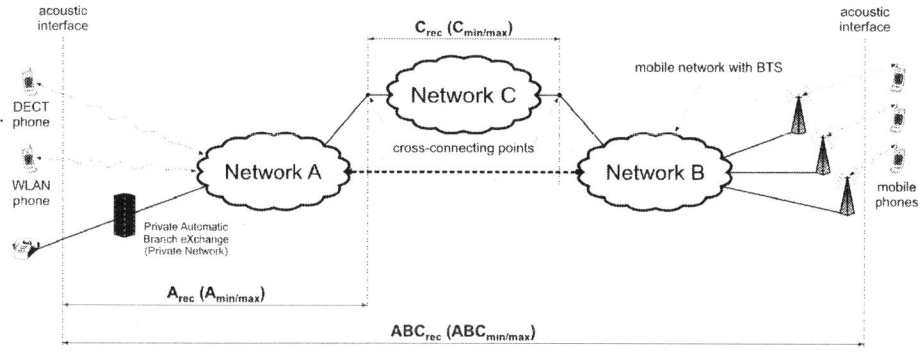

Fig. 1. The reference network configuration for telephone network plan of the transmission parameters for the Czech Republic

2.1 Recommended Values and Limits

Next overview indicates monitored parameters and their recommended and limited values. Firstly see tab. 1 for total values among acoustic interfaces of the network, marked in general ABC (via all three subnetworks). Key parameter is compliance of total R-factor higher than 50, that corresponds with total impairment factor equal to 50 (lowered by advantage factor A: the relevant value is 10 in mobile networks is, and 5 in connection with cordless terminals).

Key effect has delay in NGN networks (low value of delay close to zero (~ 0 ms), is assumed in classic networks, in practice it limits to ones ms). A delay between terminal points of network and access points is required to be in compliance with quality parameters of class 0, in the networks with packet switching according to the Recommendation ITU-T Y.1541 [14]. It is defined by IP packet delay (IPTD) at maximum of 100 ms and IP delay variation (IPDV) at maximum of 50 ms. Besides inherent delay close to session, echoes in combination with unbalanced hybrid and with acoustic coupling between earphone and microphone have very negative effect. Networks, where equally high Talker Echo Loudness Rating (TELR) at the receiving side with regard to delay of voice signal isn't guaranteed, have to be provided by suppressors of echo. Next tables 2 and 3 contain recommended and limit values for subnetworks A or B (tab. 4) and intermediate network C (tab. 5). Key parameter is compliance of particular impairment factor.

Table 1. Recommended and limit values between terminate points of public telephone network in the Czech Republic.

	Parameter		ABC_{rec}	ABC_{min}	ABC_{max}
A	Advantage Factor	-	0; 5; 10	-	-
qdu	Quantization Distortion Units	-	1 to 8	1	14
R	**Rating Factor**	-	**50 to 100**	**50**	**100**
I_{tot} - A	Impairment Factor	-	0 to 50	0	50
OLR	Overall Loudness Rating	[dB]	0 to 18	-6	29,5
T_a	Absolute Delay	[ms]	~0 to 200	~0	500
TELR	Talker Echo Loudness Rating	[dB]	Define by graph dependence on Ta		

Table 2. Recommended and limit values for terminate network A or B in the Czech Republic.

	Parameter		$A_{rec} B_{rec}$	$A_{min} B_{min}$	$A_{max} B_{max}$
A	Advantage Factor	-	0; 5; 10	-	-
qdu	Quantization Distortion Units	-	1 to 4	1	5
I_{tot} - A	**Impairment Factor**	-	**0 to 18**	**0**	**18**
RLR	Receive Loudness Rating	[dB]	-1,75 to 7,25	-4,25	13
SLR	Send Loudness Rating	[dB]	1,75 to 10,75	-1,25	16,5
T_a	Absolute Delay	[ms]	~0 to 100	~0	200
TELR	Talker Echo Loudness Rating	[dB]	Define by graph dependence on Ta		

Table 3. Recommended and limit values for transit network C in the Czech Republic.

	Parameter		C_{rec}	C_{min}	C_{max}
qdu	Quantization Distortion Units	-	0	0	4
I_{tot} - A	**Impairment Factor**	-	**0 to 14**	**0**	**14**
RLR	Receive Loudness Rating	[dB]	0	0	0
SLR	Send Loudness Rating	[dB]	0	0	0
T_a	Absolute Delay	[ms]	~0	~0	100

2.2 Using of E-model

The public telephone network plan of the transmission parameters for the Czech Republic uses the E-model for complex quality view on the network. The guidelines and planning examples of Recommendation G.108 [12] are based on the utilization of the E-Model as described in Recommendation G.107 [11]. The intent of this Recommendation is to demonstrate how the E-Model can be used in end-to-end transmission planning for a wide range of local, national, multinational and transcontinental networks. The basic equation for the Rating from Recommendation G.107 is modified to next equations:

$$R = 100 - I_{tot} + A \quad (1)$$

where A is Advantage Factor and I_{tot} is total Impairment Factor:

$$I_{tot} = I_o + I_q + I_{dte} + I_{dd} + I_{eeff} \quad (2)$$

I_o - Impairment Factor - noise and loudness rating
I_q - Impairment Factor - quantizing distortion
I_{dte} - Impairment Factor - talker echo
I_{dd} - Impairment Caused by too-long Absolute Delay
I_{eeff} - Effective Equipment Impairment Factor

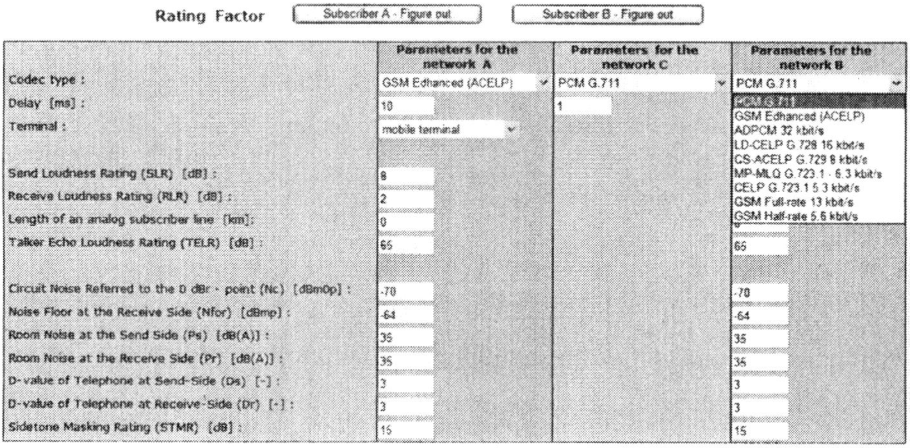

Rating Factor	Subscriber A - Figure out	Subscriber B - Figure out	
	Parameters for the network A	Parameters for the network C	Parameters for the network B
Codec type :	GSM Edhanced (ACELP)	PCM G.711	PCM G.711
Delay [ms] :	10	1	PCM G.711 / GSM Edhanced (ACELP) / ADPCM 32 kbit/s / LD-CELP G.728 16 kbit/s / CS-ACELP G.729 8 kbit/s / MP-MLQ G.723.1 - 6.3 kbit/s / CELP G.723.1 5.3 kbit/s / GSM Full-rate 13 kbit/s / GSM Half-rate 5.6 kbit/s
Terminal :	mobile terminal		
Send Loudness Rating (SLR) [dB] :	8		
Receive Loudness Rating (RLR) [dB] :	2		
Length of an analog subscriber line [km] :	0		
Talker Echo Loudness Rating (TELR) [dB] :	65		65
Circuit Noise Referred to the 0 dBr - point (Nc) [dBm0p] :	-70		-70
Noise Floor at the Receive Side (Nfor) [dBmp] :	-64		-64
Room Noise at the Send Side (Ps) [dB(A)] :	35		35
Room Noise at the Receive Side (Pr) [dB(A)] :	35		35
D-value of Telephone at Send-Side (Ds) [-] :	3		3
D-value of Telephone at Receive-Side (Dr) [-] :	3		3
Sidetone Masking Rating (STMR) [dB] :	15		15

Fig. 2. The input page of the complex R-factor calculation tool for interconnected networks A, B, C (http://matlab.feld.cvut.cz/en)

2.3 Calculation Tool

The Prague Department of Telecommunication Technology co-operated with the APVTS (Public Telecommunication Networks Provider Association) on the Public telephone network plan of the transmission parameters for the Czech Republic and produced the web application for E-model parameters calculation, which was used for optimization of basic transmission parameters and default constants of model. The basic tool calculates the R-factor from the input networks parameters. The second tool calculates the R-factor for the parameters of three co-operating networks. The input page of the complex R-factor calculation tool for interconnected networks A, B, C is shown on the Fig. 2. The calculation tools were programmed for the MATLAB Server and they are accessible on the web site http://matlab.feld.cvut.cz/en [2], [3].

3 Tandem Concatenation of Codec

Assorted combinations of codecs can occur in extreme case in the networks. For example – mobile subscriber uses codec FR (network A) through network C based on PCM and call is terminated in private network B based on codec according to the Recommendation ITU-T G.729. This tandem concatenation of more codecs isn't recommended, within codec PCM (ITU-T G.711) used in combination with any low speed codec. According to E-model the total impairment factor should add up all impairment factors of concatenated codecs. Computation model was modified in our simulation program on the basis of practical experience and results of simulations.

Table 4. Simulated values of impairment factor for single codec used in mobile networks.

Type of Codec	Equipment Impairment Factor
PCM	0
HR	29.89
FR	10.12
EFR	2.17
AMR 4,75	30.75
EFR	2.17

PCM – Pulse Code Modulation – 64 kb/s
HR – GSM Half Rate – 5,6 kb/s
FR – GSM Full Rate – 13 kb/s
EFR – GSM Enhanced Full Rate – 12,2 kb/s
AMR – GSM Adaptive Multi-Rate – from 1,85 to 12,20 kb/s

Firstly the Table 4 shows values of impairment factor, which we obtain for one codec, then the Table 5 shows values for tandem concatenation of assorted combinations of codecs. It's proved that utilization of PCM codec practically doesn't lead to impairment of resulting quality. It's impossible to say that combination of assorted codecs embodies impairment accordant to addition of particular impairment factors. Results are fundamentally better in general.

Table 5. Simulated values of impairment factor for tandem concatenation of codec.

Combination of Codec	Equipment Impairment Factor
HR - HR	38.91
FR - FR	15.12
FR - EFR	14.86
FR - HR	30.92
FR - PCM	11.69
EFR - EFR	11.15
EFR - HR	29.03
EFR - PCM	2.95
AMR - AMR	44.71
AMR - FR	36.53
AMR - EFR	36.23
AMR - HR	41.73
AMR - PCM	32.57

4 Conclusion

The Prague Department of Telecommunication Technology participation on development of the public telephone network plan of the transmission parameters for the Czech Republic was very positively evaluated by CTU (Czech Telecommunication Union) and APVTS (Public Telecommunication Networks Provider Association). The finalization of the public telephone network plan of the transmission parameters allows the QoS control for the interconnected telecommunication networks including NGN in the Czech Republic.

It is evident, that transmission delay and utilization of low speed codecs are main limiting factors lowering quality of speech transfer in next generation networks. It's necessary to adjust network parameters so that the compliant value of R-factor (the best is higher than 60) is reached, together with cooperation of different types of networks and interconnection with private networks. Quantity of network's delay is arbitrating for quality of call realized in only one network utilizing specific codec as well as type of used terminal, which specifies impairment factor A. Quality of call is good for delay up to 200 ms, but only in connection with echo cancellation process. Quality of call is unsuitable for delay over 500 ms altogether.

The best quality is in connection with utilization of codec PCM G.711 in all networks, because this codec has minimal total impairment of speech quality. More advanced coding ACELP is recommended for mobile networks, because it reaches better results than classic codec RPE-LTP.

The tandem concatenation of more codecs isn't recommended, excepting PCM codec ITU-T G.711 used in combination with arbitrary low speed codec. The tandem concatenation of codec creates lower reduction of quality than direct addition of impairments factors in E-model. It results from practical experience and results of realized simulations.

The mobile terminals are better evaluated by users (advantage factor A) because of their perception of mobility as an advantage. But it's necessary to consider communication process between mobile terminal and fixed terminal, whose users do not feel this advantage. Also, lower quality tolerance is higher for lower costs per one call. There is indeed similar quality imbalance, because user paying standard tariff requires corresponding quality, used by him, regardless from what network the call is coming.

The plan of transmission parameters of private telephone networks represents important technical, but also legal document, which influences public telephone networks and private networks connected to public telephone network providers. This plan puts a pressure on providers to serve up sufficient quality of voice communication and to do more effective networks planning. Further more it's possible to use calculation according to E-model, eventually systematic measurement in the network by suitable measurement methods providing directly R-factor, eventually MOS parameter, from which the R-factor is easily re-counted.

The problems can naturally occur in case when provider wants to measure quality in the networks through non-intrusive method, e.g. continuous monitoring in real traffic. There are distinctive deviations between results of measurements realized by subjective evaluation of quality of speech in Czech language and results of measuring by 3SQM method. It seems that this method is only limitedly applicable for Czech language and it is necessary to adjust the psychoacoustic model according to real results.

Acknowledgment. This work has been supported by the Grant Agency of the Academy of Sciences of the Czech Republic under project 1ET300750402 and Czech Technical University's grant No. CTU0715013.

References

1. Transmission Plan for New Generation Telephone Networks. Czech Telecommunication Office. Prague 2005.
2. Vodrazka, J. – Jares, P. – Hubeny, T.: Rating Factor. Simulation program on-line: Matlab Server. http://matlab.feld.cvut.cz/en/
3. Vodrazka, J. – Jares, P. – Hubeny, T.: Rating Factor for cooperative networks. Simulation program on-line: Matlab Server. http://matlab.feld.cvut.cz/en/
4. Vodrazka, J.: Transmission Plan for Public Telephone Services in NGN. In Proceedings EC-SIP-M 2005. Bratislava: The Faculty of Electrical Engineering and Information Technology of the Slovak University, 2005, s. 354-357. ISBN 80-227-2257-X.
5. Chu, Wai C.: Speech coding algorithms, Foundation and Evolution of Standardized Coders. John Wiley & Sons, Inc., Hoboken, New Jersey 2003.
6. Kuo Pei-Jeng - Omae Koji - Okajima Ichiro - Umeda Narumi: VoIP quality evaluation in Mobile wireless networks Advances in multimedia information processing. PCM 2002 IEEE Pacific Rim conference on multimedia No3, Hsinchu, 16-18 December 2002. vol. 2532, pp. 688-695, ISBN 3-540-00262-6

7. Janssen, J. - De Vleeschauwer, D. - Buchli, M. - Petit, G.: Assessing voice quality in packet-based telephony. IEEE Internet Computing. Vol. 6, no. 3, pp. 48-56. 2002
8. Rein, S. - Fitzek, F. H. P. - Reisslein, M.: Voice quality evaluation for wireless transmission with ROHC. Seventh IASTED International Conference on Internet and Multimedia Systems and Applications; Honolulu, USA; 13-15 Aug. 2003. pp. 461-466. 2003
9. Nemčík, M. - Levák, M.: An Objective Method for End-to-End Speech Quality Assessment Using PESQ Algorithm. In Digital Technologies 2006 - 3rd International Workshop [CD-ROM]. Žilina: University of Žilina, Fakulty of electrical engineering, 2006, vol. 1, ISBN 80-8070-637-9.
9. ITU-T Recommendation G.101 (1996), The transmission plan.
10. ITU-T Recommendation G.107 (1998), The E-Model, a computational model for use in transmission planning.
11. ITU-T Recommendation G.108 (1999), Application of the E-model: A planning guide.
12. ITU-T Recommendation G.109 (1999), Definition of categories of speech transmission quality.
13. ITU-T Recommendation Y.1541 (2006), Network performance objectives for IP-based services.

A Light-weight Security Protocol for RFID System

Jung-Hyun Oh, Hyun-Seok Kim, Jin-Young Choi,

Computer Theory and Formal Methods Lab,
Dept. Of Computer Science and Engineering , Korea Unversity,
5-1 Anam dong, Sungbuk gu, Seoul, 136-701, Korea,
{jhoh, hskim, choi}@formal.korea.ac.kr

Abstract. RFID is automatic object identifying technology via radio frequency. And its application areas are un-describable for its convenience and pervasiveness. However, because the communication channel between the verifier and the tag is wireless, serious privacy problems such as the data leakage and the data traceability can be occur. Without resolving these privacy problems, RFID system cannot be adapted fully in any area. Many kinds of security protocols have been proposed to resolve these problems. However, previous proposals did not satisfy security requirements and still leaved vulnerabilities. In this paper, we describe the security vulnerabilities of previous works for RFID systems. Finally, we propose a security protocol which based on one-time pad scheme using random nonce and shared secret values. The proposed protocol satisfies security requirements such as the data secrecy, data anonymity and the data authenticity between the verifier and the tag. We have proved security requirements satisfaction formally by using GNY logic.

Keywords: RFID, Security protocol, One-time pad, GNY logic

1 Introduction

RFID system is automatic object identification system using radio frequency signal. The small tags(transponders) attached to the products carries the unique information of the products and whenever the verifier(transceiver) request the product specific information tags transmit the information via radio frequency signal. The verifier relays the received information to the back-end DB whether the information is valid or not. Because of the RFID system uses the radio frequency, objects can be identified easily and quickly, and the management of the object could be efficient. Because of its merits, the attempt to apply the RFID system to many areas is in progress.

However, there are several problems that RFID system should resolve before their pervasive deployment. The problems in RFID system are privacy related problems such as *the data leakage* and *the data traceability*. These problems occur because the communication channel between the verifier and the tag is wireless. Simply eavesdropping the messages that transmitted between the verifier and the tag, the attacker can obtain the unique information of the tag, and the attacker can make private information profiles of the tag carrier. They also can track tag carrier without any authorization.

To resolve these privacy problems, security measures for the RFID system should satisfy several security requirements which are described down below.

- *Data Secrecy* means that any transmitted data between the RFID system components should not be understandable to the attacker. The tag stores the unique ID and the verifier identifies the tag by receiving the unique ID of the tag. However, if the unique ID of the tag is exposed to attackers, it could be used in identifying tag carrier's private items and making the tag carrier's private information profile.
- *Data Anonymity* means that any transmitted data between RFID system components should not be distinguishable to the attacker. Even though the attacker could not understand the meaning of the messages between the RFID system components, the attacker could track the tag or the tag carrier if the transmitted messages are fixed and being used in every session.
- *Data Authenticity* means that any transmitted data between the RFID system components should be authenticable. That is, the messages transmitted between the RFID system components should be check whether or not they are from the honest entity. If there is no measure for data authentication, the attacker would attempt to authenticate himself to the honest entities by using the previously obtained messages. And if attacker succeed in authenticating himself to the honest entities, the information such as the tag ID leakage could be possible.

The contributions of this paper are outlined as follows: First, we describe security vulnerabilities of the previous works based on three security requirements mentioned above and then we propose a light-weight security protocol based on one-time pad scheme. One-time pad scheme is proven to guarantee the perfect secrecy of the message by Shannon[8]. In our proposal, one time pad scheme with the secret values and a fresh pseudonym is applied to satisfy the three security requirements.
Second, using GNY logic[9], we proved formally the satisfaction of security requirements in our proposed protocol. In particular, GNY logic provides several notations and logical postulates that help to express the security protocol and to deduce the security requirements(goals) of the protocol logically. If the security requirements deduction of the protocol is failed, then we cannot assure that this protocol satisfies the security requirements. Because of its precise protocol expression and verification capability, GNY logic is known as one of the successful skills for formal security verification methods[17].
This paper is organized as follows: In section 2, we summarize the previous works and its security vulnerabilities. In section 3, we describe our proposal. In section 4, we present the analysis of the proposed protocol using GNY logic and discuss the module aspect of security for RFID system. Finally, the conclusion is addressed in the last section.

2 Related Works

Many proposals have been proposed to satisfy the security requirements. In this section, we have categorized the previously proposed protocols as down below.

Hash Function Based Security Protocol In [2], Weis *et al.* proposed Hash-lock protocol and the randomized hash-lock protocol. In these protocols, the tag stores the ID and the unlocking key and stay in locked state before the reader request the ID. When the reader attempts to identify the tag, the tag sends the hashed ID to the verifier. Then the reader seeks the unlocking key ID in the back-end DB with the received hashed because the tag unlock itself only when the reader sends the unlocking key. But the unlocking key is transmitted without any encryption, so the attacker can obtain the unlocking key. Therefore, *the data secrecy, the data anonymity* and *the data authenticity* are not satisfied.

In[3], Henrici *et al.* proposed the hash-based ID variation protocol which ID of the tag varies in each session with help of fresh nonce. However, the hashed ID is also sent to the reader to make unlocking key searching easier. Therefore, even though the ID of the tag is not exposed, the attacker could track the tag because the fixed hashed ID is used in every session. Therefore, *the data anonymity* is not satisfied.

In [4], Okubo *et al.* proposed the hash chain protocol which used two different one way hash functions to send the hashed ID to the reader and to update the ID after the authentication procedure is completed. However, the hashed ID should be synchronized to the back-end data-base. Therefore, the hashed ID could be used in authentication for attacker before the tag authentication session is completed. And the attacker could de-synchronize the ID between the tag and the back-end DB, the tag can be un-identifiable. Therefore, hash chain scheme does not satisfy *the data authenticity*. Moreover, this scheme gives the great burden to the DB for a single tag authentication.

Arithmetic Calculation Based Security Protocol In [5], Juel proposed minimalist cryptography using one-time pad scheme for low-cost RFID system. However, storing the triple shared keys and number of padding vectors in a single tag will cost large amount of gates to implement. Moreover, in the authentication procedure, the triple shared keys are exposed to the attacker. Therefore, the attacker could use these triple shared keys to authenticate himself to the reader just before the honest tag authentication is completed. So, this scheme does not satisfy *the data secrecy* and *the data authenticity*.

In [7], Juel *et al.* proposed a light-weight security protocol based on HB algorithm which is known to secure to both passive attack and active attack. The HB algorithm is based on hardness of the learning parity bit with noise[12]. Guessing the plaintext of the encrypted message is computationally infeasible, because guessing the plaintext is identical to solving the LPN problem. However, several papers have shown that several security vulnerabilities are still exist in the HB based protocols[13][14]. Moreover, it is open question whether the LPN problem based protocols are provably resistant to man-in-the-middle-attack or not.

In [18], Vajda *et al.* proposed light-weight security protocol using modular product such as XOR. In this scheme, the reader and tag use the secret key which is padded with random bit to authenticate each other. In [19], Defend proved that this scheme does not satisfy *the data(secret key) secrecy* and *the data authenticity* Because the fixed key is continuously used in a session before the authentication step is completed, the initially shared secret key and newly computed secret key for current session can be leaked by padding the two messages transmitted between the reader and tag.

3 Our Proposal

We have described that the previous proposed schemes did not satisfied the security requirements of the RFID system. In section, we present our protocol that satisfies the three security requirements.

Initial assumptions for the security protocol design Firstly, our scheme is focused on to design a security protocol which can be adapted into the class 1 generation 2 standard(proposed by EPC-Global) based RFID system. That is, the bit-length of a tag ID is 128. Secondly, the communication channel between the tag and verifier is wireless and between the verifier and back-end DB is, generally, wired. And, also, the hardware capability of the verifier and back-end DB is considered to be limited. Therefore, it is convenient to assume that the communication channel between the verifier and back-end DB is secure. Hence, we assumed that the verifier and the back-end DB work as one component, the verifier.

One-Time Pad Scheme Shannon has proved that the perfect secrecy of the message is guaranteed if and only if the message is padded with a fresh pseudonym which bit length is the same or longer than the message[8]. Encryption and decryption in one-time pad scheme the plaintext is combined with a random secret key, K, that is as long as the plaintext, x, and used only once. A modular addition, such as XOR, is used to combine the plaintext with the random secret key as described in (1).

$$E_K(x) = x \oplus K = D_K(x). \tag{1}$$

We applied this advantage to our protocol in encryption and decryption. In our protocol, the unique tag ID is padded with a fresh pseudonym. Because of the ID is padded with the fresh pseudonym as long as the ID, the perfect secrecy of the ID can be satisfied. Therefore, with using this scheme we can satisfy the security requirements such as *the data secrecy*.

$$E_{Pseudonym}(ID) = ID \oplus Pseudonym. \tag{2}$$

The generation of the fresh pseudonyms will be explained in detail at the next paragraph. This fresh pseudonym can be generated by both verifier and tag, and any information related to pseudonym generation will not be leaked. Therefore, the attacker could not decrypt the tag ID out of the message, $E_{Pseudonym}(ID)$, unless the attacker happened to obtain the this padded pseudonym.

Generating Padding Pseudonym To decrypt the padded ID, the verifier should know this pseudonym. The pseudonym will be generated by padding or concatenating the two fresh nonce, N_T and N_V, which are generated by the tag and the verifier ($Pseudonym = (N_T \oplus N_V)$ or $(N_T \parallel N_V)$, pseudonym generation methods depends on the bit length of the secret value and nonce). The delivery of each nonce to each other(the

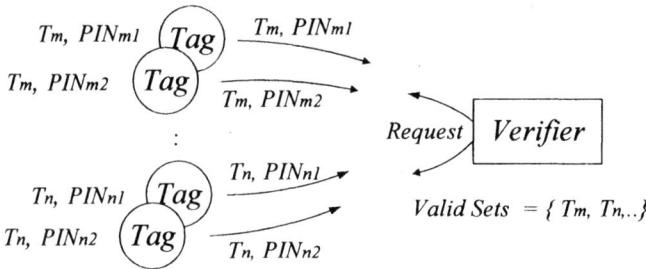

Figure. 1. The secret value set management between the different subset of RFID tags

verifier and tag) will be done by using the secret values in the secret value set which is initially shared between the verifier and the tag (Secret value set $T_x = \{S_1, S_2, \ldots, S_n\}$). The nonce N_V generated by the verifier will be delivered to the tag by padding with selected secret value S_x ($S_x \in T_x$). The nonce N_T generated by the tag will be delivered to the verifier by padding with the selected secret value S_y ($S_y \in T_x$).

$$E_{Sx}(N_V) = N_V \oplus S_x \qquad (3)$$

$$E_{Sy}(N_T) = N_T \oplus S_y \qquad (4)$$

However, a tag cannot store secret value set which contains many secret values because of its memory size or implementation cost. To make the same efficiency such as storing a number of secret values with reasonable number of secret values by considering the memory size and implementation cost, whenever the secret values are needed to deliver the pseudonym, the tag permutes the sequence of the stored secret values order and chooses necessary amount of secret values from the permuted order. For example, if stored secret value set is $T_x \rightarrow \{S_1, S_2, \ldots, S_n\}$ and if two secret values are to use in padding then available number of permuted secret value pairs are $_nP_2$. By this method, we can gain additional number of secret value orders without storing large number of secret values. The tag selects the permutation index number, *PIN*, which carries the information about the order of the permuted secret value set. And then the tag delivers the *PIN* to the verifier for reporting the verifier which secret value order should be used for nonce delivery

Secret Value Set Management   The management method of the secret values was inspired by the internet banking authentication scheme. To use internet banking, we should register our personal information in the bank and receive the secret card where a random secret value set is printed on. To use the internet banking service, the internet banking server requires you to put the specific secret values from the secret value set on the secret card for the user authentication. The secret card has several types with different secret value sets. Therefore the exposure of one type of the secret card cannot lead to the entire internet banking service failure.

Likewise, the secret value set stored in each tags should not be the same values to all tags or unique. If the shared secret value sets are the same in the tags, exposure of

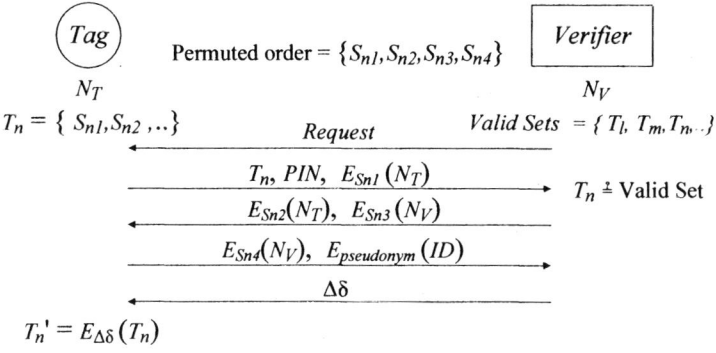

Figure. 2. The message sequence of the protocol

the one secret value set can lead us to entire information leakage of the RFID system. On the other hand, if the shared secret value set in each tag is unique, then the management of the secret value sets will be extremely difficult and this uniqueness may be used for identifying the tag. To resolve these problems, designing some subset of tags to use the same shared secret value set and the other subset of tags to use different secret value set will be the best way. Every secret value set stored in tags has label like T_x or T_y and these labels should be sent for the verifier to recognize what secret value set is being used in current session. The attacker cannot derive any information about the secret values from these labels, because the label will be assigned randomly such as, T_{123} or T_{201}. The basic idea of secret value set management between the different subset of RFID tags is depicted in Figure.1.

For *the forward secrecy*, shared secret value set will be updated at the end of the successful communication. The verifier sends the update message $\Delta\delta$ to the tag for secret value set update, and the tag update secret value set T_n to T_n' by padding the T_n with $\Delta\delta$.

$$E_{\Delta\delta}(T_n) = T_n \oplus \Delta\delta \qquad (5)$$

And the updated secret value set T_n' should be valid set which is already stored in the DB.

Protocol Steps

1. Verifier   Request the tag its unique ID
2. Tag   Permute order of T_n. Selects four secret values out of permuted order of secret values, $_nP_4$, and puts permutation order information in the *PIN*. Generates a random nonce N_T and encrypts it by padding S_{n1}. Sends messages, T_n, *PIN* and $E_{Sn1}(N_T)$ to V.
3. Verifier   Finds T_n in the *DB*. Permutes T_n with *PIN*. R decrypts the $E_{Sn1}(N_T)$ and encrypts the N_T by padding S_{n2}, to authenticate himself to T. Generates N_V and encrypts it by padding S_{n3}. Sends $E_{Sn2}(N_T)$ and $E_{Sn3}(N_V)$ to T.
4. Tag   Decrypts the $E_{Sn2}(N_T)$ to authenticate the V by verifying the N_T. If

	$D_{Sn2}(E_{Sn2}(N_T)) = N_T$ then authenticate V. T decrypts the $E_{Sn3}(N_V)$ and encrypt the N_V by padding S_{n4}, to authenticate himself to V. Generates *pseudonym* by padding two nonce N_V, N_T. Pad the ID with *pseudonym*. Sends $E_{Sn3}(N_V)$ and $E_{Pseudonym}(ID)$ to V
5. Verifier	Decrypts the $E_{Sn4}(N_V)$ to authenticate T by verifying the N_V. If $D_{Sn4}(E_{Sn4}(N_V)) = N_V$, then V authenticate T. Generates *pseudonym* for ID decryption by padding two nonce N_V, N_T. Finds ID in the *DB* and updates T information. *DB* selects the available secret value set T'_n and calculates the secret value set update message $\Delta\delta$ which can be used in updating the set T_n to T'_n Sends the $\Delta\delta$ to T
6. Tag	Updates the secret values set T_n by padding with $\Delta\delta$

4 Analysis

4.1 Security Correctness Proof

Frequently, informal and intuitive way of security verification has been used in previous proposals. However, informal and intuitive way of verification can leave design flaws and security errors undetected. For example, Needham-Schroeder protocol which was believed to be secure, and the BCY protocol which was considered to be secure in mobile communication, formal verification methods proved its security vulnerabilities[15][16].

With formal verification, we can assure that there is no undetected design flaw in system which we want to verify the security. In formal methods, there are two major methods in verification, the model checking and the theorem proving. The verification using theorem proving is logical deduction step that help to deduce the security goals of the target system. If the security goals are successfully deduced, we can assure that the target system has no design flaws.

In this section, we analyze the security correctness of our proposed protocol using theorem proving method which name is GNY logic. GNY logic is considered to be one of the successful methods for the security protocol verification[17]. With GNY logic, we have proved security goals of our proposed protocol; the tag and verifier can assure themselves that received messages are delivered from trustable agent and these messages are fresh so that they can also assure themselves that this messages are never been used before this session. Because *the data secrecy* is proved by Shannon, we are focus on proof of *the data authenticity* and *the data anonymity* satisfaction. The security verification using GNY logic involves four steps, the formalization of the protocol messages, the specification of the initial assumptions, the specification of the protocol goals and the application of the logical postulates. The precise meaning of notations and logical postulates is described in appendix A.

The formalization of the protocol messages The message with the asterisk, *, denotes that the entity who received this message did not make and send this message in the previous stage of the protocol. In the message notation, the symbol ◁ means the entity receives the message.

Message 1 : T ◁ *Request
Message 2 : V ◁ *T_n, *PIN, *$\{N_T\}_{Sn1}$
Message 3 : T ◁ *$\{N_T\}_{Sn2}$, *$\{N_V\}_{Sn3}$
Message 4 : V ◁ *$\{N_V\}_{Sn4}$, *$\{ID\}_{(NV, NT)}$
Message 5 : T ◁ *Δδ

The initial assumptions of the proposed protocol This section specifies the initial possessions and abilities of the each participant of the protocol. The message with symbol # denotes freshness of the message and the message with symbol ∅ denotes it can be recognized by the entity who receives it. The message with symbol ∋ denotes that the entity, left hand side of the symbol, possess the formula, right hand side of the symbol. The message with the arrow symbol denotes secret values described above the arrow are believed to suitable values between two entities.

$T \ni N_T$ $T \ni S_{n1}, S_{n2}, S_{n3}, S_{n4}$ $T \models \# N_T$ $T \models \varnothing N_V$
$V \ni N_V$ $V \ni S_{n1}, S_{n2}, S_{n3}, S_{n4}$ $V \models \# N_V$ $V \models \varnothing N_T$
$T \models T \xleftarrow{S_{n1}, S_{n2}, S_{n3}, S_{n4}} V$ $V \models V \xleftarrow{S_{n1}, S_{n2}, S_{n3}, S_{n4}} T$

The goals of the proposed protocol The goals of the proposed protocol are belief, ⊨, and the freshness, #, of the messages between the verifier and the tag. The belief denotes that the message is delivered from right trustable party. And the freshness denotes that the message value is not used in previous protocol session. Satisfying the freshness is important because it can make the communication party assure that the received message was not used in reply attack.

$T \models V \vdash N_V$ Tag believes verifier sent N_V
$T \models V \vdash N_T$ Tag believes verifier returned N_T
$V \models T \vdash N_T$ Verifier believes tag sent N_T
$V \models T \vdash N_V$ Verifier believes tag returned N_V
$V \models \# E_{(NV, NT)} (ID)$ Verifier believes this message is fresh
$V \models T \vdash E_{(NV, NT)} (ID)$ Verifier believes tag sent this message

The first through the fourth goal and the sixth goal related to *the data authenticity*. These goals indicate that the received messages are sent from honest agent, so these messages are trustable. The sixth goal is related to *the data anonymity*. This goal indicates that the received message is fresh, so this message is guaranteed that it was not used in previous session. If these goals are proved logically by applying the logical postulates in GNY logic, then we could assure that this protocol satisfies *the data authenticity* and *the data anonymity*.

The application of the logical postulates The security verification of our protocol will be done by goals deduction with help of the logical postulates provided by the

GNY logic. There are several postulates in GNY logic and we wrote the name of the postulates in the right-side of the messages which are used in the deduction. The first and the fifth message are removed from the following steps because message 1 and 5 are just delivering the data which are not related with the goal deduction.

Message 2 : V $\triangleleft$ *T_n, *PIN, *$\{N_T\}_{Sn1}$
             V $\triangleleft$ T_n, PIN, $\{N_T\}_{Sn1}$ /* by T1
             V $\ni$ T_n, PIN, $\{N_T\}_{Sn1}$ /* by P1
             V $\models$ T $\vdash N_T$ /*by initial assumption & I1'
             The third goal is successfully deduced

Message 3 : T $\triangleleft$ *$\{N_T\}_{Sn2}$, *$\{N_V\}_{Sn3}$
             T $\triangleleft$ $\{N_T\}_{Sn2}$, $\{N_V\}_{Sn3}$ /* by T1
             T $\ni$ $\{N_T\}_{Sn2}$, $\{N_V\}_{Sn3}$ /* by P1
             T $\models$ V $\vdash N_V$ /*by initial assumption & I1'
             The first goal is successfully deduced
             T $\models$ V $\vdash N_T$ /*by initial assumption & I1'
             The second goal is successfully deduced

Message 4 : V $\triangleleft$ *$\{N_V\}_{Sn4}$, *$\{ID\}_{(NV, NT)}$
             V $\triangleleft$ $\{N_V\}_{Sn4}$, $\{ID\}_{(NV, NT)}$
             V $\ni$ $\{N_V\}_{Sn4}$, $\{ID\}_{(NV, NT)}$
             V $\models$ T $\vdash N_V$ /*by initial assumption & I1'
             The fourth goal is successfully deduced
             V $\models$ # $\{ID\}_{(NV, NT)}$ /*by initial assumption, F1 & F2
             The fifth goal is successfully deduced
             V $\models$ T $\vdash \{ID\}_{(NV, NT)}$ /*by initial assumption & I1'
             The sixth goal is successfully deduced

The Result of the Proof *the data authenticity* : As you can see above, the tag can assures the received nonce N_V is delivered from the verifier. And the verifier also can assure that the received nonce N_T is delivered from the tag. The delivery assurance means that with this protocol *the data authenticity* is achievable. Because the nonce calculation can only be done by authenticable party who has the secret value set, the attacker cannot be authenticated to honest agent.
The data anonymity : the verifier can assure the received message is fresh which means this message is never been used before the current session. By proving the freshness of the message, the protocol can guarantee the replay attack or tag cloning is impossible by any attacker and *the data anonymity* is achievable.

4.2 Evaluation

In this section, we focus on the security module implementation cost for the passive RFID tag. The passive RFID tag is hardware constrained device so that the implementation of the complex encryption schemes such as public key encryption or the symmetric key encryption is currently very rough task. Although the complex encryption scheme equipped tag could be implemented, the tag would cost more than 5 cent. Therefore, the implementation cost should be considered very carefully before implementing the security module into the tag. According to [2], possible fabrication amount of gates for a tag within 5 cent is about 10,000 ~ 35,000 gates. Excluding the basic need for RFID tag fabrication such as antenna, IC and memory area, only 1,000 ~ 3,500 gates can be assigned for security module implementation.

To verify whether our scheme can be implemented practically in the tag or not, we made experiment on the total number of gates for our scheme with ASIC implementation.

Table 1. Total number of gates for our scheme implementation.

Bit length of data in padding	32	64	128
# of gates for XOR module	567	850	1,700
# of gates for register	928	1,875	3,713
Total gates	1,495	2,707	5,413

We have designed that the data and pseudonym is padded in parallel. Therefore, 128 XOR modules is needed, and the register which stores the 128 bit-length temporal data for padding such as the secret values, nonce or ID of a tag is also needed for 128 bit-length data padding. However, we can reduce these basic needs by reducing the bit-length of data which the padding module takes for input. For example, if we design the padding module which takes 64 bit-length data as a input then the number of XOR module for the data padding and register size for the temporal input/output data storage can be reduced almost by half. Table 1 shows the estimated total gates for our scheme by using ASIC implementation. The first row in table 1 denotes the bit-length of the input data of the one-time pad module. The fourth row denotes the estimated total number of gates for padding module based on bit-length of the input data.

Table 2. The estimated implementation gates for security modules[10].

Categories	Types	Gates
Hash Function	SHA-256	10,800
	SHA-1	8,120
	MD5	8,400
	MD4	7,350
Symmetric Enc.	AES	25,000
	Modified AES	3,595
One-time pad(our work)	XOR	1,495

In [10], Feldhofer described that the hash functions and AES based symmetric key encryption algorithms exceed at least 7,000 gates for implementation. Specifically, the implementation of the Hash function cost around 7,000 ~ 11,000 gates and the AES scheme would require 25,000 gates. Therefore, complex encryption scheme and even hash functions are not suitable in security protocol design for the RFID system because they are not satisfying the least gate limitation of the security module implementation. On the other hand, the one-time pad scheme can be implemented within 1,495 gates if we assumed that the padding module is designed to take 32 bit-length data input for padding. The total gates for our scheme are even smaller than that of the modified AES module which is proposed by Feldhofer. We have compared the total gates of the different encryption module for implementation by referring the Feldhofer's work[10] in table 2.

5 Conclusion

In designing the security protocol for RFID system to resolve privacy problems, the security requirements should be satisfied and also the implementation cost requirements as well. In this paper, we have described the previous works such as hash function based scheme and simple arithmetic calculation based scheme that failed to satisfy all the security requirements. And then, we proposed a light-weight security protocol for RFID system based on one-time pad scheme which satisfies all the security requirements and low-cost implementation requirement. The security requirements satisfaction of our protocol was presented in this paper: the data secrecy was proved by Shannon[8], and the data authenticity and the data anonymity was proved by using GNY logic. Moreover, we have showed that our protocol can be implemented with lower cost than the previous works by comparing the gates for the security module implementation.

References

1. Ari Juels, Ronald Rivest, and Michael Szydlo.: The blocker tag: Selective blocking of RFID tags for consumer privacy. In: Vijay Atluri (ed.): *Conference on Computer and Communications Security -ACM CCS*. ACM Press. Washington DC, USA (2003) 103-111
2. Stephen Weis.: Security and privacy in radio-frequency identification devices. Master thesis, Massachusetts Institute of Technology (MIT). Massachusetts, USA (2003)
3. Dirk Henrici and Paul Muller.: Hash-based enhancement of location privacy for radio frequency identification devices using varying identifiers. In: Ravi Sandhu and Roshan Thomas (eds.): *International Workshop on Pervasive Computing and Communication Security - PerSec 2004*. IEEE Computer Society. Orlando, Florida, USA (2004) 149-153
4. Miyako Ohkubo, Koutarou Suzuki, and Shingo inoshita.: Efficient hash-chain based RFID privacy protection scheme. In *International Conference on Ubiquitous Computing - Ubicomp'04*. Nottingham, England (2004)
5. Ari Juels.: Minimalist cryptography for low-cost RFID tags. In: Carlo Blundo and Stelvio Cimato (eds.): *International Conference on Security in Communication Networks - SCN 2004*. LNCS vol. 3352, Springer-Verlag. Amalfi, Italia (2004) 149-164
6. Martin Feldhofer, Sandra Dominikus, and Johannes Wolkerstorfer.: Strong authentication for RFID systems using the AES algorithm. *Workshop on Cryptographic Hardware and Embedded Systems 2004*. LNCS Vol 3156. IACR, Springer-Verlag (2004) 357-370
7. Ari Juels and Stephen Weis: Authenticating pervasive devices with human protocols. In: Victor Shoup (ed.): *Advances in Cryptology - CRYPTO 2005*. LNCS vol.3126. IACR,

Springer-Verlag. Santa Barbara, California, USA (2005) 293-308
8. C.E. Shannon: A Mathematical Theory of Communication. Bell System Technical Journal vol. 27. (1948) 379-423, 623-656
9. L. Gong, R. Needham, and R. Yahalom.: Reasoning about Belief in Cryptographic Protocols. IEEE (1990)
10. Martin Feldhofer and Christian Rechberger.: A case against currently used hash functions in RFID protocols. Printed handout of Workshop on RFID Security *2006* (2006)
11. Junichiro Saito, Jae-Cheol Ryou, and Kouichi Sakurai.: Enhancing privacy of universal re-encryption scheme for RFID tags. In: Laurence Jang, Minyi Guo, Guang Gao, and Niraj Jha, (eds.): *Embedded and UbiquitousComputing - EUC 2004*, Lecture Notes in Computer Science Vol 3207. Springer-Verlag. Aizu-Wakamatsu City, Japan (2004) 879-890
12. Hopper, N. and Blum, M.: A Secure Human-Computer Authentication Scheme. Technical Report, CMU-CS-00-139, Carnegie Mellon University (2000)
13. Selwyn Piramuthu.: HB and related light-weight authentication protocols for secure RFID tag/verifier authentication. In Collaborative Electronic Commerce Technology and Research 2006 (2006)
14. Jonathan Katz and Adam Smith.: Analyzing the HB and HB+ protocols in the "large error" case. Cryptology ePrint Archive, Report 2006/326(2006)
15. Gavin Lowe.: Breaking and Fixing the Needham Schroeder Public Key Protocol using FDR. In Tools and Algorithms for the Construction and Analysis of Systems. LNCS vol. 1055. Springer-Verlag (1996) 147-166
16. Tom Coffey, Reiner Dojen and Tomas Flanagan.: Formal verification : an imperative step in the design of security protocols. Elsevier(2003)
17. A Marithuria, R. Safavi-Naini, P. Nickolas.: Some remarks on the logic of Gong, Needham and Yahalom. Proc. of the International Computer Symposium, ROC vol.1. Hsinchu, Taiwan (1994) 303-308
18. Istvan Vajda and Levente Buttyan.: Lightweight authentication protocols for low-cost RFID tags. In *Second Workshop on Security in Ubiquitous Computing - Ubicomp 2003*. Seattle, WA, USA (2003)
19. Benessa Defend, Kevin Fu, and Ari Juels.: Cryptanalysis of two lightweight RFID authentication schemes. In *International Workshop on Pervasive Computing and communication Security - PerSec 2007.* IEEE Computer Society Press. New York, USA (2007)

Analog Digitized Data Logger with Wireless and Wired Communication Interface and RFID Features

Radek Kuchta, Radimir Vrba

Brno University of Technology, Department of Microelectronics
Udolni 53, Brno, Czech Republic
{kuchtar, vrbar}@feec.vutbr.cz

Abstract. Knowledge of temperature-time curve for a certain time interval is needed in many scientific, medical and industrial applications. In some applications, however, the recorded temperature values in monitored time should be read wirelessly. Both requirements are defined especially for the transport of biological active substances. This set was applied in case study for continuous temperature data logging during a transport of chilled electrochemical TFT sensors with applied enzymes, highly sensitive to higher temperature limit exceeding.

Keywords: RFID, wireless communication, temperature measuring, data logging.

1 Introduction

A portable data logger with RFID features was designed for applications, where portability and wireless data transfer is inevitable. Communicating reader/writer can be mounted on the wall or can be portable, too. At the beginning it was designed for temperature measuring and monitoring, but the other analog sensors can be used, too.

Two main modes of operation for a tag during temperature data logging may be remotely chosen.

Mode 1 - standard storing of data in preprogrammed regular acquisition time intervals (100 milliseconds up to 2 hours) with number of samples limited only by a data EEPROM memory size.

Mode 2 - more memory size reducing method, where only breaking lower and upper temperature limits initiates storing the date and time stamp. However, the following date and time stamp is stored only when the temperature returns back into the temperature band between lower and upper temperature predefined limits. This mode 2 corresponds with a structure of data stored in data EEPROM memory. Enhanced mode can be set when the maximum or minimum temperature between breaking and returning points of a sampling temperature course is stored, too. This is a typical example for monitoring of foodstuff transport, where the time stamp and maximum temperature after breaking the limit are proofs to define the offender who damaged the transported goods.

Wireless RFID systems generate and radiate electromagnetic waves. That's why they are legally classified as radio systems. The function of other radio services must under no circumstances be disrupted or impaired by the operation of RFID systems. For this reason, it is usually only possible to use frequency ranges that have been dedicated specifically for industrial, scientific or medical applications. These are the frequencies classified for worldwide as ISM (Industrial – Scientific – Medical), and they can also be used for RFID applications. The most important frequency ranges for RFID systems are therefore 135 kHz, 27.125 MHz, 40.68 MHz, 433.92 MHz, 869.0 MHz, 915.0 MHz, 2.45 GHz, 5.8 GHz and 24.125 GHz.

The range below 135 kHz is heavily used by other radio services because it has not been reserved as an ISM frequency range. The propagation conditions in this long wave frequency range permit the radio services that occupy this range to reach wide areas at a low technical cost. In order to prevent collisions, the future Licensing Act for Inductive Radio Systems in Europe, 220 ZV 122, will define a protected zone of between 70 and 119 kHz, which will no longer be allocated to RFID systems.

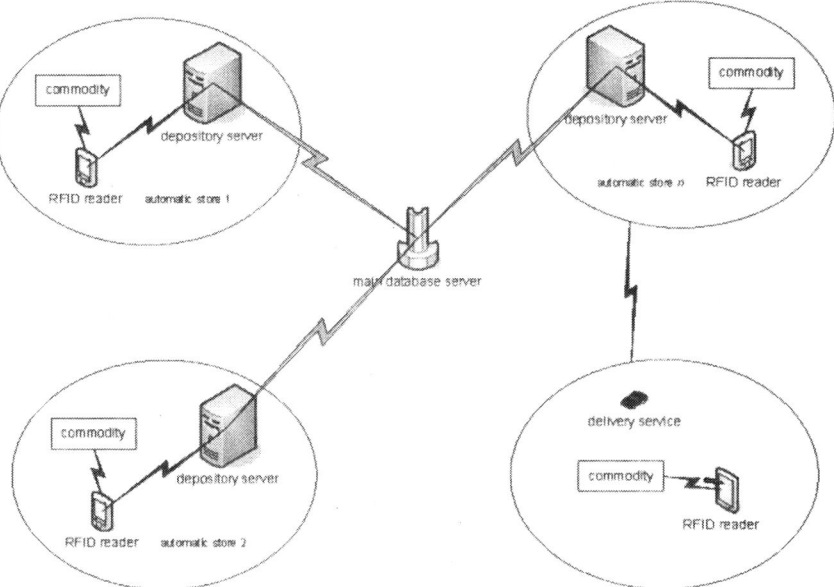

Fig. 1. Block diagram of temperature tag and reader/writer system

The main block diagram of designed tag and reader/writer system is shown in Fig. 1.

2 Preferences for frequency range below 135 kHz

This frequency range allows reaching large ranges with low cost transponders. High level of power is available to the transponder. The transponder has low power consumption due to its lower clock frequency and often sleeping a standby mode of operation.

Miniaturized transponder formats can be achieved due to the use of ferrite coils in transponder. Low absorption rate or high penetration depth in non-metallic materials and water are available due to lower frequencies. Basic block diagram is shown in Fig. 2.

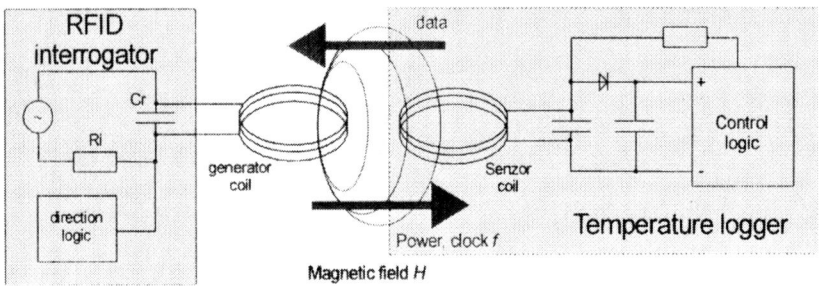

Fig. 2. Data transfer, and inductively coupled RFID interrogator and temperature data-logger block diagram

3 Communication patterns

An inductively coupled data logger comprises an electronic data-logging device and a large area coil that functions as an antenna. Inductively coupled logger is almost always operated in passive mode of data transmitting between an RFID tag and reader. This means that all the energy needed for the data transfer to the temperature logger has been provided by the reader. For this purpose, the reader's antenna coil generates a strong, high frequency electromagnetic field, which penetrates the cross-section of the coil area and the area around the coil.

4 Main data-logger features

Temperature is recorded using a temperature tag at user defined time intervals. The temperature tag can be programmed so that when the memory is full it either stops further recording or continues recording by overwriting the earliest of the previously recorded data.

Typical stored information contains: date and time stamp, temperature, temperature tag unique ID.

Recorded information can be transferred to a reader/writer and then to a PC or a PDA with wired or wireless connection to a reader/writer.

Temperature can be displayed graphically and the zoom functions allow focus on time periods where the temperature exceeds parameters.

The tag is a self-powered facility working like a wireless temperature sensor. The tag consists of a temperature probe and active part with active RFID technology. It is powered by an internal battery. The tag transmits an RF signal on demand at a pre-set time-interval. Tag life is estimated at several months depending on pre-set period of a transmission. The transmission interval can be configured via wireless connection by a reader/writer. The lifespan of the tag ends when the battery life is exhausted. Battery status can be inferred by interrogating the internal tag status value.

The lifespan of the tag can be increased by delayed turning on by the first communication attempt transmitted by the reader/writer.

5 Temperature data logger tag

This part of tag is used for temperature measuring, measured data storing and real time clock timing. Temperature is acquired in pre-set intervals. Tag can work in two basic modes as Data Logger and Out of Limit Values Logger.

- *Data Logger* - Temperature is measured in pre-set intervals. All data is stored into the internal memory. In this mode there is stored only a start time. Number of measurements depends on memory size. Data logging principle is shown in Fig. 3.
- *Out of Limit Values Logger* - If temperature is out of range, time stamp and temperature data are stored into the internal memory. Number of stored values depends on memory size. Out of Limit Values Logger principle is shown in Fig. 4.

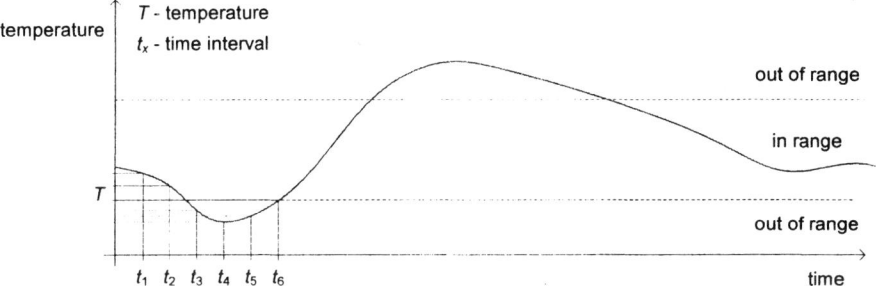

Fig. 3. Temperature data logging principle

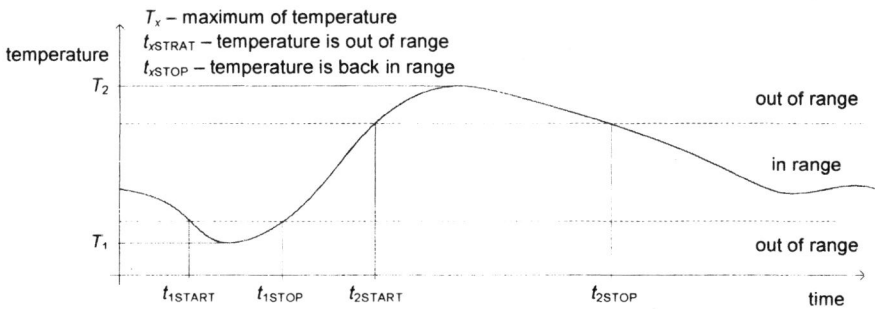

Fig. 4. Out of Limit Values Logger principle

6 Application areas

The temperature tag is designed for usage in shipping containers, dairy industry, medical applications, fuel industry, refrigerated loads, agricultural industry, refrigeration monitoring and dangerous goods areas, anywhere temperature monitoring is required. On fig. 5 are both wireless and wired types of data logger with temperature sensor.

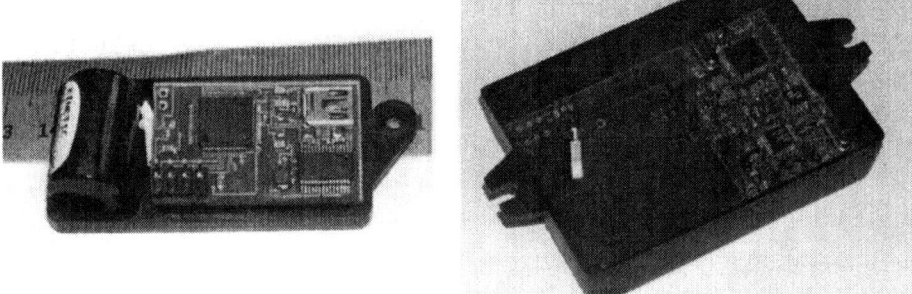

Fig. 5. Wired and wireless version of data logger

This set of applications increasingly means that temperature is a critical process and quality assurance factor for many industries.

The tag and reader/writer system is designed around a few simple ideas:
- Information is only good if it's accurate.
- Data is only useful if it's easily understood.
- Collected data is secure.
- A system is only good if it gets used.
- A system must be affordable.
- Simple setup and operation requiring minimal operator training.

7 PC based communication software

PC based communication software is used to setting up wired or wirelessly connected data logger. If wired connection is used, temperature data logger is connected directly via USB bus. In wireless connection RFID reader is necessary.

By this software tool is possible to set all selectable parameters. Setting up internal real time timer, temperature limits and work mode. One part is used for security setting.

When the connection is established all stored data is transferred to PC. In next step data can be saved to simple text file, plot to BMP picture or if the SQL server is available data is stored to the server database system. Main software window is on Fig. 6.

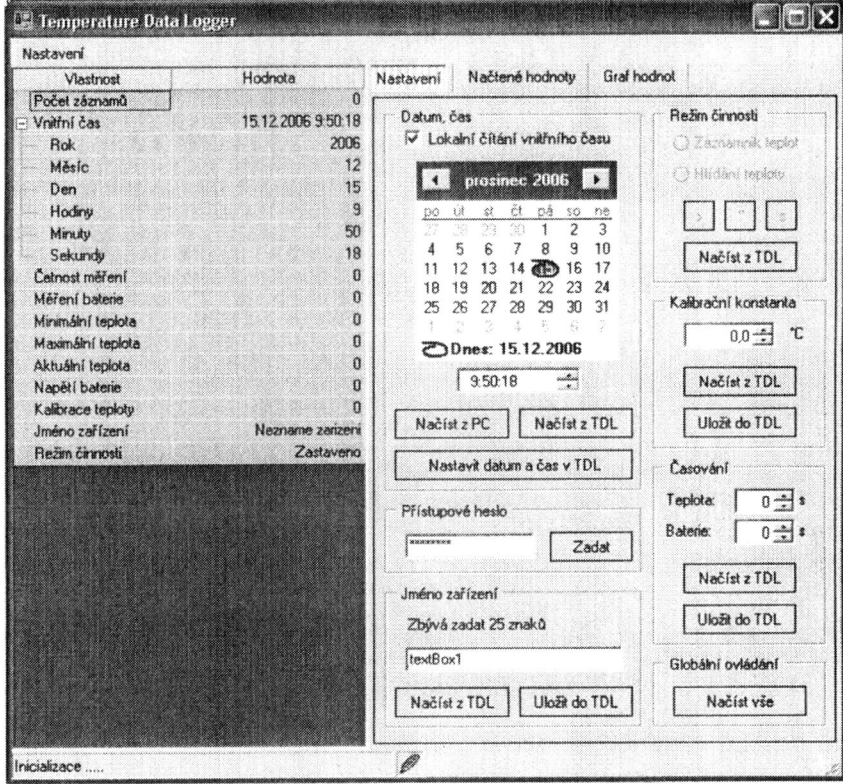

Fig. 6. PC based communication software

8 Summary

This contribution deals with a new method of portable wireless temperature data logging, which can be used for certified proof of history of temperature during monitored time interval. It fits to requirements defined by the transport of biological active substances. This set was applied for continuous temperature digital measuring and data logging during a transport of chilled electrochemical TFT sensors with applied enzymes. This type of sensors is highly sensitive and easily degradable when the temperature overcomes the maximum temperature limit.

Acknowledgement. The research has been supported by the Czech Ministry of Education in the frame of MSM0021630503 *MIKROSYN New Trends in Microelectronic Systems and Nanotechnologies*, by the Czech Science Foundation as the project GACR 102/05/0869 *ADC New Principles of Intergrated Low-Voltage and Low-Power AD Converters in Sub-Micron Technologies* and by the Ministry of Industry and Commerce of the Czech Republic in 2A-1TP1/143 *MEMS Research of New Mechatronic MEMS Structures Appropriate for Pressure Measurement.*

References

1. Data sheets and manuals on www.microchip.com.
2. Data sheets and manuals on www.rfm.com.
3. M. Ilyas and I. Mahgoub, (Eds.), *Handbook of Sensor Networks: Compact Wireless and Wired Sensing Systems*, CRC Press LLC, Boca Raton, FL, USA, 2004.
4. M. Sveda, P. Benes, R. Vrba and F. Zezulka, "Introduction to Industrial Sensor Networking," A book chapter in: M. Ilyas and I. Mahgoub, (Eds.), *Handbook of Sensor Networks: Compact Wireless and Wired Sensing Systems*, CRC Press LLC, Boca Raton, FL, USA, 2004.
5. D. Boling, "Programming Microsoft Windows CE .NET" Third edition, *Handbook of Sensor Networks: Compact Wireless and Wired Sensing Systems*, Microsoft Press, Redmond, Washington 98052-6399, USA, 2003, ISBN 978-0-7356-1884-8.

Zigbee-Based Wireless Distance Measuring Sensor System

Ondrej Sajdl[1], Jaromir Zak[1], Radimir Vrba[1]

[1] Department of Microelectronics, Brno University of Technology, FEEC,
Udolni 53, 602 00 Brno, Czech Republic
sajdl@feec.vutbr.cz

Abstract. A cost effective ZigBee-based wireless parking sensor system is presented in this article. Ultrasonic transmitter and receiver are used for distance measurement. An evaluation, control and wireless data transfer are realized by one chip - Panasonic hybrid chip PAN4450 (8-bit microcontroller + IEEE 802.15.4 radio). System consists of four independent distance sensors and one central display and control unit.

Keywords: ZigBee, IEEE 802.15.4, sensor, wireless, parking system.

1 Introduction

Increasing numbers of the cars cause a decrease of parking places especially in larger cities. It becomes more and more difficult to find parking place. Car-parking is for many drivers an uneasy task and could be potentially danger. Car crashes are very common in connection with this maneuver. Car producers are aware of this situation and more expensive and luxurious new cars are equipped with parking system very often. But there is majority of older and cheap cars which do not have such system. It is possible to buy system additionally, but there are lots of problems with installation, especially where to pull cables, how to connect system to car battery, etc. This article describes a possible cost effective wireless solution of this problem.

2 Wireless Parking Sensor System

Our parking system consists of five independent units. Four intelligent distance meters are designated for car mount at usual positions: front left (FL), front right (FR), rear left (RL) and rear right (RR) part of car fender and one central unit (CU) inside car with a 7-segment display (see
Fig. 1). The black arrows show supposed placing of the sensors into the car fender.

Please use the following format when citing this chapter:

Sajdl, O., Zak, J., Vrba, R., 2007, in IFIP International Federation for Information Processing, Volume 245, Personal Wireless Communications, eds. Simak, B., Bestak, R., Kozowska, E., (Boston: Springer), pp. 403-409.

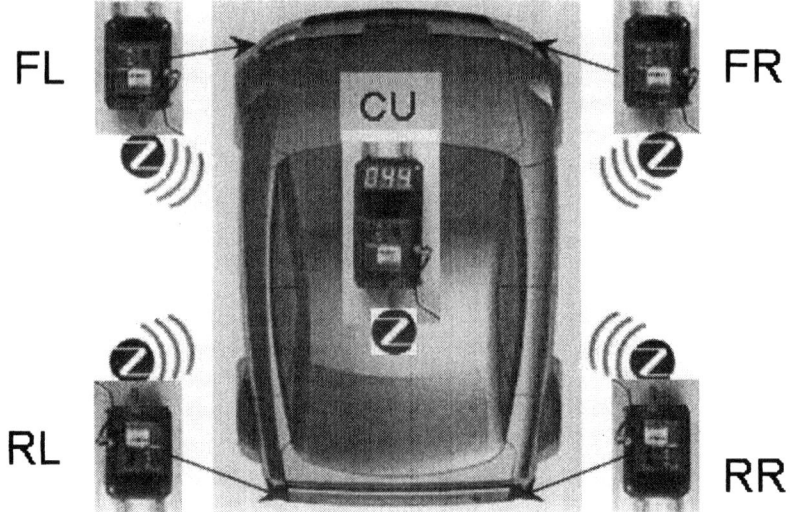

Fig. 1. Wireless parking sensor system – a car with a four wireless ultrasound distance meters (FL, FR, RL, RR) and the one central unit (CU).

2.1 Sensor Units

The sensor unit is an intelligent sensor for distance measurement with a wireless interface. It is battery powered (9 V) and it is intended for placing into the car fender (see Fig. 2).

Fig. 2. Sensor unit prototype – an antenna is on the left side, a hybrid chip with MCU and radio is in the middle and a pair of ultrasound meters (receiver, transmitter) is on the right.

Each sensor unit consists of three parts:

1. evaluation and control unit (8-bit microcontroller + software),
2. ultrasound distance meter,
3. IEEE 802.15.4 radio and antenna for wireless data transfer + software.

Evaluation and Control Unit. The heart of all units is a hybrid chip Panasonic PAN4450. It combines 8-bit Freescale HCS08 MCU and IEEE 802.15.4 radio chip Freescale MC13192. The MCU is internally connected to radio through SPI (Serial Peripheral Interface). The MCU contains all software for ultrasound distance measurement and a wireless communication driver based on a SMAC (see Wireless Data Transfer chapter). The hybrid chip, the antenna and the supporting circuits are located on a separate development board (our production) connected to other circuits (ultrasound distance meter) through a simple connector.

Ultrasound Distance Meter. The ultrasound distance measurement principle was chosen based on an extensive literature [3] [4] [5] and Internet search. We've investigated technologies like infrared light, laser, but we've found ultrasound as a good compromise between the cost versus the accuracy and reliability.

The distance meter is realized using an ultrasound transmitter-receiver pair and transmitter and receiver circuits. The transmitter sends periodically ultrasonic waves of 40 kHz for the 1 millisecond and pauses for the 62 milliseconds. The core of timing circuit is based on well-known IC NE555. The inverter is used for the drive of the ultrasonic sensor (see Fig. 3). The two inverters are connected in parallel because of the transmission electric power increase. The phase with the voltage to apply to the positive terminal and the negative terminal of the sensor has been 180 degrees shifted. It is because cutting the direct current with the capacitor, about twice of voltage of the inverter output are applied to the sensor.

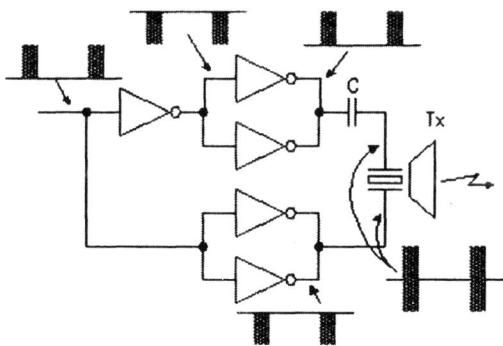

Fig. 3. Ultrasonic sensor drive circuit

The ultrasonic signal that was returned from the measurement object and received with the reception sensor is 1000 times (60 dB) of voltage amplified with the operational amplifier with two stages. It is amplified 100 times at the first stage (40 dB) and 10 times (20 dB) at the next stage (see **Fig. 4**). The output of the detection circuit is detected using the comparator. Signal is then converted into digital pulses using SR (the set and the reset) flip-flop. This circuit is the gate circuit to measure the time which is reflected with the measurement object and returns after sending out the ultrasonic wave. The set condition is the time which begins to let out the ultrasonic with the transmitter. It uses the transmission timing pulse. The reset condition is the time which detects the signal with the signal detector of the receiver circuit. That is, the time when the output of SR is in the ON condition and it becomes the time that returns after letting out the ultrasonic. Output of this SR is connected to MCU digital input. The MCU measures the duration of the pulses and computes the resulting distance.

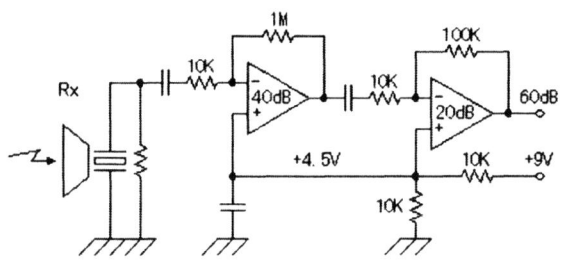

Fig. 4. Receiver circuit – signal amplification

The distance is computed based on knowledge that the speed of sound is approximately 331 m/s. This speed depends on the temperature. However, an error of approximation causes an error in distance measurement for the small distances (tens of centimeters) in maximal range of 2 centimeters in positive and negative ranges for temperature range -10 to 10 °C (see Table 1).

Table 1. The speed of sound for each temperature [2]

Temperature [°C]	Speed of sound [m/s]
-10	325.5
0	331.5
10	337.5

Our assumption is valid only for relatively small temperature ranges. It is true in the most European countries. However, this problem could be easily solved in next PCB version by adding a temperature sensor.

The distance is computed as follows:

$$d = 331 \cdot \frac{t}{2}, \quad (1)$$

where d is a distance in meters and t is a period in seconds between sending and receiving a pulse. The time must be divided by two because a reflected signal is measured.

Wireless Data Transfer. ZigBee is a new low rate wireless network standard designed for automation and control networks. The standard is aiming to be a low-cost, low-power solution for systems consisting of unsupervised groups of devices in houses, factories and offices. Expected applications for the ZigBee are building automation, security systems, remote control, remote meter reading and computer peripherals. In other words it is perfect for the presented application. A Bluetooth was also considered but it has too high power consumption and is not suitable for long-life battery powered systems. Physical radio (IEEE 802.15.4) is realized using Freescale MC13192 2.4 GHz low power transceiver.

A serious problem we've met was in the software – higher software layers of ZigBee stack are not free. The cost of ZigBee stack is a matter of thousands of dollars. But Freescale offers MAC (Medium Acces Control) layer software libraries for free. This software is called SMAC (Simple MAC). It contains API (Application Programming Interface) for complete control of MC13912 radio, including security.

Our developed software based on SMAC allows half-duplex communication with central unit (star network topology) and handles sleep modes with time synchronization. The distance sensor units are periodically awakened and it waits for possible commands from the central unit. If there is no command in a certain time frame it falls asleep. Even though software development was the most time consuming task of the whole project, we are far from implementing all ZigBee higher layers. Nevertheless we do not need a full compatibility with ZigBee standard, our solution at this stage uses "only" physical radio layer according IEEE 802.15.4, MAC layer and proprietary network and application layer implementation based on SMAC (see
Fig. 5). ZigBee compatibility has sense when there is greater number of cheap standardized ZigBee devices on the market. For example we are considering possibility to use a smartphone or a personal digital assistant (PDA) instead central unit. But there are no such devices on the market at the present time.

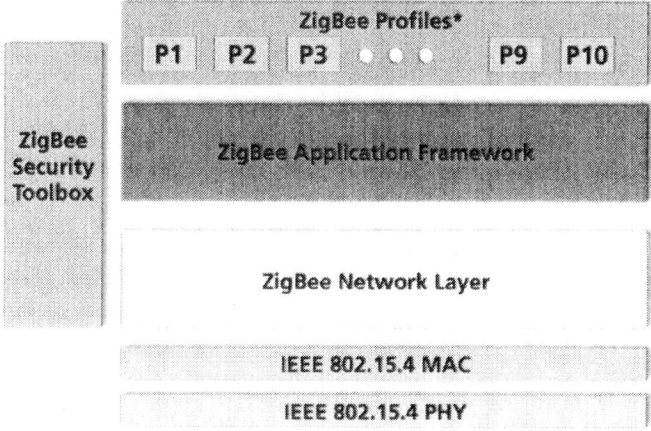

Fig. 5. ZigBee stack [1] – our system has full physical radio layer, MAC layer and proprietary implemented network and application layer

2.2 Central Unit

The central unit is very similar to sensor unit except for missing ultrasonic circuits. Additionally it has 7-segment display for displaying measured distance. The display is connected directly to the MCU's digital outputs; there are no additional circuits. The display is controlled directly by software driver stored in MCU. The central unit is also battery powered, but because of a high power consumption of a segment display it can be also connected to a car battery (12 V) through a car fuser. The display shows a value from one sensor unit and there is a button for units switching. However, the central unit receives the values from all sensor units and if the distance is smaller than predefined value (20 centimeters) the sound alarm is triggered.

A possible extension of the central unit could be taking advantage of programmable Bluetooth equipped devices - smartphones and PDAs. There is a plenty of such devices on the market and both are often used for car GPS navigation. In this case, the central unit is modified to simple ZigBee-Bluetooth converter so we can "make the best of both worlds" – ZigBee low power consumption and Bluetooth wide spread. The measurement, the alarms and the distance displaying could be solved by software on a Bluetooth device side.

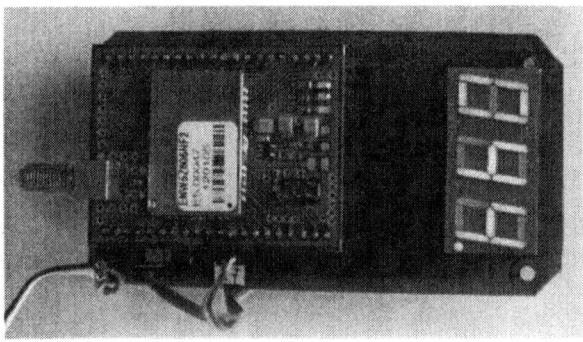

Fig. 6. Central unit prototype – similar to sensor unit with segment display; it doesn't have ultrasound circuits.

3 Conclusion

It has been presented the original design of the low power wireless parking sensor system with an extremely reduced cost. It is reliable system with quick and easy installation. The system might be easily extended. Planned support of ZigBee certified and Bluetooth equipped devices will extend system usability in future.

Acknowledgments. Research has been supported by Ministry of Education of the Czech Republic in the frame of Research Program MSM0021630503 MIKROSYN and by the Czech Science Foundation as the project GACR 102/05/0869

References

1. ZigBee Aliance: ZigBee specification.WWW: http://www.zigbee.org/
2. Wikipedia: Speed of Sound. WWW: http://en.wikipedia.org/wiki/Speed_of_sound
3. Ultrazvukový dálkoměr. A Radio (2001) 2- 5, ISSN 1212-1843.
4. Benet G.1; Blanes F.; Simo J.E.; Perez P.: Using infrared sensors for distance measurement in mobile robots. Robotics and Autonomous Systems, Volume 40, Number 4, (30 September 2002) 255-266
5. Meggitt, B. T.; Palmer, A. W.: Distance measurement using laser diodes. FOS, Volume 33 (1999) 21-31

A proposal of a Wireless Sensor Network Routing Protocol

Cláudia Barenco Abbas[1], Ricardo González[1], Nelson Cardenas[1], L. J. García Villalba[2]

[1] Universidad Simon Bolivar (USB)
Departamento de Computación y Tecnología de la Información
Apartado Postal
Caracas, 1080 Venezuela
barenco@ldc.usb.ve; rgonzalez@ldc.usb.ve; ncardenas@ldc.usb.ve

[2] Grupo de Análisis, Seguridad y Sistemas (GASS)
Departamento de Ingeniería del Software e Inteligencia Artificial-LSI
Facultad de Informática, Despacho 431
Universidad Complutense de Madrid (UCM)
C/ Profesor José García Santesmases s/n Ciudad Universitaria
28040 - Madrid, Spain
javiergv@fdi.ucm.es

Abstract. This paper presents a proposal of a routing protocol to Wireless Sensor Networks, called SHRP (*Simple Hierarchical Routing Protocol*), whose primary objective is to save battery energy. SHRP also provides both reliability and a load balance solution.

As a novel proposal, it is a proactive protocol that chooses efficient routes, by selecting just the nodes that can contribute to extend the network lifetime. In addition this protocol is also able to inform to a central point about any possible disconnection caused by a reduction of battery power or a long interference period.

The data defined by SHRP protocol can be aggregated during the sending and forwarding tasks. Redundant data are not sent, which contributes to energy saving. This protocol uses the IEEE 802.15.4 under of CC2420 radio chips and it has been implemented on TinyOS operation system.

Keywords: 802.15.4, routing protocol, wireless sensor network

1 Introduction

Monitoring and controlling activities in different systems are fundamental task for improving systems performance. Advances in micro-electronics have allowed, in a great scale, the integration of sensor, microcontroller and communication components in small footprint and low cost devices. These features are making possible the establishment of monitor and control loop in places were it was not possible or affordable until recently.

The sensor node can measure data from any physical system and sent it, usually via radio transmitter, to a command center or sink node, either directly or through a number of

---

Please use the following format when citing this chapter:

Abbas, C. B., González, R., Cardenas, N., Villalba, L. J. G., 2007, in IFIP International Federation for Information Processing, Volume 245, Personal Wireless Communications, eds. Simak, B., Bestak, R., Kozowska, E., (Boston: Springer), pp. 410-422.

communication and data concentration devices (or gateways). The decrease in the size and cost of sensors devices has increased the interest in using sets of disposable and unattended sensors. This has led to intensive research addressing with the potential of collaboration among sensors in data collecting, processing and management of the sensing activities, within the last few years.

The main goal of a WSN (*Wireless Sensor Networks*) is the transmission of data to a base station in an energy efficient. Therefore this networks are concern with lengthening battery life, reducing networks maintenance cost and providing a well connected topology unlike TCP-IP architecture. WSN's has no traditional protocol, because of they limited storage capacity, processing power and battery life, so there is some recent works that try to propose new protocols [1].

2 Related Works

Wireless Sensor Network are subject to different requirement depending on the physical environment were they are been instrumented as well as on the network characteristic. The following are some routing protocols for WSN classified according to its network architecture.

2.1 Flat networks

In flat networks all network nodes play the same role. The simplest flat protocols are based on either flooding or gossiping mechanism. Flooding involve sending message to any network node, even to destiny by sending data in any direction. This strategy is very simple to implement but it involves sending a lot of messages at the expense of large power consumption, which is not advisable in WSN. The Gossiping strategy consists in sending message, not to all nodes, but to a randomly selected subset of neighbor, this can reduce power consumption however, the arrival of message at lower transmission power cost is not warranted.

SPIN [2] (*Sensor Protocol for Information via Negotiation*) is designed to send sensed data to any node in the network reducing the amount of data messages. It uses a 3-way handshake with meta-data information to negotiate with each neighbor before transmitting data to them. This avoids the transmission of redundant data in the network.

A node 'A' intended to transmit the sensed information creates an advertising (ADV) message to send the information to its neighbor. If any neighbor 'B' wishes to receive the actual data, it sends a request for data message (REQ). Finally 'A' node sends data messages (DATA) containing actual sensor data to any node B requesting information.

In directed diffusion routing algorithm [3] a human operator's query would be transformed into an interest for named data that is diffused towards nodes in regions X or Y. Each node in the network that forwards this interest message sets up a gradient towards the source of the interest.

A reinforcement process is used to weight the gradients based on their qualities, e.g. loss ratio or hop count. When the sensors report the required information, a message is being returned along the reverse path of interest propagation towards the sink. Intermediate nodes might cache, transform, or aggregate data by combining reports from several sensors, and may direct interests based on previously cached data.

Depending on the number of nodes and the number of active interests, the utilization of the network can be very low comparable to other traditional routing approaches.

AIMRP protocol [4] organizes the network into tiers around a sink, and routes packets by progressively forwarding them to a closer tier from the sink node. This strategy do not requires coordination between the nodes which can repeatedly shut their radio modules off when not in use, independently of the nodes, while satisfying sensor-to-sink latency guarantees.

2.2 Hierarchical networks

LEACH [5] (*Low-Energy Adaptive Clustering Hierarchy*), is a hierarchical protocol built on clusters of nodes that use randomized rotation of local cluster base stations to evenly distribute the energy load among the sensors in the network. Cluster formation is a distributed task based on a subset of node randomly selected as Clusters Head (CH), which aggregate information of its cluster in order to reduce the amount of messages traveling through the network.

In this protocol operation two different phases can be identified: (i) during configuration phase CH is being elected and cluster being conformed; (ii) in transmission Phase packet is being sent from sensor nodes to a cluster head and then to a Base Station (BS) or sink node. In configuration phase CH nodes send a broadcast message to other nodes to tell them that it is a new Cluster Head. Based on signal strength each node select to which cluster it's belong to.

TEEN [6] (*Threshold sensitive Energy Efficient sensor Network*) is other hierarchical protocols for reactive networks, that respond immediately to changes in the relevant parameters. In this protocol a clusters head (CH) sends a hard threshold value and a soft one. The nodes sense their environment continuously. The first time a parameter from the attribute set reaches its hard threshold value, the node switches on its transmitter and sends its data. The nodes then transmits data in the current cluster period if the following conditions are true: the current value of the sensed attribute is greater than the hard threshold, and the current value of the sensed attribute differs from sensed value by an amount equal to or greater than the soft threshold. Both strategy looks to reduce energy spend transmitting messages.

The main drawback of this scheme is that, if the thresholds are not reached, the nodes will never communicate; the user will not get any data from the network at all and will not come to know even if all the nodes die. Thus, this scheme is not well suited for applications where the user needs to get data on a regular basis.

DIRq [7] is a protocol that decrease used power by reducing the quantity of messages sent. If a sensor has been registered and sends a value 'V1' for a desired parameter, and in the following measurement period it detect the same value or a similar one, between de (V1- δ, V1+ δ) interval then it no do send it again to sink. If sink if not receiving message of a specific node it assume that it sensed value is not been changing too much to be reported.

Flat routing algorithm such as SPIN (*Sensor Protocol Information Negotiation*) and direct diffusion strategies, seems do not have many application in wireless network sensor of small sizes, and short distances. In a simple network structure in which are not needed many hops to communicate information between sensor nodes, make no sense to use a protocol that make negotiations of complex routes of many hops.

We are interested in considering a protocol that could deal with three different aspects: battery available, number of hops and link quality to guarantee the arrival of messages in the sink node in a energy saving way. We could not find any routing protocol to WSN that use these three parameters together are at the same time be concerned about energy saving and reliability features. By all this reason we propose a new routing protocol.

3 SHRP Architecture

SHRP protocol is concerned about topology maintenance that is directed related to the reliability of data delivery. To arrange this it makes use of metrics like local battery available and link quality between neighbor nodes in choosing the best route into the sink node.

Also is concerned about energy saving as not all periodic data is sent to the sink, as is a concern that transmission is the task that more wastes energy in sensor networks [8]. SHRP protocol just sends data that have not changed from the last sensing data. The coordinators nodes can aggregate various data messages and send just one message, in this manner SHRP protocol have also an energy saving behavior.

SHRP provides a load balance scheme during the creation of best routes groups, so no always will be choose the same best route.

To be flexible and to contribute with new metrics for others routing protocols, SHRP architecture uses SP [8] (see figure 1), a specific unifying protocol of TinyOS operating system.

SP allows network protocols to choose neighbors wisely, taking into account information available at the link layer, providing a great modularity.

SHRP protocol periodically monitors the battery lifetime and link quality, cutting off from the routing table nodes that can not contribute in maintaining a well connected topology.

The proposed protocol takes care about link quality, cutting off neighbor nodes from the routing table nodes that have the average link quality indicator below a minimum threshold. This threshold is related to the IEEE LQI indicator [9].

SHRP protocol cuts off from the routing table also neighbors nodes that have the RSSI (*Received Signal Strength Indicator*) value below a minimum threshold [10].

Also, SHRP protocol cuts off nodes that do not have battery available, at least to execute what we call "minimum task cycle" [11] (see Equation 1).

Minimum Task Cycle = CCA + Sensing Task + Transmission task + Reception task + Idle Period task (1)

All the tasks of cutting off neighbor nodes, showed before, represent that if a node does not have sufficient battery or have a bad link quality, caused by interference, multipath or path loss it will not participate as a route node in the choosing of the best route.

To provide the topology maintenance SHRP protocol defines new metrics that SP does not provide. To be aware about the state of the energy of the nodes, SHRP defines a metric called "minbattrem" that keeps information about the energy available in each node along each route until the sink node.

SHRP protocol tries to choose the route that gives more reliability, this mean, the route that give the mayor energy available along all the possible routes until the sink node.

Another metric defined by SHRP protocol is called "nhops". With this metric SHRP can choose a route that takes into account the number of hops until the sink node. As the transmission is the task that spends more energy, SHRP tries to choose the route that pass through the minor number of hops, saving more energy.

Newer radios that are based on IEEE 802.15.4 standard such as CC2420 implement a parameter called link quality indicator (LQI) which is believed to be a better indicator than RSSI [2]. SHRP protocol uses the metric called "AvgLQI" that represents this parameter.

Protocol designers looking for inexpensive and agile link estimators may choose RSSI over LQI [9], but RSSI at the edge of the threshold of -87 dBm does not have a good correlation with PRR, so SHRP uses both of the metrics, LQI and RSS, in choosing the best route.

LQI and RSSI metrics are gotten from the link layer of the wireless network. In our experiments we are using the CC2420 radio chip [12].

Figure 1 shows the interaction of SHRP protocol with SP to get information about neighbor nodes and to decide about the best route into the sink node. The metrics shown in bold font are the new ones defined by SHRP protocol. In this manner, other protocol that makes use of any of these metrics will have them already available in SP module.

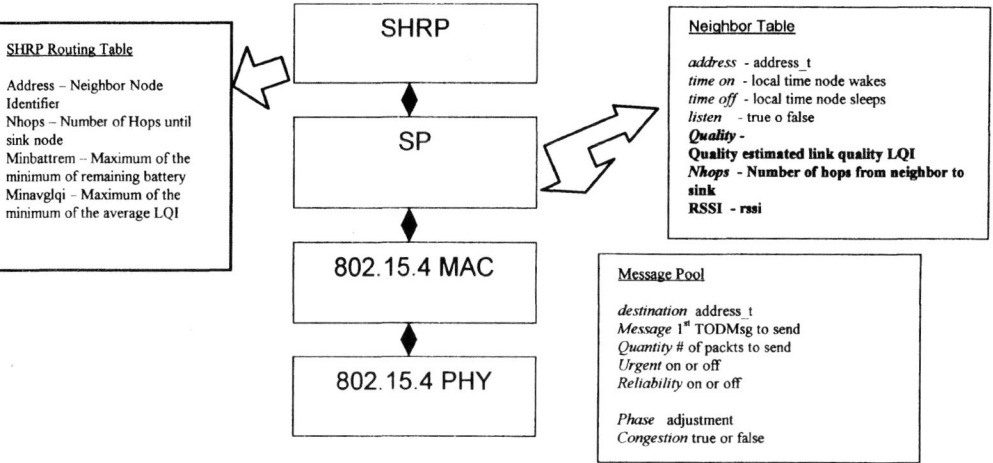

Fig. 1. Interaction between SHRP protocol and SP

SHRP protocol defines three kinds of sensing data messages: (i) periodic; (ii) alert; (iii) alarm [13].

The sink node sends the Query Message, that is explained later, with the information that the monitoring system is interested to collect periodically. Based on this message, the sensing nodes know what and when they have to sense.

The alert message is sent when the sensed data value is above an average value, specified by the monitoring system [13].

The alarm message is sent when the value of the data sensed is below a minimum or above a maximum threshold, also specified by the system [13]. This message is sent with the Urgent and Reliability bits 'on', so they have priority over any other message and the sender waits for an ACK message, to confirm that the alarm message have arrived.

4 SHRP Functional Description

There are two different types of nodes defined by SHRP protocol: sensor nodes and coordination nodes.

Sensor Node (SN): This is a node that uses one o more sensors to collect periodically a measure from a physical system, which have to be sent to its coordinator node. Between each period of collecting data the Sensor Node can sleep.

Coordinator Node (CN): This node will have two main functions: (i) route data messages coming from Sensing Node or from another neighbor Coordinator Node; (ii) aggregates messages before sending the received data, in order to decrease transmissions.

4.1 Topology Configuration

SHRP is intended to be used in a sensor network that has a static topology. The static topology can be defined after a site survey process which makes some warranty about network physical connection as any SN should communicate with at least with one CN. Nevertheless, every SN should be associated with only one CN and statically its know its CN node.

Each CN can have one or more neighbors, some of them could be reached directly from Sink. Each message sent will arrive to Sink through this CN that should be part of the best route find by SHRP protocol

During network deployment phase some politics could be taken in order to guarantee components redundancies. Even different nodes could be connected to the same sensors (if sensing device is expensive or difficult to find). If we can aggregate additional node to establish redundancy routes, this can impact network topology, increasing network lifetime. This is feasible due to the low cost of motes devices.

Redundant CNs can be used as secondary coordinators (CNsec) that must be not far from its associated primary. Each CNsec could be in sleep mode and periodically it should check if its primary CN is alive. If it does no receive any response from CNsec during a timeout time, it assumes that its primary CN has gone, so has to do an auto configuration as a primary CN.

4.2 Messages

We describe two different types of messages: data and control. Control messages are used to define the tree routes of the SHRP protocol and data messages transmits monitoring system information to the sink node.

Control Messages

To provide reliability, SHRP protocol uses battery available local CN information to choose the best route. Others parameter that SHRP uses are RSSI, LQI and number of hops into the sink node. All of them are collected in relation in a peer to peer fashion between neighbor nodes.

Number of hops is calculated based on the Query Message sent periodically by sink node. Further is showed more details about it.

RSSI is an indicator of signal strength and can be used in a simple way to determine if there is a good Reception Packet Rate (PRR). If RSSI is below some threshold, in [10] a minimum threshold of -87dbm is proposed, we can say that we have a bad quality link. It can be seen as a good estimator considering that sensibility radio use to be near of 90dbm. Although RSSI is a metric calculated in the reception of frames, it can be used as a metric to the transmission of frames, as new radios chips as CC2420 have an symmetric quality link behavior [10].

LQI is a parameter that comprises the standard IEEE-802.15.4 and is presented in CC2420 radio devices. It represents the error rate in Chips calculated in correlation with the 8 first symbols after SFD (Start Frame 2 Delimiter) [12], so that the value of LQI has a good correlation with the PRR. As LQI changes in time [9], SHRP protocol uses an average of LQI (AvgLQI) calculated within a window frame of a reasonable time.

The principal goal of SHRP protocol is to find a route that increase the network lifetime, that means maintains the connectivity at routing level, and provides a good arrival probability to the Sink Node.

Route selection can not be based only on the data of the next step, since it can lead us to select a way that offers a lower delivery guarantees in later stage. So we propose that some metrics used by the protocol, would be calculated using information from every node that will participate in the selected route.

For example in the battery available metric (battrem), each node must calculate the lower battery available for each one of the possible route to reach Sink Node and choose the greater among them (MaxMinBattAvailable). The same procedure is used with RSSI and AvgLQI metrics, calculating MaxMinRSSI and MaxMinAvgLQI, respectively.

Each CN node sends periodically to neighbor nodes a message called "Network Information Message" (NMI) (see figure 2) which includes updated data of all the metrics mentioned before. As soon as this message is received, CN node can updated its routing table (see figure 1).

The NMI message format is:

<Address, minAvgLQI, minBattrem, RSSI, AvgLQI>

SHRP protocol keeps in its routing table those nodes that show to be a good neighbor in relation to AvgLQI and RSSI metrics. To verify this feature some thresholds are defined for each metric and every CN must reach these thresholds to be considered a good neighbor.

Before sending its NMI message, the CN node does an auto evaluation of its battery available metric (battrem). It the value is below a threshold it does not send the NMI message to its neighbor, as it is not a good neighbor node to others nodes. This guarantees that when calculates the best route, CN node will have just neighbors with a minimum battery available to route messages.

In an intuitive form we can say that a good neighbor is a CN node that have enough battery to guarantee the shipment of a messages and that offer a reliable way, in terms of radio signal, to reach the Sink node.

The procedures of CN auto evaluation and the procedure of cutting off the bad neighbor are showed below.

```
            If Battrem < BattMin
                CN does not send NMI message
            Else
                For each neighbor node
                    If Neighbor(RSSI)< minRSSI
                        Cutt off the neighbor node as a neighbor
                    Else
                        If Neigbor(LQI) < minLQI
                            Cutt off the neighbor node as a neighbor
                        Endif
                    Endif
                    Mark the neighbor node as a "good" neighbor
                Endfor
            Endif
```

Some works have been made to determine reasonable values to be compared against the same metrics used by SHRP. In the case of the available battery, was done a study of battery consumption using CC2420 radio devices, in order to calculate BattMin level, that a battery should have to send a massage. It can be used to establish some comparisons in terms of energy consumption [11]. In [10] is showed a correlation between average LQI and the rate of reception of packages (PRR). This information could be used to select a reasonable MinLQI that will depend on the lower reception level that the application can reach to guarantee shipment of the package to the next node. If a node decides do not send its NMI, particularly by battery problems, it could send an special message of alert "Control Alarm Message" (CAM) to the sink node, to inform that it will leave the list of neighbors soon, and that this could cause a loose of connectivity in the network.

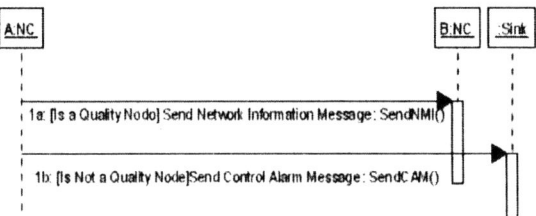

Fig.2: Messages sent to maintain the network topology

After the CN node has its good neighbor, doing an auto evaluation and cutting off bad neighbors, as explained before, it can start to calculate its best route. Each time it receives a new NMI message, it must calculate the best route.

Below is represented the algorithm to calculate the best route.

```
For each CN neighbor node
    Calculate Max(MinBattrem)
    Calculate Max(MinAvgLQI)
    Calculate(Nhops)
Endfor
Choose the neighbor that have the mayor value of Max(Min(Battrem))
If have various neighbors with same value
    Choose the neighbor that have the mayor value of Max(Min(AvgLQI))
If have various neighbors with same value
    Choose the neighbor that have the minor value of Nhops
```

The format of the routing table that uses SHRP can be shown in figure 1.

Data Messages

We have three kinds of data messages defined by SHRP protocol: periodically, warnings and alarm.

Lets suppose that a monitoring system needs to obtain information from sensors every time 't' and want to be advertised urgently it this value is below a maximum o minimum threshold. SHRP defines a periodically message, Query Message and Query Response Message, to advise sensor nodes about the monitoring system intention.

In SHRP protocol the Query Message has the following structure:

<Address, Variable, Period, Min, Max, Variance, Timestamp>

The Sink node will send these parameters through a Query message, by flooding, which guarantees that it will arrive to every SN node (see figure 3).

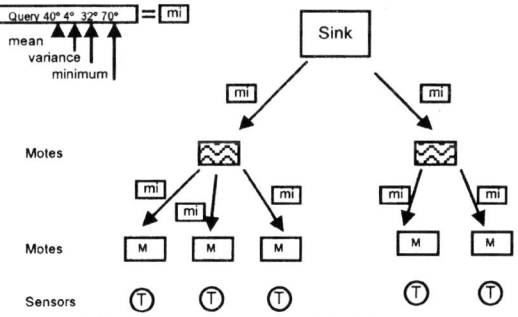

Fig.3: Query message sent by Sink Node

A sequence diagram of the Query Request Message is shown in figure 4.

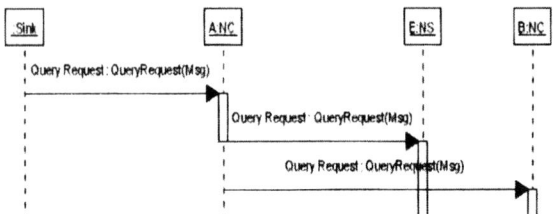

Fig. 4: Query diffusion with parameter that should be measured.

Each node SN based on the regularity of the value to be measured verifies the value sensed and determines if it has changed with respect to the last sensing. If it has changed, send a "Query Response" to its CN, which will be responsible to route the message in direction to the Sink node. If the value sensed keeps constant, the SN node will not send any message to Sink.

As a CN node cannot distinguish between the loss of messages of data of a SN or if the SN is still obtaining the same sensing values, SHRP protocol defines that at each 3 cycles of sensing, SN should send the Query Response message. In this ways the CN node can detect problems with the SN, and generates an alarm message to the central system.

The coordinator node can aggregate various data of different SN and send them as a single message reducing the number of transmissions. It can apply this same criterion in case it must route a received message of another CN.

A policy of shipment of obligatory messages each n cycles allows to detect problems and to take actions to correct the connection on the network.

Each SN node must maintain a sensing table for each of its sensors, if it has more than one sensor. The table has the following format:

$$<Variable, Period, Min, Max, Average, Variance>$$

Another kind of data messages of SHRP protocol are warning messages. SN node must compare the sensing value with the average of the measured values plus a variance (see equation 2). If the result is greater than the actual sensed value it has to send a warning message. Warning messages are used to inform about some anomaly of the sensed value, but is not a critical situation. Warning messages can be used to keep information about fluctuations in the value measured, which can mean a typical behavior but also can represent that it has some problem.

$$Last\_Sensed\_value > Average + Variance \qquad (2)$$

If measured values use to have great fluctuations, then a greater variance should be considered in order to avoid constant warning. When a CN receive a warning it can aggregate it with any message being sent to the Sink node.

The Alarms messages are generated when a measured value exceed a minimum or maximum threshold allowed by the monitoring system (see equation 3).

$$\text{Min} > \text{Last\_Sensed\_value} > \text{Max} \qquad (3)$$

The Alarm messages with the Reliability and Urgent bits of message pool table of SP configured to '1' as are urgent messages that must reach the Sink node (see figure 1).

5 Conclusion

SHRP protocol uses various quantitative parameters (battery storage, link quality and distance in number of hops up to the sink node) in choosing the best route. These parameter together contribute to have a reliable and energy saving routing protocol. This protocol is a novel proposal as it uses a mix of parameters in choosing the best route.
SHRP protocol enhances the TinyOS operating system as it defines new entries in the SP module neighboring table, so it generates greater modularity.
Also SHRP contributes with the management of the sensor networks nodes, advertising the monitoring system about critical situations of nodes in relation to battery available and link quality.
In the establishing of the link quality metric it not only uses the LQI indicator, but also RSSI indicator, what is a novel usage in the area of routing protocol.

6 Future Works

As future work we can mention that we are working in a definition of a network topology auto configuration protocol, using as a principal parameters in the choosing of the CN node by the SNs nodes, the LQI and RSSI values.
Also we are proposing a reconfiguration protocol for the secondary CN nodes, in case of the primary CN is not available any more, so contributing to have a topology fault tolerant protocol.

References

1. Al-Karaki, J.N. Kamal, A.E. "Routing Techniques in Wireless Sensor Network: a Survey" Wireless Communications, IEEE Dec. 2004 Volume: 11, Issue: 6. pag 6- 28
2. J. Kulik, W. Heinzelman, and H. Balakrishnan "Adaptive Protocols for Information Dissemination in Wireless Sensor Networks". International Conference on Mobile Computing and Networking archive. Proceedings of the 5th annual ACM/IEEE international conference on Mobile computing and networking. Seattle, Washington.
3. C. Intanagonwiwat, R. Govindan, and D. Estrin, "Directed diffusion: A scalable and robust communication paradigm for sensor networks," in 6th Annual ACM/IEEE International Conference on Mobile Computing and Networking (MobiCOM'00), Boston, MA, USA, August 2000, pp. 56–67.
4. Kulkarni S. , Iyer A and Rosenberg C. "An Address-light, Integrated MAC and Routing Protocol for Wireless Sensor Networks". April 26, 2005.
5. W. R. Heinzelman, A. Chandrakasan, and H. Balakrishnan. "Energy-Efficient Communication Protocol for Wireless Microsensor Networks". In Proc. Of Hawaiian International Conference On Systems Science, January 2000.
6. A. Manjeshwar, D. Agrawal, ."TEEN: A Routing Protocol for Enhanced Efficiency in Wireless Sensor Networks.", in WIRELESS, April 2001.
7. S. Chatterjea, S. De Luigi and P. Havinga, "DirQ: A Directed Query Dissemination Scheme for Wireless Sensor Networks", WSN2006 Wireless Sensor Networks. July, 2006.
8. J. Polastre, J. Hui, P. Levis, J. Zhao, D. Culler, S. Shenker, and I. Stoica, "A unifying link abstraction for wireless sensor networks". In *Proceedings of the Third ACM Conference on Embedded Networked Sensor Systems (SenSys 2005)*.
9. Shan L., Zhang J. et al, "ATPC: Adaptive Transmission Power Control for Wireless Sensor Networks. SenSys'06, November 2006. Bolder, Colorado, USA
10. K.Srinivasan and P. Levis. "RSSI is under appreciated". In Proceedings of the Third ACM Workshop on Embedded Networked Sensors (EmNets 2006), May 2006.
11. Polastre, J., Hill, J., and Culler, D. Versatile low power media access for wireless sensor networks. In *Second ACM Conference on Embedded Networked Sensor Systems* (2004).
12. CC2420 datasheet, http://www.chipcon.com/files/CC2420_Data_Sheet_1_4.pdf
13. Barenco C.; Gonzalez R., Cardenas N., "Proposta de um protocolo de roteamento para Redes Sem Fio de Sensores em Poços de Petróleo". I2TS'2006 - 5th International Information and Telecommunication Technologies Symposium. Cuiaba, Brasil. December 2006.

On the Accuracy of Weighted Proximity Based Localization in Wireless Sensor Networks

Peter Brida[1], Jan Duha[1], Marek Krasnovsky[1]

[1] University of Zilina, Faculty of Electrical Engineering,
Department of Telecommunications,
Univerzitna 1, 01026 Zilina, Slovakia
{Peter.Brida, Jan.Duha, Marek.Krasnovsky}@fel.uniza.sk

Abstract. Localization positioning is one of the basic problems in wide wireless sensor networks. The main objective of positioning process is to define the location of a sensor (node or device) from the relevant information obtained from reference nodes. These nodes already know their location. In many cases, these reference nodes are called beacons. The paper deals with proximity based location technique and their modifications. It does not belong to the most accurate techniques, but on the other hand it is low cost alternative to more expensive techniques. The objective of the paper is to give a survey of various proximity based location techniques performance. We analyze the influence of various design choices on the accuracy of localization techniques. The results of basic proximity based localization techniques are compared with modified proximity technique, which is proposed in this paper.

Keywords: Proximity based localization, localization error, weighted proximity based localization, wireless ad hoc sensor networks.

1 Introduction

In the near future, our surrounding will consist of large networks of wireless sensor nodes. Particularly, applications like environmental monitoring of water and soil require that these nodes be very small, light and unobtrusive. A wireless ad hoc sensor network (WASN) consists of a possibly large number of wireless devices is able to take environmental measurements (e.g. temperature, light, sound, and humidity). These sensor readings are transmitted over a wireless channel to a running application that makes decisions based on these sensor readings. Many applications have been proposed for wireless sensor networks, and many of these applications have specific quality of service (QoS) requirements that offer additional challenges to the application designers. Several applications have been envisioned for wireless sensor networks [1]. These range in scope from military applications to environment monitoring to biomedical applications.

2 Mobile Localization in Wireless Networks

The concept of localization is not limited to the geographic representation of physical location with sets of coordinates (latitude, longitude, and altitude). It is also applicable to symbolic location in a non-geographic sense such as location in time or in a virtual information space such as a data structure or the graph of a network.

Common to all notions of location is the concept that the individual locations are all relative to each other, meaning that they depend on a predefined frame of reference. This leads to a differentiation of the relative and absolute positioning [2].

When position information is used in reference to a geographic map or a global time reference, context information can be extended. An absolute position is given in respect to an inertial system and a reference point in this inertial system (see Fig. 1 a). On the other side, a relative position can only be given in respect to other points resolving the distances and the geometric configuration, e.g., the topology (see Fig. 1 b).

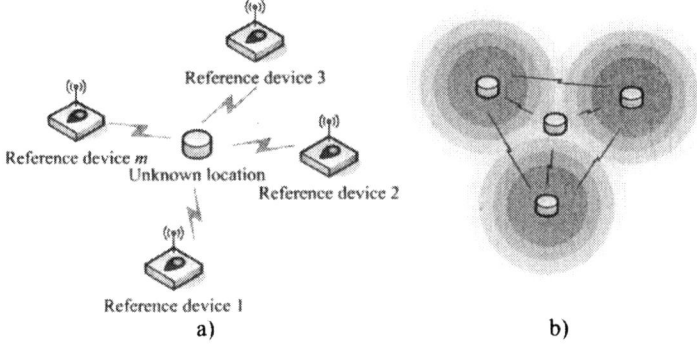

Fig. 1. a) Absolute positioning   b) Relative positioning.

When talking about physical location in the traditional way, points are usually viewed as three-dimensional coordinates $[x, y, z]$ in a Cartesian reference coordinate system.

Usually, $[x, y, z]$ coordinates by themselves are not meaningful for context-aware system services and other information needs to be associated with these position fixes. In these cases, it is important to introduce the fourth dimension - time to be able to specify where and when a certain event took place resulting in sets of $[x, y, z, t]$ for each position fix. This four-dimensional fix can then be used to put subsequent events into a context frame.

Localization techniques generally consist of three components [2]:
- Identification and data exchange.
- Measurement and data acquisition.
- Computation to derive location.

The various approaches divide these tasks differently across their system components and the data exchange between nodes is used in some cases no, or only unidirectional.

The mechanisms used for estimating location are possible to divide into two categories: range-based and range-free [3]. The range-based mechanisms are based on the indirect measurements of distance or angle between sensors. Examples of the mechanism are the following methods: received signal strength indicator (RSSI), time of arrival (ToA), time difference of arrival (TDoA), carrier phase and code measurements, ultra wide-band (UWB), ultrasound, and even visible light pulses or the angle of arrival (AoA) of a radio signal. The important thing to note is that these mentioned measurements always have errors and individual measurements are not independent of each other and are strongly influenced by the surrounding environment and the used transmission system. Because of the hardware limitations of WASN devices, solutions in range-free localization are being pursued as a cost-effective alternative to more expensive range-based approaches.

2.1 Localization in Wireless Ad hoc Sensor Networks

Wireless Ad hoc Sensor Networks (WASNs) are the special part of wireless Ad hoc networks. The localization in WASNs is a little bit different in comparison with localization in traditional wireless networks (e.g. cellular networks).

WASN is usually characterized by subjects such as vast number of devices (nodes), high overall node mobility, considerable power and resource consumption at the nodes, and moderate network sizes. WASN have brought about quite a change in the traditional connection-oriented infrastructure-dominated telecommunications world [2]. Unpredictable dynamics due to failures and changes in nodes, the environment as well as deployment in uncontrolled areas with high dynamics and possibly hostile to radio signal propagation require adaptable networking mechanisms. Other characteristics of typical WASN nodes are the limited resources available on these low power embedded systems, especially the limited transmission range and low duty cycle operation of the radio transceivers. Nodes can be reactive and are able to wake up on demand [2]. Targeted for a very long lifespan, integrated into all kinds of everyday objects and building materials, deployed once, and in many cases never collected again or decommissioned, the vast majority of nodes will form a quasistatic, multihop network topology.

The problem of localization, i.e., the determination, where a given node is physically located in a network, is a challenging one, and yet extremely crucial for many of these applications. Practical considerations such as the small size, form factor, cost and power constraints of nodes preclude the reliance on Global Positioning System (GPS) in all nodes of these networks.

2.2 Proximity Based Localization

Proximity based localization belongs to the group of range-free localization. Localization using proximity measurements is popular, when low cost takes precedence in priority over accuracy. Since, messages necessarily pass between neighbors, there is no additional bandwidth required to proximity. Proximity measurements simply report whether or not two devices are 'connected' or 'in-range'.

However, the term 'in-range' may mislead readers to believe that proximity is purely a function of geometry - whether or not two devices are separated by less than a particular distance. In fact, proximity is determined by whether or not a receiver can demodulate and decode a packet sent by a transmitter. Given the received signal and noise powers, the successful reception of a packet is a random variable. The proximity carries considerable information regarding sensor location in a binary variable. The proximity based localization has been used by numerous researchers for localization in ad hoc and wireless sensor networks [4], [5], [6] and [7].

A fixed number of reference points in the network with overlapping regions of coverage transmit periodic beacon signals. The nodes use a simple connectivity metric that is more robust to environmental vagaries, to infer proximity to a given subset of these reference points. Devices localize themselves to the centroid of their proximate reference points. The accuracy of localization is then especially dependent on the separation distance between two or more adjacent reference points and the transmission range of these reference points. In this paper we propose a weighted proximity based localization. The utilized influence of particular nodes for location estimation is increased, and on the other side the impact of other nodes is reduced. This simple enhancement leads to increasing of positioning accuracy without significant innovation.

3 System and Measurement Models

WASNs are made up of peer-to-peer links between devices (nodes). Pair-wise measurements can be made from any of these links, but only a small fraction of devices have coordinate knowledge. Thus, the measurements are made primarily between pairs of devices of which neither has known coordinates. The device with known coordinates is called a reference device. Otherwise, it is referred to as blindfolded device, since it cannot see their location. Specifically, consider a network of m reference and n blindfolded devices. The relative location problem corresponds to the estimation of blindfolded device coordinates. For simplicity, we consider the location in 2D plane. Let $[x_i; y_i]^T$ $i = 1,2,\ldots,m$ are coordinates of reference devices and $[x_j; y_j]^T$ $j = 1,2,\ldots,n$ are coordinates of blindfolded devices. Pair-wise measurements $\{RSS_{i,j}\}$ are done, where $RSS_{i,j}$ is a measurement between devices i and j. $RSS_{i,j}$ is received signal strength in device j from the device i.

3.1 Channel Model

We consider a fading channel in all measurements in this paper. The propagation path between two devices is shown in fig. 2, where RSS is received signal strength, T_x is transmitted signal strength. The influence signal attenuation is defined by: L_{LS} is signal degradation caused by large-scale propagation, L_{MS} is signal degradation caused by medium-scale propagation and finally L_{SS} is signal degradation caused by small-scale propagation. All these parameters are in [dBm]. Parameters L_{MS} and L_{SS} have the normal distribution, but their impact on the proximity based localization is negligible, because most of devices are static and these parameters are primarily changed during motion of device. Therefore, we take into consideration idealized radio model only L_{LS}. We chose this model because it is mathematically simple and easy. In this section, this model is presented. We make two assumptions in the model: the perfect spherical radio propagation and identical transmission range (power) for all devices.

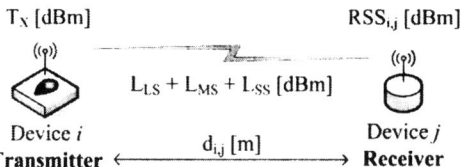

Fig. 2. Channel model.

$RSS_{i,j}$ is the measured received signal strength at device j transmitted by device i (in dBm) is Gaussian

$$RSS_{i,j}(\text{dBm}) = P_0(\text{dBm}) - 10.n_p.\log(d_{i,j}/d_0), \tag{1}$$

where $RSS_{i,j}$ is the mean power in dBm, P_0 is the received signal strength at the reference distance d_0. Typically $d_0 = 1$ meter, and P_0 is calculated by the free space path loss formula [2]. The path loss exponent n_p is a function of the environment.

3.2 Proximity Measurements

The proximity measurement S is determined based on the measured signal. $S_{i,j}$ is obtained from $RSS_{i,j}$ and it is equal to 1 if devices i and j are in range, and it is 0 if not. It is necessary to clearly define transition from status "in range" to "out of range". Therefore, we define a threshold. If the received signal strength ($RSS_{i,j}$) at j device transmitted by i device is higher as defined threshold then i devices is assumed to be in range of the j device. Thus,

$$S_{i,j} = \begin{cases} 1, & RSS_{i,j} \geq RSS_T \\ 0, & RSS_{i,j} < RSS_T \end{cases}. \tag{2}$$

3.3 Simulation Environment

The presented numerical simulations compare the performance of various solutions of proximity localization. Simulations are done for one propagation environment. The all devices are situated in square 10 m x 10 m. The reference devices locations are modeled in two cases. In the first case, the reference devices lay in expressly defined point with equally spacing, the reference devices forming raster with equally defined spacing. The raster spacing depends on the number of reference devices used in simulation. In the second case, we assume the reference devices are situated randomly (according to Gaussian distribution N(0,1)) in mentioned area. A thousand trials are performed. In each trial, the positions of reference devices and blindfolded device are generated at first. In the next step, *RSS* are calculated between each reference devices and blindfolded device based on the equation (1). Here, we use the parameters $n_p = 2$ (free space). Then, the devices are rejected which do not fulfil the threshold condition. The location coordinates of blindfolded device are determinated based on the remaining reference devices. The location coordinates can be calculated based on the following techniques (our working titles):

- Common proximity (CMP) - the estimated location is determined on the basis of the closest device location, i.e. the blindfolded device has same location coordinates as the closest device.

$$[x\_cmp_{est}; y\_cmp_{est}] = [x_{closest-device}; y_{closest-device}] \quad (3)$$

- Centroid proximity (CNP) - the position of blindfolded device is calculated as mean value of coordinates N of the closest reference devices. It is defined by the centroid of these reference devices (see equation (4)).

$$[x\_cnp_{est}; y\_cnp_{est}] = \left[\frac{1}{N}\sum_{i=1}^{N}x_i; \frac{1}{N}\sum_{i=1}^{N}y_i\right] \quad (4)$$

- Weighted proximity (WEP) - the principle of this algorithm results from centroid proximity, but the each input part (particular coordinates of devices) is individually weighted. The fundamental of WEP is the increasing of influence of closer reference devices at the expense of further devices. The mean value of coordinates obtained after the weighting gives the WEP estimate. The main benefit of the WEP algorithm is the accuracy gain by combining information contributed from multiple part inputs. Fig. 3 shows the overview of the WEP algorithm to fuse N reference devices (input parts).

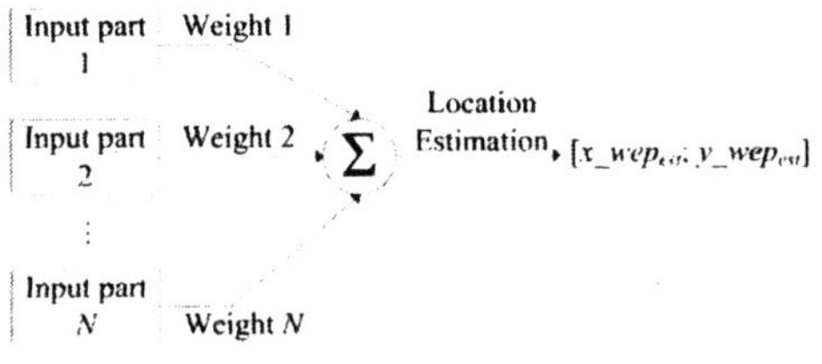

Fig. 3. Overview of the weighted proximity (WEP) algorithm.

We are interested in calculate location estimate from N available the closest reference devices, where $[x_j; y_j]^T j = 1,2,\ldots,N$ is vector of their coordinates. Then, the WEP estimate $[x\_wep_{est}; y\_wep_{est}]^T$ is written as:

$$[x\_wep_{est}; y\_wep_{est}] = \left[\sum_{i=1}^{N}x_i.w_i\left(\sum_{i=1}^{N}w_i\right)^{-1}; \sum_{i=1}^{N}y_i.w_i\left(\sum_{i=1}^{N}w_i\right)^{-1}\right], \quad (5)$$

where $[w_j]^T j = 1,2,\ldots,N$ are input weights.

The accuracy of device positioning is compared by means of *RMSE* (Root Mean Square Error)

$$RMSE = \sqrt{(x_r - x_{est})^2 + (y_r - y_{est})^2} \text{ [m]},\qquad(6)$$

where $[x_r; y_r]$ are coordinates of the real (precise) position and $[x_{est}; y_{est}]$ are coordinates of estimated position on the basis of the selected technique.

4 Simulation Results

In this section, we discuss obtained results by means of described techniques above mentioned. The results are simulated to evaluate the performance of the proximity localization in WASNs. The simulations are realized for each allocation of reference devices mentioned in previous part as following criterions:
- Impact of observed weights on the location accuracy.
- Influence of reference devices allocation on the location accuracy.
- Optimalization of number reference devices used for location estimation in all proximity techniques.
- Influence of number all reference devices used in simulation of all proximity techniques.

In the first experiment, an influence of four different weights used for weighting particular input data is examined, i.e. coordinates of reference devices. Only the three closest reference devices are weighted. It results from our preprocessing, because the accuracy does not significantly increase as the number of weighted reference devices increases.

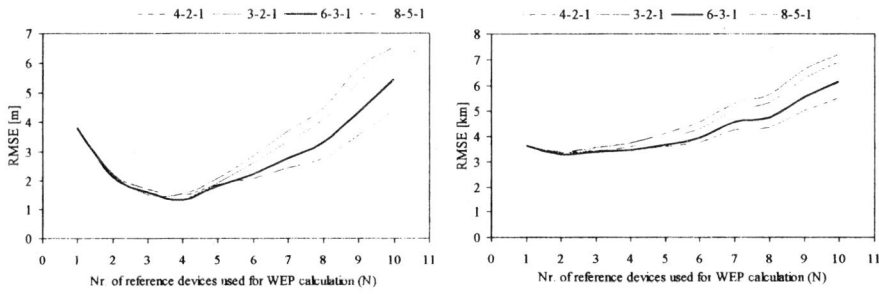

Fig. 4. RMSE [m] versus the number of reference devices used for WEP calculation ($m = 10$).

Fig. 4 is plotted for RMSE properties for $m = 10$. It can be seen two different situations from this picture: the reference devices form raster with equally defined spacing on the left and dependency random allocated reference devices on the right. The same weights are observed for both cases, but the different results are achieved. It is caused by the different allocation of reference devices. The most accurate results are drawn with bold line in all figures. Of course, the much more accurate results are achieved in the first case (reference devices form raster) in comparison with random allocation of reference devices. The most accurate results are achieved for weight {6-3-1} in both cases. The minimal localization error (RMSE) is in the case of $N = 4$ (the first case). In the second case, the minimal error is for $N = 2$. These numbers are important for next processing in particular reference devices allocation.

The above simulation is repeated for the case of $m = 100$.

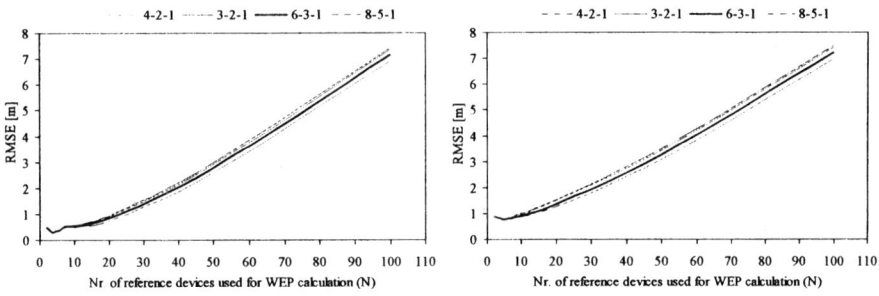

Fig. 5. RMSE [m] versus the number of reference devices used for WEP calculation ($m = 100$).

Fig. 5 depicts RMSE properties for 100 reference devices ($m = 100$). It can be seen two different situations from this picture: the reference devices form raster with equally defined spacing on the left and dependency random allocated reference devices on the right. In both cases, the same weights are observed as well as in previous situation ($m = 10$). We achieved again the different results. Of course, the much more accurate results are achieved in the first case (reference devices form raster) in comparison with random allocation of reference devices. The most accurate results are achieved for weight {6-3-1} in both cases. The minimal localization error (RMSE) is in the case of $N = 4$ (the first case). In the second case (random allocation), the minimal error is obtained for $N = 5$.

The best results are achieved for $m = 4$ for the reference devices form raster scenario. It is caused by regular allocation of the reference devices. In this case, the blindfolded device is in the square and four reference devices lay in the square edge. In the following experiments, we use the weight with the best results, e.g. {6-3-1}.

In the next part, we compare the influence of allocation of reference devices on location accuracy for all observed proximity techniques (CMP, CNP and WEP) just for $m = 10$. Because, the results are very similar and more manifestly in comparison with $m = 100$.

Fig. 6 shows the dependency of localization error on N (the number of reference devices used for calculation of CNP and WEP location estimation). Naturally, the change of N does not impact on accuracy of common proximity (dotted line). The results of CNP and WEP are similar. The number of reference devices used for estimation calculation plays important role in localization accuracy. The minimal error is interesting for us. The results confirm the fact, that weighted proximity achieves the smallest localization error in comparison of other proximity based techniques. The difference of minimal localization error is the most important factor of objectively comparison for observed techniques. In the first case, the difference is not small for $N = 4$. Therefore, WEP technique is suitable to use for estimation of location. In next simulation, only four reference devices are used for calculation of device coordinates in CNP and WEP techniques. The situation is different in the second case (random allocated reference devices). In general, we can say that the accuracy decreases as the number of reference devices used for CNP and WEP calculation (except $N = 2$) increases. The differences are smaller in comparison with the first case, but WEP is still the most accurate technique.

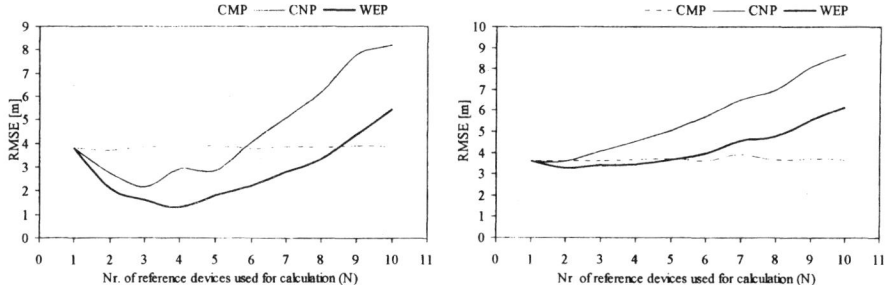

Fig. 6. RMSE [m] versus N for different allocation of reference devices.

The situation when reference devices lay in the raster is only modeled situation. Hence, we can say that the second case is more similar to reality. The positioning in the first case is much more accurate in comparison to random allocation of the reference devices. Hence, it is necessary to consider about allocation of sensors in process of implementation of WASN.

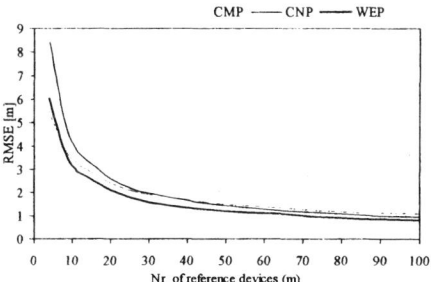

Fig. 7. RMSE [m] versus the number of reference devices.

Fig. 7 shows the dependence of RMSE on the number of reference devices m. On a basis of the obtained results we conclude that the weighted proximity technique achieves more accurate results comparing to centroid and common proximity. The RMSE is an exponential function of the number of reference devices m. Ascending value of the reference devices number means increasing of the RMSE.

5 Conclusions

We discussed three techniques for the location determination based on the proximity technology for wireless ad hoc sensor networks. The distinguished advantage of this localization technique is its simplicity and the ease of implementation. In our extensive study, we analyzed the impact of following parameters, i.e. allocation of reference devices, the number of reference devices used for location estimation and the number of all present reference devices in the observed area. The mentioned parameters were tested by means of the common, centroid and weighted proximity based localization technique. According to the results, the performance of the weighted proximity technique is better in comparison with the centroid proximity. The common proximity achieved the worst results, but this variant does not need any calculation capacity for the localization procedure. According to the results, the proposed and verified weight {6-3-1} is the most suitable for proximity based localization in wireless ad hoc sensor networks. From this study, the proximity based methods is not accurate in comparison with the sophisticated localization methods, but it is sufficient for certain non-critical applications in huge WASNs.

Acknowledgments. The work on this paper was supported by the grant VEGA 1/4065/07 of Scientific Grant Agency of the Slovak Republic.

References

1. Stojmenovič, I.: Handbook of sensor networks: algorithms and architectures, A John Wiley & Sons, 552 pages, ISBN: 978-0-471-68472-5, (2005)
2. Mohammad Ilyas, Imad Mahgoub: Handbook of sensor networks: compact wireless and wired sensing systems, CRC Press, 672 pages, ISBN 0-8493-1968-4, (2004)
3. Jayashree, L.S., Arumugam, S., Anusha, M., Hariny, A. B.: On the Accuracy of Centroid based Multilateration Procedure for Location Discovery in Wireless Sensor Networks, In proc. Wireless and Optical Communications Networks, (2006)
4. Sundaram, N., Ramanathan, P.: Connectivity based location estimation scheme for wireless ad hoc networks. In IEEE Globecom 2002, volume 1, pages 143–147, (2002)
5. Doherty, L., Pister, K. S. J., Ghaoui, L. E.: Convex position estimation in wireless sensor networks. In IEEE Infôcom, volume 3, pages 1655–1663, (2001)
6. Nagpal, R., Shrobe, H., Bachrach, J.: Organizing a global coordinate system from local information on an ad hoc sensor network. In 2nd Intl. Workshop on Inform. Proc. In Sensor Networks, (2003)
7. Niculescu, D., Nath, B.: Ad hoc positioning system. In IEEE Globecom 2001, volume 5, pages 2926–2931, (2001)
8. Patwari, N., Alfred O. Hero III: Using Proximity and Quantized RSS for Sensor Localization in Wireless Networks, In WSNA'03, San Diego, California, USA (2003)
9. Čapkun, S., Hamdi, M., Hubaux, J.-P.: GPS-free positioning in mobile ad-hoc networks. In 34th IEEE Hawaii Int. Conf. on System Sciences, (2001)

Multicast in UMTS: Adopting TCP-Friendliness

Antonios Alexiou[1,2], Christos Bouras[1,2], Andreas Papazois[2]

[1]Research Academic Computer Technology Institute, N. Kazantzaki str.,
26500 Patras, Greece
[2]Computer Engineering and Informatics Department,
University of Patras, 26500 Patras, Greece
alexiua@cti.gr, bouras@cti.gr, papazois@ceid.upatras.gr

Abstract. In this paper, we present a novel mechanism for the multicast congestion control over UMTS networks. The proposed mechanism is based on the well known TCP-Friendly Multicast Congestion Control (TFMCC) scheme. The key challenge in the design of the new scheme lies in improving the TFMCC mechanism to cope with the packet losses caused by either the temporary or the permanent degradation of the wireless channels. The proposed scheme introduces minor modifications in the UMTS nodes with respect to the computing power of the mobile terminals. Finally, our approach is implemented in the ns-2 network simulator and is evaluated it under various conditions.

Keywords: multicast, congestion control, TFMCC, TCP-friendliness, UMTS.

1 Introduction

Universal Mobile Telecommunications System (UMTS) constitutes the most prevalent standard of the 3G cellular networks. UMTS networks promise the provision of advanced services along with high data rates. Despite the high capacity that UMTS networks provide, the expected demand will certainly overcome the available resources. This is the reason why multicast transmission is one of the major goals for UMTS and 3G networks in general [1].

Multicast is an efficient method for data transmission to multiple destinations. Its advantage is that sender's data are transmitted only once over links which are shared along the paths to a targeted set of destinations. On the other hand, congestion control is a policy that regulates the source transmission rate according to the network congestion. The IP multicast, contrary to the Transmission Control Protocol (TCP), does not implement any congestion control. This means that the coexistence of multicast traffic and TCP traffic may lead to unfair use of network resources. In order to prevent this situation, the deployment of multicast congestion control is indispensable. This kind of congestion control is well-known as TCP-friendliness [2].

The adoption of a multicast congestion control in cellular networks poses an additional set of challenges which are related to the existence of wireless links and mobile terminals. In the first place, all the algorithms for congestion control treat the packet loss as a manifestation of network congestion. This assumption may not apply to networks with wireless links, in which packet loss is often induced by noise, wireless link error or reasons other than congestion [3]. Another requirement is that the mobile terminals' computing power cannot afford complicated functions. Consequently, such operations should be avoided to be held on mobile equipment.

In this paper, we present a novel mechanism for the multicast congestion control over UMTS networks. The proposed mechanism is based on the well known TFMCC scheme. TFMCC is an equation-based multicast congestion control mechanism simple enough so as to meet a prime objective for UMTS multicast services, that is scalability to applications with thousands of receivers. In our proposed mechanism, the TFMCC scheme is partly modified and extended in order to support the particularities of the UMTS Terrestrial Radio-Access Network (UTRAN). The major problem of the applicability of TFMCC over UMTS is the Current Limiting Receiver (CLR) problem. Minor modifications in the UMTS architecture are introduced by our proposed scheme.

This paper is structured as follows: Section 2 provides an overview of the work related to the scientific domain. In Section 3, we briefly present the TFMCC algorithm and describe the problem of the applicability of congestion control over the wireless access networks. Section 4 is dedicated to the proposed congestion control mechanism. Section 5 describes the simulation experiments. Finally, some concluding remarks and planned next steps are stated in Section 6.

2 Related Work

Multicast congestion control problem in fixed networks is still a domain of active research and a lot of solutions have been proposed until now. We use two distinct properties to classify the existing approaches [4], [5]:
- The rates delivered to the receivers in a session. The existing approaches fall into three categories: single-rate, multi-rate and layered.
- The place where adaptation is performed. It is either at the end systems (end-to-end service), or at the intermediate network nodes (active service).

A technical problem of major importance in multicast congestion control is scalability. When the source receives a negative feedback of congestion notification inside the network, it regulates its transmission rate. In order to avoid a feedback implosion, the majority of the researchers, like the authors of [6], suggest that the receiver of the worst congestion level should be selected as the representative. In this approach, only the representative transmits feedback information for congestion control and the number of feedbacks is limited.

In contrast to the multicast congestion control problem in fixed networks, no specific solutions and algorithms have been proposed for the variation of this problem in cellular networks. Despite radio network congestion being a widely recognized and identified problem, few relevant studies have been published. The most strongly related publication is [7]. However, this publication refers to the extended class of wireless access networks (including WLANs) and it is not well aligned with 3GPP specifications for the UMTS cellular networks.

3 Overview of Basic Concepts

3.1 TFMCC Mechanism

TFMCC is a well known equation-based multicast congestion control mechanism which is fair towards competing TCP flows, i.e. is TCP-friendly. TFMCC belongs to the class of single rate congestion control schemes and applies at the end systems (end-to-end service). Such schemes inevitably do not offer multiple transmission rates as layered schemes do. However, they are much simpler so as to offer scalability to applications with thousands of receivers.

TFMCC uses a control equation derived from a model of TCP's long term throughput to directly control the sender's transmission rate. The loss event rate and the Round-Trip Time (RTT) are the parameters that define this target throughput. Each receiver calculates its target throughput and considers it as the acceptable sending rate from the sender to itself.

TFMCC uses a feedback scheme which allows the receiver calculating the slowest transmission rate to always reach the sender. This scheme is based on the concept of the Current Limiting Receiver (CLR). The CLR is the receiver that the sender believes currently has the lowest expected throughput of the multicast group. Moreover, the TFMCC design ensures that the sender gets feedback from the receivers experiencing the worst network conditions without being overwhelmed by feedback (feedback implosion is suppressed).

For full details of TFMCC, we refer the reader to [2].

3.2 MBMS Service in UMTS

In the beginning of the current decade, the UMTS standardization body, 3G Partnership Project (3GPP), started the standardization of Multimedia Broadcast/Multicast Service (MBMS) framework. As the term Broadcast/Multicast implies, two types of service mode exist in MBMS service: the broadcast and the multicast. In the broadcast mode, data are delivered to all the receivers roaming in a specific area. On the other hand, in the multicast mode the receivers have to declare their interest for the data reception. The service then decides whether the user may receive data or not. During the rest of our analysis we will focus on the multicast mode. The multicast mode is the most complicated and also covers all the aspects of the broadcast mode.

From the physical point of view, the UMTS network architecture is organized in two domains: the User Equipment UE and the Public Land Mobile Network (PLMN). The UE is used by the subscribers to access the UMTS services while the PLMN is a network established by an operator to provide mobile telecommunications services to the public. The PLMN is further divided into two land-based infrastructures: the UMTS Terrestrial Radio-Access Network (UTRAN) and the Core Network (CN). The UTRAN handles all radio-related functionalities. The CN is responsible for maintaining subscriber data and for switching voice and data connections.

The UTRAN consists of two kinds of nodes: the first is the Radio Network Controller (RNC) and the second is the Node B. The Node B is connected to the UE via the Uu interface and to the RNC via the Iub interface. The Uu is a radio interface based on the Wideband Code Division Multiple Access (WCDMA) technology. A single RNC with all the Node Bs connected to it, is called Radio Network Subsystem (RNS).

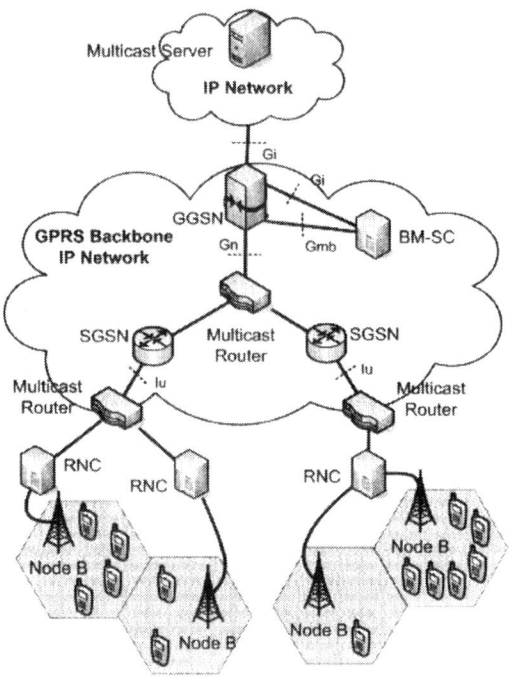

Fig. 1. MBMS architecture for UMTS using IP multicast.

The CN is logically divided into the Circuit-Switched (CS) domain and the Packet-Switched (PS) domain. The PS-domain is more relevant and therefore, in the remainder of this paper, more attention will be devoted to the PS-functionality. The PS-domain of the CN consists of two kinds of General Packet Radio Service (GPRS) Support Nodes (GSNs), namely Gateway GSN (GGSN) and Serving GSN (SGSN). The SGSN is the centerpiece of the PS-domain. It provides routing functionality, it manages a group of RNSs and it interacts with the Home Location Register (HLR) which is a database permanently storing subscribers' data. The SGSN is connected to GGSN via the Gn interface and to RNCs via the Iu interface. GGSN provides the interconnection between the UMTS network and the external Packet Data Networks (PDNs) like the Internet [1].

Fig. 1 illustrates the basic MBMS architecture for UMTS. The most significant modification of the UMTS architecture is the addition of a new node called Broadcast Multicast–Service Center (BM-SC). The BM-SC is a data source unique to MBMS. In this node the MBMS data are scheduled and interfaces are provided for the interaction with the content provider. The BM-SC may authorize and charge the content provider. At this point, it must be clarified that the data source may not originate from an external PDN, but may also originate from within the UMTS network.

In order to reduce the implementation costs, the MBMS has been designed to introduce only minor changes to existing radio and core network architectures. For simplicity reasons, in our analysis, we will consider the functionality of the BM-SC incorporated in the GGSN.

3.3 CLR Selection Problem

The traditional congestion control mechanisms treat the packet loss as manifestation of buffer overflow in the network nodes, i.e. as network congestion. Consequently, the action taken in order to resolve this situation is the reduction of the sender's transmission rate. This assumption may not apply to networks with wireless links, in which packet loss is often induced by noise, wireless link error or reasons other than network congestion. As a consequence, the network reaction should not be a drastic reduction of the sender's transmission rate [3]. The wireless-caused packet loss will eventually be resolved after the end of the fading period, without a transmission rate regulation.

Obviously, the wireless channel degradation may affect the performance of the TFMCC mechanism. If a lot of UEs participate in the multicast group, there is a high probability that at a given time, a UE suffers from fading. Eventually, the wireless-channel degradation will cause a significant and steady degradation of the performance of the TFMCC mechanism and of the multicast service. During this analysis we shall refer to this problem as the CLR selection problem.

4 The Proposed Mechanism

As we have already mentioned, the proposed mechanism follows a design very similar to that of the TFMCC scheme. Nevertheless, new functionality has been added to the existing mechanism in order to deal with the CLR selection problem.

The basic principles that govern the proposed mechanism are the following:
1. Each UE measures its packet loss rate using the packet loss history scheme of TFMCC.
2. Each Node B measures its packet loss rate. This information is written to the heading of the data packets and is then read by the UEs. This is a new functionality which combats the CLR selection problem in UMTS networks. This functionality does not exist in the TFMCC scheme and is explained below.

3. Each UE measures or estimates the RTT to the multicast server. This is achieved through an approach inherited from TFMCC. In more detail, timestamped feedback is sent to the multicast server. The server then echoes the timestamp and the corresponding UE_id in the header of a data packet. This approach causes minor traffic overhead in the network.
4. Each UE uses a control equation to calculate an acceptable sending rate from the sender back to it. The input parameters for the control equation are the loss rate and the RTT measured by the UE.
5. The feedback scheme of TFMCC is adopted. This scheme has devised a way in that the feedback from the receiver calculating the slowest transmission rate always reaches the sender. In addition, the feedback is filtered using randomized timers in order to avoid a feedback implosion.

In the proposed mechanism, the nodes located at the border between wireless and wired network (i.e. the Node Bs) have an additional responsibility. This responsibility is to provide the receivers (i.e. the UEs) with information about their measured packet loss. This means that each UE is informed by its serving Node B, of the packet loss that the Node B measures. This information is piggybacked in the data packets of a multicast session.

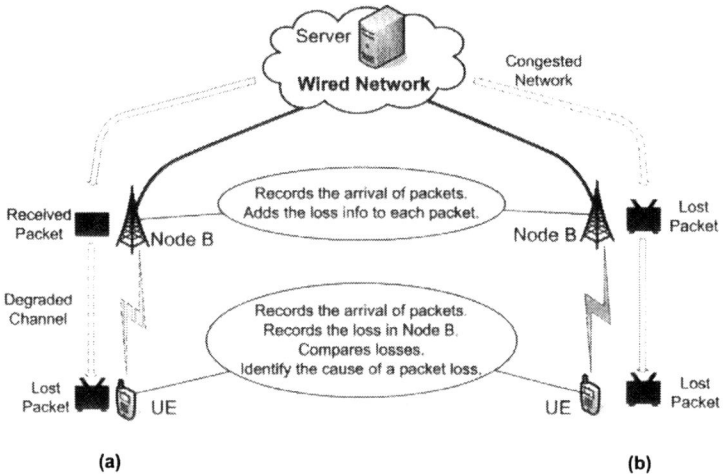

Fig. 2. Packet losses at the UE: (a) Packet loss due to wireless channel degradation. (b) Packet loss due to network congestion.

This additional functionality of the Node Bs, permits each UE to identify the reason of a packet loss. The UE compares the packet loss received from Node B with its measured packet loss. In general, the following cases are distinguished:
- When the two values differ, the UE can conclude that the reason for the difference is losses at the wireless link caused by wireless channel degradation. This kind of packet loss is not related to the network congestion and, consequently, the reduction of the transmission rate will not affect this packet loss. In this case, the

packet loss is not accounted at the CLR selection. Fig. 2(a) visualizes this functionality of the UE.
- On the other hand, when both the Node B and the UE encounter a packet loss, this packet loss is considered to be caused due to network congestion. Consequently, this kind of packet loss is taken into consideration during the CLR selection. This scenario is depicted in Fig. 2(b).

When a permanent degradation occurs on the wireless link, the buffer of the Node B will overflow and some packets will be rejected. Normally, this UE should be a CLR-candidate. In the proposed mechanism, during permanent channel degradation, the Node B counts the rejected packets as general packet losses which happened due to network congestion. These packets are taken into account by the UE during the CLR selection. This mechanism assures that this permanent degradation is not hidden from the UE. In fact, the UE is informed of the packet losses caused by buffer overflow. This functionality makes our proposed mechanism suitable not only with the CLR selection problem, but also with the permanent degradation of the wireless channel.

5 Experiments

5.1 Simulation Environment

For the verification of the proposed mechanism the ns-2 network simulator [8] along with its EURANE extension were used. The Enhanced UMTS Radio Access Network Extensions (EURANE) for ns-2 [9] comprises of extensions for the support of UMTS network functionality. Given that the ns-2 simulator does not support the multicast transmission in UMTS, we implemented the multicast packet forwarding mechanism described [10]. The next step was the installation of the TFMCC scheme. The codes used to implement and evaluate the TFMCC by the authors of [2].

Finally, the generic TFMCC was modified and extended in order to support the UMTS environment. In more detail, the implementation of the TFMCC was enhanced in order to support the functionality of the Node B and the UE as described in Section 4. The Node B implementation was modified in order to provide the UEs with information about its measured packet loss. This means that each UE is informed by its serving Node B, of the packet loss that the Node B measures. This information is piggybacked in the data packets of a multicast session. One bit in the header of the data packet is enough for the provision of this information. On the other hand, the UE implementation was modified in order to read this bit and to take the decision whether a packet loss should be accounted at the calculation of its acceptable sending rate.

5.2 TCP-Friendliness

The first aspect that we examined was the TCP-friendliness of TFMCC. In more detail, we considered the fairness of TFMCC towards the competing TCP flows when they share wired or wireless links. Below, we present the TFMCC behavior in a non-congested and in a congested UMTS network.

In the first place, fairness was analyzed using a non-congested UMTS network. We monitored the throughput over a wireless link connected a UE with its serving Node B. We supposed that UE belongs to a multicast group and receives TFMCC traffic. At the same time, this UE receives TCP traffic from an external node. Fig. 3 illustrates the throughput of TFMCC flow against that of the TCP flow.

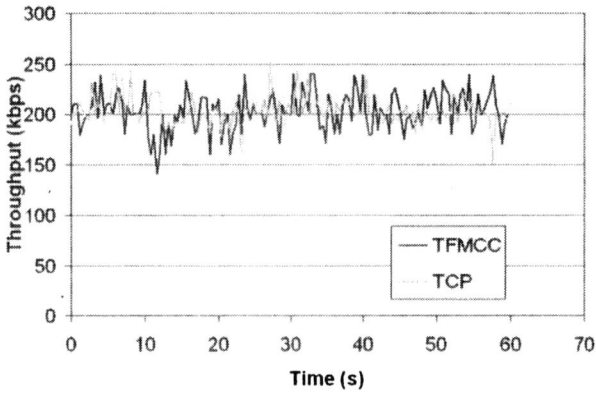

Fig. 3. TFMCC flow vs. TCP flow in a non-congested UMTS network.

Due to our initial assumption that no congestion exists, the capacity of the wireless link poses a threshold of 384kbps for the throughput of the flows towards the examined UE. As it was expected, the available bandwidth is fairly shared between the flows. Fig. 3 confirms that the average throughput of TFMCC flow closely matches the average TCP throughput.

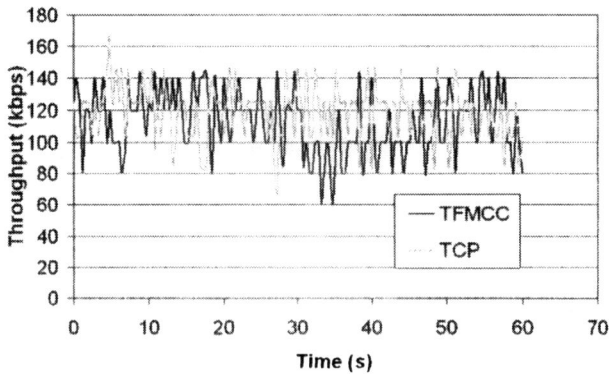

Fig. 4. TFMCC flow vs. TCP flow in a single-bottleneck UMTS network.

The next step of our experiment was to examine the fairness of the mechanism in a congested UMTS network. We considered a single-bottleneck topology and the bottleneck was applied over a link which connects an SGSN with an RNC node (Iu interface).

Fig. 4 shows the throughput of a TFMCC flow against two sample TCP flows (out of 15). The average throughput of TFMCC closely matches the average TCP throughput. Moreover, TFMCC achieves a smoother rate than the TCP.

Similar results can be obtained for many other scenarios. For example, if we suppose that congestion exists over a wired UMTS interface.

5.3 Responsiveness to Changes

An important concern in the design of congestion control protocols is their responsiveness to changes in the network conditions. This behavior was investigated using the single bottleneck topology of used in the previous experiment.

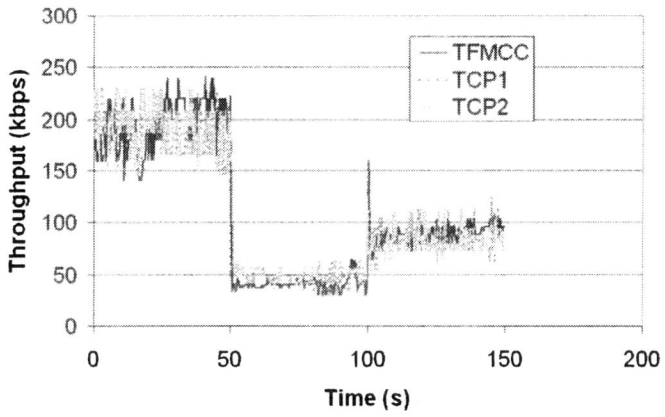

Fig. 5. Responsiveness to changes in the loss rate.

During the simulation we changed the applied loss rate of the bottleneck link. The simulation lasted 150 seconds. During this time interval three different loss rates were applied on the Iu interface. The TFMCC flow was monitored along with two TCP flows sharing the bottleneck link. The results of the simulation for the three competing flows are presented in Fig. 5.

As shown in Fig. 5, TFMCC matches closely the TCP throughput at all three loss levels. Moreover, the adaptation of the sending rate is fast enough. Actually, the simulator logs show that the UEs need 1500-2000ms after the change of the loss rate in order to adapt to the new loss rate. These figures of response time are close enough to the corresponding time of TCP (about 1000-1500 ms).

A similar simulation setting was used in order to investigate the responsiveness to changes in the RTT. The results are similar to those above. The above experiment confirms the excellent reactivity of the TFMCC to changes in congestion level of the UMTS network. Moreover, it confirms that during the application of TFMCC over the UMTS the properties and the benefits of this scheme are not affected.

5.4 Reaction to Wireless Channel Degradation

The next concern of our experiments was the evaluation of the proposed scheme when wireless-caused packet losses occur. We simulated a UMTS network and assumed a degradation of the wireless channels. In more detail, we simulated the wireless channel degradation by applying an error rate over the packets transmitted via the wireless links. We examined the proposed scheme for different number of UEs belonging in the multicast group.

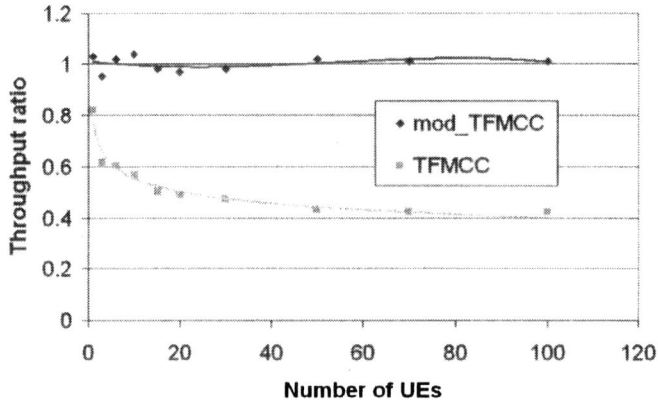

Fig. 6. Throughput of our proposed scheme vs. TFMCC when wireless packet losses occur.

In Fig. 6, our proposed scheme is referred as modified TFMCC (mod_TFMCC). On the other hand, the typical TFMCC algorithm is referred as TFMCC. The horizontal axis shows the number of the UEs belonging in the examined multicast group. Both mechanisms were examined for up to 100 UEs participating in the multicast group. The vertical axis shows the average throughput which is normalized to the corresponding TCP one. The results when 5% wireless-caused packet loss is applied are presented in Fig. 6.

The results depicted in Fig. 6 confirm the excellent behavior of our proposed scheme when wireless channel degradation occurs. The wireless-caused packet losses can be identified correctly at the UEs and be ignored at the calculation of the acceptable sending rate. This means that the CLR selection problem can be overcome and significant improvement is added on the TFMCC application over the UMTS.

5.5 Permanent Wireless Channel Degradation

The last concern of our experiments was the evaluation of the proposed scheme when wireless-caused packet losses occur in a permanent manner. We examined the behavior of the modified TFMCC when a wireless channel is permanently degraded so as to lead to buffer overflow and packet rejections in the corresponding Node B. In the simulated UMTS network of this setting we assumed a permanent degradation of the wireless link that connects a specific UE with the corresponding Node B.

In more detail, we simulated the wireless channel degradation by applying an error rate of 50% over the packets transmitted via the corrupted wireless link. In the beginning of the simulation, no wireless channel degradation occurred. After 50 seconds of simulation, we applied the error rate over the wireless channel connecting the examined UE. We monitored the changes over the throughput of the corrupted wireless link for 100 seconds. The results of our experiment are presented in Fig. 7.

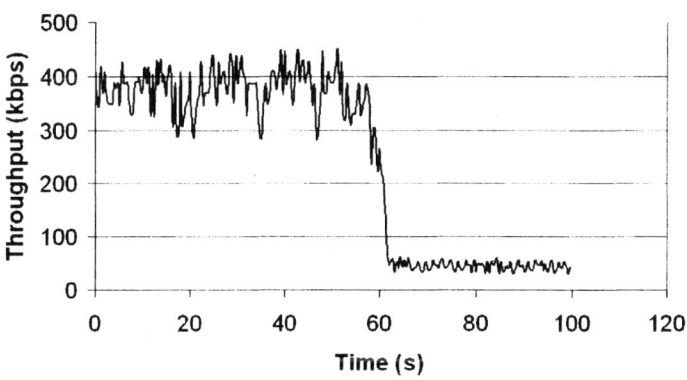

Fig. 7. Throughput of our proposed scheme vs. TFMCC when wireless packet losses occur.

The simulation results prove that our proposed scheme reacts to the permanent wireless-channel degradation. The packet losses in the Node B are considered as network congestion and are not ignored during the calculation of the acceptable sending rate in the UE. Eventually, this kind of packet losses causes reduction of the transmission rate. It took about 10 seconds to adapt to the new network conditions, but this time interval may differ according to the bit-rate of the transmission and the size of the buffer in Node B.

6 Conclusions and Future Work

We have described a congestion control scheme for the multicast transmission over UMTS. The proposed scheme was evaluated through simulation experiments and concluded that it preserves the benefits of TFMCC algorithm over the UMTS cellular network. The mechanism is fair towards competing TCP flows and has very good reactivity to changes in the congestion level.

Additionally, simulation experiments were performed in order to examine the proposed scheme against the CLR selection problem. The experiments proved that the wireless-caused packet losses can be identified correctly at the UEs, whereas the permanent wireless channel degradation is considered as network congestion.

Last but not least, the proposed scheme introduces minor modification in the UMTS architecture and respects the limited computing power of the UEs and no demanding operation is introduced in those network nodes.

The step that follows this work is the evaluation of different congestion control schemes for UMTS networks. Other multicast congestion control schemes will be investigated and modified in order to meet the UMTS requirements. Furthermore, we will try to formulate a multicast group control mechanism dedicated for the UMTS networks. In some cases, permanent wireless channel degradation may cause a large reduction to the transmission rate and eventually multicast service degradation. It will be specified under which circumstances wireless channel degradation will cause rejection of a corrupted UE from a multicast group.

References

1. Holma, H. and Toskala, A. WCDMA for UMTS: Radio Access for Third Generation Mobile Communications (3rd Edition), Wiley, 2004.
2. Widmer, J., Handley, M. Extending equation-based congestion control to multicast Applications. In ACM SIGCOMM Computer Communication Review, 31, 4 (Aug 2001), 275-285.
3. Fu, C. P. and Liew S. C. TCP Veno: TCP enhancement for transmission over wireless access networks. IEEE Journal on Selected Areas in Communications, 54, 2 (Feb 2003), 216-228.
4. Liu, J., Li, B. and Zhang, Y.-Q. Adaptive video multicast over the Internet. IEEE Multimedia, 10, 1 (Jan-Mar 2003), 22-33.
5. Vickers, B., Albuquerque, C. and Suda, T. Source-adaptive multilayered multicast algorithms for real-time video distribution. IEEE/ACM Transactions on Networking, 8, 6 (Dec 2000), 720-732.
6. Donahoo, M. J. and Ainapure, S. R. Scalable multicast representative member selection. In Proceedings of the INFOCOM'01,(Anchorage, AK, USA, Apr 22-26, 2001), 259-268.
7. Saito, T. and Yamamoto, M. Wireless-caused representative selection fluctuation problem in wireless multicast congestion control. IECE Transactions on Communications, E88-B, 7 (Jul 2005), 2819-2825.
8. The Network Simulator - ns-2. Available online at: http://www.isi.edu/nsnam/ns.
9. Ericsson Telecommunicatie B.V. User Manual for EURANE, 2005. Available online at: http://www.ti-wmc.nl/eurane/eurane_user_guide_1_6.pdf.
10. Alexiou, A., Antonellis, D., Bouras, C., Papazois, A. An efficient multicast packet delivery scheme for UMTS. In Proceedings of the MSWiM'06, (Torremolinos, Malaga, Spain, Oct 2-6, 2006), 147-150.

Simulation of Large-Scale IPTV Systems for Fixed and Mobile Networks

Radim Burget[1], Dan Komosny[1], Milan Simek[1]

[1] Department of Telecommunications,
Faculty of Electrical Engineering and Communication, UT Brno,
602 00 Brno, Czech Republic
{burgetrm, komosny}@feec.vutbr.cz, xsimek12@stud.feec.vutbr.cz

Abstract. The Internet Protocol Television has received much attention in recent years. It is a service that delivers media contents to costumers using the Real-time Transport Protocol and RTP Control Protocol. It brings many benefits for consumer and even for media distributors. IPTV is a specific application. It has only one sender and it should be able to offer its content to a huge number of receivers. As it is a new kind of service, there are still some problems to be solved. One of the most important is the Quality of Service (QoS) reporting delay and its dependence on the number of receivers in the session. The conventional RTP/RTCP architectures exceed acceptable range of the reporting interval with a relatively small number of receivers. This paper surveys the newest RTP/RTCP topologies and proposes a simulation model for further optimizations of algorithms and mathematical models for IPTV systems.

Keywords: RTP, RTCP, Real-time Transport Protocol, RTP Control protocol, Feedback, hierarchical aggregation, IPTV, Internet Protocol Television, Receiver Summary Packets, Sender reports, Receiver reports, Petri net..

1 Introduction

IPTV (Internet Protocol Television) is a service that delivers media contents (most commonly audio & video) to receivers. Each receiver (R) receives RTP data with audio and video contents and sends feedback reports, so called Receiver-Reports, in an accurate calculated interval that contains information about the Quality of Service (QoS). The calculation of this interval is described in detail in RFC 3550 [1]. Each member in this session communicates using the Any-Source Multicast in the many-to-many fashion (see Fig. 1). It is therefore really simple to inform other the members in

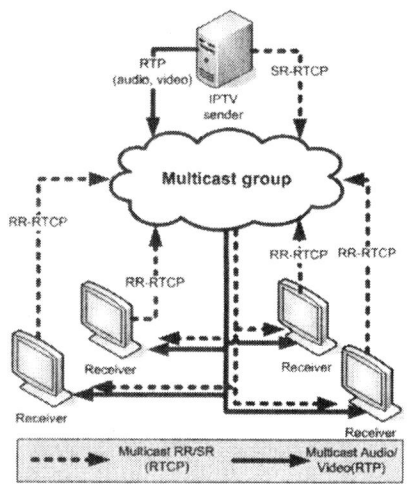

Fig. 1: IPTV session using Any-Source Multicast

Please use the following format when citing this chapter:

Burget, R., Komosny, D., Simek, M., 2007, in IFIP International Federation for Information Processing, Volume 245, Personal Wireless Communications, eds. Simak, B., Bestak, R., Kozowska, E., (Boston: Springer), pp. 445-455.

the session about the member's state. A disadvantage of the many-to-many communication is that it produces a great routing complexity, especially for huge number of receivers. Therefore a new routing protocol, the Source-Specific Multicast (SSM), has been introduced.

2 Source-Specific Multicast

A major advantage of the Source-Specific Multicast is that routing is much simpler than the older Any-Source Multicast. But it also has an unpleasant restriction. In the Source-Specific Multicast it is possible to communicate only in one-to-many fashion. Therefore the one and the only member in a whole session which is capable of sending data via multicast channel is the sender. Therefore it is not possible for any receiver to send its Receiver-Reports (RR) via the multicast channel. To bypass this restriction so called Reflection method is used. Each receiver sends its receiver reports to the sender via a unicast channel and subsequently the sender retransmits the received Receiver-Reports (RR) into the multicast channel(see Fig. 2).

As mentioned above, the interval for transmitting Receiver-Reports is computed according to equations described in RFC 3550 [1]. These folowing session parameters has an impact on the interval length: average length of packets (SR, RR), bandwidth, number of senders in the session (in case of Source-Specific multicast it equals to 1) and finally the number of members in the session.

The length of packets is limited by the protocol used, the bandwidth and the number of senders are not dynamically changing during the session time. Hence, these two could rarely be the origin of an unacceptable length of the reporting interval.

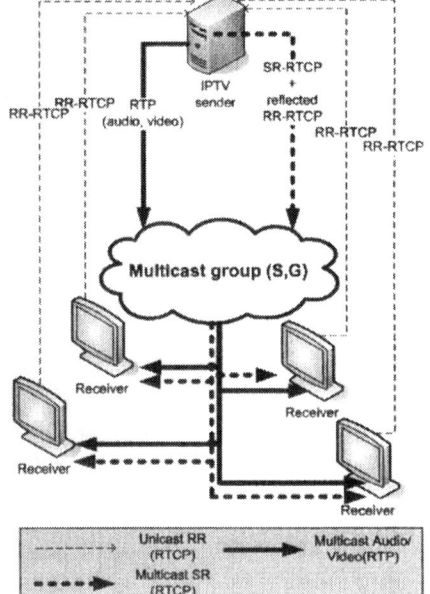

Nevertheless, if a number of members in the session is too big, the length of computed interval could be unacceptably large (minutes or even hours). If the interval exceeds some particular value, the reports might not be relevant to the current session

Fig. 2: IPTV session using Single-Source Multicast and reflecting method

state and the sender could form an inaccurate view. In this case, the optimization that sender could make according to its view could even have a negative effect on the session performance.

3 Hierarchical Aggregation

There are many ways how to reduce the interval delay. Detailed information can be found in "Real-time control protocol and its improvements for Internet Protocol Television" [2]. One of the most advanced and most promising improvements for the future use is the so-called Hierarchical aggregation (HA). This method is based on an idea where a new member type is inserted between the sender and the receivers, the so-called feedback target. Each receiver sends its reports to one of the feedback targets via a unicast channel. The feedback targets gather these reports from their related receivers and they create summaries of parameter receiver reports for each quality of service (QoS) parameter being measured. These histograms are retransmitted to the sender (using the so-called RSI packets) and it is up to the sender to decide how to optimize the audio and video contents to maximize the Quality of Service (see Fig. 3).

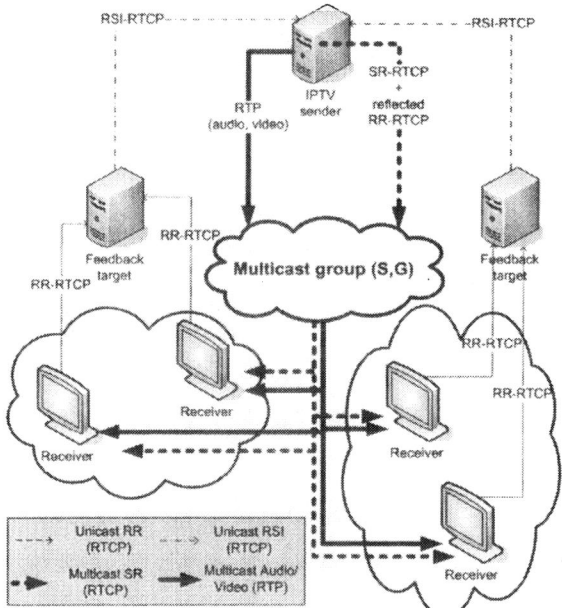

Fig. 3: IPTV session wit Single-Source Multicast using hierarchical aggregation

4 Member types in Hierarchical aggregation

In the architecture of hierarchical aggregation there are 3 types of members in the session: **sender, receiver** and **feedback target**. Each of them has a different functionality. The sender transmits audio and video data via the multicast channel, sends Sender-Reports and reflects Receiver-Reports received from receivers to the multicast group. The sender is the only member in a session which can send data into the multicast channel whereas receivers can no do that. They can only receive data from the multicast channel and therefore they have to bypass transmitting the Receiver-Reports messages using the unicast channel.

5 Data flow in Hierarchical aggregation

When we were dealing with how to verify experimentally the hierarchical aggregation model, we were asking several questions:

- How to simulate the behaviour of a network with a huge number of users (2 millions)? It would be really expensive to build an experimental network with such a huge number of nodes, routers and network devices.

- How to find out the real network conditions. The sender's view of the conditions a network according to the Receiver-Reports is almost every time affected by some diversion or a time-shift. The diversion is due to the trade-off between the effort to reduce the length of these RR packets and its precision. To monitor the the real network conditions would need an extra protocol that would communicate faster than RTCP does. Unfortunately, even this could change the network conditions and would not be accurate.

- How to visualize the behaviour of session members?

The answer to all these questions could consist in a simulation where we can relatively cost-effective by fulfil all these requirements.

The first thing we need to know is the behaviour of each member type. Each of them has a different functionality and each has different data available.

5.1 Sender data flow

The sender is the only member that can send data into the multicast channel. It sends RTP packets, Sender-Report RTCP packets (SR-RTCP) and Receiver Summary Information packets (RSI-RTCP). RTP packets are **Real-time Transport Protocol** packets that carry media contents, most commonly audio or video data. They are described in detail in RFC 3550 [1]. The **Sender Report** packets (SR-RTCP) are Real-time control protocol packets that tell each member how many packets have been sent. Each receiver makes use of it to evaluate the quality of reception and subsequently to create **Receiver-Reports** (RR-RTCP). The last one are **Receiver Summary Information** packets (RSI-RTCP). Sender creates them on the basis of information received from feedback targets and Receiver-reports received from receivers (see Fig. 4).

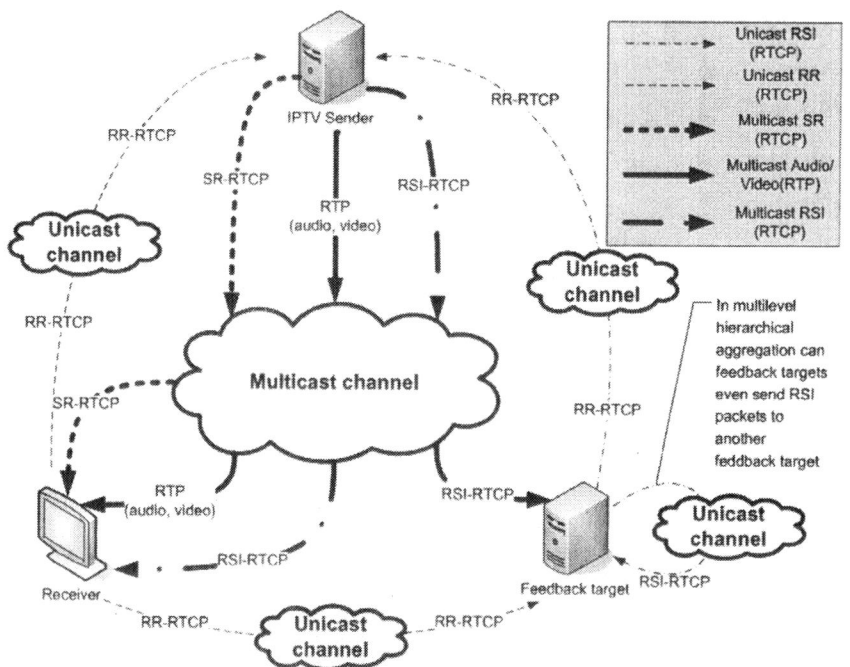

Fig. 4: Data flow between member types in Source-Specific Multicast and hierarchically aggregated session

5.1 Receiver data flow

Receivers receive RTP packets, RSI-RTCP packets and SR-RTCP packets. RSI-RTCP packets stand for Receiver Summary Information and they contain summarised statistics about network status, in particular about the sender or feedback target view of the network status of its related subgroup. And finally SR-RTCP packets that stand for Sender-report packets. They are transmitted by the sender and contain some important parameters that are used by receivers to evaluate receiver's quality of reception (see Fig. 4).

5.1 Feedback target data flow

As shown in Fig. 4, each feedback target receives **Receiver-Report** packets (RR-RTCP) from its related group of receivers. It gathers the received packets and in precalculated interval, transmits **summary information packets** (RSI-RTCP) to the sender or to other feedback targets if the level of hierarchy is greater than one. Each receiver chooses a feedback target by itself.

6 Abstract model of hierarchical aggregation

It is possible to create an experimental network with a few items in the network and verify the theoretical background on it. However an IPTV session usually has the size of millions of users in a single session. Building an experimental network of that extent would be really expensive if not impossible. This is where simulation comes to play.

6.1 Abstract sender model

The sender in an IPTV session can send via the multicast channel and receive via a unicast channel. It sends Sender-Reports (SR-RTCP), Receiver Summarization (RSI-RTCP) packets and RTP packets with audio and video contents. And it receives Receiver Reports (RR-RTCP) from receivers and Receiver Summarization (RSI-RTCP) packets from feedback targets.

The corresponding model is designed using the Petri-net model. Firing the transition will start asynchronous processes that send sender report packets, receiver summary packets, and audio and video data (RTP packets). On the other hand, it receives Receiver-Reports from the receiver and Receiver-Summary packets from feedback targets (see Fig. 4 and Fig. 5).

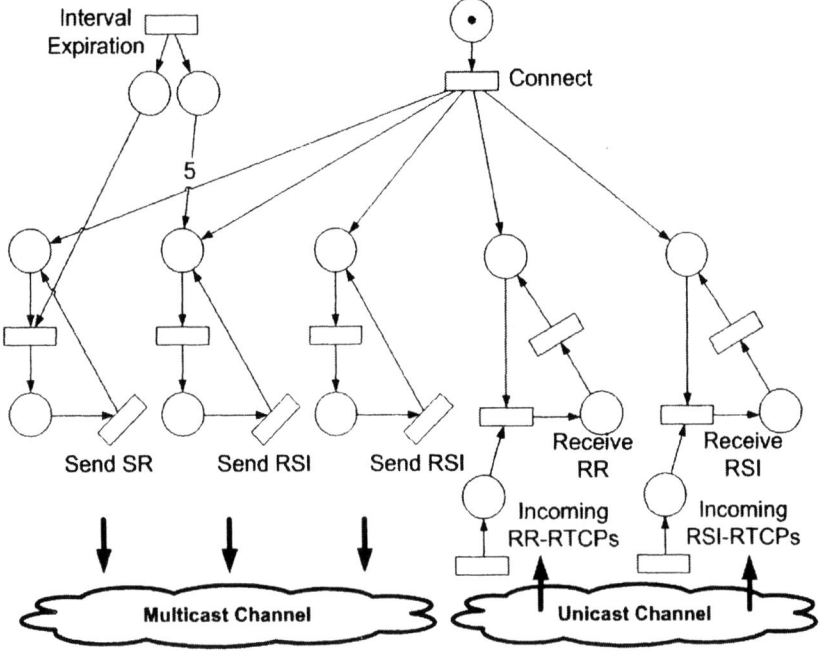

Fig 5: Petri-net model of sender behavior

6.2 Abstract receiver model

The receiver can send Receiver-Report (RR-RTCP) packets via a unicast channel and receive RTP data and Receiver Summarization (RSI-RTCP) packets via the multicast channel. The corresponding model is depicted in Fig. 5.

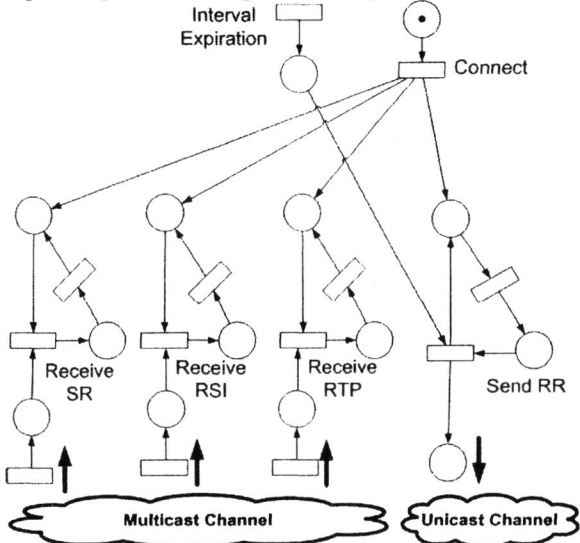

Fig 6: Petri-net model of receiver behavior.

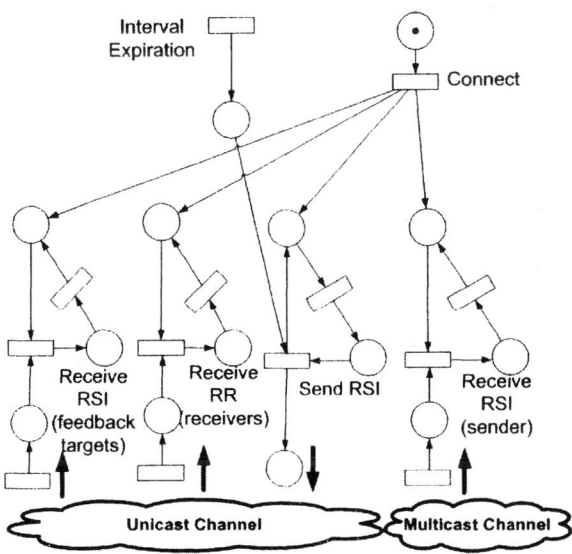

Fig 7: Petri-net model of feedback target behavior

6.3 Abstract feedback target model

The feedback target can send packets using a unicast channel and receive from both the unicast and the multicast channel. It sends Receiver Summarization (RSI-RTCP) packets to the other feedback targets. It receives Receiver Reports (RR-RTCP) from a related group of receivers, and Receiver Summarization (RSI-RTCP) packets from the other feedback targets from a unicast channel. And finally, it receives Receiver Summarization (RSI-RTCP) packets from sender. Thus it knows the number of members in a session, which is necessary to calculate the reporting interval T_d.

The corresponding model is depicted in Fig. 7.

7 Application for IPTV simulation

The simulation model allows obtaining results for large numbers of receivers. However, it lacks the real-network conditions. For that purpose, the IPTV client/server application has been developed. The main application goal is to broadcast IPTV via a real network and to evaluate the feedback interval sent from receivers to the source. For that purpose, two multimedia sessions were implemented for audio and video transmissions. The application creates a histogram and monitors the actual number of receivers. The IPTV server and client use separate streams with associated ports as depicted in Figure 8.

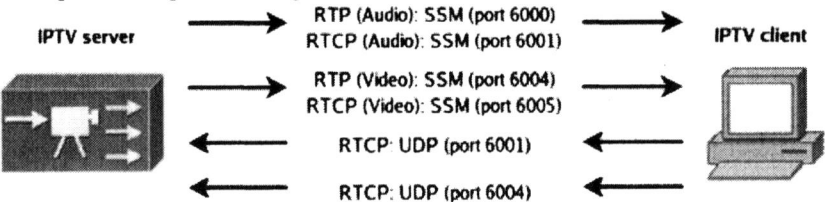

Figure 8: IPTV server media streams

The IPTV server works with one multicast socket and one unicast socket. The multicast socket is used for transmitting the media stream encapsulated in RTP packets and also for sending the SR-RTCP and RSI-RTCP packets to all receivers. The unicast socket is used for receiving RR-RTCP and RSI-RTCP packets which come from receivers and summarization nodes respectively. The receiver also works with two ports associated with RTP and RTCP packets. Besides the receivers and the sender, a new type of member was introduced – the feedback target. The feedback target can behave as a summarization node for a specific group of receivers. The implementation overview of the IPTV receiver is shown in Figure 9 using UML notation. The application has also a graphical interface to show detailed information about simulation outcomes, see Figure 11. It can show every transmitted and received packet. Furthermore it displays a chart with information about the reported number of receivers and senders in a session and the value of an RR-RTCP packet transmission interval.

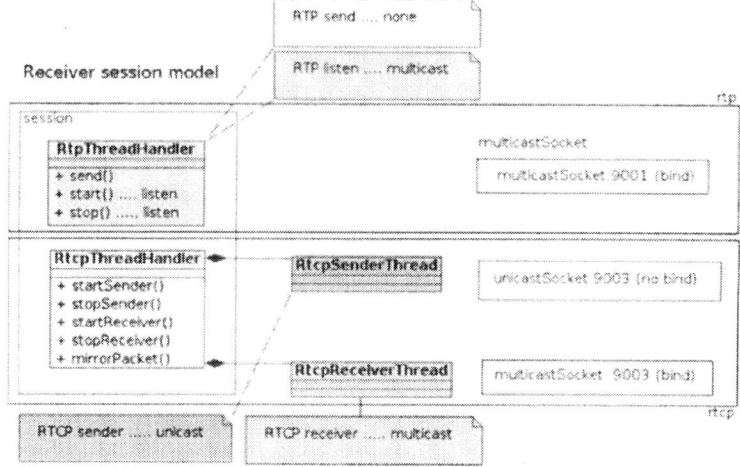

Fig 9: IPTV client structure

The large number of IPTV receivers is allowed due to multiple instances of IPTV application. A selected number of virtual IPTV clients behave as summarization servers for specified groups. The software has been developed in order to run multiple instances with minimal consumption of PC resources (memory, CPU). For example, a group of IPTV receivers can be simulated by one virtual IPTV client with the respective bandwidth consumption and generating multiple RR packets according to equation). Simulation results are gathered at the IPTV server and stored in the text files for further processing using additional software.

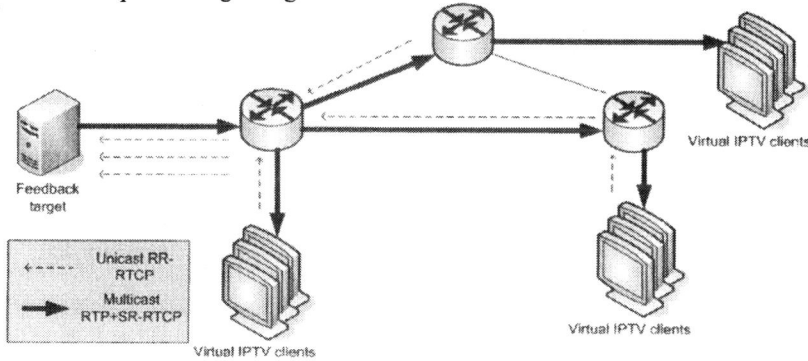

Figure 10: Experimental network overview

The application is used in an experimental network, whose structure is shown in Figure 9. The routers used create domains for a set of session members. The routers support the necessary routing protocols to implement an SSM session - PIM-SSM (Protocol Independent Multicast) [An Overview of Source-Specific Multicast (SSM)] and IGMPv3 (Internet Group Management Protocol) [Using Internet Group Management Protocol Version 3 (IGMPv3)]. However, in order to provide real conditions, the session members could be assigned to hierarchical tree groups with no relation to the router position. Also, the IPTV client/server application could cooperate with commercial IPTV solutions. This provides for future analysis of IPTV broadcasting from the experimental network. The IPTV solution used in the experimental network covers three components Cisco IP/TV 3442 Broadcast Server, Cisco Content Engine CE-566A (including IPTV Program Manager) and IPTV Viewer.

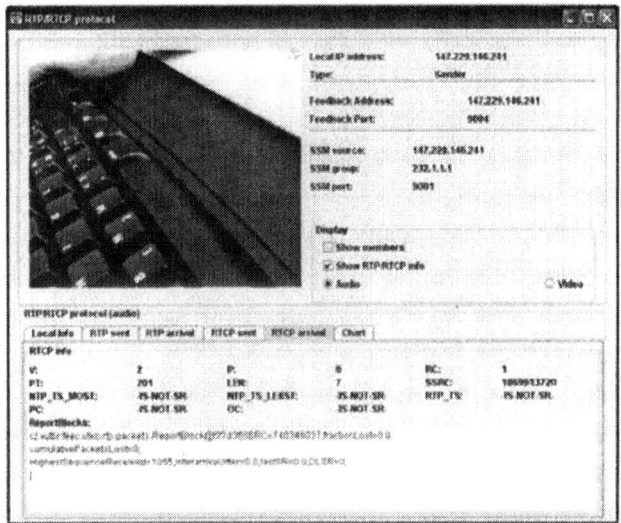

Figure 11: IPTV client graphical interface

Conclusion

The Internet is currently used for the distribution of classical broadcasting services such as TV (IPTV). The reason could be seen in the great number of possible subscribers and more features enabling the control of media transmission. Our research deals with the feedback transmission within IPTV sessions. The feedback is usually used by high-layer protocols to control and monitor the session behaviour. The research is mainly focused on the hierarchical feedback aggregation. The algorithm significantly reduces the feedback transmission interval sent in the receiver-to-source communication.

The paper gives an overview of the feedback interval transmission problem and proposes a simulation model of feedback transmission in large-scale IPTV sessions. The simulation model was proposed using Petri-net model and was implemented using a JAVA discrete event simulation library. However, this kind of simulation faces problems arising from neglecting some aspects of network properties. Therefore the IPTV client/server was developed and used in an experimental network. The required number of IPTV receivers is created using virtual clients. Special logging files are created to store the simulation results, mainly at the IPTV server side. Also the application is able to cooperate with the IPTV Cisco Broadcast Server. This feature allows future deployment of the IPTV client/server application in a large real networks for further simulations and testing of new feedback transmission algorithms.

Acknowledgments. This work was supported by the Grant Agency of the Czech Republic - project No. 102/07/1012.

References

1. H. Schulzrinne, S. Casner, R. Frederick, V. Jacobson, "RTP: Transport Protocol for Real-time Applications", RFC 3550, July 2003, ftp://ftp.rfc-editor.org/in-notes/rfc3550.txt
2. R. Burget, D. Komosny, "Improvements for RTCP Feddback for IPTV", GESTS, July 2006.
3. Chesterfield, J., E. Schooler, J. Ott, "RTCP Extensions for Single-Source Multicast Sessions with Unicast Feedback," Work in progress, October 2004.
4. SCHULZRINNE, H., CASNER, S., FREDERICK, R.. RTP Profile for Audio and Video Conferences with Minimal Control, Request for Comments 3551, Internet Engineering Task Force, 2003.
5. ROSENBERG, J, SCHULZRINNE, H. Timer reconsideration for enhanced RTP scalability, Proceedings of Seventeenth Annual Joint Conference of the IEEE Computer and Communications Societies, IEEE, 1998.
6. BHATTACHARYYA, S.. An Overview of Source-Specific Multicast (SSM), Request for Comments 3569, Internet Engineering Task Force, 2003.
7. Chesterfield, J.; Schooler, E.M.. An extensible RTCP control framework for large multimedia distributions. Network Computing and Applications, 2003. NCA 2003.
8. KOMOSNY, D., NOVOTNY, V. Analysis of bandwidth redistribution algorithm for single source multicast In Proceedings of the Sixth International Network Conference. Sixth International Network Conference. United Kingdom: University of Plymouth, 2006, s. 45 - 52, ISBN 1-84102-157-1
9. NOVOTNY, V., KOMOSNY, D. Optimization of Large-Scale RTCP Feedback Reporting in ICWMC 2007. ICWMC 2007 - The Third International Conference on Wireless and Mobile Communications. Guadeloupe, 2007, ISBN: 0-7695-2796-5
10. N. Baldo, U. Horn, M. Kampmann, F. Hartung "RTCP feedback based transmission rate control for 3G wireless multimedia streaming", 15th IEEE International Symposium on Personal, Indoor and Mobile Radio Communications, 2004 - PIMRC 2004, Volume: 3, page(s): 1817- 1821, ISBN: 0-7803-8523-3, 2004
11. Randa El-Marakby, David Hutchison "A Scalability Scheme for the Realtime Control Protocol". Proceedings of the IFIP TC-6 Eigth International Conference on High Performance Networking, Pages: 153 - 168, Kluwer, B.V, ISBN:0-412-84660-8, 1998
12. M. Castro, P. Druschel, A. Kermarrec, A. Rowstron "A large-scale and decentralized application-level multicast infrastructure", IEEE Journal on Selected Areas in Communications, vol. 20, no. 8, pp. 1489-1499, 2002
13. R. El-Marakby, D. Hutchison, "Scalability Improvement of the Real-Time Control Protocol (RTCP) Leading to Management Facilities in the Internet," iscc, p. 125, Third IEEE Symposium on Computers & Communications, 1998.

Multicast Feedback Control Protocol for Hierarchical Aggregation in Fixed and Mobile Networks

Dan Komosny, Radim Burget

Dept. of Telecommunications, Brno University of Technology
Brno, Czech Republic
{komosny, burgetrm}@feec.vutbr.cz

Abstract. For large-scale multimedia distributions, multicast is the preferred method of communication. ASM (Any Source Multicast) and SSM (Source-Specific Multicast) are the two types of multicast used. ASM is designed for either many-to-many or one-to-many communication. SSM is derived from ASM. SSM is used when only one session member is allowed to send data. An example use of SSM could be an IPTV broadcasting system over fixed or mobile network. The paper deals with describing hierarchical aggregation for feedback transmission in SSM. For the purpose of hierarchical aggregation, multicast receivers are organized into a tree structure. We present a tree structure consisting of end and summarization nodes. End nodes act as multicast receivers and summarization nodes perform feedback aggregation. The proposed MFCP (Multicast Feedback Control Protocol) is used to establish the tree structure and to exchange signalization needed for the feedback hierarchical aggregation.

Keywords: multicast, feedback, hierarchical aggregation, tree, feedback target, IPTV

1 Introduction

ASM (Any Source Multicast) and SSM (Source-Specific Multicast) are the two types of multicast used in IP-based networks [1] [2]. SSM is expected to cover all types of multimedia sessions with many receivers and only one source, such as IPTV broadcasting. The feedback could be transmitted with either ASM or SSM. The feedback which comes from receivers is used by the media source, for example, for the parameterization of a multicast forward error correction (FEC) algorithm or the tuning of audio suppression algorithms. The feedback transmitted usually contains information about the synchronization of transmitted media (audio, video), packet loss, packet delays, and jitter. RTP (Real-time Transport Protocol) and the accompanying RTCP (Real-time Control Protocol) [3] [4] are typically used for multimedia real-time transmissions. RTP is designed to transmit the multimedia (video/audio) whereas RTPC is used for the feedback transmission. Two main types of packets are used within RTCP - SR (Sender Report) transmitted from the source to receivers, and RR (Receiver Report) transmitted from receivers to the source. For the purpose of communication quality monitoring, the RR and SR packets can be also distributed to end nodes not actually involved in the multimedia reception, i.e. to a dedicated monitoring application. H.323 and SIP (Session Initiation Protocol) [5] are some of the architectures working on the RTP/RTCP protocol stack.

An SSM session is described by the multicast group address and the source unicast address. SSM is much simpler than ASM as regards the protocol complexity. Unlike ASM, there is no need to deploy complex routing trees for bidirectional communication among all participants. Therefore, SSM is more suitable for large-scale conferences than ASM. However, SSM lacks the support for communication among session members (i.e. many-to-many). Therefore, RR packets cannot be transmitted directly via multicast. The existing solutions employ unicast connections from receivers to the source and a summarization method [6] [7] is used to distribute the feedback data back from the source to the receivers via multicast. The method is based on aggregating the received data from RR packets in the source. When the aggregation is finished, a summary packet called RSI (Receiver Summary Information) is assembled and sent to all receivers. In addition, the aggregated values can be compressed up to a factor of 16. The compression significance grows when there are large sessions. The RSI packet is sent from the source together with the sender SR packets.

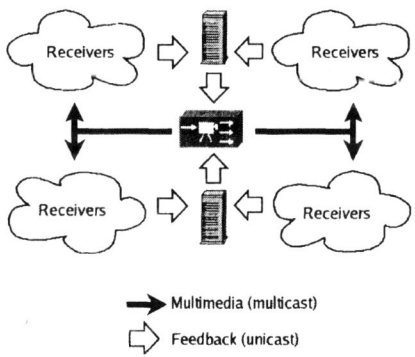

Fig. 1. Feedback hierarchical aggregation

2 Tree Structure for Hierarchical Aggregation

With the hierarchical aggregation, some nodes behave as summarization servers for a group of receivers, see Fig. 1. Members from a group report their feedback to the summarization node and the node puts the feedback received into a single summary. This summary is sent using RSI packets. Summarization nodes are organized hierarchically, thus a node produces a summary for another node of higher level, all the way up to the source. It is supposed that a summarization node is a dedicated server not involved in the media reception.

For the feedback aggregation, we assume the structure of a tree consisting of both end and summarization nodes. The summarization node highest in the tree hierarchy is the multicast source. On other levels, summarization nodes are presented except the lowest level, which consists of end nodes only. A set of end nodes represent multicast receivers of the media being transmitted via one-to-many multicast. The tree structure is depicted in Fig. 2. The feedback is transmitted from end nodes using RR packets to the summarization node higher in the tree branch. The summarization node aggregates the received feedback values into one RSI packet as described above. The RSI packet is then sent to the next higher summarization node. Note that a higher summarization node aggregates values only from the received RSI packets. Finally, the highest summarization node in the tree receives the summary feedback from all end nodes in a session. Then the highest summarization node (also the multicast source) sends the summary feedback back to all session members via one-to-many multicast. In addition the multicast source sends its own feedback to all session members in the SR packets.

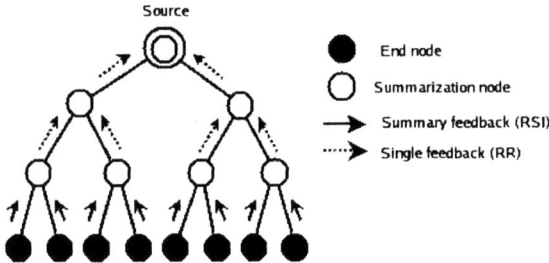

Fig. 2. Tree structure for feedback transmission using hierarchical aggregation

The tree structure should be formed in a way that keeps the round-trip delay in feedback transmission RI_{SS} as low as possible. The session feedback reporting interval could be expressed as

$$RI_{SS} = RI_{RR} + RI_{RSI} \times I, \qquad (1)$$

where I is the number of tree levels, RI_{RR} is the reporting interval of RR packets and RI_{RSI} is the reporting interval of RSI packets provided that RI_{RSI} is of a constant value trough tree levels $i=0,1,...,I-1$. In order to keep the reporting interval as low as possible the reporting interval values should be substituted with RI_{min}, which is the lowest possible reporting interval (5 seconds). The reporting interval for RR packets is identified as

$$RI_{RR} = \frac{PS_{RR} \times n_{gend}}{0.75 \times BW_{RTCP}}, \quad (2)$$

and for RSI packets as

$$RI_{RSI} = RI_{RSI}(i) = \frac{PS_{RSI}(i) \times n_{gsmr}}{0.75 \times BW_{RTCP}}, \quad (3)$$

where $i \geq 0$ is the tree level, PS_{RR} is the size of the RR packet, n_{gend} is the number of end nodes below a summarization node, BW_{RTCP} is the session bandwidth used RTCP packet transmission (5% of the total allowed session bandwidth BW_{SS}), PS_{RSI} is the size of the RSI packet and n_{gsmr} is the number of summarization nodes in a group on a tree level i. In order to assure that the statement $PS_{RSI}=PS_{RSI}(i)$ from equation (3) holds, we need to keep the size of the RSI packet at the same value through all tree levels where aggregation is done, i.e. for $I \geq i \geq 0$. The RSI packet size can be calculated as

$$PS_{RSI} = RSI_{fix} + \sum_{k=1}^{K}(RBL_{fix} + RBL_{data}(k)), \quad (4)$$

where RSI_{fix} is the fixed part size of RSI packet, K is the number of report blocks, RBL_{fix} is the fixed part size of report blocks, and $RBL_{data}(k)$ is the variable part size of report block k. For more information about the use of report blocks see [7]. Classic headers for IP, UDP, RSI can be used to calculate the fixed part size of the RSI packet. Then, we can identify the variable part size of the report block RBL_{data} as

$$RBL_{data} = DBN \times DBL, \quad (5)$$

where DBN is the number of distribution buckets (also see [7]), and DBL is the size of the distribution bucket. For the purpose of identifying the bucket size, we need to consider the worst case i.e. all end nodes reports feedback values which belong in one bucket. In other words, we need to have enough bits to express the number of all end nodes in a session. Therefore, the worst-case bucket size DBL is

$$DBL(i) = \log_2(n_{gend} \times n_{gsmr}^{(I-i-1)}), \quad (6)$$

where $DBL(i)$ is the bucket size on tree level i. It can be seen from the equation above that on the higher tree levels, the distribution bucket size is growing (more end nodes are involved in aggregation). However, we need to keep the RSI packet size of a constant value as we proposed above. To assure this, we use the multiplicative factor MF defined in [8]. Utilizing the following equation

$$DBL = DBL(i) = \log_2(\frac{n_{gend} \times n_{gsmr}^{(I-i-1)}}{D(i)}) \quad (7)$$

we are able to calculate the divisor D for the specific tree level i as

$$D(i) = n_{gsmr}^{(I-i-1)} \tag{8}$$

and with the definition $D = 2^{MF}$ in [8], we are able to identify the multiplicative factor MF for each tree level as

$$MF(i) = \log_2 n_{gsmr}^{(I-i-1)}. \tag{9}$$

Now, the number of end nodes in a group n_{gend} and the number of summarization nodes in a group n_{gsmr} can be identified for the purpose of a tree establishment. Utilizing equation (2), we can express n_{gend} as

$$n_{gend} = \frac{RI_{min} \times 0.75 \times BW_{RTCP}}{PS_{RR}}, \tag{10}$$

and similarly, utilizing equation (3), we can identify the n_{gsmr} as

$$n_{gsmr} = \frac{RI_{min} \times 0.75 \times BW_{RTCP}}{PS_{RSI}}, \tag{11}$$

where RI_{min} is the lowest possible reporting interval (5 seconds).

3 Multicast Receivers Clustering into Tree Groups

Using a defined number of group members n_{gend} and n_{gsmr} from equations (10) (11) only to set the tree structure is not adequate in terms of data transmission in IP networks. In order to achieve a proper routing performance, session members should also be organized in groups considering the relative distance between them and the group summarization node. This relative distance should be kept as small as possible. In large-scale multimedia distributions, the uncontrolled members partitioning into groups could lead to an inefficient IP-level routing. For example, for European IPTV broadcasting, it could happen that a member located in Russia reports its feedback to a summarization node located in Spain, whereas the next higher summarization node is situated in Italy, see Fig. 3. This leads to a difference in the feedback-level (or application-level) overlaying topology and the IP-level underlying topology. For more information about the problem see paper [9]. In a case involving European-scale or even word-scale IPTV broadcasting with feedback transmission, this could seriously degrade the overall network performance.

The intuitive solution of the feedback-level routing problem is to integrate the exact IP-level routing topology into the tree establishment algorithm. This solution is however quite tricky since a close cooperation with all involved routing protocols is required. Instead, we think of another known solution that uses the network latency to find the appropriate closest nodes see [10]. The algorithm called "binning" partitions nodes into bins and nodes within a bin are thought to be relatively close. Results presented in [10] show, that only approximate tree structure information offered by the binning algorithm allows significant routing complexity improvements. The algorithm is based on a set of landmark nodes (LM) placed on the Internet in a certain way. Then, a node evaluates its round-trip time (RTT) to these landmarks in a specified period of time. On the basis of the measured RTTs, nodes join bins as follows: Each node creates a vector consisting of landmarks ordered by increasing RTTs. The resulting landmark vector then defines a bin, and nodes with the same vector belong to the same bin. Furthermore, nodes can be partitioned into bins using the vector similarity. An important feature of this algorithm is that nodes assign themselves into bins without any communication with other nodes. This is ideal for large-scale multimedia sessions where communication among all nodes would be harmful to the network bandwidth load.

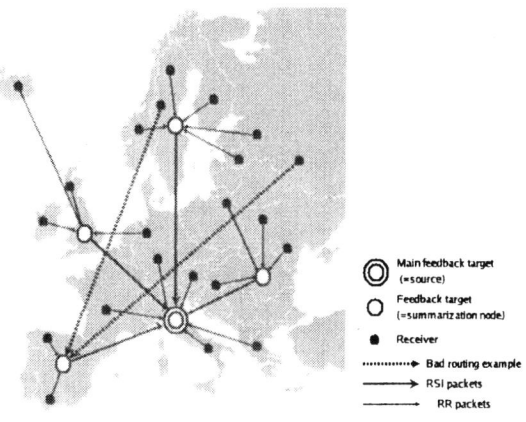

Fig. 3. Feedback transmission example in an European-scale IPTV broadcasting session

4 Multicast Feedback Control Protocol for Hierarchical Aggregation

To organize a tree structure for feedback hierarchical aggregation, every session member should be assigned to a previously defined group, which is represented by the summarization node (also feedback target). In other words, the source communicates with all session members to tell them to which summarization node they should send their feedback to. Since only a one-way multicast channel is provided from the source to session members, this information cannot be addressed to a particular member. Establishing new unicast connections from the source to every session member would greatly increase the network traffic. Therefore, we prefer solution based on a messages sent by the source to all session members, using the existing multicast channel.

For that purpose, we have proposed the multicast feedback control protocol for hierarchical aggregation (MFCP). Fig. 4 shows the protocol position in the protocol hierarchy. Note, that the protocol exploits both unicast and multicast (SSM) communication. The protocol should be considered as an enhancement to the standard protocol RTP/RTCP protocol stack.

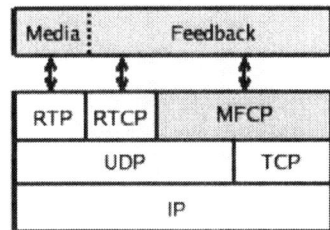

Fig. 4. Multicast feedback control protocol

The key idea used in the protocol is as follows: Let us suppose that the sender is aware of landmark RTT vectors of session members, including both summarization nodes and end nodes. The vector is formed as a list of landmarks starting with the landmark with the lowest RTT value measured. To assure that the source knows all vectors, each session member sends its vector to the sender periodically. We will discuss below how members perform this and how they obtain the landmark IP addresses for the purpose of RTT measurement. As the source is informed about all landmark RTT vectors, it is able to establish the tree structure in terms of relative distance between summarization nodes and group members. The next thing is to involve the required number of group members n_{gend} and n_{gsmr} into the tree structure. The source could meet this requirement by selecting suitable summarization nodes to form an available set of dedicated servers. The servers' availability should be managed by the service provider. After summarization nodes selection, the source calculates the number of group members belonging to a summarization node and compares it with the required values n_{gend} and n_{gsmr} from equation (10) and (11) respectively. Also, the ratio between the significance of relative distance and the number of group members could also be set at the source. In this way, the required tree structure is found. If the source cannot achieve this by selecting summarization nodes from the available set, it could also use fake landmark vectors of summarization nodes. Finally, the source sends the resulting tree structure using a MFCP message to all multicast members in order to put the tree structure in place.

For the purpose of data transmission via the multicast channel, we have proposed a general message shown in Fig. 5. The message consists of the following fields: version of the protocol (4bits), padding bit to signalize the use of padding bytes at the end of the packet (1bit), reserved bits for the future use (11bits), packet type (8bits), length in 32-bit words including the header (8bits) and SSRC value (32bits) carrying the synchronization source identifier of the originator of the packet within a RTP/RTCP session. Using the SSRC value, we are able to create different tree structures for each synchronization source. This could be useful, for example, if we are interested only in feedback on particular stream within a session.

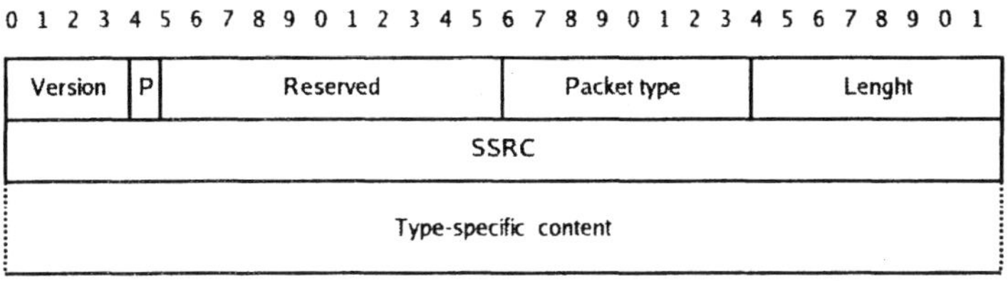

```
 0 1 2 3 4 5 6 7 8 9 0 1 2 3 4 5 6 7 8 9 0 1 2 3 4 5 6 7 8 9 0 1
```

Version	P	Reserved	Packet type	Lenght
SSRC				
Type-specific content				

Fig. 5. General multicast feedback control packet (GMFC)

4.1 Transmission from Source to Receivers

The simplest way to transmit information about the required tree structure to the receivers is to format the data as a session member IP address and also its related summarization node IP addresses, which a session member should send feedback to. The IP address of a session member is needed since we use a multicast channel and we are not able to address a packet carrying this information to a particular member. However, provided that the source knows the landmark RTT vectors of all multicast members, instead of sending data containing two IP addresses, it can contain only IP addresses of the selected summarization nodes and their calculated (or fake) landmark RTT vectors. When a multicast member receives this data, it compares its own measured landmark RTT vector with the list of vectors provided and finds the closest summarization node. Then a session member joins the group by sending its feedback to this summarization node. In order to avoid the transmission of landmarks IP addresses in a vector, which would produce a great amount of data, these IP address could be replaced by a landmark ID number. The relation between ID and IP addressed of landmarks are sent to receivers in the landmark packet (LM). The packet also provides a list of landmarks IP addresses for receivers to evaluate the RTT vector. The landmark ID is set according to its position in the list, so the first landmark has a ID=0, second ID=1 and so on. The LM packet (Fig. 6) consists of the following fields: general message header with payload type=1, sequence number (32bit) which increments by one for each LM packet sent and the list of landmarks IP addresses. The purpose of the sequence number is to identify the current ID list for receivers. When a receiver receives the feedback target packet (FT) carrying information about the tree structure, it also includes the sequence number of the LM packet with the information for landmark IDs transformation to corresponding IP addresses. If the sequence number received from LM packet is different from the number received in FT packet, an error occurred during the communication and the receiver should start to send its feedback directly to the source.

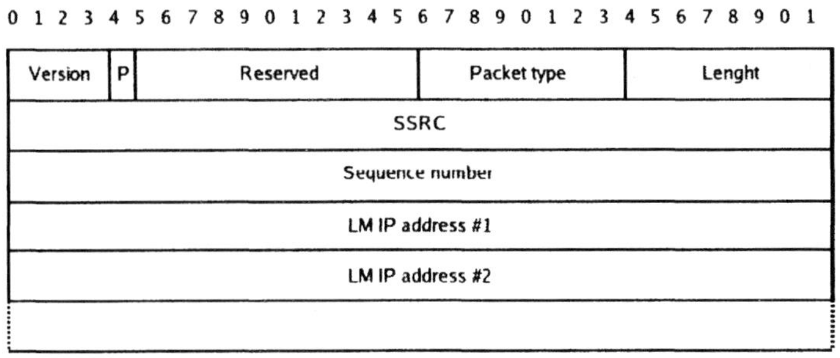

0 1 2 3 4 5 6 7 8 9 0 1 2 3 4 5 6 7 8 9 0 1 2 3 4 5 6 7 8 9 0 1

Version	P	Reserved	Packet type	Lenght
SSRC				
Sequence number				
LM IP address #1				
LM IP address #2				
⋮				

Fig. 6. Landmark packet (LM)

The FT is shown in Fig. 7. The message consists of the following fields: general header with payload type=2, sequence number of LM packet (32bits) which is used to identify the current list of ID for landmark IP addresses, first group feedback target (=summarization node) IP address (32bits), feedback target port (16bits), group size (32bits) allowing a receiver to calculate the proper RR packet transmission interval RI_{RR} from equation (2), vector specification (10bits) and length of landmark ID vector list in 32-bit words (6bits). The purpose of the vector specification field is to set the vector accuracy. The landmark vector could be reduced to only the several most significant values, i.e. to landmarks closest to the session member. The vector size then depends on accuracy required when identifying a session member position. Because the tree specification data can be too large to be sent in one FT packet, the information should be encapsulated into several packets.

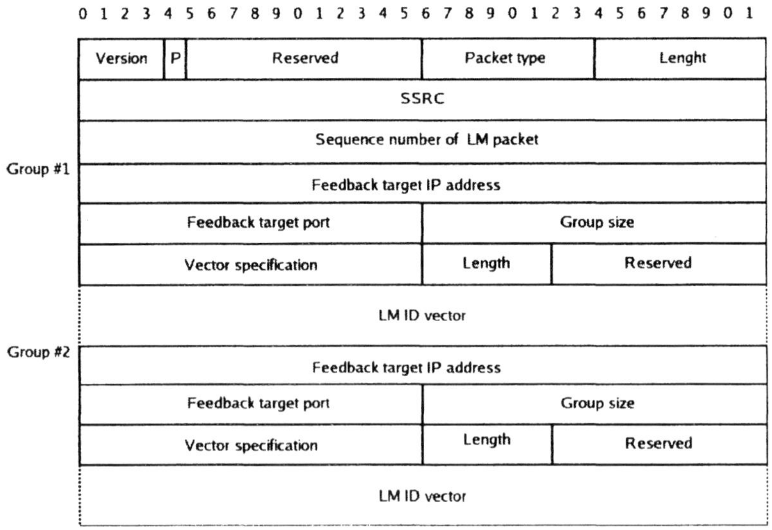

	0 1 2 3 4 5 6 7 8 9 0 1 2 3 4 5 6 7 8 9 0 1 2 3 4 5 6 7 8 9 0 1				
	Version	P	Reserved	Packet type	Lenght
	SSRC				
	Sequence number of LM packet				
Group #1	Feedback target IP address				
	Feedback target port			Group size	
	Vector specification			Length	Reserved
	LM ID vector				
Group #2	Feedback target IP address				
	Feedback target port			Group size	
	Vector specification			Length	Reserved
	LM ID vector				

Fig. 7. Feedback target packet (FT)

4.2 Transmission form Receivers to Source

Every session receiver sends its evaluated landmark IDs vector to the sender periodically in the landmark vector packet (LMV) shown in Fig. 8. The LMV packet is sent using a unicast connection. The sequence number included is the number receiver from the last LM packet. As mentioned above, this allows the source to check whether a receiver use the current set of landmarks and their corresponding IDs.

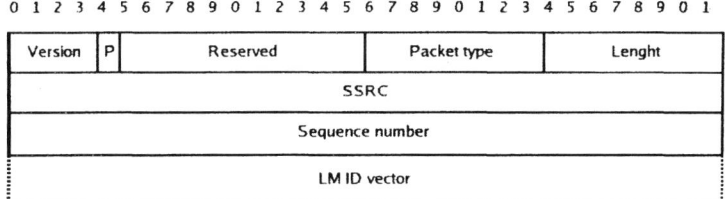

Fig. 8. Landmark vector packet (LMV)

4.3 Initiation of Feedback Transmission

This section describes how a new receiver initiates the feedback transmission in a previously established multicast session, for example using SDP (Session Description Protocol) [11]. When a new receiver joins a session, it starts to transmit its feedback directly to the source, i.e. it immediately belongs to the group where the source behaves as a summarization node. This scenario means that no immediate communication is required. This feature is quite important in case of massive session joining, for example, when an interesting IPTV program is beginning to start broadcast. After a specific time which depends on the LM packet transmission period, the new receiver will receive information about available landmark set and their corresponding IDs, see Fig. 9. Using this information, the receiver evaluates its landmark IDs vector and sends it to the sender in the LMV packet. After another time interval, the receiver will receive the FT packet with information identifying a new summarization node to send feedback to. Then the process continues by sending the LM, FT and LMV packet in specific periods of time. The LMV packet periodic transmission allows the source to check whether the receiver is still participating in the session.

5 Conclusion and Future Work

The most advanced method known for SSM feedback transmission is hierarchical aggregation, which uses a tree consisting end and summation nodes. A summarization nodes acts as a feedback target for a group of end nodes. For the purpose of end nodes clustering into groups, we use the binning algorithm. The algorithm partitions end nodes into bins and end nodes within a bin are thought to be relatively close. The binning algorithm uses a set of landmark nodes placed on the network and each end node evaluates RTT values of these landmarks. Then, end nodes are assigned to bins on the basis of a vector consisting of landmarks ordered by increasing RTT values and nodes from a bin send their feedback to the closest summarization node.

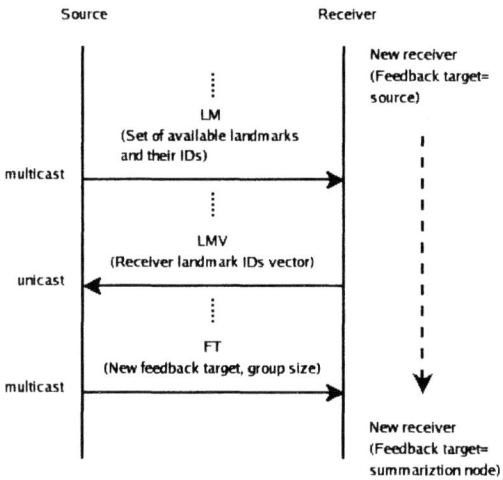

Fig. 9. New receiver joins a session

In this paper, we proposed the MFCP protocol which is used to establish and manage a tree structure for hierarchical aggregation. The main idea of the MFCP protocol is that end and summarization nodes send their landmark vectors to the source periodically. A vector consists of a list of landmark IDs starting with the landmark with the lowest RTT value. When the source knows all landmark vectors, it can calculate the required tree structure for hierarchical aggregation. The tree structure establishment covers two conditions as follows: 1) It has to meet the identified number of end nodes in groups n_{gend} and the identified number of summation nodes in groups n_{gsmr} 2) It has to assure low routing complexity for the feedback transmission. The source finds the tree structure by changing summarization nodes from an available set of dedicated servers. If this method does not work, it can create a fake vector of selected summarization nodes. Then the source sends the calculated tree structure information to receivers.

The MFCP protocol is simple since it uses only three packet types (LM-landmark packet, LMV-landmark vector packet and FT-feedback target packet). The protocol is scalable in terms of massive number of receivers joining a session since no immediate communication is needed. The convergence time for changes in the tree structure is strongly affected by transmission intervals of MFCP packets. Small transmission intervals could cause unnecessary network load. For the purpose of proper interval settings, we plan to test the protocol in the PlanetLab network. A similar RTT measurement method among PlaneLab nodes is presented in [12]. As this network consists of over 700 hundred nodes spread throughout the world, the test results should give an accurate overview of the MFCP protocol world-scale use.

Acknowledgement. This work was supported by the Academy of Sciences of the Czech Republic project 1ET301710510.

References

1. HOLBROOK H., CAIN B. Source-Specific Multicast for IP. Request for Comments: 4607. Internet Engineering Task Force. 2004.
2. BHATTACHARYYA, S. An Overview of Source-Specific Multicast (SSM). Request for Comments 3569. Internet Engineering Task Force. 2003.
3. SCHULZRINNE H., CASNER S., FREDERICK R., JACOBSON V. RTP: A Transport Protocol for Real-Time Applications. Request for Comments 3550. Internet Engineering Task Force. 2003.
4. SCHULZRINNE H., CASNER S., FREDERICK R. RTP Profile for Audio and Video Conferences with Minimal Control. Request for Comments 3551. Internet Engineering Task Force. 2003.
5. HANDLEY M., SCHULZRINNE H., SCHOOLER E., ROSENBERG J. SIP: Session Initiation Protocol. Request for Comments 2543. Internet Engineering Task Force. 1999.
6. CHESTERFIELD J., SCHOOLER E. An Extensible RTCP Control Framework for Large Multimedia Distributions. Proceedings of the Second IEEE International Symposium on Network Computing and Applications. IEEE Computer Society. 2003.
7. CHESTERFIELD J., SCHOOLER E., OTT J. RTCP Extensions for Single-Source Multicast Sessions with Unicast Feedback. Internet Draft, work in progress. Internet Engineering Task Force. 2007.
8. CHESTERFIELD, J., SCHOOLER, E., OTT, J. RTCP Extensions for Single-Source Multicast Sessions with Unicast Feedback. Internet Draft, work in progress. Internet Engineering Task Force. 2004.
9. CASTRO M., DRUSCHEL P., KERMARREC A. , ROWSTRON A. A large-scale and decentralized application-level multicast infrastructure. IEEE Journal on Selected Areas in Communications. IEEE. 2002.
10. RATNASAMY S, HANDLEY M, KARP R, SHENKER S Topologically-Aware Overlay Construction and Server Selection. Proceedings of 21rd Annual Joint Conference of the IEEE Computer and Communications Societies. IEEE. 2002.
11. HANDLEY M., JACOBSON V., PERKINS C. SDP: Session Description Protocol. Request for Comments 4566. Internet Engineering Task Force. 2006.
12. TANG L., CHEN Y., FEI L., ZHANG H., JUN L. Empirical Study on the Evolution of PlanetLab. Sixth International Conference on Networking - ICN 07. IEEE. 2007

Experiences of Any Source and Source Specific Multicast Implementation in Experimental Network

Milan Simek[1], Radim Burget[1] and Dan Komosny[1]

[1] Department of Telecommunications
Brno University of Technology
Brno, Czech Republic
simek.mil@phd.feec.vutbr.cz, {burget,komosny}@feec.vutbr.cz

Abstract. Any Source Multicast (ASM) called Internet Standard Multicast (ISM) and its extension, Specific Source Multicast (SSM), are two known multicast technologies for delivering a data flow to a great number of customers over IP networks without cumulative bandwidth consumption. ASM technology is suitable for the many-to-many delivering model such as videoconferencing, where more sources are sending the traffic to one multicast group. SSM technology was developed as a solution for one-to-many delivering model such as the Internet Protocol Television (IPTV), where the receiver exactly defines the source of the multicast traffic. ASM and SSM can co-operate together on the network if specific conditions are complied with. The paper deals with describing the experiences and the measurement results from the implementation of these two multicast technologies in an experimental network.

Keywords: Any Source Multicast, Specific Source Multicast, Protocol Independent Multicast, Experimental Network, Multicast Routing Table

1. Introduction

One of the current solution for the traditional IP multicast service routing is the Protocol Independent Multicast with Sparse Mode (PIM-SM) together with Rendezvous Points (RPs). The ASM technology uses the Shared Path Tree (to deliver the data flow towards the memberships from the RP) and traditionally Internet Group Management Protocol version 2 (IGMPv2) (for membership advertisement). For the identification of the Shared Path Tree the notation (*,G), called "star comma G" is used, where the letter G represents the multicast group address. This service consists of the delivery of IP datagrams from any source to a group of receivers called the multicast host group.

In SSM, the delivery of datagrams is based on (S,G) channels. The traffic for one (S,G) channel consists of datagrams with an IP unicast source address S and the multicast group address G as the IP destination address. In SSM, no RP is needed, because the data are delivered over the Shortest Path Tree (SPT), i.e. the shortest path from the source toward the receiver. Different from ISM is the use of the IGMPv3 protocol, which is the standard track protocol used for hosts to signal multicast group membership to routers. Would be members must subscribe and unsubscribe to (S,G) channels to receive or not to receive traffic from the specific sources. PIM-SSM, which is an extension of the standard PIM-SM is used in this case.

The purpose of this article is to describe the different principles in the building of the distributive tree and behaviour of the multicast routing tables in the ASM and the SSM in the proposed experimental network.

2. Experimental multicast network

Cisco IP/TV Release 5.1 was chosen as a suitable tool for creating the experimental network. IP/TV [2] is a network-based application delivering on-demand or scheduled programs to an unlimited number of users over any IP-based local network. IP/TV supports a lot of audio and video formats, including MP3, Advanced Audio Codec-Low Complexity (AAC-LC), H.261, Moving Picture Experts Group-1 (MPEG-1), MPEG-2, and MPEG-4 (Simple Profile). IP/TV Program Manager, IP/TV Server and IP/TV Viewer are the three main components that form the experimental multicast network [5].

2.1. IP/TV Program Manager

The Program Manager (PM) is a Linux based application running on the Cisco Content Engine 566 hardware. The PM concerns about programs on IP/TV servers and allows configuring and maintaining the programs, channels, records, and the File Transfer Protocol (FTP) relation among IP/TV servers. The administrative interface is available over web-explorer with the JavaScript support [5].

2.2. IP/TV Server

The IP/TV Server (ITS) is a part of the Cisco IP/TV 3400 Series Server hardware appliance or it is available as software. The ITS runs on the Windows 2000 Server and Windows NT 4.0. It is managed by the PM and makes it possible to multicast scheduled programs to the network. The ITS reports the important information about the actual session such as the use of multicast group addresses, the ports, the audio and video codecs, the TTL and duration time, and so on [5].

2.3. IP/TV Viewer

The IP/TV Viewer (ITV) is the software part of the IP/TV Solution. The ITV serves for the high quality reception of the video stream. The list of the programs is offered represented by means of the Session Description Protocol (SDP) [5].

2.4. Experimental network topology

The experimental multicast network contains a Content Engine 566, IP/TV 3400 Series Server and three pieces of the Cisco 1800 Router. Because of the requirement to use the Shared Path Tree with Rendezvous Point for ASM, the ring topology was chosen (Fig. 1 shows the network component arrangement). The receivers are connected to the Vlan interfaces of the routers.

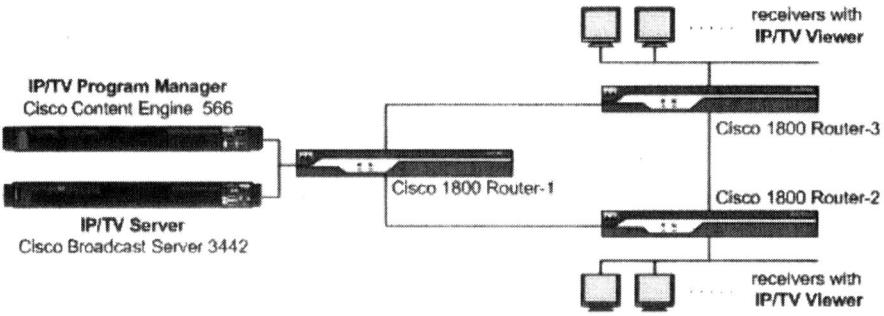

Fig. 1. Arrangement of the network components

3. Building of Distributive Trees

The distributive trees are a complex of paths from the source over all multicast routers to the receivers of the specific multicast groups. In accordance with the path building mechanism, the distributive trees are divided into Shared Path Tree for ASM technology and Shortest (Source) Path Tree for SSM.

3.1. Shared Path Tree

To build the Shared Path Tree, the Protocol Independent Multicast with Sparse Mode (PIM-SM) is used. One specific router in the network creates the root of this shared tree. This root is known as the Rendezvous Point (RP). The would be sources have no information about interested receivers location and send the traffic only upstream toward to RP that takes care of delivering the packets to the receivers. For one multicast group, only one specific RP can be defined. When the first packet is received over the Shared Path Tree, the Receiver's Designated Router (RDR - the router closest to the receiver) finds out the unicast IP address and makes a request for the traffic to be delivered over the Shortest Path Tree. This process is called the SPT-Switchover. The SPT-Switchover is performed if the rate of traffic flowing down the Shared Path Tree exceeds the SPT-Threshold [4] set for the group. The SPT-Threshold is the bitrate on the shared tree (default value is 0kbps – it means immediately switching to the SPT). At the moment of the SPT establishment, the traffic from the RP is cancelled. The notation of the Shared Path Tree is (*,G). Fig. 2 describes the process of building the distributive tree for ASM technology, with the one interested receiver connected to router 2.

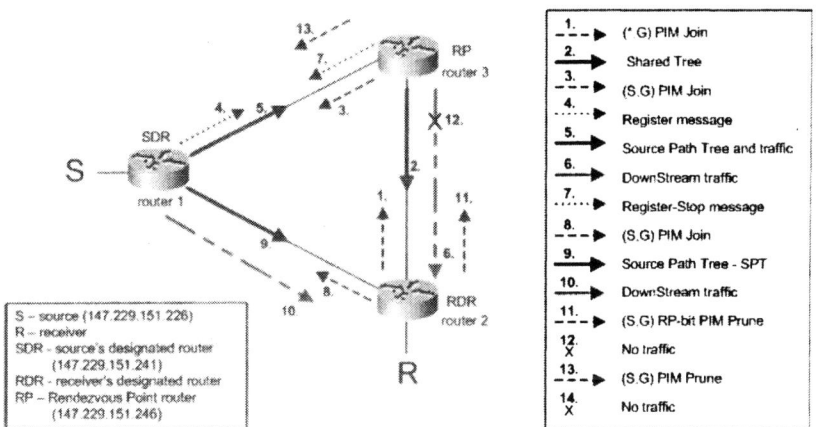

Fig. 2 Building the distributive tree in Any Source Multicast

3.2. Source Path Tree

In SSM, the traffic delivery is based on (S,G) channels. If the receiver wants to receive the traffic from the specific source, it is necessary to use the multicast routing protocol enabling the SPT to be created immediately after the receiver request. This issue is solved by means of the Source Specific Protocol Independent Multicast PIM-SSM using the PIM sparse-mode functionality to create an SPT between the receiver and the source without RP support. The receiver subscribes the reception from the specific (S,G) channel by means of the IGMPv3 protocol that allows defining the specific source for the multicast group. The SPT building process is shown in Figure 3.

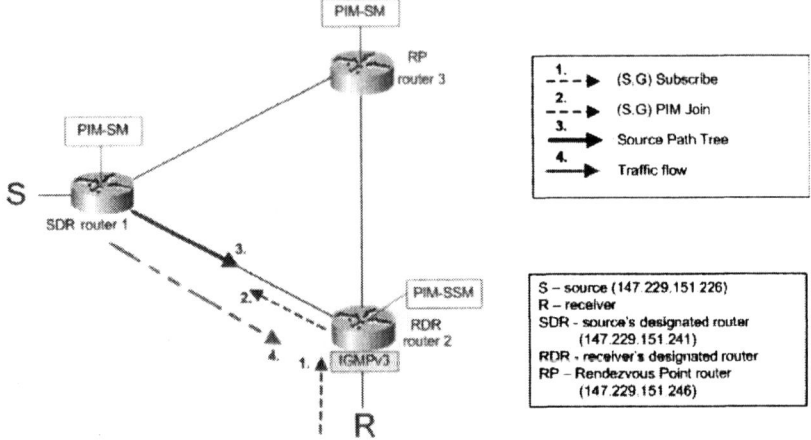

Fig. 3 Building the distributive tree in Source Specific Multicast with the PIM-SSM and IGMPv3 protocol.

4. Configuration of ASM and SSM in the Experimental Network

The detailed steps of the configuration basic IP multicast are described in [2]. The basic common setting for ASM and SSM consists in enabling multicast routing on all the routers used together with the setting of the PIM-SM protocol on every interface that uses the IP multicast. The ASM with the PIM-SM requires the RP for building the Shared Path Tree. For the settings of the RP in the network the single static RP configuration was used. This way of setting is sufficient for the experimental purposes.

4.1. Configuring ASM with a Single Static RP

It is necessary to perform the `ip rp-address x.x.x.x [access-list]` command on any router in the experimental network (x.x.x.x specifies the IP address of the RP). This access-list defines the range of the IP multicast addresses for which the RP will be used. The access-list should be assigned from the 224.0.0.0 to 239.255.255.255 range with the exception of the reserved address. These reserved address for the multicast are described in [5]. For automating the RP announcement process Auto-Rendezvous Point (Auto-RP) should be used or the Bootstrap router should be used to discover and announce RP-set information for each group. For the display of the known RPs in the network the `show ip pim rp` command is useful.

4.2. SSM Configuration

The configuration and management of SSM in the network are a great advantage of this technology because no RP is required for delivering traffic from the source to the multicast group. The network or rather routers need not maintain the information about active sources. The configuration of the SSM must be performed only on the routers, where the receivers are interested in the traffic reception from the specific (S,G) channel. These routers need to support two specific protocols for the SSM implementation: PIM-SSM and IGMPv3. It is necessary to check the version of the Cisco Internetwork Operating System (IOS) on the requested routers. IGMPv3 is supported in Cisco IOS starting with 12.0(15)S, 12.1(8)E [6]. The current IOS version of Cisco 1800 routers used in the experimental network is 12.4(6) T3.

To use SSM in the experimental network it is neecesarry to designate IP multicast address ranges which are to be used with SSM. The Internet Assigned Number Authority (IANA) has designated the IP multicast address range 232/8 [3].

There are some restrictions on the use of the IGMPv3 protocol. The client application used for the traffic received from the specific (S,G) channel needs the IGMPv3 protocol support in the operating system. In the Microsoft OS, the IGMPv3 is supported in the Windows XP, Windows 2003 and Windows Vista versions. For the older versions (Windows 95, NT 4.0 SP3, 98, NT 4.0 SP4,ME a 2000) with the IGMPv1,2 support, it is possible to use the Cisco solution URL Rendezvous Directory (URD) or IGMPv3lite protocol [4]. In the UNIX OS with the kernel version starting with 2.4.22, the protocol is supported.

5. Multicast Routing Tables

The multicast routing tables (MRT) keep and periodically update the information about the (*,G) and (S,G) states.

5.1. ASM Multicast Routing Table

When the receiver sends the PIM-Join (IGMP record) upstream toward the RP, the record about the (*,G) state is added to MRT of the Last Hop router. If this router has no limitation in the SPT-Threshold value, the record about the (S,G) states is added immediately. Table 1 describes the record in the MRT for the 224.3.8.8 multicast group. Two entries are maintained and each of them has two timers. The literature [4] says: „Uptime indicates per interface how long in hours, minutes, and seconds the entry has been in the MRT and Expires indicates per interface how long in hours, minutes, and seconds until the entry will be removed from the multicast routing table". Expires time of the (*,G) state is stopped (never expires). The expire timer for the (S,G) state counts down from 3:00 min and it is updated every 10 seconds. The SJC flag in the (*,G) state indicates: S-Sparse mode, J - indicates that the rate of traffic flowing down the shared tree is exceeds the SPT-Threshold set for the group (default 0 kbps), C – a member of the multicast group is present on the directly connected interface [4]. The JT flag in the (S,G) state indicates: J – mentioned above,

T – packets are received over the SPT. The RPF nbr abbreviation describes the IP address of the upstream router to the source [4].

Table 1. ASM multicast routing table of the Last Hop router

Action	Receiver is joined to the group		
	Source is active		Source is inactive
State	(*,G)	(S,G)	(*,G)
	(*, 224.3.8.8)	(147.229.151.226, 224.3.8.8)	(*, 224.3.8.8)
Timers (Uptime/Expires)	00:14:06/**stopped**	00:03:26/00:02:51	00:17:36/**00:02:13**
RP	147.229.151.246	----	147.229.151.246
Flags	SJC	JT	SJC
Incoming Interface	FastEthernet1	FastEthernet 0	FastEthernet1
RPF nbr	147.229.151.246	147.229.151.241	147.229.151.246
Outgoing Interface	Vlan1	Vlan1	Vlan1
State/Mode	Forward/Sparse	Forward/Sparse	Forward/Sparse
Timers (Uptime/Expires)	00:00:29/00:02:30	00:00:29/00:02:30	00:04:00/**00:02:13**

If the server stops sending traffic, the entry referring to the (S,G) state is deleted after 3 minutes and only the (*,G) entry is maintained as long as the receiver is requesting receipt. The entry is then updated every 60 seconds (in accordance with the IGMP Query Interval).

5.2. SSM multicast routing table

Since in the SSM there is no mechanism to announce to a receiver that the source is not sending the traffic to the multicast group, the (S,G) state is maintained in the MRT together with the Source Path Tree from the receiver to the source as long as the receiver sends the (S,G) subscriptions. The SSM multicast routing table is described in the Table 2. If the receiver is sending the (S,G) subscription and the source is sending traffic to the multicast group, the entry in the MRT is updated every 10 seconds. If the source becomes inactive, the entry in the MRT is updated every 60 seconds. The sTI flag indicates: s – indicates that a multicast group is within the SSM range of IP address, T – indicates that packets have been received on the source path tree, I – indicates that an (S,G) entry was created by an (S,G) report [4].

Table 2. SSM multicast routing table of the Last Hop router

Action	Receiver is joined to the group	
	Source is active	Source is inactive
State	(S,G) (147.229.151.226, 232.3.10.10)	(S,G) (147.229.151.226, 232.3.10.10)
Timers (Uptime/Expires)	00:03:26/00:02:51	00:17:36/**00:02:26**
RP	----	----
Flags	sTI	sTI
Incoming Interface RPF nbr	FastEthernet 0 147.229.151.241	FastEthernet 0 147.229.151.246
Outgoing Interface State/Mode Timers (Uptime/Expires)	Vlan1 Forward/Sparse 00:00:29/00:02:30	Vlan1 Forward/Sparse 00:04:00/**00:02:26**

6. Benefits of SSM using

In the SSM, there is no problem with address allocation. If the transmission is identified from the source unicast IP address and the multicast group IP address of the 232/8 range, this final couple of addresses is unique too. It does not matter that two sources (S1 and S2 e.g.) will send the traffic to one multicast group, thus for the (S1,G) and (S2,G) transmissions the network will maintain two different distributive trees. As, the transmission is identified with the unique source address, it offers more effective defense against the Denial of Service Attack (DoS). The measurement of the DoS attacks in the ASM and SSM will be performed in the future. The easy configuration and management is another benefit of the SSM too.

7. Conclusion

In this paper are presented matters concerning the ASM and the SSM implementation in the experimental network. Using the configuration of the SSM in the experimental network, it was proved that implementation of this technology in the network where the ASM exists is easier because only the Last Hop router needs to be configured for the PIM-SSM and IGMPv3 protocols used but the IGMPv3 protocol is used only in the latter operation systems. An examination of the multicast routing tables has revealed that the SSM has no mechanism for notifying the receivers that the source has stopped sending the traffic to the multicast group. Thus the multicast routing table keeps the information about the (S,G) state as long as the receivers request the receipt of that channel. The multicast routing tables thus keeps a lot of needless information. In ASM, the entry of the (S,G) state is deleted after 3 minutes when the source becomes inactive.

The SSM is ideal for internet broadcast applications such as the audio/video transmission in real time. Customers need to get ready this technology still has not still sufficient support in the internet community. For the realization of videoconferencing and for playing online games a many-to-many delivering model is necessary that is fully supported by the ASM technology.

Extensions to the experimental network and subsequent measurements are planned in the future.

Acknowledgement. This work was supported by the Grant Agency of the Czech Republic - project No. 102/07/1012.

References

1. Bhattacharya, S.; An Overview of Source Specific Multicast (SSM). IETF RFC 3569, 08/1989, ftp://ftp.rfc-editor.org/in-notes/rfc3569.txt
2. Cisco System, Inc.: Configuring Basic IP Multicast. Cisco Systems, Inc., 1992-2007. www.cisco.com
3. Thaler, D. Handley, M., Estrin, D.: The Internet Multicast Address Allocation Architecture. RFC 2908, September 2000
4. Cisco System, Inc.: Source Specific Multicast. Cisco Systems, Inc., 1992-2007. www.cisco.com
5. Cisco System, Inc. Cisco IP/TV Broadcast Server User Guide, Release 5.1. Cisco Systems, Inc., 1992-2007, www.cisco.com
6. Cisco System, Inc. IGMP Version 3. Cisco Systems, Inc., 1992-2007, www.cisco.com

Image compression in digital video broadcasting

Kamil Bodecek, Vit Novotny, Milan Brezina

Faculty of Electrical Engineering and Communication,
Brno University of Technology
Purkynova 118, 612 00 Brno, Czech Republic
kamil.bodecek@phd.feec.vutbr.cz, novotnyv@feec.vutbr.cz, imb@email.cz

Abstract. Images form integral part of interactive services broadcasted in digital television. Natural images compression is always trade-off between visual quality and file size. Annoying artifacts arise in highly compressed images. The scope of this article is to investigate the possibilities of image compression enhancement for interactive services. The aim is to decrease size of the image file in such way that the following post-processing compensates the drop of the visual quality. A low-complex post-processing technique is evaluated. Proposed solution is suitable for decreasing downloading time of the images in digital video broadcasting services.

Keywords: mhp, dvb, jpeg, post-processing.

1 Introduction

In recent days there is a fast growing market in digital television. Together with digitalization, the interactive services have become available. With the term Interactive Digital Television (IDTV) the experts refer to TV with interactive content and/or digital enhancements. IDTV combines traditional TV watching with new interactive digital applications that may be developed to run on TV. Digital applications are broadcasted within a TV video stream by broadcasters and retrieved by Set Top Boxes (STBs) that execute them. These applications may be complementary to TV programs (e.g., voting systems, electronic program guides), or even completely uncorrelated to the program being watched by the user, thus representing new services exploitable while the TV program goes on (e.g., games, news, t-learning, t-commerce, delivery of Web contents).

2 Interactive Digital Television

Digital television in Europe is covered by Digital Video Broadcasting (DVB) project which defines global standards for the delivery of digital television and data services [1]. DVB was extended with generic, common application programming interface to enable interactive applications to be downloaded from DVB broadcast networks and run on STB receivers. The applications and the running environment of the STB are standardized by Multimedia Home Platform (MHP) which defines a generic interface between interactive digital applications and the terminals (STB, integrated digital TV) [2].

2.1 MHP interactive applications and data

An MHP application is an interactive application written in Java programming language. These applications are called "Xlets". The application model is very similar to the Applets from Web pages in that they are loaded and run by a life cycle manager, residing on the STB. DVB has adopted Java language and has created a lightweight version called DVB-J. Typical application design scenario is to separate application executive code and data content (text, images...), which is transported independently.

2.2 Broadcast carousel - delivery of interactive data in DVB

DVB networks are based on the MPEG-2 (Motion Picture Experts Group) standard [3]. MPEG-2 transport stream (TS) is form by multiple elementary streams containing video, audio, interactive data and service information, see Fig.1. Although MPEG-2 provides a means of transporting the Java application along with the audio-visual content, there is a problem in that the viewer may change channel and want to run Java program at any point in the transmission. Unless a user elect to view that program from the beginning, the application would already have been broadcasted. The result is that the STB and the viewer would have missed it.

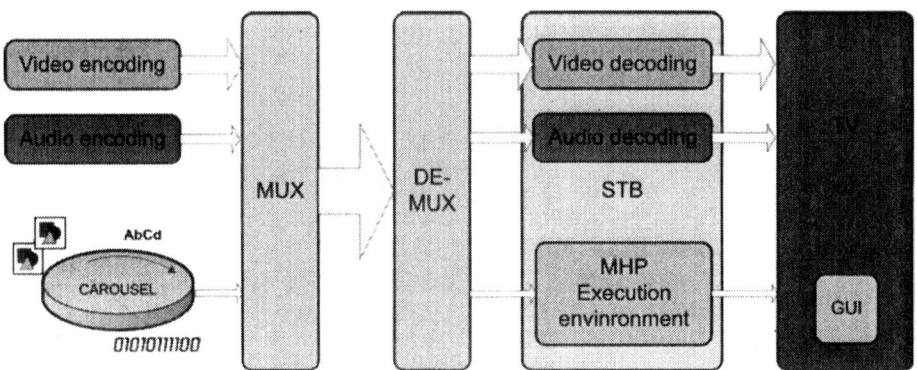

Fig. 1. Audiovisual and interactive data delivery in DVB.

The solution is in sending it over and over again so that the STB can pick up and assemble a complete application at any time during the transmission. This is provided by broadcast carousel (Digital Storage Media - Command and Control DSM-CC) - it keeps playing the same application around and around [4].

The packetized application is then continuously multiplexed with the audiovisual content for transmission. User can now join the transmission at any time and still have access to the interactive TV application. But it must be kept in mind that users can wait a full lap in the worst case to access a specific file unless the file is repeated during the lap. The longest access time for a file is defined by the bandwidth of the carousel and the size of the data in the carousel [5].

3 Images in DVB-MHP applications

Because of the visual character of the IDTV MHP applications, the images play important role in the application design and information transmission. The amount of the images in context with total size of the application and its data is very dependent on the type of the application. Images usually have large file size in comparison with the application code or text content.

The one of the goals of this paper is to investigate the influence of the perceived image quality of lossy compressed images in DVB-MHP applications.

3.1 Image compression in DVB-MHP

In MHP the image formats are JPEG and PNG as default for pictures. Because GIF format decoder implementation is left on the STB manufacturer so it is not recommended to use it for compressed pictures. When presenting Web pages on the STB using Web browser client the decoding of the GIF images should be possible. Besides these well-known image formats there is a possibility to use MPEG I-frames for compressed images.

JPEG
The most known image compression format in MHP applications is JPEG (Joint Photographic Expert Group). Officially, the JPEG corresponds to the ISO/IEC international standard 10928-1, a digital compression and coding of continuous-tone still images or to the ITU-T Recommendation T.81 [6].

JPEG is used for lossy compression of the natural images. It is a representative of the transform-based image codecs. Compression ratio is in range from 0.25 to 5 bit/pixel. Visual quality and compression efficiency is content dependent. Especially in low bit rates, the typical compression artifacts as blockiness and ringing became noticeable.

PNG and GIF
GIF (Graphics Interchange Format) is an 8-bit-per-pixel bitmap image format that was introduced by CompuServe company. It is lossless with exception that only 256 colors can be used. This limitation makes the GIF format unsuitable for reproducing color photographs and other images with continuous color, but it is well-suited for more simple images such as graphics or logos with solid areas of color [7].

PNG (Portable Network Graphics) is a bitmapped image format that also employs lossless data compression as GIF does. PNG was created to improve and replace the GIF format, as an image-file format not requiring a patent license. PNG supports palette-based (palettes of 24-bit RGB colors) or greyscale or RGB images [8]. Because these formats are lossless they are out of scope of this article.

MPEG I-frames and video drips

Besides familiar image formats the MHP receiver can use MPEG I-frames to display a full-screen image (such as a background image) or video "drips". In both cases, there is a need to be aware of a few limitations. Some receivers may use the hardware MPEG decoder for decoding I-frames and video drips. Decoding those image formats may disrupt any broadcasted video that is playing. This may be a short glitch while a frame is decoded, or it may last for the entire time the image is displayed, depending on the receiver.

Still images in the background layer will normally be MPEG I-frames. It is a MPEG-2 video sequence with only one frame. Video drip is a new content format that is pretty much unique to the digital TV world. The main aim of this format is to provide a memory-efficient way for displaying several similar images. Basically, it is a very short piece of MPEG-2 stream. The first thing in the file is an MPEG-2 I-frame that can be decoded and presented to the user. This is followed by one or more P frames, which are then decoded based on the preceding I-frame. This allows the decoder to update a static image in a very memory-efficient way [9].

3.2 Degradation of decoded image

Low bit rate image coding is essential for many visual communication applications. When bit rates become low, most compression algorithms yield visually annoying artifacts that highly degrade the perceptual quality of image and video data. When compressed at the same bit rate, images with more details usually degrade more than those with fewer details. The coding bit rate is another important factor that determines quality. In lossy compression, there is a trade-off between the bit rate and the resulting distortion.

The blocking effect is the most noticeable artifact associated with JPEG and MPEG compression standard. These blocking artifacts are square blocks with the edges aligned with the 8x8 regions processed via the discrete cosine transform block. Blockiness arises since each block is encoded without considering the correlation between adjacent blocks [10]. JPEG operates in spectral domain, trying to represent the image as a sum of smooth oscillating waves. Spectral domain is appropriate for capturing relatively smooth color gradients, but not particularly appropriate for capturing edges, so the "ringing" artifacts occur.

Generally speaking, the ringing effect occurs in all coding schemes that involve quantization in the frequency domain, see Fig.2.

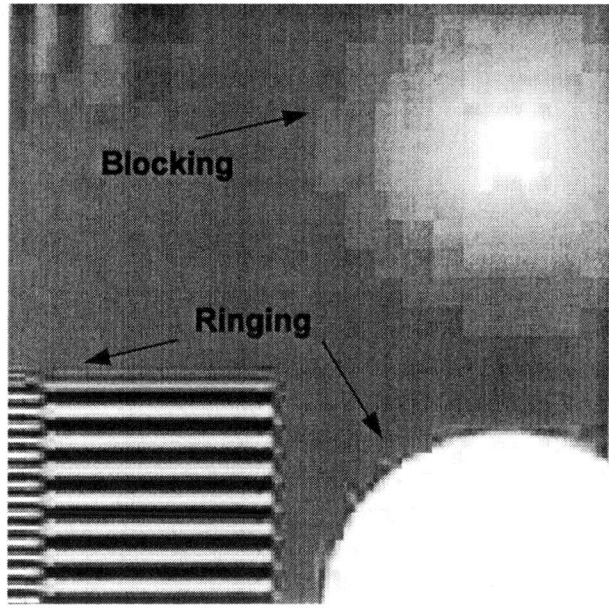

Fig. 2. Artifacts in JPEG image.

4 Transmission efficiency improvements

The second goal of this article is the evaluation of the proposals for image compression and/or transmission efficiency improvement. Firstly, it must be analyzed the multiplexed TS. For example, typical DVB-Terrestrial broadcasters' scenario is to have in one multiplex with 22 Mbits/s 4 or 5 audiovisual services with additional data for Electronic Program Guide (EPG). The multiplex also contains MHP applications in broadcast carousel. Because the MHP applications are supplementary services complementary to broadcasted audiovisual service the carousel bit rate is quite small, in range from 100kbits/s to 1Mbits/s. The size of the carousel data should not be too large. Long downloading time of the requested data is very annoying for the consumers. It could be expected that the waiting time should not be longer than 10 to 20 s. Therefore the maximum size of the carousel is 2Mbits to 20Mbits.

MPEG I-frame and JPEG is the only standardized solution in MHP for natural images coding.

4.1 Image post-processing

The investigated idea is that the broadcasted JPEG images are transmitted with lower quality and the STB employs sophisticated post-processing method for image enhancement. The file size of the image is reduced so the amount of the broadcasted images is higher or the same number of images has smaller file size and downloading time of the images is shorter. The first requirement that must be accomplished is that the total time from user request for the image to the finished downloading, decoding and displaying is longer then total time of downloading time of more compressed image, decoding, post-processing and displaying, i.e.

$$L_1 + D + V \geq L_2 + D + P + V , \qquad (1)$$

where D is decoding time, V is displaying time, P is post-processing time, L_1 and L_2 are downloading times. Therefore

$$L_1 \geq L_2 + P , \qquad (2)$$

because decoding and displaying time is the same for both cases.

4.2 Overview of post-processing techniques

The aim of post-processing is to reduce coding artifacts, like blockiness and ringing, whereby the objective and mainly subjective quality is improved. Post-processing methods can be divided into two directions:
- Algorithms based on image enhancement, it means subjective improving visual quality.
- Algorithms based on image restoration, it means objective recovering image as close to original as possible.

Another division can be done according to the compression used for images. Many post-processing methods have been developed for removing ringing and blocking artifacts in JPEG images. Existing methods for image enhancement JPEG compressed images are discussed. Recovering methods will be introduced very briefly.

Searching boundaries in JPEG images
Blockiness reduction using wavelet subband decomposition
This method uses wavelet-based decomposition which locates the blocky noise on the predetermined block boundaries of their corresponding subbands. The coefficients with high value in detail subbands are reduced by a linear minimum mean square error filter, which exploits the signal and noise characteristics [11].

Blockiness reduction employing spatial filtering of the image
Many approaches have been developed to reduce blockiness and ringing utilizing the local image characteristics and filtering methods driven by quantization parameters.

Blocking artifacts detection is based on variance differences within each 8 x 8 block. If the block boundary is recognized the 1D spatial low-pass filter is applied across the boundary. Then each pixel is classified according to a local variance value into three groups: smooth region, texture region and edge region. All pixels in block, where at least one pixel is classified as an edge pixel, are filtered by an adaptive fuzzy filter [12], [13].

A spatial filtering performed by an adaptive filters controlled by quantization parameters are also very often used. These filters can be linear or nonlinear. It must be noticed that nonlinear filter reduces the blockiness more effectively. A quantization parameter is important parameter for artifact reduction, because a new pixel value should not exceed the interval delimited by a quantization step.

Usually a 1D spatial filter is used for reducing the block artifacts and 2D spatial filter for reducing ringing artifacts [14].

Iterative methods for image restoration

Several classical image restoration techniques, including projection onto convex sets (POCS) and maximum a posteriori (MAP) restoration have been used for minimizing the compression artifacts. These methods are mostly very complex due their iterative execution.

Method using weighted combinations of shifted transforms is developed for deringing and deblocking [15]. A shifted transform is applied to the input image, then modified by a non-linear denoising point transformation and then inverse transform to the shifted transform is applied.

An N-point DCT is employed to obtain the local image characteristics and a 2N-point DCT is employed to obtain the global one. The comparison between N-point and 2N-point DCT coefficients makes the possibility to detect the high frequency components, which correspond with the block artifacts [16].

Proposed algorithm

All post-processing methods are computationally extensive. One of the low-complexity denoising methods has been introduced in [17]. This method simply re-applies JPEG algorithm to the spatially shifted versions of the already compressed images and forms an average. Although very simple, the effectiveness is at least comparative to nonlinear filtering methods, POCS and wavelet denoising methods. In that method, the forward and backward JPEG coding is performed 63 times and the result is average of these computations and received image. Only non-invertible portions of JPEG (DCT and quantization) are participated in the algorithm for speed-up the processing.

In our experiments the simplified post-processing version of this algorithm has been evaluated. We have used only 3 shifted forward and backward JPEG coding as depicted on Fig.3. The drop of the improvement is trade-off between computational complexity and visual enhancement as can be seen on Table 1.

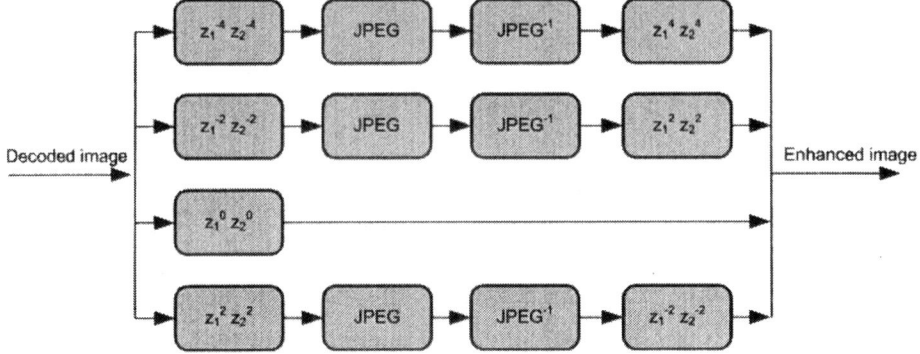

Fig. 3. Proposed algorithm of reduced re-application of JPEG.

Table 1. PSNR comparison for Lena image.

JPEG PSNR [dB]	PSNR improvement [dB]	
	JPEG [16]	Proposed
26,65	1,17	0,83
29,74	1,01	0,74
32,34	0,65	0,64

5 Post-processing time requirements analysis

Analysis of the computational load of the processor in the STB is based on assumptions defined in Eq.2. As a possible model of the carousel we have established broadcast carousel with bit rate of 200 kb/s. Lena image has been used in 3 different spatial resolutions and 1 cropped version of Bike image in greyscale. Quality factor 15 has been set in JPEG compression. After proposed post-processing, the new less compressed image has been created with objective quality equal to the original JPEG image. The size differences between original JPEG and Equal JPEG give the maximum available post-processing time which is dependent on the carousel bit rate as can be seen in Table 2. Visual comparison shows that there is some loss of sharpness in post-processed image in comparison with equal objective quality JPEG, but the subjective perceived quality has been improved, see Fig.4.

Table 2. Analysis of post-processing time

Width	px	720	256	512	512
Height	px	576	256	256	512
Pixels	-	414 720	65 536	131 072	262 144
PAL size	%	100,0	15,8	31,6	63,2
Decoded	dB	26,93	30,59	31,46	32,50
Bit rate	b/p	0,58	0,42	0,34	0,26
Size	B	30 315	3 420	5 549	9 531
Processed	dB	27,37	31,15	32,04	33,13
Equal JPEG	dB	27,45	31,20	32,10	33,20
Size	B	33 948	3 832	6 222	10 617
Size saving	B	3 633	412	673	1 086
Carousel bit rate	kb/s	200			
Delay	s	10			
Processing time	s	0,142	0,016	0,026	0,042

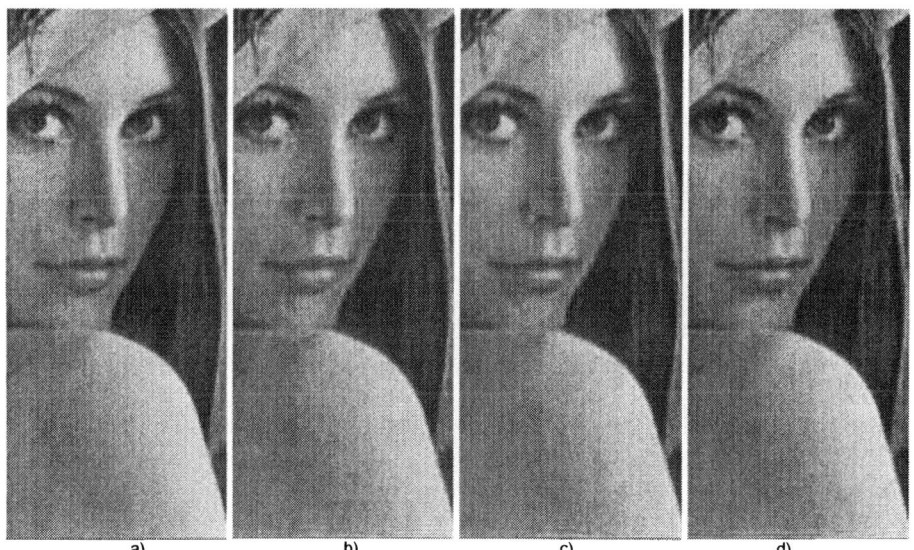

a) b) c) d)

Fig. 4. Post-processing of the Lena image: a) Original, b) Compressed (PSNR = 32,50 dB), c) Post-processed (PSNR = 33,13 dB), d) Equal objective quality JPEG without post-processing (PSNR = 32,30 dB).

6 Conclusions

This paper presented image compressions standardized for DVB-MHP applications. In lossy compression there is a trade-off between visual quality and file size. Compression artifacts occur in highly compressed images. Short overview of post-processing techniques is described and low-complex post-processing technique based on re-application of the JPEG algorithm proposed. The PSNR improvement is more then 0,6 dB. Although there is some loss of sharpness in post-processed image in comparison with equal objective quality JPEG, the subjective perceived quality has been improved. Post-processing time speed-up and enhancement efficiency is still in subject of further research.

Acknowledgement. The research has been supported by Research Program No. MSM0021630513 and Academy of Science Project No. 1ET301710510.

References

1. DVB project website, URL: http://www.dvb.org.
2. Digital video broadcasting (DVB), Multimedia Home Platform (MHP) specification 1.0.3, ETSI Doc. No. ES 201 812 V1.1.1, 2003.
3. ISO/IEC 13818: Generic coding of moving pictures and associated audio information.
4. ISO/IEC 13818-6:1998, Generic coding of moving pictures and associated audio information; Part 6: Extensions for digital storage media command and control, 1998.
5. Jones, J.: DVB-MHP/JavaTV Data Transport Mechanisms. In Proc. Fortieth International Conference on Technology of Object-Oriented Languages and Systems (TOOLS Pacific 2002), Sydney, Australia, pp. 115-121, Volume 10, 2002.
6. Wallace G. K.: The JPEG still picture compression standard, In Communications of the ACM, pp. 30 – 44, Volume 34, Issue 4, 1991.
7. Cover Sheet for the GIF89a Specification, URL:http://www.w3.org/Graphics/GIF/spec-gif89a.txt.
8. Portable Network Graphics (PNG) Specification (Second Edition) Information technology - Computer graphics and image processing -Portable Network Graphics (PNG): Functional specification. ISO/IEC 15948:2003 (E), URL: http://www.w3.org/TR/PNG/.
9. Morris, S., Chaigneau, A. S.: Interactive TV Standards: A Guide to MHP, OCAP, and JavaTV. Focal Press, 2005.
10. Shen M.Y.; Kuo C.-.C.J.: Review of Postprocessing Techniques for Compression Artifact Removal, In Journal of Visual Communication and Image Representation, pp. 2-14, Volume 9, Issue 1, 1998.
11. Huyk Choi and Taejeong Kim: Blocking-Artifact Reduction in Block-Coded Images Using Wavelet-based Subband Decomposition, In IEEE Transactions on Circuit and Systems for Video Technology, pp. 801 - 805, Volume 10, 2000.
12. Hao-Song Kong, Yao Nie, Vetro, A., Huifang Sun and Barner, K.E.: Adaptive fuzzy post-filtering for highly compressed video, In Proceedings of International Conference on Image Processing ICIP '04, 24-27 Oct. 2004, Volume 3, pp. 1803 - 1806, 2004.
13. Hao-Song Kong, Vetro, A., Huifang Sun: Edge map guided adaptive post-filter for blocking and ringing artifacts removal, In Proceedings of International Symposium on Circuits and Systems, 2004. ISCAS '04, 23-26 May 2004, Volume 3, pp. 929 – 932, 2004.

14. Hoon Paek, Rin-Chul Kim and Sang-Uk Lee: A DCT-Based Spatially Adaptive Post-Processing Technique to Reduce the Blocking Artifacts in Transform Coded Images, In IEEE Transactions on Circuit and Systems for Video Technology, pp. 36 - 41, Volume 10, 2000.
15. Samadani, R.; Sundararajan, A.; Said, A.: Deringing and deblocking DCT compression artifacts with efficient shifted transforms, In Proceedings of International Conference on Image Processing ICIP '04, 24-27 Oct. 2004, Volume 3, pp. 1799 - 1802, 2004.
16. Hoon Paek and Sang-Uk Lee: On the POCS-based Postprocessing technique to Reduce the Blocking Artifacts in Transform Coded Images, In IEEE Transactions on Circuit and Systems for Video Technology, pp. 358 - 367, Volume 8, June 1998.
17. Nosratinia, A.: Denoising JPEG images by re-application of JPEG, In IEEE Second Workshop on Multimedia Signal Processing, 7-9 Dec 1998, pp. 611-615, 1998.

Fast lifting wavelet transform and its implementation in Java

Jan Maly [1], Pavel Rajmic[1]

[1] Brno University of Technology,
Faculty of Electrical Engineering and Communication,
Department of Telecomunications,
Purkynova 118, 61200 Brno, Czech Republic
xmalyj05@stud.feec.vutbr.cz, rajmic@feec.vutbr.cz

Abstract. Fast lifting wavelet transform is a technique which replaces standard discrete wavelet transform used in computation of wavelet coefficients. The idea of lifting comes from the lifting scheme, a method used in wavelet design. The standard method relies on convolution of the original signal with FIR filter structures. Fast lifting scheme basically breaks up the original filters into a series of smaller structures, providing a very sophisticated and versatile algorithm that is up to 50 % faster than the standard way with no extra memory requirements. This paper discusses an implementation of this algorithm in Java language, comparing both speed and efficiency of standard and fast lifting wavelet transform for CDF 9/7 filters, which are used in lossy image compression in JPEG2000 standard. Java has been chosen for its platform independent character and easy integration in mobile devices..

Keywords: Fast, lifting, wavelet, transform, convolution, java, CDF

1 Introduction

The *discrete wavelet transform* has lately gained significance in many signal applications, including analysis and compression. Lossy image compression standards based on DWT alow us to benefit from much better compression efficiency and many interesting features, such as progressive codec behavior or supression of blocking artifacts. Anyway, these depend heavily on a choice of wavelet coefficients coding (typical representatives are vector-based methods - Embedded Zerotree Wavelet [3] and Set Partitioning in Hierarchical Tress [4] and scalar EBCOT – used in JPEG2000 standard). All of these features need a versatile and fast DWT coder and decoder; without that component in the coding chain, no efficient implementation could ever be proposed. This problem is much more relevant in any application which lacks computation power and need to use as little memory, as possible. A typical representative of such an application is a mobile device – such as a standard today's cell phone.

Although it is generally possible to improve existing hardware architecture to contain support for DWT directly (specific realizations were proposed for this purpose, exploiting many interesting hardware features such as parallel processing -

[1]), sometimes it is cheaper and more desirable to implement this feature by software. Mobile devices do integrate support for Java™ Virtual Machine, making it easy to write platform-independent programs. In the following paper, we will analyze the efficiency of CDF 9/7 filter fast lifting wavelet transform realization and compare it to standard convolutional approach. But first we need to focus on the theory beyond the lifting scheme.

2 Discrete wavelet transform

Proposed by [5], discrete wavelet transform consists of analysis (wavelet coefficients computation) and reconstruction (signal re-assembly) stage. Both stages incorporate filtering with two adjacent FIR structures, the high pass (h) and low-pass (l) filters. Analysis filters are formed by a pair ($\bar{h}$, $\bar{g}$), synthesis (reconstruction) filters are described by a pair (h, g). Every set of coefficients created by analysis is a subject of subsampling (removing even results), so every stage produces exactly half the amount of source data. This process meets the Nyquist's rule, as long as half of the frequencies have been discarded in every filtering step. At the synthesis stage, zeros must be inserted consequently before filtering steps (Fig.1). If within this scheme perfect reconstruction is archieved (which is an essential condition for successful compression), we call the filter pair (h, g) *complementary*.

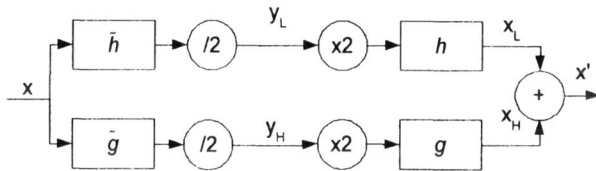

Fig. 1: DWT stages description. x is the input signal, y stands for wavelet coefficients, x' is the reconstructed signal. /2 a x2 blocks represent the subsampler and upsampler, respectively.

2.1 Perfect reconstruction, polyphase matrices

The perfect reconstruction is archieved by satisfying the following set of conditions [6]:

$$h(z)\widetilde{h}(z^{-1}) - g(z)\widetilde{g}(z^{-1}) = 2$$
$$h(z)\widetilde{h}(-z^{-1}) + g(z)\widetilde{g}(-z^{-1}) = 0 \quad (1)$$

To understand the problem more, we need to define a polyphase representation of the filters defined in the form of polyphase matrix, a 2×2 transform matrix that treats the odd and even samples independently. Polyphase matrix for synthesis is defined as

$$P(z) = \begin{bmatrix} h_e(z) & g_e(z) \\ h_o(z) & g_o(z) \end{bmatrix}, \qquad (2)$$

where h_e and h_o are defined as even and odd samples of h, that is

$$h(z) = h_e(z^2) + z^{-1} h_o(z^2). \qquad (3)$$

The same principle as (3) applies to g_e and g_o coefficients. We can also define analysis polyphase matrix in a very simmilar way:

$$\widetilde{P}(z) = \begin{bmatrix} \widetilde{h}_e(z) & \widetilde{g}_e(z) \\ \widetilde{h}_o(z) & \widetilde{g}_o(z) \end{bmatrix}. \qquad (4)$$

For a perfect reconstruction, $P(z)$ and $\widetilde{P}(z)$ must satisfy the following:

$$P(z)\widetilde{P}(z^{-1})^T = I \qquad (5)$$

Based on the above formulations, discrete wavelet transform can be described in terms of polyphase matrices as

$$\begin{bmatrix} y_L(z) \\ y_H(z) \end{bmatrix} = \widetilde{P}(z) \begin{bmatrix} x_e(z) \\ z^{-1} x_o(z) \end{bmatrix} \qquad (6)$$

for analysis and

$$\begin{bmatrix} x_e(z) \\ z^{-1} x_o(z) \end{bmatrix} = P(z) \begin{bmatrix} y_L(z) \\ y_H(z) \end{bmatrix} \qquad (7)$$

for synthesis.

3 Lifting scheme

The idea of lifting is based on the following: If a filter pair (h, g) is complementary, then for every filter s the pair (h', g), where

$$h'(z) = h(z) + s(z^2) \cdot g(z) \qquad (8)$$

is complementary, too. This rule also applies symetrically as

$$g'(z) = g(z) + t(z^2) \cdot h(z). \qquad (9)$$

The principle can be reversed, so that we can explicitly say: if the filterbanks (h, g) and (h', g) allow for perfect reconstruction, then there exists an unique filter s satisfying (8). Each such transform is called a *lifting step* – because what we perform is lifting the values of one particular subband with the help of the other.

In the language of polyphase matrices, lifting step described by (8) produces new polyphase matrix $\widetilde{P}^{new}(z)$, which is defined (based on (2) - [5]) as

$$\widetilde{P}^{new}(z) = \begin{bmatrix} 1 & \widetilde{s}(z) \\ 0 & 1 \end{bmatrix} \widetilde{P}(z). \qquad (10)$$

Because we have lifted the low-pass subband with the help of the high-pass subband, this step is called *primal lifting* or *update step*. By taking (9) into consideration we can define

$$\widetilde{P}^{new}(z) = \begin{bmatrix} 1 & 0 \\ \widetilde{t}(z) & 1 \end{bmatrix} \widetilde{P}(z). \qquad (11)$$

By lifting the high-pass subband with the help of the low-pass one, we have just made *dual lifting* or *predict step* (prediction of the odd samples from the even samples).

Lifting factorization [6] is a generally defined process that is used to factorize complementary wavelet filter pair into a series of lifting steps. By computing greatest common divisor (*gcd*) of even and odd filter values using the Euclidean algorithm, we obtain a structure consisting of subsequent pairs of primal and dual lifting steps:

$$\widetilde{P}(z) = \left\{ \prod_{i=1}^{m} \begin{bmatrix} 1 & \widetilde{s}_i(z) \\ 0 & 1 \end{bmatrix} \begin{bmatrix} 1 & 0 \\ \widetilde{t}_i(z) & 1 \end{bmatrix} \right\} \begin{bmatrix} K & 0 \\ 0 & 1/K \end{bmatrix} \qquad (12)$$

where $\widetilde{s}_i(z)$ and $\widetilde{t}_i(z)$ are Laurent polynomials computed for each step i by the *gcd* algorithm as the resulting quotient (with the division remainder being zero). In practice, these are usually first or second order polynomials (one to three-tap FIR filters). K and $1/K$ act as resulting stream scaling constants and can be omitted if we accept the fact of resulting coefficients being scaled.

4 CDF 9/7 lifting implementation

The most efficient CDF (Cohen-Daubechies-Feauveau wavelet) 9/7-tap lifting factorization is as follows [6] (only corresponding step factors s_i and t_i from (12) are listed):

$$s_1(z) = a(1+z^{-1})$$
$$t_1(z) = b(1+z)$$
$$s_2(z) = c(1+z^{-1})$$
$$t_2(z) = d(1+z)$$
(13)

where a=-1.586134342, b=-0.05298011854, c=0.8829110762 and d=0.4435068522. Corresponding scaling factor is K=1.149604398. Constants a, b, c, d and K are derived from the factorization process in order to keep the final structure being capable of perfect reconstruction. The result is in the form of Laurent polynomials with degree being 1 - as proposed in [7], every pair of symmetric filters with at least dissimilar leghts (with difference in length being at least 2) may be factored in lifting steps of this form.

4.1 Data dependency diagram

Fig. 2: Data dependency diagram for 4-stage lifting factorization of CDF 9/7 filters analysis.

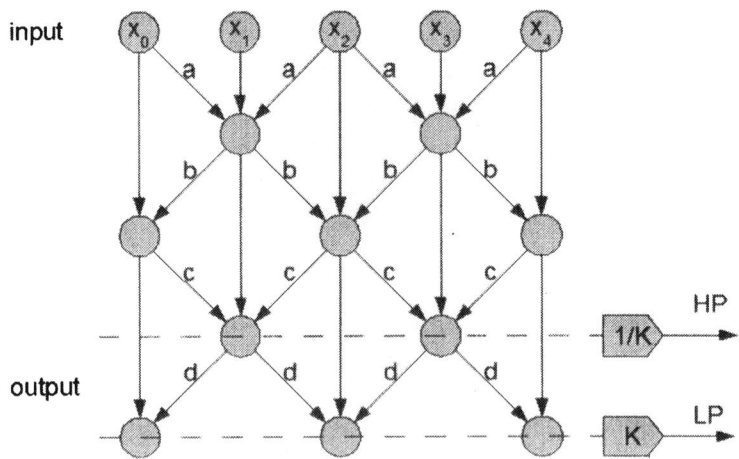

Every circle under the input stage represents an accumulator, arrows with a,b,c or d specify multiplication with the respective constant.

To explain the situation more in detail, we define a *data dependency diagram*, which focuses on data flow in the analyzing (synthesizing) structure (Fig.2). The data dependency diagram has an interesting property – at any point of operation no extra memory is needed to store data dedicated for next step computation – making the whole transform possible to be done "in place".

When compared with classic approach via convolution, for a 9-tap filter (worst case), not only we need to perform 9 multiplications and summarize them to generate one resluting sample, we also have to actually store 8 previous coefficients from the input stream for the next sample to be processed. Therefore, savings of the fast lifting approach are evident.

4.2 Boundary handling in general

Because the filtering process is applied to finite signals, the coding system must handle boundaries with special approach to avoid discontinuity-based errors. A general solution to this problem is to extend the signal. Extending is done *periodically* at both boundaries. When using CDF 9/7 filterbank, we use the "whole-sample" symmetric extension (WSS), which is described in the following simple example [1]:

Example: Consider a signal ABCDEFGH. For odd-length filters (which is a case of CDF 9/7) we can extend the signal (underlined) as

$$HGFEDCB\underline{ABCDEFGH}GFEDCBA$$

For even-length filters, a very simmilar way is proposed, known as HSS ("half-sample" symmetric extension), with the only difference being a duplication of the boundary item into the extended part, too (mirror symmetry).

4.3 Fast lifting boundary handling

When we discuss the boundary handling of fast lifting approach [2], we usually adopt the point of view that treats every lifting step as a separate subband transform. That means, extension of the signal is defined due to the nature of the corresponding Laurent polynomial in (12). Because in the case of CDF 9/7 the degree of this polynomials is always 1 (odd-length filters), we use HSS extension with one sample.

While observing the data dependency diagram on (Fig.2), we can deduce that no extra memory will be needed as we only have to estimate the value from the previous step symetrically to left or right, depending on the location of signal boundary (Fig.3). This leads to additional savings when compared to convolutional approach, where extra memory (or conditional algorithmic solution, which needs no extra memory, but on the contrary slows down the computation) is needed to extend the original signal – and in the case of CDF 9/7 the extension is quite noticeable.

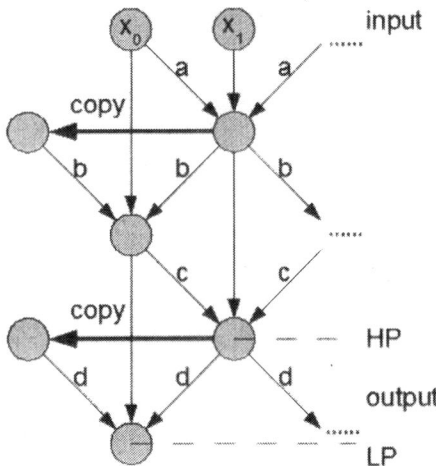

Fig.3: Data dependency diagram for fast lifting CDF 9/7 approach, with signal boundaries taken into consideration.

5 Java Implementation

The proposed Java™ implementation is based on two classes, one for convolutional-approach dwt, which has been written for testing purposes, and one for lifting-based dwt. Both classes offer the same interface – with two static methods analyze() and synthesize(), whose parameter is always the source data to be transformed to the resulting output. As we intend to use this lifting-based coder on tasks of lossy image compression in the future, we must declare all the input and output memory space as double precision floating point numbers1.

Convolution-based coder is basically a periodically extended standard convolution of the source signal with the corresponding filter of the CDF 9/7 set. It uses extra memory buffer with extended source signal, as this solution is computationally most efficient. The only drawback is that we need to process the resulting array to gain only the relevant coefficients, which slows the process down.

Lifting-based approach has no need to allocate extra memory space. It consists of four loops, that process the entire signal by the mean of each step's factorized filter. The signal is then scaled by the corresponding constant. As we use the "in-place" memory approach, source array is modified directly in the four steps. As visible from data dependency diagram (Fig.2), odd values represent the high-pass output, while

---

[1] It is indeed possible to use integers – for example in the case of lossless compression mode used by JPEG2000 format with LeGall 5,3 filterbank [2].

even values stand for the low-pass output. Thus, we need to rearrange the resulting array to get whichever format suitable for our needs.

Boundary handling is solved by the algorithm as proposed in the previous chapter, that means predicting values out of bounds by estimating it from 1-sample HSS extension. This solution is simple and results in perfect reconstruction (with the exception of precision-based floating point errors).

Sample: A source code used to compute two adjacent lifting steps (predict and update) is listed here. x represents the full source data vector and n is length of the vector.

```
a=-1.586134342;
for (i=1;i<n-2;i+=2) {
   x[i]+=a*(x[i-1]+x[i+1]);
}
x[n-1]+=2*a*x[n-2];          // boundary handling

a=-0.05298011854;
for (i=2;i<n;i+=2) {
   x[i]+=a*(x[i-1]+x[i+1]);
}
x[0]+=2*a*x[1];              // boundary handling
```

To test the two coders, we have analyzed a very simple bicubic signal into wavelet coefficients, then synthesized them back to gain source signal and measured the elapsed time. Resulting graph is on (Fig.4).

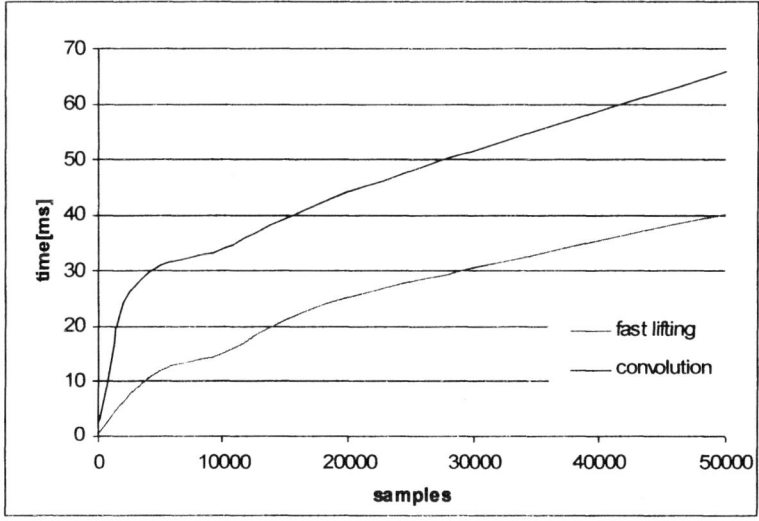

Fig.4: Dependency of elapsed computational time on the number of processed samples for fast lifting and convolution approach. Tested on PC with 2GHz AMD Athlon CPU with Java J2SE JDK1.6.0 on Windows.

6 Conclusion

The fast lifting-based approach for the discrete wavelet transform presents a very efficient way of computing wavelet coefficients. We have successfully created a Java^{TM} implementation of this coding technique for the case of CDF 9/7 filterbanks, which is a starting point of proposing lossy image compression format based on DWT, that will make use of this platform.

Testing of the algorithm has confirmed it to be superior when compared to the standard convolutional approach, both in terms of computational speed and memory efficiency. The algorithm is also much simpler to implement than the standard convolutional way, because we don't need to perform manipulation on the source signal in order to archieve boundary extension and cutoff (after the transform).

Although being very simple to use in one-dimensional processing, this algorithm needs to be analyzed in order to be used in 2D-DWT. Standard way of processing 2D data is to apply the transform row-wise (to produce L and H subbands) and then column-wise on the results to produce four more subbands LL, LH, HL and HH. This forces us to store the results in corresponding format in memory in order to archieve a standard dyadic-tile wavelet decompostion.

Acknowledgments. The paper was prepared within the framework of No. 102/06/P407 and No. 102/07/1303 projects of the Grant Agency of the Czech Republic and No. 1ET301710509 project of the Czech Academy of Sciences.

References

1. Acharya T.: JPEG 2000 Standard For Image Compression: concepts, algorithms and VLSI architectures, Wiley-Interscience (2005)

2. Taubman D.S., Marcelin M.W.: JPEG 2000 – Image compression fundamentals, standards and practice, Kluwer Academic Publishers (2002)

3. Shapiro, J. M.: Embedded Image Coding Using Zerotrees Of Wavelet Coefficinet,. IEEE Transactions on Signal Processing, Vol. 41, No. 12 (1993)

4. Said A., Pearlman W.A.: A New Fast and Efficient Image Codec Based on Set Partitioning in Hierarchical Trees, IEEE Transactions on Circuits and Systems for Video Technology, vol. 6 (1996)

5. Mallat, Stéphane: A Wavelet Tour of Signal Processing, 2nd edition, Academic Press (1999)

6. Daubechies I., Sweldens W.: Factoring wavelet transforms into lifting schemes, The J. of Fourier Analysis and Applications, Vol.4, pp. 247-269 (1998)

7. Vetterli,M., Le Gall D.: Perfect reconstruction FIR filter banks: Some properities and factorizations. IEEE Trans. Acoust. Speech and Sig. Proc., 37(7):1057-1071 (1989)

New watermarking scheme for colour image

Petr Cika[1]

[1] Department of Telecommunications,
Brno University of Technology,
Purkynova 118, 612 00 Brno, Czech Republic
cika@feec.vutbr.cz

Abstract. This paper deals with a new watermarking scheme in the time domain. The method tested follows up on a method that was presented in [1]. Compared with the method from [1] there are some changes in the new one, which improve the detection and extraction process. The watermark is embedded into the red and green components of a colour image. This has, on the one hand, an effect on the quality of watermarked image; on the other hand it increases the possibility of extracting the watermark even from very modified images.

Keywords: spatial watermark, encryption, watermarking system

1 Introduction

Most of the multimedia data are presently saved in the digital form. The possibilities of long-time archiving and copying without loss of quality belong to the big advantages of data saved in this form. But such data also have some disadvantages. One of them is the necessity of using compression; another disadvantage is, for example, data authentication. For the protection of multimedia data against theft, modification and for data authentication new methods are sought.
Nowadays, two basic possibilities exist for securing multimedia data: encryption and watermarking. By encryption the original multimedia data are modified and made unreadable for attackers or plagiarists. These encrypted audio or video data make a correct playback without the knowledge of decryption method and key impossible. The opposite situation occurs in the case of watermarking. The watermarking methods are designed such that they hide the information into the original multimedia data. During image or video viewing or audio listening the secret information is imperceptible. The watermark is uncovered only for authentication.
The multimedia data can be watermarked in the time domain, frequency domain or parametric domain. in most time- domain watermarking methods the watermark data are embedded into the least significant bits (LSB). The control sum of all image elements is embedded into the LSB, for example. In [2] the watermark is embedded near the object boundaries. It is very easy to remove the watermark, for example, by means of compression. In [3] a watermarking method is described that chooses randomly n-pairs of image pixels (a_i,b_i) and increases the a_i by one, whereas it

decreases b_i by one. The detection of this watermark is performed by comparing the sum of differences a_i and b_i. The assumed result is $2n$. The next example is watermarking which modifies the luminance blocks of the image [4]. In this method the block selection is very important. Single blocks are classified as hard, progressive and noise contrast blocks. The pixels in the blocks are divided into 2 zones: zone 1 and zone 2. Each of these is split into two categories, A and B. The insertion of bits can be described by the following equations:

If the embedded bit equals 0 $m1B^* - m1A^* = L; m2B^* - m2A^* = L,$
if the embedded bit equals 1 $m1A^* - m1B^* = L; m2A^* - m2B^* = L,$

where $m1A$, $m1B$, $m2A$ and $m2B$ are the average values of the luminance component after the bit insertion, and L is the insertion depth. The watermark block scheme using the MD5 hash function is described in [5]. In [6] watermarking schemes are presented that are based on the image properties. A new view of colour image watermarking is described in [1]. This method uses the green component of the RGB space and modifies it for watermarking. This method was used for the proposal of the present new watermarking scheme.

2 New watermarking scheme

Some information from [1] was applied when creating the new watermarking scheme. The watermark data bit size must equal the number of 8x8 blocks in the original image. The original image data are summed by the ex-or operation with a pseudo-randomly generated sequence of the same length as the watermark data. In the next step, each bit of this new bit sequence is embedded into the 8x8 block of the red and the blue components of the original image. On the user side it is possible to decode the watermark via using the original data, the embedded watermark and the pseudo-random sequence. The algorithm is described in Figures 2.1 and 2.2.

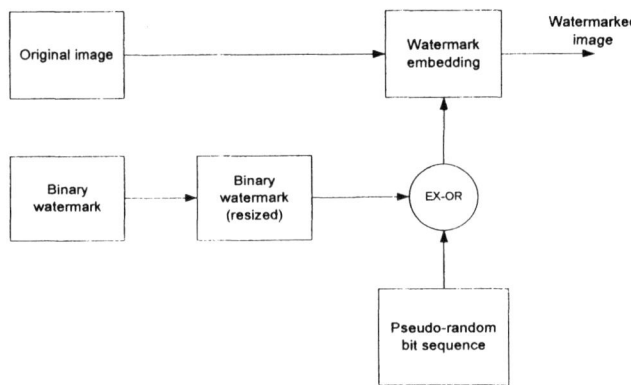

Fig 1. Watermark embedding

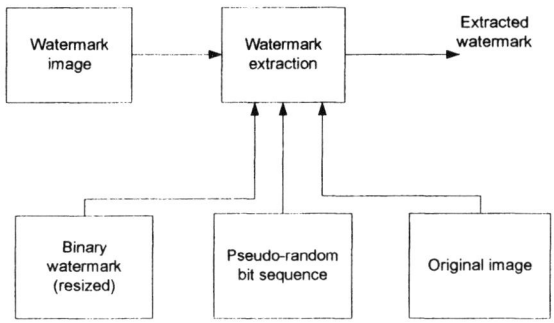

Fig 2. Watermark extraction

2.1 Watermark embedding

In the proposed algorithm the watermark is embedded into the 8x8 blocks of the image. The image is divided into the 8x8 blocks in the first step. The watermark is then resized according to the number of 8x8 blocks of the original image. During the next step, a pseudo-random bit sequence is generated. The length of this sequence equals the number of 8x8 blocks of the original image. The watermark bits are ex-or summed with the pseudo-randomly generated bit sequence. Each bit of the new sequence is embedded into the 8x8 blocks of the original image. For the insertion the blue and the red components of the image were chosen. The whole method for watermark insertion is described by the following equations.

If $W = 1$:
For all pixels of the 8x8 block

$$I_W = I_0 + k, \qquad (1)$$

if $W = 0$:
For all pixels of the 8x8 block

$$I_W = I_0 - k, \qquad (2)$$

where I_0 are the original image values, I_W are the watermarked image values, and k is the insertion depth. This mechanism is described in [1], but there the watermark is embedded only into the green component. The advantage of embedding the watermark into both (the red and the blue) components can only be observed at watermark extraction. From these modified blocks a watermark image of the same size as the original image is composed.

2.2 Watermark extraction

For a correct function, the algorithm for watermark extraction needs the original image, the watermarked image, the original watermark, and the pseudo-random bit sequence. The whole process begins with watermark detection. The watermarked and the original images are divided into 8x8 blocks. Subsequently, the blocks of the individual colour components in the watermarked and the original images are compared. For each block of the selected colour component the following operations are executed:

- Choosing two parameters, P_0 and P_1, those determine the probability of the occurrence of 0 or 1.
- Each pixel of the 8x8 blocks of the original image I_0 is compared with the same pixels of the watermarked image block I_W.
- If $I_0 > I_P$ then $P_1 = P_1 + 1/64$.
- If $I_0 \leq I_P$ then $P_0 = P_0 + 1/64$.
- the decoded bit of one block is
    - 1, if $P_1 > P_0$,
    - 0, if $P_1 \leq P_0$.

In the above way the sequence of the extracted watermark is gradually obtained. This sequence is ex-or summed with the pseudo-random sequence. The extracted watermark, which is later compared with the original watermark, is the result. As two colour components are used for watermark embedding, a new, probability-based method was developed. During watermark extraction we get two different watermarks from the watermarked image (the red and the green components). The original watermark, C, is available too. For the extraction we developed the following rules:

- if $0.3A + 0.3B + 0.3C > 0.5$, then the resultant bit is 1,
- If $0.3A + 0.3B + 0.3C \leq 0.5$, then the resultant bit is 0.

3 Quality parameters

The Peak Signal to Noise Ratio (PSNR) was used for testing the quality of the embedded image. The final PSNR value is expressed in the case colour images by the equation [7]

$$PSNR = 10\log_{10}\left(\frac{255^2}{\frac{MSE(R)+MSE(G)+MSE(B)}{3}}\right), \qquad (3)$$

where MSE is Mean Square Error defined by the equation [7]

$$MSE(x) = \frac{1}{MN}\sum_{m=0}^{M-1}\sum_{n=0}^{N-1}[x(m,n)-x'(m,n)]^2, \qquad (4)$$

where M, N define the image size, x is the pixel value of the original image, and x' is the pixel value of the watermarked image.
The Normalized Cross Correlation function was used for quality testing of the extracted watermark. NCC is defined with the equation [7]

$$NCC = \frac{\sum_{i=0}^{I-1}\sum_{j=0}^{J-1}W_{ij}W'_{ij}}{\sum_{i=0}^{I-1}\sum_{j=0}^{J-1}[W_{ij}]^2}, \qquad (5)$$

where I, J define the size of the embedded watermark and W, W' define the original and extracted watermark bits.

4 Results of new watermarking scheme testing

The proposed algorithm was tested for the robustness to the following modifications:
- JPEG compression
- Image rotation
- Image resizing
- Image cropping

The Lenna colour image, size 512x512 pixels, was chosen for testing (Fig3a). The binary watermark is in Figure 3b. The insertion depth $k = 4$ was chosen for testing.

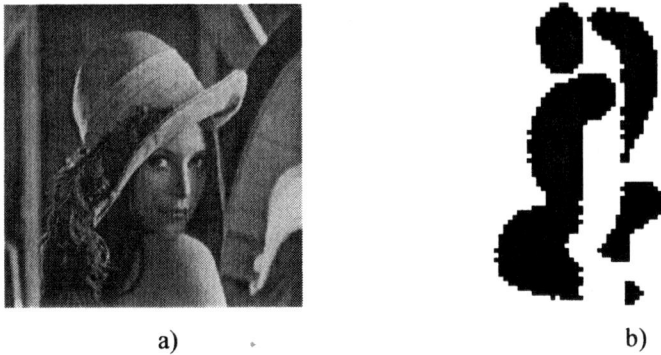

Fig 3. a) Original image - Lenna, b) Embedded watermark

Figures 4a and 4b show the results after applying the JPEG compression to the watermarked image. Figure 4c is the extracted watermark from the JPEG compressed image with a compression factor of 60. The similarity with the original watermark in Figure 3b is strong.

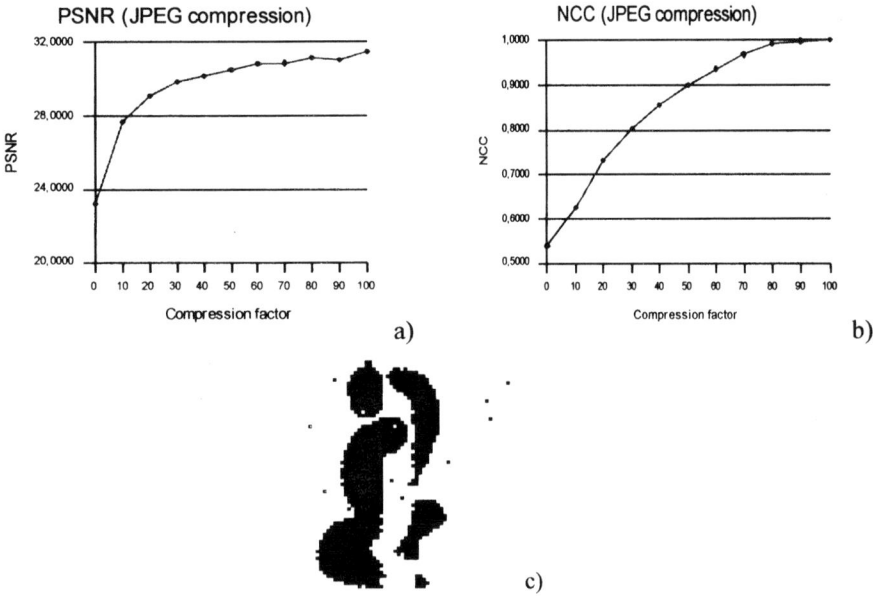

Fig 4. JPEG compression a) Extracted image PSNR , b) Watermark NCC, c) Extracted watermark (JPEG $q=60$)

Figures 5a and 5b show the results after resizing the original image. The original image size is 512x512 pixels. This size was multiplied by the values 0.1, 0.2, ..., 1.7.

Above a resize factor of 0.6 the extracted watermark is almost the same as the original one.

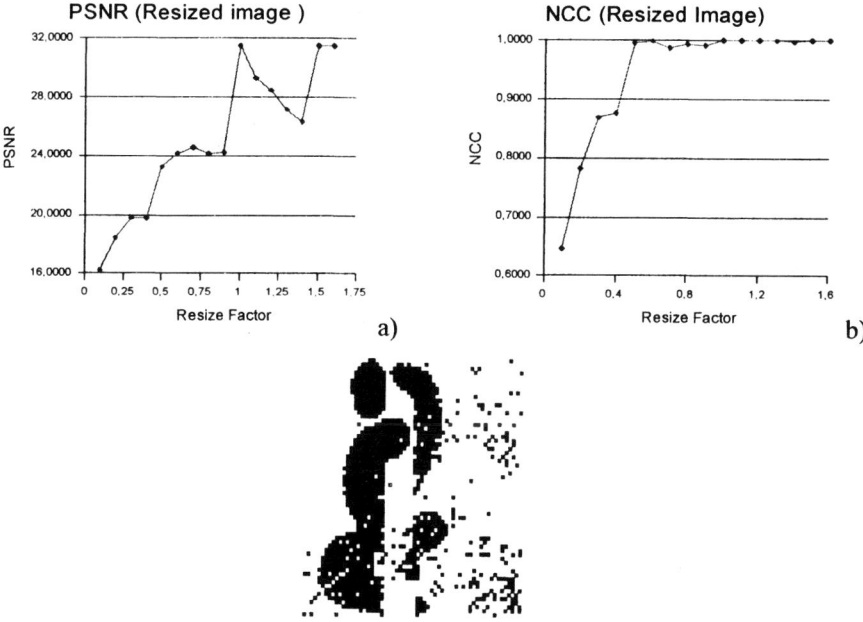

Fig 5. Resized image a) Extracted image PSNR , b) Watermark NCC, c) Extracted watermark (resized 0.4x)

Figure 6 shows the NCC value of the watermark after image rotation. Rotation has not had any effect on the extracted watermark.

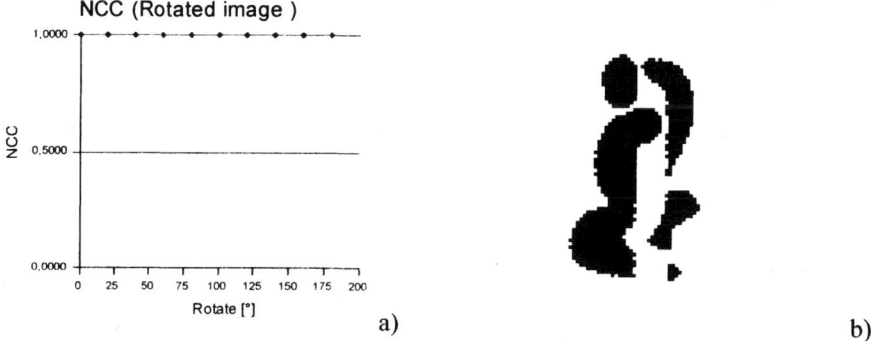

Fig 6. Rotated image a) Watermark NCC, b) Extracted watermark (rotation = 35°)

If the image is modified by cropping, the NCC value of the extracted watermark is very low. This method is not robust enough to resist image cropping, because the watermark is embedded in the all image. When we cropped some a part of an image, we lost the part of the embedded watermark.

5 Conclusion

This paper describes a watermarking technique in the spatial domain. The algorithm is shown to be robust as regards image rotation, image resizing and JPEG compression, but it is not suitable to image cropping. Other modifications have not been tested yet. The PSNR and NCC values, which were obtained from individual examples of watermarked image modifications, are shown in the graphs in the fourth chapter. The algorithm is in the blue print stage. The new improvements, such as forward error correction codes or encryption mechanisms, will be added. Then this algorithm will yield much better results.

Acknowledgments. Research described in this paper was financially supported by the university development fund FRVS No. 993/2007 and by the research programme MSM 0021630513 "Advanced Electronic Communication systems and technologies (ELCOM)".

References

1. Verma, B., Jain, S., Agarwal, D., Phadikar, A. A New Color Image Watermarking Scheme. INFOCOMP Journal of Computer Science. 2006. ISSN 1807-4545. Accepted in April 2006
2. Macq, B., Quisquater, J. Cryptology for digital TV broadcasting. Proceeding of the IEEE. 1995, vol. 83, p. 944-957 ISSN 0018-9219.
3. Bender, W., Gruhl, D. Mormoto, N., Lu, A. Techniques for data hiding. IBM Systems Journal. 1996, vol. 35, no. 3, p. 313-336, ISSN 0018-8670
4. Darmstaedter, V., Delaigle, J., Quisquater , J., Benoit, M. Low-cost spatial watermarking. Computer&Graphics. 1998, vol. 33, no. 4, p. 417-424. ISSN 0097-8493
5. Wong, P., Memon, N. Secret and public key image watermarking schemes for imageauthentication and ownership verification. IEEE Transactions on image processing. 2001, vol. 10, no. 10, ISSN 1057-7149
6. Arnold, M., Schmucker, S., Wolthusen, D. Techniques and Applications of Digital Watermarking and Content Protection. Norwood: Artech House, inc., 2003. 274 pages. ISBN 1-58053-111-39
7. Min, W. Multimedia Data Hidding. Doctoral thesis. 2001 Princeton University

Data hiding error concealment for JPEG2000 images

Milan Brezina, Kamil Bodecek, Milan Brezina,

Faculty of Electrical Engineering and Communication,
Brno University of Technology
Purkynova 118, 612 00 Brno, Czech Republic
imb@email.cz , kamil.bodecek@phd.feec.vutbr.cz , brezina.milan@email.cz

Abstract. This paper presents an error concealment technique for image transmission in JPEG2000 for the lowest frequency coefficients. The proposed method uses the layer structure that is a feature of the JPEG2000. The most significant layer is hidden in place of the lowest layer of the bit stream. After transmission this embedded layer is used for error concealment. The bit stream encoded using the proposed method has the same data structure as the standard JPEG2000. Experiments show the effectiveness of these algorithms.

Keywords: Error concealment, data hiding, JPEG2000, layer structure.

1 Introduction

Multimedia communication through wireless and broadcast network is becoming increasingly important. Since these networks transmit data with a very high bit error rate and packet losses, some regions of the transmitted data may not be decoded. Error resilience issue has become a necessity. Error concealment (EC) is an effective error resilience method which combats against the effects of residual errors staying in the bit stream after transmission.

The standard JPEG2000 includes some error resilience tools such as entropy coding, packet markers and also data partitioning. These tools are used to detect and locate possible errors, and also to resynchronize the decoding process. They can minimize the impact of errors in images, but are only applied at the source coding stage. The errors during transmission can still cause the loss of some wavelet coefficients [3]. Any EC method has not been standardized for JPEG2000 yet. Only JPEG2000 VM 7.2 proposes to replace the lost wavelet coefficients by zeros, which will affect the image quality in a certain extent. New efficient EC techniques for JPEG2000 are necessary. Some of the proposed EC methods for JPEG2000 actually aimed at the study of wavelet-based coding [2], [3], [4] and [5]. In a recent work [6], the authors proposed to use the data hiding to facilitate EC, which employed the layer structure of a JPEG2000 bit stream. The method required that the length of the least significant

Please use the following format when citing this chapter:

Brezina, M., Bodecek, K., Brezina, M., 2007, in IFIP International Federation for Information Processing, Volume 245, Personal Wireless Communications, eds. Simak, B., Bestak, R., Kozowska, E., (Boston: Springer), pp. 505-513.

layer must be the same as that of the most significant layer. Fig. 1 shows the case for 20 layers. Our proposed method uses linear bit distribution which causes that all layers in the bit stream have nearly same length as required in [6] as well. It means that each layer has the same quality increment. Kakadu software employed in this work uses logarithmical distribution in general which is not suitable and the distribution has to be add set up manually. The last geometrical distribution is shown as an example when the goal bit rate is divided by 2 in each step for layer. It means last significant layer includes 50% of quality increment, the following 25% and so on.

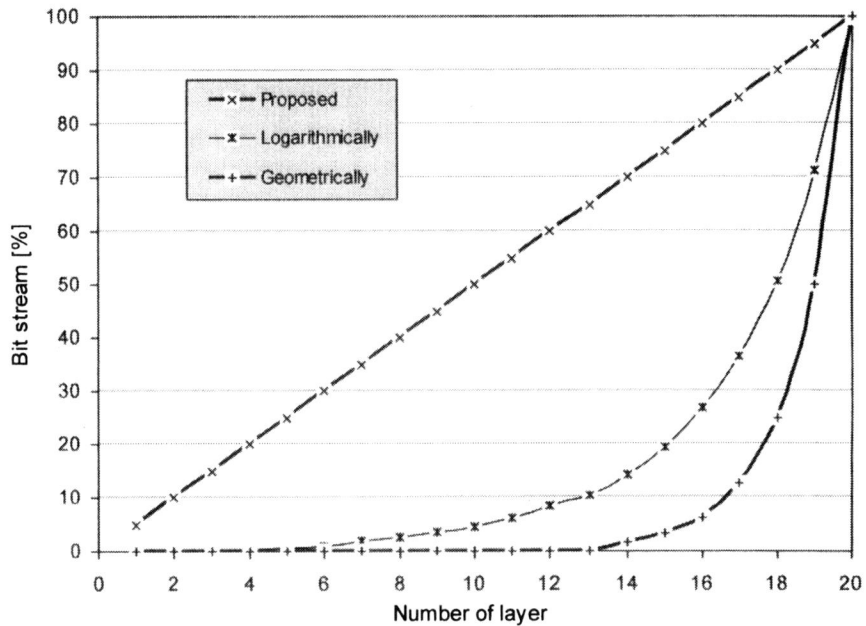

Fig. 1. Bit distribution in layers

This paper is organized as follows. Section 2 briefly reviews the features of the JPEG2000 encoder and the general concept of exploiting the data hiding to facilitate error concealment. Section 3 describes the proposed method in the details. Following experimental results are presented in Section 4. Section 5 is a conclusion of this paper.

2 JPEG2000 encoder

The JPEG2000 standard has a lot of useable features which were described in many previous papers of many authors. The most important of them are error resilient tools and scalabilities. Scalable coding means that image can be simultaneously available for decoding at a variety of qualities or resolutions. The lowest units of JPEG2000 code stream is a packet. Packets contain bit data from a certain layer, component, resolution, and precinct of one tile (see Fig. 2). The order in which these packets are arranged is called as the progression order [2]. The standard defines four different progression orders. The most used is RLCP or LRCP. For example the Layer – Resolution – Component – Position progression is defined as the interleaving of the packets in the following order:

$$\text{for each } l=0,...,N-1$$

$$\text{for each } r=0,...,R$$

$$\text{for each } i=0,...,I$$

$$\text{for each } k=0,...,K$$

packet for layer l, resolution r, component i and precinct k, where N is number of layers, R is maximum number of decomposition levels, I is number of component and K is number of precincts. Obviously by encoding, the number of components and precincts are equal number one.

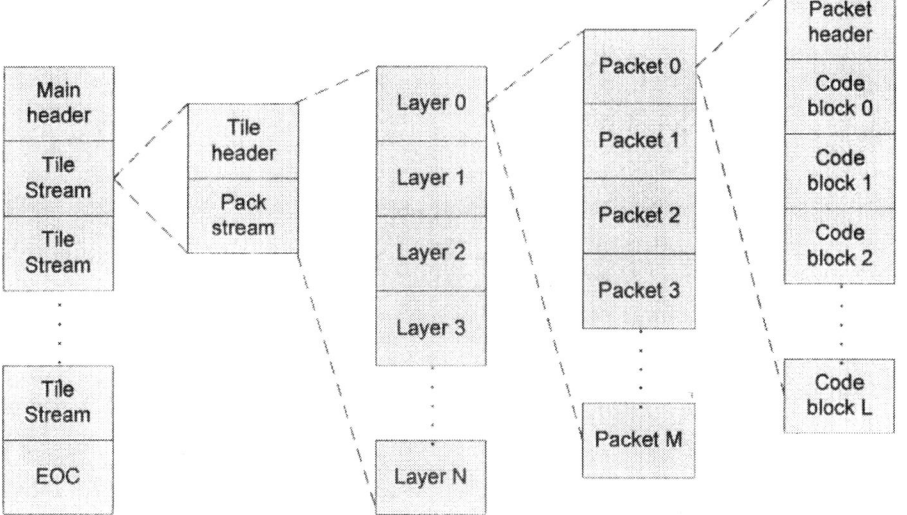

Fig. 2. JPEG2000 code stream structure

In addition, the JPEG2000 decoder can generate an image with a lower bit rate by using the upper layers or by truncating the lower layers of the bit stream. It means that an image with an arbitrary bit rate can be decoded from one JPEG2000 bit stream in the decoder side. It should be noted that the number of layers and the length of each layer can be assigned arbitrarily in the JPEG2000 encoding process [6]. Note the number of generated packets is expressed as a multiplication of the number of layers by the number of resolution levels. Since the size of a packet is determined automatically, we cannot assign an arbitrary size.

As it was mentioned above the JPEG2000 standard has error resilience tools which combat against destructive impact of transmission errors. Error resilience is achieved at the entropy coding and packet level. At the entropy coding level, to localize random and burst errors, the arithmetic coder can be terminated after every coding pass. This allows the decoder to continue the decoding process if errors are detected. At the packet level, the situation is very similar. The decoder can detect and localize errors in the packets. In case of data loss, the result cannot be achieved by using standard error resilience tools. In particular, when the significant layer or layers are lost, there is significant image distortion.

3 Proposed method

First at all a layer structure has to be formed for scalability of a bit stream to use this method. Figure 3 shows a block diagram of a JPEG2000 extended coder which hides the highest layer in the lowest layer. Each layer with a specific data size is generated using the standard JPEG2000 encoder. After that the least significant layer is replaced by the most significant layer (see Figure 4).

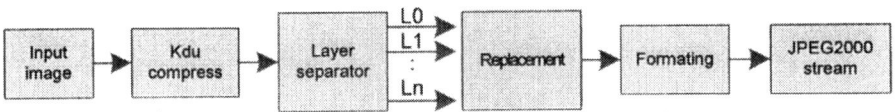

Fig. 3. Block diagram of extended JPEG2000 coder

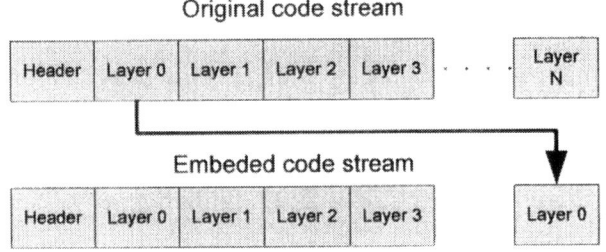

Fig. 4. Proposed encoder extension

Generally it works as follows; the most significant layer (Layer 0) is duplicated. One of duplicates is placed instead of the highest layer second one is placed as a first layer in the bit stream. The length of the least and the most significant layers has to be the same to do mentioned process. The length of the bit stream with the embedded data is therefore the same as that of the original file. If the encoded bit stream with embedded data has the same structure as the original stream, a standard decoder tools can decode this stream. In addition we can use also the JPEG2000 error resilience tools. Finally the least significant layer to inhibit the image deterioration caused by the data hiding, that data is set to zero [7].

The errors mainly are burst errors in the general wireless channel or in Internet, so the probability that the error position of the first packets is the same as that of the hidden data is very small. The reason why we duplicate the most significant layer is that the most important data (low wavelet coefficients) are included in the most significant layer, in our case it is layer 0. Wavelet coefficients are separated into the bit-planes from MSB to LSB. In this way if the upper bit-plane coefficients included in the upper significant layer are affected by errors, the other coefficients of lower bit-plane(s) are getting to be meaningless. Therefore error concealment of these coefficients is desired. Figure 5 shows how the mentioned method reduces the goal PSNR of much known testing pictures in comparison to the images with only the most significant layer 0. You can see differences between using Layer 0 and Layer Max in particular pictures. But differences between picture with all layers (Layer Max) and pictures with hiding data (Max -1) is varying from 0.1 dB to 0.3 dB.

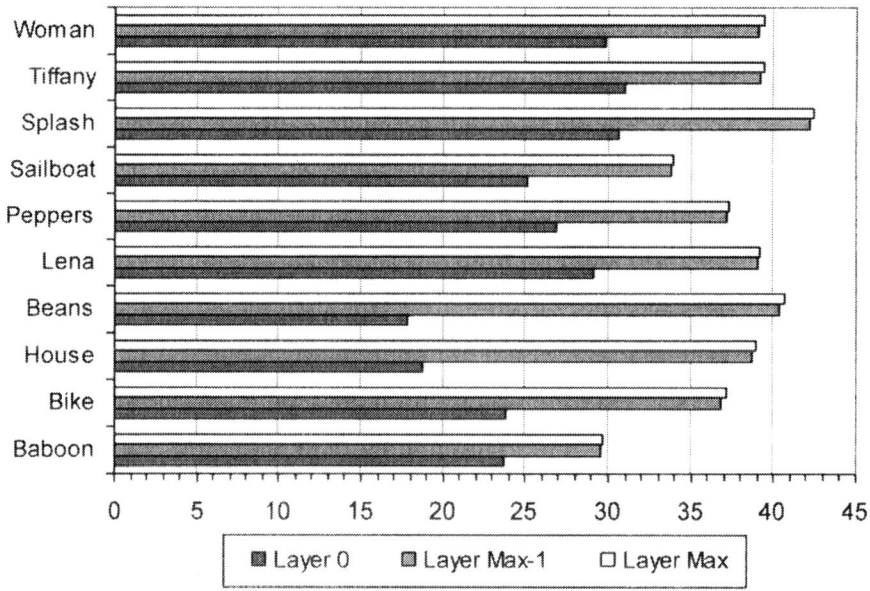

Fig. 5. PSNR of test pictures for 1 bpp

Figures 6 and 7 show a block diagram of an extended JPEG2000 decoder and error concealment technique. In the concealment process, errors are concealed using embedded data. The embedded layer is compared with the most significant layer. When this layer is affected by errors, the data of this layer are concealed by the hidden data of embedded layer. There are two main types of problems; random errors and packet losses. Random errors can be localized in each code-block and coding pass. Therefore these errors are concealed using the code-block and pass data of hidden layer. On the other side packet losses are detected using packet header and resynchronization symbols. Packet losses from the most significant layer are replaced by packets extracted from the hidden layer.

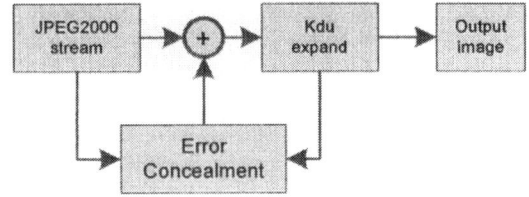

Fig. 6. Block diagram of extended JPEG2000 decoder

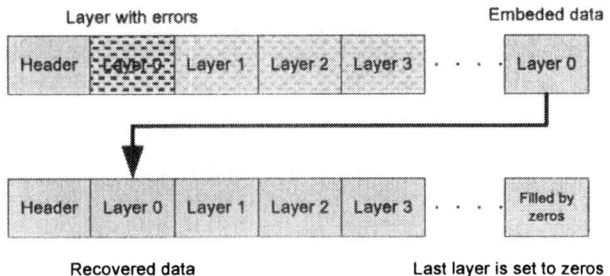

Fig. 7. Proposed error concealment process

4 Experimental results

We use simulations to establish the performance of the proposed method. The method was applied to a 10 standard black and white and color images in original resolution. Image coding was done using 5-level composition process with reversible wavelet filters. The target bit rate was 0.5 bit/pixel. A bit stream was consisted of 10 layers. Note the most significant layer is layer 0. Image quality was measured in terms of the peak signal to noise ratio (PSNR) between the original and the decoded image.

4.1 Images without errors

The image distortion caused by hiding layer 0 in layer 9 of a bit stream is shown here. The original compressed image without hidden data is in Figure 8a and its PSNR is 32.8 dB. Figure 8b shows image with the only layer 0 and its PSNR is 23.9 dB. The last image in Figure 8c shows image with hiding data in the least significant layer 9 with PSNR 32.6 dB. Additionally the hidden data is truncated from the bit stream during the decoding process. As these figure showed, the proposed method can hide certain amount of data without apparent deterioration. Embedded layer makes this deterioration about 0.2 dB. This means that the image deterioration caused by hiding data can be easily ignored. Using more number of layers and higher bit rate the embedded deterioration is going to be meaningless.

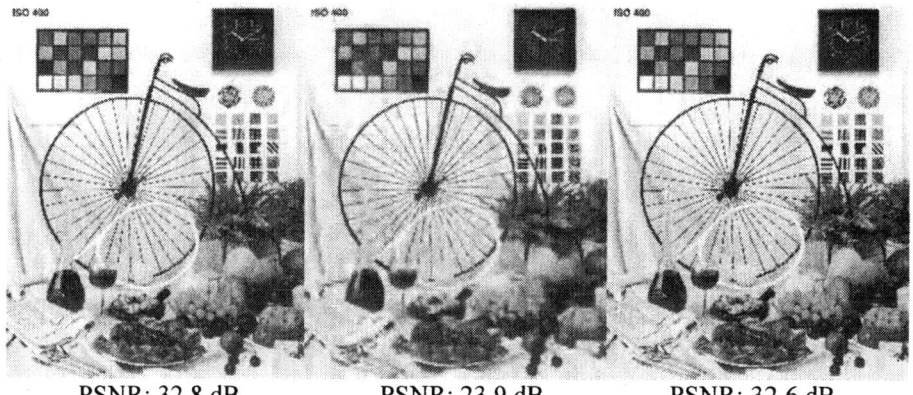

PSNR: 32.8 dB   PSNR: 23.9 dB   PSNR: 32.6 dB

Fig. 8. Visual quality of JPEG2000 images (a) original, (b) only Layer 0, (c) Layer N-1

4.2 Images with errors

In this section we can compare the quality of the proposed error concealment in relation to burst errors. It is assumed that errors are detected perfectly by using error resilient tools. The errors are generated randomly and independently at a rate of 10^{-3}. Figures 9 and 10 show images decoded from erroneous stream without and with using the JPEG2000 error resilient tools, respectively. Right parts of figures show the concealed images decoded by the proposed method. These figures show that the proposed method improves the visual image quality as well as its PSNR. Moreover, using JPEG2000 error resilient tools, we obtain better goal image quality after using error concealment even the PSNR difference between image with and without using error concealment is little bit smaller than without using these tools [8], [9], [10].

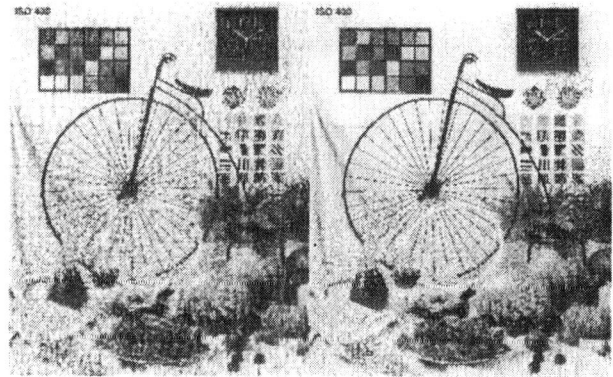

PSNR: 16.17 dB   PSNR: 21.71 dB

Fig. 9. Images decoded without JPEG2000 error resilient tools

PSNR: 20.33 dB 　　　　PSNR: 24.51 dB

Fig. 10. Images decoded with JPEG2000 error resilient tools

Note that it is best to do this kind of experiments on larger images. Obviously used only Lena (512 x 512) image is extremely old. Modern image compression applications more often than not involve images with at least a few mage-pixels. Lena and the other USC images were scanned from magazines in the 1980's using a scanner whose color channels were not even properly aligned.

5 Conclusion

We have presented an efficient error concealment method which is proposed aiming at the recovery of the missing visual information in the transmission of JPEG2000 images over error prone channels. This method is based on the features of JPEG2000 standard at the layer level of a bit stream. Error concealment is implemented by adopting the data hiding technique.

The proposed method is available for any error correcting codes. Thus, under a higher error rate a more powerful code can be used for the error concealment. Moreover it should be noted that a bit stream produced by the proposed method with using a code has the same data structure as a standard JPEG2000. Simulation results showed that hiding data had only a small effect on image quality in comparison with the quality of the proposed error concealment method.

Acknowledgment

This article was created with aid of project FRVŠ 1363/2007.

References

1. ISO/IEC 15444-1, JPEG2000 Part 1 final draft international standard, Aug. 2000.
2. Taubman, D., Marcelin, M. W.: JPEG2000 Image compression fundamentals, standard and practice, Kluwer academic publisher (2001)
3. Moccagatta, I., Soudagar, S., Liang, J., Chen, H.: Errorresilient coding in JPEG-2000 and MPEG-4, IEEE Journal on Selected Areas in Communications, (2000) vol. 18, pp. 899-914
4. Shirani, S., Kossentini, F., Ward, R.: Error concealment methods: A comparative study, In IEEE Canadian Konference on electrical and computer engineering, Edmonton, Canada, (1999) pp. 835-840
5. Atzori, L., Dodoña, S., Giusto, D., D.: Error recovery in JPEG2000 image transmission, in Proc. IEEE ICASSP, Salt Lake City, USA, (2001) vol. 3, pp. 1733-1736
6. Kurosaki, M., Kiya, H., Error concealment using layer structure for JPEG2000 images, in Proc. 15th Workshop on Circuits and Systems in Karuizawa of IEICE, Japanese, pp.77-82, (2002)
7. Lee, P., G., Chen, L., G.: Bit-plane error recovery via Gross subband for image transmission in JPEG2000, IEEE International Conference on Multimedia and Expo, Lausanne pp. 149-152, (2002).
8. Brezina, M.,: Transmission of still images JPEG2000, Telecommunication and Signal Processing, Brno: Brno University of Technology (2005)
9. Brezina, M.,: JPEG2000 transmission over lossy channel, Research in Telecommunication Technology, Brno: Brno University of Technology (2005)
10. Březina, M.: Wavelet coding for mobile applications, Research in Telecommunication Technology, Brno: Brno University of Technology (2005)

Optimized discrete wavelet transform to real-time digital signal processing

Jan Vlach[1], Pavel Rajmic[1],
Jiri Prinosil[1], Josef Vyoral[1], Ivan Mica[1]

[1] Brno University of Technology, Faculty Electrical Engineering and Communication,
Department of Telecomunications, Purkynova 118,
612 00 Brno, Czech Republic
jan.vlach@phd.feec.vutbr.cz, rajmic@feec.vutbr.cz,
{xprino01,xvyora01,xmicai00}@stud.feec.vutbr.cz

Abstract. In this paper, we propose optimized method of discrete wavelet transform. There is many use of wavelet transform in digital signal processing (compression, wireless sensor networks, etc.). In those fields, it is necessary to have digital signal processing as fast as it possible. The new segmented discrete wavelet transform (SegWT) has been developed to process in real-time. It is possible to process the signal part-by-part with low memory costs by the new method. In the paper, the principle and benefits if the segmented wavelet transform is explained.

1 Introduction

If we use wavelet transform in real applications we handle signals of finite length. It is not usually possible to process whole signal at a time, we must process the signal segment-by-segment. To calculation of sufficient amount of no redundant wavelet coefficients, it is necessary to know the signal behind the segment borders. Generally in this case, we determine signal behaviors behind the borders. Typical examples are extension techniques like zero-padding, smooth, symmetric, asymmetric and period extension. Zero-padding assumes samples outside the segment boundary are zero [1], [2]; periodic extension assumes that the signal is periodic [1], [2]. Symmetric extension assumes the signal is reflected at the segment boundaries [1]–[3]. These basic types of extension cause signal distortion at the boundaries. The sort of distortion depends on estimation range of signal specification behind boundaries. The more level of wavelet decomposition is chosen the much amount of distortion is caused. Such distortion happens to be unacceptable.

The way how to circumvent problem is adoption the segment techniques based upon overlap-save and overlap-add methods. In the case of fast discrete wavelet transform [5], we use overlap-save (OSC) and overlap-add convolution (OAC). Conditions are more complicated with increasing level of wavelet decomposition. These problems solved new method SegWT, which is optimized to discrete wavelet transform for segment processing with various segment lengths. It has a great potential application also in cases when it is necessary to process a long signal off-line and no sufficient memory capability is available. It is then possible to use this new method for equivalent segment wise processing of the signal and thus save the storage place.

2 Discrete-time wavelet transform

In digital signal processing we use finite discrete (or discrete-time) wavelet transform, abbreviated DTWT, which can be represented by an orthogonal matrix $\mathbf{W}$ of size $n \times n$ [7].

Let $\mathbf{x}$ be a vector of length n. Its wavelet transform is vector $\mathbf{y}$, obtained as $\mathbf{y} = \mathbf{W}\mathbf{x}$. Due to the ortogonality of $\mathbf{W}$, the inverse wavelet transform is $\mathbf{x} = = \mathbf{W}^{-1}\mathbf{y} = \mathbf{W}^T\mathbf{y}$.

In fact, instead of multiplying vector $\mathbf{x}$ by the matrix $\mathbf{W}$, more effective Mallat's pyramid algorithm [5] is used for computing the transform. Each step of this algorithm corresponds to:
1. extending the input vector
2. filtering this vector by specific low-pass and high-pass filters
3. cropping the central part of the results
4. decimation the results

The coefficients from low-pass branch are called "approximations" and those from the high-pass branch are called "details". We can repeat this single transformation step with the approximations standing for the input. The number of such repetitions, d, is called transformation depth. Scheme of one step of this algorithm is depicted in Fig. 1. This way the input is divided into number of frequency subbands [1].

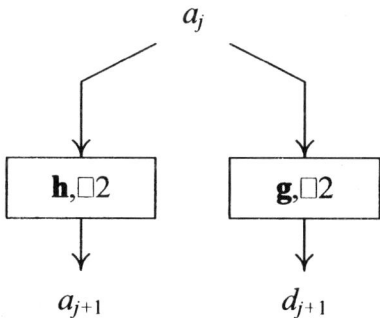

Fig. 1. One step of the forward wavelet transform – decomposition into details and approximations.

3 Segmented wavelet transform

The task for the segmented wavelet transform based on wavelet over-lap save convolution (WOSC) and wavelet overlap-add convolution (WOAC) techniques [6], SegWT, is naturally to allow signal processing by its segment, so that in this manner we get the same result (same wavelet coefficients) as in the ordinary DTWT case. In this problem, the transform depth d, wavelet filter length m and the segment length s play a crucial role.

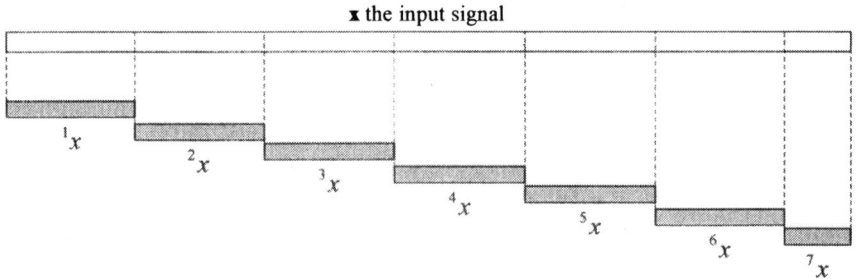

Fig. 2. Segmentation of the input signal. The last segment can be shorter than the others.

Derivation of the SegWT algorithm requires a very detailed knowledge of the DTWT and IDTWT [8] algorithms. Thanks to this it is possible to deduce fairly sophisticated rules how to handle the signal segments. We found out that, in dependence on d, m, s it is necessary to extend every segment from left by exact number of samples from the preceding segment and from right by another number of samples from the subsequent one. Fig. 3 illustrates the principle of segment extending.

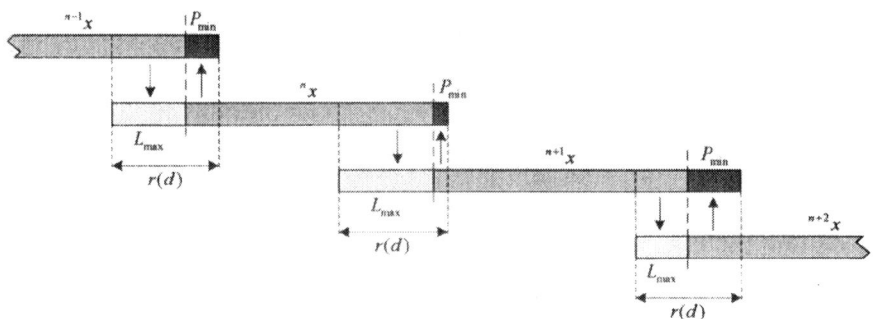

Fig. 3. Scheme of extending input segments. Each segment has to be extended different length form left and right (L_{max}, R_{min}) and the length can also differ from segment to segment. The sum of actual L_{max} and previous R_{min} is always constant ($r(d)$).

After using segmented forward wavelet transform which includes extending, cropping and decimation steps we gain wavelet coefficients ready to application process. Then we must processed coefficients reconstruct to source form. Segmented inverse wavelet transform includes similar step like in wavelet decomposition. The hardest part is to add overlapped parts of neighboring segments. The example of WOAC technique is show on Fig. 4.

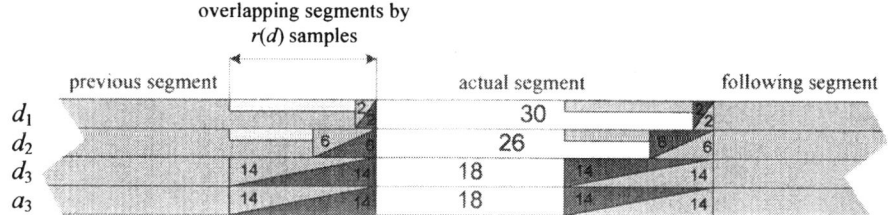

Fig. 4. Example of segmented wavelet reconstruction, show as superposition of all details and aproximations.

4 Experiments

To testing and checking the method we have used MATLAB, because of comfortable implementation the algorithm, also for availability build-in functions like fast convolution algorithm using FFT, downsampling, upsampling, etc. Testing experiment was based on comparison wavelet reconstruction process (SegIWT) of new method with standard IDTWT algorithm. Indeed, it was necessary to adjust the standard algorithm with the view to comparison, otherwise results would be incomparable.

In order to get valuation the new algorithm we defined a quantity, called *percentage velocity gain* (Δt), which indicates how much faster SegIWT process is in comparison to IDTWT algorithm. For example, $\Delta t = 50\%$ means, it is possible to process whole signal in two units of time with new method and in three units of time with standard algorithm; in other words, new algorithm is three over two times faster. The percentage velocity gain is defined

$$\Delta t = \frac{t_{old}}{t_{new}} \cdot 100 - 100 \quad [\%], \qquad (1)$$

where t_{old} is time necessary to signal reconstruction with IDTWT and t_{new} is time necessary to signal reconstruction with SegIWT. Simulation parameters were chosen:
– decomposition depth $d = 4$,
– Daubechies wavelet type 4, filter length $m = 8$,
– vector length in depth level $x_d = 4$

These parameters were chosen with respect to acceptable simulation duration and also for showing model behavior of new method.

4.1 Dependence on signal length

First experiment was percentage velocity gain dependence on whole signal length (number of samples).

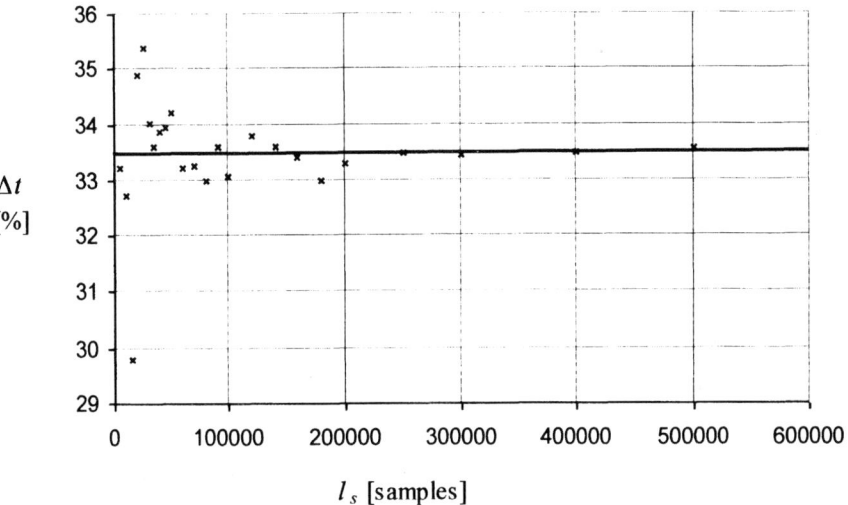

Fig. 5. Percentage velocity gain Δt in dependence on whole signal length l_s. It is evident that mean value of Δt is not dependent on l_s. The mean value of Δt is circa 33% for above-mentioned experiment. Considerable variation of Δt values is caused by short simulation time for short length l_s. These durations are not possible to gauge precisely.

4.2 Dependence on decomposition depth

This experiment was based on changing decomposition depth in signal with constant length l_s = 100 000 samples (Fig. 6). When we set simulation parameters in order to with increasing decomposition depth the number of segments is constant, we obtain the similar dependence (Fig. 7). The descent of Δt is caused by immense time cost to overlap segments (lots of segments are being overlapped). If we choose the decomposition depth greater than 10, the time cost of the new algorithm is approximately the same. Indeed, it is important to notice the IDTWT standard algorithm needs to read much more data from the computer memory by same time cost which means if the access to the memory is slow it could affect simulation results markedly.

In those experiments in which there are another filters, the results are almost the same like in Fig. 6. With increasing filter length the length of overlapped parts also increases. However, this not affects the simulation process as well, because wavelet filters length is very short in comparison to segment length.

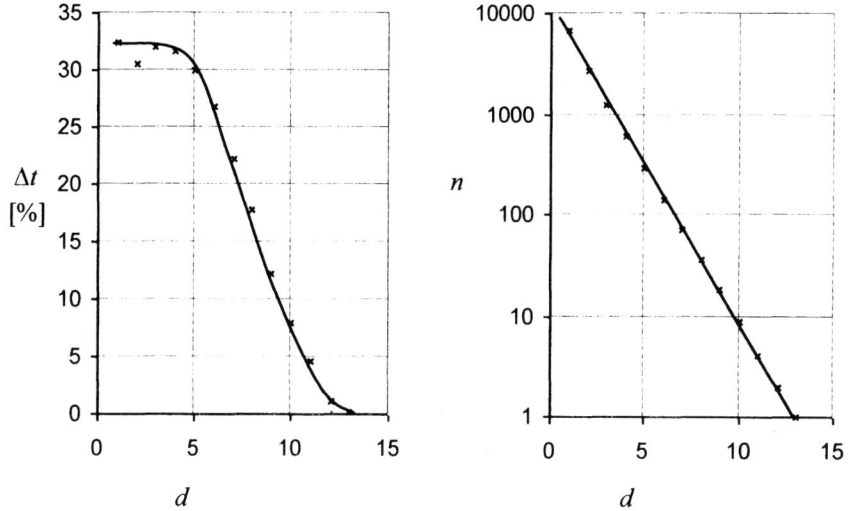

Fig. 5. Percentage velocity gain Δt in dependence on d the constant decomposition depth. The value n is the number of segments.

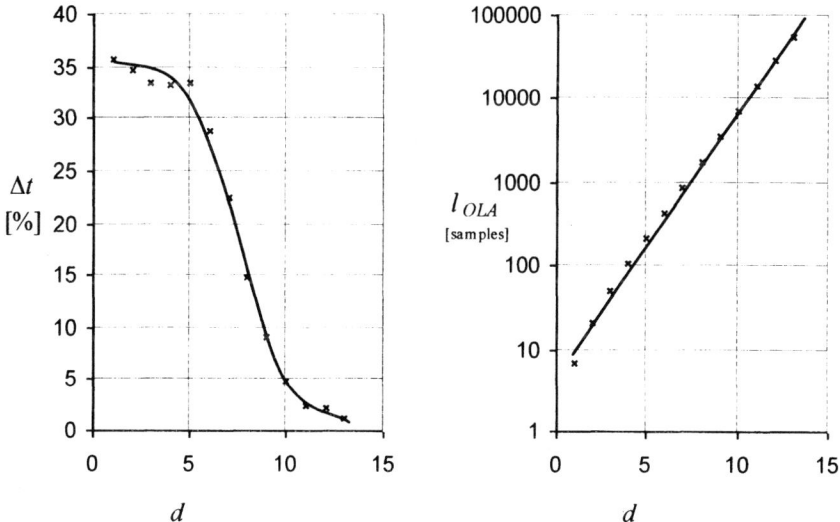

Fig. 6. Percentage velocity gain Δt in dependence on d the decomposition depth with constant number of segments. The value l_{OLA} is the number of overlapped samples.

5 Conclusions

In the paper, a novel SegWT method was described. The method can be used in real-time applications, in which the signal segment-by-segment processing is required. As it was proved in presented experiments the computational time was decreased up to about 33% in comparison to standard fast wavelet transform algorithm. Thus, the method would be suitable e.g. for implementation on digital processors and the range of applications of the new algorithms is very wide – from noise cancelation in speech signals by using tresholding of wavelet coefficients to image processing (compression JPEG2000, pattern recognition). These areas are two typically subjects of telecommunications research.

Acknowledgements. The paper was prepared within the framework of No. 102/06/P407 and No. 102/07/1303 projects of the Grant Agency of the Czech Republic and No. 1ET301710509 project of the Czech Academy of Sciences.

References

1. Strang, G., Nguyen, T.: Wavelets and Filter Banks. Cambridge, MA: Wellesley-Cambridge.(1996)
2. Mertins, A.: Signal Analysis Wavelets, Filter Banks, Time–Frequency Transforms and Applications. New York: Wiley (1999)
3. Bradley, J. N., Brislawn, C. M., Faber, V.: Reflected boundary conditions for multirate filter banks, In Proc. IEEE Int. Symp. Time–Frequency and Time–Scale Analysis. (1992)
4. Muramatsu, S., Kiya, H.: Extended overlap-add and -save methods for multirate signal processing," IEEE Trans. Signal Processing, Vol. 45. (1997) 2376–2380
5. Mallat, S.: A Wavelet Tour of Signal Processing. 2nd edition, Academic Press. (1999)
6. Nealand, J. H., Bradley, A. B., Lech, M.: Overlap-Save Convolution Applied to Wavelet Analysis. IEEE Signal Processing Letters, Vol. 15, 2. (2003)
7. Vidakoci, B.: Statistical Modeling by Wavelets (Wiley Series in Probability and Statistics), John Wiley & Sons, New York. (1999)
8. Misiti, M., Misiti, Y., Oppenheim, G.; Poggi J.-M.: Wavelet Toolbox User's Guide, The MathWorks, Inc. (2001)

Enhanced estimation of power spectral density of noise using the wavelet transform

Petr Sysel[1] and Zdenek Smékal[1]

Department of Telecommunications, Brno University of Technology,
Purkynova 118, 612 00 Brno, Czech Republic,
sysel@feec.vutbr.cz

Abstract. In practice, different methods for enhancing speech hidden in noise are used but none of the available methods is universal; it is always designed for only a certain type of interference that is to be suppressed. Since enhancing speech masked in noise is of fundamental significance for further speech signal processing (subsequent recognition of speaker or type of language, compression, processing for transmission or storing, etc.), it is necessary to find a reliable method that would work even under considerable interference and will be modifiable for different types of interference and noise. The methods known to date can basically be divided into two large groups: single-channel methods and multi-channel methods. The basic problem of these methods lies in a rapid and precise method for estimating noise, on which the quality of enhancement method depends. If the noise is of stationary or quasi-stationary nature, its determination brings further difficulties. A method is proposed in the article for enhancing the estimation of power spectral density of noise using the wavelet transform.

Keywords: *Speech Enhancement, Power Spectral Density, Wavelet Transform Thresholding.*

1 Introduction

The wavelet analysis is a certain alternative to the Fourier representation for the analysis of short-term stationary real signals such as speech that is degraded by noise. If the noise is of non-stationary nature, then the greatest problem consists in estimating its power spectral density with sufficient frequency resolution. Two types of estimating the power spectral density are known: *non-parametric methods* and *parametric methods*. The best-known non-parametric methods include the Barlett method of periodogram averaging [1], the Blackman and Tukey method of periodogram smoothing [2], and the Welch method of averaging the modified periodograms [3]. Although the three methods have similar properties, the Welch method is the most widely used. These methods are called non-parametric because the parameters of the data being processed are not sought in advance. To yield a good estimate of the power spectral density the methods require the application of a long recording of data (at least 10^4 samples).

The periodogram is defined as follows:

$$P_{xx}\left(e^{j2\pi f}\right) = \frac{1}{N}\left|\sum_{n=0}^{N-1} x[n]e^{-j2\pi fn}\right|^2 = \frac{1}{N}\left|X\left(e^{j2\pi f}\right)\right|^2. \tag{1}$$

The function $X\left(e^{j2\pi f}\right)$ is the Fourier transform of discrete signal $x[n]$. It can be shown that the periodogram is an asymptotically unbiassed estimate but its variance does not decrease towards zero for $N \to \infty$. This is to say that the periodogram itself is not a consistent estimate [4]. To be able to use the FFT, we must choose the discrete frequency values:

$$P_{xx}\left[e^{j2\pi f_k}\right] = \frac{1}{N}\left|\sum_{n=0}^{N-1} x[n]e^{-j2\pi f_k n}\right|^2 = \frac{1}{N}\left|\sum_{n=0}^{N-1} x[n]e^{-jk\frac{2\pi}{N}n}\right|^2, \tag{2}$$

$$k = 0, 1, \ldots, N-1.$$

The periodogram is calculated at N frequency points f_k. For the comparison of the properties of non-parametric methods, the quality factor was proposed:

$$Q = \frac{\left\{E\left(P_{xx}\left[e^{j2\pi f_k}\right]\right)\right\}^2}{\text{var}\left(P_{xx}\left[e^{j2\pi f_k}\right]\right)}, \tag{3}$$

$$k = 0, 1, \ldots, N-1.$$

A comparison of the non-parametric methods is given in Table 1. All the three methods yield consistent estimates of power spectral density. In the Bartlett method a rectangular window is used whose width of the main lobe in the frequency response when the maximum value drops by 3 dB is $\Delta f = 0.9/M$, where M is the length of partial sequences. In the Welch method a triangular window is used whose width of the main lobe in the frequency response when the maximum value drops by 3 dB is $\Delta f = 0.28/M$, where M is again the length of partial sequences and their overlap is 50 %. In the Blackman-Tukey method, too, the triangular window is used and $\Delta f = 0.64/M$. As can be seen, in the Blackman-Tukey and the Welch methods the quality factor is higher than in the case of the Bartlett method. The differences are small though. What is important is that Q increases with increasing length of data N. This means that if for a defined value Q we want to increase the frequency resolution of estimate Δf, we must increase N, i.e. we need more data.

The main drawback of non-parametric methods is the fact that they assume zero values of the autocorrelation estimate $r[m]$ for $m \geq N$. This assumption limits the frequency resolution of the periodogram. It is further assumed that the signal is periodic with period N, which is not true either. Since we have at our disposal only a sequence of finite length, there is aliasing in the spectrum.

There are, of course, also other methods that can extrapolate the values of autocorrelation for $m \geq N$. On the basis of the data analysed the parameters of the model are estimated (that is why they are called parametric methods), and using the model the properties of power spectral density are determined. Three

Method of PSD estimation	Q	Number of complex multiplications
Bartlett [1]	$1.11\Delta f N$	$\frac{N}{2} \log_2 \frac{0.9}{\Delta f}$
Blackman-Tukey [2]	$2.34\Delta f N$	$N \log_2 \frac{1.28}{\Delta f}$
Welch [3]	$1.39\Delta f N$	$N \log_2 \frac{1.28}{\Delta f}$

Table 1. Comparison of the quality of non-parametric methods according to [4].

types of model are known: *AR (Auto Regressive)*, *MA (Moving Average)*, and *ARMA*. From the three models, the AR model is the most frequently used. This is because it is well-suited to represent a spectrum with narrow peaks and then the Yule-Walker equation can be used to calculate the model coefficients. Wold [5] derived a theorem that says that any random process of the type of ARMA or MA can be represented uniquely using an AR model of infinite order.

The noise of blender was chosen for the comparison of different methods for estimating power spectral density. Blender noise is no typical random process since in addition to the random signal due to the friction between the blender content and blender parts it also contains periodic components due to rotor rotation. This noise was chosen because this type of noise is often encountered in speech enhancement. The majority of household appliances (vacuum cleaner, hair drier, etc.) and workshop machines (drilling, grinding, sawing and other machines) produce a similar type of noise. Fig. 1 gives a comparison of the estimates of power spectral density of blender noise obtained using a periodogram (full line) and a 15th order AR model (dashed line). As stated above, the periodogram is an asymptotically unbiased estimate but its variance may cause an error of as much as 100 % [6]. Moreover, it is inconsistent since it does not decrease with increasing signal length. This is shown in Fig. 1 by the large variance of values given by the full line. By contrast, the AR model provides a smoothed estimate of power spectral density. In the calculation, however, we encounter difficulties when estimating the order of the model. If a very low order of the model is used, we only obtain an estimate of the trend of power spectral density and lose the details, which in this case represent the maxima of the harmonics of periodic interference. If, on the contrary, the order of AR model is high, we obtain statistically unstable estimates with a large amount of false details in the spectrum. The variance of such an estimate will be similar to the estimate obtained using the periodogram. The choice of the order of the model is an important part of the estimation and depends on the statistical properties of the signal being processed. These properties are, however, in most cases also only estimated. For a short stationary signal the value of the order of the model can be chosen in the range:

$$0.05N \geq M \geq 0.2N, \qquad (4)$$

where N is the signal length, and M is the order of the model [6]. For segments whose length $N = 200$ samples the minimum order is $M = 10$. But for signals formed by a mixture of harmonic signals and noise this estimate mostly fails.

An estimate obtained using the periodogram can be made more precise via averaging the modified periodograms, which we obtain by dividing the signal into segments and weighting the latter by a weighting sequence. This approach is used in the Welch method for estimating the power spectral density [3]. In Fig. 2 we can see a comparison of the estimates of power spectral density obtained using the Welch method (with 13 segments weighted) and the AR model of 15th order. It is evident from the Figure that in comparison with the estimate obtained using the periodogram the Welch method yields a smoothed estimate, whose variance, moreover, decreases with increasing number of averaged segments (consistent estimate). Dividing the signal into segments naturally results in a reduced frequency resolution of the estimate, which shows up in less pronounced maxima that represent the components of periodic interference. In addition, in the case of non-stationary noise, periodograms may be averaged for segments that include interference of different statistical properties and thus also of different power spectral densities. This further reduces the accuracy of estimating the power spectral density.

2 Enhancing the estimate of power spectral density, using the wavelet transform

To enhance the estimate of power spectral density the wavelet transform can also be used. In [7] the non-parametric estimate of the logarithm of power spectral density is made more precise using the wavelet transform. With this method, statistically significant components of the estimate are obtained from thresholding different levels of wavelet decomposition and thus its non-negative values are ensured. The input signal is interfered with by additive non-Gaussian noise, and the wavelet coefficients of additive noise are assumed to be independent of the wavelet coefficients of power spectral density of useful signal. For the processing, type Daubechies wavelets of the 1st, 4th, 6th and 8th orders and type coiflet wavelets of the 2nd and 3rd orders were used. The method does not assume any preliminary knowledge of the type of noise. In [8] the authors endeavour to find a better estimate of power spectral density than the periodogram logarithm itself. To do this, they use the Welch method of modified periodograms, when they first average K periodograms that have first been multiplied by the spectral window. In [9] the method of wavelet transform thresholding is used to estimate noise, and this estimate is used to enhance speech by the spectral subtraction method.

Consider a stationary random process $x[n]$, which has a defined logarithm of power spectral density $\ln G_{xx}\left(e^{j2\pi f}\right)$, $|f| \leq 0.5$. As the function $\ln G_{xx}\left(e^{j2\pi f}\right)$ is periodic by frequency f, it can be expanded into a discrete Fourier series [4]:

$$\ln G_{xx}\left(e^{j2\pi f}\right) = \sum_{m=-\infty}^{\infty} v[m] e^{-j2\pi fm}. \qquad (5)$$

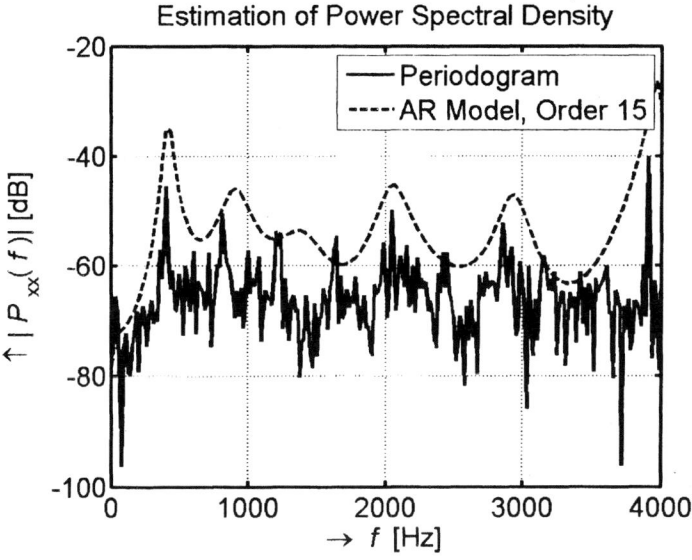

Fig. 1. Estimation of power spectral density of blender noise, using the periodogram and the 15th order AR model

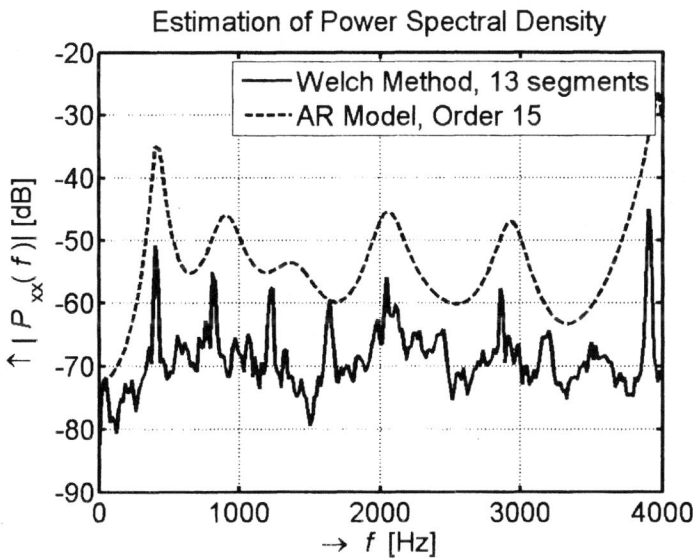

Fig. 2. Estimation of power spectral density of blender noise, using the Welch method of averaging modified periodograms for 13 segments, and the 15th order AR model.

Assuming that $G_{xx}\left(e^{j2\pi f}\right)$ is a real and even function by f, it holds that $v[m] = v[-m]$. The coefficients of the discrete Fourier series are:

$$v[m] = \int_{-0.5}^{0.5} \left\{\ln G_{xx}\left(e^{j2\pi f}\right)\right\} e^{j2\pi fm} df, \qquad (6)$$
$$m = 0, \pm 1, \pm 2, \ldots.$$

Discrete Fourier series coefficients $v[m]$ are cepstral coefficients and the sequence $v[m]$ is the cepstrum of autocorrelation sequence $\gamma_{xx}[m]$, where the Wiener-Khinchine relation holds:

$$G_{xx}\left[e^{j2\pi f}\right] = \sum_{m=-\infty}^{\infty} \gamma_{xx}[m] e^{-j2\pi fm}, \qquad (7)$$

where it holds $\gamma_{xx}m] = E\left(x^*[n]x[n+m]\right)$, where the symbol $*$ denotes a complex conjugate.

The estimate of autocorrelation sequence equals:

$$r_{xx}[m] = \frac{1}{2N+1} \sum_{n=-N}^{N} x^*[n]x[n+m], \qquad (8)$$

and it holds:

$$\gamma_{xx}[m] = \lim_{N\to\infty} r_{xx}[m]. \qquad (9)$$

The inverse equation to equation (5) has the form:

$$G_{xx}\left(e^{j2\pi f}\right) = e^{\left(\sum_{m=-\infty}^{\infty} v[m] e^{-j2\pi fm}\right)} = \sigma_w^2 H\left(e^{j2\pi f}\right) H\left(e^{-j2\pi f}\right), \qquad (10)$$

where $\sigma_w^2 = e^{v[0]}$ is the variance of white noise sequence $w[n]$. The transfer function $H(z)$, $z = e^{j2\pi f}$, is the causal part of discrete Fourier series (10) and $H(z^{-1})$ its non-causal part. In case the AR model is used, it holds for the transfer function:

$$H(z) = \frac{1}{A(z)} = \frac{1}{1 + \sum_{i=1}^{p} a_j z^{-j}}. \qquad (11)$$

The power spectral density will be obtained as follows:

$$G_{xx}\left(e^{j2\pi f}\right) = \sigma_w^2 \left|H\left(e^{j2\pi f}\right)\right|^2, \qquad (12)$$

where σ_w^2 is the variance of white noise sequence $w[n]$, for which it holds:

$$\sigma_w^2 = E\left(|w[n]|^2\right). \qquad (13)$$

The periodogram $P_{xx}\left(e^{j2\pi f}\right)$, as the estimate of power spectral density of noise $G_{xx}\left(e^{j2\pi f}\right)$, is defined as follows:

$$P_{xx}\left(e^{j2\pi f}\right) = \sum_{m=-(N-1)}^{N-1} r_{xx}[m]e^{-j2\pi fm} = \frac{1}{N}\left|\sum_{n=0}^{N-1} x[n]e^{-j2\pi fn}\right|^2. \quad (14)$$

Using the periodogram we can determine the power spectral density as:

$$G_{xx}\left(e^{j2\pi f}\right) = \ln G_{xx}\left(e^{j2\pi f}\right) + \epsilon\left(e^{j2\pi f}\right) + \gamma. \quad (15)$$

If we assume noise to be a Gaussian random process, then the logarithm of periodogram can be modelled as:

$$\ln P_{xx}\left(e^{j2\pi f}\right) = \ln G_{xx}\left(e^{j2\pi f}\right) + \epsilon\left(e^{j2\pi f}\right) + \gamma, \quad (16)$$

where $\epsilon\left(e^{j2\pi f}\right)$ is a random process with probability distribution χ_2^2 with two degrees of freedom, and $\gamma \approx 0.57721$ is the Euler-Mascheroni constant [7, 11]. The random process $\epsilon\left(e^{j2\pi f}\right)$, which is responsible for the periodogram variance, can be removed via thresholding the wavelet transform coefficients. The coefficients of the discrete wavelet transform with discrete time $C_{j,m}[k]$ will be calculated according to the relation:

$$C_{j,m}[k] = \sum_{k=0}^{N-1} \left(\ln P_{xx}[k] - \gamma\right) \psi_{j,m}[k], \quad (17)$$

where $P_{xx}[k]$ are samples of the periodogram of the implementation of a random process of length $2N = 2^{M+1}$ obtained by the discrete Fourier transform, and the base functions $\psi_{j,m}[k]$ are derived by a time shift $j = 0, 1, \ldots, 2^{m-1}$ and dilatation with the scale $m = 0, 1, \ldots, \log_2 N$ of a single mother function $\psi[k]$ according to the relation:

$$\psi_{j,m}[k] = \frac{1}{\sqrt{2^m}}\psi\left[\frac{k}{2^m} - j\right]. \quad (18)$$

The transform is linear and therefore the coefficients will represent the sum of the coefficients representing the sought power spectral density $g_{j,m}[k]$ and the coefficients representing the noise $e_{j,m}[k]$:

$$C_{j,m}[k] = g_{j,m}[k] + e_{j,m}[k]. \quad (19)$$

As the random processes $\epsilon[k]$ are independent and the transform is orthogonal, the coefficients $e_{j,m}[k]$ will be non-correlated. At the same time, however, they are not independent since their probability distribution is independent of the shift j but is dependent on the scale m. But according to the central limiting theorem their probability distribution converges with increasing $m \to \infty$ to the Gaussian normal distribution. The coefficients are modified via soft thresholding according to the relation:

$$C_{j,m}^{(s)}[k] = \text{sgn}\left(C_{j,m}[k]\right)\max\left(0, |C_{j,m}[k]| - \lambda\right), \quad (20)$$

where λ is the threshold value chosen. After that the smoothed estimate of power spectral density of noise $\hat{G}_{xx}[k]$ is obtained (via the inverse discrete wavelet transform) in the form:

$$\ln \hat{G}_{xx}[k] = \frac{1}{N} \sum_{m=0}^{M} \sum_{j=0}^{2^m-1} C_{j,m}^{(s)}[k]\psi_{j,m}[k], \qquad (21)$$

$$k = 0, 1, \ldots, N-1.$$

With regard to the asymptotically normal distribution of coefficients $e_{j,m}[k]$ the threshold value λ can be determined using the universal thresholding proposed by Johnstone, et al., [10] in the form:

$$\lambda = \sigma\sqrt{2\log N}, \qquad (22)$$

where σ is the standard deviation of noise, and N is the length of data. In [11] an optimum determination of the threshold in dependence on the scale is proposed. If the scale is large, then the threshold equals:

$$\lambda_m = \alpha_m \ln N, \qquad (23)$$

where the constants α_m are given in Table 2 and N is the length of data. If the scale is small, $m << M - 1$, the threshold will be determined according to the relation:

$$\lambda_m = \frac{\pi}{\sqrt{3}}\sqrt{\ln N}. \qquad (24)$$

The blender noise, which was used in testing the estimations using the AR model, the Welch method and the periodogram, was also used in the estimation via thresholding the coefficients of the wavelet transform of the periodogram. A comparison of this estimation with that using the AR model is shown in Fig. 3. It is evident that the estimate using the AR model is more smoothed but has a smaller frequency resolution than the estimate obtained via thresholding the

scale m	α_m	scale m	α_m
$M-1$	1.29	$M-6$	0.54
$M-2$	1.09	$M-7$	0.46
$M-3$	0.92	$M-8$	0.39
$M-4$	0.77	$M-9$	0.32
$M-5$	0.66	$M-10$	0.27

Table 2. Values of constant α_m for determining the threshold when thresholding the wavelet transform coefficients.

wavelet transform coefficients. On the contrary, the estimate obtained via thresholding the wavelet transform coefficients is less smoothed but the pronounced peaks in the power spectrum, which represent the harmonic components of the blender motor interference, are localized in the frequency more easily than in the case of AR models. This is also evident from a comparison of the method of thresholding the wavelet transform coefficients with the Welch method of thresholding modified periodograms in Fig. 4.

3 Experimental results

The proposed method of power spectral density estimation was tested on speech enhancement by spectral subtraction [12] of actual recordings of speech signal interfered with by different types of noise. The speech signal was interfered with by the noise of blender and vacuum cleaner, which have the character of wideband noise almost approximating white noise. In the testing, the speech signal was also exposed to interference by noise from a drilling machine and a Ford Transit, which on the contrary have the character of narrow-band noise. In view of the fact that the recordings were made in a real environment and it was impossible to obtain pure speech signal without interference, the estimation of the quality was performed using the signal-to-noise ratio determined segmentwise according to the relation:

$$\text{SNR}_\text{seg} = 10 \log_{10} \frac{R_s}{R_\nu}. \qquad (25)$$

The power of signal R_s is determined from a segment containing the speech signal while the power of interference R_ν is estimated from a segment that does not contain the speech signal.

A noisy speech signal was enhanced using the proposed modified method of spectral subtraction. For comparison, the speech was also enhanced by the method of spectral subtraction with the power spectral density of interference being estimated using the Welch method. For enhancement, the RASTA method was also applied. The values of SNR of the original signal and of the signal reconstructed by the individual methods are given in Table 3.

Compared to the original spectral subtraction method and the RASTA method there was a marked improvement in the SNR in the case of interference of wideband character - vacuum cleaner or blender. Less good results are obtained in the case of noise of narrow-band character, where additionally much depends on the position of narrow-band noise. If noise is in the same frequency band as speech, e.g. drilling machine, the modified method of spectral subtraction exhibits an improvement of SNR which is only a little lower than the improvement in the original method of spectral subtraction or the RASTA method. If the position of narrow-band noise is outside the speech frequency band (low-frequency noise of the Ford Transit engine), then the modified method is comparable with the other methods.

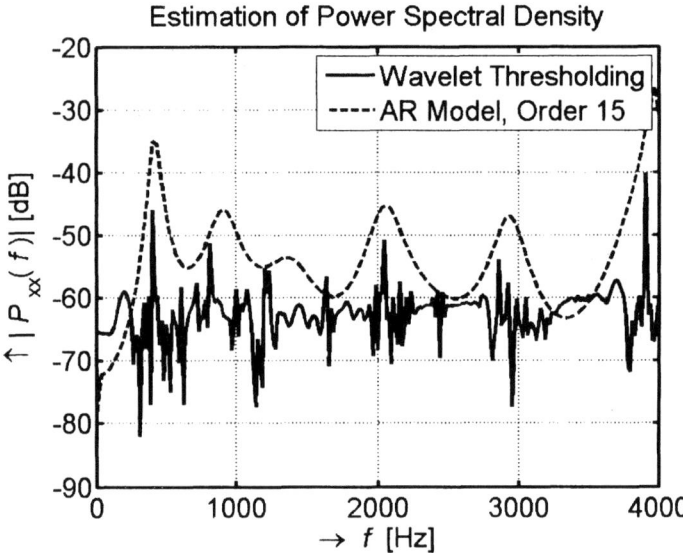

Fig. 3. Estimation of power spectral density of blender noise, via thresholding the wavelet transform coefficients of periodogram and via the AR model of 15th order.

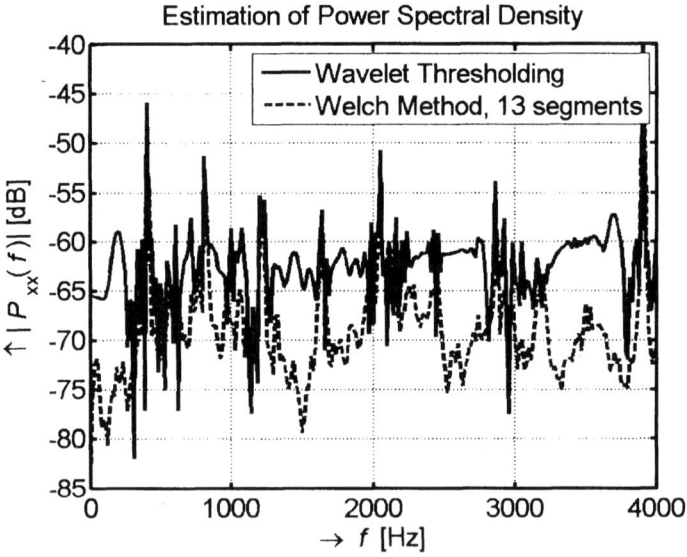

Fig. 4. Estimation of power spectral density of blender noise, via thresholding the wavelet transform coefficients of periodogram and via the Welch method of averaging modified periodograms for 13 segments.

Estimation of signal-to-noise ratio SNR_{seg} [dB]				
Method Type of noise	Original signal	Modified method	Spectral subtraction	RASTA method
Vacuum cleaner	12	22	16	16
Blender	5	17	12	13
Drill	0	13	17	14
Ford Transit	3	20	18	17

Table 3. Estimation of signal-to-noise ratio for four different types of noise and three different methods of speech signal enhancement.

4 Conclusions

A new method for enhancing the estimation of power spectral density of noise via thresholding the wavelet transform coefficients was proposed. It led to a reduced variance in the estimate of power spectral density of noise. The smoothed estimate was then used to enhance the speech signal hidden in noise, using the spectral subtraction method. Thanks to the lower variance in the estimate of power spectral density of noise there was also a lower variance in the estimate of the power spectrum of speech signal and a reduction in the signal reconstruction error. Judging by the experiments made, the method is in the first place suitable for noise that is of wideband nature - e.g. shower, vacuum cleaner, and the like. Listening tests showed that thanks to the lower noise variance the musical noise was suppressed which was the by-product of the existing method of spectral subtraction. The estimation of power spectral density of noise via thresholding the wavelet transform coefficients can also be used in other single-channel methods of speech enhancement.

Acknowledgements

This work was supported within the framework of project No 102/07/1303 of the Grant Agency of the Czech Republic and the National Research Project "Information Society" No 1ET301710509.

References

1. Bartlett, M.S.: Smoothing Periodograms from Time Series with Continuous Spectra. Nature (London), Vol. 161 (May 1948), 686-687.

2. Blackman, R. B., Tukey, J. W.: The Measurement of Power Spectra. Dover, New York (1958).
3. Welch, P. D.: The Use of Fast Fourier Transform for the Estimation of Power Spectra: A Method of Time Averaging over Short Modified Periodograms. In IEEE Trans. Audio and Electroacoustics, Vol. AU-15 (June 1967), 70-73.
4. Proakis, J.G., Manolakis, D.G.: Digital Signal Processing–Principles, Algorithms and Applications. Third Edition. Prentice Hall, New Jersey (2006).
5. Wold, H.: Study in the Analysis of Stationary Time Series. Reprinted by Almqvist & Wiksell, Stockholm (1954).
6. Uhlíř, J., Sovka, P.: Digital Signal Processing. Prague: CVUT Publishing(1995). (In Czech)
7. Moulin, P.: Wavelet Thresholding Techniques for Power Spectrum Estimation. In IEEE Transactions on Signal Processing. Vol. 42, No 11 (1994), 3126-3136.
8. Walden, A.T., Percival, D.B., McCoy E.J.: Spectrum Estimation by Wavelet Thresholding of Multitaper Estimators. In IEEE Transactions on Signal Processing. Vol. 46, No. 2 (1998), 3153-3165.
9. Sysel, P.: Wiener Filtering with Spectrum Estimation by Wavelet Transform. In Proceedings of International Conference on "Trends in Communications". Bratislava, Slovakia:(2001), 471-474.
10. Donoho, D. L., Johnstone, L. M., Kerkyacharian, G., Picard, D.: Wavelet shrinkage: asymptotic (with discussion). J. R. Statist. Soc. B, Vol. 57, (1995), 301-369.
11. Vidakovic, B.: Statistical Modeling by Wavelets. John Wiley & Sons, Inc. Publication, New York (1999).
12. Benesty, J., Makino, S., Chen, J.: Speech Enhancement. Springer, Berllin (2005).

Face detection in image with complex background

Jiri Prinosil [1], Jan Vlach [1]

[1] Brno University of Technology, Faculty Electrical Engineering and Communication,
Department of Telecommunications,
Purkynova 118, 612 00 Brno, Czech Republic
xprino01@stud.feec.vutbr.cz, jan.vlach@phd.feec.vutbr.cz

Abstract. This paper deals with an approach to detect face from image with complex background. That can be use in many areas (eg. security systems, biometrics, telecommunications, etc.).In the first part there is presented brief overview of common used method for face localization in image. On the basis of comparison of its properties there was developed appropriate combination of several methods with particular improvements. The goal was to find suitable accuracy to speed rate. Methods were implemented in MATLAB and accuracy rate was tested on Georgia Tech face database.

1 Introduction

In recent years the price of computer and forcing-off techniques decreases, on the other side the quality improves, which faces to its massive expansion. Computers must be able to interact with surrounding environment very often, especially with a man. These reasons result in increasing interest in the field of pattern and object recognition with the use of computers. Perhaps, the most energy was devoted to human faces recognition. Mainly for the reason this interest is useful to many spheres (biometrics, sight systems, user interface, content image coding), also for the complexity of all problems with many ambiguous solving. In telecommunication it can be used in video-calls, where the scanning device positioning is controlled on the base of a head motion tracking. Other application can be in video-calls too, where only the face is transmitted and the own background is added by user. There are more other fields in communication where the face localization can be used (java applications in mobile phones, security systems in wireless communication), All of this lead to more comfortable and more attractive use of wireless networks.

Face detection is first step in systems for face recognition and its pertinent analysis. All next works in the process of recognition depends on the rate of success of detection. Human face is very heterogeneous because of the form and skin color. Detection algorithm faces to problems of different lighting scenes, angle of scan camera, head turning and physiognomy. Complexity of the task is proportional to numberless quantity of solutions that arose during research. Methods derive benefit from knowledge and progress from many scientific regions like image processing, probability and statistics, matrix arithmetic, AI, fuzzy logic or spectral analysis.

In present days there are many ways to detect face in an image. On the bases of approach we can divided these methods into four main classes [1].

- **Knowledge-based methods** – The face is determined by defined rules, which describe a "typical face." Usually these rules describe relations among face features. This method requires very precise localization and description of particular features, which leads to necessity of using complicate and robust algorithms. On that ground these methods in principle do not reach demanded results.
- **Feature invariant approaches** – detection pursuant to generally valid human faces features, which do change to light conditions or head rotation. Such features can be facial features (eyes, nose, etc.), facial texture or skin color.
- **Template matching methods** – searching on the basis of correlation image with preset pattern of either the whole face or its part. A drawback is necessity to create and to keep saved individual pattern in memory. It is mostly necessary to set up these patterns manually, which is very tough and time-consuming.
- **Appearance-based methods** – human face models are obtained by learning from learning set, which contains variety of face patterns. Detection deals with comparing of image segment with the model. There is necessary to create face model again, but the learning is generally semiautomatic. Methods based upon this approach reach very good results.

From see above mentioned reasons the most perspective methods for our work are invariant feature methods.

1.1 Feature invariant approaches

A common denominator of these methods is presumption, that there are some features that generally may describe the face independently on its form or background parameters. The methods differ at that way, how to define invariable features and how it is possible to find them in the image. In several methods the invariable features are assumpted as a geometric form of each face component part as eyes, eyebrows or mouth. With extraction of these parts by edge detectors and creation of statistical models of their relations, it is then check presence of a face in image.

Face features

These techniques are focused directly on searching or facial feature eyes, mouth, nose etc. Various math transformations are used to highlight these features with respect to the rest of face or entire image. Among these techniques belong e.g. edge detection and other filtering methods [2]. More advanced algorithms derive benefit from knowledge of distance of one part from another or from the others. Further these algorithms use probability where we can find particular part and probability where we cannot find it. An example could be the simple fact, that the classifier wouldn't evaluate tested object as the face in that case, in which we can find lips among the eyes and nose. This method fails especially if we find the face in picture with complex environment (high rate of edges and other structural elements).

Face textures
Human faces have distinct structure, which differs from other objects in image. We usually separate skin, hair and other objects in image. Color information is used to detect face texture [3]. Although this method is fast, it is used rather in more complicated implementations.

Skin color
Searching for skin-like color regions in color image has become very much effective and used in many applications [4]. The advantage of using this method is rate, above all; farther minimal sensitivity to change luminous conditions, insensitiveness on size and face angular displacement, age, or sex of subject.

1.2 Skin detection

Basic principle of all methods coming-out from observation, that the human color skin participate explicit compact subspace in colored space [5]. By the limitations of the space it is possible to model human color skin and then classify the color of each single pixel in image. Skin color classification can be divided into three steps:
- suitable color space selection,
- skin color model selection and the accuracy of this model,
- determination of the criteria for classification.

Suitable color space selection
In the field of colorimetry, imaging and image processing, the research has brought many color spaces with various properties. Form and solidity of skin color subspace differs in each model. Suitable color space selection is primary target for many classification methods.

On behalf of fundamental color space used in image processing we can mark color *RGB* space (Fig. 1, RGB cube is identical with RGB space). The color is defined by three fundamental colors composition (red, blue and green) with various intensities.

PAL the television standard and JPEG image compress algorithm use YCbCr color space. It is transformed RGB model, where *Y* represents luminance components; *Cb* and *Cr* are chrominance components (Fig. 2, cube RGB is transformed into particular hexahedron). There is a simple relation between both models:

$$\begin{bmatrix} Y \\ Cb \\ Cr \end{bmatrix} = \begin{bmatrix} 16 \\ 128 \\ 128 \end{bmatrix} + \begin{bmatrix} 65.481 & 128.553 & 24.966 \\ -37.797 & -74.203 & 112.000 \\ 112.000 & -93.786 & -18.214 \end{bmatrix} \cdot \begin{bmatrix} R \\ G \\ B \end{bmatrix} \quad (1)$$

Simplicity and the fact that many formats result from this space assured great popularity for classification systems.

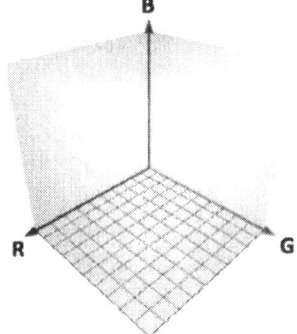 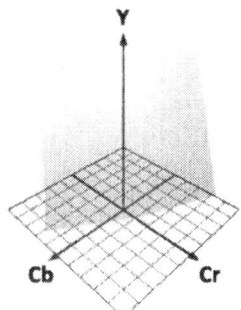

Fig. 1. RGB the color space.   **Fig. 2.** YCbCr the color space.

Skin color model selection and the accuracy of this model

The principle of color segmentation in face detection in scene is section individual pixel accordance with its color value into blocks of skin-like pixels and not-skin-like pixels [6]. At the first, it is necessary first to define the color skin area. Explicitly, it is possible to express the color skin area by defined rules, on the basis of distribution skin-like pattern in color space.

The distribution is obtained experimentally by comparison different testing patterns (Fig. 3). The best benefit is the easy implementation, thereby very good classification rate. Quality of results very depends upon chosen model and accuracy of rules, which it is necessity determine experimentally.

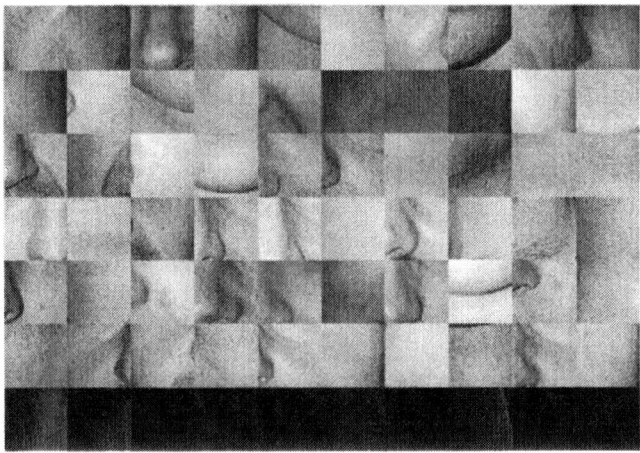

Fig. 3. Example of skin-like pattern.

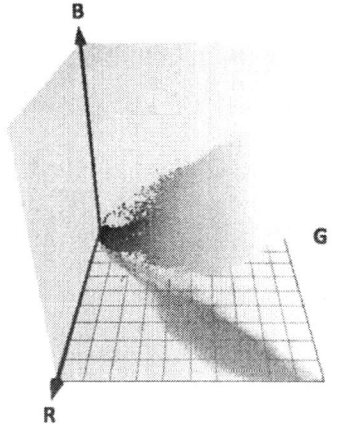 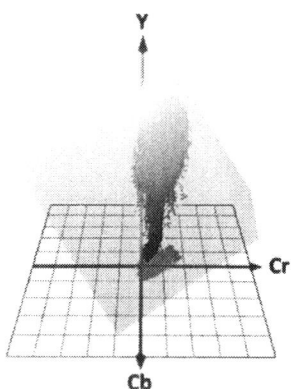

Fig. 4. Skin-like subspace in RGB model.   **Fig. 5.** Skin-like subspace in YCbCr model.

Non-parametric models of skin-like color distribution
Keynote is to find out skin-like color distribution from training set without explicit expression of color subspace by defined rules. Instead of that it is mostly designed so-called stochastic map – for particular point of discrete color space it is associated probability of that the pixel's color is skin-like color.

Parametric models
Parametric models try to describe the form of skin-like color distribution with suitable mathematical model. Parameters of the model are then obtained from training set. Skin-like color subspace has in most case elliptic form and normal distribution approximately.

2 Automatic face localization algorithms

On the occasion of relation of the algorithm we went out from presumption that there could occur unlimited amount of faces with different angular displacement and various distances from picture camera. But for localization, we consider those only, in which eyes and mouth are visible and which attain at least minimal size (approximately 20×20 pixels). The localization can be divides into two parts. The first part presumable face candidates are looked for and in second there are eliminated inappropriate candidates on the basis of founds eyes and lip positions.

2.1 Face parts

In this part we search presumptive face region candidates by the use of skin-like color segmentation, wavelet transform and morphologic operations. For skin-like color segmentation we use already created database binary model, which is later converted to stochastic representation.

Binary pattern database

In our case we first assembled sufficient number of different image skin patterns (totally over 400 patterns), which were scanned by three different cameras from various facial parts of 50 people. From this, we created database, which is represented by three-dimension matrix of 256×256×256 size, where on index of each matrix element represents corresponding color value of RGB color space and is set either on 1 (given value coincides into skin-like color area) or 0 (given value does not coincide into skin-like color area). Now we are already able to create binary map on the basis of comparing each pixel of tested picture with this database. This map determinates which pixel from tested picture coincides into skin-like color area. The main advantage of using the database, which describes skin-like color region, in comparison with mathematical model [7] is better image processing rate, but although the bigger memory requirements (it is necessary to have saved database in memory, about 16 MBits) too.

However according to [8] and practical tests, it has been realized, that the RGB color space is not really good for face detection. In practice it is mostly used different color spaces like YCbCr, HSV, RG, etc. For this project it has been chosen YCbCr space as an optimal color space. According to (1) we performed from RGB to YCbCr transform and then operate herewith color space.

Stochastic pattern database

However we can use the binary database only in conditions that pixels accordant with skin-like color are able to find rarely in the regions out of the face. These pixels also do not flock together. That conditions it is possible satisfy only if the scanned scenes are created artificially, which is practically useless. On this account, we modify the binary matrix onto stochastic matrix with respect to percept occurrence of each color patterns. Further, we then consider skin-like color of face if it is homogenous to the considerable measurement. It means the color value changes gradually only. From here, we proceed on next adjustment of our application. For dispatching of the following processing, we would not proceed pixel-by-pixel, but block-by-block, one of 3×3 pixels. For every single pixel of appropriate block we determine probability if the value coincides to skin-like color in accordance with stochastic matrix. On the base of comparison first and second static stochastic moment from all pixels in given block, we assign to this block either the one (high mean value, low scatter – block coincides into skin-like color) or the zero (low mean value, high scatter – clock does not coincide into skin-like color). We retrieved binary map again this way, which corresponds to appearance of skin-like color regions. Indeed, this map has one third size only compared to previous map and the percent occurrence of not-skin-like color pixels coincide into skin-like color region is reduced distinctively. The comparison both binary and stochastic map is in Fig. 6.

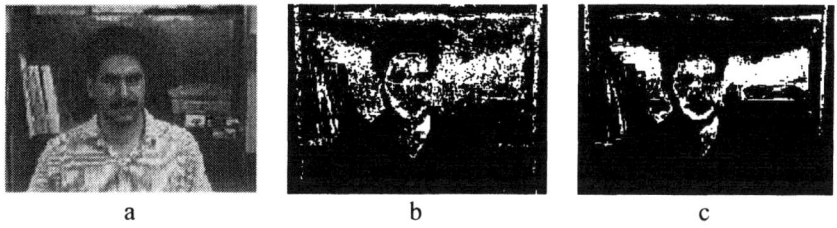

Fig. 6. a) original image, b) binary mask in accordance to binary matrix, c) binary mask in accordance to stochastic matrix.

Notice that the amount of not skin-like color pixels is reduced on the other hand amount of skin-like color pixels is reduced too. It is therefore necessary to perform filtration by low-pass filter (for example blurring).

Wavelet transform

In spite of using stochastic matrix there is significant amount of not skin-like color pixels. To better reduction we treat luminance components of the image to wavelet transform. From presumption there is expressive brightness transition in vertical direction in the area of eyes, lips, etc, we use detailed coefficients in vertical direction and first level of decomposition only. Transitions in horizontal direction are insignificant (especially if the face melts to background). We use Daubeschi wavelet type 2. The high coefficient value means sharp transitions (eyes, lips, etc.). To cover larger face area, we use low-pass filter again. However this time, we go out of presumption that it is most important to distend the coefficients more vertically than horizontally (see face anthropology) and therefore we use special averaged out function of sinusoid (Fig. 7). This field of coefficient we adjust on binary map size and multiply them with this map.

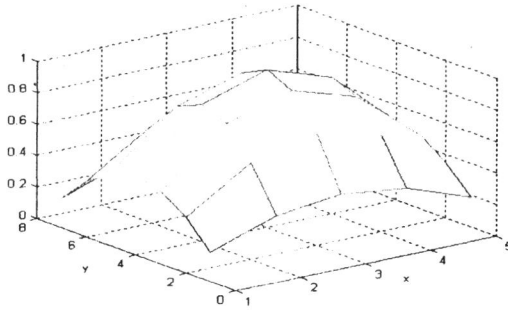

Fig. 7. Impulse response of averaged out function of wavelet coefficients.

Morphologic operations

By the use of wavelet transform we reduced all areas of pixels, which were homogenous in vertical direction. Nevertheless even in the binary map could be still pixels, which do not belong to face area and satisfy all determine criteria so far. Now, we perform series of morphologic operation closing and opening (2) and (3) [9], which result in removing standalone pixels and groupage of other pixels into compact units.

$$X \bullet B = (X \oplus B) \ominus B, \quad (2)$$

$$X \circ B = (X \ominus B) \oplus B, \quad (3)$$

where $\oplus$ means dilatation morphologic operation (4) and $\ominus$ erosion morphologic operation (5).

$$X \oplus B = \{p \in \varepsilon^2 : p = x+b, x \in X, b \in B\}, \quad (4)$$

$$X \ominus B = \{p \in \varepsilon^2 : p+b \in X, \forall b \in B\}, \quad (5)$$

Then we spatially describe these units by the use of simple edge detector and suitable tracker algorithm. In our case we described them with rectangular tetragons with $x_{1i}, y_{1i}, x_{2i}, y_{2i}$, where i means order of tetragon and x_1, x_2, y_1, y_2 are initial and end coordinates in horizontal/vertical direction. On the basis of size rate of each side we discard area, which not satisfy form of face [10]. Further we discard such areas, which are too small to provide fair information. Fig. 8 shows all binary masks after averaging, after multiplication with wavelet coefficients and after morphologic operations with suitable face candidate selection.

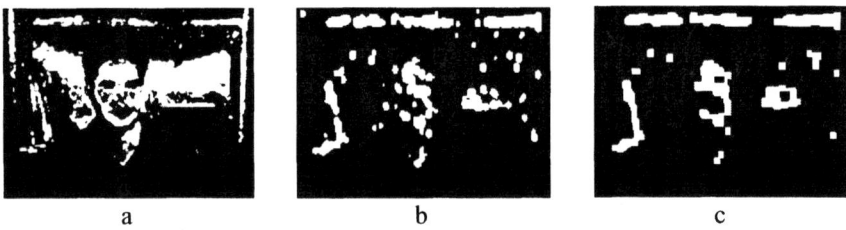

Fig. 8. a) averaged out binary mask, b) multiplied with averaged out wavelet coefficients, c) after morphologic operations with suitable face qualifiers selection.

2.2 Face features localization

Indeed, how we see in Fig. 8c, we still have not discarded all areas, which are not skin-like color areas. So we perform last verification test for elected candidates, and it is location the position of mouth and eyes. To solve the problem we accede by the use of so-called color map of each qualifier region. First we focus on lips localization, because this face part is invariable against head turning and pertinent structural face components (beards, glasses, etc.).

Mouth localization

In first part, we perform FLD the color transformation [11] (6), which comes out from presumption that color of lips composition consists of high levels of red and low levels of blue of color space RGB. Unfortunately this color composition varies in various ways in different people and with FLD we are not able to involve all possibilities. From this reason we establish *rg* color transformation [12] (7), which would support FLD transformation in some cases.

$$FLD = [-0.289 \quad 0.379 \quad 0.038] \cdot \begin{bmatrix} R \\ G \\ B \end{bmatrix}, \qquad (6)$$

$$r = \frac{R}{(R+G+B)} \quad g = \frac{G}{(R+G+B)}, \qquad (7)$$

We use only *g* component, because *r* component does not appeared to be good for this, because of redundation and low dispersion. Results are two similar color map of mouth (Fig. 9, we can see very low values in corresponding areas for both cases). Drawback is possibility of occurrence those background parts with very high value of red and low value of blue (for example red scarf about neck, etc.). We use that fact we keep at one's disposal averaged out binary mask of supposed face, thus we process that parts only, which are concerned in this binary mask. Further, there could be that case, in which the red color is in the face in larger quantity (e.g. strongly reddish face), therefore we use already performed wavelet transform in vertical direction (lips appear like sharp transition in vertical direction, especially if opened) and then multiply the result with these maps. Now we transform each map onto binary map by tresholding (thresholds are obtained pursuant to histograms of each color map.). Then we merge these binary maps by (8) and then apply *mean shift* algorithm [13]. The result of this transform is coordinates of area with the best probability of lips appearance.

$$b = \frac{1}{4} \cdot (b_1 + b_2) + \frac{3}{4} \cdot b_1 \cdot b_2. \qquad (8)$$

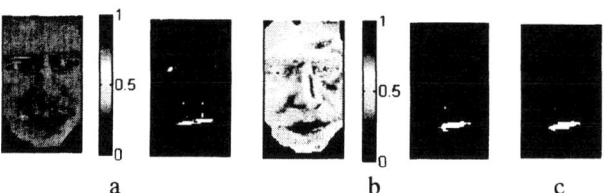

Fig. 9. a) *g* the color map and its binary image,
b) FLD the binary map and its binary image, c) result of (8)

Eyes localization

By eyes localization we use similar attempts like by the lips localization, however we use only one color map [14] (9). For the reason pupil of the eye and white of the eye colors are very similar in most people. Example of eye color map is in Fig. 10 (areas answer to eyes reach of high level values.). Here again we use averaged out face binary mask because of disturbing background influences (e.g. white shirt collar) and apply multiplication with wavelet coefficients (just in case of closed eye, when there is a influence of eyebrow and eyelashes). And again, we transform color map onto binary mask by the help of tresholding (thresholds are obtained from histogram) and by using mean shifting algorithm we localize eyes position (it is necessary take in team lips position here, because teeth could reach also very high level values in eye color map).

$$b_{eye} = \frac{1}{3} \cdot C_b^2 + (256 - C_r)^2 + \frac{C_b}{C_r}. \tag{9}$$

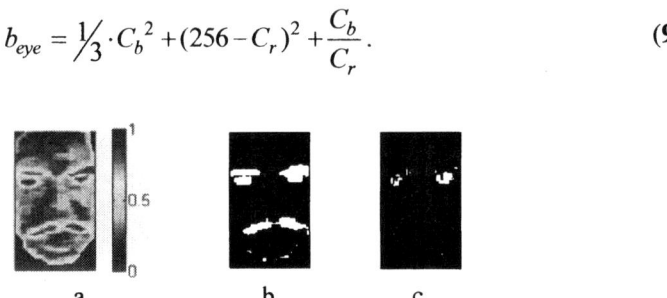

Fig. 10. a) color eye map, b) binary image, c) binary image multiplied with wavelet coefficients.

2.3 Face regions verification

On the base of mutual positions and distances among lips, both eyes with a respect to vertical and horizontal size supposed face we determine whether it is face or not [10]. In the case of agreeable the face coordinates are saved including the size and eyes and lips positions for next appropriate process. The face is included into result image. In opposite case the qualifier areas are reprobate. All qualifier areas are preceded with this detection by sequel. The final result is in Fig. 11 (green dots mean approximate eyes and lips positions).

Fig. 11. a) original image, b) detected face with eyes and lips higlight.

3 Conclusion

For benchmarking developed algorithm, we have chosen *Georgia Tech face database* [15], which has 750 shots of 50 subjects (men and women in different ages). The advantage compared to others is that it includes complex backgrounds in combination with skin-like color background. This presents challenge for many face localization algorithms. Another advantage is subject diversity, which are various head turning, emotional expression, lightning and structural face accessories (beard, glasses, etc.). We achieved 93,4% rate of success, which is very good result with regard to complexity of the database. Main cause of unsuccessful localization was especially insufficient suppression of color influence of background in the area of face with wavelet coefficients. It would be good to find better way to separation in next work in the future. To other problem was present of dense beard, in which lisp was not included to facial area and face candidate had to be then denied. This would be prevented by vertical expansion of area in lips localization even except candidates face area.

Acknowledgements. The paper was prepared within the framework of No. 102/07/1303 project of the Grant Agency of the Czech Republic and No. 1ET301710509 project of the Czech Academy of Sciences. The paper has been supported by the Ministry of Education, Youth and Sport of the Czech Republic project FRVS 972/2007/G1.

References

1. Yang, M.-H.: Detecting Faces in Images: A Survey. IEEE Transactions on pattern analysis and macgine intelligence, Vol. 24, 1. (2002)
2. Yow, K. C., Cipolla, R.: Feature-Based Human Face Detection, Image and Vision Computing, Vol. 15. (1997)
3. Augusteijn, M. F., Skujca, T.L.: Identification of Human Faces through Texture-Based Feature Recognition and Neural Network Technology, Proc. IEEE Conf. Neural Networks. (1993)
4. Jones, M. J., Rehg, J.M.: Statistical Color Models with Application to Skin Detection, Proc. IEEE Conf. Computer Vision and Pattern Recognition, Vol. 1. (1999)
5. Graf, H. P., Chen, T., Petajan, E., Cosatto, E.: Locating Faces and Facial Parts, Proc. First Int'l Workshop Automatic Face and Gesture Recognition. (1995)
6. Singh, S. K., Chauhan, D. S., Vatsa, M., Singh, R.: A Robust Skin Color Based Face Detection Algorithm, Tamkang Journal of Science and Engineering, Vol. 6. (2003)
7. Garcia, C., Zikos, G., Tziritas, G.: Face Detection in Color Images using Wavelet Packet Analysis, IEEE Int. Conf. on Multimedia Computing and Systems. (1999)
8. Kawato, S., Ohya, J.: Automatic Skin-color Distribution Extraction for Face Detection and Tracking, ICSP2000: The 5th Int. Conf. on Signal Processing, Beijin, China. (2000)
9. Boomgaard, Balen: Image Transforms Using Bitmapped Binary Images, Computer Vision, Graphics, and Image Processing: Graphical Models and Image Processing. (1992).
10. Young, J. W.: Head and Face Anthropometry of Adult U.S. Civilians, U. S. Government Printing Office. (1993)

11. Chaloupka, J.: Rozpoznávání akustického signálu řeči s podporou vizuální informace, disertační práce Technická univerzita v Liberci. (2005)
12. Stern, H., Efros, B.: Adaptive Color Space Switching for Face Tracking in Multi-Colored Lighting Environments, Proceedings of the fifth IEEE International Conference on Automatic Face and Gesture Recognition. (2002)
13. Toth, D., Stuke, I., Wagner, A., Aach, T.: Detection of Moving Shadows using Mean Shift Clustering and a Significance Test, Proceedings of the 17th International Conference on Pattern Recognition. (2004)
14. Campadelli, P., Cusmai, F., Lanzarotti, R.: A Color-Based Method for Face Detection, Proceedings of the International Symposium on Telecommunications (IST2003), 186-190. (2003)
15. Nefian, A., V.: Georgia Tech face database, http://www.anefian.com/face_reco.htm

RF Pure Current-Mode Filters using Current Mirrors and Inverters

Jan Jerabek, Kamil Vrba

Department of Telecommunications, Brno University of Technology,
Purkynova 118, 612 00 Brno, Czech Republic
xjerab08@stud.feec.vutbr.cz, vrbak@feec.vutbr.cz

Abstract. In this paper the design method leading to the radio-frequency (RF) filters working in the pure current mode is shown. New active elements called current mirrors and inverters (CMI) are presented, a generalized variant (GCMI) is used for the design purposes, and a universal version (UCMI) is fitted to the filtering-circuit final solution. Initial autonomous circuits are derived from the full admittance network, which is defined. In this paper, a few autonomous circuits with five passive elements are presented. An illustration of the filter design with independent quality factor adjustment with respect to the characteristic frequency is given. Results obtained from the measurement of samples are shown.

Keywords: pure current mode, frequency filter, current mirror, current inverter, CMI, GCMI

1 Introduction

Frequency filters and all other common electrical circuits traditionally work in the voltage mode, which means that circuit input and output values are expressed in the voltages. Of course, there is also a current flowing in the circuit, but its value does not carry any useful information. The operational amplifier (OPA) is often used as an active element, when the voltage mode is mentioned. Despite the sundry innovations effected, this circuit cannot today meet all the filter-design requirements.

The continual progress in the integrated-circuit manufacturing technologies is, among others, accompanied by decreasing the values of supply voltage and power consumption, and thus naturally also by decreasing the voltage being processed (at the same time). It causes an undesirable decrease of signal-to-noise ratio. This ratio is a little more advantageous when we use a current as a carrier of information. Also, a better bandwidth can be obtained when the circuit operates in the current mode. These are some of the reasons leading to the present increase in frequency filters working in the current mode [1] to [6].

There are sometimes mixed-working elements in place of active elements (e.g. OTA amplifiers or Current Conveyors), but the aim of this paper is to deal with the pure current-mode filter design, which means that both the active element and the filter itself are working in the current mode.

We have a number of ways how to design a current-mode frequency filter. The adjoint network method can be regarded as the simplest one, but with the least applications in practice. This method is based on the transformation of the voltage-mode filtering circuit into the current-mode circuit [7]. A method that has been developed in great detail is the method employing synthetic elements and impedance converters forming high-order immittances [8]. From other methods we should mention the signal flow graph method [9] and, last but not least, the autonomous circuit method [10], which is used for the design procedure starting from the full admittance network and which is shown in this paper.

2 Current Mirrors and Inverters

The filtering circuit works in the pure current mode when all of its input and output values are currents and, at the same time, the active element also works only with current values. This means that current inputs and current outputs are the only acceptable terminals of proposed active elements. For the generalization of design procedures, an active element GCMI (Generalized Current Mirrors and Inverters) [11] was defined in our department. This element is suitable for the new filtering circuit design according to all the design methods mentioned in the introduction. The circuit symbol of the GCMI is shown in Fig. 1.

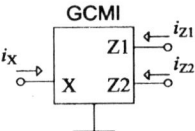

Fig. 1. Symbol of the GCMI circuit used in the schematics

This active element is described by two equations, $i_{Z1} = a \cdot i_X$, $i_{Z2} = b \cdot i_X$, where a, b are the current transfer coefficients taking the values -1 or 1 only. From these equations it is evident that we can define CMI circuits with four possible combinations of output current orientation when we use both outputs of the GCMI and generally also four other variants if we use just one of the outputs. All these variants are summarized in Table 1.

Such a generalized circuit (GCMI) is used in the design procedure of new filtering circuits, and when we come to the realization we use a particular CMI variant, made by universal CMI, see below.

Table 1. All possible variants resulting from the GCMI definition

Circuit labelling	CMI+/+	CMI-/-	CMI+/-	CMI-/+	CM_1	CI_1	CM_2	CI_2
Value of coefficient a	+1	-1	+1	-1	+1	-1	-	-
Value of coefficient b	+1	-1	-1	+1	-	-	+1	-1

From the application point of view it is advantageous to have both outputs of the CMI having the same current-flow orientation, so that both inner paths in the CMI circuit mentioned will have approximately the same phase shift. After the basic research was done, it turned out that for the filter realization it was advantageous to have all CMI variants available. Thus the idea of developing a universal current unit came into being that would be capable of realizing these four CMI variants. This circuit (the UCMI) is indicated in Fig. 2.

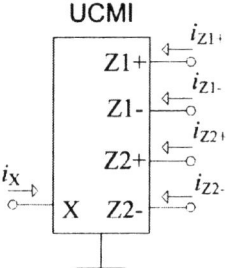

Fig. 2. Circuit symbol of the UCMI used in the schematics

The X terminal represents current input while terminals Z_{1+} and Z_{2+}, or Z_{1-} and Z_{2-} are current outputs with positive or negative current transfer from terminal X. It is clearly evident from the following equations describing the circuit behavior:

$$i_{Z1+} = i_{Z2+} = i_X, \; i_{Z1-} = i_{Z2-} = -i_X . \tag{1}$$

Proposing the internal structure of this pure-current element was the subject of completed research, the first fabricated samples produced in cooperation with AMI Semiconductor will be ready later this year (2007). To be able to experimentally verify the CMI-application functionality sooner, a universal current conveyor UCC-N1B [12] made also in AMI Semiconductor was employed. It contains all terminals necessary for the UCMI realization but it also has voltage inputs, which have to be grounded for proper UCMI implementation. Although the UCC circuit was originally not designed for this kind of usage and its bandwidth is only about 30 MHz, it did not show in any markedly negative way in the verification of the newly designed filters. Of course, we can expect that the commencement of production of the UCMI, which will be internally simpler than the UCC, will bring us the widening of frequency bandwidth up to 100 MHz.

3   Autonomous Circuit Design Method

The autonomous circuit filter design method [10] seems to be really advantageous. This method consists of a few steps:
- An autonomous circuit containing only general active elements (GCMI) and general passive elements (admittances) is proposed. We usually proceed

intuitively or follow some analogies. A systematic method of autonomous circuit design is introduced in chapter 4 of this paper.
- The characteristic equation (CE) of autonomous circuit is calculated [10].
- CE is simplified by choosing suitable transfer coefficients. We also have to ensure that all the terms are positive in view of the condition of filter stability.
- Passive elements are concretized (admittances are replaced by capacitors or resistors) in such a way that appropriate order of the CE is obtained.
- Input and output positions are appointed, while respecting that current sources can only excite the nodes of a circuit and that current response can be monitored in the branches.
- The transfer functions found are determined.

4 Design of Autonomous Circuits

The first point of the procedure described above is the most sophisticated task and that is why we tried to algorithmize it [11]. As already mentioned, we usually also proceed intuitively, but it is much more favorable to start from a general full admittance network (Fig. 3) connected between two GCMI elements, in case we want to obtain second-order filters. Four passive elements are often enough for a particular solution and it is further possible to realize more than one filter type within one circuit (referred to as the multifunction filter) [11]. In this paper, autonomous circuits with five passive elements are discussed. The main advantage of such circuits is the potential independence of the quality factor on the characteristic frequency and hence these filters are suitable for tuning.

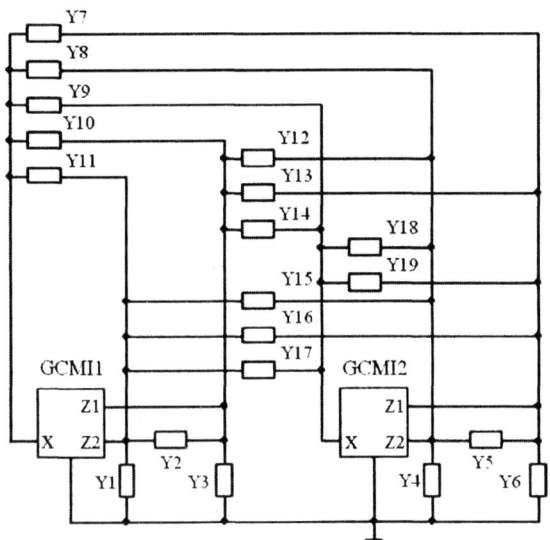

Fig. 3. Improved full admittance network connected to two GCMI

As can be seen from Fig. 3, the improved full admittance network is made up of 19 admittances, so the number of admittances is the same as in [11], but same changes have been made. Based on later findings, the admittance connected between the X terminal of the GCMI and the ground has been removed while an admittance has been added between two outputs (Z1 and Z2) of the GCMI, which holds for both GCMI placed in the network. Like in [11], it should be noted that each of the admittances should theoretically be replaced by a resistor, capacitor, their parallel combination, short-circuit or disconnected, which will be applied below.

A large number of autonomous circuits can be derived from this full admittance network, which is beyond the scope of this paper and that is why only a few autonomous circuits with five passive elements will be shown, see Table 2. The left side of the characteristic equation is presented on the right side of every autonomous circuit depiction. As is well known, the left side of the CE will appear in the denominator of the transfer function of every filter derived from the autonomous circuit.

Table 2. Autonomous circuits with five passive elements

No.	Autonomous circuit	Left side of the characteristic equation
1.		$Y_1Y_3 + Y_1Y_4 + b_2Y_1Y_4 - a_2b_1Y_1Y_4 +$ $+ Y_3Y_5 + Y_2Y_3 + Y_4Y_5 + b_2Y_4Y_5 +$ $+ Y_2Y_4 + b_2Y_2Y_4$
2.		$Y_2Y_3 + Y_3Y_5 + Y_1Y_3 - a_1a_2Y_1Y_3 +$ $+ Y_2Y_4 + b_2Y_2Y_4 + Y_4Y_5 +$ $+ b_2Y_4Y_5 + Y_1Y_4 + b_2Y_1Y_4 -$ $- a_1a_2Y_1Y_4 - a_2b_1Y_1Y_4$
3.		$Y_1Y_5 + Y_2Y_5 + Y_3Y_5 + Y_4Y_5 +$ $+ a_1Y_3Y_5 + Y_1Y_4 + Y_2Y_4 + Y_3Y_4 +$ $+ a_1Y_3Y_4 + b_2Y_2Y_4$

Table 2. Autonomous circuits with five passive elements - continued

No.	Autonomous circuit	Left side of the characteristic equation
4.		$Y_2Y_3 + Y_3Y_4 + a_1Y_3Y_4 - $ $- b_1b_2Y_3Y_4 + Y_3Y_5 + Y_1Y_2 +$ $+ Y_1Y_4 + a_1Y_1Y_4 + Y_1Y_5$
5.		$Y_3Y_4 + a_2Y_3Y_4 + Y_1Y_4 + a_1Y_1Y_4 +$ $+ a_2Y_1Y_4 + a_1a_2Y_1Y_4 + Y_2Y_4 +$ $+ a_2Y_2Y_4 + Y_1Y_5 + a_1Y_1Y_5 +$ $+ Y_2Y_5 + Y_3Y_5$
6.		$Y_2Y_5 + Y_4Y_5 + b_2Y_4Y_5 + Y_1Y_2 +$ $+ a_1Y_1Y_2 + Y_1Y_4 + b_2Y_1Y_4 +$ $+ a_1Y_1Y_4 + a_1b_2Y_1Y_4 - a_2b_1Y_1Y_4 +$ $+ Y_2Y_3 + Y_3Y_4 + b_2Y_3Y_4$
7.		$Y_3Y_4 + a_2Y_3Y_4 + b_2Y_3Y_4 + Y_2Y_4 +$ $+ a_2Y_2Y_4 + Y_1Y_4 + a_1Y_1Y_4 +$ $+ a_2Y_1Y_4 + a_1a_2Y_1Y_4 + Y_3Y_5 +$ $+ b_2Y_3Y_5 + Y_2Y_5 + Y_1Y_5 + a_1Y_1Y_5$
8.		$Y_2Y_3 + Y_2Y_5 + a_2Y_2Y_5 + Y_2Y_4 +$ $+ Y_1Y_2 + a_1Y_1Y_2 + Y_3Y_5 + a_2Y_3Y_5 +$ $+ Y_3Y_4 + Y_1Y_3 + a_1Y_1Y_3 + b_1Y_1Y_3$

5 Example of the Frequency Filter Design

The frequency filter design will be shown on autonomous circuit No. 1 from Table 2. Its characteristic equation is

$$D = Y_1Y_3 + Y_1Y_4 + b_2Y_1Y_4 - a_2b_1Y_1Y_4 + Y_3Y_5 + Y_2Y_3 + Y_4Y_5 +$$
$$+ b_2Y_4Y_5 + Y_2Y_4 + b_2Y_2Y_4 = 0 \ . \tag{2}$$

Choosing the products of coefficients $a_2b_1 = -1$ and the value of coefficient $b_2 = -1$ we obtain a simplification of CE (2) and ensure that all admittance products in the CE will be positive, which is one of the conditions required for filter stability. Equation (2) will change to

$$D = Y_1Y_3 + Y_1Y_4 + Y_3Y_5 + Y_2Y_3 \ . \tag{3}$$

Following the steps mentioned in chapter 3 we now come to the passive elements choice. We always have more than one option of selecting which of the admittances will be replaced by capacitor and which of them will be replaced by resistor. When we want to obtain a second-order filter, Table 3 lists all suitable variants.

Table 3. All possible variants of passive elements selection

Variant	Y_1	Y_2	Y_3	Y_4	Y_5
A	pC_1	G_1	G_2	pC_2	G_3
B	G_1	G_2	pC_1	G_3	pC_2
C	G_1	pC_1	pC_2	G_2	G_3

If we choose, for example, variant B, CE (3) will change to

$$D = p^2C_1C_2 + pC_1(G_1 + G_2) + G_1G_3 \ , \tag{4}$$

where the complex variable $p = j\omega$. The current-mode multifunction frequency filter designed according to variant B is shown in Fig. 4. At the same time, the exciting current sources connected to the nodes and all possible current responses are shown.

There are lots of possible transfer functions in this circuit, so it is advantageous to summarize them all for greater lucidity, see Table 4. From this Table, it is much better evident how to choose the remaining transfer coefficients appropriately and how to obtain the required filtering transfer functions (low pass, band pass and high pass frequency filters at least). There are only numerators of transfer functions in Table 4; denominators are equal to the CE (4).

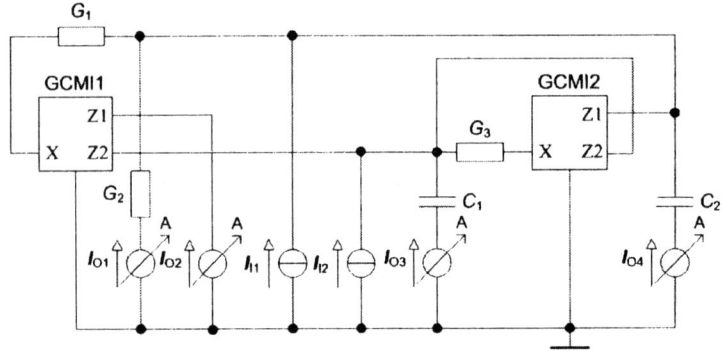

Fig. 4. Frequency filter designed for current mode

Table 4. Numerators of all potential transfer functions

		Current source	
		I_{I1}	I_{I2}
Current response	I_{O1}	$-pG_2C_1$	$a_2G_3G_2$
	I_{O2}	$a_1pG_1C_1$	$-a_1a_2G_1G_3$
	I_{O3}	$b_1pG_1C_1$	$-p(G_1C_1+ G_2C_1) - p^2C_1C_2$
	I_{O4}	$-p^2C_1C_2$	$a_2pG_3C_2$

In case of current-mode filter, the filtering functions can also be obtained by adding two or more output current responses. The remaining non-defined transfer coefficients of the GCMI have to be chosen with respect to the two previously mentioned conditions and in such a way that most functions are obtained. Table 5 contains all possible versions of the transfer coefficient selection.

Table 5. All possible versions of transfer coefficient selection

Version	a_1	b_1	$GCMI_1$	a_2	b_2	$GCMI_2$
1.	1	-1	+/-	1	-1	+/-
2.	1	1	+/+	-1	-1	-/-
3.	-1	-1	-/-	1	-1	+/-
4.	-1	1	-/+	-1	-1	-/-

The third version in Table 5 appears to be the best because it enables us to implement two identical inverting band pass filters by adding up two output currents. The first filter by adding up $I_{O1} + I_{O2}$ and second one by adding up $I_{O1} + I_{O3}$, while regarding I_{I1} as the current source. Thus if we choose the third version of coefficient selection, we obtain the following second-order transfer functions:

$$K_{HP} = \frac{I_{O4}}{I_{I1}} = -\frac{p^2 C_1 C_2}{D}, \quad (5)$$

$$K_{BP1} = \frac{I_{O1} + I_{O2}}{I_{I1}} = -\frac{pC_1(G_1 + G_2)}{D}, \quad (6)$$

$$K_{BP2} = \frac{I_{O1} + I_{O3}}{I_{I1}} = -\frac{pC_1(G_1 + G_2)}{D}, \quad (7)$$

$$K_{LP} = \frac{I_{O2}}{I_{I2}} = \frac{G_1 G_3}{D}. \quad (8)$$

From (5) to (8) it is evident that going down from the top, we are concerned with an inverting high pass (HP) filter, two identical inverting band pass (BP1 and BP2) filters and a low pass (LP) filter.

The final solution of the multifunction filter working in the pure current mode is presented in Fig. 5. Both types of the GCMI chosen have been implemented by the UCC-N1B circuits to be able to verify the correctness of the whole design process.

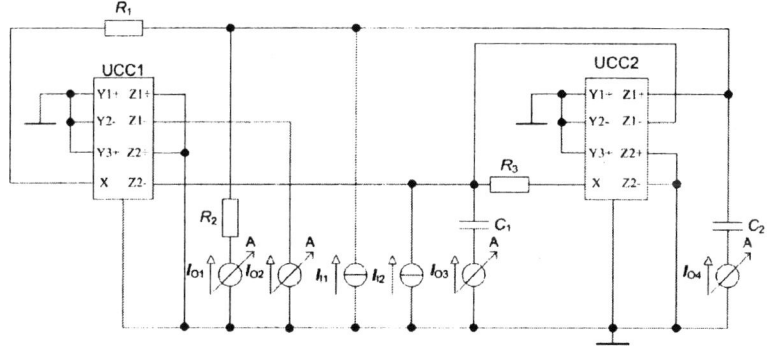

Fig. 5. Multifunction second-order filter working in the current mode with UCC

The characteristic frequency and quality factor formulas can be derived from CE (2) for all filter kinds given and they are

$$f_0 = \frac{1}{2\pi}\sqrt{\frac{1}{C_1 C_2 R_1 R_3}}, \quad (9)$$

$$Q = \sqrt{\frac{C_2 R_1}{C_1 R_3}} \frac{R_2}{R_2 + R_1}. \quad (10)$$

When comparing eq. (9) and eq. (10), we can notice that the quality factor can be adjusted independently of the characteristic frequency by the value of resistor R_2. The values of capacitors have been chosen as $C_1 = C_2 = 68$ pF, the characteristic frequency equal to $f_0 = 2$ MHz, and resistor $R_1 = 1.8$ kΩ, as an example of a particular solution. From all these selected values, the value of R_3 has been calculated, $R_3 = (C_1 C_2 R_1 4\pi^2 f_0^2)^{-1} \approx 820$ Ω. We can then tune the quality factor of this multifunction filter by the value of the R_2 resistor. Three suitable values of the R_2 resistor have been chosen (1 kΩ, 2.7 kΩ, 22 kΩ) to obtain three quality factor values (0.53, 1.00, 1.37).

The common circuit analyzer unit, used for measuring the voltage-mode circuits, was complemented with a voltage-current converter to obtain current excitation for the filter input. It was implemented by an operational transconductance amplifier (OTA) integrated in the OPA860 circuit. The reverse current-voltage conversion was also realized in a similar way, in order to transform the output current of the filter into the analyzer (voltage) input [13]. The frequency bandwidth of such converters is approximately 100 MHz, which is absolutely sufficient for design verification. The measured transfer characteristics of such a multifunction filter are depicted in Fig. 6 to Fig. 9.

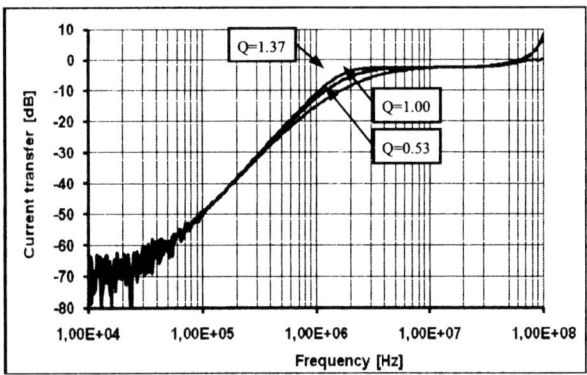

Fig. 6. Measured characteristics of HP filters

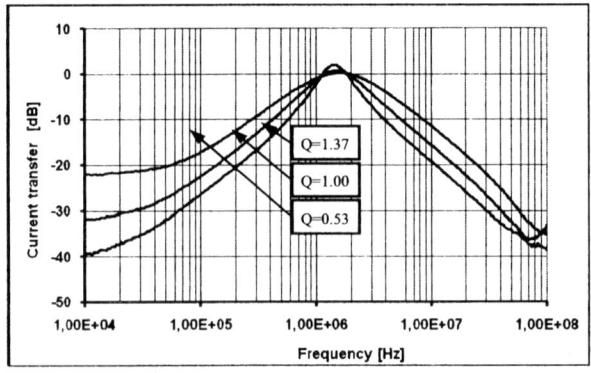

Fig. 7. Measured characteristics of BP1 filters

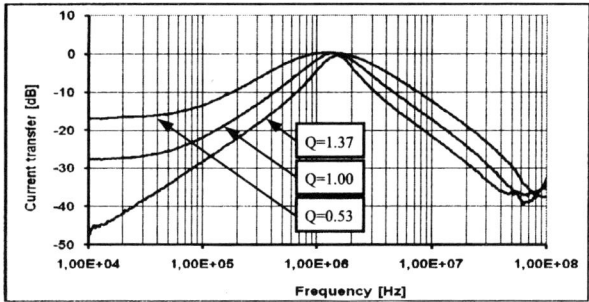

Fig. 8. Measured characteristics of BP2 filters

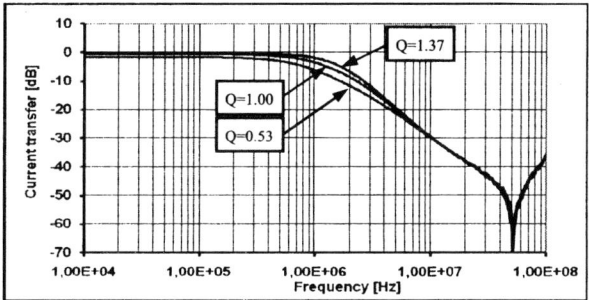

Fig. 9. Measured characteristics of LP filters

The characteristics obtained for all the filters have a lower quality factor than it was expected. The characteristic frequency is also a little bit lower, about 1.5 MHz. The pattern of frequency curves at frequencies of over 30 MHz is damaged by parasitic effects present on the sample boards and by the real features of the UCC circuit. Theoretically similar band passes have different characteristics, which can be attributed to the divergent circuit and also to the board disposition. Measured characteristics performance is absolutely sufficient for design verification, which was the objective of the measurement. Filters based on the GCMI circuit can be assessed overall as good.

6 Conclusion

The paper describes the design process of autonomous circuits with five passive elements starting from a general full admittance network connected between two generalized active elements GCMI. These autonomous circuits are suitable not only for the frequency filter design but also for oscillators working in the current mode, which however was not part of this paper. We limited ourselves to autonomous circuits with five passive elements, but the procedure is similar also when looking for autonomous circuits with another number of passive elements.

The whole design procedure was demonstrated on one of the autonomous circuits, leading to a second-order multifunction frequency filter working in the pure current mode and providing the possibility of the quality factor being independent of the characteristic frequency tuning. The circuit characteristics were subsequently verified experimentally, the evaluation of results can be found in chapter 5.

The active element (UCC) was used only to enable an experimental verification of the CMI-application functionality. As mentioned before, the first fabricated samples of the UCMI circuit produced in cooperation with AMI Semiconductor should be ready by the end of this year (2007). The commencement of production of the UCMI, which will be internally simpler than the UCC, is expected to widen the frequency bandwidth up to the 100 MHz and also to improve other features.

Acknowledgments. The research into new active elements is supported by The Czech Science Foundation, project No. 102/06/1383 and by research project of the Ministry of Education No. 2B06111.

References

1. Tomazou, C., Lidgey, F. J., Haigh, D. G.: Analog IC design: the current-mode approach. Institution of Electrical Engineers, London, 1996.
2. Gunes, E., Anday, F.: Realization of voltage and current-mode transfer functions using unity gain cells. International Journal of Electronics, Vol. 83, No. 2, pp. 209-213, 1997.
3. Salama, K.: Continuous time universal filters using unity gain cells. International Journal of Electronics and Communication, Vol. 56, No. 2, pp. 1-4, 2002.
4. Palmisano, G., Pennisi, S.: A novel CMOS current-mode power amplifier. 2nd IEEE CAS Region Workshop on Analog and Mixed Design, pp. 83-86, Baveno, 1997.
5. Alzaher, H. A., Ismail, M.: Current-mode universal filter using unity gain cells. Electronics Letters, Vol. 35, No. 25, pp. 2198-2200, 1999.
6. Palmisano, G., Pennisi, S.: Solution for CMOS current amplifiers with high-drive output stages. IEEE Transactions on Circuits and Systems – II Analog and digital signal processing, Vol. 47, No. 10, pp. 988-997, 2000.
7. Cajka, J., Dostal, T., Vrba, K.: Transformations providing adjoint networks working in the current mode. Elektrorevue.cz, No. 23, 2000, (in Czech).
8. Gubek, T., Biolek, D.: New approaches to gm-C filters using non-cascade synthesis and CDTA elements. ElectronicsLetters.com, No. 3, 2004.
9. Biolek, D.: Novel signal flow graphs of current conveyors. 38th MWSCAS, Rio de Janeiro, Brazil, pp. 1058-1061, 1995.
10. Koton, J., Vrba, K.: Method for designing frequency filters using universal current conveyors. International Transaction on Computer Science and Engineering, Vol. 13, No. 1, pp. 144-154, 2005.
11. Jerabek, J., Vrba, K.: Design of High-Frequency Filters Working in the Pure Current Mode with CMI. The Second International Conference on Systems, ICONS 2007, Martinique, 2007.
12. Sponar, R., Vrba, K.: Measurements and behavioral modeling of modern conveyors. International Journal of Computer Science and Network Security, Vol. 6, No. 3A, pp. 57-65, 2006.
13. Jerabek, J., Vrba, K.: U/I and I/U converters using current conveyors for measurement of circuits in current mode. In Proceedings of RTT'06, Part II, pp. 125-128, Czech Republic, 2006, (in Czech).

Multifunction RF Filters Using OTA

Norbert Herencsar[1], Kamil Vrba[1]

[1] Dept. of Telecommunications,
Faculty of Electrical Engineering and Communication,
Brno University of Technology, Purkynova 118, 612 00 Brno, Czech Republic
herencsar@phd.feec.vutbr.cz, vrbak@feec.vutbr.cz

Abstract. In this paper the procedure of RF filter design using two active elements OTA is shown. The general method of filter design using OTA amplifiers, when the fully connected feedback network was designed around the two OTA elements, is shown. Autonomous circuits were gradually picked out from this general arrangement, using two to five passive elements. These autonomous circuits then serve as an initial circuit in designing different types of frequency filters. On a selected autonomous circuit, all the procedure of multifunction filter design is shown. The properties of the proposed second–order multifunction filter were subjected to an AC analysis in the PSpice software [8], a generalized sensitivity analysis and experimental measurement.

Keywords: OTA, frequency filter, autonomous circuit, general admittance network, sensitivity analysis.

1 Introduction

In the last few years a number of new active elements for designing RF filters working in the current or the voltage mode were developed. Primarily these are CFA (Current Feedback Amplifier), OTA (Operational Transconductance Amplifier), BOTA (Balanced Output OTA), MOTA (Multiple Output OTA), active element working in the mixed mode CDBA (Current Differencing Buffered Amplifier), CC (Current Conveyor), VC (Voltage Conveyor), COA (Current Operational Amplifier) or pure current element CMI (Current Mirror and Inverter). Numerous scientific papers and publications dealing with RF filters using these active elements have been presented [3], [4]. Unfortunately, most of these elements are only on the theoretical level or are currently being developed. Commercially available amplifiers are only OTA and BOTA amplifiers, e.g. MAX43/MAX436 by MAXIM–Dallas Semiconductor [9], LM13600/LM13700 by National Semiconductor or LT1228 by Linear Technology [10].

This paper is focused on the design procedure of RF filters with two transconductance amplifiers, which are referred to as OTA amplifiers. These filters are often appropriate for signal processing of high–speed data communication systems, in cable modems or in hard–drive communication interfaces.

2 Transconductance amplifiers

Operational transconductance amplifiers with a single output (OTA) were made commercially available for the first time in 1969 by RCA. The first publications with OTA came out in 1985, when R. L. Geiger and S. E. Sánchez presented to the general public the new CMOS OTA architectures and new circuits of frequency filters [2] with this new active element. The schematic symbol of OTA is shown in Fig. 1a. For tuned RF filters it is appropriate to use an OTA element with variable values of transconductance g_m set by the control current I_{SET} (Fig. 1b).

a)   b)

Fig. 1. Schematic symbol of OTA: a) with constant transconductance, b) with variable transconductance.

The ideal OTA is a voltage–controlled current source characterized by transconductance g_m. The output current of the OTA element is given by the following equation:

$$I_0 = -g_m(U_p - U_n), \qquad (1)$$

where U_p and U_n are the voltages on non–inverting and inverting inputs of OTA with respect to the ground. The ideal OTA amplifier is characterized by a finite, frequency–independent transconductance g_m while its input and output impedances are theoretically infinite. An important feature of OTAs is the possibility of driving the transconductance g_m by the control current I_{SET}.

Currently, the OTA elements are supplied on the market by many manufacturers. A commercially available OTA element, as mentioned above, is the circuit LT1228 (Linear Technology) [10] or MAX435 (MAXIM–Dallas Semiconductor) [9], which is a high–speed wideband transconductance amplifier (WTA) with high–impedance inputs and output. Due to its unique performance features, it is suitable for a wide variety of applications such as high–speed instrumentation amplifiers, wideband, high–speed RF filters, and high–speed differential line driver and receiver applications. In this paper we deal with the filtering applications of OTA amplifiers.

3 General design of autonomous circuits with two OTA elements

In our workplace, the method of generalized frequency filter design, when the initial circuit is a full–admittance network connected to active elements [5], [6], was developed. The same procedure was used in the case of OTA elements. The basic general autonomous circuit is shown in Fig. 2.

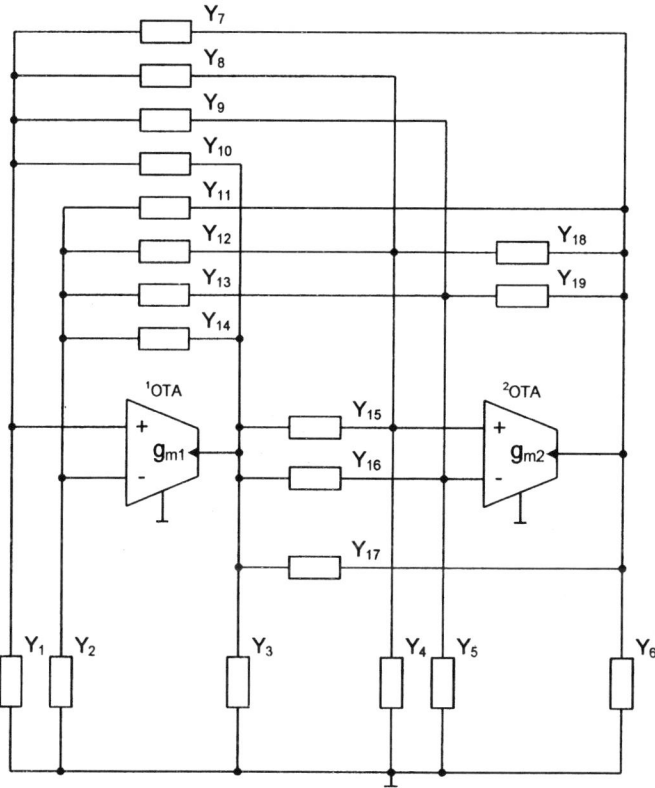

Fig. 2. Full admittance network connected to two OTA elements.

Simpler autonomous circuits were gradually picked out from this general arrangement. We tried to have the passive elements grounded with one terminal, because such elements are easier to implement in integrated circuits. The selected autonomous circuits presented in Table 1 have three or four admittances. In Table 1 the characteristic equations are also given, which describe the behaviour of autonomous circuit. Only their left side is mentioned, the right side is always equal to zero. The characteristic equations were obtained using the SNAP software [7].

Table 1. Designed autonomous circuits and their characteristic equations.

No.	Autonomous circuits and their left side of characteristic equations
1	$Y_1Y_2 + Y_2Y_3 + g_{m1}g_{m2}$
2	$Y_1Y_2Y_3 + g_{m1}Y_2Y_3 + g_{m2}Y_1Y_3 + g_{m1}g_{m2}Y_3$
3	$Y_1Y_2Y_3 + Y_1Y_2Y_4 + Y_2Y_3Y_4 + g_{m2}Y_2Y_3 + g_{m1}g_{m2}Y_3$
4	$Y_1Y_2Y_4 + Y_2Y_3Y_4 + g_{m2}Y_2Y_4 + g_{m1}g_{m2}Y_4$

4 Filter design

For the verification of the correct function of filters with OTA elements autonomous circuit No. 1, Table 1, has been selected. The characteristic equation of this circuit is:

$$D = Y_1 Y_2 + Y_2 Y_3 + g_{m1} g_{m2} = 0. \tag{2}$$

Due to choosing the passive elements $Y_1 = G_1$, $Y_2 = pC_1$, $Y_3 = pC_2$ equation (2) changes to a form which satisfies the feasibility conditions of the frequency filter

$$D\,p = p^2 C_1 C_2 + p\, G_1 C_1 + g_{m1} g_{m2}, \tag{3}$$

where $p = j\omega$ is the complex variable. The frequency filter designed in the voltage mode (VM) is shown in Fig. 3.

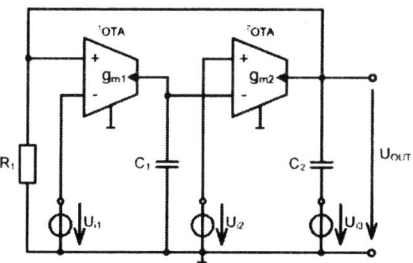

Fig. 3. Designed filter working in VM.

The complex voltage transfer functions of the designed filter with driving voltages U_{i1}, U_{in2} and U_{in3} have the form:

$$K_{LP} = \frac{U_{out}}{U_{i1}} = \frac{g_{m1} g_{m2}}{D\,p}, \tag{4}$$

$$K_{BP} = \frac{U_{out}}{U_{i2}} = -\frac{pC_1 g_{m2}}{D\,p}, \tag{5}$$

$$K_{HP} = \frac{U_{out}}{U_{i3}} = \frac{p^2 C_1 C_2}{D\,p}, \tag{6}$$

$$K_{BR} = \frac{U_{out}}{U_{i1} + U_{i3}} = \frac{p^2 C_1 C_2 + g_{m1} g_{m2}}{D\,p}, \tag{7}$$

$$K_{AP} = \frac{U_{out}}{U_{i1} + U_{i2} + U_{i3}} = \frac{p^2 C_1 C_2 - pC_1 g_{m2} + g_{m1} g_{m2}}{D\,p}. \tag{8}$$

In the voltage mode the circuit according to Fig. 3 can be used as a low–pass (4), band–pass (5), high–pass (6), band–reject (7) or an all–pass (8) filter.

The characteristic frequency of all filters is:

$$\omega_0 = \sqrt{\frac{g_{m1}g_{m2}}{C_1 C_2}} \,. \tag{9}$$

The quality factor Q_0 of all filters is given by the following equation:

$$Q_0 = \frac{1}{G_1}\sqrt{\frac{C_2 g_{m1} g_{m2}}{C_1}} \,. \tag{10}$$

For the required values Q_0 and ω_0, and the selected values C_1 and C_2 we can determine other parameters necessary for the design:

$$g_{m1} = \omega_0 Q_0 C_1, \qquad G_1 = g_{m2} = \frac{\omega_0 C_2}{Q_0} \,. \tag{10}, (11)$$

The filter example was designed for the tunable range of characteristic frequency $f_0 \approx$ 1 MHz to 10 MHz, the range of controlling the quality factor Q_0 = 0.5 to 5 based on the Butterworth approximation [1]. Capacitors C_1 = 220 pF and C_2 = 110 pF were chosen. Other parameters are within these limits: $g_{m1,2}$ = 1 mA/V to 10 mA/V, and R_1 = 100 Ω to 1 kΩ. Commercially available transconductance amplifiers (OTA) are the MAX436 or LT1228 circuits. Their company models were used in the computer simulation in the PSpice software [8]. The simulation results for the voltage–mode second–order multifunction filter are shown in Fig. 4. The characteristics of a filter using LT1228 and MAX436 are compared with the ideal models of OTA amplifiers. From the results it is evident that when using MAX436 the results are close to the ideal curves. The possibility of tuning the characteristic frequency and the quality factor is shown in Fig. 5. For a second–order low–pass filter using LT1228 and MAX436 the two cut–off frequencies and the mid frequency of tuning $f_0 \approx$ 1 MHz to 10 MHz and Q_0 = 0.5 to 5 are shown. From the results it is again evident that for filters working at a higher frequency it is preferable to use the MAX436 elements. A specific filter solution is shown in chapter six.

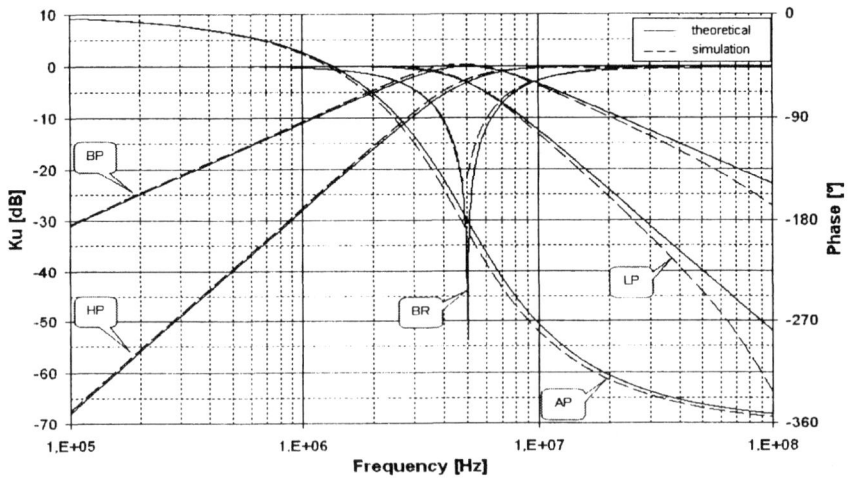

a)

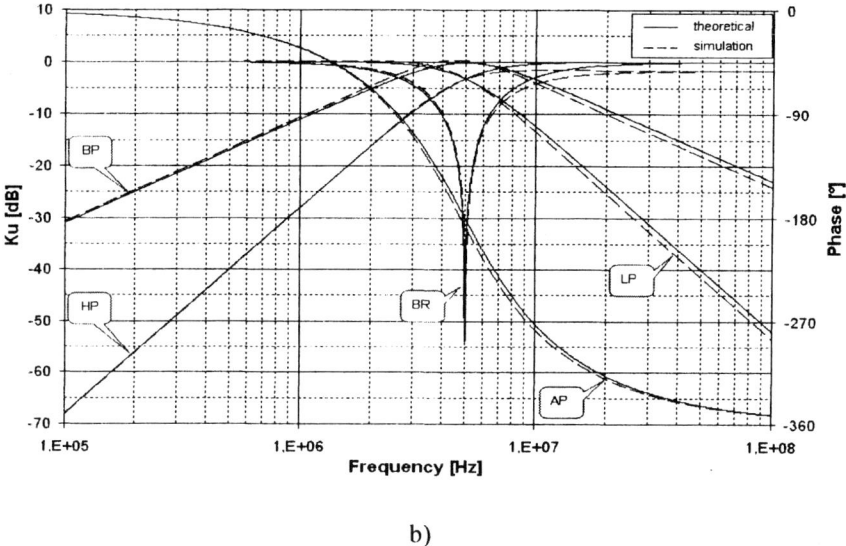

b)

Fig. 4. Frequency characteristics of multifunction filter shown in Fig. 3 using: a) LT1228 and b) MAX436.

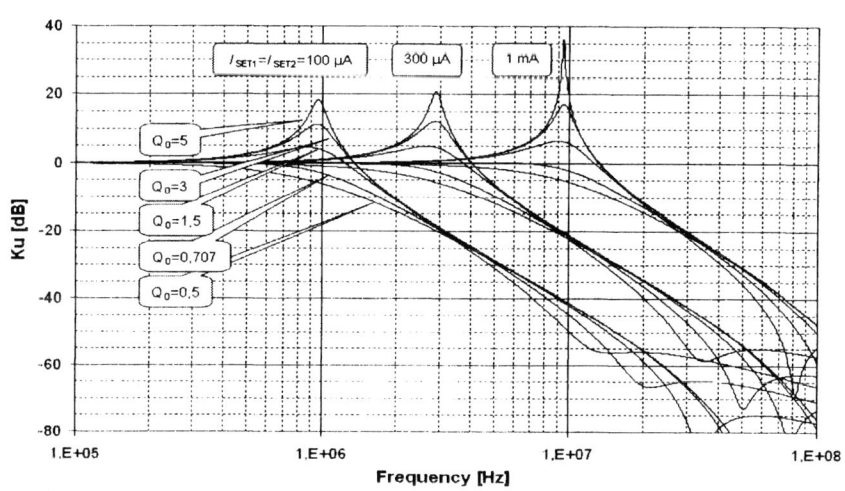

a)

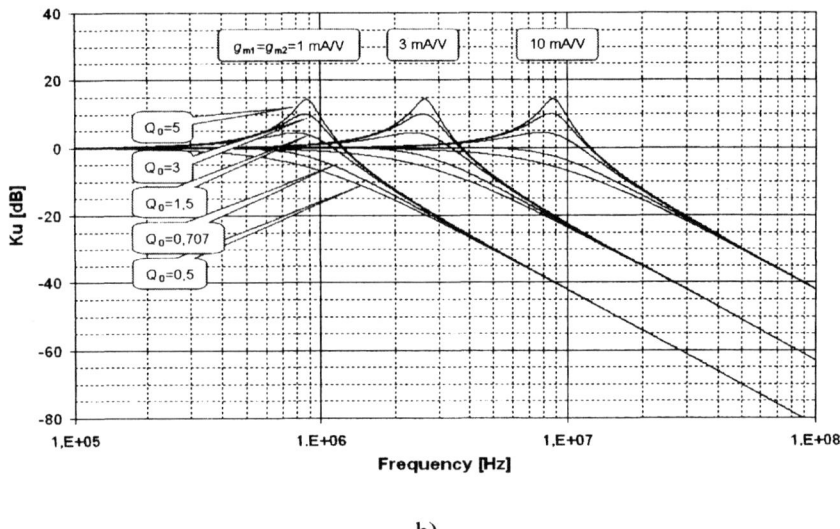

b)

Fig. 5. Frequency characteristics of tuned low–pass filter shown in Fig. 3 using: a) LT1228 and b) MAX436.

5 Sensitivity function of multifunction filter

All filters have in their denominator the same left side of characteristic equation (3) of the circuit function and thus the characteristic frequency; the quality factor Q_0 and the relevant relative sensitivities are identical for all types of filters [3]. For the circuit in Fig. 3 the particular relative sensitivities of characteristic frequency ω_0 of all filters are given by the following equation:

$$-S_{C_1}^{\omega_0} = -S_{C_2}^{\omega_0} = S_{g_{m1}}^{\omega_0} = S_{g_{m2}}^{\omega_0} = \frac{1}{2}, \qquad S_{G_1}^{\omega_0} = 0,$$

The relative sensitivities of the quality factor Q_0 are:

$$-S_{C_1}^{Q_0} = S_{C_2}^{Q_0} = S_{g_{m1}}^{Q_0} = S_{g_{m2}}^{Q_0} = \frac{1}{2}, \qquad S_{G_1}^{\omega_0} = -1.$$

From the results it is evident that the sensitivities of the circuit are low. Potential deviations, which occur in manufacturing or due to changing temperature, can be compensated by changing the values of transconductance g_m by the control current I_{SET}.

In frequency filters the greatest emphasis is put on the overall waveform of the frequency characteristics, and that is why we focus on the effects of parameter tolerances of circuits on overall frequency characteristics. The generalized sensitivity function [3] was introduced for this purpose. This function is complex and frequency–dependent, and it gives us overall information about how the transmission will be affected by tolerances of single circuit parameters throughout the frequency band. When we want know how the frequency characteristic is most affected, it is good to define the worst–case relative sensitivity of transmission modulus, which informs about the worst combinations of circuit parameter tolerances. Partial generalized sensitivity functions of filters to active and passive elements are:

$$^{LP}S_{G_1}^{K_u} = -\frac{pC_1G_1}{p^2C_1C_2 + pC_1G_1 + g_{m1}g_{m2}}, \tag{12}$$

$$^{LP}S_{g_{m1}}^{K_u} = {}^{LP}S_{g_{m2}}^{K_u} = -{}^{LP}S_{C_1}^{K_u} = -\frac{p^2C_1C_2 + pC_1G_1}{p^2C_1C_2 + pC_1G_1 + g_{m1}g_{m2}}, \tag{13}$$

$$^{LP}S_{C_2}^{K_u} = -\frac{p^2C_1C_2}{p^2C_1C_2 + pC_1G_1 + g_{m1}g_{m2}}, \tag{14}$$

$$^{BP}S_{G_1}^{K_u} = -\frac{pC_1G_1}{p^2C_1C_2 + pC_1G_1 + g_{m1}g_{m2}}, \tag{15}$$

$$-{}^{BP}S_{g_{m1}}^{K_u} = {}^{BP}S_{C_1}^{K_u} = \frac{g_{m1}g_{m2}}{p^2C_1C_2 + pC_1G_1 + g_{m1}g_{m2}}, \tag{16}$$

$$^{BP}S_{g_{m2}}^{K_u} = \frac{p^2C_1C_2 + pC_1G_1}{p^2C_1C_2 + pC_1G_1 + g_{m1}g_{m2}}, \tag{17}$$

$$^{BP}S_{C_2}^{K_u} = -\frac{p^2C_1C_2}{p^2C_1C_2 + pC_1G_1 + g_{m1}g_{m2}}, \tag{18}$$

$$^{HP}S_{G_1}^{K_u} = -\frac{pC_1G_1}{p^2C_1C_2 + pC_1G_1 + g_{m1}g_{m2}}, \tag{19}$$

$$^{HP}S_{g_{m1}}^{K_u} = {}^{HP}S_{g_{m2}}^{K_u} = -\frac{g_{m1}g_{m2}}{p^2C_1C_2 + pC_1G_1 + g_{m1}g_{m2}}, \tag{20}$$

$$^{HP}S_{C_1}^{K_u} = \frac{g_{m1}g_{m2}}{p^2C_1C_2 + pC_1G_1 + g_{m1}g_{m2}}, \tag{21}$$

$$^{HP}S_{C_2}^{K_u} = \frac{pC_1G_1 + g_{m1}g_{m2}}{p^2C_1C_2 + pC_1G_1 + g_{m1}g_{m2}}. \tag{22}$$

For easier interpretation, the generalized sensitivity functions were expressed graphically. The characteristic frequency $f_0 \approx 5$ MHz and quality factor $Q_0 = 0.707$ based on the Butterworth approximation ($b_1 = 1.4142$, $b_2 = 1$) [1] were chosen. For component parameters the recommendation is: capacitor values $C_1 = 220$ pF, $C_2 = 110$ pF, admittance and transconductance values are $G_1 = g_{m1} = g_{m2} \approx 4.9$ mA/V.

The partial generalized sensitivities of transmission are shown in Fig. 3. The relative sensitivities of filter transmission are on the left side, on the right side are semi–relative sensitivities of the argument. From the graphs it was found that the sensitivities of low–pass and high–pass filters in the transmission band were zero, which could be of advantage.

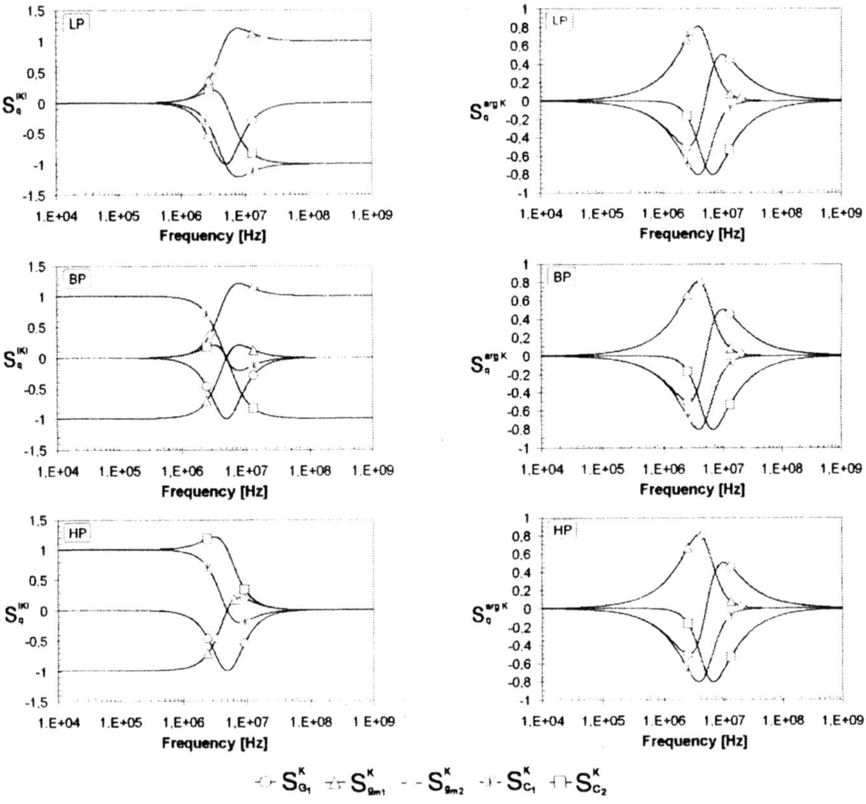

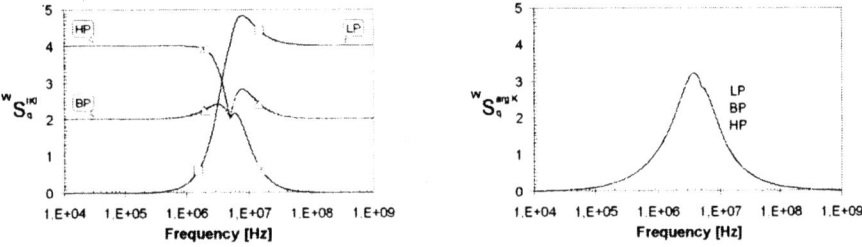

Fig. 6. Partial generalized sensitivities of filter based on circuit in Fig. 3.

The worst–case global relative sensitivity of transmission modulus and the semi–relative sensitivity of the filter transmission argument are shown in Fig. 7.

Fig. 7. The worst–case global relative sensitivities of modulus and semi–relative sensitivities of argument for multifunction filter based on circuit in Fig. 3.

The semi–relative sensitivity of transmission argument is the same for all filters. The worst–case sensitivity of the band–pass is approximately identical throughout the frequency band. The sensitivity of high–pass filter in the stop–band is high but in the transmission band it decreases to zero. Similarly, the sensitivity of the low–pass in the stop–band is worse but in the transmission band it is zero.

6 Experimental results

The band–reject filter was chosen for the realization. To set the transconductances of OTA amplifiers the AD5258 [11] digital potentiometers were used. Capacitors C_1 = 220 pF and C_2 = 110 pF, resistor R_1 = 200 Ω and transconductances $g_{m1,2}$ = 4,9 mA/V were used in the realization of the RF filter. A simplified schematic of the system measured is in Fig. 8.

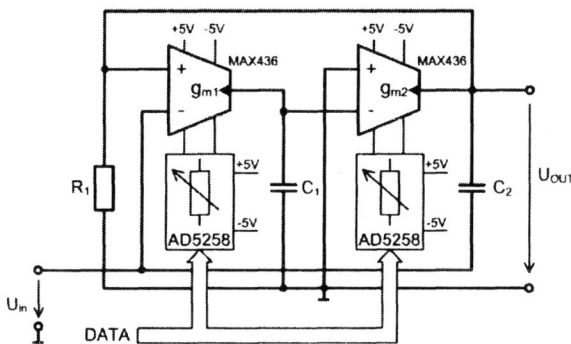

Fig. 8. Schematic of measured band–reject.

The measurement was carried out with an HP3589A network/spectrum analyzer connected to computer via the GPIB bus system. The frequency response measured for the second–order band–reject frequency filter is given in Fig. 9. The results of measurement are matches to simulations.

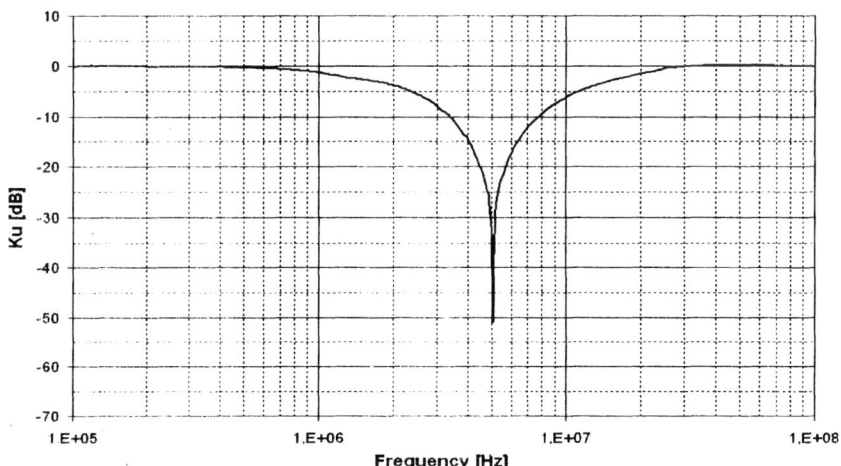

Fig. 9. Frequency response measured for second–order band–reject filter.

7 Conclusion

In this paper a general method for designing new circuits of frequency filters with two OTA elements is presented. Usually, when proposing autonomous circuits we proceed intuitively. The method of connecting a general network of passive elements to two OTAs is shown. Simpler autonomous circuits were gradually eliminated from this general arrangement. An example of the procedure of frequency filter design with two OTA elements is shown. A specific filter solution was simulated in the PSpice software [8], MAX436 [9] and LT1228 [10] company models were used. A generalized sensitivity analysis is shown. The frequency filter designed is expected to be used in the signal processing of high–speed data communication systems, in cable modems or in hard–drive communication interfaces. The band–reject filter was realized experimentally using a MAX436 [9].

Acknowledgment. The paper has been supported by the Czech Science Foundation project GACR 102/06/1383 and Ministry of Education, Youth and Sport of the Czech Republic project No. 1ET301710508.

References

1. Tietze, U., Schenk, Ch.: Halbleiter – Schaltungtechnik, Springer, 12 ed., 2002. ISBN 3-540-42849-6.
2. Geiger, R. L., Sánchez, S. E.: Active Filter Design Using Operational Transconductance Amplifiers: A Tutorial, IEEE Circuits and Devices Magazine, Vol. 1, pp. 20–32, 1985.
3. Chen, W. K.: The Circuits and Filters Handbook. CRC Press, Inc. Boca Raton, Florida, 1995.
4. Deliyannis, T., Sun, Y., Fidler, J. K.: Continuous–Time Active Filter Design, CRC Press, Boca Raton, 1999, 443 pages, ISBN 0-8493-2573-0.
5. Koton, J., Vrba, K.: Method for Designing Frequency Filters using Universal Current Conveyors, International Transactions on Computer Science and Engineering, 2005, Vol.13, No.1, pp. 144–154.
6. Herencsár, N., Vrba, K.: Method for Designing Frequency Filters Using BOTA. In Proceedings of the Second International Conference on Systems, ICONS 2007. Sainte-Luce, Martinique: IEEE Computer Society, 2007, pp. 1–6. ISBN: 0-7695-2807-4.
7. Biolek, D., Kolka, Z.: SNAP – symbolic, semisymbolic, and numerical analysis of electronic circuits. URL http://snap.webpark.cz/indexa.html
8. OrCAD, Inc. OrCad PSpice User's Guide [pdf online file] USA: OrCAD, 1998. http://www.electronics–lab.com/downloads/schematic /013/ . 436 pages.
9. MAX435/MAX436 – Wideband Transconductance Amplifier with Differential Output. Datasheet, MAXIM – Dallas Semiconductor, 1993.
10. LT1228 – 100MHz Current Feedback Amplifier with DC Gain Control. Datasheets, Linear Technology, 1994.
11. AD5258 – Nonvolatile, $I^2C^®$–Compatible 64–Position, Digital Potentiometer, Analog Devices, 2007.

New Multifunctional Frequency Filter Working in Current-mode

Jaroslav Koton, Kamil Vrba

Brno University of Technology, Dept. of Telecommunications
Purkynova 118, 612 00 Brno
JaroslavKoton@phd.feec.vutbr.cz, vrbak@feec.vutbr.cz

Abstract. Increasing demands on circuits with low supply voltage and low power consumption lead to the realization of analogue circuits working in the current-mode. In this paper a new topology is presented of multifunctional frequency filter SIMO (Single Input Multiple Output) working the in current-mode. The proposed circuit is realized using four second-generation current conveyors and six passive elements. A high value of quality factor can be achieved, changeable independently of the characteristic frequency. The results of sensitivity and tolerance analyses of the proposed filter are given. The properties of the designed filter were also experimentally verified.

Key words: signal processing, frequency filters, current conveyor, current-mode, autonomous circuit

1 Introduction

Currently research and development in the area of designing linear circuit structures is focused on the application of new active elements, such as current conveyors [1], voltage conveyors [2] or transadmittance amplifiers [3]. More often we can meet with these elements in structures working in the current-mode because of their wider frequency bandwidth. Another advantage of current-mode circuits is their higher dynamic range. Using circuits working in the voltage-mode the demand of sufficient signal to noise ratio cannot be fulfilled in low supply-voltage applications. The value of supply voltage has not such an influence on the dynamic range in current-mode circuits, as in voltage-mode circuits, and this is their main advantage.

New circuits working in the current-mode can be designed using the method of adjoint transformation to a voltage-mode prototype [4]. Although this design method is relatively fast, it does not solve the problem of decreasing supply voltage, because the active elements stay the same, mostly if operational amplifiers are used. It is better to base the design of new circuits realizing frequency filters working in the current-mode on seeking a suitable autonomous circuit, which can be extended by other active and passive elements [5], [6], [7].

2 Frequency filter design

For the design of new circuits current conveyors were used. Although these active elements are not currently industrially produced, they can be found as a part of some types of current feedback amplifiers, such as AD844, or elements labeled as OPA860 and OPA861. The current conveyor of these elements is the second-generation current conveyor CCII+. In our department we use the universal current conveyor UCC-N1B, which was developed in cooperation with the AMI Semiconductor Centre in Brno. By a suitable connection of the branches of this integrated circuit all three-port and some more-port first class, that is with a single port X, current conveyors can be realized.

For the design it is suitable to use the generalized current conveyor (Fig. 1a). The relation between branch currents and voltages is given by equations

$$u_X = a \cdot u_Y, \quad i_Y = b \cdot i_X, \quad i_Z = c \cdot i_X, \tag{1}$$

where a, b, and c are voltage or current gains between individual ports. Choosing the value of these parameters $a \in \{-1;1\}$, $b \in \{-1;0;1\}$, $c \in \{-1;1\}$ a specific type of current conveyor is determined, which has to be used for the realization itself of the proposed circuit.

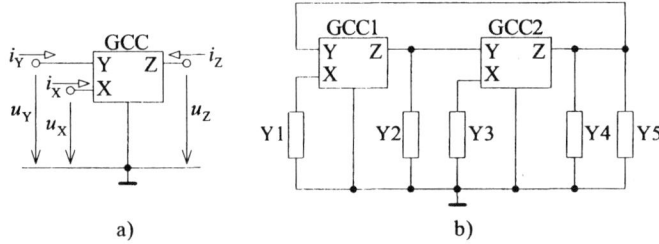

Fig. 1. a) Generalized three-port current conveyor, b) autonomous circuit with two GCCs

As an initial structure for further design the autonomous circuit in Fig. 1b [7] was chosen. This circuit is described by its general characteristic equation

$$CE = a_1 b_1 a_2 b_2 Y_1 Y_3 - a_1 b_1 Y_1 Y_2 - a_2 b_2 (Y_3 Y_4 + Y_3 Y_5) - a_1 c_1 a_2 c_2 Y_1 Y_3 + Y_2 Y_4 + Y_2 Y_5 = 0. \tag{2}$$

The following analysis of this circuit is focused on the realization of filters that will enable us to change the quality factor independently of the characteristic frequency. To make the realization itself as simple as possible, we will not consider any type of current conveyor as in [7], but only second-generation current conveyors, i. e. $a_1 = a_2 = 1$, $b_1 = b_2 = 0$.

Choosing the product $c_1 c_2 = -1$ equation (2) will simplify to the form

$$CE = Y_1 Y_3 + Y_2 Y_4 + Y_2 Y_5 = 0. \tag{3}$$

In this case all filters derived from this autonomous circuit will fulfill the condition of stability. If the character of passive elements is chosen as follows $Y_1 = G_1$,

$Y_2 = pC_1$, $Y_3 = G_2$, and $Y_4 = pC_2+G_3$ (Fig. 2) then (3) will change to a form that agrees with the conditions of second-order frequency filters design

$$CE = p^2 C_1 C_2 + pC_1 G_3 + G_1 G_2 = 0. \quad (4)$$

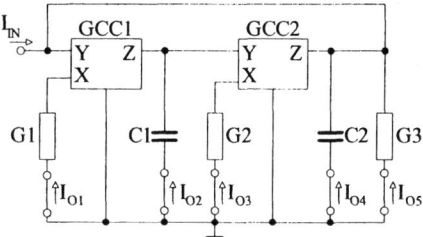

Fig. 2. Multifunctional filter working in current-mode with two CC

Possible general current transfer functions of the circuit in Fig. 2 are

$$K_{BP1} = \frac{I_{O1}}{I_{IN}} = -\frac{pC_1 G_1}{CE}, K_{BP2} = \frac{I_{O2}}{I_{IN}} = -\frac{c_1 pC_1 G_1}{CE}, K_{LP} = \frac{I_{O3}}{I_{IN}} = -\frac{c_1 G_1 G_2}{CE}, \quad (5a,b,c)$$

$$K_{HP} = \frac{I_{O4}}{I_{IN}} = -\frac{p^2 C_1 C_2}{CE}, K_{BP3} = \frac{I_{O5}}{I_{IN}} = -\frac{pC_1 G_3}{CE} \quad (5d,e)$$

$$K_{BR} = \frac{I_{O3} + I_{O4}}{I_{IN}} = -\frac{p^2 C_1 C_2 + c_1 G_1 G_2}{CE} \quad (5f)$$

The proposed circuit can be used for the realization of low-pass, (5c), high-pass (5d), band-pass (5a, b, e), and band-reject (5f) frequency filters. By respecting the condition $c_1 c_2 = -1$ in some cases it is possible to realize either inverting or non-inverting transfer function choosing the value of coefficients c_1 and c_2.

According to (4) the quality factor Q and radian frequency ω_0 are given by

$$Q = \sqrt{\frac{C_2}{C_1}} \frac{\sqrt{G_1 G_2}}{G_3}, \quad \omega_0 = \sqrt{\frac{G_1 G_2}{C_1 C_2}}. \quad (6)$$

Relative sensitivities of these parameters to individual passive elements are

$$S_{R\,C1}^Q = -S_{R\,C2}^Q = -S_{R\,G1}^Q = -S_{R\,G2}^Q = -\frac{1}{2}, \quad S_{R\,G3}^Q = -1, \quad (7)$$

$$S_{R\,C1}^{\omega_0} = S_{R\,C2}^{\omega_0} = -S_{R\,G1}^{\omega_0} = -S_{R\,G2}^{\omega_0} = -\frac{1}{2}, \quad S_{R\,G3}^{\omega_0} = 0. \quad (8)$$

If some of the passive elements C_1, C_2, G_1 or G_2 changes by 1%, the quality factor Q and radian frequency ω_0 will change by 0.5% or by -0.5%. The highest sensitivity of the quality factor is to the change of the admittance G_3.

Using the circuit in Fig. 2 for frequency filter design, problems can appear if a high value of quality factor is required. The analyses show that in that case it is suitable to use a low impedance of resistor R_3. This requirement can be fulfilled if the resistors R_1 and R_2 are also of low impedance. Their values cannot be arbitrarily low, because if the impedance connected to a current port X is close to or even smaller than the input impedance of the port X, the circuit leaves its functionality at high frequencies. The value of the input impedance of the port X is a very limiting factor of current conveyors and its reduction is the main problem currently solved by many researches [8–10].

The influence of the parasitic impedance of the port X can be suppressed by the modification of the circuit in Fig. 1b) by extending it with other active and passive elements. The advantage is that the resistors connected to the port X can a have higher value and hence the parasitic properties of the current conveyor used will not show so much.

In Fig. 3a) the modified structure of the autonomous circuit with four generalized current conveyors and six passive elements is presented.

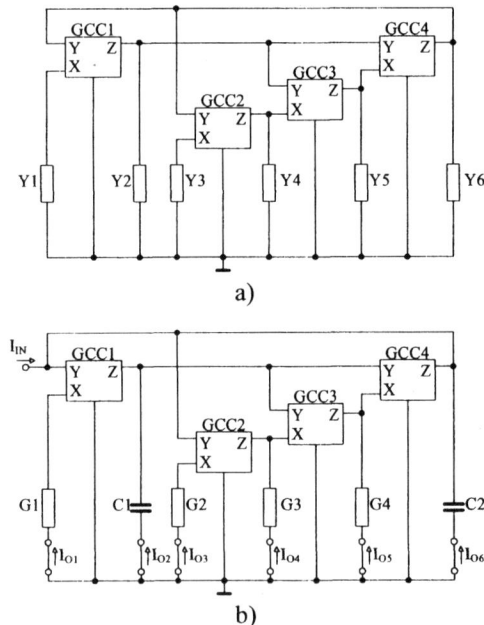

Fig. 3. a) Modified autonomous circuit with four GCC, b) multifunctional filter working in current-mode

The design of the circuit starts with the general characteristic equation, as in the previous case. In the further design we will also use only second-generation current conveyors. The characteristic equation of the autonomous circuit in Fig. 3a is given by

$$CE = c_1c_3c_4Y_1Y_4 - c_1c_4Y_1Y_5 - c_2c_3c_4Y_2Y_3 + Y_2Y_6 = 0. \quad (9)$$

Choosing the value of the products of the coefficients $c_1c_3c_4 = 1$, $c_1c_4 = -1$, and $c_2c_3c_4 = -1$ and the character of passive elements $Y_1 = G_1$, $Y_2 = \mathbf{p}C_1$, $Y_3 = G_2$, $Y_4 = G_3$, $Y_5 = G_4$, and $Y_6 = \mathbf{p}C_2$ it is possible to determine the type of the current transfer functions realizable by the circuit designed in Fig. 3b

$$\mathbf{K}_{BP1} = \frac{\mathbf{I}_{O1}}{\mathbf{I}_{IN}} = \frac{-\mathbf{p}C_1G_1}{CE}, \quad \mathbf{K}_{BP2} = \frac{\mathbf{I}_{O2}}{\mathbf{I}_{IN}} = \frac{-c_1\mathbf{p}C_1G_1}{CE}, \quad \mathbf{K}_{BP3} = \frac{\mathbf{I}_{O3}}{\mathbf{I}_{IN}} = \frac{-\mathbf{p}C_1G_2}{CE}, \quad (10\text{a,b,c})$$

$$\mathbf{K}_{LP1} = \frac{\mathbf{I}_{O4}}{\mathbf{I}_{IN}} = \frac{-c_1G_1G_3}{CE}, \quad \mathbf{K}_{LP2} = \frac{\mathbf{I}_{O5}}{\mathbf{I}_{IN}} = \frac{-c_1G_1G_4}{CE}, \quad \mathbf{K}_{HP} = \frac{\mathbf{I}_{O6}}{\mathbf{I}_{IN}} = \frac{-\mathbf{p}^2C_1C_2}{CE}, \quad (10\text{d,e,f})$$

$$\mathbf{K}_{LP3} = \frac{\mathbf{I}_{O4} + \mathbf{I}_{O5}}{\mathbf{I}_{IN}} = \frac{-c_1G_1(G_3 + G_4)}{CE}, \quad (10\text{g})$$

$$\mathbf{K}_{BR} = \frac{\mathbf{I}_{O4} + \mathbf{I}_{O5} + \mathbf{I}_{O6}}{\mathbf{I}_{IN}} = -\frac{\mathbf{p}^2C_1C_2 + c_1G_1(G_3 + G_4)}{CE}, \quad (10\text{h})$$

where $CE = \mathbf{p}^2C_1C_2 + \mathbf{p}C_1G_2 + G_1(G_3 + G_4)$. This circuit can be used as high-pass (10f), low-pass (10d, e, g), band-pass (10a, b, c) or band-reject (10h) frequency filter. In the case of transfers (10b), (10d), (10e), and (10g) the choice of the coefficient c_1 will realize either the inverting or the non-inverting type of the transfer function of the filter. However, it is necessary to respect the conditions of the products of the coefficients c_1 to c_4.

The quality factor Q and radian frequency ω_0 are given as follows

$$Q = \sqrt{\frac{C_2}{C_1}} \frac{\sqrt{G_1(G_3 + G_4)}}{G_2}, \quad \omega_0 = \sqrt{\frac{G_1(G_3 + G_4)}{C_1C_2}}. \quad (11\text{a,b})$$

The relative sensitivities of the quality factor and radian frequency to passive elements are

$$S_{R\,C1}^{Q} = -S_{R\,C2}^{Q} = -S_{R\,G1}^{Q} = -\frac{1}{2}, \quad S_{R\,G2}^{Q} = -1, \quad (12\text{a})$$

$$S_{R\,G3}^{Q} = \frac{1}{2}\frac{G_3}{G_3 + G_4}, \quad S_{R\,G4}^{Q} = \frac{1}{2}\frac{G_4}{G_3 + G_4}, \quad (12\text{b})$$

$$S_{R\,C1}^{\omega_0} = S_{R\,C2}^{\omega_0} = -S_{R\,G1}^{\omega_0} = -\frac{1}{2}, \quad S_{R\,G2}^{\omega_0} = 0, \quad (13\text{a})$$

$$S_{R_{G3}}^{\omega_0} = \frac{1}{2}\frac{G_3}{G_3+G_4}, \quad S_{R_{G4}}^{\omega_0} = \frac{1}{2}\frac{G_4}{G_3+G_4}. \qquad (13b)$$

The values of the relative sensitivities are analogous to those of the initial autonomous circuit. Moreover, the values of admittances G_3 and G_4 can be optimized here such that their influence on the properties of this circuit is minimized.

3 Simulations

Using the OrCAD – PSpice simulation program the behaviour of the proposed circuit was analyzed. The universal current conveyor UCC-N1B was used as the active element. For simulations the values of coefficients $c_1 = c_3 = -1$, $c_2 = c_4 = 1$ were chosen, which fulfill the conditions for circuit stability.

If we choose $R_1 = R_3 = R_4 = R$ and $C_1 = C_2 = C$, then using (11) the values of resistors are given can be calculated for the required values of ω_0 and Q

$$R = \frac{\sqrt{2}}{\omega_0 C}, \quad R_2 = \frac{QR}{\sqrt{2}}. \qquad (14a,b)$$

The magnitudes of current transfer functions of the multifunctional frequency filter analyzed are given in Fig. 4. The value of the characteristic frequency considered is $f_0 = 4.5$ MHz and of the quality factor is $Q = 20$. The results are given for transfer functions (10c), (10f), and (10g).

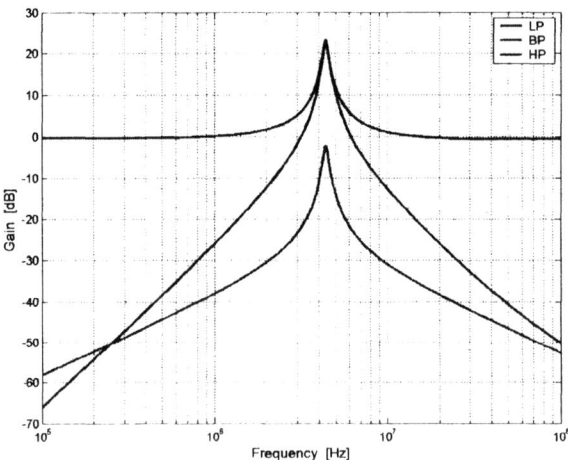

Fig. 4. Magnitudes of current transfer functions, $f_0 = 4.5$ MHz, $Q = 20$

The real properties of the active elements used cause that the quality factor is not as high as required. The current value according to simulations is just about $Q = 14$.

Also the characteristic frequency f_0 has decreased. However, it can be said that the behaviour of the designed circuit is very satisfactory. The sensitivity of the circuit to the change of passive elements is expressed by the histogram in Fig. 5. It shows the characteristic frequency shift if resistors and capacitors with 5% tolerance are used.

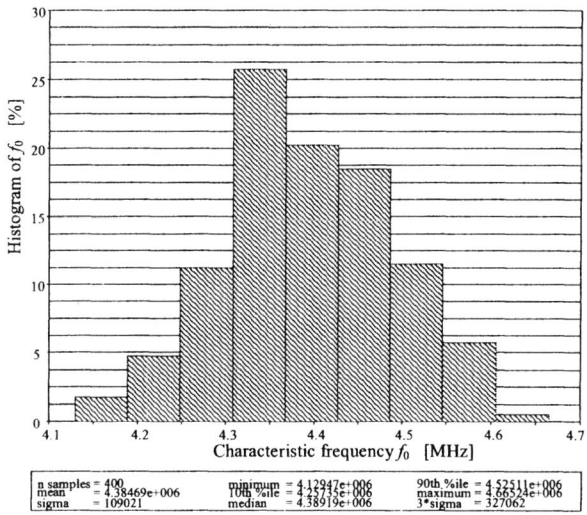

Fig. 5. Histogram of the characteristic frequency f_0 of the proposed circuit

4 Experimental results

The real behaviour of the designed multifunctional frequency filter was also experimentally verified. The value of characteristic frequency f_0 is 4.5MHz. The measurements were performed for the values 147 Ω, 210 Ω, 2.94 kΩ, and 14.7 kΩ of the resistor R_2, which according to (11a) agrees with the values 0.5, 0.707, 10, and 50 of the quality factor Q. The results for individual current transfers are shown in Fig. 6. The transfer functions measured were (10c), (10f), a (10g).

The value of the quality factor Q is not as high as the theoretical one according to (11a), which is caused by real the properties of the circuit elements used. The best results were achieved for the low-pass frequency filter (Fig. 6c). As regards on the shape of the magnitude function, the worst of all behaviours is that of the high-pass filter, where in the frequency area of 100 MHz a local maximum of the magnitude is formed. This can be already caused by the parasitic properties of the wiring on the PCB board. The magnitude of the transfer function realizing the band-pass frequency filter has at low frequencies a lower attenuation than expected, which is caused by the non-zero impedance of the port X [11]. This feature mostly shows in the low values of quality factor, where the impedance of resistor R_2 is close to the impedance of the port X.

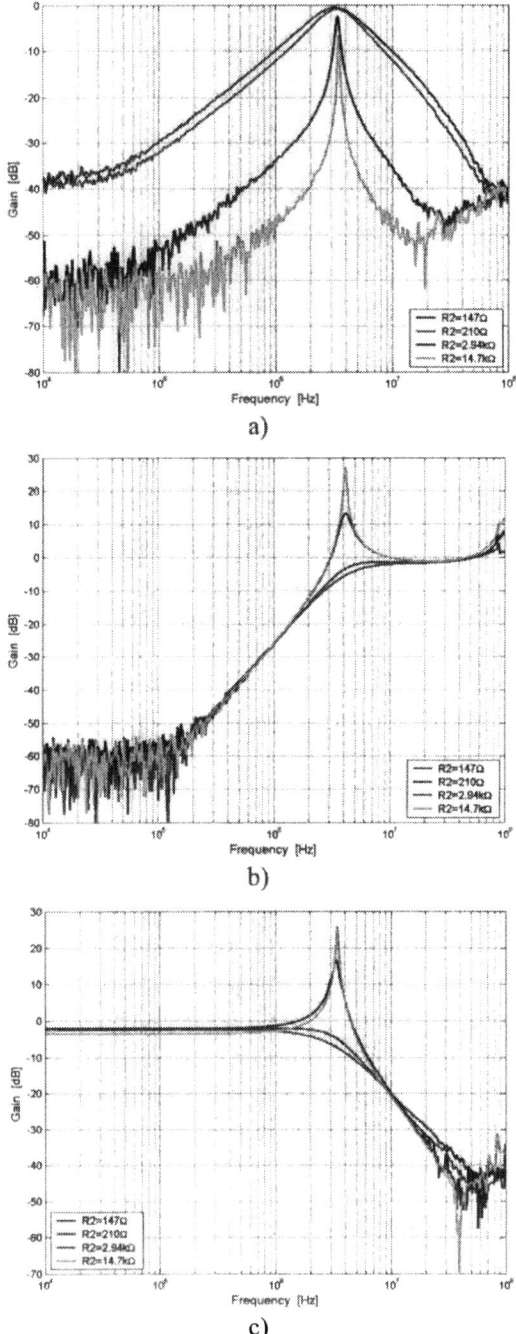

Fig. 6. Measured magnitudes of transfer functions a) band-, b) high-, c) low-pass

5 Conclusion

In this paper a new filtering circuit structure was proposed, which starts from an autonomous circuit. Current conveyors were used as active elements. Using this circuit it is possible to realize a second-order low-, high-, band-pass and band-reject frequency filter working in the current-mode. The circuit presented enables changing the quality factor Q independently of the characteristic frequency f_0 using a single passive element. The behaviour of the designed structure was not only verified by simulations, but it was also practically realized.

Acknowledgement

The research of the current-mode using modern active elements is supported by the science project of Ministry for education MSM 0021630513 and by the Czech Science Foundation, project No. 102/06/1383.

References

[1] H. M. Hassan, A. M. Soliman: „Novel Accurate Wideband CMOS Current Conveyor", *Frequenz*, 2006, Vol. 60, No. 11-12, pp. 234-236.
[2] K. N. Salama et al.: „Parasitic-Capacitance Insensitive Voltage-Mode MOSFET-C Filters Using Differential Current Voltage Conveyor", Circuits Systems Signal Process, 2001, Vol. 20, No. 1, pp. 1-16.
[3] Y. Sun, C. Hill, A. Szczepanski: „Large Dynamic Range High Frequency Fully Differential CMOS Transconductance Amplifier", Analog Integrated Circuits and Signal Processing, 2003, No. 34, pp. 247-255.
[4] G. W. Roberts, A. S. Sedra: "All Current-Mode Frequency Selective Circuits," Electronics Letters, 1989, Vol. 25, No. 12, pp. 759-760.
[5] J. Koton, K. Vrba, Method for Designing Frequency Filters using Universal Current Conveyors, *International Transaction on Computer Science and Engineering*, ISSN 1738-6438, 2005, Vol. 13, pp. 144-154.
[6] J. Koton, K. Vrba, P. Hanak, Frequency Filter with Current Conveyors for Signal Processing of Data-Buses Working in the Current-mode, International Conference on Networking, ICN 2006, Morne, 2006
[7] K. Vrba, J. Cajka: "Application of the General Four-port Second-kind Current Conveyor for Universal Filter Design", Electronic Journal for Engineering Technology, Vol. 5, No. 1, 2003, ISSN 1523-9926.
[8] F. Seguin, A. Fabre, New Sekond Generation Current Conveyor with Reduced Parasitic Resistance and Bandpass Filter Application, IEEE Transaction on Circuits and Systems – I, Vol. 48, No. 6, 2001, pp. 781-785.
[9] H. M. Hassan, A. M. Soliman, Novel Accurate Wideband CMOS Current Conveyor, Frequenz 60, 11-12, 2006, pp. 234-236.
[10] S. B. Salem et al., A High Performance CMOS CCII and High Frequency Applications, Analog Integr. Circ. Sig. Process, Springer Science Business Media, 2006.
[11] H. Schmidt, G. S. Moschytz, Fundamental Frequency Limitations in Current-Mode Sallen-Key Filters, *Proceedings of the ISCAS*, Monterey, California, May 31–June 3, vol. 1, pp. 57–60, 1998.

Second-Order Multifunction Filters with Current Operational Amplifiers

David Kubanek and Kamil Vrba

Department of Telecommunications, Faculty of Electrical Engineering and Communication,
Brno University of Technology, Purkynova 118, 612 00 Brno, Czech Republic
{kubanek, vrbak}@feec.vutbr.cz

Abstract. Novel second-order multifunction filters operating in current mode are described. Their structure is based on three current operational amplifiers connected in a feedback loop. These active elements are dual to the well-known operational amplifiers and are suitable for high-frequency signal processing. The designed filters offer low-pass, band-pass and high-pass transfer functions simultaneously. Computer simulation is carried out to prove the functionality of a filter variant. The influence of real properties of active elements is analyzed.

Keywords: active filter, current mode, current operational amplifier

1 Introduction

The demands on modern signal processing systems are still stricter. Particularly multimedia and other broadband signals must be processed by high-frequency electronic circuits. The development of new sub-micron integrated circuit technologies leads to the supply voltage lowering of and thus reducing the levels of processed voltage signals. The signal-to-noise ratio (SNR) decreases as well and that is why designers of recent electronic circuits increasingly use current signals to carry information. The SNR is better in current-mode circuits and their frequency bandwidth usually also increases.

Frequency filtering is one of the most often used operations in signal processing and frequency filters belong to the most widely used electronic circuits. This paper presents a design procedure of several second-order filter variants that employ current operational amplifiers as active building blocks, operate in current mode, and provide multiple transfer functions simultaneously.

2 Current operational amplifier

We have chosen the current operational amplifier (COA, also known as TCOA - True Current Operational Amplifier) for the analog filter design. It is a relatively new building block for active circuits and several its variants were described in [1], [2], [3], [4], and [5] together with their possible internal structures designed in MOS or bipolar technologies. COA is a dual element to the well-known voltage feedback operational amplifier. It has a differential current input and infinite current gain. COA can be advantageously implemented in current mode active circuits, which process high-frequency analog signals. COA is usually connected in the negative feedback loop, which provides the equality of its input currents. The variant of COA with differential input and single output (DISO) as shown in Fig. 1 will be suitable for our design.

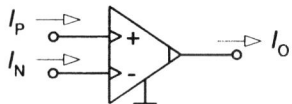

Fig. 1. Schematic symbol of COA

3 Filter design

The basic circuit for the filter design is shown in Fig. 2. It consists of three COAs connected in a global negative feedback loop. Each amplifier also has its local feedback loop formed by a passive one-port element and the circuit has three current inputs and three current outputs [6].

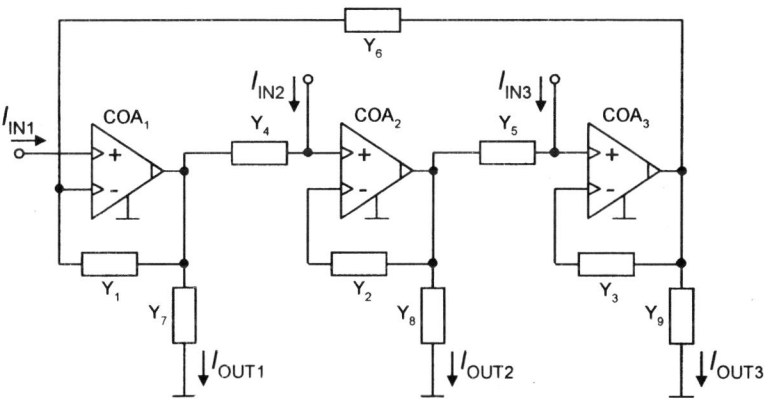

Fig. 2. The chosen general circuit with COAs

The proposed general network has the following characteristic polynomial (CP):

$$D = Y_1 Y_2 Y_3 + Y_4 Y_5 Y_6 . \tag{1}$$

Here Y_i (i = 1, 2, 3, 4, 5, 6) are admittances of one-port elements used in the network in question. Let us divide these elements into two groups: the first one includes elements with admittances Y_1, Y_2, Y_3, the second one the elements with admittances Y_4, Y_5, Y_6. If arbitrary two admittances in the first group are sC_1 and sC_2

and the remaining admittance equals G_1, and if arbitrary one admittance in the second group is chosen as (sC_3+G_4) and the remaining admittances are then denoted as G_2 and G_3, then we obtain the CP in the form:

$$D(s) = s^2 C_1 C_2 G_1 + s C_3 G_2 G_3 + G_2 G_3 G_4. \tag{2}$$

We get the same CP if we interchange the groups above. If we skip the identical variants that arise due to the cyclic permutation, we can summarize the possible types of admittances in Tab. 1. We obtained 6 circuits (denoted by A, B, … F), which are suitable for our purpose.

Table 1. Possible values of admittances

Variant	Y_1	Y_2	Y_3	Y_4	Y_5	Y_6
A	sC_1	sC_2	G_1	sC_3+G_4	G_2	G_3
B	G_1	sC_2	sC_1	sC_3+G_4	G_2	G_3
C	sC_1	G_1	sC_2	sC_3+G_4	G_2	G_3
D	sC_3+G_4	G_2	G_3	sC_1	sC_2	G_1
E	sC_3+G_4	G_2	G_3	G_1	sC_2	sC_1
F	sC_3+G_4	G_2	G_3	sC_1	G_1	sC_2

All the possible current transfer functions of the network depicted in Fig. 2 have the form as seen in Tab. 2.

Table 2. Transfer functions of the network in Fig. 2

$\dfrac{I_{OUT1}}{I_{IN1}} = \dfrac{Y_2 Y_3 Y_7}{D}$	$\dfrac{I_{OUT2}}{I_{IN1}} = \dfrac{Y_3 Y_4 Y_8}{D}$	$\dfrac{I_{OUT3}}{I_{IN1}} = \dfrac{Y_4 Y_5 Y_9}{D}$
$\dfrac{I_{OUT1}}{I_{IN2}} = -\dfrac{Y_5 Y_6 Y_7}{D}$	$\dfrac{I_{OUT2}}{I_{IN2}} = \dfrac{Y_1 Y_3 Y_8}{D}$	$\dfrac{I_{OUT3}}{I_{IN2}} = \dfrac{Y_1 Y_5 Y_9}{D}$
$\dfrac{I_{OUT1}}{I_{IN3}} = -\dfrac{Y_2 Y_6 Y_7}{D}$	$\dfrac{I_{OUT2}}{I_{IN3}} = -\dfrac{Y_4 Y_6 Y_8}{D}$	$\dfrac{I_{OUT3}}{I_{IN3}} = \dfrac{Y_1 Y_2 Y_9}{D}$

It is evident that for each of 6 circuit variants (see Tab. 1) 9 current transfer functions can be calculated, i.e. a total of 54 transfer functions. Let us choose the admittances at the outputs conductive: $Y_7 = G_{OUT1}$, $Y_8 = G_{OUT2}$, $Y_9 = G_{OUT3}$. Using the computed functions we can select 4 networks, which can be used as *multifunction filters*. It means that they operate as low-pass (LP), band-pass (BP) and high-pass (HP) filters simultaneously. All these networks have one input and three outputs or three inputs and one output. They are summarized with their transfer functions in Tab. 3.

Table 3. Transfer functions of multifunction filters

Variant (Tab. 1)	Transfer functions		
	LP	BP	HP
B	$\dfrac{I_{OUT1}}{I_{IN2}} = -\dfrac{G_2 G_3 G_{OUT1}}{D(s)}$	$\dfrac{I_{OUT1}}{I_{IN3}} = -\dfrac{sC_1 G_3 G_{OUT1}}{D(s)}$	$\dfrac{I_{OUT1}}{I_{IN1}} = \dfrac{s^2 C_1 C_2 G_{OUT1}}{D(s)}$
C	$\dfrac{I_{OUT1}}{I_{IN2}} = -\dfrac{G_2 G_3 G_{OUT1}}{D(s)}$	$\dfrac{I_{OUT3}}{I_{IN2}} = \dfrac{sC_1 G_2 G_{OUT3}}{D(s)}$	$\dfrac{I_{OUT2}}{I_{IN2}} = \dfrac{s^2 C_1 C_2 G_{OUT2}}{D(s)}$
D	$\dfrac{I_{OUT1}}{I_{IN1}} = \dfrac{G_2 G_3 G_{OUT1}}{D(s)}$	$\dfrac{I_{OUT2}}{I_{IN1}} = \dfrac{sC_1 G_3 G_{OUT2}}{D(s)}$	$\dfrac{I_{OUT3}}{I_{IN1}} = \dfrac{s^2 C_1 C_2 G_{OUT3}}{D(s)}$
E	$\dfrac{I_{OUT1}}{I_{IN1}} = \dfrac{G_2 G_3 G_{OUT1}}{D(s)}$	$\dfrac{I_{OUT1}}{I_{IN3}} = -\dfrac{sC_1 G_3 G_{OUT1}}{D(s)}$	$\dfrac{I_{OUT1}}{I_{IN2}} = -\dfrac{s^2 C_1 C_2 G_{OUT1}}{D(s)}$

4 Influence of COA real properties on filter characteristics

Let us choose the variant C as an example for further analyses. A frequency-dependent model of COA as shown in Fig. 3 will be used for PSpice simulation. This model includes parasitic properties that can be expected in a real COA (unfortunately, no COA has been commercially manufactured yet). The controlled sources CCCS and VCCS in the model are frequency independent. The terminals + and − have internal series resistances of 5 Ω. A shunt resistance of 10 MΩ and a capacitance of 5 pF are connected to the high-impedance node and determine the amplifier bandwidth. The simulated magnitude frequency characteristics of the designed multifunction filter are depicted in Fig. 4.

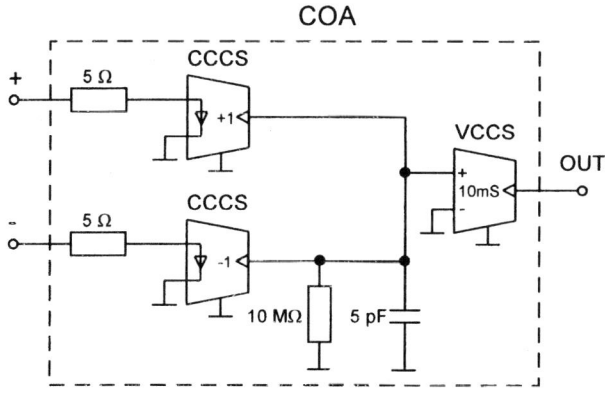

Fig. 3. The COA model used

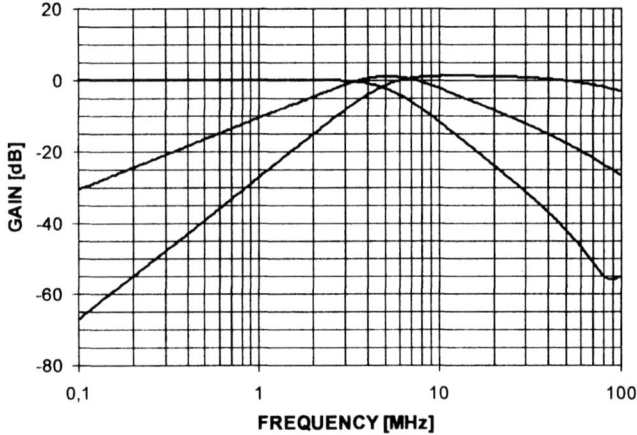

Fig. 4. The magnitude frequency characteristic of the designed filter

The magnitude frequency characteristics are correct up to ca 30 MHz. High-pass filter has a parasitic attenuation above this frequency, which is caused by non-zero input resistances and particularly by non-zero capacitance of the internal node in the COA model. This capacitance causes additional attenuation also at the low-pass and band-pass characteristics at high frequencies. The internal capacitance also causes the parasitic zero of the low-pass filter (apparent at about 90 MHz in Fig. 4) and its higher values shift the zero to lower frequencies.

The resistance of the high-impedance node has an influence only on the band-pass and high-pass characteristic. It causes finite and constant attenuation at low frequencies, which are out of range of Fig. 4. Finite attenuation appears also at high-frequency parts of the low-pass and band-pass characteristics which are above 100 MHz and thus also not apparent in Fig. 4. Causes of these effects will be investigated in detail in chapter 4.1.

We are looking for the replacement of COA by a commercially available element and if it is possible, we will present practically measured characteristics in the conference presentation.

4.1 Analysis of finite attenuation in stop-band

As mentioned in the last chapter, the real COA properties included in the model in Fig. 3 cause finite and constant attenuation in stop-bands of all LP, BP, and HP filters. Let us find relations for the transfer values in stop-bands. The parameters of filter passive elements and the COA input resistance R_{IN} can be expressed as a product or a ratio of geometric means C, G, and coefficients m, n, p, z as shown in Tab. 4. This representation leads to the independent relations for the ideal pole frequency ω_{pi} and the ideal pole quality factor Q_{pi} of the filter, see (3).

Table 4. Substitution of filter component parameters

C_1	C_2	C_3	G_1	G_2	G_3	G_4	G_{OUT1}	G_{OUT2}	G_{OUT3}	R_{IN}
mC	C/m	nC	G	G	G/p	pG	pG	G	$nG/(mp)$	$1/(zG)$

$$\omega_{pi} = \frac{G}{C}; \quad Q_{pi} = \frac{p}{n}. \tag{3}$$

With the substitution according to Tab. 4, it is easy to express the limits of transfer magnitudes in the stop-bands. They are summarized in Tab. 5, where A_0 is the low-frequency gain of COA, which is equal to the product of the resistance of the high-impedance node and the transconductance of VCCS.

Table 5. Limits of transfers in stop-bands

P	B		H
$\lim\limits_{f \to \infty} K = \dfrac{p^2}{2pz + p^2 + 3}$	$\lim\limits_{f \to 0} K = \dfrac{n}{A_0 m}$	$\lim\limits_{f \to \infty} K = \dfrac{n}{2p^2 z^2 m}$	$\lim\limits_{f \to 0} K = \dfrac{1}{A_0^2}$

The values of z and A_0 can be assumed much higher than m, n, p in most cases. Then the high-frequency transfer of LP filter is indirect proportional to z and thus it increases proportionally to the COA input resistance. The relation for the high-frequency transfer of BP has also z in the denominator, but this transfer does not increase with the COA input resistance so dramatically thanks to the square. The low-frequency transfers of BP and HP filters increase with reducing the low-frequency gain of COA, whereas the HP transfer increases less dramatically due to the square in the relation.

5 Conclusions

New second-order multifunction filters are described in the paper. Four unique variants were derived from a general basic circuit. They can operate as low-pass, band-pass and high-pass filters simultaneously. One of the variants was chosen for the computer simulation with frequency dependent COA model. It was proved that the filter operates correctly also at relatively high frequencies. It was investigated how the real properties of COA affect the functionality of the filter. The non-zero input resistances and internal node capacitance of the COA model mostly influence the characteristics.

Acknowledgment. The paper has been prepared as a part of the solution of the Czech Republic Grant Agency projects 102/06/1383 and 102/07/P353.

References

1. Tomazou, C., Lidgey, F. J., Haigh D. G.: Analogue IC design: the current-mode approach. Peter Peregrinus Ltd, 1990
2. Bruun, E.: A differential-input, differential-output current mode operational amplifier. Int. J. Electronics, 1991, Vol. 71, No. 6, pp. 1047-1056
3. Kaulgerg, T.: A CMOS Current-Mode Operational Amplifier. Int. Journal of Solid-State Circuits, Vol. 28, No. 7, July 1993, pp. 849-852
4. Mucha, I.: Towards a true current operational amplifier. Proceedings of the ISCAS'94 Conference, London, May 30-June 2, 1994, Vol. 5, pp. 389-392
5. Nagasaku, T., Hyogo, A., Sekine, K.: A Synthesis of a Novel Current-Mode Operational Amplifier. Analog Integrated Circuits and Signal Processing, 11, 1996, pp. 183-185
6. Cajka, J., Vrba, K., Kubanek, D.: Novel multifunction second–order filters which are electronically tunable. Proceedings of the Radioelektronika Conference, Brno, May 6-7, 2003, pp. 21-24
7. Ghausi, M. S., Laker, K. R.: Modern filter design. Prentice Hall, New Jersey, 1981

Continuous – Time Active Filter Design Using Signal – Flow Graphs

Martin Minarcik, Kamil Vrba

Department of Telecommunications, Faculty of Electrical Engineering and Communication,
Brno University of Technology,
Purkynova 118, 612 00 Brno, Czech Republic
xminar08@stud.feec.vutbr.cz, vrbak@feec.vutbr.cz

Abstract. This article deals with active frequency filter design using signal-flow graphs. The procedure of multifunctional circuit design that can realize more types of frequency filters is shown. To design a new circuit the Mason – Coates graphs with undirected self-loops have been used. The voltage conveyors whose properties are dual to the properties of the well-known current conveyors have been used as the active element.

Keywords: frequency filter design, multifunctional frequency filters, signal-flow graps.

1 Introduction

The voltage feedback amplifier is the most often used part in the area of signal processing. Nowadays, new active elements such as transconductance amplifiers, transadmittance amplifiers, voltage and current conveyors come to be used. The utilization of circuits with the new active elements offers many advantages compared to structures with the voltage feedback amplifiers. The internal structures of modern active elements are realized with the help of simple building blocks and that is why the signal with a wide frequency range can processed. A big group of new active elements is formed by current and voltage conveyors. At our department the signal – flow graph method has been used to design new circuits with voltage conveyors.

Signal – flow graphs have been used in many areas of engineering. In the area of signal processing the signal flow – graphs were used especially as an effective instrument for a quick analysis of linear circuits. Nowadays this problem is solved mainly by computer technology and suitable computer programmes so that signal – flow graphs are rarely used in signal processing today [1], [2]. However they are a suitable aid in the design of new circuits using nontraditional parts.

2 Signal – flow graph of voltage conveyor

In 1999 one type of voltage conveyor was presented [3]. The other types or generations of voltage conveyors have been developed in our department [4]. The voltage conveyor properties were derived from the current conveyor properties by interchanging the currents and the voltages in the describing equations. The basic types of voltage conveyors are three – port active elements with one high – impedance input **x**, low – impedance input **y**, and voltage output **z**. Twelve types of voltage conveyors can be defined according to the way voltages and currents are transferred between separate ports of the voltage conveyor. For their common definition the generalized voltage conveyor – GVC (Fig.1.) can be used.

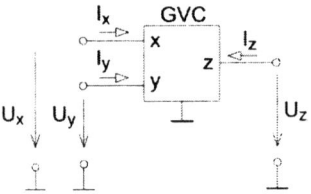

Fig. 1. Schematic representation of the GVC

The currents and voltages relations between GVC ports are described by the following matrix equation

$$\begin{pmatrix} I_x \\ U_y \\ U_z \end{pmatrix} = \begin{pmatrix} 0 & \alpha & 0 \\ \beta & 0 & 0 \\ \gamma & 0 & 0 \end{pmatrix} \cdot \begin{pmatrix} U_x \\ I_y \\ I_z \end{pmatrix}, \qquad (1)$$

where α, β, γ are general coefficients. These coefficients can take the values $\alpha \in \{-1,1\}$, $\beta \in \{-1,0,1\}$, $\gamma \in \{-1,1\}$. Different types of voltage conveyors can be defined by combining the transfer coefficients (Table.1).

Table 1. Definition of three-port voltage conveyors.

type	α	β	γ	type	α	β	γ
VCI+	1	1	1	IVCI+	-1	1	1
VCI-	1	1	-1	IVCI-	-1	1	-1
VCII+	1	0	1	IVCII+	-1	0	1
VCII-	1	0	-1	IVCII-	-1	0	-1
VCIII+	1	-1	1	IVCIII+	-1	-1	1
VCIII-	1	-1	-1	IVCIII-	-1	-1	-1

Other types of voltage conveyors were derived from the multiple – port current conveyors [5], [6]. For the practical realization of circuits using voltage conveyors a universal voltage conveyor – UVC [7] that can realize all types of three – port and multiple – port voltage conveyors has been designed in our department.

For the graphic description of voltage conveyor and the design of new circuit the Mason – Coates graphs with undirected self-loops are used. The description of linear circuits by signal – flow graphs is based on the graphic representation of mathematical formulas of circuits. The linear circuit can be described by a system of n linear algebraic equations

$$\mathbf{X} = \mathbf{AY}, \tag{2}$$

where **Y** is the column matrix of the known parameters $y_1, y_2, ..., y_n$, symbol **X** stands for the column matrix of unknown parameters $x_1, x_2, ..., x_n$. Symbol **A** denotes the square matrix of coefficients a_{ij} ($i,j = 1,2,...n$). Equation (2) can be displayed in the graph in many ways that lead to various types of graphs [1]. The Mason – Coates graphs with undirected self-loops are the graphical representation of equation (2) formulated in the form

$$a_{ii}x_i = y_i - \sum_{\substack{j=1 \\ j \neq i}}^{n} a_{ij}x_j . \quad (i = 1, 2, ..., n). \tag{3}$$

The system of equations (2) can be displayed in the graph as follows. We display all parameters in the graph by ringlets that are called *nodes*. The relations between parameters are displayed by oriented lines that are called *branches*. The transfer of each branch is given by coefficient a_{ij}. Coefficients a_{ii} on the left side of equation (3) represents transfer of *undirected self-loops* which belong to the nodes x_i. In signal – flow graphs representing linear circuits we can start from any system of independent equations (impedances, admittances). For circuits with voltage conveyors it is suitable to start from the system of admittance equations. The branches in a graph will correspond to node voltages in the circuit. The transfers of oriented branches will be given by the sum of admittances between a given pair of nodes while the transfers of undirected self-loops will be equal to the sum of admittances connected to given node. However, the voltage conveyors are irregular active elements. Because of this their signal – flow graphs must be created with the help of hybrid equations [8]. The GVC element can be described by the system of equations

$$Y_x V_x = I_x, \tag{4}$$

$$Y_y V_y = I_y, \tag{5}$$

$$Y_z V_z = I_z, \tag{6}$$

$$1.I_x = \alpha I_y, \tag{7}$$

$$0.I_y = \beta V_x - V_y, \tag{8}$$

$$0.I_z = \gamma V_x - V_z. \tag{9}$$

Equations (4) to (6) describe the properties of the voltage conveyor from the viewpoint of external circuit. Because of this they are written in the form of (3) and correspond to node voltages on GVC terminals. The transfers of the self – loops of these nodes are given by the sum of admittances connected to the given node. Equations (7) to (9) describe the voltage conveyor properties and belong to the internal nodes of the GVC signal – flow graph. The final GVC signal – flow graph is shown in Fig.2.

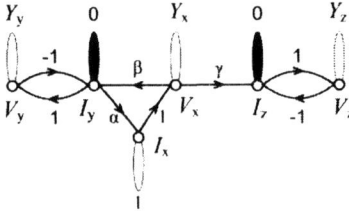

Fig. 2. Full signal – flow graph of the GVC

The graph in Fig.2 describes all properties of the GVC element. However, it is not suitable to design new circuits. This graph can be simplified by removing node I_x. This node will be used only in the case that the current I_x is the output value of the designed circuit. By removing this node from the graph only the transfer between node V_x and internal node I_y is changed. Another modification of the GVC graph is based on the idea that the voltage output has an ideally infinite impedance. Because of this, the output port voltage does not depend on the loading impedance. According to this idea the output terminal can be represented in the graph only by one voltage node. The self-loop transfer of this node is at all times equal to one. From the viewpoint of external circuit the branches can only get out of the node V_z. This property can be with advantage used in the design of new circuits. The simplified GVC signal – flow graph is shown in Fig.3.

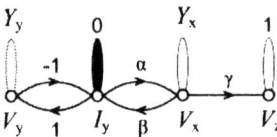

Fig. 3. Simplified signal – flow grapf of the GVC

Just as the voltage output properties can be used in the graph, the properties of the current input which has ideally zero impedance can be used too. In the graph it is possible to display the current input by one current node with self-loop transfer always equal to one. Conversely the branches from an external circuit can only enter into the current node. In the case of voltage conveyors this modification can be used only in second – generation voltage conveyors ($\beta=0$). A simplified graph of the second – generation voltage conveyors is shown in Fig.4.

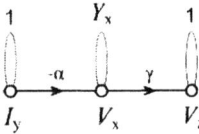

Fig. 4. Simplified graph of second – generation voltage conveyors

3 Frequency filter design using signal flow graphs

The signal – flow graphs can be used to design general circuits with the required transfer function in the frequency filter design. For this purpose Mason's rule for evaluating circuit quantities from a signal – flow graph is used [2].

$$\frac{x_i}{y_j} = \frac{1}{\Delta}\sum_{k=1}^{n} P_k \Delta_k , \qquad (10)$$

where
- x_i is output quantity
- y_j is input quantity
- P_k is the transfer of the kth forward path from input node (x_{in}) to output node (x_{out})
- Δ is the product of all self-loops
    - (sum of loop gain products of all possible sets of nontouching oriented loops taken two at a time)
    + (sum of loop gain products of all possible sets of nontouching oriented loops taken three at a time)
    ....
- Δ_k is the value of Δ for that portion of the graph which does not touch the kth forward path.

The graphs of generalized elements that have the possibility of choosing the transfer coefficients can be used with advantage to design circuits with voltage conveyors. However, the biggest advantage in the design of new circuits consists in the properties of the nodes V_x (I_y). From the signal – flow graph theory it is known that by adding a general admittance between a pair of nodes (a, b) two oriented branches originate. One from node a to node b and another in the opposite direction. The self-loop transfers of these nodes are changed too. If the general admittance is added between any node in the graph and the node V_z (I_y) only one branch originates and the self-loop transfer of the node V_z (I_y) is not changed.

The application of signal – flow graphs will be shown in the multifunctional frequency filter design. Multifunctional circuits are multiple – port networks realizing more types of transfer functions. A multifunctional circuit with one output and more inputs will be designed. Various types of transfer function will be realized by changing the input port without changing the topology and passive elements of the circuit.

From Mason's Rule it is evident that by changing the input node only the transfer function numerator is changed. The expression for the transfer function denominator is given by the determinant of the whole graph and it is the same for all transfer functions. So the multifunctional filter design can be divided into two parts. In the first part the basic graph structure will be designed which the required change of the transfer function numerator is reached. In the second part the oriented loops will be created that determine the graph determinant.

The basic multifunctional filter graph structure (Fig.5) has been designed using two simplified voltage conveyor graphs (Fig.4).

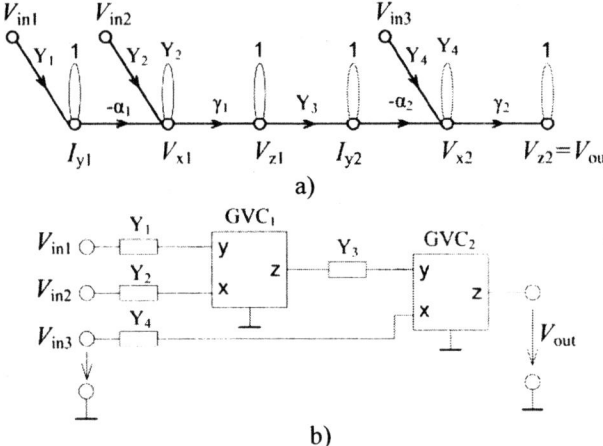

Fig. 5. The basic structure of multifunctional circuit: a) signal – flow graph b) circuit realization

From the graph in Fig.5 it is evident that by changing the input node the transfers of forward paths between these nodes and the output node are changed. The transfers of forward paths correspond to transfer function numerators and can be expressed by the equations

$$P_{in1} = \alpha_1 \alpha_2 \gamma_1 \gamma_2 Y_1 Y_3, \tag{11}$$

$$P_{in2} \Delta_{in2} = -\alpha_2 \gamma_1 \gamma_2 Y_2 Y_3, \tag{12}$$

$$P_{in3} \Delta_{in3} = \gamma_2 Y_2 Y_4. \tag{13}$$

The next step in the design of the general multifunctional circuit structure is creating oriented loops in the graph. There must be a sufficient number of oriented loops in the graph in order that in the final realization with concrete passive elements the stability criterion will be fulfilled. Oriented loops can be created by adding new branches between any pairs of graph nodes. However, this can give a risk to undesirable forward paths between the input nodes and the output node in the graph which will lead to a change transfer in functions numerators. The specific properties of the nodes I_y and V_z can be used to create oriented loops. Several possibilities of realizing oriented loops in the graph are shown in Fig.5. The determinants of the graphs are always depicted. Oriented loops can be designed by using voltage conveyors graphs too (Fig.5c, d)

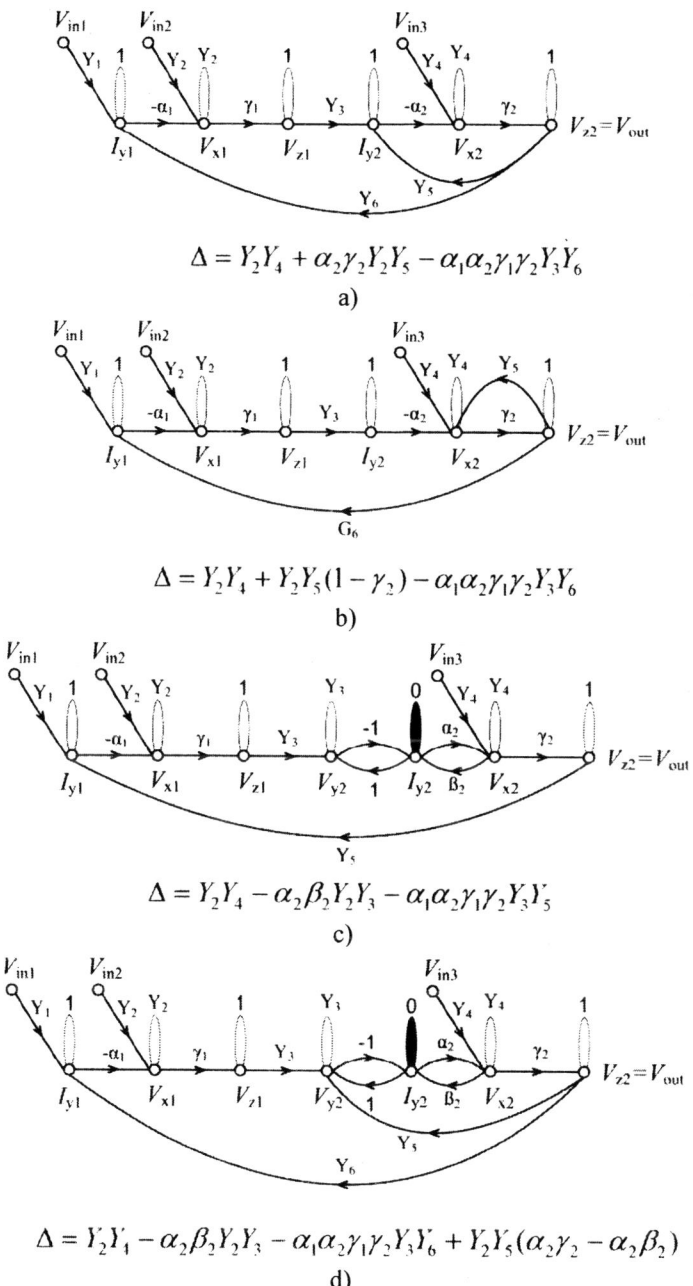

Fig. 6. General graphs of multifunctional circuits

The graph in Fig.5c has been chosen for the next design. By choosing particular transfer coefficients the graph determinant has been modified such that all the members of the determinant have the positive sign. In this way the final circuit stability has been achieved. Possible combinations of the concrete transfer coefficients are given in Table 2.

Table 2. Choice of the transfer coefficients

variant	α_1	β_1	γ_1	α_2	β_2	γ_2	GVC_1	GVC_2
1	-1	0	1	1	-1	1	IVCII+	VCIII+
2	-1	0	-1	1	-1	-1	IVCII-	VCIII-
3	1	0	-1	1	-1	1	VCII-	VCIII+
4	1	0	1	1	-1	-1	VCII+	VCIII-
5	1	0	1	-1	1	1	VCII+	IVCI+
6	1	0	-1	-1	1	-1	VCII-	IVCI-
7	-1	0	1	-1	1	-1	IVCII+	IVCI-
8	-1	0	-1	-1	1	1	IVCII-	IVCI+

The last step in the multifunctional frequency filter design is choosing concrete passive elements. Several possibilities are offered again here. For example $Y_1 = 1/R_1$, $Y_2 = 1/R_2$, $Y_3 = pC_1$, $Y_4 = pC_2$, $Y_5 = 1/R_3$ can be chosen. The final realization of the multifunctional frequency filter is shown in Fig.6.

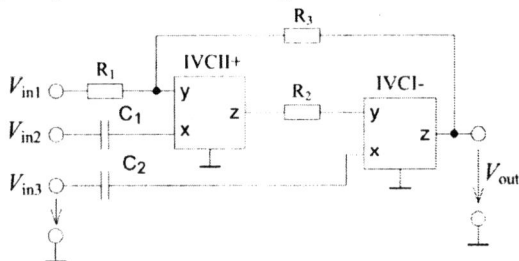

Fig. 7. Designed multifunctional frequency filter

The circuit in Fig.6 realizes frequency filters of the type of low pass, band pass and high pass with transfer functions (15) to (17):

$$K_{V,LP} = \frac{V_{out}}{V_{in1}} = -\frac{R_3/R_1}{1 + pC_1R_3 + p^2C_1C_2R_2R_3} \quad (15)$$

$$K_{V,BP} = \frac{V_{out}}{V_{in2}} = -\frac{pC_1R_3}{1 + pC_1R_3 + p^2C_1C_2R_2R_3} \quad (16)$$

$$K_{V,HP} = \frac{V_{out}}{V_{in3}} = -\frac{p^2C_1C_2R_2R_3}{1 + pC_1R_3 + p^2C_1C_2R_2R_3} \quad (17)$$

By the connection of inputs V_{in1} a V_{in3} the band rejection can be realized too (18).

$$K_{V,BR} = \frac{V_{out}}{V_{in1} + V_{in3}} = -\frac{R_3/R_1 + p^2 C_1 C_2 R_2 R_3}{1 + pC_1 R_3 + p^2 C_1 C_2 R_2 R_3}. \quad (18)$$

4 Simulations

Using the OrCad – Pspice simulation program the multifunctional frequency filter was analyzed. The universal voltage conveyor was used as an active element. The frequency filter was designed for characteristic frequency ω_o = 1 MHz and quality factor Q = 0,707 (Butterworth approximation). Analyzed amplitude frequency responses of the designed multifunctional filter are shown in Fig. 8.

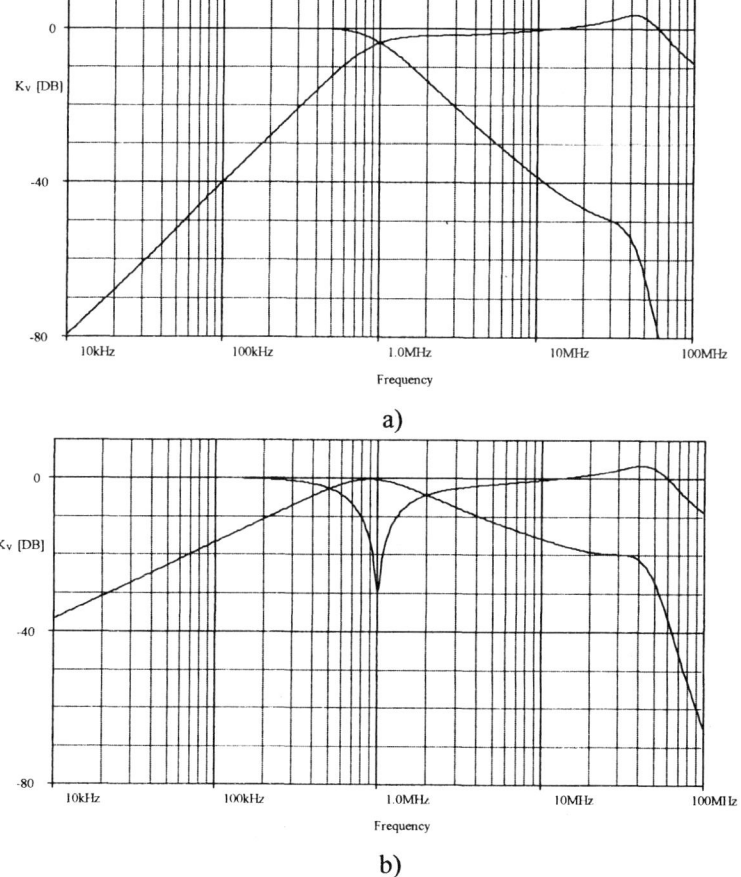

Fig. 8. Analyzed magnitudes of transfer functions a) low – pass, high – pass b) band – pass, band – rejection

5 Conclusion

The method of active frequency filter design with voltage conveyors using signal – flow graphs was introduced in this paper. The multifunctional circuit that can realize low pass, band pass, high pass and band rejection filters was designed using this method. The same procedure of the circuit design can be used for the other types of active elements.

Acknowledgments. The research into new active elements is supported by The Grant Agency of the Czech Republic, project No. 102/06/1383.

References

1. Wai-Kai Chen.: Circuits and Filters Handbook 2nd Edition. CRC Press, London (2003)
2. Biolek, D., Biolkova,V. Novel Signal Flow Graphs of Current Conveyors. In Proceedings of the International Conference on 38th MWSCAS. Rio de Janeiro (1995) 1058-1061
3. Acar, C., Ozoguz, S.: A new versatile building block: current differencing buffered amplifier suitable for analog signal - processing filters. Microelectronics Journal, Vol. 30 (1999) 157–160
4. Becvar, D.: Voltage Conveyors. Elektrorevue.cz, No. 51, 2001, (in Czech).
5. Vrba, K., Cajka, J. Universal network using DVCC elements for filter realization. In Proceedings of Conf. Telekomunikace 98, Brno (1998) 136-137
6. Vrba, K., Cajka, J.: Universal filter design using the general four-port current conveyor. In Proceedings of Conf. TSP 2000, Brno (2000) 16-19
7. Minarcik, M., Vrba, K.: Low-output and high-input impedance frequency filters using universal voltage conveyor for High-Speed Data Communication Systems. In Proc of Int. Conf. on Networing ICN 2006, Mauritius (2006)
8. Biolek, D., Cajka, J.,-Biolkova, V.: Modeling of current and voltage conveyors by flow graph technique. IASTED, Mexico (2003) 140–145

The Design of Optical Routes Applications

Martin Kyselak[1] and Miloslav Filka[1]

[1] Department of Telecommunications, Purkynova 118, Brno
Faculty of Electrical Engineering and Communication
Brno University of Technology;
martinkyselak@phd.feec.vutbr.cz; filka@feec.vutbr.cz

Abstract. Contemporary optical fibers can deal with almost all of the unfavorable effects which are known these days. They have sufficiently low specific slump, they can handle a slump caused by OH- ions, they can restrain the multimode effect and finally they can compensate a chromatic dispersion. But there is one problem, which the present science can't solve and this problem is the Polarization Mode Dispersion (PMD). This effect is a restricting factor of high-speed long-distance optical routes. The capacity increase of the existing optical routes is more often realized by multiplexing methods. Using more wavelengths makes multiplication of the optical fiber transfer rate possible. It evokes not only the advancement of fiber letting (known as "Dark Fiber" service), but more often also a single wavelength letting.

Keywords: optical fibre, design, application, pmd, polarization mode dispersion.

1 Introduction

Contemporary demand for data services is not a realistic demand anymore because of capabilities of currently existing optical routes. Services like Video on Demand, videoconference, real-time audio and video streams, large-file transmissions and other multimedia services lead to an increasing demand for transmission capacity. The quick transport layer is not established only on high-quality end-point equipment with sufficiency throughput, but it is necessary to ensure an equivalent physical layer by optical fiber routes.

2 Interface Description

The project documentation, including DGD values analysis along the optical fiber, is very important in order to build up a new optical route. The paper deals with efforts to develop the new complex application set to demonstrate the attenuation values and chromatic dispersion values. Further the applications, which can be able to simulate and calculate the probable PMD value by various ways in MATLAB environment.

OptView Application – is considered as an overall route overview. OptView will be a complex application which can depict the behavior of the basic values along the route. Fundamental features include the selection of: optical components, route parameters, modification possibilities, libraries reading, component selections and also fully automatic route protocol creating and of course creating and printing of attenuation and dispersion values graphs. OptView application is using a simple linear mathematics to determine the particular attenuations and to determine a total value of chromatic dispersion. The PMD calculation will be made by unsophisticated sum of partition squares bellow the square root. The preview is shown in the fig. 1.

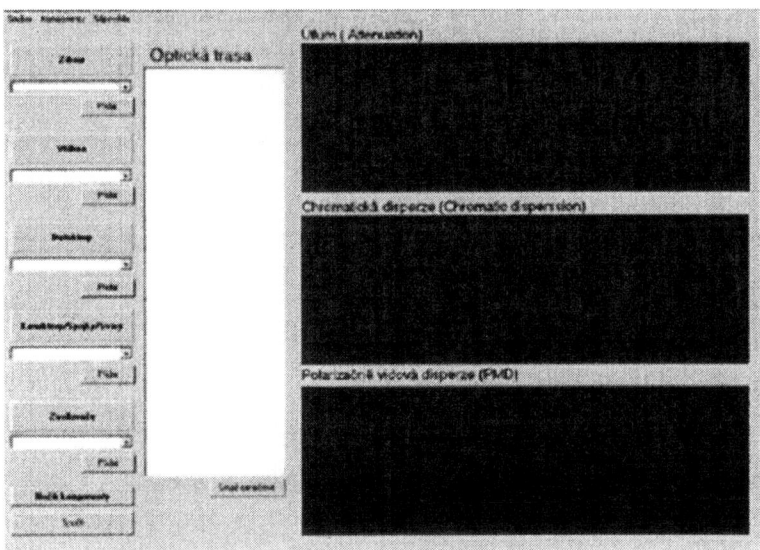

Fig. 1. The OptSim preview

MATLAB simulation process is being prepared for both of the polarized waves. It is a complicated three-dimensional simulator with the "mode delay regulation" option and "watch the vector sum" option. The result of the simulation will serve for the transparent depiction of complicated physical principles of light polarization results, which are not really easy to grasp mathematically and present them in an easy way. Thanks to the option of "mode delay regulation", the simulator will be fabulous to estimate the PMD effect on the multi-channel optical transmissions, especially on the dense wavelength multiplexing optical systems - DWDM. As an aid in simulation program development in Matlab environment, a 30-day trial version of OptSim application was provided by Safibra Company. The connection diagram is shown in the fig. 2.

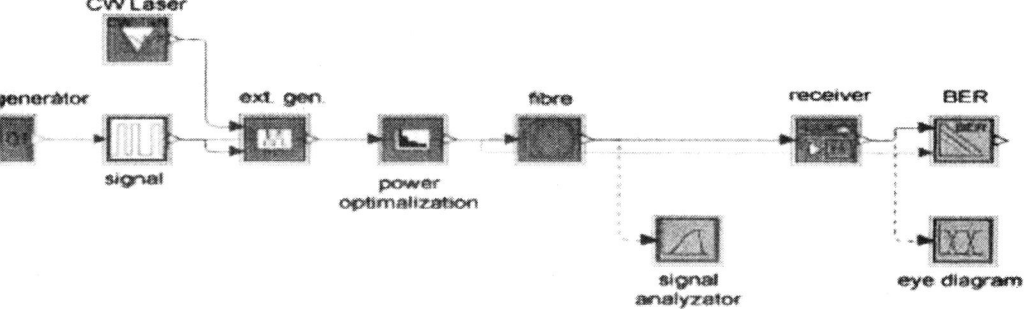

Fig. 2 The OptView connection diagram

Thanks to a sophisticated function, which can draw an "eye diagram", application outputs were used in order to compare it to the results from simulation that is currently being developed. For individual optical route input parameters, three "eye diagrams" have been generated. Two most important of them are shown in the figure 3 and 4.

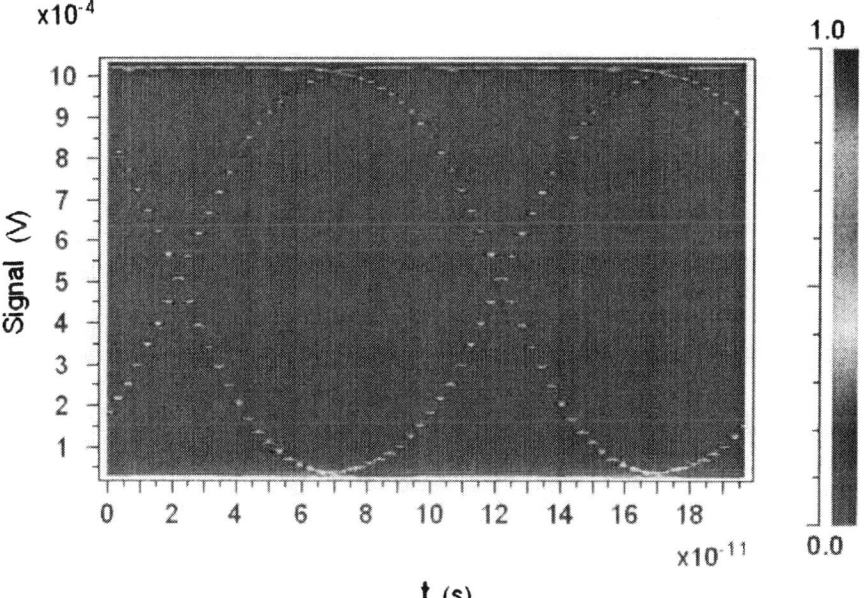

Fig. 3 Eye Diagram, 10 Gbps

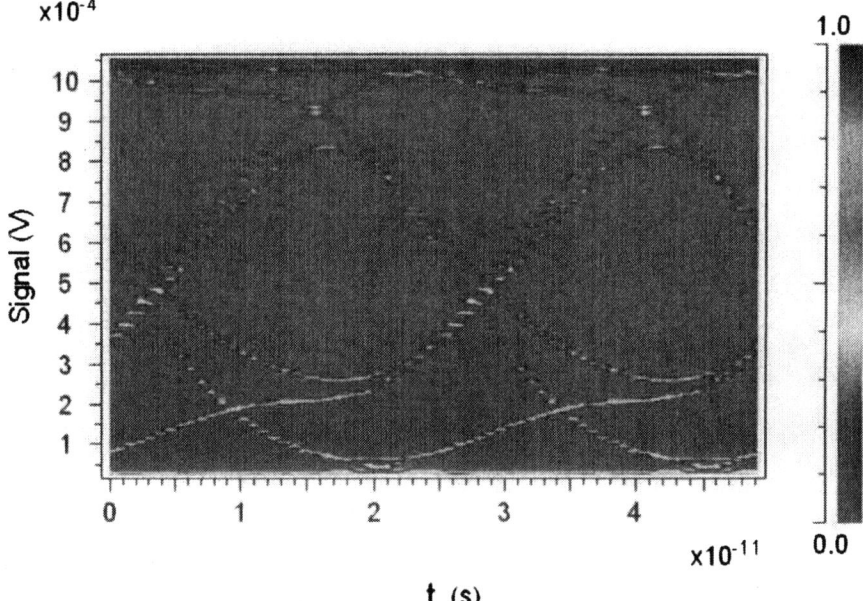

Fig. 4 Eye Diagram, 40 Gbps

Obviously clear depiction is caused by intentional omission of all influences on the optical transmission, except PMD. Following optical paths were used for the simulation: length 50 km, fiber G.652.d, attenuation and chromatic dispersion - zero, PMD = 0,1 ps/sqrt km. Bit rate 10 Gb/s for the fig. 3 and 40 Gb/s for the fig. 4.

There is only a block diagram prepared for the simulation of compensation technique in MATLAB environment. The block diagram is shown in the figure 5. The first output from MATLAB application is shown in the figure 6.

Simulation Application Block Diagram

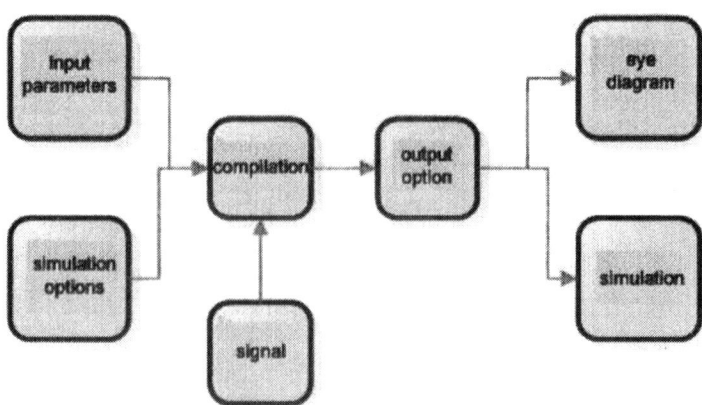

Fig. 5 The Simulation Application Block Diagram

The application capable of simulating both of the polarization planes of optical ray going through the optical fiber is being completed. Similar principle has been used as for compensation techniques simulator. One of the simulation moments is shown in the figure 6. Instantaneous sum of vectors with different input parameters option can be seen here.

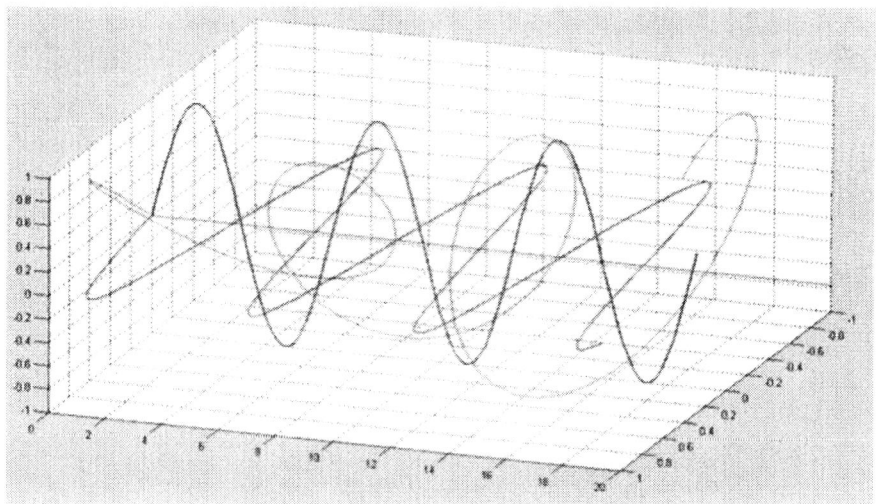

Fig. 6 The Simulation Application Output

The output will be curves of all polarization planes in both channels. This application deals with the transmission of power among particular channels. In the figure 7 there is a vector showing a sum of both of the polarization (reciprocally delayed) planes in time "t". The application calculates using the Maxwell distribution way and the probability of single DGD values occurrence [1]. Amplitudes are (for this example) non-dimensional values, since the values from the measuring are not used for the simulation, but the random values.

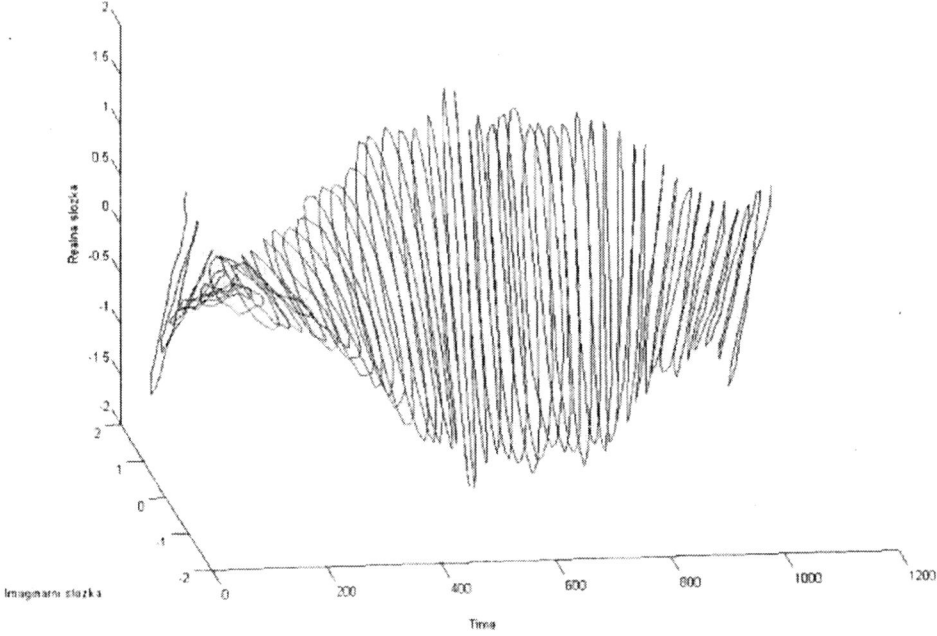

Fig. 7 The Vector Sum in the Time t

3 Conclusion

All the applications mentioned above can be used to obtain any accessible data about existing or planned optical paths. These application results will lead to an effective judgment of polarization mode dispersion resultant value. The partial results will lead to the frailest place of the paths locating in light of dispersion and will help with planning really high-capacity long-distance optical paths. Simulation results will also be used to determine the final values and polarization factors, and with advantages it can be used to assess the causes of each influence and precede them effectively. Appreciable advantage is transforming a mathematical machinery of optical physics (optic polarization) to the transparent graphical illustrations with "input parameters choice" option.

Final results and application outputs will be compared to each other and investigated in the future. It will be possible to presume the causes of aggravated parameters and to localize the place with the largest portion of total PMD value. Finally there will be a comparison of theoretically and practically obtained values - gained by measuring on existing paths. The equipment FTB-5500B and POTDR (Polarization Optical Time Domain Reflectometer) will be used for measuring. By a mutual comparison of calculated and measured values, it will be possible to evaluate the successfulness of the whole work. In case of equivalence of partial results and total results, the developed applications might be used as a professional part of design for high-speed multi-channel optical paths designers and developers.

References

[1] Galtarossa A., Menyuk C.R.: *Polarization Mode Dispersion*, Springer, Padova, ISBN 0-387-23193-5
[2] Saleh, B. E. A., Teich, M. C.: *Základy fotoniky II, Matfyzpress*, Praha, 1994, ISBN 80-85863-01-4
[3] Fischer, S., Randel, K., Petermann, J.K. *PMD outage probabilities of optical fiber transmission systems employing bit-to-bit alternate polarization*, IEEE Photonics Technology Letters, Volume: 17, Issue: 8, pp. 1647-1649, August 2005.
[4] Martin Hájek: *Zkušenosti s měřením polarizační vidové disperze (PMD) jednovidových optických kabelových tras*, OPTICKÉ KOMUNIKACE, Praha 2002
[5] Damask J. N.: *Polarization Optics in Telecommunications*, Springer, New York 2004 ISBN 0-387-22493-9
[6] KYSELÁK, M. *The Optimalization of the High-Speed Optical Networks*, Wave and Quantum Aspects of Contemporary Optics, XV Czech-Polish-Slovak Optical Conference. Technical University Liberec, Czech Republic, 2006. s. 75 (1 s.). ISBN: 80-86742-13-X.
[7] KYSELÁK, M. *Způsoby řešení polarizační vidové disperze u stávajících optických tras*, Optické komunikace 2006. Praha: Agentura Action M, 2006. s. 85-174. ISBN: 80-86742-16-4.
[8] KYSELÁK, M. FILKA, M. KOVÁŘ, P. Novell *Approach to the Solution of Optical Fibre Dispersion Effects*, Telecommunications and Signal Processing TSP - 2006. Brno, 2006. s. 25-28. ISBN: 80-214-3226-8.
[9] ITU-T Recommendation G.652: *Characteristics of a single-mode optical fibre cable*. ITU-T, April 1997

Measurement and therapeutical system based on Universal Serial Bus

Lukas Palko[1]

Department of Telecommunications
Faculty of Electrical Engineering and Communication
Brno University of Technology, Czech Republic[1]
lucas.palko@enjoy-rampl.com

Abstract. The paper presents complex treatment and measurement system based on USB and its enhancements. The article is divided into two parts. The first one deals with up-to-date USB enhancements which are USB On-The-Go, USB Test & Measurement Class, Wireless USB, PictBridge, PoweredUSB, Inter-Chip USB, Battery charging specification. The second part presents the practical usage for measurement and therapeutical system.

Keywords: USB, USB On-The-Go, OTG, USB Test & Measurement Class, USBTMC, Wireless USB, WUSB, PictBridge, PoweredUSB, Retail USB, USB PlusPower, USB +Power, Inter-Chip USB, Battery charging specification.

1 Introduction

The theme of this paper is to find the communication solution for the research into the possibility of skin therapy by the digitized form of magnetotherapy and phototherapy.

The digitization approach enables an easy control of batching exactly determined amounts of magnetic and electromagnetic (luminous) energy in space and time into the treated environment as well as the transmission of necessary control and diagnostic information between the therapist and the patient. Developing this method has had impacts that were not quite anticipated. There is supposed a controlled treatment of damaged and otherwise degraded cells of skin (burns, bumps, infection, etc.).

In view of the known contraindications in both magnetotherapy and phototherapy (laserotherapy), interest has concentrated on the joint action of planar pulsed magnetic and luminous fields penetrating into small depths of human skin. This medical technology is based on generating a combination of the two types of field close above the skin, according to a computation algorithm given in advance.

Here are the research aims:
- development of a laboratory type of therapeutic device that would allow research into digitally controlled planar pulsed magnetic and luminous fields,
- studying the effect of this combined field on biological materials, especially on human skin, with simultaneous search for suitable chemotherapeutics and their application possibilities,

- preparation of a user-friendly form of the controlling device software, inclusive of keeping records of therapist's patients included in the experimental program,
- complementing the therapeutic device with an imaging equipment, with the possibility of the therapist demarcating the surface under treatment directly on the computer display, controlling the device from a distance to facilitate an interactive out-patient treatment as registered in the program.

The measurement and therapeutical system contains a stand-alone therapeutic device for dermatology and cosmetics (one of the research aims), medical diagnostics apparatuses (ultrasound scanner, skin surface roughness diagnostics, skin parameters analyzer – amount of sebum, hydration, skin elasticity, water loss survey, skin conductivity) and additional utilities for better patient monitoring (PDA, digital camera). The whole system is intended to find out effects of the dermatology therapeutic device.

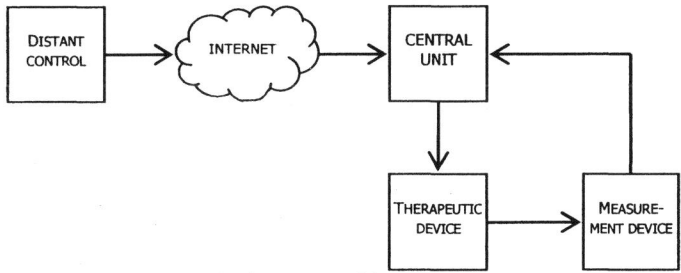

Fig. 1. Measurement and therapeutical system architecture

As the most perspective communication interface was chosen USB with its enhancements. Next chapter gives a brief survey of USB basics and its enhancements, which are supposed to use in the project.

2 USB enhancements

The very first idea of USB was to create uniform interface for different computer peripheral units. This point was completed and even more it was broken through. The USB become the most popular PC interface and it practically pushed away other interfaces. This huge popularity is given by unique combination of proprieties:
- plug-and-play feature – user friendly, no troubles with plugging or unplugging, can be connected to a PC "on-the-fly" without need for restart,
- high transfer rates up to 480 Mbps – sufficient for most of applications,
- supply "on board" – VBUS contains DC +5V power source with current drain up to 500mA,
- plenty of connected devices – to one host can be connected up to 127 peripherals,
- serial communication – the count of wires was reduced down to 4 (including supply), the cables are easy to manipulate with.

It is obvious that all of these features are not exactly the best, but the combination seems to be excellently balanced. There exists a FireWire Standard, which offers higher transfer speed and it is used for multimedia signal transfers, but it legs behind the USB in other features. Here are shown the USB enhancements which push this phenomenon beyond the state of art [1], [2].

2.1 USB On-The-Go (OTG)

The USB OTG was created for making the current USB interface available for smaller portable devices (PDA, digital camera, mobile phone, MP3 player, etc.). A basic goal is apart from smaller connectors an ability to mutually communicate one-to-one without master computer. USB OTG became very important standard for uniforming different manufacture approaches and it guarantees compatibility across platforms. The interface independency offers for example a contacts synchronization of a mobile phone and a PDA, a connection between GPS and PDA, a song exchange between two MP3 players, etc. Implementation of existing USB interface offers a wide bandwidth and so a transfer rate up to 480 Mbps. This high rate allows mass data transfer.

The classic USB uses architecture of a master and a slave. A USB host (PC) acts as the master and a USB peripheral (USB Device) acts as the slave. Only the USB host can control the configuration and data transfers over the link and initiate data transfers. Against that is USB OTG, which brings new possibility. The USB-OTG device is able to initiate the session, control the connection and exchange host and peripheral roles between each other. There are two special protocols:
- Session Request Protocol (SRP) allows both devices (host and peripheral) to enable or disable the active status of the communication link. It is very important for power consumption control of portable devices.
- Host Negotiation Protocol (HNP) brings possibility to exchange the host and the peripheral roles. Devices, which have implemented HNP, are called OTG dual-role devices. Here is point, that both communicating devices can initiate the data transfer.

The USB OTG is backward compatible with common USB. When the OTG device is connected to USB (no OTG), it acts as a common USB host or peripheral. The same effect of loosing OTG capabilities will arose after connection of USB hub, all OTG devices will become common USB devices (one of these will be a host a the others will be peripherals) [1], [3], [8].

2.2 Wireless USB (WUSB)

The basic idea of WUSB is to combine the speed and user-friendly technology with the wireless comfort. The range is supposed to the short-range category alike a Bluetooth technology. It is designed as a substitution for all devices that are now connected by common USB (printers, scanners, MP3 players, hard discs, game controllers, etc.). It used the WiMedia Alliance's Ultra-WideBand (UWB) common radio platform which offers transmission rates up to 480 Mbps at 3 meters and up to 110 Mbps at 10 meters.

The radio operates between 3.1 and 10.6 GHz and it can be adjusted according to a local regulatory policy in different countries around the world. A maximal count of

devices remains on 127, but there is no longer need for hubs – all devices are connected directly to the host. There became new types of adapters:
- Device Wire Adapter (DWA) for existing USB devices to connect to the wireless host,
- Host Wire Adapter (HWA) for personal computer to gain a wireless host capability. It can be connected externally to desktop common USB ports or to Card-Reader of laptop.

Apart from that, there is one additional feature - Dual-Role Devices (DRDs) support. It is an approach to a USB-OTG with limited capabilities. There is given a chance for a typical device to become a host for specified activity. For example a digital camera is device for a desktop but even more it can control a direct image transfer to a printer [3], [5], [8].

2.3 USB Test & Measurement Class (USBTMC)

This class is intended for automated measurement systems to interconnect a central control unit (usual computer) and measurement devices. It becomes a modern alternative to current interfaces GPIB (IEEE-488) and RS-232. The USBTMC development is based on Virtual Instrument Software Architecture (VISA), which is supported by GPIB and also RS-232. The USBTMC is a result of a collective effort of companies joined in VISA and USB Implementers Forum (USB IF) community.

USBTMC is able to control signals of measurement devices and accept their requests for interruption as well. Physical layer of USBTMC is equal to classic USB, thus the user comfort remains with all advantages:
- number of connected devices up to 127,
- high speed up to 480 Mbps,
- plug-and-play feature.

USBTMC can be used as a GPIB-USB converter. It enables to connect older GPIB measurement instruments to common PC via the USB. A bit rate is adapted to GPIB capabilities and it can be up to 8 Mbps (or 64 Mbps for an extension version). The converter acts for an operating system as a virtual GPIB port and therefore there are no additional software needs. Current GPIB measurement software detects a device as a standard GPIB [1], [2].

2.4 Inter-Chip USB

The Inter-Chip USB (IC_USB) is defined in a supplement to the USB 2.0 specifications for inter chip communication up to 10 cm distance. IC_USB specifications define a USB physical layer adapted for embedded devices. The main change to a USB standard is definition of multiple operating voltage classes (1.0 V, 1.2 V, 1.5 V, 1.8 V and 3.0 V) which facilitate migration to planned higher-density IC technology. The features are:
- all IC_USB products (host, peripheral) are USB products able to use USB compliant software,
- any peripheral device may by supplied by one of power-supply voltage selected from mentioned voltage classes,

- Full-speed (12 Mbps) support is required for all IC_USB products, Low-speed (1.5 Mbps) support is required only for hosts, for peripherals it is optional and High-speed (480 Mbps) is reserved for future study.

This specification is supported by all of Europe's major mobile network operators and it will be probably adopted by Multi Media Card Association (MMCA) into a MMC I/O specification. Another usage is for microcontroller technology [3], [7].

2.5 PictBridge

PictBridge is a standard related to the USB for direct printing. It was developed by Camera & Imaging Products Association (CIPA) and the aim was to print images directly from a digital camera (DSC) to a printer without a need for additional control equipment (computer). It is an open standard, but it cannot be published or revealed to the third party without a permission of CIPA. Printers have usually USB port type A and portable devices have usually USB port type Mini-B. There is no need for additional drivers and control software and all settings depend on devices firmware. PictBridge offers these features:
- printing one image displayed on the DSC or determined selection of more then one,
- automatic printing of images using the Digital Print Order format (DPOF) specification.

PictBridge is expanded standard and it is supported by most of producers of digital cameras, camcorders and also printers [6], [8].

2.6 PoweredUSB

The USB has it own supply distribution (VBUS), where is available voltage +5 V DC and current up to 100 mA (500 mA after announcement) for each USB-hub port. It can be sufficient for smaller devices like mice, keyboards, web cams, etc., but when the requirements exceed this limitation, there is a need for external power source and the advantage of real plug-and-play is devalued.

The PoweredUSB (also know as Retail USB, USB PlusPower, USB +Power) brings extra power lines to the common USB, which can fulfill the higher requirements and it also remains full compatibility to common USB. There became a new robust connector with 4 additional pins. This standard was developed by IBM, NCR, and FCI/Berg.

There are three key position locations on host connector. It determines which of these voltages is available in the connector:
- +5 V DC, amperage up to 6 A (power up to 30 W),
- +12 V DC, amperage up to 6 A (power up to 72 W),
- +24 V DC, amperage up to 6 A (power up to 144 W).

There is special requirement for the used cable, because of need for high transmitting rate and capability to supply up to 6 A on mentioned voltages. The recommended maximum length is 4 meters. Hot-Plugging (Hot-Unplugging) feature known from

USB is supported by a construction of connectors, where ground pins are front-ended the voltage ones. This feature partially suppresses contact arcing into the connector, in spite of it, the devices should be designed to avoid connector degradation or electronic component failure that could occur [4].

2.7 Battery charging specification

This specification was developed by the USB Implementers Forum Battery Charging workgroup for defining a standard way to draw electrical current from PCs. In addition, it allows PCs to control levels of current, in order to shorten the charging time of connected portable devices. The USB port is currently used for charging batteries in peripheral devices without any closer description and the trend among mobile devices to utilize USB for charging is growing. The new specification allows multiple types of portable devices to charge from USB ports thus it eliminates the need to carry multiple chargers for different kinds of devices.

The main benefit is definition of mechanism for portable devices of power source detection. If the source is wall charger or high current host or high current hub, the system allows immediately draw currents transcending 500 mA barrier. If the source is PC or hub, peripheral devices are limited up to 100 mA or 500 mA according to USB specification. The group also comes up with a provision in the new specs to allow portable devices with dead batteries to recharge at 100 mA even if they are not able to power up [3].

3 Application

All of these additions to the USB are intended to use in measurement and therapeutical system:

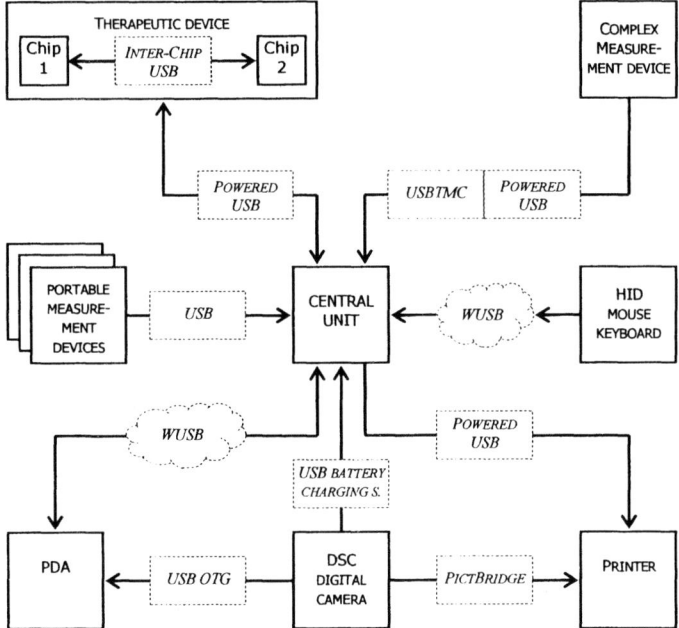

Fig. 2. Interface block diagram of the measurement and therapeutical system.

The core of the system is the central unit which is implemented into a high performance personal computer. It is operated wirelessly by WUSB keyboard and mouse and it has these key functions:
- controlling of the therapeutic device with a safety feedback,
- acquisition and processing of measured data,
- internet connection to a distant control center (not shown on the figure 2).

The therapeutic device is connected to the central unit via the PoweredUSB, which ensures both power supply and data link. Data link is used for the device control with a safety feedback. The power supply need is fulfilled by voltage +12V.

Inside the device are placed integrated microcontrollers. Their data exchange is via Inter-Chip USB specification.A complex measurement device is a GPIB diagnostic apparatus, which uses USBTMC transducer, for a measurement data transfer simplification. Power supply is provided by the PoweredUSB link with +24V.

A block mentioned as portable measurement devices means a set of simple medical diagnostic apparatuses for skin parameters measurement. Most of them are equipped by the common USB and the rest has a RS-232 output. These are supplemented by converters to the USB. PDA and a digital camera (DSC) are used for better patient monitoring out of the central control unit. The PDA uses wireless connection (WUSB) to the central unit. The DSC can be either connected to the PDA by the USB on-to-go or directly to the central unit, when it can use the USB battery charging specification for both images downloading and a battery charging. The last is an optional printer which can be powered by the PoweredUSB from the central unit and it can be used for a direct images printing from the DSC with the aid of the PictBridge enhancement.

3 Conclusions

There were designed three generations of the therapeutic device for dermatology and cosmetics. The block diagram of the actual measurement and therapeutical system with the third generation device is showed on the figure 2. Among the features of the third generation therapeutic device belongs a simultaneous planar pulsed magnetic and luminous field generating, dynamic intensity field application, a surface that is acted on demarcation and application visualization on the computer display for the operator to have an idea of its dynamic course.

The USB as a communication interface was found suitable for the system by virtue of its versatility and easy of use in all enhancements.

The results, obtained during the research into the possibility of skin therapy by the digitized magnetotherapy and phototherapy, were presented at the first international conference on the Digital Society ICDS 2007 in Guadeloupe, French Caribbean [9].

References

1. Palko, L.: Modern trends of serial interface USB (in Czech). Article in electrotechnic magazine Elektrorevue, 2005, ISSN 1213-1539, article number 05025,
2. Palko, L.: Universal Serial Bus (in Czech). Article in technical magazine Automa (12/7), 2006, ISSN 1210-9592,
3. USB Implementers Forum. [citation 2007-05-13]. Available on http://www.usb.org
4. PoweredUSB.org, [citation 2007-05-13]. Available on http://www.poweredusb.org
5. Cypress webpage, [citation 2007-05-22]. Available on http://www.cypress.com
6. PictBridge website, (Camera & Imaging Products Association) [citation 2007-05-23]. Available on http://www.cipa.jp/pictbridge
7. Léonard, Y.: MMC on the verge of adopting USB. [citation 2007-05-23]. Available on http://www.smartcardstrends.com
8. Wikipedia, the Free Encyclopedia. [citation 2007-05-13]. Available on http://www.wikipedia.org
9. Resl, V., Leba, M., Lojek, A., Číž, M., Hyršl, P., Palko, L., Rampl, I.,: Digitized biophysical therapy of skin. ICDS 2007 [including workshop TELEMED 2007]. Guadeloupe, French Caribbean, 2007, ISBN 0-7695-2760-42007

Transmitting Conditions of Cable Tree

Vaclav Krepelka[1] and Miloslav Filka[2]

[1] FITCE CZ, Mahenova 2, Brno
vaclav.krepelka@iol.cz
[2] Department of Telecommunications, Purkynova 118, Brno
Faculty of Electrical Engineering and Communication
Brno University of Technology
filka@feec.vutbr.cz

Abstract. Research and development of Spectrum Administration for metallic access networks analyzed possibilities of matrix representation for transmission circumstances. Some interesting remarks were gained as a side output of this work.

Keywords: Cable tree, signal spectrum function, noise spectrum function, transmitting function, PWD.

1 Cable tree

Key term **Cable tree** is defined in the *last mile* topology. **Cable tree** itself consists of sequence coupled cable sections concentrated in site of Main distribution frame in common trunk cable, see Fig. 1.
Note: Cable tree may be equipped by optical network units ONU, where optical signal is transformed into electrical and vice versa respectively; and which is in ONU-MDF received or transmitted into/from pairs of tided metallic cable.

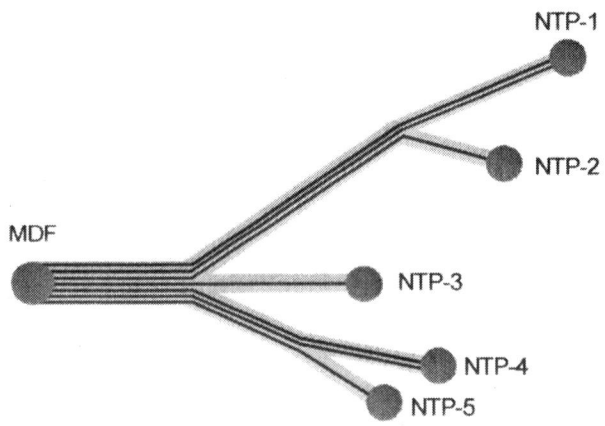

Fig. 1. Basic architecture of cable tree

2 Analytical Performance of Useful Signals & Disturbing Noises

Circumstances of mutual influences upon symmetric metallic cable are performed in Fig. 2:

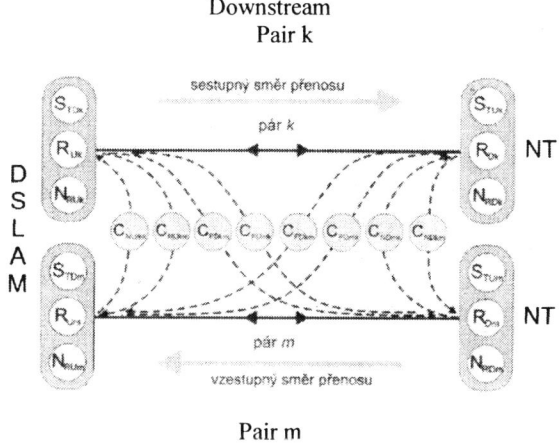

Fig. 2: Signals a Couplings between two Pairs

The simplest generalizable case of bidirectional transmission alongside two symmetric pairs is performed here. These pairs are terminated in exchange site in v DSLAM, in subscriber site in two network terminations. Three transmission characteristics belong to each of four here performed terminating points:

- Transmitted spectrum of signals $S(\omega)$ or maybe $S(f)$
- Non wanted disturbing noise $N(\omega)$
- Receiver sensitivity $R(\omega)$

as well as inter pair crosstalk couplings
- Near end crosstalk (NEXT) $C_N(\omega)$
- Far end crosstalk (FEXT) $C_F(\omega)$

Indexes **k** and **m** define mutual position of pairs;
indexes **D** and **U** discriminate downstream and upstream transmission directions;
indexes **T** and **R** assign appropriate characteristics to transmission and receive *of* signal.

Transmitting circumstances in site of subscriber termination, *in Fig. 1 pointed NT* are expressed by following equation:

$$S_{RDk}(\omega) = T_{CTk}(\omega)\, S_{TDk}(\omega) \tag{1a}$$

Equally in site of exchange termination *(DSLAM)* is valid:

$$S_{RUk}(\omega) = T_{CTk}(\omega)\, S_{TUk}(\omega) \tag{1b}$$

where T_{CT} is a vector of cable tree transmitting functions

Noise circumstances in site of NT may be described by following equation:

$$N_{RDk}(\omega) = S_{TDm}(\omega)\, C_{FDmk}(\omega) + S_{TUm}(\omega)\, C_{NDmk}(\omega) \qquad (2a)$$

as well as dually in site exchange termination DSLAM

$$N_{RUk}(\omega) = S_{TDm}(\omega)\, C_{NUmk}(\omega) + S_{TUm}(\omega)\, C_{FUmk}(\omega) \qquad (2b)$$

Such circumstances are only exceptional in reality; maybe only in last cable section led from subscriber cabinet supposing one-quad subscriber cable and both pairs exploited by broadband service. All couplings in real cable tree are multiplied and combined many times. We are able to express these circumstances by matrixes. These matrixes will be of equal order to number of pairs in trunk cable of cable tree. This number of pairs may reach hundreds as well as exceed thousand in metropolitan networks. Composition of such matrix exceeds real possibility of calculation.

Note: Variable ω will be again left out in following text

The equation 1a is able to be performed by product of diagonal matrix:

$$\begin{bmatrix} S_{RD1} & 0 & 0 & 0 \\ 0 & S_{RD2} & \ldots & \ldots \\ \ldots & \ldots & \ldots & 0 \\ 0 & \ldots & 0 & S_{RDp} \end{bmatrix} = \begin{bmatrix} T_{CT1} & 0 & \ldots & 0 \\ 0 & T_{CTp} & \ldots & \ldots \\ \ldots & \ldots & \ldots & 0 \\ 0 & \ldots & 0 & T_{CTp} \end{bmatrix} * \begin{bmatrix} S_{TD1} & 0 & \ldots & 0 \\ 0 & S_{TD2} & \ldots & \ldots \\ \ldots & \ldots & \ldots & 0 \\ 0 & 0 & 0 & S_{TDp} \end{bmatrix} \qquad (3a)$$

Equally equation 1b:

$$\begin{bmatrix} S_{RU1} & 0 & 0 & 0 \\ 0 & S_{RU2} & \ldots & \ldots \\ \ldots & \ldots & \ldots & 0 \\ 0 & \ldots & 0 & S_{RUp} \end{bmatrix} = \begin{bmatrix} T_{CT1} & 0 & \ldots & 0 \\ 0 & T_{CTp} & \ldots & \ldots \\ \ldots & \ldots & \ldots & 0 \\ 0 & \ldots & 0 & T_{CTp} \end{bmatrix} * \begin{bmatrix} S_{TU1} & 0 & \ldots & 0 \\ 0 & S_{TU2} & \ldots & \ldots \\ \ldots & \ldots & \ldots & 0 \\ 0 & 0 & 0 & S_{TUp} \end{bmatrix} \qquad (3b)$$

The equation 2a performing noise circumstances by matrixes of pth order for p pairs of cable tree trunk cable as a product of left diagonal signal matrix multiplied by right crosstalk matrix with zero diagonal. Resulting noise matrix is again with zero diagonal in accordance with multiplying rule for matrixes.

$$\begin{bmatrix} 0 & N_{RD12} & \cdots & N_{RD1p} \\ N_{RD21} & 0 & \cdots & \cdots \\ \cdots & \cdots & 0 & N_{RD(p-1)p} \\ N_{RDp1} & \cdots & N_{RDp(p-1)} & 0 \end{bmatrix} =$$

$$\begin{bmatrix} S_{TD1} & 0 & \cdots & 0 \\ 0 & S_{TD2} & \cdots & \cdots \\ \cdots & \cdots & \cdots & 0 \\ 0 & 0 & 0 & S_{TDp} \end{bmatrix} \star \begin{bmatrix} 0 & G_{FD12} & \cdots & G_{FD1p} \\ G_{FD21} & 0 & \cdots & \cdots \\ \cdots & \cdots & 0 & G_{FD(p-1)p} \\ G_{FDp1} & \cdots & G_{FDp(p-1)} & 0 \end{bmatrix} +$$

$$+ \begin{bmatrix} S_{TU1} & 0 & \cdots & 0 \\ 0 & S_{TU2} & \cdots & \cdots \\ \cdots & \cdots & \cdots & 0 \\ 0 & 0 & 0 & S_{TUp} \end{bmatrix} \star \begin{bmatrix} 0 & G_{NU12} & \cdots & G_{NU1p} \\ G_{NU21} & 0 & \cdots & \cdots \\ \cdots & \cdots & 0 & G_{NU(p-1)p} \\ G_{NUp1} & \cdots & G_{NUp(p-1)} & 0 \end{bmatrix} \quad (4a)$$

The equation 2b may be performed dually as follows:

$$\begin{bmatrix} 0 & N_{RU12} & \cdots & N_{RU1p} \\ N_{RU21} & 0 & \cdots & \cdots \\ \cdots & \cdots & 0 & N_{RU(p-1)p} \\ N_{RUp1} & \cdots & N_{RUp(p-1)} & 0 \end{bmatrix} =$$

$$\begin{bmatrix} S_{TD1} & 0 & \cdots & 0 \\ 0 & S_{TD2} & \cdots & \cdots \\ \cdots & \cdots & \cdots & 0 \\ 0 & 0 & 0 & S_{TDp} \end{bmatrix} \star \begin{bmatrix} 0 & G_{NU12} & \cdots & G_{NU1p} \\ G_{NU21} & 0 & \cdots & \cdots \\ \cdots & \cdots & 0 & G_{NU(p-1)p} \\ G_{NUp1} & \cdots & G_{NUp(p-1)} & 0 \end{bmatrix} +$$

$$\begin{bmatrix} S_{TU1} & 0 & \cdots & 0 \\ 0 & S_{TU2} & \cdots & \cdots \\ \cdots & \cdots & \cdots & 0 \\ 0 & 0 & 0 & S_{TUp} \end{bmatrix} \star \begin{bmatrix} 0 & G_{FU12} & \cdots & G_{FU1p} \\ G_{FU21} & 0 & \cdots & \cdots \\ \cdots & \cdots & 0 & G_{FU(p-1)p} \\ G_{FUp1} & \cdots & G_{FUp(p-1)} & 0 \end{bmatrix} \quad (4b)$$

Note: Due to the fact of invalidity commutative law for multiplying of matrixes, this sequence should be abided. On the other hand **change of this obligatory sequence leads to same phenomenon as by changing disturbing and disturbed lines, called as crosstalk measurement „straight" and „across" or marked = and x.**

3   The Condition for Information Recoverability

The couple of equations 1 and 2 is simplified for next consideration, again missing variable ω for useful signal in downstream as well as upstream:

$$S_{Ri} = T_{CT} S_{Ti} \quad \text{where } i = D, U \tag{5}$$

and analogous noise circumstances in site of subscriber exchange:

$$N_{RD} = S_{TD}\, G_{FD} + S_{TU}\, G_{ND} \tag{6a}$$

and again dually in site of network termination.

$$N_{RU} = S_{TD}\, G_{NU} + S_{TU}\, G_{FU} \tag{6b}$$

The condition for information transmission secured in the simplest configuration as to Fig. 2 in sufficient quality may be performed for down(up)stream as follows:

$$N_{RD(U)k}(\omega) \leq R_{D(U)k}(\omega) \quad \text{for } \omega \ni \Omega \tag{7}$$

These inequalities may be overwritten for cable tree generally:

$$N_{RD} \leq R_D \tag{8a}$$

$$N_{RU} \leq R_U \tag{8b}$$

Induction of both inequalities 8 into couple of equations 6 enables determination for highest assigned signals, there define masks of Power spectral density of transmitted signals:

$$S_{TD} \leq \frac{1}{G_{NU} - \dfrac{G_{FU} G_{FD}}{G_{FU}}} \left(R_U - \frac{G_{NU} R_D}{G_{ND}} \right) \quad \text{for } \omega \ni \Omega \tag{9a}$$

and

$$S_{TU} \leq \frac{1}{G_{ND} - \dfrac{G_{FU}G_{FD}}{G_{NU}}} \left(R_D - \frac{G_{FD}R_U}{G_{NU}} \right) \text{ for } \omega \ni \Omega \qquad (9b)$$

Inequality system 9 may be altered by other constants produced by legislative measures of non-technical character. Matrixes of higher order are proposed to be calculated generally; therefore the volume of necessary calculations may exceed realizable measures.

List of symbols:

G_{FD} — matrix of transmitting functions of FEXT in downstream
G_{FU} — matrix of transmitting functions of FEXT in upstream
G_{ND} — matrix of transmitting functions of NEXT in downstream
G_{NU} — matrix of transmitting functions of NEXT in upstream
N_{RD} — noise spectrum vector in site of receivers in downstream
N_{RU} — noise spectrum vector in site of receivers in upstream
R_D — receiver sensitivity vector in downstream
R_U — receiver sensitivity vector in upstream
S_{RD} — signal spectrum vector in site of receivers in downstream
S_{RU} — signal spectrum vector in site of receivers in upstream
S_{TD} — signal spectrum vector in site of transmitters in downstream
S_{TU} — signal spectrum vector in site of transmitters in upstream
T_{CT} — transmitting functions vector of cable tree
ω — radian frequency
Ω — frequency interval (transmitted band)

References

[1] Křepelka, V.: Doctoral Theses, VUT FEKT ÚTKO, Brno, 2005

[2] Meninger, M.: Úvod do správy spektra metalické sítě, Telekomunikace 1/03, Praha, ISSN 0040-2591

Non-linear circuits with CCII+/- current conveyors

Jiri Misurec

Department of Telecommunications, Brno University of Technology
Purkynova 118, 612 00 Brno, Czech Republic
misurec@feec.vutbr.cz

Abstract. In the area of analog techniques and primary processing of analog signals in the last decade, some authors have focused on circuits with current or voltage conveyors. Most of them have concentrated on filters with current conveyors, their design and properties, different connections, sensitivity analysis, etc. The present paper is devoted to the basic theoretical description of non-linear circuits with CCII+/- current conveyors in non-filter applications. The basic connections of circuits with current conveyors are chosen, in which a non-linear three-pole is considered, and the functional relations of these connections are established. The applications of non-linear circuits given in the paper are half-wave and full-wave precise rectifiers, which are an analogy to precise measuring rectifiers with operational amplifiers. Rectifiers with current conveyors operating in the current mode can exhibit some positive properties. Only the basic connections are given and the basic functional computer simulations are made here. The active element chosen for the practical rectifier realization was an appropriately connected OTA amplifier. In conclusion, some measurement results are given.

Keywords: current conveyor, non-linear circuits, rectifiers.

1 Introduction

CCII+/- conveyors are active elements that form a numerous group of functional blocks, which realize unit transfers of current and voltage (with either positive or negative polarity) between individual gates. The description of these elements is sufficiently known and can be found in many publications, for example in [1], [2]. Current conveyors enable the design of circuits operating in the voltage, current or hybrid mode. Fig. 1 gives a schematic symbol that describes the conveyor relations and its ideal model with controlled sources of a second-generation current conveyor (CCCS – Current-Controlled Current Source, VCVS – Voltage-Controlled Voltage Source).

The practical availability of these elements is currently poor; and in experimental work elements are utilized which include in their internal structure the CCII element in some form. These are, for example, the AD844 and OPA660 amplifiers. The following amplifiers can also be used: the MAX435 and MAX436 OTA and BOTA (Balanced Operational Transconductance Amplifier) amplifiers by Maxim [3], and the LM13600 amplifiers by National Semiconductor [4].

In the following, only the CCII+ element will be considered, for which the respective relations will be derived. For the other elements, the relations are similar. In the basic connections with CCII+ the relations between input and output signals are linear. Fig. 2 gives the basic connections that represent the non-inverting voltage amplifier, inverting voltage amplifier and inverting current amplifier. For simplification, we will consider the above amplifications to be ideal. In these basic connections the relations between input and output signals are linear.

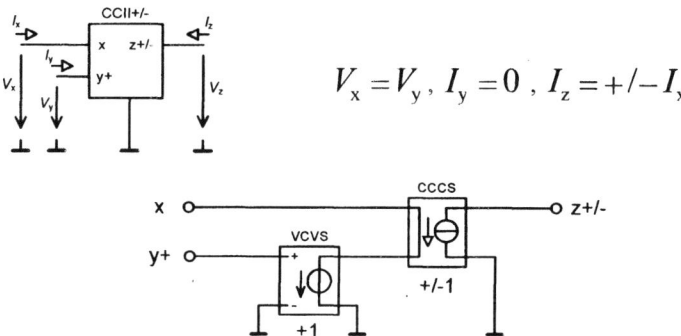

$$V_x = V_y, \ I_y = 0, \ I_z = +/- I_x$$

Fig. 1. Definition of CCII+/− current conveyor

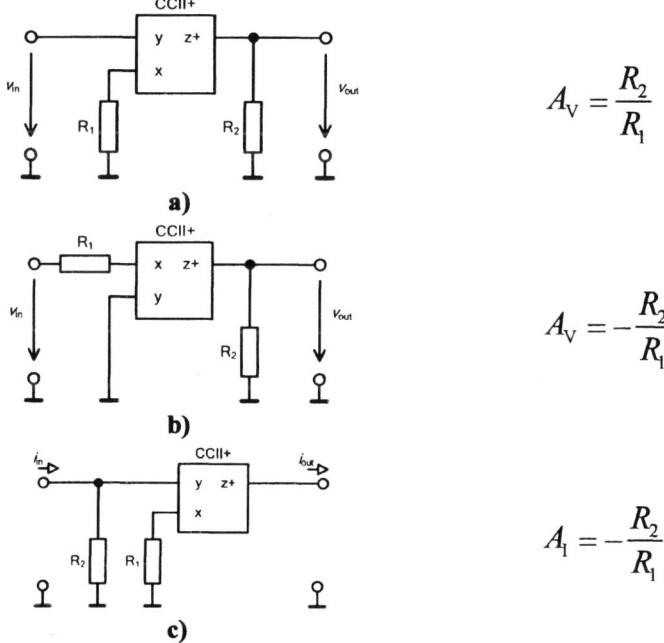

$$A_V = \frac{R_2}{R_1}$$

a)

$$A_V = -\frac{R_2}{R_1}$$

b)

$$A_I = -\frac{R_2}{R_1}$$

c)

Fig. 2. Basic connections of amplifiers with current conveyors: a) non-inverting voltage amplifier, b) inverting voltage amplifier, c) inverting current amplifier

2 Non-linear amplifiers with CCII+ current conveyor

If instead of linear elements we use non-linear elements, the circuit amplification will be non-linear, and the dependence relation between the input and the output signal will be also non-linear.

In technical practice there are a number of non-linear elements controlled by an electric quantity, namely a voltage or a current. These elements thus have three or more poles by which they are connected into a circuit. When analyzing circuits with CCII+ we are interested not only in the output circuit of these elements but also in the input circuit where the control signal is acting, since the control signal source and the non-linear controlled element influence each other. From the basic connections with current conveyors given in Fig. 2 it follows almost immediately that the controlled non-linear element is a non-linear three-pole resistance.

Let us define a non-linear resistance three-pole by currents and voltages as given in Fig. 3. Between the currents i_A, i_B, and i_C flowing into the non-linear three-pole (NTP) and the voltages on three-pole terminals v_A, v_B, and v_C with respect to the common potential the following functional relations hold.

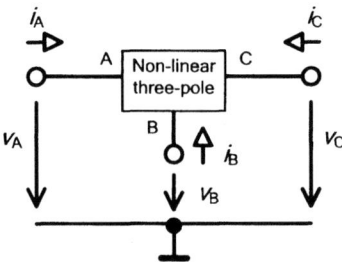

Fig. 3. Non-linear three-pole, designation and definition

$$i_A = r_A(v_A, v_B, v_C) \;, \tag{2.1}$$
$$i_B = r_B(v_A, v_B, v_C) \;, \tag{2.2}$$
$$i_C = r_C(v_A, v_B, v_C) \;. \tag{2.3}$$

Since the three-pole can also be viewed as a node, Kirchhoff's law also holds,
$$i_A + i_B + i_C = 0 \tag{2.4}$$

The three-pole under consideration can also be described by the characteristics
$$v_A = g_A(i_A, i_B, i_C) \;, \tag{2.5}$$

$$v_B = g_B(i_A, i_B, i_C), \quad (2.6)$$

$$v_C = g_C(i_A, i_B, i_C), \quad (2.7)$$

and the description by hybrid characteristics is also generally known. These characteristics can, for example, be in the form

$$i_A = h_A(v_A, i_B, v_C), \quad (2.8)$$

$$v_B = h_B(v_A, i_B, v_C), \quad (2.9)$$

$$i_C = h_C(v_A, i_B, v_C). \quad (2.10)$$

If the resistance element in the above connections with current conveyors is replaced by a non-linear resistance three-pole from Fig. 3, a number of circuits will be obtained whose transfer characteristic will depend on the properties of this three-pole. When solving these circuits, it is necessary to know the appropriate characteristics describing the non-linear three-pole.

Connecting the non-linear three-pole in place of resistor R_1 in Fig. 2a) will yield the situation given in Fig. 4.

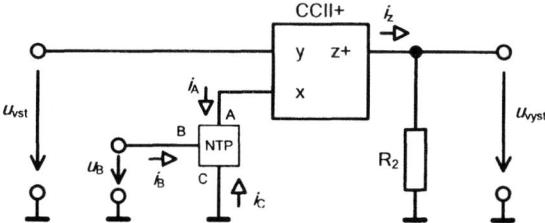

Fig. 4. Non-linear resistance three-pole in the input part of non-inverting amplifier with CCII+

If we consider the relations holding for the CCII+ current conveyor, then $v_A = v_{in}$, $i_A = i_x$, $i_z = i_x$, and $v_{out} = v_{in}R_2/R_1 = i_A R_2 = i_z R_2$; further we consider $v_C = 0$. Then the relations describing the three-pole will be of the form

$$i_A = r_A(v_A, v_B), \quad (2.11)$$

$$i_B = r_B(v_A, v_B), \quad (2.12)$$

$$i_C = r_C(v_A, v_B). \quad (2.13)$$

It can be seen from Fig. 4 that there can be two cases of driving the circuit. The three-pole can be driven either by voltage or by current, and the input terminal "y" can only be driven by voltage since it is the voltage input that is concerned here. For the first case let us consider that ideal voltage sources are connected to the B terminal of the three-pole and the conveyor. In that case it is of advantage to start from the knowledge of the characteristic

$$i_A = r_A(v_A, v_B). \quad (2.14)$$

Then the amplifier output voltage is determined.

$$V_{out} = R_2 i_z = R_2 i_A = R_2 r_A(v_A, v_B) \ . \tag{2.15}$$

Another possible case is that an ideal current source is connected to the B terminal of non-linear three-pole. In this case it is of greater advantage to exploit the hybrid characteristic of non-linear three-pole

$$i_A = h_A(v_A, i_B) \ , \tag{2.16}$$

and, similarly, the amplifier output voltage will be

$$V_{out} = R_2 i_z = R_2 i_A = R_2 h_A(v_A, i_B) \ . \tag{2.17}$$

It can be seen from the relations given above that the output voltage of the circuit under consideration depends directly on the respective characteristic of the non-linear three-pole.

Yet another connection of amplifier with current conveyor is the inverting voltage amplifier given in Fig. 2b). In the input circuit we replace the resistor R_1 as shown in Fig. 5. From the description of current conveyor it is evident that $v_C = 0$, and the description of NTP is again given by the relations (2.11), (2.12), (2.13) or by their hybrid equivalents.

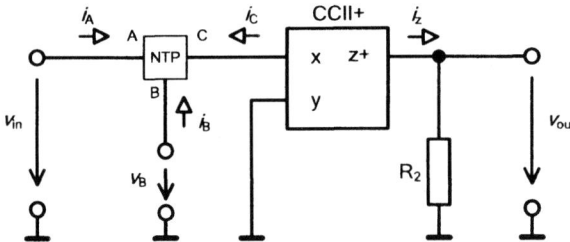

Fig. 5. Non-linear resistance three-pole in the "input circuit" of inverting amplifier

In this case there will be a total of four possible sources being connected to terminals A and B. It is understood that the sources are connected between the respective terminal and the common ground terminal.

  A) Ideal voltage sources are connected to the A and B terminals. Then we start from the knowledge of the characteristic

$$i_C = r_C(v_A, v_B) \ . \tag{2.18}$$

Then the amplifier output voltage is determined.

$$V_{out} = -R_2 i_z = -R_2 i_C = -R_2 r_C(v_A, v_B) \ . \tag{2.19}$$

  B) If a voltage source is connected to terminal A and a current source to terminal B, the tree-pole will be described using the hybrid characteristic

$$i_C = h_C(v_A, i_B) \ , \tag{2.20}$$

and then the amplifier output voltage will be determined.

$$V_{out} = -R_2 i_z = -R_2 i_C = -R_2 h_C(v_A, i_B) \quad . \tag{2.21}$$

C) If a current source is connected to terminal A and a voltage source to terminal B, then the hybrid characteristic is

$$i_C = l_C(i_A, v_B) \quad , \tag{2.22}$$

and the amplifier output voltage is

$$V_{out} = -R_2 i_z = -R_2 i_C = -R_2 l_C(i_A, v_B) \quad . \tag{2.23}$$

D) In the case that the two sources are current sources, and if we consider the validity of Kirchhoff's law (2.4), then for the output voltage it holds

$$V_{out} = -R_2 i_z = -R_2 i_C = R_2(i_A + i_B) \quad . \tag{2.24}$$

It follows from the above relations that in cases A), B), and C) the output voltage of the inverting amplifier under consideration depends directly on the non-linear three-pole characteristic. In case D) the output voltage does not depend on the three-pole properties. The operation described by relation (2.24) can be realized in a simpler way, without using the non-linear three-pole.

In the inverting current amplifier with current amplifier connected as in Fig. 2c) the non-linear three-pole replaces resistor R_2, as indicated in Fig. 6.

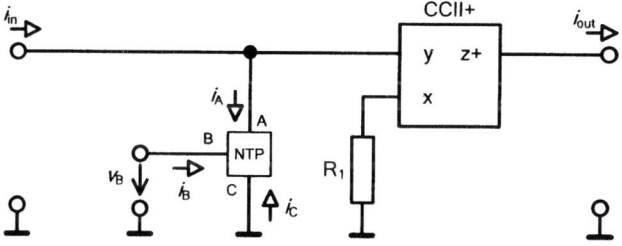

Fig. 6. Non-linear resistance three-pole in inverting current amplifier

Let us consider that an ideal voltage source is connected to the B terminal of the tree-pole and that the current $i_B = 0$. Then the characteristic of non-linear three-pole is

$$v_A = g_A(i_A, i_C) \quad , \tag{2.25}$$

and the output current of current amplifier will be determined as

$$i_{out} = -\frac{v_A}{R_1} = -\frac{g_A(i_A, i_C)}{R_1} \tag{2.26}$$

There is no use in considering a current source connected to terminal B, the reason being the same as in point D) of the preceding case.

Similar relations can also be found for CCII− current conveyors.

3 Application of circuits with non-linear current-conveyor amplifier

The following section focuses on selected applications of non-linear amplifiers using current conveyors. The processing of the positive and the negative part of the signal being amplified is separate, as made possible by complementary non-linear structures. The latter are the basis for creating rectifier circuits with conversion characteristics approximating ideal characteristics approximated by a broken line. These circuits are known as "operational rectifiers", "absolute value rectifier", etc. which have some importance in measuring techniques in particular. Use is made, above all, of high-precision half-wave or full-wave rectifiers, various kinds of clippers or function converters. The subject of the present paper is primarily half-wave and full-wave rectifiers.

3.1 Half-wave rectifier

The connection of fast inverting half-wave rectifier using a CCII- current conveyor is shown in Fig. 7. The output current i_z of the terminal "z" is given by the relation

$$i_Z = -\frac{u_{in}}{R_1}, \qquad (3.1)$$

and the rectifier output voltage is then given by the relation

$$V_{out} = i_Z \cdot R_2 = -\frac{R_2}{R_1} \cdot v_{in}. \qquad (3.2)$$

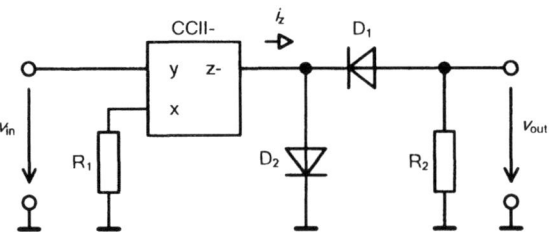

Fig. 7. Inverting half-wave rectifier with CCII–

The operation of half-wave rectifier was simulated by an idealized model of CCII– current conveyor. The voltage applied to the output was of sine waveform, amplitude v_{in} = 10 V, frequency f = 100 kHz, resistance values $R_1 = R_2 = 100\ \Omega$, with models of the Schottky diodes 1PS70SB40 being used. Results of the computer simulation are given in Fig. 8, it can be seen that the circuit implements the function of inverting half-wave rectifier.

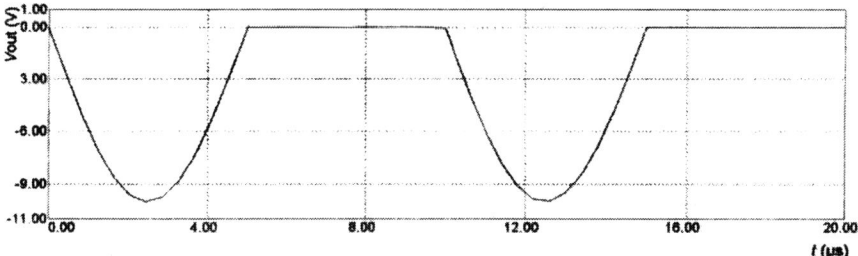

Fig. 8. Inverting half-wave rectifier with CCII–, $f = 100$ kHz

3.2 Full-wave rectifier

A full-wave rectifier with current conveyors is given in Fig 9 [5], [6]. The two CCII+ conveyors form a difference amplifier with current-to-voltage conversion on the output resistor R_2 such that with positive values of input signal the output current values are given by the relation

$$i_z = -\frac{v_{in}}{R_1} \qquad (3.3)$$

The output current i_z flows from the output terminal "z" 1CCII+ through resistor R_2, which has the same value as R_1. Diodes D_4 and D_2 are on, and the voltage on the output is $v_{out} = v_{in}$.

With negative values of input signal, diodes D_3 and D_1 are on. The output current of ^{2}CCII+ conveyor flows again through resistor R_2 and it again holds $v_{out} = v_{in}$. The magnitude of voltage transfer is given by the resistance ratio R_2/R_1.

In the rectifier, the fast Schottky diodes are expected to be used in order to obtain a high operating frequency. Voltage V_x serves to suitably set the operating mode of the diodes.

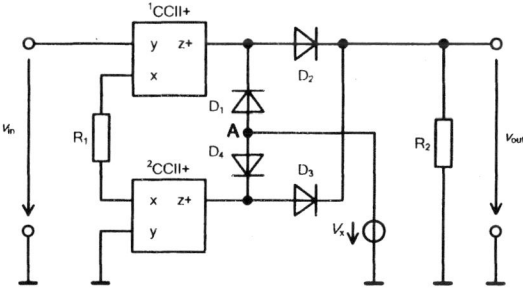

Fig. 9. Full-wave rectifier with CCII+

In the computer simulation of the above connection in the time domain the model of AD844 circuit was used, which is part of the Microcap program library. The diodes used were the Schottky diodes IPS70SB40, $R_1 = R_2 = 100$ kHz. The simulation was conducted for two voltages, $V_x = 0$ V, and $V_x = 1$ V, for the frequency $f = 100$ kHz, and a harmonic input signal amplitude of 100 mV. The simulation results are given in Fig. 10.

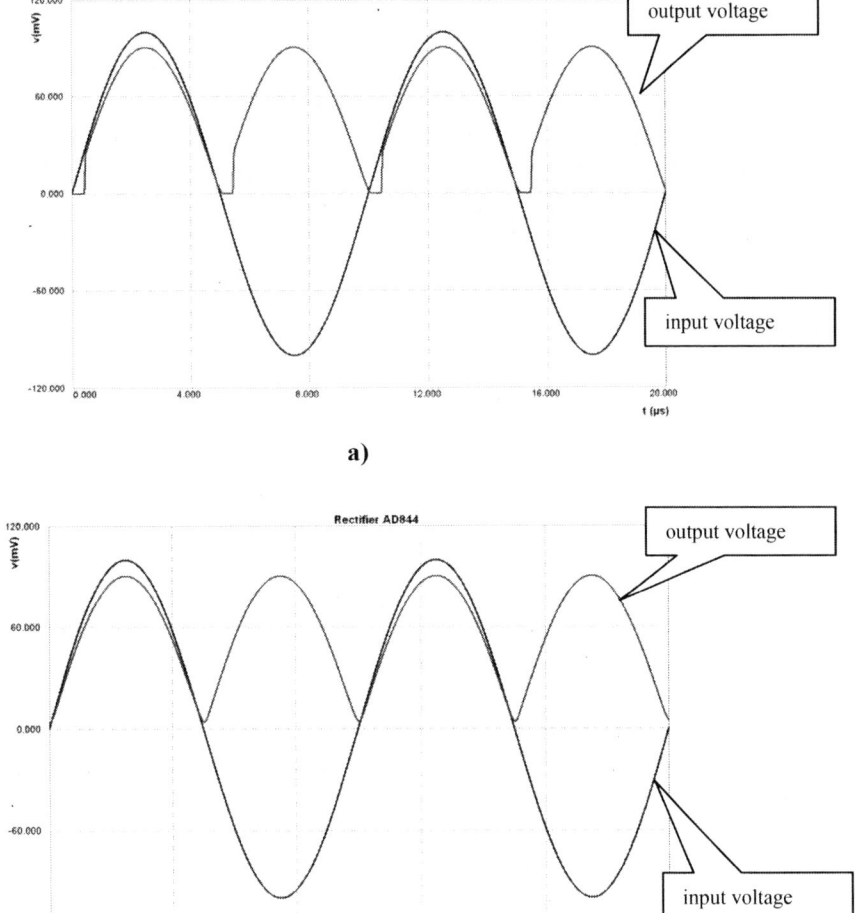

Fig. 10. Full-wave rectifier with CCII+, a) $f = 100$ kHz, $V_x = 0$ V, b) $f = 100$ kHz, $V_x = 1$ V

It is obvious from the simulation results that the magnitude of voltage V_x plays a role here. The negative effect of voltage V_x increases at higher frequencies. The voltage on the anodes of diodes D_1 and D_4, that is to say at point A, is influenced by the low impedance (in computer simulation it is zero) of the auxiliary source of voltage V_x while on the diodes there is a small voltage. When these diodes are subsequently connected to the terminal "z", the output voltage is zero. The magnitude of auxiliary voltage needs to be made balanced depending on the input voltage decrease on the diodes.

4 Practical rectifier realization

The operation of rectifiers was tested experimentally. In the specimen realized, a MAX435 operational transconductance amplifier (OTA) was utilized. With this element, the transfer conductance depends on the control current by means of which the element conductance can be changed. In Fig. 11 the connection of rectifiers with a MAX435 circuit is shown. The setting of the transfer conductance is not critical and therefore it is not given in the schematic.

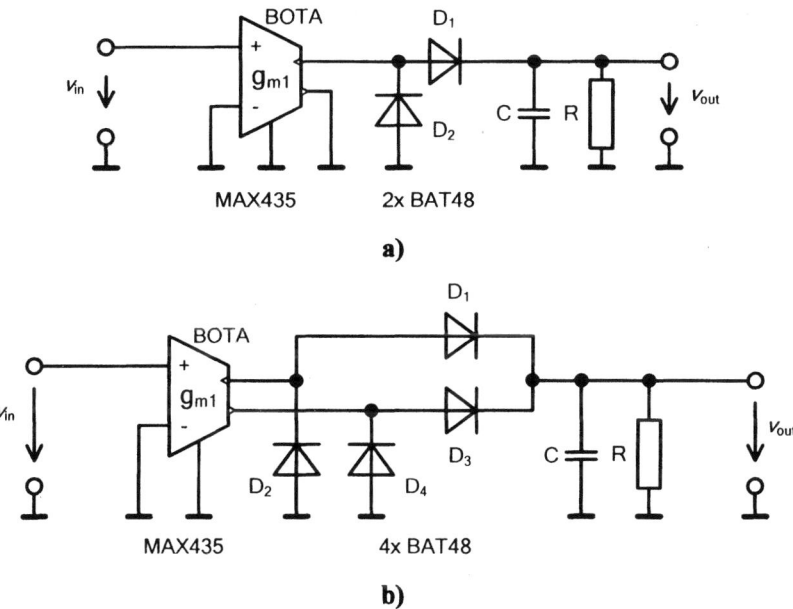

Fig. 11. Ideal connection of rectifiers for experimental verification, a) half-wave rectifier, b) full-wave rectifier

The waveforms measured for the half-wave rectifier are given in Fig. 12. The rectifier was measured at frequency of 5 MHz. The waveforms measured for the full-wave rectifier are similarly given in Fig. 13.

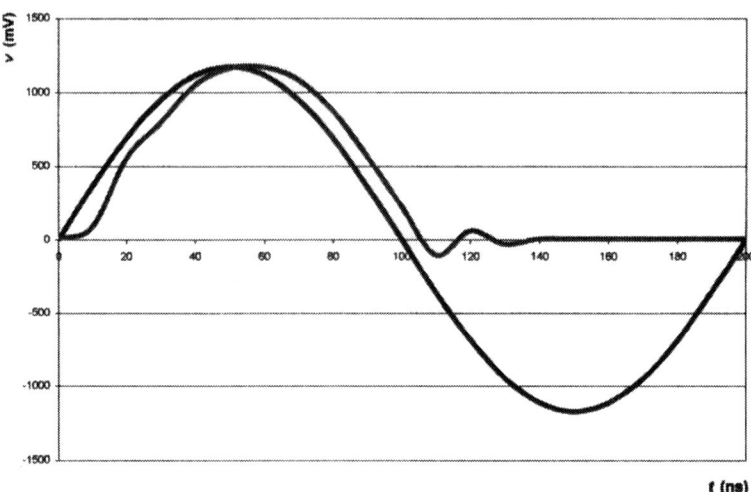

Fig. 12. Time behaviours measured for half-wave rectifier with OTA (MAX435), $f = 5$ MHz

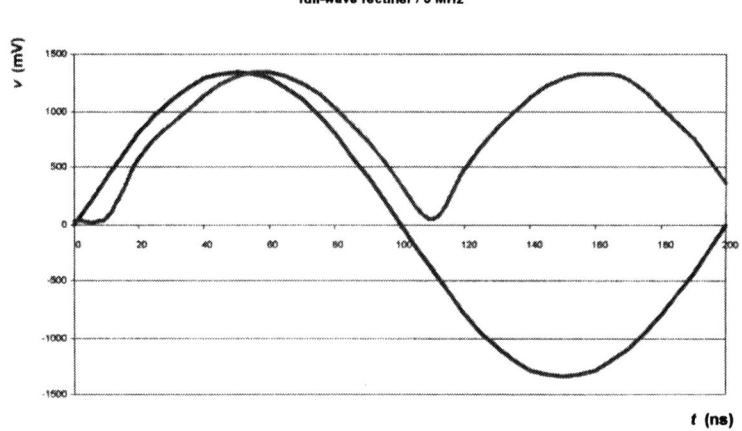

Fig. 13. Time behaviours measured for full-wave rectifier with OTA (MAX435) $f = 5$ MHz

5 Conclusion

The paper is focused on problems of non-linear elements in circuits with current conveyors. A non-linear three-pole is considered and the functional relations of selected basic connections are determined. The application of non-linear elements is verified on the connections of half-wave and full-wave rectifiers. Only the basic connections are given and the basic functional computer simulations are conducted. An experimental verification of the rectifiers was performed using an OTA element. The paper provides a theoretical foundation for the solution of non-linear circuits with current conveyors. The scope of the paper does not allow a more in-depth analysis of the rectifier solution. Practical measurements were only performed to verify the functionality of circuits. It will be necessary to focus on obtainable properties and to analyse also other circuit structures.

References

1. SEDRA, A., and SMITH, K.C. The current conveyor: A new circuit building block. Proc. IEEE, Vol.56, pp. 1368-1369, Aug. 1968.
2. SEDRA, A.S., and SMITH, K.C. A second generation current conveyor and its application. IEEE Trans., 1970, CT-17, pp. 132-134.
3. MAX435/MAX436 - Wideband Transconductance Amplifiers. Datasheet, Maxim, 1993. URL: http://pdfserv.maxim-ic.com/en/ds/MAX435-MAX436.pdf.
4. LM13600 - Dual Operational Transconductance Amplifiers with Linearizing Diodes and Buffers. Datasheet, National Semiconductor, 1998. URL:http://cache.national.com/ds/LM/LM13600.pdf.
5. GRIGORESCU, L. Precision rectifier. The annuals of Dunarea de Jos, 2003. University of Galati, pp. 55-57, URL http://thefibannals.home.ro/anale-fib-2003-16.pdf, ISSN 1224-5615.
6. BIOLEK, D., BIOLKOVÁ, V., KOLKA, Z. AC analysis of operational rectifiers via conventional circuit simulators. Brno, Fakulta elektrotechniky a komunikačních technologií VUT, 2004, URL http://user.unob.cz/biolek/veda/articles/Tenerife04.pdf.

Field Programmable Mixed-Signal Arrays (FPMA) Using Versatile Current/Voltage Conveyor Structures

Jiri Stehlik, Daniel Becvar

Brno University of Technology, Faculty of Electrical Engineering and Communication
Dept. of Microelectronics, Udolni 53, CZ-60200 Brno, Czech Republic
{stehlik, becvar}@feec.vutbr.cz

Abstract. Field Programmable Mixed-Signal Arrays (FPAA) designs are reviewed and standard building blocks described with respect to circuit parameters and design limitations. The second-generation current conveyor is introduced as an analog building block with properties similar to those of an operational amplifier, and which has the potential for high frequency operation. A current conveyor can be implemented in an area similar to that of a simple operational amplifier. Potentially greater bandwidths are achieved, while using CMOS technology instead of bipolar technology and test circuit design is proposed.

Keywords: Field Programmable Analog Array, Field Programmable Mixed-Signals Array, Programmable Universal Current Conveyor.

1 Introduction

FPAA designs are used in continuous-time or discrete-time modes. Discrete-time designs are programmed in terms of capacitor ratios but have limitations of lower bandwidths. Continuous-time circuits extend operating range to higher bandwidths, and offer no need of input/output anti-aliasing filters for applications in real-time signal processing, but they usually have higher distortion and require complex programming.

Published FPAAs designed at the building block level have used op-amps as their basic active element. For higher frequency bandwidths they could potentially be obtained by specifying higher power dissipation. However, in order to obtain significant gain from a simple op-amp-based amplifier, a designer must settle for significantly lower frequencies.

There are three potential options. The first would be to design an op-amp with greater frequency response. However, such an op-amp would occupy much greater die area and consume more supplying power. A second option would be to use a programmable compensation capacitor [1], where op-amp bandwidth can be extended if a high gain is desired. A third option is to use a block different than the operational amplifier, of which one possibility is the second-generation current conveyor [2].

2 Filed Programmable Analog Array Designs

A field-programmable analog array is an integrated circuit, which can be configured to implement analog functions using a set of configurable analog blocks (CAB) and a programmable interconnection network [3], and is programmed using on-chip memories.

An early conceptual FPAA design by Sivilotti [4] consists of CABs designed at the transistor level, and the interconnection network is based on a tree structure. Its target application was for the prototyping of analog neural networks. A fully-differential continuous-time CMOS design based on operational amplifiers and a modification to the Czarnul four MOSFET transconductors [5]. Its target application is for signal processing applications in the audio range, IC test results were presented for biquad filter, squaring, rectifier and VCO circuits. The CAB contains an op-amp and switchable feedback capacitors, and can also be used to implement a comparator by turning off the compensation capacitor.

Zetex Semiconductors Ltd. has introduced the Totally Reconfigurable Analog Circuit TRAC which includes 20 CABs, organized in two rows of 10 CABs, each capable of implementing one of the eight following functions: log, anti-log, non-inverting pass, addition, negating pass, op-amp, half-wave rectification, and off. Topological programming is implemented by turning CABs off, and by external wiring of the pins. Architecture is shown in Fig. 1.

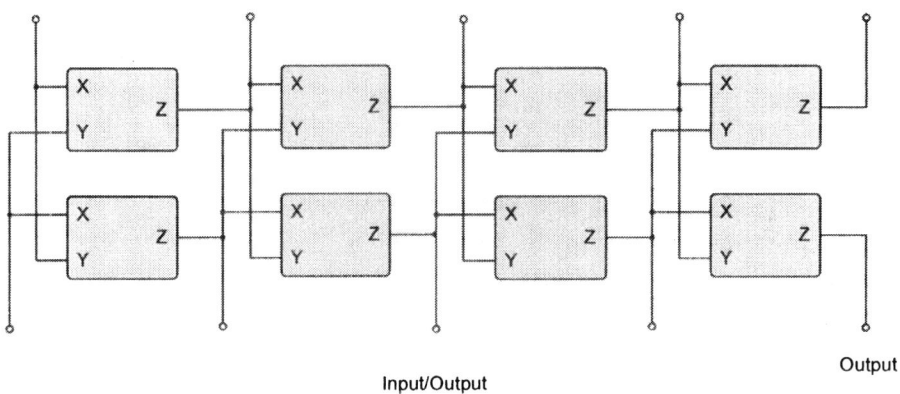

Fig. 1. TRAC Array Architecture

3 Configurable analog block

CAB (Configurable Analog Block) is used in an FPAA design as a principal block. Functions of a CAB include comparison, differentiation, amplification, integration, addition, subtraction, multiplication, log, anti-log. Current conveyor based implementation of basic CAB is shown in Fig. 2. Circuit is capable to realize amplification, first-order filtering functions, and with supplemented switchable diodes on the X and Z nodes log and anti-log functions could be realized, too.

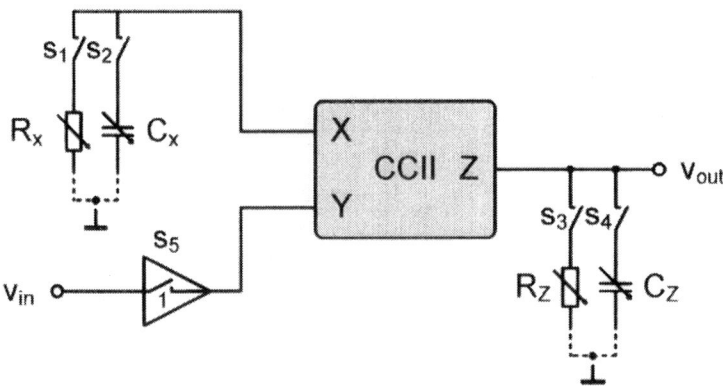

Fig. 2. Configurable Analog Block

The CAB consists of a second generation current conveyor, two transconductors, two programmable capacitors, and a buffer. Programmable resistors are realized with the transconductors. The function of the CAB is described in Tab. 1.

Table 1. Realization of CAB function with second generation current conveyor

Function	Configurable Analog Block				
	S1	S2	S3	S4	S5
Amplifier	1	0	1	0	1
Integrator	1	0	0	1	1
Lossy integrator	1	1	1	1	0

4 Programmable Universal Current Conveyor – PUCC

Programmable Universal Current Conveyor (PUUC) is a special type of the Universal Current Conveyor (UCC) [6] developed for FPAA. The idea is depicted in Fig. 3.

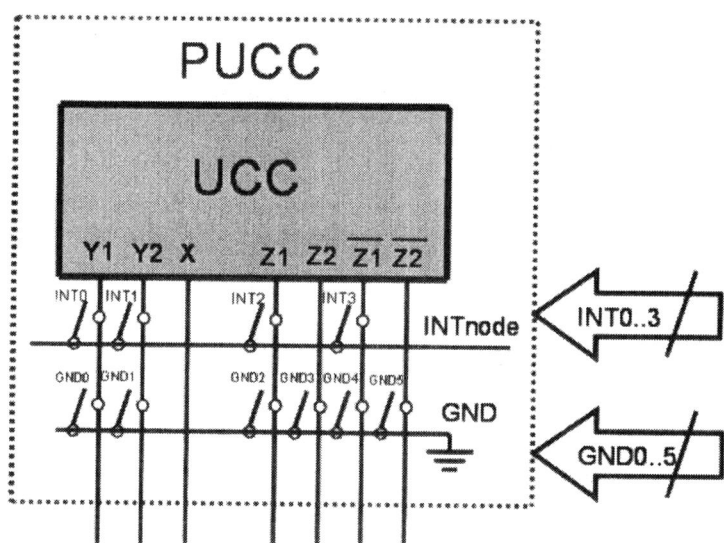

Fig. 3. Universal Current Conveyor

PUCC uses a standard UCC with 8 switches controlled by 2 independent buses (4b INT bus, 6b GND bus). It is possible to reconfigure UCC into 18 different simple current conveyors (with only one Y input) using these buses (Tab. 2). Moreover, the UCC structure offers current conveyors with fully differential voltage input. When using UCC (PUCC), the current conveyors with differential input could be use in very simple way.

Table 2. Reconfiguration of UCC

PUCC	INT BUS (4b)						GND BUS (6b)					
	Y1	Y2	Z1	Z2	notZ1	notZ2	Y1	Y2	Z1	Z2	notZ1	notZ2
CC type	INT0	INT1	INT2		INT3		GND0	GND1	GND2	GND3	GND4	GND5
CCI-	1	0	1		0		0	1	0	1	0	1
CCI+	1	0	1		0		0	1	0	0	1	1
CCI+/-	1	0	1		0		0	1	0	0	0	1
CCII-	0	0	0		0		0	1	1	1	0	1
CCII+	0	0	0		0		0	1	0	1	1	1
CCII+/-	0	0	0		0		0	1	0	1	0	1
CCIII-	1	0	0		1		0	1	1	1	0	0
CCIII+	1	0	0		1		0	1	0	1	0	1
CCIII+/-	1	0	0		1		0	1	0	1	0	0
ICCI-	0	1	1	0	0	0	1	0	0	1	0	1
ICCI+	0	1	1	0	0	0	1	0	0	0	1	1
ICCI+/-	0	1	1	0	0	0	1	0	0	0	0	1
ICCII-	0	0	0	0	0	0	1	0	1	1	0	1
ICCII+	0	0	0	0	0	0	1	0	0	1	1	1
ICCII+/-	0	0	0	0	0	0	1	0	0	1	0	1
ICCIII-	0	1	0	0	1	0	1	0	1	1	0	0
ICCIII+	0	1	0	0	1	0	1	0	0	1	0	1
ICCIII+/-	0	1	0	0	1	0	1	0	0	1	0	0

The proposed FPAA test chip structure is shown in Fig. 5. The FPAA structure (consists of 4 CABs) on the right side is similar to architecture of TRAC device [7]. The structure uses PUCC is on the left side. Test chip offers the possibility of architecture comparison. The PUCC structure promises huge versatility, but on the other hand the complexity of the PUCC block are probably the disadvantage if it is compared with simple CCII structure (more parasitics from switches, more control logic). Next chapter presents a current conveyor-based FPAA architecture.

5 Current Conveyor Implementation of the Zetex TRAC

Previously introduced Zetex TRAC is a bipolar design achieving 4 MHz bandwidth. Its noteworthy feature is the absence of switches in the signal paths, resulting from the hardwired interconnection network. This can boost performance by limiting the parasitics in the routing and can result in greater linearity.

A CMOS CCU-based TRAC design includes CABs which are arranged in a manner as depicted in Fig. 4.

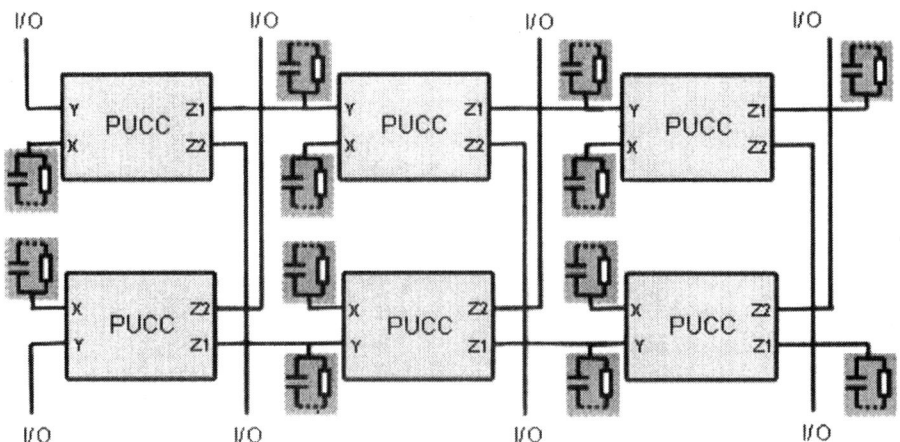

Fig. 4. Proposed FPPA TRAC based architecture using PUCC

Hardwired connections are used to route voltage signals between subsequent CABs. The leftmost pins act as inputs and the rightmost pins act as outputs. The intermediate pins can be configured both as inputs and outputs. To configure a pin as an input, the previous CAB whose output is connected to the pin, should be turned off. This is accomplished by turning off the CCII's bias currents, as shown in Fig. 4. This approach is similar to that proposed for the Praemont et al architecture [8], but could result in a significant waste of silicon area if too many CABs should be turned off.

Proposed FPAA test chip

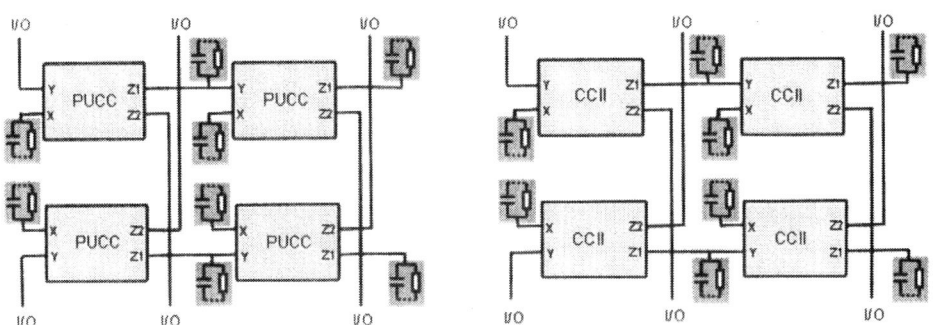

Fig. 5. Proposed test chip

6 Fully-Differential Implementation

All current conveyor blocks presented thus far have employed single-ended signaling. However, fully-differential signaling has advantages in terms of immunity to common-mode noise. A potential fully-differential current conveyor amplifier block is depicted in Fig. 6, and includes two CCII, and two resistors. Here the current I_x is equal to $(V_{in+}-V_{in-})/R_X$.

That current then flows across R_Z, producing an output $(V_{out+}-V_{out-})=(R_Z/R_X)(V_{in+}-V_{out-})$.

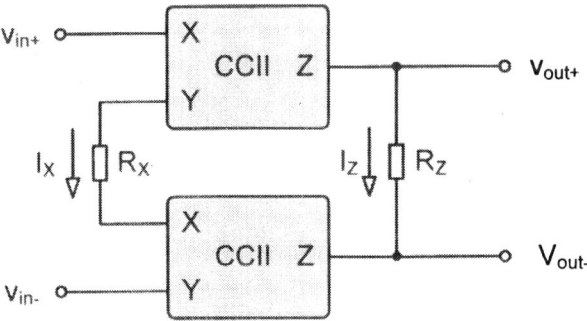

Fig. 6. Fully-Differential CCII Amplifier

7 Universal current conveyor

PUCC can be used as basic building block for design of Fully Differential Amplifier. Realized differential inputs are implemented in Universal Current Conveyor [6].

8 Filter design using FPAA

Suitable autonomous circuit was our starting point when designing the RC-active networks with current conveyors. For example, a simple circuit as shown in Fig. 7a could be chosen. It contains two general current conveyors, GCC, and five grounded passive one-port elements (resistors or capacitors) characterized by their admittances. The realization of this autonomous circuit using proposed FPAA is shown in Fig. 7b.

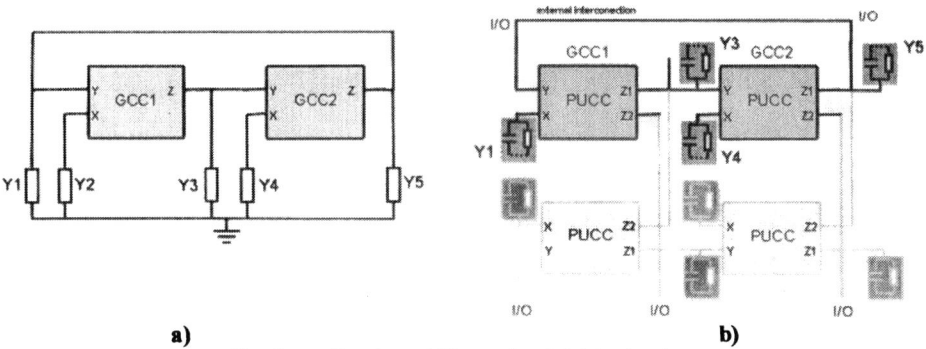

Fig. 7. Realization of filter using FPAA circuits

The universal filter design procedures using autonomous circuit (Fig. 7) were published in [9], [10], [11]. There was shown in those papers that this circuit works as multifunctional filter, moreover it could operate in both modes – voltage and current ones. Filter is determined by used types of current conveyor (instead of GCC1 and GCC2) and types of passive elements. Using proposed PUUC based FPAA for realizations of these filters brings huge versatility and design freedom for designers.

9 Summary

This paper has presented a design for a field-programmable analog array, with architecture similar to that of a commercially available FPAA. The advantage of the new method is that potentially greater bandwidths could be achieved, while using CMOS technology instead of bipolar one. Also, a fully-differential configurable analog block, which uses PUCC, has been presented.

10 Conclusion

A field-programmable analog array consists of configurable analog blocks, interconnections, as well as memories capable to configure the array into useful analog circuits. FPAA circuits in the literature were reviewed and discussed obstacles in the development of a high-frequency FPAA. Among these main role was played by the low bandwidth of the simple operational amplifier.

The second-generation current conveyor was introduced, an analog building block with properties similar to those of an operational amplifier. It has the potential for higher frequency operation. A current conveyor can be implemented in an area similar to that of a simple operational amplifier. Also, reasons were stated as to why a current conveyor is preferable over a simple operational amplifier.

Low Power Supply Voltage

With the increase in the use of mixed-signal techniques, there will be compulsion to design analog circuits with lower power supply voltages. This is already evident in many deep submicron designs which use a single 3.3 V power supply (or even lower one) [12], [13], [14]. A current conveyor providing rail to rail swings at low power supplies should be developed.

Performance Limitations

Performance limitations of FPAAs were described. Some performance limitations included the speed of op-amps, transconductors, and comparators. By using current conveyors rather than simple op-amps, the performance of FPAAs could be increased to higher bandwidths. However, current conveyors as they stand, as well as existing transconductors, will not allow an extension of CMOS FPAAs to RF bandwidths. For that purpose, new architectures will need to be studied.

FPAA performance parameters other than bandwidth should be examined and improved upon, as well. Such parameters should include versatility and linearity.

Acknowledgement. This research has been supported by Ministry of Education of the Czech Republic in the frame of Research Program MSM0021630503 *MIKROSYN*, by the contract GAAV 1QS201710508, and by the FRVS 2353/2007 project.

References

[1] Gray, P.R., and Meyer, R.G., "Analysis and Design of Analog Integrated Circuits," Second Edition, John Wiley and Sons, 1984.

[2] Sedra, A.S., and Smith, K.C. "A second generation current conveyor and its application", IEEE Trans., 1970, CT-17, pp. 132-134

[3] Lee, K.F.E., and Gulak, P.G., "Field Programmable Analogue Array Based on MOSFET Transconductors," Electronics Letters, vol. 28, no. 1, pages 28-29, January 2, 1992.

[4] Sivilotti, M.A., "A Dynamically Configurable Architecture for Prototyping Analog Circuits," Advanced Research in VLSI: Proceedings of the Decennial Caltech Conference on VLSI, 1988, MIT Press, 1988.

[5] Dupuie, S.T., and Ismail, M., "High Frequency CMOS Transconductors," Analogue IC Design: The Current-Mode Approach, Peter Peregrinus, London, pages 181-238, 1990.

[6] D. Becvar, K. Vrba, R. Vrba: "Universal current conveyor: a novel helpful active building block. Proc. ICT 2000 (International Conference on Telecommunications), Acapulco, México, 2000, pp. 216-219, ISBN 968-36-7762-2.
[7] Zetex Semiconductors Ltd., http://www.zetex.com/trac.
[8] Prémont, C., Grisel, R., Abouchi, N., and Chante, J.-P., "Current-Conveyor Based Field Programmable Analog Array," 1996 Midwest Symposium on Circuits and Systems, Ames, Iowa, August 1996.
[9] Vrba, K., Čajka, J.: Universal filters using universal current conveyors. Proc. Int. Conf. RTT 2000 (Research in Telecommunication Technology), 2000, pp. 168-171.
[10] Vrba, K., Čajka, J.: Universal filter design using the general four-port current conveyor. Proc. Int. Conf. TSP 2000, FEI VUT Brno, 2000, pp. 16-19.
[11] Vrba, K., Čajka, J., Dostál, T.: How to design universal filters with the aid of general current conveyors. Proc. Int. Conf. RADIOELEKTRONIKA 2001, 2001, UREL VUT Brno, pp. 34-37.
[12] Ferri, G., and Sansen, W., "A 1.3 V Op-Amp in Standard 0.7pm CMOS with Konstant gm and Rail-to-Rail Input and Output Stages," ISSCC'96, pages 382-383, February 1996.
[13] Baschirotto, A., and Castello, R., "A 1V 1.8MHz CMOS Switched-Opamp SC Filter with Rail-to-Rail Output Swing," ISSCC'97, pages 58-59, February 1997.
[14] Dedic, I.J., Amos, N.C., King, M.J., Schofield, W.G., and Kemp, A.K., "A 16b 100 kSample/s 2.7V 25mW ADC/DSP/DAC-Based Analog Signal Processor in 0.8pm CMOS," ISSCC'97, pages 96-97, February 1997.

Modeling and Design of a Novel Integrated Band-Pass Sigma-Delta Modulator

Lukas Fujcik[1], Jiri Haze[1], Radimir Vrba[1], Jiri Forejtek[1], Pavel Zavoral[1], Roman Prokop[1], Linus Michaeli[2]

[1] Dept. of Microelectronics, FEEC, Brno University of Technology
Brno, Czech Republic, fujcik@feec.vutbr.cz
[2] Dept. of Electronis and Multimedia Communications
FEI, Technical University of Kosice, Kosice, Slovak Republic

Abstract. The paper deals with a bandpass sigma-delta modulator (BP SDM), which is used for conversion of signal from capacitive pressure sensor. This approach is absolutely new and unique, because this kind of modulator is utilized only for wireless and video applications. The main advantage of BP SDM is due to its defined band. That is why it is resistant against offsets of its sub-circuits. Another important advantage is low power consumption, since the BP SDM digitizes only narrow band instead of whole Nyquist band with similar dynamic range. The paper shows basic ideas of this approach and simulation results. The main stages are implemented in switched-capacitor (SC) technique. The paper presents two possibilities how to design PLL block. The first one is conventional approach and the other one is with digital sigma-delta modulator as generator of input harmonic signal.

Keywords: Band-pass sigma-delta modulator, capacitive pressure sensor, phase locked loop

1 Introduction

Band pass sigma-delta modulators (BP SDMs [1], [2], [3]) are well suited for direct conversion of the digitally modulated signals (QAM, PSK) from the frequency to the digital output. Once the RF signal is digitized, most of the signal processing tasks like channel filtering, demodulation etc. can be easily done in the digital domain with high degree of programmability. Induced noise and high sampling frequencies requires corresponding electronic technology as it is used in implementation for GPS/GSM communication systems.
The binary flow from the BP SDM output is down converted by digital multiplier and in the LP digital filter transformed in the digital number. The whole structure represents band pass sigma-delta analog to digital converter (BP SD ADC). Coherency between input signal fin and clock frequency ($f_s = 4f_{in}$) in the BP SD ADC makes converter phase sensitive.

Authors introduce a new application of BP SDM as the processing circuit from the capacitive pressure sensor with the direct digital outputs representing real and imaginary components of the input vector. The simple BP SDM first generation was designed for integrated sensor system using CMOS 07 technology. The defined narrow band is advantage against offsets of modulator sub circuits such as OPA, delay stage and summator. The modulator is tuned at 62.5 kHz sampling frequency. It processes the signal with central frequency f_c = 15 625 Hz. The paper shows comparison of proposed ideal and real BP SDM.

2 The BP SDM topology

Fig. 1 shows block diagram of BP SDM [4]. It contains band-pass filter (BPF), an N-bit quantizer and a digital to analog converter (DAC), which is connected in a loop. The BPF can be synthesized by cascading two or more second-order biquadrate filters or resonators, which must have a sharp transfer function and well-defined resonance at f_n. These resonators may be implemented as a discrete-time filter using either SC or SI techniques or they may be implemented as a continuous time filter.

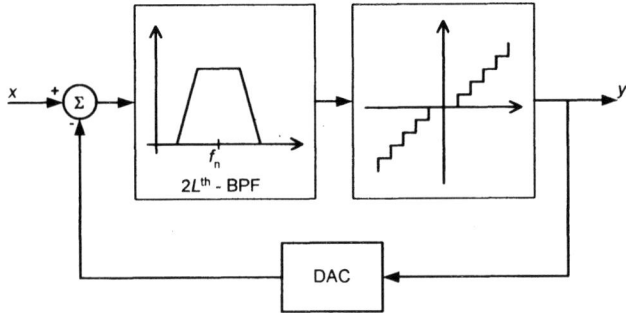

Fig. 1. Basic block diagram of a BP SDM

Let consider $2L^{th}$-order BPF composed of a cascade of L resonators with a DT transfer function given by

$$H_R(z) = \frac{1}{(1-z^{-1}z_n)(1-z^{-1}z_n^*)} \quad (1)$$

where z_n and z_n^* are the conjugate-complex poles of $H_R(z)$.
The output of modulator in Z-domain, assuming quantization error is

$$Y(z) = S_{TF}(z)X(z) + N_{TF}(z)E_q(z) \quad (2)$$

The signal transfer function (*STF*) and noise transfer function (*NTF*) are as follows

$$S_{TF}(z) = \frac{[N_R(z)]^L}{[N_R(z)+(1-z^{-1}z_n)(1-z^{-1}z_n^*)]^L} \quad (3)$$

$$N_{TF}(z) = \frac{[(1-z^{-1}z_n)(1-z^{-1}z_n^*)]^L}{[N_R(z)+(1-z^{-1}z_n)(1-z^{-1}z_n^*)]^L} \quad (4)$$

The output transfer characteristics for the first generation BP SDM utilized for processing circuit is

$$Y(z) = X(z).z^2 + E_q(z)(1+z^2) \quad (5)$$

Power density after noise shaping in the spectral domain is

$$\varepsilon_{NS}^2(f) = \varepsilon_q \frac{4}{f_S}\left(1+\cos\frac{4\pi f}{f_S}\right) \quad (6)$$

The Fig. 2 shows the capacitive pressure sensor where one branch is sensing branch and another is reference branch.

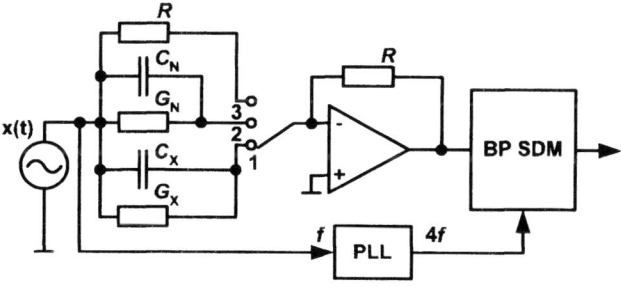

Fig. 2. Measurement chain with BP SDM

While the pressure influences the capacity ΔC_X the humidity impacts the conductivity ΔG_X of the sensing branch. Pre-processing circuit with BP SDM converts it after synchronized down-conversion into its real and imaginary part. The binary flows are converted into digital number in two digital low pass (D-LP) filters. The digital resolution is proportional to the time window of the D-LP filter. The BP SDM is controlled coherently with the input source frequency. The digital output values are determined by the quantized vectors in all three switch positions as follows

$$U_1 = RU[j\omega(C_N + \Delta C_X) + G_N + \Delta G_X] \quad (7)$$
$$U_2 = RU[j\omega C_N + G_N]$$
$$U_3 = RU$$

The phase shift of the processing block is suppressed by the subtracting operation. The difference of the digital values from the output of the LP filter in the position 1 and 2 normalized to the value measured in the position 3 is expressed by the formula. Here measured changes of capacity and resistance are obtained

$$\frac{U_1 - U_2}{U_3} = R(j\omega\Delta C_X + \Delta G_X) \quad (8)$$

The BP SDM is tuned on 62,5 kHz of sampling frequency as mentioned, it means that central frequency is

$$f_c = \frac{f_s}{4} = \frac{62500}{4} = 15625 \, Hz \quad (9)$$

Formula (6) shows the minimal value of the noise shaped power density around the central frequency f_c.

3 Design of auxiliary stages

This chapter describes the design procedure of generator of harmonic signal and phase locked loop (PLL).

3.1 Generator of input harmonic signal utilizing sigma-delta modulation

The digital design has been used for simplicity of chip realization. The sine wave signal is generated by means of digital sigma-delta modulation. The output from sigma-delta modulator is connected to analog low pass filter. The output sine wave signal is obtained after high frequency filtering. The input clock frequency of digital part is 4 MHz. This frequency is common for all digital stages. The digital part is proposed using VHDL language.
The whole system consists of four blocks as can be seen from Fig. 3.

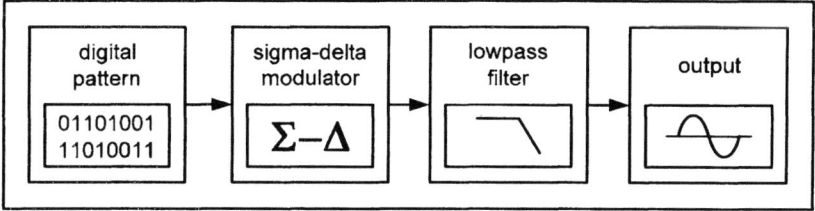

Fig. 3. Generator block diagram

Thanks to this condition, the design is very flexible and each part could be changed very easily. The first stage of this proposal is generator of digital image of output sine wave signal, which is described in VHDL language.

The core of the whole generator consists of sigma-delta modulator. The first order modulator is used for simplicity of design. The modulator (Fig. 4) includes summator and subtractor. The input and output signals are connected on it. The difference (sum) is connected to the register. The value in this register is increased at each clock pulse. The register output is compared in comparator zero reference value. In case when register value is higher than zero, the comparator reverse into logic high, in other cases is zero. The comparator output is modulated sine wave signal. It could be filtered and analog signal could be obtained. The output is directly connected to the analog low-pass filter.

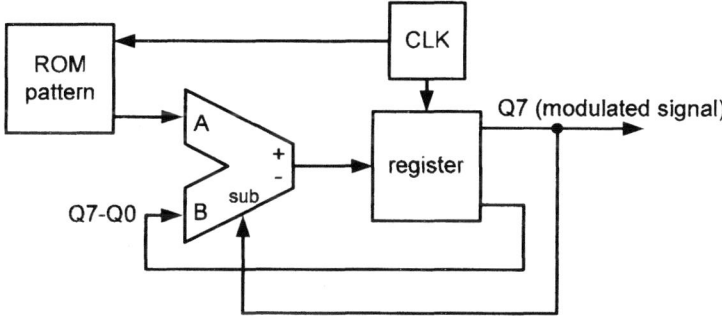

Fig. 4. The block diagram of sigma-delta modulator

The modulator realization is shown on Fig. 4. This whole part (from look-up memory ROM till output) is programmed in VHDL. The comparator has been replaced by MSB of register for simplicity, because the decision level of comparator has been set at binary number 64, which is zero reference value. Thanks to this modification, the design is sign-less.

The last stage is low-pass filter. It is only analog part of the whole proposal. This filter has been designed separately. It has been chosen third order active Butterworth RC filter with cut-off frequency of 15 625 Hz. The filter makes analog signal from modulated signal.

The digital stages have been designed in VHDL language in Xilinx ISE WebPack design environment. The simulation (Fig. 5) proceed in simulation program ModelSim Xilinx Edition. The main aim was to obtain as simple control logic as possible because of chip area. The design has been tested and finalized in FPGA Spartan-3.

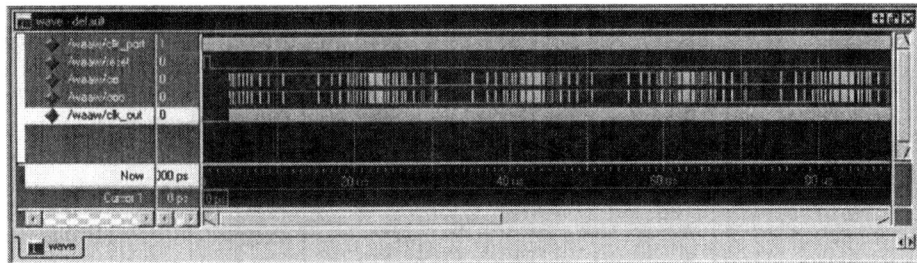

Fig. 5. The simulation of digital stage function

The analog part is low-pass filter. First the simple RC filter has been used and its function has been verified. Then the third order RC filter has been utilized with exact frequency of 15 625 Hz. The 4 MHz frequency of pulses at modulator output is obtained. Therefore this filter should be allowed to separate sine wave signal from digital noise at the output sufficiently. Filter consists of one fast OPA.

3.2 Design of PLL

The PLL stage is designed utilizing well-known structures, which is shown on Fig. 6. The input sinewave signal is converted by means of comparator onto pulses. Than phase detector, which is bipolar charge pump compares phase between these pulses and signal generated by VCO (Voltage Controlled Oscillator).

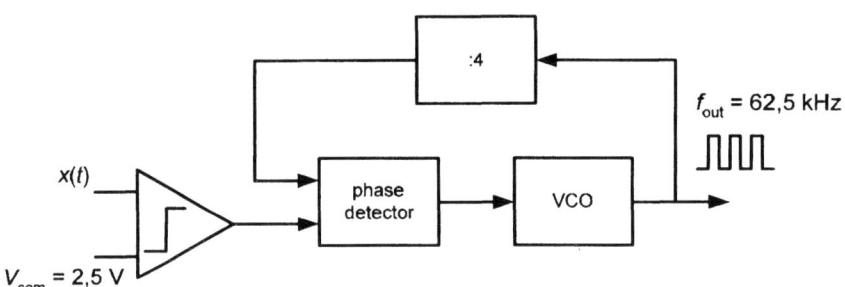

Fig. 6. Block diagram of designed PLL

The VCO is designed as starving ring oscillator with current controlled invertors as shown in Fig. 7.

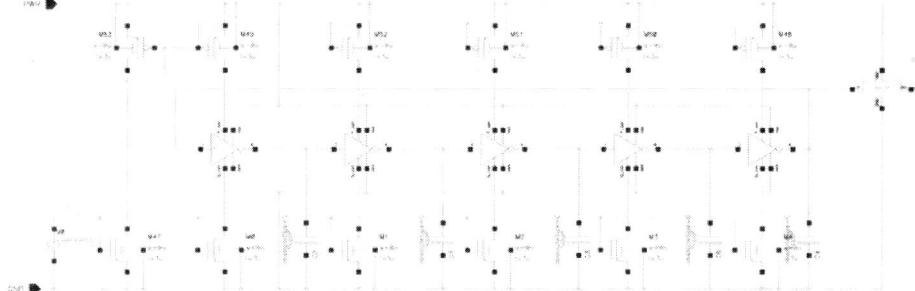

Fig. 7. Schematic diagram of VCO

The VCO is designed for 25 µA driving current at control voltage 1.3 V. The ring oscillator consists of 5 controlled invertors. Since the output frequency has to be very low, there are used capacitors to produce the time delay.

The phase detector (PD) detects the phase difference between the reference signal and the feedback signal from the VCO and frequency divider. Note that, although the PD of a PLL can be an analog multiplier, an exclusive-or (XOR) gate or a J-K flip-flop, etc, for a frequency synthesizer we always use the charge-pump PLL with a tri-state phase-frequency detector (PFD) that also detects frequency errors.

4 Modeling of PLL with generator of input harmonic signal

The PLL with generator of harmonic signal utilizing sigma-delta modulation was designed and simulated in MATLAB SIMULINK and ModelSim. Model of generator of input harmonic signal and additional synchronizing digital logic are described in VHDL language. Output signal of generator is filtered by low pass filter. This filtered signal is led to comparator. Comparator output is feedback signal to VHDL model of generator. This feedback signal is represented as synchronization signal. Proposed model of PLL with generator of harmonic signal utilizing sigma-delta modulation is shown in Fig. 8.

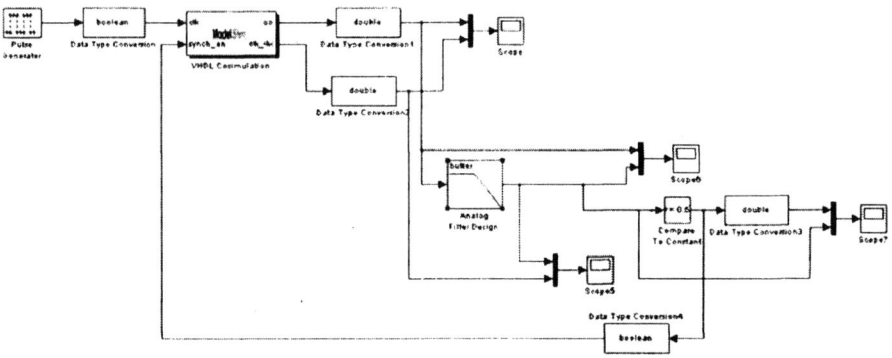

Fig. 8. Model of PLL with generator of input harmonic signal utilizing sigma-delta modulation

Matlab simulation results of PLL with generator of harmonic signal utilizing sigma-delta modulation is shown on Fig. 9. Modelsim simulation results of PLL with generator of harmonic signal utilizing sigma-delta modulation is shown on fig. 10.

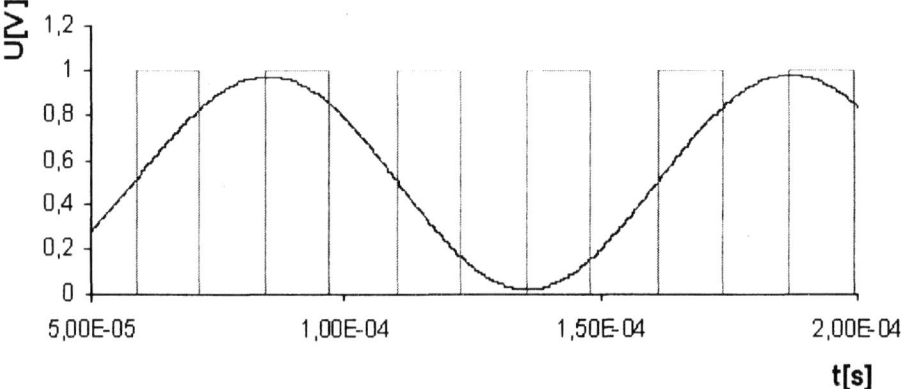

Fig. 9. Matlab simulation results of PLL with generator of harmonic signal utilizing sigma-delta modulation

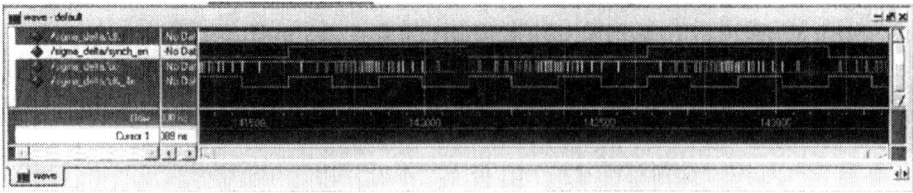

Fig. 10. ModelSim simulation results of PLL with generator of harmonic signal utilizing sigma-delta modulation

5 Design of band-pass modulator in Cadence

The switching of the inputs is synchronized with internal switching of modulator since it uses blocks with SC technique.

The ideal BP SDM consists of input S/H circuit, summator, four delay circuits, comparator and DAC connected in closed loop. There are several addition stages mainly D flip-flop and XOR logic gate. These stages are utilized to satisfy clock synchronization with driving clock. It is important to note that ideal BP SDM uses ideal OPAs (voltage controlled voltage source with very high gain), ideal capacitors and it has no offsets, noise sources etc.

The behavior of modulator has been simulated with input signal frequency 15625 Hz with sweeped amplitude from 0 to 1 V. The computed values are average values of logic 1 and logic 0 (represented as 5 V and 0 V respectively). It means that number on y axis is ratio between logic levels appropriate to input amplitude. The most important output is XOR output.

It can be seen that this output is quite linear and appropriate to input amplitude. The beginning of the curves is affected by start-up phase of modulator and should not be considered. The real BP SDM circuitry is depicted on Fig. 11.

Fig. 11. Block diagram of real BP SDM

The delay and S/H circuits are designed in SC technique. Comparator is proposed with latch. The DAC is simple 1-bit circuit. This modulator has been designed and simulated in Cadence software using CMOS 07 technology. The power supply is 5 V. The systematic offset of the basic loop is 140 µV.

Since this is the first realization of that kind of chip for capacitive pressure sensors, there are many auxiliary pins used for testing measurement. The aim of this arrangement is to obtain maximum information about behaviour of each stage of BP SDM. The most critical parts are blocks using switched-capacitor approach, because there are many nonidealities and error sources, mainly clock feedthrough and noise. Table 1. lists all pins, which will be led out of the chip for testing.

Tab. 1. Expected list of pins for 24-DIL package

Pin no.	Name	Description	Function
1		not used	
2	VSS	input	supply of digital part
3	VSSA	input	supply of analog part
4	VDDA	input	supply of analog part
5	OUT_SH	output	output of S/H circuit
6	IN	input	analog input
7	REZ_IN	output	output of summator
8	DAC_OUT	output	output of DAC
9		not used	
10	REZ_OUT	output	output of delay circuits
11	AGND	input	analog ground
12	NREF	input	negative reference voltage of DAC
13		not used	
14	PREF	input	positive reference voltage of DAC
15	PH2_n	output	output of nonoverlapping clock PH2_n
16	PH2	output	output of nonoverlapping clock PH2
17	PH1_n	output	output of nonoverlapping clock PH1_n
18	PH1	output	output of nonoverlapping clock PH1
19	OUT_D	output	output of D flip-flop
20	OUT_XOR	output	output of XOR (most important)
21		not used	
22	CLK	input	input of X-tal clock
23	KOMP_OUT	output	output of comparator
24	VDD	input	supply of digital part

The average outputs of real BP SDM depending on magnitude of analog input signal (range from 0 to 1 V) are shown on Fig. 12. It can be seen, that output of XOR is mostly linear, which is expected result.

Consequently the output plots of real BP SDM correspond with ideal one, so the proposed 1st generation BP SDM for pressure sensing works correctly.

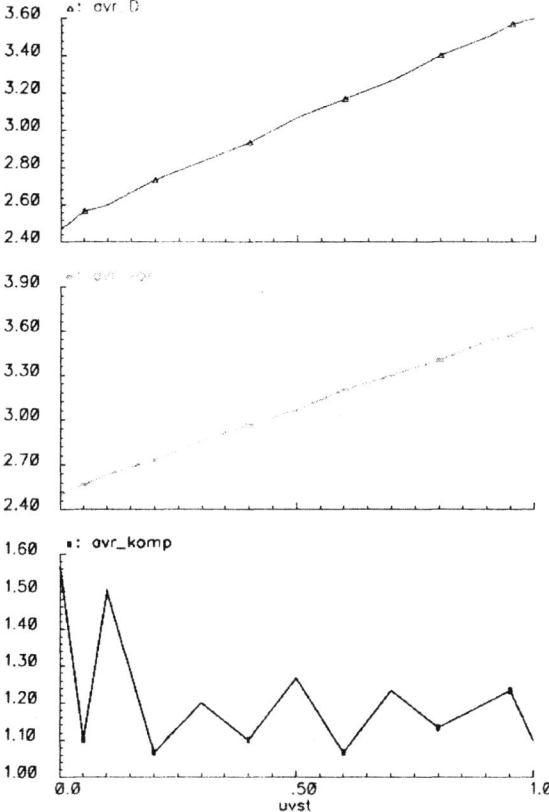

Fig. 12. Output plots of real BP SDM

6 Conclusions

The BP SDMs are well suited for wireless applications. This paper shows another way how to use its advantages. Authors designed 1st generation BP SDM as preprocessing block for capacitive pressure sensing. The behavior of real modulator has been verified and compared. Two possibilities of PLL design are presented in this paper. Both solutions will be designed on ASIC. The first one is conventional approach which is shown on fig. 6. The second one uses digital sigma-delta modulator as generator of input harmonic signal. This part with additional synchronizing digital logic is described by VHDL language. This digital part will be implemented on chip by using Synthesis and Place&Route tools. Matlab modeling of PLL with digital sigma-delta modulator is presented in this paper. Moreover, the test results will serve to the chip redesign targeted on the improvement of the conversion accuracy and the reduction of the power consumption.

Acknowledgments

The research has been supported by the Czech Ministry of Education in the frame of Research Program MSM0021630503 *MICROSYN*, by the Czech Grant Agency as the project GA102/03/0619 *Smart Microsensors and Microsystems for Measurement and Regulation* and project GA102/05/0869.

References

1. Ong, A., K., Wooley, B., A. A Two-Path Bandpass SD Modulátor for Digital IF Extraction at 20 MHz, IEEE Journal of Solid-State Circuits, vol. 32, No. 12, December, 1997
2. Lao, CH., Leong, H., Au, K., Mok, K., U, S., Martins, R., P. A 10.7-MHz Bandpass Sigma-Delta Modulator Using Double-Delay Single-opamp SC Resonator with Double-Sampling,
3. Tabatabaei, A., Wooley, B., A. A Two-Path Bandpass Sigma–Delta Modulátor with Extended Noise Shaping, IEEE Journal of Solid-State Circuits, vol. 35, No. 12, December, 2000
4. Rodrígez-Vázquez, A., Madeiro, F., Janssens, E. CMOS Telecom Data Converters, Springer Verlag, 2004, 375 pages, ISBN 1402075464
5. Shu, K., Sánchez-Sinenco, E., CMOS PLL Synthesizers, Analysis and Design, Springer Science, 2005, 232 pages, ISBN 0-387-23669-4

Composite Materials for Electromagnetic Interference Shielding

Pavel Steffan, Jiri Stehlik, Radimir Vrba

Brno University of Technology, Dept. of Microelectronics
Udolni 53, CZ-60200 Brno, Czech Republic
{steffan, stehlik, vrba}@feec.vutbr.cz

Abstract. This paper demonstrates that the addition of chemical agents and carbon fibers to cement can greatly enhance the shielding effectiveness of the concrete. In addition to improving the shielding effectiveness, carbon fibers and chemical agents enhance the tensile and flexural strengths significantly. As both carbon fibers and steel fibers are electrically conductive, both can be added to cement to enhance the shielding effectiveness, but steel fibers tend to rust whereas carbon fibers are chemically stable and inert.

1 Introduction

Electromagnetic interference (EMI) shielding refers to the reflection and/or absorption of electromagnetic radiation by a material, which thereby acts as a shield against the penetration of the radiation through the shield. As electromagnetic radiation, particularly that at high frequencies (e.g. radio waves, such as those emanating from cellular phones) tend to interfere with electronics (e.g. computers), EMI shielding of both electronics and radiation source is needed and is increasingly required around the world. The importance of EMI shielding relates to the high demand of today's society on the reliability of electronics and the rapid growth of radio frequency radiation sources [1].

EMI shielding is to be distinguished from magnetic shielding, which refers to the shielding of magnetic fields at low frequencies (e.g. 50 Hz). Materials for EMI shielding are different from those for magnetic shielding. EMI shielding is a rapidly growing application of carbon materials, especially discontinuous carbon fibers. This review addresses carbon materials for EMI shielding, including non-structural and structural composites, colloidal graphite, as well as EMI gasket materials.

2 Mechanisms of shielding

The primary mechanism of EMI shielding is usually reflection. For reflection of the radiation by the shield, the shield must have mobile charge carriers (electrons or holes) which interact with the electromagnetic fields in the radiation. As a result, the shield tends to be electrically conducting, although a high conductivity is not required. For example, a volume resistivity of the order of 1 [Ωcm] is typically sufficient. However, electrical conductivity is not the scientific criterion for shielding, as conduction requires connectivity in the conduction path (percolation in case of a composite material containing a conductive filler), whereas shielding does not. Although shielding does not require connectivity, it is enhanced by connectivity. Metals are by far the most common materials for EMI shielding. They function mainly by reflection due to the free electrons in them. Metal sheets are bulky, so metal coatings made by electroplating, electroless plating or vacuum deposition are commonly used for shielding. The coating may be on bulk materials, fibers or particles. Coatings tend to suffer from their poor wear or scratch resistance [1].

A secondary mechanism of EMI shielding is usually absorption. For significant absorption of the radiation by the shield, the shield should have electric and/or magnetic dipoles which interact with the electromagnetic fields in the radiation. The electric dipoles may be provided by $BaTiO_3$ or other materials having a high value of dielectric constant. The magnetic dipoles may be provided by Fe_3O_4 or other materials having a high value of the magnetic permeability, which may be enhanced by reducing the number of magnetic domain walls through the use of a multilayer of magnetic films. The absorption loss is a function of the product $\sigma_r \mu_r$, whereas the reflection loss is a function of the ratio σ_r/μ_r, where σ_r is the electrical conductivity relative to copper and μ_r is the relative magnetic permeability. Silver, copper, gold and aluminum are excellent for reflection, due their high conductivity. Superpermalloy and mumetal are excellent for absorption, due to their high magnetic permeability. The reflection loss decreases with increasing frequency, whereas the absorption loss increases with increasing frequency [1].

3 Composite materials for shielding

Due to the skin effect, a composite material having conductive filler with a small unit size of the filler is more effective than one having conductive filler with a large unit size of the filler. For effective use of the entire cross-section of a filler unit for shielding, the unit size of the filler should be comparable to or less than the skin depth. Therefore, a filler of unit size 1 μm or less is typically preferred, though such a small unit size is not commonly available for most fillers and the dispersion of the filler is more difficult when the filler unit size decreases.

Electrically conducting polymers [2] are becoming increasingly available, but they are not common and tend to be poor in the process ability and mechanical properties. Nevertheless, electrically conducting polymers do not require conductive filler in order to provide shielding, so that they may be used with or without filler. In the presence of conductive filler, an electrically conducting polymer matrix has the added advantage of being able to electrically connect the filler units that do not touch one another, thereby enhancing the connectivity. Cement is slightly conducting, so the use of a cement matrix also allows the conductive filler units in the composite to be electrically connected, even when the filler units do not touch one another. Thus, cement–matrix composites have higher shielding effectiveness than corresponding polymer–matrix composites in which the polymer matrix is insulating. A shielding effectiveness of 40 dB at 1 GHz has been attained in a cement–matrix composite containing just 1.5 vol. % discontinuous 0.1 μm diameter carbon filaments. Moreover, cement is less expensive than polymers and cement–matrix composites are useful for the shielding of rooms in a building [3]. Similarly, carbon is a superior matrix than polymers for shielding due to its conductivity, but carbon matrix composites are expensive [1].

4 Results

Tab. 1 gives the shielding effectiveness at 1.0, 1.5, and 2.0 GHz for nine types of cement mortars (for example, electromagnetic attenuation at 1.5 GHz frequency increased from 0.5 dB for plain cement to 10.2 dB for the same thickness of disc (3.6 mm) with chemical agents and short carbon fibers in the amount of 0.5 % by weight of the cement). Comparison of Rows 1 and 2 of Tab. 2 shows that the use of chemical agents (even without carbon fibers) enhances the shielding effectiveness substantially. This is consistent with the fact that the presence of these chemical agents reduces the electrical resistivity of the cement. However, an even larger enhancement can be obtained by the further addition of carbon fibers, as shown by the comparison of Rows 1, 2 and 3. The use of chemical agents and 0.5 % fibers gives a shielding effectiveness comparable to that obtained by the use of no chemical agents and 1 % fibers, as shown by comparing Rows 3 and 4. Furthermore, comparison of Rows 4, 6, 8 and 9 and of Rows 3, 5 and 7 shows that the shielding effectiveness increases monotonically with increasing fiber content. The trends are similar for all three frequencies [4].

Tab. 1.: Shielding effectiveness of cement mortars [3]

No.	Material	Attenuation [dB]			Thickness
		1.0 GHz	1.5 GHz	2.0 GHz	[mm]
1.	Plain cement	0.4	0.5	1.5	3.6
2.	Cement + chemical agents	3.7	3.7	7.3	4.0
3.	Cement + chemical agents +0.5 % fibres	9.4	10.2	11.7	3.6
4.	Cement + 1 % fibres	10.2	9.8	15.8	3.8
5.	Cement + chemical agents + 1 % fibres	14.8	12.3	18.5	3.8
6.	Cement + 2 % fibres	1.5	15.2	21.8	3.9
7.	Cement + chemical agents + 2 % fibres	15.6	13.7	19.6	3.9
8.	Cement + 3 % fibres	19.2	16.8	23.8	4.1
9.	Cement + 4 % fibres	21.1	18.6	25.1	3.9

5 Conclusions

Short carbon fibers (as low as 0.5% by weight of cement or 0.21% by volume of cement mortar) and chemical agents (triethanolamine, sodium sulphate and potassium aluminium sulphate) are effective in increasing the electromagnetic interference shielding effectiveness of cement mortar to about 10 dB or more in the frequency range 1.0 to 2.0 GHz for a mortar thickness of 4 mm. This degree of shielding effectiveness is sufficient for the construction of electromagnetic interference shielded structures. A small carbon fibre content is desirable for material cost saving and ease of dispersing the fibers.

Acknowledgment

Funding for this work was provided by the Ministry of Industry and Commerce under the contract FT-TA3/027 and Czech Ministry of Education in the frame of Research Program MSM0021630503 MIKROSYN New Trends in Microelectronic Systems and Nanotechnologies.

References

1. Chung D.D.L.: Electromagnetic interference shielding effectiveness of carbon materials. Elesevier Science Ltd.: Pergamon (2000), Carbon 279-285
2. Xiangcheng Luo, Chung D.D.L. Electromagnetic interference shielding using continuous carbon-fiber carbon-matrix and polymer-matrix composites, Elesevier: Composites (1999)
3. Jeng-Maw Chiou, Qijun Zheng and D.D.L. Chung, "Electromagnetic Interference Shielding by Carbon Fiber Reinforced Cement", Composites 20(4), (1989) 379-381
4. Nikkanen, P. On the Electrical Propertis of Concrete and Their Applications. Valtion Teknillinen Tiedotus, Sarja III, Rakennus 60:1962, 77pp, In Finnish in English Summary
5. Vrba, R., Stehlík, J., Šteffan, P. Multifunkční kompozity mimořádných vlastností na bázi anorganických nanosložek. Výzkumná zpráva projektu MPO ČR FT-TA3/027: Multifunkční kompozity mimořádných vlastností na bázi anorganických nanosložek (2006)

Single Chip Potentiostat Measurement System

Pavel Steffan, Radimir Vrba

Brno University of Technology, Dept. of Microelectronics
Udolni 53, CZ-60200 Brno, Czech Republic
{steffan, vrba}@feec.vutbr.cz

Abstract. The paper describes new approach of potentiostat for very low sensor current measurements. Recent potentiostats were used as measurement devices which were standalone and sensors were connected by a screened wire. These concepts are inconvenient for the presented very low current measurements. The new approach is based on the potentiostat circuitry integrated on a chip that should be directly contacted to a Thick Film Technology (TFT) sensor.

1 Introduction

The new approach of the amperometric sensor measurement is given. It solves the problem of measurement of low currents in order of nA for electrochemical biosensors. However the sensitivity of integrated measurement system, which processes biosensor output currents, has to be in order of pA. There are leakage currents and noise constraints which could be avoided by on-chip solution. The on-chip solution allows direct chip connection on the basic ceramic plate of biosensors. It means the idea is similar to hybrid integrated circuit. Main advantage of this approach is keeping sensitive low potential and low current ASIC electronic circuitry placed just on the TFT sensor apart of control and supply electronics of portable device. Additional advantages of the sensors fabricated using TFT are low cost, small dimensions, good reproducibility, chemical, mechanical and electrical properties of electrodes and well accessible and ecological fabrication process.

The electrochemical applications are widespread. The preparation of microsensor brings issues of large instrumentation connected with long leads. This set is sensitive to noise induced into measuring path. It is very important to minimize the measuring path to decrease noise because of very small current and voltage measured on microsensor. The potentiostat design and construction is discussed from the era of electron tubes discovery. Modern techniques and problems of design with modern circuitry as operating amplifiers are well described e.g. in [1], [2]. Then the requirements to built-up portable instrumentation brought a need to minimize power consumption [3]. Modern design utilizes CMOS technology for microchip fabrication [4]. The very simple potentiostat was integrated in this work where circuitry was presented with basic three operating amplifiers [5], or additionally supplemented with digital part for signal pre-processing [4], working at narrow range of measured currents.

To reach the highest possible sensitivity of the potentiostat, there was necessary to create a new approach of measurements. To minimize noise and other parasitic parameters, the integrated potentiostat placed directly on the body of sensor (see Fig. 4) was developed, designed and manufactured.

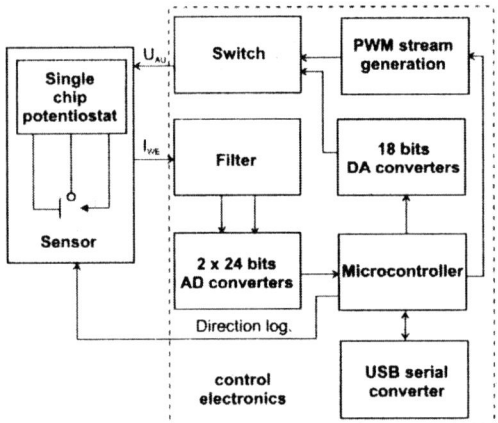

Fig. 1. New approach of potentiostat

The output sensor signals reflect current and potential of the working electrode. The potential is controlled by a programmable triangle voltage source. The required period of this signal supplying triangle is switchable in the range of 2 s to 60 min. A high linearity of the output signal is a condition for a precise measurement. The previous system solved this problem either by using fully analog integrator [6] (it had problems with large linear capacitor, period switching, small currents) or by counters and DA converters with filtered output (problems with linearity, not suitable for on chip systems). Our novel solution of the triangle voltage source is based on the generation of a digital PWM (Pulse Wide Modulation) stream, which means that the voltage is equal to the requested output voltage and following low pass filtration. The mean value of the stream must be accurately defined in time with the triangle shape.

2 Sensor

The sensor is located on a corundum ceramic base. On this surface the working, the reference and the auxiliary electrodes are applied. The working and the auxiliary electrodes are made of gold and the reference one is made of silver in the standard product. At the end of the sensor there is a contacting field which is connected with the active part by the silver conducting paths covered by a dielectric protection layer.

The solution of the issues during measurements is to miniaturize the circuitry of the instrumentation. Taking measurements by potentiometric method can be done only with potentiostat which is the most important circuitry in electrochemical analysis. The potentiostat should work at wide dynamic range of 8 orders; it means the current ranges 10 pA to 1 mA. The total current range has to be dividend into partial ranges because very small output voltage of converter current/voltage is reached for the smallest current. The current subranges changing keep good S/N ration within total range. The subranges are adjusted by controlled resistances in feed-back loop. The measurement of current has to automatically change the subrange according to the value of the output voltage. The subrange control should not make high glitches to measured signal. Glitches can cause significant error in measurement because of influencing electrode reaction.

The micropotentiostat was designed with circuits shown in simplified Fig. 6. All parts were integrated together with digital part for subrange control and correction constant stored in PROM with 48 bits (see topology in Fig. 2). The memory is downloaded by top system which is providing and controlling the measuring method to know the system parameters which are formed by micropotentiostat and electrode system and to do an automatic correction of measured signal. The microchip of the micropotentiostat is presented in Fig. 6. The microchip is mounted on thick-film sensor and connected by aluminum wire bonding (see Fig. 3 and 4).

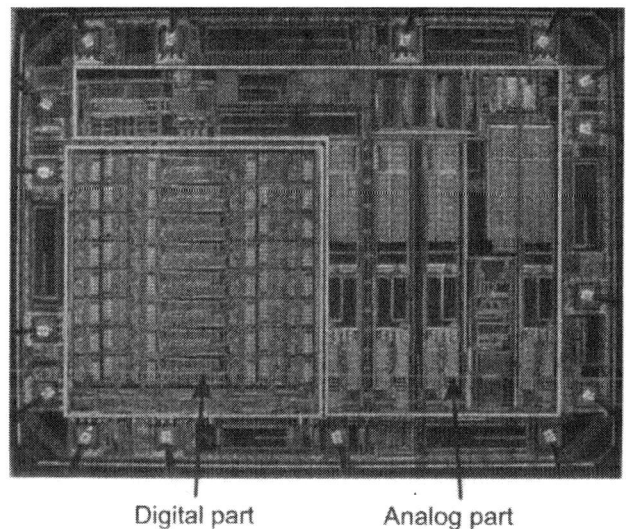

Fig. 2. Topology of designed microchip

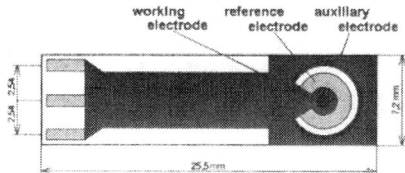

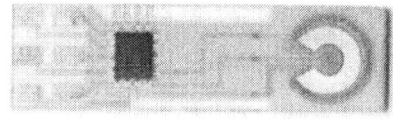

Fig. 3. Standard sensor   **Fig. 4.** Modified sensor

3 Chip Layout

Chip was designed in technology AMIS CMOS 0,7 μm. The integrated measurement system has two sections. Analog section is designed to control the potential on RE and to measure the current which flows through the sensor. This part is very important to settle final accuracy of measurement because this system measures very low currents in subranges from 1 nA to 1 mA. The measured signal is amplified and converted into current (Fig. 5, 6). This signal features high absolute resistance against electric and magnetic interference. The simplified circuitry of the analog part is presented in Fig 6. The chip is powered by 5 V DC. The internal analog ground potential is at + 2,5 V level related to external analog ground, i.e. the chip internal supply span is ± 2,5 V to internal analog ground.

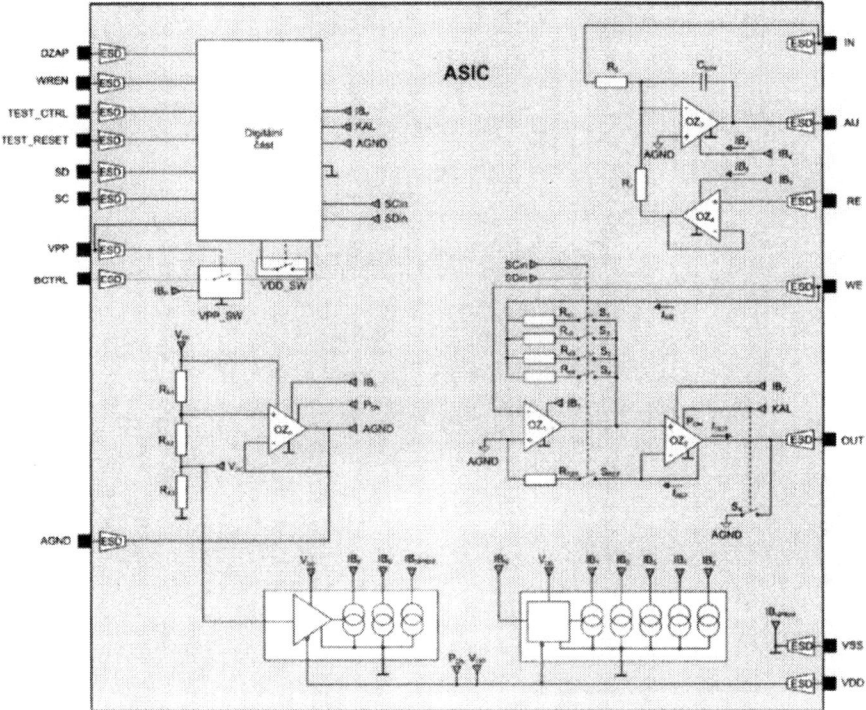

Fig. 5. The chip block diagram

Parameters of the type and series of the sensor are stored in the digital part of the chip. This part includes PROM memory, control logic, multiplexers and shift registers.

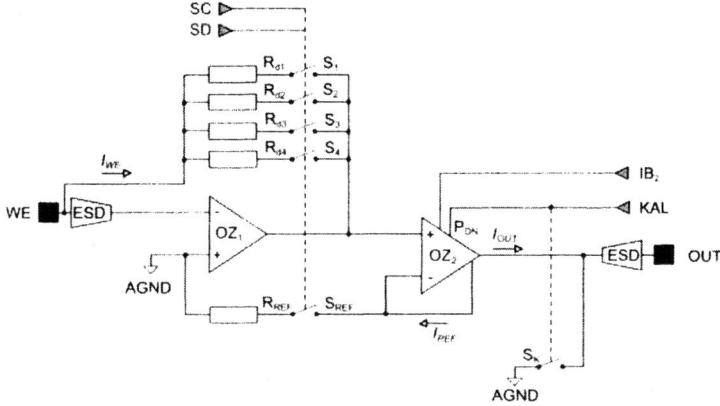

Fig. 6. Simplified circuit of analog part

4 Results

After the measurement system was properly checked (see Fig. 10) and tested, it was used for sensor measuring by the cyclic voltampere method. The intelligent measuring sensor was plunged into the 10 ml of 10 mM ferro-ferrikyanidu liquid. The results of the measuring are drawn in Fig. 11. These results were compared with measurement results of professional analyzer Sycopel AEW 2-10. Correct function of the proposed ASIC was proved.

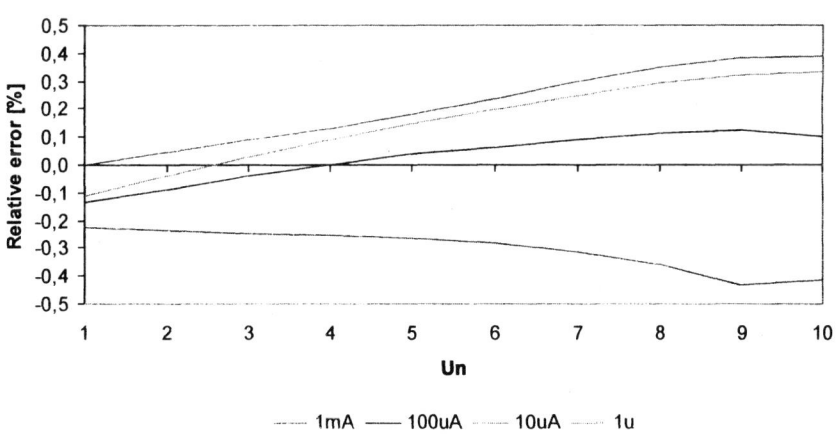

Fig. 7. Relative linearity error

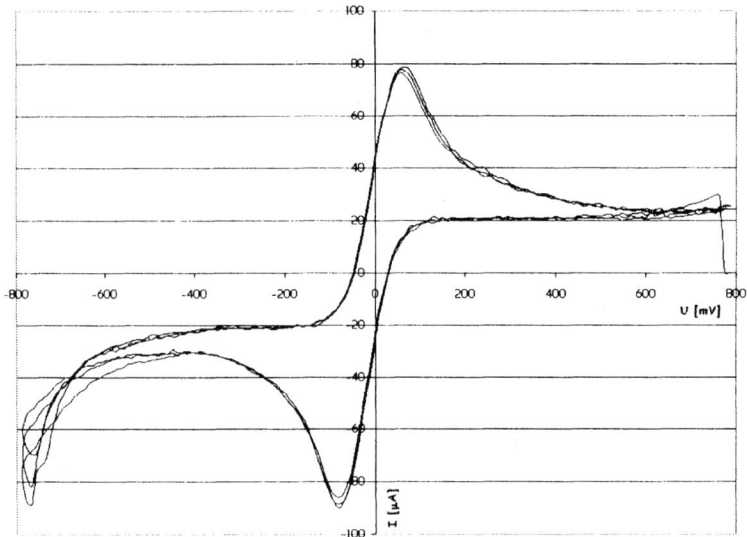

Fig. 8. The resulting plot of measured data

5 Conclusions

The new approach of the amperometric sensor measurement is given. It solves the problem of measurement of low currents (in order of nA) from electrochemical biosensors. However the sensitivity of integrated measurement system, which process output currents from biosensor, has to be in order of pA. There are leakage currents and noise constraints which could be avoided by on-chip solution. The on-chip approach allows direct chip connection on the basic ceramic plate of biosensors. It means the idea is similar to hybrid integrated circuit. Main advantage of this solution is separating sensitive low potential and low current ASIC electronic circuitry placed just on the TFT sensor apart of control and supply electronics of portable device. Additional advantages of the sensors fabricated using TFT are low cost, small dimensions, good reproducibility, chemical, mechanical and electrical properties of electrodes and well accessible and ecological fabrication process.

This paper focuses to deal with the design of a mixed signal ASIC single chip measurement system for embedded digital sensing of TFT biosensors. This ASIC monolithic measurement system is embedded on thick film amperometric sensors. The sensor applies fused alumina substrate processed by thick film technology. The ASIC prototype prepared in CMOS 0.7 µm technology contains mixed signal analog measurement block and programmable digital calibration PROM memory located on the same chip. The digital part of the chip of the smart sensor contains 48 OTP (One Time Programmable) cells that are supposed to store calibration data for the precision measurement. The full system is controlled by embedded software. PC and programmer communicate with the biosensor embedded ASIC chip via a USB interface.

Acknowledgment. Funding for this work was provided by the Ministry of Industry and Commerce under the contract FT-TA/050 and Czech Ministry of Education in the frame of Research Program MSM0021630503 MIKROSYN New Trends in Microelectronic Systems and Nanotechnologies.

References

1. Potentiostats. Bank Elektronic – Inteligent controls GmbH, Clausthal – Zellerfeld
2. C. E. D. Chidsey, Lab B. Amperometry: Electrochemistry with a 3-Electrode Potentiostat. Chem 174: Instrumental and Physical Principles of Chemical Measurements, (2002), pp. 1-17.
3. Richard J. Reay, Anthony F. Flannery, Christopher W. Storment, Matthew D. Steinberg, Christopher R. Lowe, A micropower amperometric potentiostat. Sensors and Actuators B 97 (2004), pp.284–289.
7. Michael Bollerott, Low Power Single-Chip CMOS Potentiostat. Fraunhofer Institut Mikroelektronische Schaltungen und Systeme, Duisburg, http://www.ims.fhg.de, (1999)
5. Samuel P. Kounaves, Gregory T.A. Kovacs, Microfabricated electrochemical analysis system for heavy metal detection. Sensors and Actuators B 34 (1996), pp.450- 455.
6. Prokop, R. - Vrba, R. – Skočdopole, M. – Fujcik, L.: A Slow Triangle Generator for Smart Sensors, EDS (2003), ISBN 80-214-2452-4

Portable and Precise Measurements with Interdigital Electrodes at Wide Range of Conductivity

Jaromir Hubalek, Radek Kuchta

Brno University of Technology, Dept. of Microelectronics
Udolni 53, 60200 Brno, Czech Republic
{hubalek, kuchtar}@feec.vutbr.cz

Abstract. A planar structure of the electrodes employed in conductivity measurements is utilizable at small range of conductivity measurement because of small sizes, planar ordering and technology which significantly influence the impedance behavior. The paper presents new approach in the field of precise electrolytic conductivity measurements with planar sensors. This novel measuring method was suggested for measurement using thick-film screen-printed interdigital electrodes (IDEs). Correction characteristics over a wide range of specific conductivities were determined from an interface impedance characterization of the IDEs. A local minimum of the imaginary part of the interface impedance is used for corrections to get linear responses. The method is seeking the optimal frequency for determination of specific conductivity of the solution by IDEs in order to achieve a highly accurate response. The method takes precise conductivity measurements in concentration ranges from 10^{-6} M to 1 M without cell replacement.

Keywords: interdigital electrodes, accuracy, conductometry, thick-film sensors.

1 Introduction

The impedance spectroscopy is well known method to study interfaces solution-electrode. It is very important in corrosion science [1]. In the area of sensors it helps us understand as the ions reacts with the electrode, to determine their concentration. The ion membrane of biosensor can be characterized with this method as well [2]. Generally the impedance characterisation is tool to find the working conditions for conductometry. This knowledge increase significance when the electrode system is much distanced from ideal theory conditions where large macroelectrodes and high frequency for measurement are utilized because of the interface capacitance close to zero and only solution resistance (conductivity) measuring [1]. In a case of miniaturise electrodes as IDEs, additionally that are created in a planar form, the impedance spectroscopy at wide range of conditions has to be applied. The sizes, spherical currents, porosity or/and roughness and the edge effect can absolutely change impedance behaviour and make the accurate conductivity determination very difficult. A parasitic capacitance of the planar electrode structure with the substrate precludes high frequency measurements. A work of Sansen's team was focused on the conductivity error determination using comb-like structure [3]. Their results were verified over a concentration range from 0.1 M to 0.2 M NaCl solution. They defined that the cell constant depends on the correction factor and a ratio of the electrode width and interelectrode distance. Optimal IDEs design was found to decrease error of measurement [4]. The double layers of microelectrodes, planar structure of IDEs, roughness significantly impact the total capacitance behaviour of the electrode interface impedance [5], [6], [7]. The impedance range can be suppressed at low and high frequencies which results to narrow band for the conductivity measurement. Therefore the method of precise conductivity measurement based on elimination of capacitance from measurements was discovered built-up on bipolar pulse technique [8]. However this technique also does not depress the dependence of the cell constant on IDES structure at wide range of specific conductivities. The correction factor of our thick-film planar Pt IDEs strongly depends on measured conductivity [9]. The surface of the thick-film electrodes is very rough and porous which causes the electrode surface to be much larger than the thin-film equivalent. The sensitivity of the sensor can be higher in comparison with geometric sizes. Also, the capacity of the electrodes should be higher.

The electrode impedance theory described in [1], [10] can be compared with the interface impedance of the measured conductometric sensor. The interface impedance can be calculated from a cell constant determined by cross-sectional area A and interelectrode distance d over specific conductivity κ of the solution. Because geometric sizes do not represent the real cell constant, it is usually calibrated in a known solution.

The measurements took in local minimum of an imaginary part of the interface impedance correspond very well with measurement taken at high frequencies with large macroelectrodes [9]. In this point, at the ridge frequency, the capacitive part of the impedance interface is smallest and the resistance is approximately the sum of the solution resistance (R_S) and charge transfer resistance (R_{CT}). The R_{CT} is constant if the measuring voltage is smaller than 100 mV [11], but it depends proportionately on concentration as well as R_S [10]. Therefore, we can express that

$$R_S + R_{CT} \approx R_S = f(\kappa). \qquad (1)$$

A relative error of the interface impedance measurement can be determined at the minimum capacitive part by

$$\Delta \equiv \left| \frac{Z_{measured} - R_s}{R_s} \right| = \frac{R_{CT}}{R_s} = \frac{1}{K_{cor(\kappa)}}. \qquad (2)$$

$K_{cor(\kappa)}$ is a cell constant corrected for measured conductivity at the ridge frequency, where its value can be very close to constant. The parameter R_{CT} is factored into the $K_{cor(\kappa)}$ according to (1).

2 Experimental

Thick-film interdigitated electrodes IDEs were fabricated from Pt paste (Dupont) screen-printed on alumina substrate and finally covered by over glaze paste. The geometry of the sensors under test has had these parameters: width of fingers is 0.5 mm, length is 6.5 mm, interelectrode distance is 0.5 mm the electrodes are formed by 10 fingers. The impedance characterization was performed using Impedance analyser Agilent 4284. The sensing element is shown in Fig. 1. New instrumentation was designed, built-up on PCB and software based on our method that performs the measurement was created. The instrumentation was tested with the same sensing element.

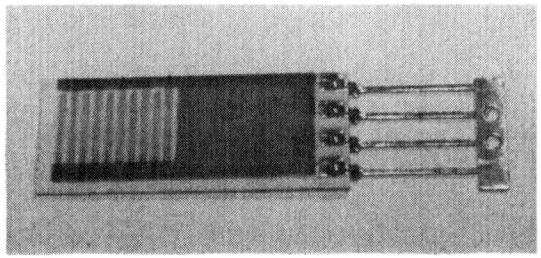

Fig. 1. Thick-film screen printed sensor

The sensors were fabricated by deposition Pt thick-film comb structure with 5 fingers of every one electrode. Non-electrode parts were covered with overglass paste. An interface impedance characterisation of the thick-film sensor has been performed with an impedance analyser (Agilent 4284). The sensor, through testing, gives us better insight into sensor behavior.

3 Results and discussion

3.1 Sensor characterization

The frequency characteristics were found to depend strongly on the specific conductivity or solution concentration respectively. This dependence can be shown on logarithmic axis as plotted in Fig. 2.

Measured characteristics can be processed to a 3D plane graph as shown in Fig. 3. The measuring frequency and the conductivity specify the correction factor for the cell constant correction. It is clear that the frequency has to be changed as the conductivity is changed to get a very small dependence of the cell constant over a wide range of measurements.

Our experimental results confirm the behaviour, where only values at the ridge frequencies were taken within the minimum of the capacitance for the wide conductivity range of KCl solution [7]. The ridge frequency characteristic can be expressed in a complex plane as a power function of the real and the imaginary part of the interface impedance. The real part of the interface impedance has an approximately linear relation with frequency and conductivity. Extracted relations are expressed in (3) and (4).

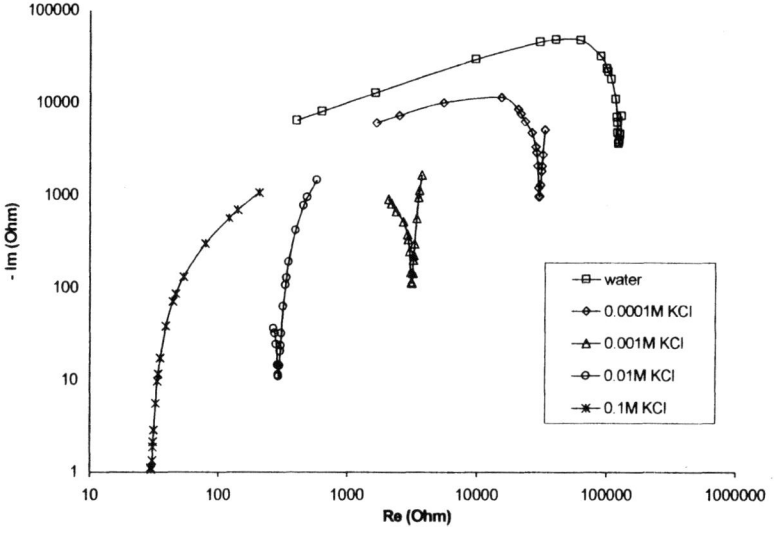

Fig. 2. Nyquist plots of impedance characteristics of the sensor at wide range of the conductivity

$$Re = C_1 \cdot f^{-\alpha}, \tag{3}$$

$$Re = C_2 \cdot \kappa^{-\beta}. \tag{4}$$

The fraction α/β is approximately equal to 1. It is clear that if the real part of the impedance depends on the ridge frequency and conductivity, the relation between both variables can be derived. The frequency in the minima of the capacitance (the ridge frequency) was put into a resistance relation (5).

$$c_2 \cdot f^{-\alpha} = c_1 \cdot \kappa^{-\beta}. \tag{5}$$

Then the equation for calculation of measuring frequency at minima of the capacitance can be derived into (6). The new equation describes the relation between measuring frequency and specific conductivity. The relation is linear (Fig. 4).

$$f = \left(\frac{c_1}{c_2}\right)^{-t} \kappa^{\nu/t} = c_C \cdot \kappa + c_{OFF} \tag{6}$$

The constants C_1, C_2 have to be determined experimentally for each sensor. We found that the measurement has to be taken at the ridge frequency where the capacitance is minimal (Fig. 2) and real value of the impedance is invariable on a cell constant dependency on the measured conductivity or concentration respectively and this real value very well proximate to the solution conductivity value.

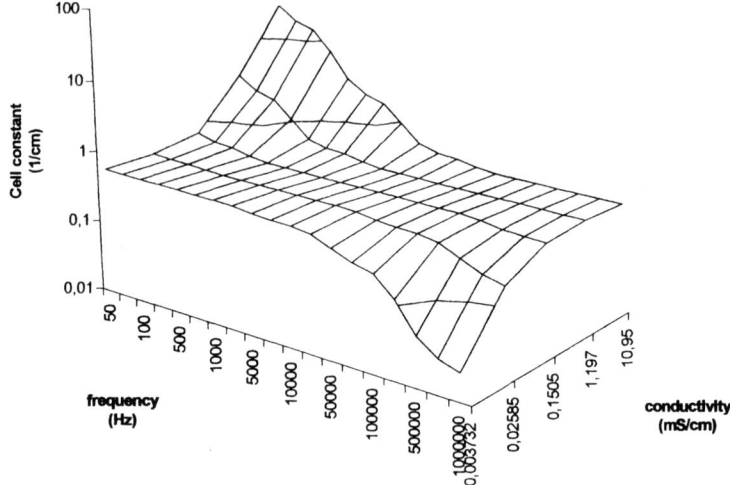

Fig. 3. 3D characteristic of thick-film planar sensor

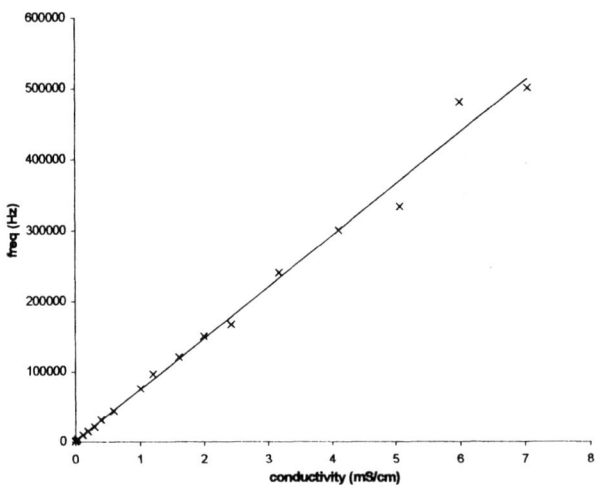

Fig. 4. Frequency dependence on measured conductivity at ridge frequency

As it was described above, the cell constant also depends on the measured conductivity. Because the interface impedance is measured, the cell constant should be derived from this impedance. An extracted equation is

$$K_{cell} = -c \cdot Ln(Z_{min}) + d \: . \qquad (7)$$

The constants c,d have to be determined experimentally. This correction equation can increase accuracy of the measurement. When the influence of the conductivity on the cell constant is very strong, the correction of the cell constant should be performed according to (6).

3.2 Suggested measuring method

Our novel method provides measurements exactly at the ridge frequency [9] corresponding to the measured conductivity value. The idea of measuring method appears from the cell constant dependency (see fig.4) where lowest change of the cell constant can be reached if the measuring frequency changes in respect of measured specific conductivity because of corresponding with the ridge frequency. We observed that it can be simple achieved with iterative algorithm. The responded conductivity is calculated from the cell constant corrected at this ridge frequency. Because the measured conductivity is not known at the start of the measurement, subsequent approximations to the ridge frequency and to the corresponding cell constant have to be performed in a few repeated measuring steps. This iterative method of the measurement can be demonstrated clearly in Fig. 5.

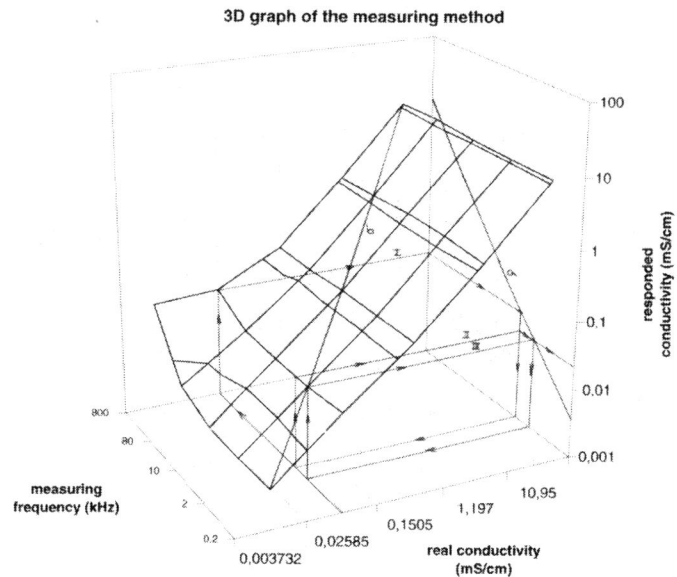

Fig. 5. Iterative steps of measurements

The 3D plot demonstrates the measured and the responded conductivity at different ridge frequencies and the corresponding cell constant of our planar sensor. If a solution with a conductivity value 0.02585 mS/cm is measured at the start frequency 800 kHz, the iteration steps go from front of x-axis to the back, where measuring frequency is 800 kHz. Then the steps continue in direction of arrows to determine the responded conductivity. In the second step, a new measuring frequency is extrapolated from curve (a) to do the new measurement. This curve (b) is a 3D representation of (a). Steps are repeated until the responded conductivity is equal to the measured real conductivity (step III in Fig. 5). The method quickly finds the correct ridge frequency for a measured conductivity.

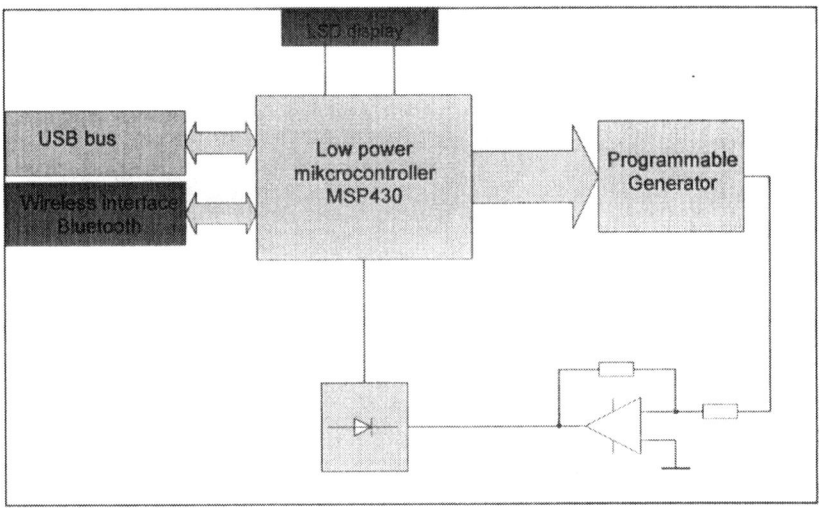

Fig. 6. Designed system providing method of automatic tuning of measuring frequency

This iterative method converges very quickly to the ridge frequency for the specific conductivity as show preliminary results in Tab.1. The measurements have done with Agilent impedance analyser, where the method was applied, and they helped us to determine maximal error to 2% at the range from 1 µS/cm to 100 mS/cm. This error was decreased below 1% by cell constant correction according to measured conductivity using (6). The responded conductivity is calculated from the cell constant corrected to the ridge frequency.

The method was implemented into new instrumentation. The microcontroller based system automatically providing the iterative method was designed. The block schematic is presented in Fig. 6. A fine tuneable oscillator connected with microcontroller including 16 bit A/D sigma-delta converter and analog circuitry including RMS to DC converter and low pass filter of 2nd order was utilized. All system is built-up as a portable battery supplied instrument. The system can be easily programmed through USB bus and data can be collected also by Bluetooth.

Table 1. Measured values during iterative steps of measurements

It.	Calcul. Freq (kHz)	Meas. Freq (kHz)	Re (kOhm)	Im (Ohm)	Conduct. (mS/cm)
1		1000	1.598	-589.2	0.1444
2	11.243	11.3636	2.1375	-77.81	0.1148
3	9.0871	8.9286	2.136	-77.03	0.1149
4	9.093	9.2307	2.128	-76.75	0.1154

It.	Calcul. Freq (Hz)	Meas. Freq (Hz)	Re (Ohm)	Im (Ohm)	Conduct. (mS/cm)
5	9.1248	9.2307	2.128	-76.75	0.1154
		500	235	-123	0.9397
2	69132	68571.4	187	-6.25	1.3303
3	97569	96000	186.4	-5.94	1.339
4	98244	100000	185.9	-5.92	1.34
5	98491	100000	185.9	-5.92	1.34

The test instrumentation for precise specific conductivity performing our iterative method (see Fig. 7) was employed for comparison measurement with analyser Agilent. Our first results confirm validity of the method with accuracy 1% at wide

range of the specific conductivity without cell replacement.

Fig. 7. Final design of the instrumentation with display

4 Conclusion

The suggested method can improve the electrolytic conductivity measurement of the chemical sensor with planar screen-printed IDEs. Results show that a high accuracy of the measurement with the developed method can be achieved. The method was implemented into new designed instrumentation which performs the specific conductivity measurement automatically to reduce measuring time. The system can be connected to computer.

The elimination of the cell replacement with the maintenance of the accuracy over a wide range of the conductivity with our portable instrumentation using IDEs is the most important result of our work.

Acknowledgement. This research has been supported by Grant Agency of the Academy of Sciences of the Czech Republic under the contract GAAV 1QS201710508 Impedimetric Chemical Microsensors with Nanostructured Electrode Surface, and by the Czech Ministry of Education within the framework of Research Program MSM0021630503 MIKROSYN New Trends in Microelectronic Systems and Nanotechnologies. We thank to Petr Novak for his work in PCB design and mounting.

References

1. Digby D. MacDonald Application of Electrochemical Impedance Spectroscopy in Electrochemistry and Corrosion Science. Toronto, John Wiley & Sons, Inc., 1991.
2. Kell, D.B. The principles and potential of electrical admittance spectroscopy: an introduction. Biosensors – Fundamentals and Applications. Oxford University press, 1987.
3. Paul Jacobs, Jan Suls and Willy Sansen: Performance of planar differential-conductivity sensor for urea. *Sensors and Actuators B*, 1994, Vol. 20, pp. 193-198.
4. P. Jacobs, A. Varlan, W. Sansen. Design optimisation of planar electrolytic conductivity sensors. *Medical & Biological Engineering & Computing*, 1995, 33, pp. 802-810.
5. Norman F. Sheppard, Robert C. Tucker and Christine Wu. Electrical Conductivity Measurements Using Microbabricated Interdigitated Electrodes. *Analytical Chemistry*, 1993, Vol. 65, pp.199-1202.
6. W. Laureyn, D. Nelis, P. Van Gerwen, K. Baert, L. Hermans, R. Mahnee, J.-J. Pireaux, G. Maes. Nanoscaled interdigitated titanium electrodes for impedimetric biosensing. *Sensors and Actuators B*, 2000, Vol. 68, pp. 360-370.
7. Adam E. Cohen, Roderick R. Kunz. Large-area interdigitated array microelectrodes for electrochemicl sensing. *Sensors and Actuators B*, 2000, Vol. 62, pp. 23-29.
8. D.E. Johnson and C.G. Enke. Bipolar Pulse Technique for Fast Conductance Measurements. *Analytical Chemistry*, Vol.42, no.3, March 1970, pp. 329-335.
9. Hubálek, J., Krejčí, J., Correction Factors of IDES for Precise Conductivity Measurements, *Sensors and Actuator B*, 2002, Vol. 91, no.1-3, pp. 46-51.
10. Rieger, P.H.: *Electrochemistry.* Prentice-Hall, Inc., A Division of Simon & Schuster, Englewood Cliffs, New Jersey, 1987.
11. Moussavi, M., Schwan, E.T., Sun, H.H., Harmonic distortion caused by electrode polarisation. Med. & Biol. Eng. & Comput., 1994, 32, 121-25.

New Architecture of Network Elements

Vladislav Skorpil, Martin Kral

Department of Telecommunications,
Brno University of Technology,
Purkynova 118, 612 00 Brno, Czech Republic
skorpil@feec.vutbr.cz,

Abstract: This paper describes design and computer simulation of a new architecture of a node active network element, based on artificial neural network technology with the support of priority processing for different connection types. As an example of a network element was selected switching area. This network element with optimized switching area is able to transfer large data quantity with minimum delay. Architecture of network element, that contains artificial neural network for optimized priority switching is described in this paper. It describes implementation of neural network in control process for data units switching. The programming language MATLAB 7.0 was used for software simulation. Network elements with new architecture, which uses a neural network, as well as intimated simulation, are suitable for working for example in personal wireless network communication systems.

Keywords: Neural network, Switch, Switching area, Network element, Neuron

1 Introduction

The switching-over is the basic function of the active network elements that work above physical layer of OSI model. The function of switch consists at that, so data incoming on inputs must be transported to target output and it what fastest. The speed can be limited by blocking, it is caused by situation when data flow from two or more inputs directs to one output. It is necessary in this case for everyone data flow to reserve fair output allocation. Basic communication protocols for computer networks, like protocol IP or Ethernet, does not contain any mechanism for controlling of fair or priority communication channel reserving and they use queue of FIFO type. As modern multimedia applications would such mechanism often need, the possibilities how to guarantee this mechanism are searched very hardly in present. The common quality of these solutions is, that it is assumed the identification of single data or groups of data flows and priority allocation, on basis that the switching process is made.

Please use the following format when citing this chapter:

Skorpil, V., Kral, M., 2007, in IFIP International Federation for Information Processing, Volume 245, Personal Wireless Communications, eds. Simak, B., Bestak, R., Kozowska, E., (Boston: Springer), pp. 669-678.

2 Model of the switch making use of artificial neural network

The basic model of switch-over switching area consists of several partial blocks. The block diagram of the switching area is illustrated on the Fig.1. The received data units are first storage into input buffer. As data units contain target address, it is possible to determine over which output port they leave switching system. Except information about target port are detected also information about data flow whereto the data unit belongs to. Mentioned information are assembled from of all data units retained in input buffers and serve as input data for neural network.

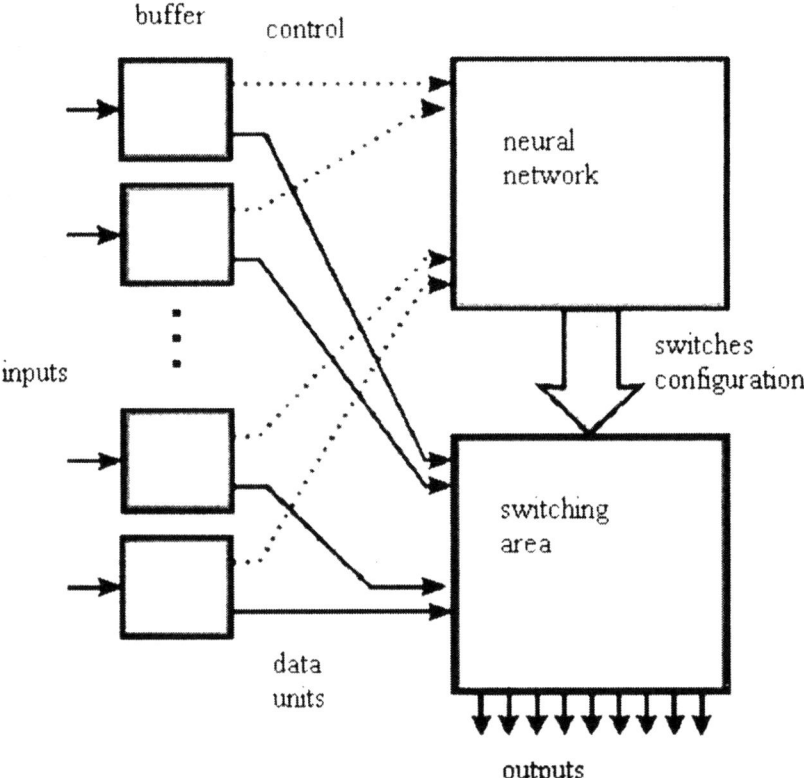

Fig. 1 Block diagram of switching area utilizing neural network

In the event of, so the frames in the port's buffer have such target addresses, so to everyone's output direction exists at most one requirement, it is possible directly put together N-dimensional vector, formed from priorities of single requirements. The first component of the vector includes the frame's priority oriented to the first output, the second one includes the frame's priority oriented to the second output and so one.

The vector must always contain concrete values. In the event of, so no requirement is to output port, have to be on relevant position of the vector the value corresponding to the lowest priority.

In the event of, so more frames in buffer are headed to the same output and it either with the same or with the different priority, have to exist way how to determine, which concrete one will be chosen for optimization process. In the event of different priority the frame with the highest priority is chosen of course. As far as the data units have the same priority, another algorithm for selection must be implemented.

It is possible from vectors of priority since single input ports to create the NxN dimensional matrix, that shall contain input conditions, to be necessary for optimize. Like answer on this input data the neural network will generate explicit configuration of switches, on basis that the switches in the switching area will be set. Generated configuration will be in the form of NxN dimensional matrix, so-called configuration matrix. Demand on generated result is to be connected always one input and one output. This condition will be projected to the configuration matrix that is generated by neural network thus, that at every row and at every column will be exactly one active output, i.e. the value is one. The other will be non-active, i.e. the value is zero. Past setting of switches it can come to own transmission of data units from input buffer units to outputs of system.

It was already states, that setting of the switches has to be optimal in light of the matrix of priority. To the configuration matrix, obtained as a result that is generated by the neural network, we can look like to a template. By the help of this template, it is possible to choose elements from arbitrary NxN dimensional matrix. **P** so, that from the matrix **P** are selected only elements, which stands on the position, where the configuration matrix has the value of one. That way we obtain from the matrix **P N** selected elements by the help of the template. When we apply this template to the matrix of priority obtained from the input buffers, we are able to determine the priority of elements, that just will be transported. We can then formulate the optimization exercise so, in order to be summation of priorities minimal. But at the same time, how it was already noted above, we have to guarantee in every row and column of matrix only one value of one. This condition is not completely accorded to the optimization travelling-salesman problem, but it is very close to it.

3 The basic scheme of the element

We think over the single-stage switching area, which has three inputs and three outputs, it is switch on the Fig.2. The switching area is realized on the cross-bar switch, i.e. in the described case the switching area with 9 switching points. We can connect arbitrary input to arbitrary output. However in one time moment t we can transfer information from one input only to one output in terms of only one's switch.

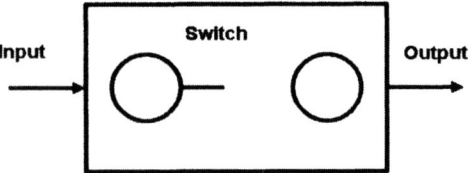

Fig.2 Switch

We create the switching area by in a way, that the switches will be shape partial positions, generally m-n dimensional matrix, in described case 3x3 dimensional one. We do not speculate about multiplexing of inputs or outputs, every switching point of the switching area presents one switch (Fig.3). The multiplexing has meaning for generally m-n dimensional switching area, main benefits have economic character, saving of a material and of a place.

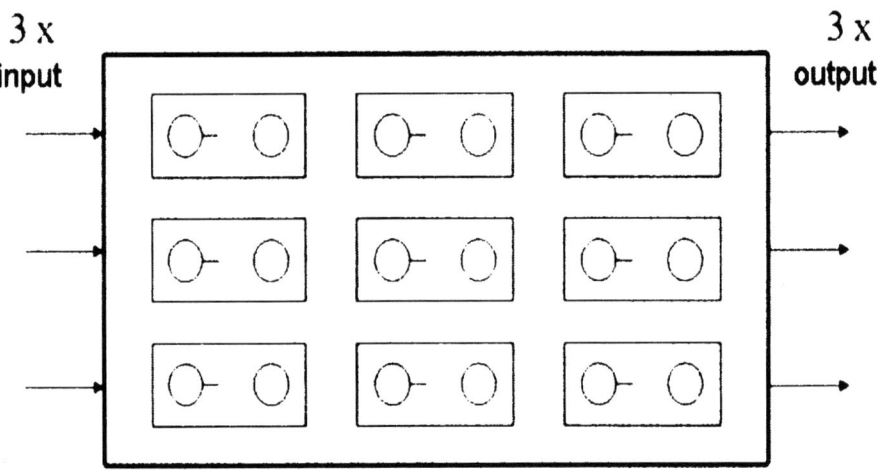

Fig.3 Switching area

The function of switching area is, in order to data unit, which reveals on the input, was transported to relevant output of the area and continues further to a receiver. As far as we would introduce data units into switching area only from one input and other inputs would be cancelled, function of all area would be greatly unused and efficiency would be small. That is why we want, to appear in the same instant of time data units on every input, which have to be transported pass through switching area. One quite serious problem is here. It can become, so on two and more inputs in the same time moment data units appear, which are addressed to the same output. Then the question arrives, which one of data have priority of processing and for which of them is not real-time transmission critical. This question is solved by the priority processing method.

We return back to the switching area. We introduce the simplest network about so much message sources (data), how much it has inputs in the node and so much receivers, how much it has outputs. We should have only one node in the network. So that we prove to deliver the right data units to the correct receiver, we have to set up addressing. The simple two-bits addressing word, how in the event of input, that way also of output of the switching area, is used in our example (Fig.4). The first input has address 01, the second one 10 and the third one 11. The first output has address 01, the second one 10 and the third one 11. The resolution, whether or not it walks about input or output address, is given by virtue of the data frame header, where it is stated, whether it walks about source or target address. The priority is stated also in the header. The frame structure is on the Fig.5,

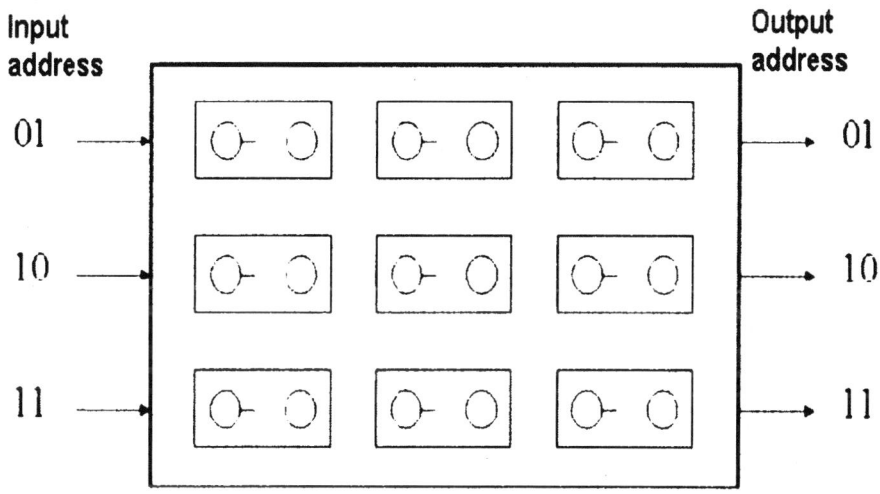

Fig.4 Switching area with addressing

2 bits	2 bits		
Target address	Source address	4 bits Priority	16 bits Data

Fig.5 Frame structure

4 Control of switching area

The switching area would not be operational without control. The control is here realized by control matrix, which has dimension of switching area and every element of this matrix control one switching point. It is matrix with the dimension 3x3, in described case.

$$\mathbf{C} = \begin{pmatrix} x & x & x \\ x & x & x \\ x & x & x \end{pmatrix} \quad (1)$$

Indexing of switching area according to control matrix $\mathbf{C}$ is simple. The switching point, which switches the first input, i.e. with the address 01, to the first output (address 01) corresponds to the matrix element with the index 11 (C_{11}). And farther, the switching point which switches the first input to the second output will be operating by the element of control matrix with the index 12 (C_{12}), etc. The elements of the matrix put on binary value 0,1. The value of logical 1 means bracing of the switching point and thereby the way through the switching area is closed and data units from the appropriate input are transported to the given output. The switching point, controlled by the appropriate element of matrix about the value of logical 0, will be in a sleep position, the switch does not close and it means, so the data units from appropriate input were routed to other output. On the Fig. 6 we can see the particular graphic example. The outputs of switching area are for simpler notion removed from the right side to low one, because on the original picture is not clear, which switching point corresponds to which output. The control matrix will be given by elements:

$$\mathbf{C} = \begin{pmatrix} 0 & 1 & 0 \\ 1 & 0 & 0 \\ 0 & 0 & 1 \end{pmatrix} \quad (2)$$

It is given, at this structure of control matrix, so logical value 1 of the matrix element C_{12} switches switching point, which connects the first input to the second output. Further the logical value 1 of the matrix element C_{21} switches switching point which connects the second with the first output and last the logical value 1 of the matrix element C_{33} switches switching point which connects the third input with the third output.

Personal Wireless Communications 675

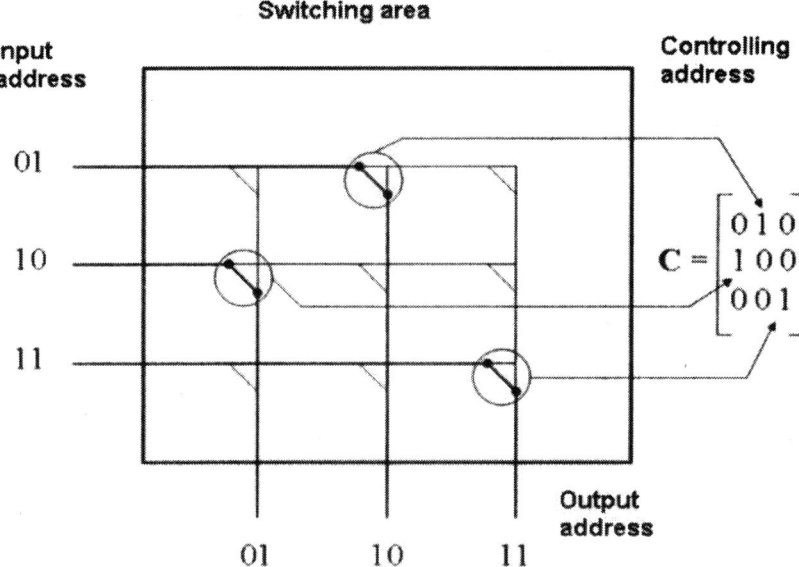

Fig.6 Switching area controlled by control matrix

5 Scheme of neurons examining priority selection

The most important part of the whole network element is a block, which examines priority selection of frames. This frames will be subsequently sent to the switching area. The block by virtue of certain information evaluates the most acceptable or rather the most important frame from the possible selection and gives up advice for sending it. Just this block works on the principle of neural network. The neurons practise this important priority selection. The scheme of neurons examining priority selection in on the Fig.7.

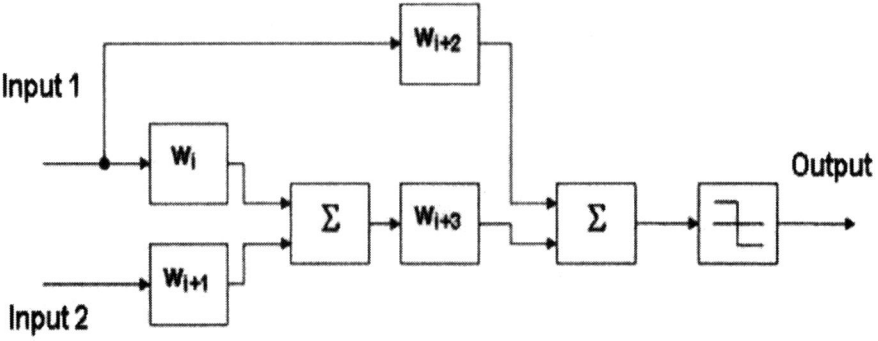

Fig.7 Scheme of neurons examining priority selection

The block, that examines priority selection, contains from two neurons. This block selects within one output of switching area if you like of one output address. For our actual switching area, and consequently 3 possible output addresses, 3 these blocks will be used. The values of priority of single frames, which ask for input to the switching area, are set to the input of the block. The frames have one sharing, all frames have the same final address as well as different source address. It comes to this, that frames from different directions are routing to the one particular output. So that we decide, which frame to choose, we have to compare their priority. At the beginning we take priority from frames, which are on the first two inputs, during which time the priority of frame, which applies approach to the first input of switching area, will allocate to the input 1. The first neuron these two priority sums and numbered value sends to the second input of the second neuron. On the first input of the second neuron we send priority from the input 1. The result is sent to the block of non-linear function after numbered of both values to the summarize block of the second neuron. It decides, whether on the output will be value log0 or log1 and all process repeats. The value of output controls partly the assignment of priority to the inputs one and two as well as the associated matrix of priority **B**.

6 Conclusion

The aim of this research was assembling of theoretical materials and knowledge concerning to the problems of network elements and artificial neural networks and then design and subsequently computer simulation of a new architecture of node active network element. This network element will be founded on the technology of artificial neural networks and will support priority processing of different types of connection. The optimization of switching area of these active elements is very actual. In the paper is stated the architecture of network element, which includes neural network for optimizing of priority switching. It was used programme language MATLAB 7.0 for the software simulation. The idea of the implementation of neural network at control process of switching of data units is indicated at simulation. It is also during the simulation demonstrated, to what purpose is every bloc in the element used.

At last it is possible to claim, so target of the research were realized. It was designed a simulated network element, containing in the process of control of switching area artificial neural network. The element further switches single data units making provision for priority. The next research will be expansion of the number of inputs and outputs of the network element.

Acknowledgement. This research was supported by the grants:
No 102/07/1503 Advanced Optimizing the design of Communication Systems via Neural Networks. The Grant Agency of the Czech Republic (GACR)
Grant 1884/2007/F1/a "Innovation of computer networks participation in high-speed communication" (grant of the Czech Ministry of Education, Youth and Sports)
Grant 1889/2007/F1/a „ Repair of digital exchanges education in the course Access and Transport Networks" (grant of the Czech Ministry of Education, Youth and Sports)
No MSM 0021630513 Research design of Brno University of Technology " Electronic communication systems and new generation of technology (ELKOM)"
2E06034 Lifetime education and professional preparation in the field of telematics, teleinformatics and transport telematics (grant of the Czech Ministry of Education, Youth and Sports)

References

1. Gupta,M. and Liang,J. *Static and dynamic neural network*, John Wiley & Sons, Inc., New Jersey, 2003
2. Jirsík, V. and Hráček,P. Neural networks, experts systems and speech recognition., BUT Brno 2004
3. Novák,M. Artificial nearal networks and applications. C. H. Beck, Praha,1998
4. Marťán, P. Visualization of selected methods of machine learning, MU, 2002
5. Novotný, V. Network architecture. BUT, Brno, 2002
6. Molnár,K. Aplication of neural networks in high-speed active network elements. BUT, Brno, 2002
7. Minsky,M. and Papert, S. *Perceptrons*, MIT Press, 1969
8. Hassoun,M.H. *Fundamentals of artifical neural network*, MIT Press, 1995
9. Síma J. and Neruda, R. Theoretical questions of neural networks. Matfyzpress, Praha, 1996
10. Fausett,L. Fundamentals of Neural Networks, Architecture, Algorithms and Applications, Prentice Hall, 1994

11. Kosko,B. Neural Networks and Fuzzy Systems, Prentice-Hall International, Inc., London, 1992
12. Molnar,K. Switching area of ATM switch controlled by neural network. BUT, Brno 1999
13. Rinipe, J. and Euliano, R.N. and Lefebvre, W.C. Neural and Adaptive Systems, John Wiley & Sons, Inc., USA, 1999
14. Krbilová, I. and Vestenický, P. Use of Intelligent Network Services in Proc. of ITS. RTT , CTU Prague, 2004
15. Král,M. Design of new network element architecture. BUT, Brno 2006
16. Vestenický, M. and Vestenický, P. Evolutionary Algorithms in Design of Switched Capacitors Circuit in Proc. of International Workshop „Digital Technologies 2004". Slovak Electrical Society and University of Zilina, pp.34-37, 2004
17. Vestenický,P. The Prediction Properties of Kalman Filter in Proceedings of TRANSCOM '95. UT&C, Zilina, pp. 243-248, 1995
18. Krbilová, I. and Vestenický,P.: Forgalomszabályozásés szolgáltatásminöség" Magyar távközlés 7, Vol.. 6/96, pp. 32-33, 1996
19. Bubeníková, E. and Vestenický, P. Principles Of The Intranet Information System Creation in Proc. of ELEKTRO'99 Conference Proceedings, section Information & Safety Systems. University of Žilina, pp. 77-81, 1999
20. Vestenický, P. The Functions of ATM Interfaces in Proc. of DDECS '97 Conference Proceedings. VSB Technical University, Ostrava, pp. 186-191, 1997
21. Vestenický, P. Optimization of Selected RFID System Parameters in Proc. of. AEEE 3, Vol. 2, pp. 113-114, 2004

Author Index

Alexiou, Antonios, 433
Almenárez, Florina, 304

Baba, Kensuke, 230
Barenco Abbas, Cláudia, 410
Becvar, Daniel, 628
Becvar, Zdenek, 107
Bestak, Robert, 75, 99
Bodecek, Kamil, 477, 505
Bouras, Christos, 433
Brezina, Milan, 477, 505
Brida, Peter, 423
Burget, Radim, 445, 456, 468

Canales, María, 1, 13
Cardenas, Nelson, 410
Carrasco, Loren, 264
Casilari, Eduardo, 63
Choi, Jin-Young, 384
Choi, Yong Seouk, 217
Cika, Petr, 497
Cuevas, Antonio, 294
Cvrk, Lubomir, 355

Díaz-Sánchez, Daniel, 304
Delicado, Francisco M., 87
Delicado, Jesús, 87
Dostal, Otto, 316
Duha, Jan, 423

Einsiedler, Hans, 294

Femenias, Guillem, 184, 264
Filka, Miloslav, 595, 610
Forejtek, Jiri, 637
Fujcik, Lukas, 637

Gállego, José Ramón, 1, 13
García Villalba, L. J., 410
García-de-la-Nava, Jorge, 63
Gaugue, Alain, 241
González, Ricardo, 410
González-Cañete, Francisco J., 63
Gozdecki, Janusz, 25

Haze, Jiri, 637

Herencsar, Norbert, 557
Hernández, Ángela, 133
Hernández–Solana, Ángela, 1
Hernández-Solana, Ángela, 13
Hopjan, Miroslav, 325
Hubalek, Jaromir, 660
Hubeny, Tomas, 332

Jerabek, Jan, 545

Kadlec, Jaroslav, 349
Khara, Sibaram, 206
Kim, Hongsoog, 123
Kim, Hyun-Seok, 384
Kim, Junsik, 123
Kim, Kyung Soo, 217
Kim, Nam, 123, 217
Klajbor, Tomasz, 145
Komosny, Dan, 445, 456, 468
Koton, Jaroslav, 569
Kotuliak, Ivan, 174
Kovar, Petr, 277, 341
Kozlowska, Ewa, 115
Kral, Martin, 669
Krasnovsky, Marek, 423
Krepelka, Vaclav, 610
Krichl, Vit, 252
Kubanek, David, 341, 578
Kuchta, Radek, 349, 396, 660
Kyselak, Martin, 595

Lanza, Jorge, 38
Liebe, Christophe, 241
Lubacz, Jozef, 196

Mach, Pavel, 75
Maly, Jan, 488
Marín, Andrés, 304
Matejka, Juraj, 174
Mica, Ivan, 514
Michaeli, Linus, 637
Mikoczy, Eugen, 174
Minarcik, Martin, 585
Misurec, Jiri, 616
Molnar, Karol, 277, 367
Moreno, Jose I., 294

Muños, Luis, 38

Nakamura, Toru, 230
Natkaniec, Marek, 25
Novotny, Vit, 285, 477

Ogier, Jean-Marc, 241
Oh, Jung-Hyun, 384
Oh, Sangchul, 123
Orozco-Barbosa, Luis, 87

Pérez, David, 133
Palko, Lukas, 602
Pani, Chandi, 157
Papazois, Andreas, 433
Park, Namhoon, 123
Podhradsky, Pavol, 174
Pravda, Ivan, 376
Prinosil, Jiri, 514, 533
Prokop, Roman, 637

Rajmic, Pavel, 488, 514
Ramis, Jaume, 264
Riera-Palou, Felip, 184, 264
Rudinsky, Jan, 169

Sánchez, Luis, 38
Saha Misra, Iti, 157, 206
Saha, Debashis, 206
Sajdl, Ondrej, 403
Sargento, Susana, 25
Shklyaeva, Anna, 341

Simek, Milan, 445, 468
Sisma, Ondrej, 241
Skorpil, Vladislav, 54, 669
Slavicek, Karel, 316
Smékal, Zdenek, 521
Stastny, Jiri, 54
Steffan, Pavel, 649, 653
Stehlik, Jiri, 628, 649
Svoboda, Pavel, 285
Sysel, Petr, 521
Szczypiorski, Krzysztof, 196

Triviño-Cabrera, Alicia, 63

Ulvan, Ardian, 99

Valdovinos, Antonio, 1, 13, 133
Valenzuela, José Luis, 133
Vlach, Jan, 514, 533
Vodrazka, Jiri, 332, 376
Vranova, Zuzana, 325
Vrba, Kami, 578
Vrba, Kamil, 545, 557, 569, 585
Vrba, Radimir, 349, 396, 403, 637, 649, 653
Vrba, Vit, 355
Vyoral, Josef, 514

Yamasaki, Tomomi, 230
Yasuura, Hiroto, 230

Zak, Jaromir, 403
Zavoral, Pavel, 637

Printed in the USA